PROCEEDINGS OF THE INTERNATIONAL INSTITUTE OF SPACE LAW

2008

51st Colloquium on the Law of Outer Space

24th IAA–IISL Scientific-Legal Roundtable

(29 September–3 October 2008, Glasgow, Scotland)

IISL–ECSL Symposium held on the occasion of the 47th Session of the Legal Subcommittee of UNCOPUOS

(31 March–1 April 2008, Vienna, Austria)

3rd Eilene Galloway Symposium on Critical Issues in Space

(11 December 2008, Washington, D.C., USA)

17th Manfred Lachs Space Law Moot Court Competition 2008

(2 October 2008, Glasgow, Scotland)

and other materials

AIAA

The World's Forum for Aerospace Leadership

Published and Distributed by

AMERICAN INSTITUTE OF AERONAUTICS AND ASTRONAUTICS

Editor:

Corinne M. Contant Jorgenson

Washington, D.C., USA

ISBN-13: 978-1-60086-717-0
ISBN-10: 1-60086-717-0

Printed in the U.S.A.

TABLE OF CONTENTS

3. WEAPONISATION OF OUTER SPACE IN THE LIGHT OF THE OUTER SPACE TREATY

4. LEGAL ASPECTS OF NATURAL NEAR EARTH OBJECTS (NEOs)

5. OTHER LEGAL MATTERS

6. GLASGOW SESSION REPORTS

24th IAA–IISL SCIENTIFIC-LEGAL ROUNDTABLE

United Nations

2008 IISL–ECSL SPACE LAW SYMPOSIUM HELD ON THE OCCASION OF THE 47th SESSION OF THE LEGAL SUBCOMMITTEE OF UNCOPUOS, VIENNA, AUSTRIA

The 3rd Eilene M. Galloway Symposium On Critical Issues In Space Law, Washington D.C., December 2008

17th Manfred Lachs Space Law Moot Court Competition

The aim is to optimize the functioning of the Institute and to clarify and update the language of its rules. It will also be proposed to slightly increase the number of directors, to better reflect the required geographical and professional spread.

In March 2009, the IISL attended the meeting of the UNCOPUOS Legal Subcommittee for the first time as independent observer, and it was with great pride that we occupied the seat behind our very own nameplate. IISL members are most welcome to attend the COPUOS meetings as observers, and are invited to contact the Secretary for this purpose. We are currently also fulfilling the requirement to obtain consultative status with ECOSOC, which is an obligation for all COPUOS observers. This process may take up to three years.

The website launched last year is continuously expanded and improved (see www.iislweb.org). The "IISL groups" on Facebook and LinkedIn continue to grow, and prove to be useful communication and networking tools.

The proposals of the IISL Directorate of Studies to improve the format of our annual Colloquia will be implemented for the first time in Daejeon in October 2009, and will see the inaugural Nandasiri Jasentuliyana Keynote Lecture on Space Law delivered by Prof. Vereshchetin, Honorary Board Member and Former Member of the ICJ in The Hague. Daejeon will also celebrate the first "Young Scholars" session.

In addition to an assistant secretary, the Board has agreed to appoint an assistant treasurer, and Ms. Axelle Cartier agreed to serve in this capacity for two years. Dr. Gérardine Goh will hand over the task of assistant secretary to Mr. Karl Abendschein in October. The Board is most grateful for Dr. Goh's work since she was appointed in 2007. A report about the Board elections and new members is published separately.

3. Cooperation

Cooperation with our sister organisations, the International Astronautical Federation (IAF) and the International Academy of Astronautics (IAA), but also with the Committee on Space Research (COSPAR) continue to be intensified. This is essential because interaction between scientists, engineers and lawyers is of the utmost importance in today's world of space activities. Thus, several members of the IISL actively participate in and contribute to various committees, commissions and working groups of these organisations. The Board will prepare an overview of these for information of the members, who are welcome to contribute in areas of their interest and expertise.

4. IISL Activities

In *September/October 2008*, the Colloquium in Glasgow, Scotland, was the first one at which I had the honour to preside. As usual, five sessions were held with many good papers that are published in this volume. Sessions dealt with: (1) Private international law regarding space activities; (2) the 40th anniversary of the Rescue Agreement: looking ahead; (3) Weaponization of outer space in the light of the Outer Space Treaty; (4) Legal aspects of natural Near Earth Objects (NEOs); and (5) Other legal matters.

The 23rd Scientific/Legal Roundtable addressed the topic "Paper Satellites—Problems of Policy, Regulation and Economics". For the first time the Roundtable was opened by IAA and IISL officials (Dr. J.M. Contant and myself), in order to raise the profile of this important gathering. It is our firm intention to continue our efforts to increase awareness about this important meeting of scientists and lawyers.

Another successful IISL Plenary Event on "Real space, real law, real progress" had been coordinated by Mr. Chris Stott, and was moderated by Mr. Bob Richards. It was again a good mix of theory and practice, and the audience participated actively.

On *11 December 2008*, the Institute co-organized with the National Center for Remote Sensing, Air & Space Law at the University of Mississippi School of Law and a new partner, Arianespace Inc., the third "Eilene M. Galloway Symposium on Critical Issues in Space Law", a one day interdisciplinary meeting at the Cosmos Club in Washington, D.C. The symposium addressed "Art. VI of the Outer Space Treaty: Issues and Implementation" and involved lawyers, scientists, space agency representatives and industry. We are very happy that Mr. Clay Mowry, president of Arianespace Inc., is willing to support this event, as his involvement will ensure broader access to space industry and policy-makers.

In December 2009, this event will be organised for the fourth time, again at the Cosmos Club, sadly without Mrs. Galloway to guide us. The event will not be the same without her, but hopefully we can continue to hold it in her honour and addressing topics that lay close to her heart.

5. IISL Awards

In 2008, the Board of Directors gave a Lifetime Achievement Award to its Board Member Prof. Francis Lyall of the University of Aberdeen, Scotland, recognizing his outstanding, life-long educational, scholarly, managerial, and cooperative contributions to the development of international space law.

Unfortunately the Prof. Dr. I.H.Ph. Diederiks-Verschoor Award and Prize (honouring the best paper presented at an IISL Colloquium by a young author) could not be given in 2008, mainly because there had been too few submissions.

6. Manfred Lachs Moot Court Competition

The 17th Manfred Lachs Space Law Moot Court Competition was held during the Glasgow IISL Colloquium. Preliminary rounds were held at regional level in Europe, North America, and the Asia Pacific region. The Finals were judged by Judges Koroma and Tomka of the International Court of Justice. Prof. Lyall capably replaced Judge Owada who was prevented to attend because of urgent Court matters.

The final took place at the beautiful Glasgow City Chambers, and the stunning converted church Orán Mór was the location for the elegant and animated dinner, attended by around 170 guests. The Local Organising Committee and various local sponsors had made excellent arrangements for these events, for which we are most grateful. Special mention must be made of EADS Astrium, which generously sponsored the fabulous dinner. Various local universities provided support for the ICJ Judges, in exchange for which, as is becoming customary, the judges gave a lecture for academic staff and students.

A report on the competition, the case and the finalists' briefs are published in this volume. The competition will be held for the 18th time in Daejeon, Korea, in October 2009.

7. United Nations

Another symposium was organized by IISL with ECSL during the Legal Sub-Committee of COPUOS in March 2008, on "Legal Implications of Space Applications for Global Climate Change". The two-day symposium was chaired by Ambassador Jankowitsch and Prof. Marchisio.

The IISL again submitted and presented a Report on the work of the Institute, as well as its annual contribution to the UN publication on "Highlights in Space".

Several IISL members as usual attended the annual meetings of COPUOS and its subsidiary bodies.

8. Words of thanks

In the last, but not least important section of this Introduction I wish to express my thanks for the continued support by the members of the Institute. Thanks to their dedication, the work is not only carried out efficiently, timely and expertly, but, perhaps most importantly, in a spirit of cooperation, friendship and collegiality that has marked the IISL since its creation.

My first words of thanks this year go to my fellow members of the IISL Board of Directors, whom I have called upon quite often in the past year, asking for their advice and guidance. They were perhaps not used to receiving so many calls for advice and help, but they have never faltered in giving me valuable support and feedback on numerous matters. I believe that the team spirit among the Board has benefitted from this increased communication, and I fully intend to continue calling on them in the coming year.

I want to thank Corinne Jorgenson for her efforts in combining motherhood with the IISL Secretariat work. She has managed to keep her head above water during this year in a remarkable way; I am confident that the coming year will be easier, not only because her sons are starting to grow (which might in fact make things less easy), but especially because she has now experienced a full year of IISL activities and knows first hand what it's all about. Corinne, I hope you will enjoy the combination of IISL work and being a young mother as much as I have!

Special thanks are again expressed to the Judges of the International Court of Justice, who have continued to honour us by judging the finals of the Manfred Lachs Space Law Moot Court Competition.

Likewise, the continuing support of NASA, ESA/ECSL and JAXA must be put on record again. Their support enables a new generation of space lawyers to attend their (in most cases) first IISL Colloquium, and I think you will all agree that no one ever forgets their first IISL Colloquium!

The Local Organizing Committee and Institute members overseeing the arrangements for IISL activities in Glasgow should be congratulated and thanked for the outstanding arrangements they made for the Colloquium, Moot Court and Annual Dinner. Thanks and congratulations to Prof. Francis Lyall, Prof. Lesley-Jane Smith, Ms. Joanne Wheeler and Ms. Suszann Parry for their hard work.

I am also very grateful to the following persons:

- Prof. Jonathan Galloway, Dr. Peter van Fenema and Prof. Toshio Kosuge, members of the nominating committee;
- Prof. Stephan Hobe, Prof. Ram Jakhu and Prof. Steven Freeland, members of the Directorate of Studies;
- Dr. Andrei Terekhov, Chairman of the committee that prepares the report on the Status of International Space Treaties;
- Dr. Sylvia Ospina, for preparing the Institute's contribution to the UN report on Highlights in Space, and for leading the effort to review the IISL Statutes and Bye-laws, in close cooperation with Prof. Vladimir Kopal, Prof. Francis Lyall and Prof. Frans von der Dunk, whose hard work is also greatly appreciated;

- Mr. Chris Stott and Mr. Bob Richards, for another successful IISL Plenary Event in Glasgow;
- Dr. Skip Smith, Mr. Ricky Lee and Ms. Melanie Vincent for organizing the regional rounds of the Moot Court;
- Mr. Ian Awford, Dr. Peter van Fenema, Dr. Martha Mejia-Kaiser, Dr. Larry Martinez, Dr. Fabio Tronchetti, Dr. Yun Zhao, Prof. Stephan Hobe, Mr. KRS Murthi and Prof. Maureen Williams for judging the briefs and the semi-finals of the Moot Court;
- Mr. Ricky Lee for writing the Problem of the 17th Moot Court;
- Mr. Scott Hatton, for being such a responsive and capable webmaster;
- And all members of the Institute who assisted with its work in numerous other ways and whose names are not mentioned above.

I look forward to my next year as President, and intend to continue my efforts to enhance the relevance of the Institute by undertaking new initiatives and strengthening existing activities.

The Hague, May 2009

IISL Board of Directors 2008 - 2009

INTERNATIONAL INSTITUTE OF SPACE LAW

94BIS AVENUE DE SUFFREN
75015 PARIS – FRANCE
www.iislweb.org

TEL.: (33) 1 45 67 42 60
FAX: (33) 1 42 73 21 20
info@iislweb.org

Statement of the Board of Directors of the International Institute of Space Law (IISL)*

In 2004, the Board of Directors of the IISL, an international non-governmental organization, issued a statement relating to the issue of 'property rights' in outer space. The statement can be found on the website of the IISL, at http://www.iislweb.org/publications.html

In view of recent misleading views and discussions on this subject in the press, the Board considers that it is appropriate to further clarify a number of salient points as follows:

International Law establishes a number of unambiguous principles, according to which the exploration and use of outer space, including the Moon and other celestial bodies, is permitted for the benefit of mankind, but any purported attempt to claim ownership of any part of outer space, including the Moon and other celestial bodies, or authorization of such claims by national legislation, is forbidden as following from the explicit prohibition of appropriation, and consequently is prohibited and unlawful. Since there is no territorial jurisdiction in outer space or on celestial bodies, there can be no private ownership of parts thereof, as this would presuppose the existence of a territorial sovereign competent to confer such titles of ownership.

The current international legal regime is binding both on States and, through the precise wording of Article VI of the Outer Space Treaty of 1967, which has been ratified by 100 countries, including all the space-faring countries, also on non-governmental entities, i.e. individuals, legal persons and private companies. The clear goal of such a regime is to preserve outer space, including the Moon and other celestial bodies, for the exploration and use of all mankind, not only for those States and private enterprises that are capable of doing so at any particular time.

At present, international space legislation does not include detailed provisions with regard to the exploitation of natural resources of outer space, the Moon and other celestial bodies, although it does set down a general framework for the conduct of all space activities, including those of private persons and companies, with respect to such natural resources.

The IISL is of the opinion that a specific legal regime for the exploitation of such resources should be elaborated through the United Nations, on the basis of present international space law, for the purposes of clarity and legal certainty in the near future. The IISL will continue to play an active role in any such discussions as they develop.

22 March 2009

---

* The Board of Directors of the IISL comprises specialists in space law from all continents and various backgrounds, including past and present chairmen of the UN Committee on the Peaceful Uses of Outer Space (UNCOPUOS) and its Legal Subcommittee. This Statement is based on a proposal from the members of the IISL Directorate of Studies. The views expressed in this Statement represent a consensus among the Members of the IISL Board of Directors acting in their personal capacity, and do not necessarily reflect the views of any entities with which they may be affiliated.
For more information about the IISL, please visit www.iislweb.org or contact info@iislweb.org

INTERNATIONAL INSTITUTE OF SPACE LAW

94BIS AVENUE DE SUFFREN
75015 PARIS – FRANCE
www.iislweb.org

TEL.: (33) 1 45 67 42 60
Fax: (33) 1 42 73 21 20
info@iislweb.org

Standing Committee on the Status of International Agreements Relating to Activities in Outer Space

Chairman:
Andrei D. Terekhov
Deputy Director
General Legal Division
Office of Legal Affairs
United Nations

ANNUAL REPORT – 2008

Background

At the meeting of the IISL Board of Directors held at Brighton, United Kingdom, in October 1987, the Board decided to create a standing committee of the Institute to prepare and submit at each annual Colloquium a report on the status of agreements relating to activities in outer space.

Institute members who can contribute substantively to the work of the committee are invited to serve on it. Membership is open to any interested IISL member. The committee members help collect, review and verify data concerning selected space agreements. Agreements of general global applicability or regional applicability are selected for reporting. Bilateral agreements are not included in the report. In general, the agreements included in the present report, and their status, are believed to be valid as of 1 January 2008.

Coded entries are used in the table as follows:

R = ratification, accession or succession
S = signature only
D = declaration of acceptance of rights and obligations

When no entry is made in a column opposite a state's name, that state has not signed that agreement, or is not a party to it, or that state has withdrawn from the agreement. Changes from last year are highlighted by "*****".**

At the meeting of the IISL Board of Directors held at Paris, France, in April 1998, the Board decided that the report should also contain basic data concerning Declarations of Principles Relating to Activities in Outer Space adopted by the General Assembly of the United Nations.

COMMITTEE MEMBERSHIP

Prof. Dr. K.-H. Böckstiegel
Institut für Luft- und Weltraumrecht
University of Cologne
Albertus-Magnus-Platz 5000
Cologne, 41 GERMANY

Prof. Carl Q. Christol
5500 Calle Real C-124
Santa Barbara, CA 93111
USA

Prof. Dr. Aldo Armando Cocca
Lavalle 1527, piso 10
1048 Buenos Aires
ARGENTINA

Prof. Dr. Isabella Diederiks-Verschoor
President Emeritus, IISL
Leestraat 43, 3743 EH Baarn
NETHERLANDS

Dr. Stephen E. Doyle
3431 Bridget Brae
Shingle Springs, CA 95682
USA

Dr. Eilene Galloway
4612 29th Place, NW
Washington, D.C. 20008
USA

Prof. Dr. Vladimir Kopal
IAF General Counsel
Vidlicová 2200
16000 Prague 6 - Dejvice
CZECH REPUBLIC

Mr. Gabriel Lafferranderie
Le Clos Salibert
3 allée des Cèdres
78860 Saint Nom La Breteche
FRANCE

Prof. Francis Lyall
Department of Public Law
University of Aberdeen
Aberdeen, AB243UB
SCOTLAND

Mr. David Sagar
Flat 4, 51 Cadogan Gardens
London SW3 2 TH
UNITED KINGDOM

Dr. A. D. Terekhov (Chairman)
Office of Legal Affairs
Room 3430P, UN Secretariat
New York, NY 10017 USA

Mr. Leo A. Webb
4041 Valley Oaks Drive
Clinton, IA 52732-9467
USA

AGREEMENTS INCLUDED

1. **1963 NTBT** - **Treaty Banning Nuclear Weapon Tests in the Atmosphere, in Outer Space and Under Water; done at Moscow on 5 August 1963; entered into force on 10 October 1963. (14 UST 1313; TIAS 5433; 480 UNTS 43)**
   http://fletcher.tufts.edu/multi/texts/BH454.txt

2. **1967 OST** - **Treaty on Principles Governing the Activities of States in the Exploration and Use of Outer Space, including the Moon and Other Celestial Bodies; done at Washington, London and Moscow on 27 January 1967; entered into force on 10 October 1967. (18 UST 2410; TIAS 6347; 610 UNTS 205)**
   http://www.unoosa.org/oosa/en/SpaceLaw/outerspt.html

3. **1968 ARRA** - **Agreement on the Rescue of Astronauts, the Return of Astronauts and the Return of Objects Launched into Outer Space; done at Washington, London and Moscow on 22 April 1968; entered into force on 3 December 1968. (19 UST 7570; TIAS 6599; 672 UNTS 119)**
   http://www.unoosa.org/oosa/en/SpaceLaw/rescue.html

4. **1972 LIAB** - **Convention on International Liability for Damage Caused by Space Objects; done at Washington, London and Moscow on 29 March 1972; entered into force on 1 September 1972. (24 UST 2389; TIAS 7762; 961 UNTS 187)**
   http://www.unoosa.org/oosa/en/SpaceLaw/liability.html

5. **1975 REG** - **Convention on Registration of Objects Launched into Outer Space; done at New York on 14 January 1975; entered into force on 15 September 1976. (28 UST 695; TIAS 8480; 1023 UNTS 15)**
   http://www.unoosa.org/oosa/en/SORegister/regist.html

6. **1979 MOON** - **Agreement Governing the Activities of States on the Moon and Other Celestial Bodies; done at New York on 18 December 1979; entered into force on 11 July 1984. (18 ILM 1434; UN doc. A/RES/34/68 of 14 December 1979, Annex)**
   http://www.unoosa.org/oosa/en/SpaceLaw/moon.html

7. **1971 ITSO** - **Agreement Relating to the International Telecommunications Satellite Organization (ITSO) with annexes; done at Washington on 20 August 1971; entered into force on 12 February 1973. (23 UST 3813, 4091; TIAS 7532; 1220 UNTS 1220; 10 ILM 909, 946)**
The current agreement entered into force on 30 November 2004, but was provisionally applied from 18 July 2001. It may be found at–
http://www.itso.int/php_docs/tpl1_itso.php?dc=agreement
The 1971 agreement, with annexes and Operating Agreement, may be found at–
http://fletcher.tufts.edu/multi/texts/BH585.txt

8. **1976 IMSO** - **Convention on the International Mobile Satellite Organization (IMSO) with annex; done at London on 3 September 1976; entered into force on 16 July 1979. (31 UST 1, 135; TIAS 9605; 1143 UNTS 105; 15 ILM 233, 1051)**
The current agreement entered into force on 31 July 2001, but was provisionally applied from 15 April 1999. It may be found at–
http://www.imso.org/Basic_Documents.asp
The 1976 agreement, as amended 1985, with Protocol (1981), may be found at–
http://fletcher.tufts.edu/multi/texts/BH688.txt

9. **1971 INTR** - **Agreement on the Establishment of the "Intersputnik" International System and Organization of Space Communications; done at Moscow on 15 November 1971; entered into force on 12 July 1972. (862 UNTS 3)**
The Protocol on Amendments to the Agreement (4 November 2002), and the current Operating Agreement (4 February 2003) may be found at–
http://intersputnik.com/press_centre/downloads/
Not all the Parties to the 1971 Agreement have yet ratified the new Intersputnik Agreements. The 1971 agreement, with annexes, may be found at–
http://intersputnik.com/press_centre/downloads/

10. **1976 INTC** - **Agreement on Cooperation in the Exploration and Use of Outer Space for Peaceful Purposes (INTERCOSMOS); done at Moscow on 13 July 1976; entered into force on 25 March 1977. (16 ILM 1)**
http://www.jaxa.jp/library/space_law/chapter_2/index_e.html
The document served from this website omits the words "and Use" in the title and is incorrectly dated "1973" instead of "1976" in the note following the title.

11. 1976 ARBS - The Agreement of the Arab Corporation for Space Communications (ARABSAT); done at Cairo on Wednesday, 14 Rabi Al Akhar 1396 H., corresponding to April 14, 1976; entered into force on July 16, 1976. (Space Law and Related Documents, U.S. Senate, 101st Congress, 2nd Session, 395 (1990))
http://www.jaxa.jp/library/space_law/chapter_2/index_e.html

12. 1975 ESA - Convention for the Establishment of a European Space Agency (ESA) with annexes; done at Paris on 30 May 1975; entered into force on 30 October 1980. (14 ILM 864)
http://www.esa.int/convention/

13. 1982 EUTL - Convention Establishing the European Telecommunications Satellite Organization (EUTELSAT); done at Paris on 15 July 1982; entered into force on 1 September 1985. (XI Annals of Air and Space Law 416 (1986); UK Misc. No. 4, Cmnd. 9154 (1984))
The current agreement entered into force 28 November 2002; it may be found at–
http://www.assemblee-nationale.fr/11/pdf/projets/pl2173.pdf
The 1982 agreement, with annexes, may be found at–
http://www.jaxa.jp/library/space_law/chapter_3/index_e.html

14. 1983 EUMT - Convention for the Establishment of a European Organization for the Exploitation of Meteorological Satellites (EUMETSAT); done at Geneva on 24 May 1983; entered into force on 19 June 1986. (Federal Republic of Germany Bundesgesetzblatt, Jahrgang 1987, Teil 11, p. 256 (1987). This Convention has been published in the national bulletins of the ratifying States)
http://www.jaxa.jp/library/space_law/chapter_3/index_e.html

15. 1974 BRUS - Convention Relating to the Distribution of Programme-carrying Signals Transmitted by Satellite; done at Brussels on 21 May 1974; entered into force on 25 August 1979. (1144 UNTS 3)
http://www.jaxa.jp/library/space_law/chapter_2/index_e.html

Treaty Status Table as of January 1, 2008

TREATY ---- STATE	(1) 1963 NTBT	(2) 1967 OST	(3) 1968 ARRA	(4) 1972 LIAB	(5) 1975 REG	(6) 1979 MOON	(7) 1971 ITSO	(8) 1976 IMSO	(9) 1971 INTR	(10) 1976 INTC	(11) 1976 ARBS	(12) 1975 ESA	(13) 1982 EUTL	(14) 1983 EUMT	(15) 1974 BRUS
Afghanistan	R	R					R		R						
Albania													R		
Algeria	S	R		S	R		R	R			R				
Andorra													R		
Angola							R								
Antigua and Barbuda	R	R	R	R	R										
Argentina	R	R	R	R	R		R	R							S
Armenia	R						R						R		R
Australia	R	R	R	R	R	R	R	R							R
Austria	R	R	R	R	R	R	R					R	R	R	R
Azerbaijan							R		R				R		
Bahamas	R	R	R				R	R							
Bahrain							R	R			R				*R*
Bangladesh	R	R					R	R							
Barbados		R	R				R								
Belarus	R	R	R	R	R			R	R				R		
Belgium	R	R	R	R	R	R	R	R				R	R	R	S
Benin	R	R		R			R								
Bhutan	R						R								
Bolivia	R	S	S				R								
Bosnia and Herzegovina	R		R	R			R	R					R		R
Botswana	R	S	R	R			R								
Brazil	R	R	R	R	R		R	R							S
Brunei Darussalam							R	R							
Bulgaria	R	R	R	R	R		R	R	R	R			R		
Burkina Faso	S	R					R								
Burundi	S	S		S	S		S								
Cambodia				S											
Cameroon	S	S	R				R	R							
Canada	R	R	R	R	R		R	R				(1)			
Cape Verde	R						R								
Central African Republic	R	S		S			R								
Chad	R						R								
Chile	R	R	R	R	R	R	R	R							
China		R	R	R	R		R	R							
Colombia	R	S	S	S			R	R							
Comoros							R	R							
Congo			S				R								
Costa Rica	R		S	S			R	R							R
Cote d'Ivoire	R						R								S
Croatia	R						R	R					R	*R*	R
Cuba		R	R	R	R		R	R	R	R					
Cyprus	R	R	R	R	R		R	R					R		S

Note: (1) Canada has a Cooperation Agreement with ESA, but is not a Member.

Treaty Status Table as of January 1, 2008

TREATY ---- STATE	(1) 1963 NTBT	(2) 1967 OST	(3) 1968 ARRA	(4) 1972 LIAB	(5) 1975 REG	(6) 1979 MOON	(7) 1971 ITSO	(8) 1976 IMSO	(9) 1971 INTR	(10) 1976 INTC	(11) 1976 ARBS	(12) 1975 ESA	(13) 1982 EUTL	(14) 1983 EUMT	(15) 1974 BRUS
Czech Republic	R	R	R	R	R		R	R	R	R			R		
Dem. People's Rep. of Korea							R		R						
Dem. Rep. of the Congo (formerly - Zaire)	R	S	S	S			R								
Denmark	R	R	R	R	R		R	R				R	R	R	
Djibouti											R				
Dominican Republic	R	R	S	R			R								
Ecuador	R	R	R	R			R								
Egypt	R	R	R	S			R	R			R				
El Salvador	R	R	R	S			R								
Equatorial Guinea	R	R					R								
Estonia													R		
Ethiopia	S	S					R								
Fiji	R	R	R	R			R								
Finland	R	R	R	R			R	R				R	R	R	
France		R	R	R	R	S	R	R				R	R	R	S
Gabon	R		R	R			R	R							
Gambia	R	S	R	S			R								
Georgia							R		R				R		
Germany	R	R	R	R	R		R	R	R			R	R	R	R
Ghana	R	S	S	S			R	R							
Greece	R	R	R	R	R		R	R				R	R	R	R
Guatemala	R			S		S	R								
Guinea							R								
Guinea-Bissau	R	R	R				R								
Guyana		S	R												
Haiti	S	S	S	S			R								
Holy See		S					R						R		
Honduras	R	S		S			R								*R*
Hungary	R	R	R	R	R		R	R	R	R			R		
Iceland	R	R	R	S			R	R					R		
India	R	R	R	R	R	S	R	R	R						
Indonesia	R	R	R	R	R		R	R							
Iran (Islamic Republic of)	R	S	R	R	S		R	R							
Iraq	R	R	R	R			R	R			R				
Ireland	R	R	R	R			R					R	R	R	
Israel	R	R	R	R			R	R							S
Italy	R	R	R	R	R		R	R				R	R	R	R
Jamaica	R	R	S				R								R
Japan	R	R	R	R	R		R	R							
Jordan	R	S	S	S			R				R				
Kazakhstan		R	R	R	R	R	R		R				R		
Kenya	R	R		R			R	R							R
Kuwait	R	R	R	R			R	R			R				
Kyrgyzstan							R		R						
Lao People's Dem. Republic	R	R	R	R					R						

Treaty Status Table as of January 1, 2008

TREATY ---- STATE	(1) 1963 NTBT	(2) 1967 OST	(3) 1968 ARRA	(4) 1972 LIAB	(5) 1975 REG	(6) 1979 MOON	(7) 1971 ITSO	(8) 1976 IMSO	(9) 1971 INTR	(10) 1976 INTC	(11) 1976 ARBS	(12) 1975 ESA	(13) 1982 EUTL	(14) 1983 EUMT	(15) 1974 BRUS
Latvia								R					R		
Lebanon	R	R	R	S	R	R	R	R			R				S
Lesotho		S	S												
Liberia	R							R							
Libyan Arab Jamahiriya	R	R					R	R			R				
Liechtenstein				R	R		R						R		
Lithuania													R		
Luxembourg	R	*R*	S	R			R					R	R	R	
Madagascar	R	R	R				R								
Malawi	R						R								
Malaysia	R	S	S				R	R							
Maldives			R												
Mali	S	R		R			R								
Malta	R		S	R			R	R					R		
Marshall Islands								R							
Mauritania	R						R				R				
Mauritius	R	R	R				R	R							
Mexico	R	R	R	R	R	R	R	R							R
Micronesia (Fed. States of)							R								
Monaco			S				R	R					R		
Mongolia	R	R	R	R	R		R		R	R					
Montenegro	*R*		*R*	*R*	R			*R*							R
Morocco	R	R	R	R		R	R	R			R				R
Mozambique							R	R							
Myanmar	R	R	S												
Nepal	R	R	R	S			R								
Netherlands	R	R	R	R	R	R	R	R				R	R	R	
New Zealand	R	R	R	R			R	R							
Nicaragua	R	S	S	S	S		R		R						R
Niger	R	R	R	R	R		R								
Nigeria	R	R	R	R			R	R							
Norway	R	R	R	R	R		R	R				R	R	R	
Oman				S			R	R			R				
Pakistan	R	R	R	R	R	R	R	R							
Panama	R	S		R			R	R							R
Papua New Guinea	R	R	R	R			R								
Paraguay	S						R								
Peru	R	R	R	R	R	R	R	R							R
Philippines	R	S	S	S		R	R	R							
Poland	R	R	R	R	R		R	R	R	R			R		
Portugal	S	R	R				R	R				R	R	R	R
Qatar				R			R	R			R				
Republic of Korea	R	R	R	R	R		R	R							
Republic of Moldova													R		
Romania	R	R	R	R		S	R	R	R	R			R		
Russian Federation	R	R	R	R	R		R	R	R	R			R		R

Treaty Status Table as of January 1, 2008

TREATY ---- STATE	(1) 1963 NTBT	(2) 1967 OST	(3) 1968 ARRA	(4) 1972 LIAB	(5) 1975 REG	(6) 1979 MOON	(7) 1971 ITSO	(8) 1976 IMSO	(9) 1971 INTR	(10) 1976 INTC	(11) 1976 ARBS	(12) 1975 ESA	(13) 1982 EUTL	(14) 1983 EUMT	(15) 1974 BRUS
Rwanda	R	S	S	S			R								R
Samoa	R														
San Marino	R	R	R										R		
Saudi Arabia		R		R			R	R			R				
Senegal	R		S	R			R	R							S
Serbia and Montenegro	R	S	R	R	R		R	R					R		R
Seychelles	R	R	R	R	R										
Sierra Leone	R	R	S	S											
Singapore	R	R	R	R	S		R	R							R
Slovakia	R	R	R	R	R			R		R			R	*R*	
Slovenia	R		R	R									R		R
Somalia	S	S	S				R				R				
South Africa	R	R	R	S			R	R							
Spain	R	R	R	R	R		R	R				R	R	R	S
Sri Lanka	R	R		R			R	R							
St. Vincent and Grenadines		R	R	R	R										
Sudan	R						R				R				
Suriname	R														
Swaziland	R		R				R								
Sweden	R	R	R	R	R		R	R				R	R	R	
Switzerland	R	R	R	R	R		R	R				R	R	R	R
Syrian Arab Republic	R	R	R	R			R		R		R				
Tajikistan							R		R						
Tanzania							R	R							
Thailand	R	R	R				R								
The former Yugoslav Republic of Macedonia							*R*						R		R
Togo	R	R		R			R								R
Tonga	R	R	R					R							
Trinidad and Tobago	R	S		R			R								R
Tunisia	R	R	R	R			R	R			R				
Turkey	R	R	R	*R*	R		R	R					R	R	
Turkmenistan									R						
Uganda	R	R					R								
Ukraine	R	R	R	R	R			R	R				R		
United Arab Emirates		R		R	R		R	R			R				
United Kingdom	R	R	R	R	R		R	R				R	R	R	
United Rep. of Tanzania	R			S			R								
United States of America	R	R	R	R	R		R	R							R
Uruguay	R	R	R	R	R	R	R								
Uzbekistan							R								
Venezuela	R	R	S	R			R	*R*							
Viet Nam		R					R	R	R	R					*R*
Yemen	R	R	R				R		R		R				
Zambia	R	R	R	R			R								
Zimbabwe							R								

Treaty Status Table as of January 1, 2008
~ Totals for 179 States ~

TREATY ---- STATE	(1) 1963 NTBT	(2) 1967 OST	(3) 1968 ARRA	(4) 1972 LIAB	(5) 1975 REG	(6) 1979 MOON	(7) 1971 ITSO	(8) 1976 IMSO	(9) 1971 INTR	(10) 1976 INTC	(11) 1976 ARBS	(12) 1975 ESA	(13) 1982 EUTL	(14) 1983 EUMT	(15) 1974 BRUS
"Old" Ratifications	124	98	89	83	51	13	148	88	25	10	20	17	49	18	28
"New" Ratifications	1	1	1	2			1	2						2	3
Total Ratifications	**125**	**99**	**90**	**85**	**51**	**13**	**149**	**90**	**25**	**10**	**20**	**17**	**49**	**20**	**31**
"Old" Signatures	10	25	23	25	4	4	1								10
"New" Signatures															
Total Signatures	**10**	**25**	**23**	**25**	**4**	**4**	**1**								**10**

~ Changes in status since the 2007 Report ~

Agreements for which status has changed:	(1) 1963 NTBT	(2) 1967 OST	(3) 1968 ARRA	(4) 1972 LIAB			(7) 1971 ITSO	(8) 1976 IMSO						(14) 1983 EUMT	(15) 1974 BRUS
Bahrain							R	R			R				*R*
Croatia	R						R	R					R	*R*	R
Honduras	R	S		S			R								*R*
Luxembourg	R	*R*	S	R			R					R	R	R	
Montenegro	*R*		*R*	*R*	R			*R*							R
Slovakia	R	R	R	R	R			R		R			R	*R*	
The former Yugoslav Republic of Macedonia							*R*						R		R
Turkey	R	R	R	*R*	R		R	R					R	R	
Venezuela	R	R	S	R			R	*R*							
Viet Nam		R					R	R	R	R					*R*

Treaty Status Table as of January 1, 2008, for certain IGOs

TREATY *(See Note below.)* ---- ORGANIZATION	(3) 1968 ARRA	(4) 1972 LIAB	(5) 1975 REG	(6) 1979 MOON
European Space Agency (ESA)	D	D	D	
European Org. for the Exploitation of Meteorological Satellites (EUMETSAT)	D	D	D	
European Telecommunications Satellite Org. (EUTELSAT)		D		
Totals for 3 Organizations:	**2**	**3**	**2**	

Note: The Committee believes that only four of the agreements covered in the report (1968 ARRA, 1972 LIAB, 1975 REG and 1979 MOON) provide for a declaration of acceptance of rights and obligations ("D") by IGOs.

Agreements for which status has changed
1999 - 2007

TREATY ---- STATE	(1) 1963 NTBT	(2) 1967 OST	(3) 1968 ARRA	(4) 1972 LIAB	(5) 1975 REG	(6) 1979 MOON	(7) 1971 ITSO	(8) 1976 IMSO	(9) 1971 INTR	(10) 1976 INTC	(11) 1976 ARBS	(12) 1975 ESA	(13) 1982 EUTL	(14) 1983 EUMT	(15) 1974 BRUS
~ Changes in status for 1999 ~															
		(2)	(3)	(4)	(5)			(8)							(15)
Indonesia	R	S	*R*	R	R		R	R							
Jamaica	R	R	S				R								*R*
Liechtenstein				R	*R*		R						R		
Morocco	R	R	R	R		R	R	*R*			R				R
St. Vincent and Grenadines		*R*	*R*	*R*	R										
United Arab Emirates					*R*		R	R			R				
~ Changes in status for 2000 ~															
			(3)		(5)	(6)	(7)		(9)			(12)	(13)		(15)
Dem. People's Rep. of Korea							*R*		R						
India	R	R	R	R	R	S	R	R	*R*						
Kazakhstan		R	R	R	*R*	*R*	R		R				R		
Portugal	S		R				R	R				*R*	R	R	R
Rwanda	R	S	S	S			R								*R*
Spain	R	R	*R*	R	R		R	R				R	R	R	S
The former Yugoslav Republic of Macedonia													*R*		R
Ukraine	R	R	R	R	R			R	*R*				R		
~ Changes in status for 2001 ~															
		(2)		(4)			(7)								
Cuba		R	R	R	R		*R*	R	R	R					
Gambia	R	S	R	S			*R*								
Georgia							*R*		R				R		
Guinea-Bissau	R	R	R				*R*								
Portugal	S	*R*	R				R	R				R	R	R	R
United Arab Emirates		*R*		*R*	R		R	R			R				
~ Changes in status for 2002 ~															
				(4)	(5)			(8)						(14)	(15)
Comoros							R	*R*							
Greece	R	R	R	R	*R*		R	R					R	R	R
Luxembourg	R	S	S	R			R						R	*R*	
Peru	R	R	R	*R*	R	S	R	R							R
Togo	R	R		R			R								*R*
~ Changes in status for 2003 ~															
						(6)		(8)							
Belgium	R	R	R	R	R	*R*	R	R				R	R	R	S
Tonga	R	R	R					*R*							

Agreements for which status has changed
1999 - 2007

TREATY ---- STATE	(1) 1963 NTBT	(2) 1967 OST	(3) 1968 ARRA	(4) 1972 LIAB	(5) 1975 REG	(6) 1979 MOON	(7) 1971 ITSO	(8) 1976 IMSO	(9) 1971 INTR	(10) 1976 INTC	(11) 1976 ARBS	(12) 1975 ESA	(13) 1982 EUTL	(14) 1983 EUMT	(15) 1974 BRUS
~ Changes in status for 2004 ~															
									(9)			(12)			(15)
Azerbaijan							R		*R*				R		
Greece	R	R	R	R	R		R	R				*R*	R	R	R
Luxembourg	R	S	S	R			R					*R*	R	R	
Singapore	R	R	R	R	S		R	R							*R*
~ Changes in status for 2005 ~															
		(2)		(4)	(5)	(6)									
Brazil	R	R	R	R	*R*		R	R							S
Indonesia	R	*R*	R	R	R		R	R							
Italy	R	R	R	R	*R*		R	R				R	R	R	R
Nigeria	R	R	R	*R*			R	R							
Peru	R	R	R	R	R	*R*	R	R							R
~ Changes in status for 2006 ~															
			(3)		(5)	(6)									(15)
Algeria	S	R		S	*R*		R	R			R				
Lebanon	R	R	R	S	*R*	*R*	R	R			R				S
Montenegro					*R*										*R*
Turkey	R	R	*R*		*R*		R	R					R	R	
~ Changes in status for 2007 ~															
	(1)	(2)	(3)	(4)			(7)	(8)						(14)	(15)
Bahrain							R	R			R				*R*
Croatia	R						R	R					R	*R*	R
Honduras	R	S		S			R								*R*
Luxembourg	R	*R*	S	R			R					R	R	R	
Montenegro	*R*		*R*	*R*	R			*R*							R
Slovakia	R	R	R	R	R			R		R			R	*R*	
The former Yugoslav Republic of Macedonia							*R*						R		R
Turkey	R	R	R	*R*	R		R	R					R	R	
Venezuela	R	R	S	R			R	*R*							
Viet Nam		R					R	R	R	R					*R*

Agreements for which status has changed
Totals for 1999 - 2007

TREATY ---- **STATE**	**(1) 1963 NTBT**	**(2) 1967 OST**	**(3) 1968 ARRA**	**(4) 1972 LIAB**	**(5) 1975 REG**	**(6) 1979 MOON**	**(7) 1971 ITSO**	**(8) 1976 IMSO**	**(9) 1971 INTR**	**(10) 1976 INTC**	**(11) 1976 ARBS**	**(12) 1975 ESA**	**(13) 1982 EUTL**	**(14) 1983 EUMT**	**(15) 1974 BRUS**
Ratifications															
Ratifications at 1-1-99	124	94	85	79	41	9	143	85	22	10	20	14	48	17	23
Ratifications added	1	5	5	6	10	4	6	5	3			3	1	3	8
Total Ratifications	**125**	**99**	**90**	**85**	**51**	**13**	**149**	**90**	**25**	**10**	**20**	**17**	**49**	**20**	**31**
Signatures															
Signatures at 1-1-99	10	27	24	26	4	5	1								10
Signatures added															
Signatures completed by ratification *(See Notes below)*		-2	-1	-1		-1									
Total Signatures	**10**	**25**	**23**	**25**	**4**	**4**	**1**								**10**
Notes: **STATE** and Year Action Reported															
Indonesia		2005													
Luxembourg		2007													
Peru				2002		2005									
Turkey			2006												

Declarations of Principles Relating to Activities in Outer Space adopted by the General Assembly of the United Nations

1. **Declaration of Legal Principles Governing the Activities of States in the Exploration and Use of Outer Space (General Assembly resolution 1962 (XVIII)). Adopted without vote on 13 December 1963.**
http://www.unoosa.org/oosa/SpaceLaw/gares/html/gares_18_1962.html

2. **Principles Governing the Use by States of Artificial Earth Satellites for International Direct Television Broadcasting (General Assembly resolution 37/92).** http://www.un.org/documents/ga/res/37/a37r092.htm
Adopted by 107 votes to 13, with 13 abstentions, on 10 December 1982 as follows (voting results are reproduced from UN document A/37/PV.100 of 17 December 1982):

> **In favour: Afghanistan, Algeria, Argentina, Bahrain, Bangladesh, Barbados, Benin, Bhutan, Bolivia, Botswana, Brazil, Bulgaria, Burma, Botswana, Burundi, Byelorussian Soviet Socialist Republic, Central African Republic, Chad, Chili, China, Colombia, Comoros, Congo, Cuba, Cyprus, Czechoslovakia, Democratic Kampuchea, Democratic Yemen, Djibouti, Dominican Republic, Ecuador, Egypt, El Salvador, Ethiopia, Fiji, Gabon, Gambia, German Democratic Republic, Ghana, Guyana, Haiti, Honduras, Hungary, India, Indonesia, Iran (Islamic Republic of), Iraq, Jamaica, Jordan, Kenya, Kuwait, Lao People's Democratic Republic, Liberia, Libyan Arab Jamahiriya, Madagascar, Malaysia, Maldives, Mali, Malta, Mauritania, Mauritius, Mexico, Mongolia, Mozambique, Nepal, Nicaragua, Niger, Nigeria, Oman, Pakistan, Panama, Papua New Guinea, Peru, Philippines, Poland, Qatar, Romania, Rwanda, Sao Tome and Principe, Saudi Arabia, Senegal, Sierra Leone, Singapore, Solomon Islands, Somalia, Sri Lanka, Sudan, Suriname, Syrian Arab Republic, Thailand, Togo, Trinidad and Tobago, Tunisia, Turkey, Uganda, Ukrainian Soviet Socialist Republic, Union of Soviet Socialist Republics, United Arab Emirates, United Republic of Cameroon, United Republic of Tanzania, Upper Volta, Uruguay, Venezuela, Viet Nam, Yemen, Yugoslavia, Zaire, Zambia.**
>
> **Against: Belgium, Denmark, Germany, Federal Republic of, Iceland, Israel, Italy, Japan, Luxembourg, Netherlands, Norway, Spain, United Kingdom of Great Britain and Northern Ireland, United States of America.**
>
> **Abstaining: Australia, Austria, Canada, Finland, France, Greece, Ireland, Lebanon, Malawi, Morocco, New Zealand, Portugal, Sweden.**

3. **Principles Relating to Remote Sensing of the Earth from Outer Space (General Assembly resolution 41/65). Adopted without vote on 3 December 1986.** http://www.un.org/documents/ga/res/41/a41r065.htm
The document served from this website omits the word "Outer" in the title.

4. **Principles Relevant to the Use of Nuclear Power Sources in Outer Space (General Assembly resolution 47/68). Adopted without vote on 14 December 1992.** http://www.un.org/documents/ga/res/47/a47r068.htm

5. **Declaration on International Cooperation in the Exploration and Use of Outer Space for the Benefit and in the Interests of All States, Taking into Particular Account the Needs of Developing Countries (General Assembly resolution 51/122). Adopted without vote on 13 December 1996.**
http://www.un.org/documents/ga/res/51/a51r122.htm

Note: A copy of this report can be downloaded from the Institute's web site–
http://www.iislweb.org/committee.html

2008/09 BOARD OF DIRECTORS ELECTIONS AND NEW MEMBERS

1. Board of Directors Elections

The result of the Board elections in Glasgow was as follows:

Prof. Dr. Elisabeth Back Impallomeni, Prof. Dr. Stephan Hobe, Prof. Dr. Vladimir Kopal, Prof. Jose Monserrat, Filho and Prof. Dr. Frans von der Dunk were all re-elected as per the advice of the Nominating Committee, in light of their commitment to continue serving the Institute, its missions and its members.
Prof. Kopal was also re-elected as Vice-President, and Prof. Hobe was elected as the new Treasurer, replacing Prof. von der Dunk.

2. Members Elections

The following institutional and individual members were elected in the fall of 2008 and spring of 2009:

INSTITUTIONAL:

Romanian Space Agency
M.-I. Piso, V. Pop, I. Simion
Str. Mendeleev 21-25, sector 1
Bucharest 010362, Romania
marius-ioan.piso@rosa.ro

SES ASTRA
J. Purvis, R. Roelandt
Chateau de Betzdorf
L-6815 Luxembourg
anja.palm@ses.com

INDIVIDUAL:

CROWTHER, Prof. Richard
STFC
Harwell Science and Innovation Campus
Chilton, Oxfordshire, OX11 0QX, UK
Richard.crowther@stfc.ac.uk

DETTER DE LUPIS FRANKOPAN, Prof. Dr. Countess Ingrid
Château de Verrey
F-21690 Verrey, France
idf@frankopan.net

JIMENEZ MONROY, Ms. Cynthia
Frankenslag 350
The Hague, 2582 JB, The Netherlands
jm.cynthia@gmail.com

JOLLY, Ms. Claire
98 rue Bobillot
75013 Paris, France
Claire.jolly@oecd.org

KONAR-STEENBERG, Prof. Mehmet K.
William Mitchell College of Law
875 Summit Avenue
St. Paul, MN 55105, USA
mehmet.konar-steenberg@wmitchell.edu

KOPLOW, Mr. David
600 New Jersey Avenue, NW
Washington, DC, 20001, USA
koplow@law.georgetown.edu

LANGSTON, Ms. Sara
1410 N. Scott Street #265
Arlington, VA 22209, USA
langstonsara@gmail.com

MANTL, Dr. Leopold W.
12, Rue des Boers
1040 Brussels, Belgium
Leopold.MANTL@ec.europa.eu

MARBOE, Dr. Irmgard
Faculty of Law, University of Vienna
Schottenbastei 10-16
A-1010 Wien, Austria
irmgard.marboe@univie.ac.at

MASTROIANNI, Mr. Giovanni
Street S. P. 336 n.15
81013 Caiazzo, Italy
giovanni.mastroianni@libero.it

McCORMICK, Dr. Patricia
585 Manoogian Hall
Detroit, MI 48201, USA
pmccormak@gmail.com

RENDLEMAN, Mr. James
975 War Eagle Drive N
Colorado Springs, CO 80919, USA
napatarheel@hotmail.com

RICHER, Mr. Joseph
206 North Patrick Street
Alexandria, Virginia, 22314, USA
joe.richer@gmail.com

RICHTER, Mr. Werner
17 Avenue de Saxe
75007 Paris, France
Werner.richter@dlr.de

SCHOENMAKER, Ms. Annelie
13 rue de la Louviere
78120 Rambouillet, France
annelie.schoenmaker@esa.int

VICENTE, Captain Esteban Pacha
99 City Road
London, ECIY IAX, UK
Esteban_pacha@imso.org

WICHT, Mr. Anthony
20 Coronga Crescent
Killara NSW 2071, Australia
anthony.wicht@gmail.com

WILSON, Mr. Ernst
2525 B Peartree Road
Elizabeth City, NC 27909, USA
ewilson@mail.ecsu.edu

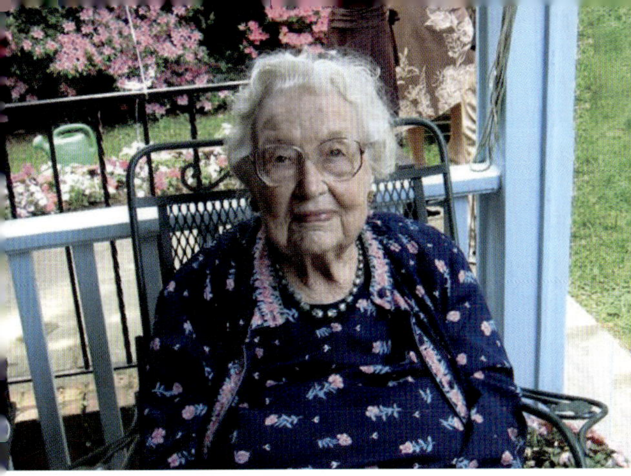

Eilene Marie Galloway
1906–2009

Eilene Marie Galloway passed away on 2 May 2009, two days before her 103rd birthday. She was instrumental in creating the International Institute of Space Law (IISL) and was its Vice President from 1967 to 1979, after which she served as its Honorary Director.

She received the IISL's Andrew Haley Gold Medal in 1968, and the Lifetime Achievement Award in 1990.

In her name, the IISL created the *Eilene Galloway Award for Best Written Brief* in the Manfred Lachs space law moot court competition from 2000 onwards, and since 2006, the *Eilene Galloway Symposium on Critical Issues in Space Law* has been organized in Washington, D.C.

The Institute availed itself of her valuable advice until her passing. As IISL Emeritus President Prof. Isabella Diederiks-Verschoor wrote in her tribute to Eilene Galloway in the 2006 IISL Proceedings: "She has lived her life of distinguished service to the U.S. and to the Space World and has been an inspiration to us all, and above all a reliable friend."

Below follow a biography of Eilene Galloway and a compilation of tributes written by members of the space law community on the IISL website, to honour this grand pioneer of space law who has been a true inspiration and friend to so many of us.

BIOGRAPHY

Eilene Galloway has been an influential force in the development of domestic and international space policy and law since the birth of the Space Age. When *Sputnik* was launched on October 4, 1957, she was working as a Senior Specialist in International Relations for the Legislative Reference Service (later the Congressional Research Service) at the Library of Congress. (Her career at the Library began in 1941 as an abstracts editor. In 1951, she became an analyst in national defense; by 1957 she had risen to specialist in national defense; and in 1966, she became a senior specialist, the highest rank possible in the Service). She was immediately seconded by Senate Majority Leader Lyndon Johnson and Senator Richard Russell (chairman of the Armed Services Committee) to work with them in formulating America's response to this dramatic event. From that day, she has worked tirelessly with the United States Congress (both the House and Senate) and as a consultant to executive branch agencies (NASA, FCC, State Department) in assessing alternatives and developing solutions for U.S. and international space policy and law for the exploration and utilization of outer space. Among her many achievements was helping to write the National Aeronautics and Space Act, which created NASA and remains today as the foundation for U.S. space policy.

In 1958, Sen. Johnson sent Eilene to represent the United States at a meeting of the International Court of Justice in The Hague, where she gave a speech on "The Community of Law and Science." She was also helpful in establishing the United Nations Committee on the Peaceful Uses of Outer Space (UNCOPUOS), where she represented the United States at many meetings. She was Editor of *Space Law Senate Symposium* (577 pages), 1958. The United Nations invited her to speak at its 1995 Congress on Public International Law about "Space Law Role of the United Nations: Organization and Management."

After *Sputnik*, Rep. John W. McCormack, Majority Leader of the House of Representatives, asked Eilene for advice on whether the House should have a Space Committee. She recommended establishing one, and he became chairman of the Select Committee on Astronautics and Space Exploration. The House Committee was first to complete space recommendations, which would include the creation of a national space agency. When President Eisenhower sent Rep. McCormack the text of NASA's organization, McCormack again asked Eilene for advice. She proposed that NASA be an Administration rather than an Agency. Mr. McCormack made the change immediately, enabling the new National Aeronautics and Space Administration to plan and coordinate across federal agencies the variety of activities involved in the development and uses of outer space.

Dr. Galloway was born Eilene Marie Slack in Kansas City, Missouri, on May 4, 1906. She was the only child of Joseph Locke Slack and Lottie Rose (Harris) Slack. Her father joined the Marines in 1915, and her mother raised her alone. In 1923, Eilene graduated from Westport High School in Kansas City, Missouri, where she was twice captain of the debate team. She received a full scholarship to Washington University in St. Louis, attending from 1923 to 1925. In 1928, she received her bachelor of arts degree in political science from Swarthmore College. She received honorary doctorates from Swarthmore (1992) and Lake Forest College (1990). She was a member of Phi Beta Kappa and Kappa Alpha Theta. On December 24, 1924, Eilene married George Barnes Galloway, who would later become a foremost expert on the workings of Congress. They had two sons, David and Jonathan. The Galloways moved to the District of Columbia in 1931.

During the Great Depression, Eilene became assistant to Aubrey Williams, assistant administrator of the Federal Emergency Relief Administration (FERA). Overseeing adult education programs nationwide, she also developed specific programs for the District of Columbia, primarily to address adult illiteracy (1934-5). She met with Eugene Meyer, publisher of the *Washington Post*, to ask that the paper distribute copies of a guide to adult education that she had compiled for FERA (*The Emergency Education Program and College Student Aid Program of the Federal Emergency Relief Administration*). Meyer initially demurred, responding that he would need some indication of demand for the publication. Eilene countered by securing orders for more than 1,600 copies, and Meyer complied with her request. The guide was so popular that it was reprinted, and ultimately the *Post* distributed some 175,000 copies.

Dr. Galloway was a founding member of the International Institute of Space Law (IISL). Up to her death, she continued to study and write about space law and policy, authoring numerous papers, speeches, and opinion pieces. She opposed the weaponization of space and consistently and persuasively argued that the 1967 Outer Space Treaty has enabled the world community to maintain peace in space and should be preserved and upheld. She constantly affirmed the core principle of the 1967 UN Treaty on Principles Governing the Activities of States in the Exploration and Use of Outer Space, Including the Moon and Other Celestial Bodies (the Outer Space Treaty), stated in Article II: "Outer Space, including the moon and other celestial bodies, is not subject to national appropriation by claim of sovereignty, by means of use or occupation, or by any other means."

Dr. Galloway was the recipient of many awards and honors. She was the first woman elected Honorary Fellow of the American Institute of Aeronautics and Astronautics (2006), the organization's highest honor. She also was a fellow of the American Astronautical Society (1996), an honorary director of the International Institute of Space Law, and a Trustee Emeritus of the International Academy of Astronautics. She was the first recipient of a Lifetime Achievement Award from Women in Aerospace (1987) and was honored as well with the NASA Public Service Award and Gold Medal (1984), granted for her achievements in advising Congress about outer space and her service to international organizations including the UN regarding international space law.

Eilene was fearless. When she was six, her mother enrolled her in the Georgie Brown Dramatic School in Kansas City, Missouri, where she learned to be confident in front of audiences. She was one of a group of girls from the School who performed a patriotic song and dance for Theodore Roosevelt at the Muhlbach Hotel in Kansas City, when he was campaigning for the U.S. presidency in 1912. Throughout her life, she was not afraid to take on any task that came her way. She always found a way to turn crises and problems into opportunities. In times of stress, she turned to her favorite book, the *Meditations of Marcus Aurelius*, for guidance.

Eilene was an expert story-teller, sharing with wit and a wonderful sense of humor tales of her experiences as a child, in the Great Depression, her work with powerful members of Congress, her travels around the world, and especially her work as a space law and policy analyst. She had a wonderful sense of humor. Once she arrived in Germany to begin a tour of countries before meeting a deadline in England. At the U.S. Army Office there, a colonel pulled down a large wall map of Europe and told her, "You cannot visit that many countries before your deadline," Eilene pulled out her pocket map and said, "The reason you think I cannot do it is because you have such a large map. I have a little map, which makes it possible for me to meet my deadline."

TRIBUTES TO EILENE GALLOWAY

In my position as Chairman of the UN Committee on the Peaceful Uses of Outer Space, COPUOS and on behalf of all states members and permanent observers, I would like to express our condolences to Professor Eilene Galloway´s family. I was fortunate to meet her for a very brief moment just enough to see in her brilliant eyes the soul of a profound loving person to whom we have an immense debt of gratitude.

—***Ambassador Ciro Arévalo-Yepes***, *UN COPUOS Chairman*

During my whole tenure of the UN Committee on the Peaceful Uses of Outer Space (and of course thereafter) Eilene Galloway was a constant source of inspiration and motivation and was certainly always first amongst those who guided and helped our work. It was a particular pleasure for myself and for the members of the Committee when she participated in our work. Her devotion to the idea of the rule of law in Outer Space is at the root of many of the advances that space law has seen over the years and she will also be an important and helpful point of reference for future space legislators.

—***Peter Jankowitsch***, *Past Chair, COPUOS*

Dr. Galloway's contribution to international law, and international space law in particular, has been a remarkable achievement. Her fruitful and dedicated work in this area, as well as training and encouraging of young professionals, earned her world-wide recognition and respect from specialists all over the world. Her presence will be sorely missed.

—***Mazlan Othman***, *Director, Office for Outer Space Affairs (OOSA), United Nations*

Eilene was indeed one of the great International space lawyers. I had the privilege, although on limited opportunities, to listen to her present on various topics while I was at the United Nations Office for Outer Space Affairs. However, it was during group discussion periods that her profound knowledge of the geopolitical situations and compromises that led the international space community to agreements on key aspects of the international space legal regime came through best. It was also in those discussions that her steadfast internationalism came through. Although sad that Eilene has now left us, we are very fortunate to inherit the fruit of her work of many decades.

—***Sergio Camacho***, *Former Director, United Nations Office for Outer Space Affairs*

Eilene Galloway was an outstanding expert in space law. Moreover she was a very close friend during many years. She will be missed badly, but her contributions to space law remain.

—***Isabella Diederiks-Verschoor***, *IISL President Emeritus*

It is with a great sense of personal loss that I learned of the demise of a giant in the field of space law and policy, Dr. Eilene Marie Galloway. I have had the great honor and privilege of knowing and working with her closely, since mid 1960s, when she first began to be involved in the work of the United Nations Committee on the Peaceful Uses of Outer Space. I believe, she had traveled to the United Nations on a previous occasion, when President Eisenhower proposed to the UN General Assembly, the establishment of a United Nations Agency for space activities. Naturally, the Soviet Union, which had just launched the first Sputnik opposed the proposal at the time, and ironically, much later, during the UNISPACE discussions, the Russians made the same proposal, only to be opposed by the United States. Mrs. Galloway was a true internationalist and her major preoccupation was to ensure that space and space resources are used for peaceful activities. It was her abiding belief that in order to achieve that objective, strong international cooperation was indispensable. In pursuit of her interest in international cooperation she authored, in the 1960s, the first two senate publications relating to space law and policy. They were of great assistance to diplomats and scholars involved in space treaty making and were the documents most consulted during those negotiations. In recognition of her commitment to international cooperation and her knowledge of space law and policy, she was highly respected by diplomats negotiating in the COPUOS. In the early years at the height of the cold war, when US and Soviet delegates hardly exchanged courtesies, she was treated with great respect, and more importantly, with genuine friendship by the Soviet delegates. The world has now lost a champion for peaceful uses and international cooperation, themes that were highlighted in all her writings.

She was a pioneer in the field of space law and policy, a founder of the International Institute of Space Law in 1957, when first steps were taken to establish it. She shepherded it through several decades as a member of the Board of Directors, Vice President and Honorary Director, and remained as the primary spirit of the organization till her death. She never failed to point out the importance of building the Institute's work on solid scientific and technical basis. More often than not this was a theme that laced her writings. It is then no surprise that she was a very active member of the IAF and equally active academician of IAA. For many years, she served as chairman of the Social Sciences Section of the Academy and member of its Board of Trustees, always promoting scientific and legal interaction between the Academy and the Institute.

For some reason, I was fortunate enough to be taken under her wings in my early career and when she relinquished the chair of the Social Sciences Section of IAA, she chose me to be recommended to succeed her, and later again, to take her place in the IAA Board of Trustees. In 1993, following the death of Judge Manfred Lachs, who was then the President of IISL, it was Dr. Galloway who approached me and said that I should succeed Judge Lachs. It was her advice that prompted me to let my name be placed for election, though I was discouraged by others with advice not to do so in my own interest, as I was running into a very difficult campaign. In all my work relating to the Institute and the Academy, I never failed to consult her, and she provided me with her wise and valuable advice and stood behind me in taking the important work of the Institute and the Academy forward.

The international space community has lost a pioneer and a giant in the field of space and I have lost a valuable mentor and friend whose wisdom, warmth, and humor I will always miss.

—***Nandi Jasentuliyana***, *IISL President Emeritus, and former Deputy Director-General and Director Office of Outer Space Affairs, United Nations*

My fondest memory of Eilene is when I met her at my very first IISL Colloquium in Innsbruck in 1986. We, a bunch of young space lawyers, were so happy and proud she would have coffee with us; we could hardly believe it! I can still remember how excited we were, and a bit nervous too, to meet all those great space lawyers whose books we had been studying to prepare our first paper. Eilene was encouraging and interested, and really made us feel good. We had our picture taken, and some time later she mailed me a copy, I was amazed she did that. When I became secretary of the IISL a few years later, she was immediately in favour of my becoming Secretary with a big "S" (i.e. member of the Board of Directors) a year later, in Montreal (1991). Then in 2007, when Nandi Jasentuliyana decided to step down and suggested that I become the next IISL President, Eilene was yet again fully supportive and helped pave the way for my election. Her beautifully handwritten letters filled with crisp analysis will remain with me. So will her annual Christmas cards with the "space window" of Washington National Cathedral that she always wrote, until this last Christmas. That first encounter in the hills of Austria was the beginning of a beautiful friendship; I cherish her support and guidance during all these years and will do my best to live up to her standards and expectations.

—***Tanja Masson-Zwaan***, *IISL President*

Eilene Galloway had a significant impact on the evolution of the International Academy of Astronautics (IAA) the past 40 years. This time span represents almost the entire existence of this organization as the IAA is about to celebrate its 50th birthday in 2010.

Eilene was first elected as a Corresponding Member of the Academy in 1968 and full Member in 1977. Her election was in Section 3, covering Life Sciences and Social Sciences that also included space law. There was a restricted number of Academicians at that time: in 1968 the total of Corresponding Members was 56 for the entire world, and in 1977 the total of full Members was only 63, in her section.

As a result of all Eilene's key support to the IAA, she was nominated as a Trustee (1982) and Chair of the new Section 4 Social Sciences (1983 to 1987). Under the new statutes she achieved one of the highest scores ever, showing an international consensus on her skill and experience. Upon her request, as she was just over 80 years old, she asked to be released from the task of the Section Chair, but agreed to stay as a Trustee for a few more terms. She was a very special person and despite her age she stayed as a Trustee for two more terms until she was 87, making this contribution very unique in the Academy.

In 1986, Eilene was granted the IAA Theodore von Kármán Award, the most prestigious award of the Academy. In 1999, she received the IAA Social Sciences Award to recognize her contribution to the building of that particular section (Section 4).

She was one of the founders of the IAA-IISL Scientific-Legal Roundtable which was the very first place for discussions between scientists, engineers and lawyers. This was very important to her to create a structure for a permanent dialogue between the two communities. Next year, we will celebrate the 25th Roundtable.

We will miss her greatly and she was a unique model for those who feel old after 45! She proved that pro-active intellect is the only way to preserve individuals from degeneration. In addition, she was a sweet lady with both immense culture and a great sense of humor. It has been a privilege to work with her and I will miss her friendship.

—***Dr. Jean-Michel Contant***, *Secretary General, International Academy of Astronautics*

It is with profound grief that we have learned of the death of Mrs. Eilene Galloway. Through her contributions to space law and policy and her efforts to promote international cooperation among space-faring countries, she has helped shape the first half century of space age. Mrs. Galloway has also enriched the International Astronautical Federation and, in so doing, helped establish the International Institute of Space Law and support the International Academy of Astronautics. I would like to kindly ask you to convey our condolences to her family and to the space law community.

—***Prof. Dr. Berndt Feuerbacher***, *President, International Astronautical Federation*

I would like to express my sincere condolences, on behalf of the entire COSPAR community, to Dr. Galloway's family and to the fraternity of space lawyers represented in the IISL for the recent passing of Eilene Galloway.

The importance of her contribution to the international community of lawyers, engineers, and scientists working in the domain of space is well known. She and her work are very highly respected throughout this community and profession, and her great contribution is part of the legacy of the first 50 years of the space age.

—***Roger Bonnet***, *COSPAR President*

As a very long-time friend of Eilene these words are offered in her memory. Her creativity and steadiness of purpose in the entire area of outer space activity made her a world figure. Her good and faithful friends extend throughout the universe based on the enormous respect held for her and her significant contributions. Eilene was a living proof that a superbly qualified person can have a meaningful impact on a national government, on its very highest leaders, and on international organizations and institutions. Her's was a life well lived. It is with sadness that we must acknowledge she is gone—but not forgotten. For me I have lost a very special friend.

—***Carl Q. Christol***

Dear Jonathan, This is to express all my sadness in learning that your dear and beloved Mother faded away at the eve of her birthday. I remember you always at her side whenever she attended our IISL meetings and you taught us the love and respect we all owe to our parents for their lifetime. Her wisdom and the active contribution she gave to the progressive development of space law ever since its start will remain a cornerstone in the conquest of outer space by humankind.

—***Prof. Dr. Elisabeth Back Impallomeni***

I was honored to have known and admired Eilene Galloway. Our destinies coincided for over half a century during which I admired her dedication to space law and her exceptional contributions to space legislation and research. In 1976, I was appointed Director of the Institute of Air and Space Law by McGill University and I consider the foundation of a special publication to be named the *Annals of Air and Space Law (Annals)*. Many times when I was at a crossroad, I would meet Eilene at her favourite Cosmos Club to have her precious advice and enlightenment. I considered her the only person who could give me a decisive answer as to the potential success of such an endeavour as the *Annals*. Not only she encouraged me to publish the *Annals* if possible the same year but she sent me an excellent leading article of the first issue. Later she continued enriching many other volumes with her visionary articles (Vol. I, III, IV, V, VII and IX). Years later, in 1996, I received Social Sciences Book Award from the International Academy of Astronautics "for outstanding achievement as evidenced by the publication of the *Annals of Air and Space Law*". It was more than well deserved that, on the occasion of Eilene's 100th birthday, the Institute of Air and Space Law dedicate a volume to her (2006 – Vol. XXXI). Eilene will be sorely missed but her ideas will continue to be carried out by people she so proudly inspired. Rest in peace, Eilene.

—***Nicolas Mateesco Matte***

Eilene Galloway was a great jurist and an outstanding person. Her contribution to the establishment and development of the law of outer space cannot be overstated. I had the privilege of knowing Eilene for over 20 years and always admired her wisdom, tremendous expertise and cheerful personality, including a great sense of humour. I remember several years ago, when she was in her 90s, in response to my question as to how she was feeling, Eilene said: "Well, I have some pain in my knees, but below and above them I feel fine." Eilene Galloway will be sorely missed.

—***Andrei Terekhov***, *Executive Director, Office of Administration of Justice, United Nations*

I was sorry to learn about the decease of Eilene. I respect her many achievements. It is an honour to count myself as her friend and colleague.

—***Lubos Perek***, *Czech Republic*

I express my deep condolences to Jonathan and the entire Galloway family on the recent passing of Dr. Eilene Galloway. I met her, for the first time, in 1979 at the Institute of Air and Space Law in Montreal, where I was doing my doctoral studies in space law. A few days after that, as promised, she sent me a copy of the US Senate Hearings on the Outer Space Treaty, which is an indispensable source for a proper space law research. Since then Eilene has been my mentor. It was a great pleasure for us in the Institute of Air and Space Law to dedicate to her the recent issue of our Annals of Air and Space Law as well as our 2006 International and Interdisciplinary Workshop on Policy and Law Relating to Outer Space Resources. I met Eilene, the last time, in December 2008 when I visited her at her residence in Washington, D.C., and had long and very interesting talk about our favourite passion; i.e. space law. I will always cherish the great memories of my meetings and conversations over the last thirty years with Eilene, who had since 1957 tirelessly advocated and promoted actively inter-national cooperation and peaceful uses of outer space.

—***Ram Jakhu***, *the Institute of Air and Space Law, Montreal, Canada*

It is with my deepest concern, sadness, and condolence that I have learned about the death of Eilene Galloway. She was one of the founders of the International Institute of Space Law, back in 1961. But above that, she was editor, I may say creator, of the U.S. Senate publication "Legal Problems of Space Exploration". This was an opus that really opened the door for international Space Law studies. Everyone should have read—or read even now—her excellent introduction on p. XIff of that book. I have sent my personal expression of condolence to her son, the distinguished Professor and Space Lawyer Jonathan Galloway. Eilene Galloway will always be remembered by all of us Space Lawyers, and her colleagues and friends.

—***Ernst Fasan***, *Director, HC of IISL*

I first met Eilene Galloway in 1989 at the 32nd IISL Colloquium in Malaga-Torremolinos. I already knew her writing, but to meet one so influential in the development of space law was a privilege. It was also interesting for her mind was so active, and her sense of humour infectious. She was also encouraging to a re-entrant into the space community, encouragement which continued down through the years including lunch at the Cosmos Club or at her home when I visited Washington. Like so many I am glad to have known her.

—***Francis Lyall***, *Scotland*

With the passing away of Eilene Galloway space law has lost its Grandmother, as is very clear from all the testimonies to her key role and enduring enthusiasm in promoting the cause of space law, and through it, of the peaceful development of space. Who does not remember the accident of her finger being crushed to the bone by accident during an IAC, where she insisted on being back at the IISL colloquium as soon as the hospital had managed to get her hand bandaged appropriately—she was well over ninety years old by then! In addition, her support of young and not-so-young space lawyers has been evidenced amply in those testimonies, as was of course her sharp intelligence—lest anybody could have been fooled by the knitting needles she sometimes brought to space law meetings. One side of Eilene that may not have been brought to the fore in all of the above that much is her unique wittiness, combining a sharpness of mind with a modest and warm understanding of humanity also at the personal level. I had the privilege to sit next to her at both gala dinners during what must probably have been the last occasion Eilene was able (see the famous statement on the knee quoted by Andrej Terekhov) to join us for an IAC and IISL Colloquium. In Houston in 2002, she told me a number of wonderful stories, like the one in which she was made, in recognition of her many achievements for the U.S. space endeavour, Admiral of the Navy—of Texas, which did not have its own real navy since many years, so that indeed it was an honorary title only, which she told me with a twinkle in her eye she greatly regretted. But the best one, the one that probably captures most of her personality, was the following. In the early days of developing a U.S. space policy Eilene, as the one given the task to work on such a policy, was summoned by then-Senate majority leader Lyndon Johnson to fly to the UK for the purpose of briefing that closest of U.S. allies on these developments, as these had after all many military- and security-related aspects to them. In fact, a special plane was chartered for picking her up and flying her immediately across the Atlantic for the purpose. Upon the late night arrival in the UK, immediately after the plane had come to a standstill on the tarmac the pilot came up to Eilene and said: "Ma'am, we have received an invitation for a late dinner by our hosts, would you please join us?" Eilene begged to be excused on account of the long flight, and indicated she preferred to stay on the plane, get some proper sleep and prepare for next morning's sessions. The pilot, however, then started to shuffle somewhat uneasily, and when Eilene asked what the matter was, explained in an embarrassed tone: "Well, ma'am, you see, by regulation it is only allowed to charter a special plane like this one for someone with the rank of a four-star general, so Lyndon Johnson just had you promoted to a four-star general for the purpose. Which makes you the highest officer on board, which in turn means that if you don't leave the plane, nobody gets to leave the plane..." Upon which Eilene, whom of course could not find it in her heart to refuse the crew their chance to see the London nightlife, graciously agreed to join for dinner. I feel it is in this spirit that we should remember Eilene: a four-star general and Admiral of the Texan Navy in terms of intellect and perspicacity, needle-knitting a safety-net of legal rules and obligations for peaceful space exploration and use, and at the same time endowed with the modest and warm humanity to see the humour and relativity of it all—and the graciousness to give the crew their due.

—***Frans von der Dunk**, Harvey and Susan Perlman Alumni/Othmer Professor of Space Law, University of Nebraska-Lincoln*

When I first met Eilene, just after completing my first law degree at the University of Buenos Aires, I was thrilled. I had heard so much about her from Ambassador Aldo A. Cocca and Professor Manuel A. Ferrer when I was a student in Argentina. In later years, back in England, I had the pride and joy of corresponding with Eilene—never missing a Christmas greeting from her—and of working together at a number of international meetings on the law of outer space. Buenos Aires, Mar del Plata, Vienna, Brussels, Munich, Brighton are so full of nostalgic memories linked to Eilene. Over the years, and particularly in the framework of IISL Colloquia, we learnt many lessons from her, not only on space matters but on many other aspects of academic life. And life itself.

Her very simplicity was impressive. I shall ever remember, in more recent years, a summer afternoon in Vienna during UNISPACE III having tea with Eilene and my daughter Viviane—who also had the privilege of meeting her. All three of us were enjoying tea with the famous viennese pastries and chatting about fashion trends, a favourite topic we later resumed at the Cologne Golf Club on the occasion of a Colloquium organised by our friend Karl H. Böckstiegel in May 2001. That evening Eilene was wearing a turquoise dress with matching bracelet and earrings. She looked splendid and sparkled as we entered deeper analysis of the world of fashion.

These mementos are meant as a modest tribute to a very admirable lady. We shall miss her for ever.

—***Maureen Williams***

I first met Eilene at the Innsbruck IISL Colloquium in 1986, which I was able to attend quite by chance. It was also the occasion of my first presentation of a paper on space law and telecommunications satellites, and boy! was I nervous, especially because the only people that had a copy of my paper were Eilene and Dr. Isabella Diederiks-Verschoor. They kindly let me have my paper, so that I could make my presentation. (I returned it to them at the end of the Colloquium). I moved to Washington, D.C. a few months later, and lived not far from Eilene, but I was in such awe of this wonderful lady that I didn't dare call her. A couple of years later, I finally got the courage to call her, and made an appointment to go see her shortly before the World Space Congress in 1992. She was so gracious and gentle, that I felt badly for myself, and silly, for not having called her before. After that first invitation to tea at her home, we became friends, and would have tea and even dinner occasionally. One day she called me, saying that she was "so agitated" (one of her pet words) and had to answer someone's letter, but didn't want to wait for it to go out by regular mail. I suggested that she get a FAX machine, an idea that intrigued her. I told her I could take her to look at them, to see if she would like to buy one. So off we went to the Office Depot near her house, and 10 minutes later, we came out with a new FAX machine for Eilene. I installed it for her, and she was able to respond to the letter, and get over her agitation in no time... I have so many wonderful memories of Eilene, and of her gentle but acute sense of humour! One time, she showed me a letter an organization had sent her, asking her if she would be its Secretary. She declined the invitation, replying that they should get someone younger for the position, someone in their early 80s... (Eilene was about 87 at the time). Like so many people, I am deeply indebted to

Eilene for her helping me grow in knowledge of space law; she was always willing to discuss events or issues that I may have been writing about, and share her learned opinions on them. I could never repay her willingness to mentor me, and so many others, but if we, in turn, can be as open and helpful to younger space lawyers, perhaps this is the best way of repaying Eilene, and ensuring that her legacy continues. I visited Eilene in March, and took her some yellow flowers. She often mentioned that every house should have a room painted in yellow, so that on gray days, it would seem that the sun was still shining. She loved the yellow flowers, and had them placed where she could see them from her bed. We spoke about space law, about space debris, about some of her accomplishments, as well as more mundane matters. What a privilege to have been able to spend two hours with her! I kissed her good-bye, and inside me, I said farewell to the matriarch of space law. She is now with other celestial bodies, in that infinite cosmos that she cared so much about. God bless Eilene.

—Sylvia Ospina

Much of what I am saying here, at this sad time, I had a happy occasion to say in my message to the 2nd Eilene M. Galloway Symposium, when Eilene was still with us. I had the privilege of enjoying Eilene's friendship for many decades. For the first time, we met at the International Astronautical Congress in Stockholm back in 1960. Since then, we frequently met and worked together at the annual colloquia on the law of outer space, meetings of the IISL Board, and other occasions. Once I had the pleasure to receive Eilene and show her around the famous Peace Palace in The Hague and the Museum of the International Court of Justice housed therein. Like so many other colleagues, I always admired her vast knowledge and experience, unquenchable interest in all aspects of life, her sense of humor. One may ask how it is possible that mutual respect and sincere friendship connected two persons with such different backgrounds, who had lived most of their lives in States ideologically and militarily opposed to each other. I would think that an answer to this question can be found in the most insightful paper presented by Jonathan Galloway at the 1st Eilene M. Galloway Symposium. In his paper Jonathan spoke of "the abiding concerns and dreams" of his mother in the field of space law and of her having been "a fervent advocate" of the international civil space programs of the United States of America. Eilene Galloway's lasting and passionate devotion to the cause of international space cooperation and to the development of legal regulation of space activity has forever gained her respect and affection of the world space community. I would like to finish this tribute to the memory of my dear friend with the words from one of the last New Year Greetings I received from Eilene. In her characteristic calligraphic handwriting, Eilene has written, among other things: "I think of the years since we met in Sweden and all that has happened in Outer space. So wonderful we have had peace from Outer space to the earth for some 47 years. And now hope this will continue".

—Vladlen S. Vereshchetin

I am honored to have known and experienced the intelligence and friendship of Eilene. Everyone knows or should have known that part of Eilene that was her public side. I was honored and privileged to have received her loving friendship when it most was needed. When I lost my cat of twenty years, Eilene wrote to me a long and touching letter about the cat she had as a girl. At a very young age she recognized the many sides of her cat and gave her eight names. These were very sophisticated names and showed the brilliance of a child at a young age. To me, it showed me the side of Eilene that was friendship and love. She had tolerance and desire for intelligent conversations and writings. Likewise, she had the gift of giving loving friendship when I most needed it, as well as others. A very funny thing happened when I sent a photo from the Ranch. Eilene received it and immediately wrote back. She thought what a beautiful picture of me in such a gorgeous dress. Then she looked more closely and thought what skinny legs I had: Then she realized that she had received and had been looking at a photo of one of our PEACOCKS in full bloom. I could see her laughing with her beautiful smile and eyes. Then her distant laugh hit Les and me. We laughed together the miles apart. I really did feel complimented that she would think I was as beautiful as a peacock, but the skinny legs were too much. People throughout the world will miss Eilene. Jonathan, to you and the rest of your family I send my condolences, knowing that nothing will fill the void you are feeling. In many ways I feel fulfillment in having known several sides of Eilene and will continue to miss her at unexpected times. Our memories of you will keep Eilene alive in each of us and as such, she will continue to guide us through difficult times in space law and in other personal ways.

—Patricia Margaret Sterns

It was with great sadness that we learned of Eilene's passing. She was one of the first friends I made in the Institute, and she was very supportive and helpful over the years. Eilene made a significant and lasting contribution to the preservation of outer space for peaceful purposes, which she termed a tangible benefit of space law. I extend condolences to Jonathan and the entire family. We will all miss Eilene.

—***Les Tennen***

Eilene was a dear friend and I miss her very much. I think she had two qualities that made her truly remarkable. First was her enthusiasm for the space program, which never waned over all these decades. She was already 51 years old when Sputnik was launched and had a rich and fulfilling career as a national defense expert for the U.S. Congress. But that day in October 1957 changed everything. She became an instant "space cadet" filled with fascination about the potential of the space program to help bring peace on Earth through international cooperation and with determination that space itself not become a theatre of war. Second was her mentoring of so many people around the world who were interested in the legal and policy aspects of space. Her legacy is not only her direct contributions to the law that created NASA and her many studies and papers that helped guide the development of space policy, but the large number of her "pupils" who now can carry on her work.

—***Marcia Smith****, Washington, D.C.*

I met Eilene when I was developing a paper for the Bangalore IAF/IISL meeting. Since her name appeared in so many citations, I called her and she invited me to her home, as she has invited so many others. Over the years, her warmth, intellectual generosity, and idealism bound me to her for life. More than anything, I will miss the best friend a person could have.

—***JoAnn Clayton Townsend****, Washington, D.C.*

I first met Eilene Galloway when I was just beginning my work on space law. She graciously welcomed me into her home and we talked for a very long time. I was impressed by how genuine she was and that she, who had accomplished so much, was willing to spend time with a novice in the field. My condolences to Jonathan and to the rest of the family.

—***Colleen M. Driscoll, Ph.D.****, Director, The Kurtz Institute of Peacemaking*

I have known Prof. Eilene Galloway ever since my research in space law started in 1964 or so. Prof. Galloway has been an outstanding scholar-philosopher who guided the development of space law in the formative stage. I have had the privilege to work upon her great contributions that provide torchlight to present and future generations to use outer space law for human benefits, and for promoting a peaceful world order. Her passing away is a great loss to the world community.

—***Prof. Saligram Bhatt****, former Professor of Space Law, JNU, New Delhi, India*

I met Ms. Galloway in 1979 when I was a Fellow at the Smithsonian Institution in Washington, D.C. She provided a lot of support and ideas for my thesis on Transfer of Space Technology. I enjoyed very much our lunches at the Cosmos Club and the salad with cherry "bomb" tomatoes that would seem to explode when we tried to eat them... and the visits to her house, where we exchange ideas on how to move the raccoon living in her chimney. I think that you never die when you have planted so many seeds....

—*With Love,* ***Valnora Leister***

Eilene Galloway sponsored my admission in to IISL a bit over 20 years ago! During the inaugural class of the International Space University in 1988, I wrote a paper on the Common Heritage of Mankind, and the development of lunar bases. When I left Boston after graduation, I returned to Washington, D.C. My paper was selected for the Student section of the AIAA to be held in Bangalore, India. I wanted to make sure that my paper was vigorously reviewed and tested before I fulfilled this great honour and dream of delivering my first international paper. Eilene took the time to thoughtfully and thoroughly review my paper and my arguments. Some of her comments were minor. Her biggest issue was that she felt that I should be much more optimistic about the role of international organization in the development of the rules of conduct for outer space. But what made this encounter so special and so transformational? Eilene was exceptionally intelligent, knowledgeable, thorough and yes she was detail-oriented. But most striking was how passionate and idealistic she was about this great field of space law. She was one of the mothers of this great new movement. And the reason why she took the time out to meet with an overly enthusiastic student, was because she cared. She cared about the space law movement, and she cared about bringing new blood into the movement. She did not do this lightly, as a favour or to be polite (and she was wonderfully polite) but she did this because she really wanted to raise the overall level and the standards of this exhilarating endeavour of space.... When I look into the night skies, I will think of the remarkable impact that Eilene had on my colleagues and me, and I will remain daunted at the stratospheric standards, and the civility that she extended to all of us. Godspeed!

—***Michael Potter***

Je présente mes condoléances les plus chaleureuses à toute la famille d'Eilene Galloway. Elle a été et demeurera toujours un phare doctrinal pour moi comme pour beaucoup de juristes. La lecture de ses réflexions et ses efforts pour développer le droit spatial a été pour moi un exemple remarquable. Parce que ses idées continueront de vivre et de fructifier, Eilene Galloway ne nous quitte pas vraiment.

—***Prof. Mireille Couston*** *(France)*

I am so sorry to hear of the demise of one of our very distinguished and active members, Dr. Eilene Galloway. She was a real pillar in many areas of international law, particularly, Space Law. She will be missed a lot. May her gentle soul rest in perfect peace, Amen. Best,

—***Chris N. Okeke***

Eilene was a towering figure in space culture. On a personal note, she was the one in 1984 who asked me to start organising in the international astronautical congress for IAA the first art and literature sessions where artists as professionals could present side by side with their scientific and engineering peers, leading to the establishment of IAA Commission VI on Space Activities and Society and now to ITACCUS at IAF. I am truly saddened tonight but at the same time joyous at acknowledging her crucial role in establishing a space culture on earth.

—***Roger Malina***

Requiescat in pace.

—***Zeldine O'Brien***

I got to know Eilene only in March 2009 when it was my great honour to interview her on behalf of the International Astronautical Federation. Submitting my questions via Corinne Jorgenson of the IISL, I received Eilene's responses back in beautiful handwritten script. She had taken such care with the answers in what proved to be her final interview; I was truly astonished. Reading other tributes I am sad not to have known Dr Galloway for longer—she was obviously a very special person and dear friend to a number of people. However I am very privileged to have had the one chance to get to know her.

—***Scott Hatton***

I would like to express my sincerest condolences.

—***Christian Bruenner***

Eilene was the spur for all us when we came to study outer space law and always the landmark and emulation in the University career. Dear Jonathan, we are close to you for this sad loss of the mother-teacher.

—***Gabriella Catalano Sgrosso***

Today, in Uruguay, the Mother's Day is celebrated. I think that, it is not by chance that I will refer to Mrs. Eilene Galloway, who added, to her condition of biological mother, the other one, most universal, of being a creator and precursor of the law and policy to be applied to outer space. And, like all mothers, she followed with lucidity and persistence the development of those favourable structures, pursuing cooperation and peace. Similar to all great teachers, she guided us with total generosity in our first steps in these disciplines, as Eugene Pepin, Isabella Diederiks-Verschoor, Aldo Armando Cocca and Carl Q. Christol did it too, only to mention the precursors.... The "passing away" of Eilene, allows me to appeal to memories (re-cordis) to revive in my heart, my first encounter with her. In 1982, I was invited by the Government of the United States to make a trip of specialization and exchange in space sciences. During a month, I travelled across the country, from East to West, with a stop at Cape Kennedy, in order to attend the launching of the STS Columbia. In my short stay in Washington, I fulfilled two essential commitments: my recognition to NASA for its collaboration with our Research Center and my visit to my Teacher's home. I arrived in time for the "five o'clock tea" with a bouquet. But in fact, she dazzled me with the flowers she cultivated in her home and essentially with those that, invisible to the eyes, had the significance and permanence of the essential. She talked me about her son Jonathan, whom I met in Innsbruck in 1986, making both of us, presentations about Space Law education. Without concealing her satisfaction, she also proposed me to visit the Library of Congress, so I could see the registries about my budding bibliography. Historians of our remembered Teacher affirm that, from her youth, she was an assiduous reader of the *Meditations of Marco Aurelio*, that Roman emperor and stoic philosopher (121–180 A.C). I take one of them, which can well synthesize the work of Eilene: "A single one is the light of the sun, although walls, forests and uncountable impediments obstruct it; unique is the common substance although divided in multiple natures and particular limitations. One is the intelligent soul although it seems to be divided".

—***Prof. Dr. Eduardo D. Gaggero Montevideo***

On behalf of the Centro de Investigación y Difusión Aeronáutico-Espacial I want to express our great sorrow for the loss of the distinguished and appreciated Dr. Galloway, who was Corresponding Member of CIDA-E since 1985.

—***Marta Gaggero***

I feel deeply sorry for the passing away of Dr. Galloway. It was a good memory for me to meet her every year in IISL annual meeting around the world. She certainly contributed to the present success of IISL for many years. I hope that she will look at us from the outer space from now on.

—***Hashimoto**, Yasuaki, Japan*

With Eilene Galloway I lost a dear friend and mentor: we first met more than three decades ago in the Committee on the Peaceful Uses of Outer Space where I had just started my work as the legal adviser of my Delegation (Germany). I learnt from Eilene's experience, political foresight and criticism and enjoyed her friendship until the end of her life. I deeply miss her wisdom, amity and encouragement.

—***Marietta Benkö**, Institute of Air and Space Law, Cologne University*

I first met Eilene in the early 1970s when I started looking for a job in or close to Space Law, after earning my LL.M. in International Law at GWU in Washington, D.C. She had great advice, such as to keep things short if you want them to be read in Washington as no one has any time to read anything (I had just shown her a very long article I was working on). It was good advice and I have followed it since. Over the years I would look forward to meeting her at the IAF/IISL and other conferences around the world. She was always an inspiration due to her dedication to the study of Space Law and to her helping of students and fellow professionals with ideas and encouragement. She also helped me to obtain publication of one of my first law review articles. My sympathy goes out to Jonathan and the entire family. It was a life well spent!

—***George Paul Sloup***

I was fortunate to watch a pre-recorded interview of Eilene. Her contribution to the NASA Act and space law in general will be fondly remembered. May her soul rest in peace.

—***V.S. Mani**, India*

Dear Jonathan, This is to express my condolences with the passing away of your beloved mother and the woman who did so much for the progressive development of space law. Her life-long work in this field of law and her devotion and enthusiasm will be an example for many others. I hope this will help you in coping with the grief of missing her as a mother and not in the last place as an inspiring colleague.

—***Hanneke van Traa-Engelman**, The Netherlands*

I met Dr. Eilene M. Galloway for the first time in Oslo, Norway, during the IAF Congress/IISL Colloquium in 1995. I have been keeping in touch with her by telephone, because she was more than a mother to me—kind, caring, tender-hearted and inspiring. Her Christmas cards brought joy and happiness to the recipients—reminding them that they were always in her thoughts. She was a gift and a blessing from the Lord to all of us. May her precious and noble soul rest in peace. Amen.

—***Maurice N. Andem***

Dr. Eilene Galloway was a pioneer in space law and a wonderful inspiration to us all. In my capacity as head of the Legal and Business Support Department (Agency Affairs) of the German Aerospace Center, I had the opportunity to see the direct impact of her enduring legacy on international space law. Dr. Galloway will be missed. I send my condolences to her family, also on behalf of the colleagues in my Department.

—***Dr. Bernhard Schmidt-Tedd**, Legal and Business Support Department (Agency Affairs), German Aerospace Center (DLR)*

It is with deep sadness that I heard of the passing away of Dr. Eilene Galloway. The occasions I had the honour to meet her, even if they were few, were most impressive and most encouraging. In the summer of 2000 I had sent her within the work of our research some materials of the Project 2001 Workshop on the International Space Station. She thanked with one of her wonderfully worded letters, saying—in her very gracious way—that when the package arrived she had felt "as if it were Christmas" and that the material was exactly what she needed for her work in the NASA Space Flight Advisory Committee. Of course, it was reverse: Eilene Galloway had created Christmas in July by her unexpected, gentle words! My condolences go to Dr. Jonathan Galloway, her family and all those who loved her.

—***Susanne Reif***

The IISL Board of Directors at the offices of ESPI in Vienna, March 2009.

IISL Honorary Directors Dr. Fasan and Prof. Priyatna Abdurrasyid at the annual IISL/ECSL Space Law Symposium organised for the members of the UNCOPUOS Legal Subcommittee, Vienna, March 2009.

Annual IISL/ECSL Space Law Symposium in Vienna.

Group photo taken at the Third Eilene Galloway Event on Critical Issues in Space Law, held on 11 December 2008 in Washington, D.C., organised by the IISL, the National Center for Remote Sensing, Air, and Space Law of the University of Mississippi, and Arianespace, Inc.

two finalist teams of the Manfred Lachs Moot Court Competition with the Judges, H.E. Judge Abdul Koroma (ICJ), H.E. Judge Peter Tomka (ICJ) and Professor Francis Lyall iversity of Aberdeen), with IISL President Tanja Masson-Zwaan, at the City Chambers in Glasgow.

L President Tanja Masson-Zwaan with board members Ricky Lee, Patricia Sterns, and lie Tennen.

IISL Director Prof. Francis Lyall, recipient of the IISL Lifetime Achievement Award, with Dr. Gerardine Goh.

ests at the IISL Annual Dinner, Òran Mór, Glasgow.

Ms. Zeldine O'Brien, Dr. Berndt Schmidd-Tedt, Ms. Agnes Mellot, and IISL Treasurer Prof. Stephan Hobe.

51st Colloquium on the Law of Outer Space, Glasgow, Scotland

51st Colloquium on the Law of Outer Space, Glasgow, Scotland

Session 1:
PRIVATE INTERNATIONAL LAW REGARDING SPACE ACTIVITIES

Chairmen:

Francis Lyall

Lesley Jane Smith

Rapporteur:

C. Doldirina

IAC-08-E8.1.1

51st INTERNATIONAL COLLOQUIUM ON THE LAW OF OUTER SPACE (IISL) (E8.)
Private International Law Regarding Space Activities (1.)

REGARDING FORMATION OF THE INTERNATIONAL SPACE PRIVATE LAW

Mariam Yuzbashyan

Moscow State Institute of International Relations (University), Moscow, Russia,
E-mail: m_you@mail.ru

ABSTRACT

For several years international lawyers specializing in the field of international space law have been discussing formation of international space private law. This discussion has been also supported by a number of countries' official delegations at UN COPUOS and its Legal Subcommittee. To date this issue has not been comprehensively analyzed.

This paper raises the issue of formation of international space private law; gives an analysis of its legal sources, principal institutions and evaluates the perspectives and tendencies of its future development. In light of the growing tendencies towards globalization, commercialization and privatization, international space private law could provide adequate legal regulation for private space activities.

The author defines international space private law as a body of legal norms comprising substantive law norms and rules of conflict of laws regulating connected with space activity property and personal non-property relations, complicated with "foreign element". International space private law is characterized by its specific subject of legal regulation; and methods of legal regulation such as: substantive law method, method of conflict of laws, international and national methods of legal regulation.

Principal existing and potential legal sources of the international space private law comprise:

• The 2001 Cape Town Convention on International Interests as Applied to Space Assets (the title in accordance with the Art. II of the Preliminary Draft Space Assets Protocol) where is emphasized the primary character of UN space treaties;

• National space legislation (its growing role can not be denied but it would not be right to rely only on civil and national methods in commercial space activity regulation);

• Space law cases (for those states where they are applicable).

This paper also contains a brief analysis of legal status of subjects of the international space private law, of problems of space activity licensing, responsibility of private legal entities, space insurance and other actual issues. Finally the author analyzes the perspectives and tendencies of development of the international space private law. The author hopes that rather soon these ideas will evolve from "imagination" to reality as the objectives, principals and legal background for it already exist, even if there is still much to be done.

FULL TEXT

INTRODUCTION

By the end of the XX century significant changes took place in the world cosmonautics. At the present time space activity is characterized by the growing tendencies towards privatization, commercialization and globalization. As Prof. Dr. Yuri M. Kolosov notes "the private sector literally "digs in" the outer space making it closer to the practical needs of a human being and at the same time creating new legal problems to be resolved by states in the near future"[1].

Contemporary international space law allows space activity of non-governmental entities provided that such activity is authorized and supervised by the appropriate State – Party to the Treaty on Principles Governing the Activity of States in the Exploration and Use of Outer Space, Including the Moon and Other Celestial Bodies (1967) (hereinafter referred to as "Outer Space Treaty") for assuring that national activities are carried out in conformity with the international space law provisions. States Parties to the Outer Space Treaty bear international responsibility for national activities in outer space, including the Moon and other celestial bodies, whether such activities are carried on by governmental agencies or non-governmental entities[2].

In view of the growing tendencies of space activity privatization and commercialization an update of the existing space law provisions is required. Private entities' activity at the international telecommunication services market in laissez faire conditions and development of national space legislation in a number of countries predetermine formation of the international space private law[3]. Moreover, this tendency was distinguished earlier. In 1990 Prof. O Kunc expressed opinion that international private space law would be formed just as international private maritime law had been formed[4].

I. INTERNATIONAL SPACE PRIVATE LAW (DEFINITION, SUBJECT AND METHODS OF LEGAL REGULATION)

It should be noted that the notion "international space private law" is not strict and so far relative. Unfortunately, this issue has not been comprehensively analyzed.

The author of this paper defines ***international space private law*** as *a body of legal norms comprising substantive law norms and rules of conflict of laws regulating connected with space activity property and personal non-property relations, complicated with "foreign element"*[5].

In official documentation *space activity* is defined as exploration of outer space and celestial bodies and their use in practical purposes effected with space technology. This kind of activity is not limited by the sphere of outer space. It can also expand to Earth provided that it is organically connected either with launch of a space object and its operation or with reentry of this object.[6]. Still there is no definition for *commercial space activity* in space law; however it could be determined as activity concerning sale, purchase and exchange or commercial use of space goods and services.

International space private law is characterized by its specific ***subject of legal regulation*** comprising connected with space activity property and personal non-property relations complicated with "foreign element". "Foreign element" may be distinguished by subject of relation, object of relation or juridical act, on grounds of which legal relationship arises, changes or terminates.

Subject of legal regulation requires specific ***methods of legal regulation.*** In this case applicable are the following methods: substantive law method, method of conflict of laws, international and national methods of legal regulation.

Existence of these objective preconditions (specific subject requiring specific methods of legal regulation) is enough to raise the issue of formation of a new branch of law – international space private law that could be able to fill in the actual "vacuum" in regulation of commercial space activities.

II. OVERVIEW OF THE LEGAL SOURCES OF THE INTERNATIONAL SPACE PRIVATE LAW

As any other branch of law international space private law has its own legal sources. Existing and potential legal sources of international space private law both at international and national level are analyzed hereafter.

1. The Cape Town Convention on International Interests in Mobile Equipment and the Preliminary Draft Protocol on Matters specific to Space Assets[7].

Convention on International Interests in Mobile Equipment (hereinafter referred to as "Convention") was prepared in frames of the International Institute for the Unification of Private Law (hereinafter referred to as "UNIDROIT"), opened to signature at the diplomatic Conference held in Cape Town on 16 November 2001, entered into force on 1 March 2006[8].

The aim of the Convention is to increase the efficiency of financing high-value "mobile equipment" (aircraft objects, space objects (assets), railway rolling stock, etc.), because such equipment moves from jurisdiction to jurisdiction, and because not all jurisdictions provide equivalent recognition of creditor's rights, creditors face higher risks and this increases the cost of obtaining credit. The Convention establishes internationally applicable legal regimen for security, title-retention and leasing interests: this will reduce the risks faced by creditors and thereby reduce the costs of financing high value mobile equipment. Financiers will be able to assure themselves that their proprietary interests in a financed asset are superior to all potential competing claims against that asset, and upon default will be able to promptly realize the value of that asset.[9] In course of preparation of the Convention it was decided to divide the rules to be prepared in those general rules applicable to all types of high-value mobile equipment and those special rules needed to adapt the general rules to the specific characteristics of each type of equipment: the general rules were to be placed in the Convention and the equipment specific rules were to be placed in separate protocols applying to each such type of equipment. The both documents (the Convention and the corresponding Protocol) shall be used and interpreted in integrity;

validity of these documents being impossible in case the Convention is ratified without a protocol and vice versa.

The first developed protocol that was opened to signature and entered into force on the same date with the Convention is the Protocol to the Convention on International Interests in Mobile Equipment on Matters specific to Aircraft Equipment (Aircraft Protocol)[10]. The Luxembourg Protocol to the Convention on International Interests in Mobile Equipment on Matters specific to Railway Rolling Stock (The Rail Protocol) was opened to signature at the Diplomatic Conference held in Luxembourg on 23 February 2007 and has not yet entered into force[11].

Another protocol dealing with the application of the Convention to *space assets* is also under development. This protocol will address the difficult task of applying the benefits of the Convention to space assets which are increasingly being financed by private sector investors rather than by governments. The preliminary draft Protocol to the Convention on International Interests in Mobile Equipment on Matters specific to Space Assets (hereinafter referred to as "draft Protocol")[12] is currently under consideration by an inter-governmental negotiation process which includes representation by private sector financiers and the space industry. In accordance with the Article XXIV of the draft Protocol this Protocol will enter into force on the first day of the month following the expiration of three months after the date of the deposit of the [fifth][13] instrument of ratification, acceptance, approval or accession, between the States which have deposited such instruments.

Important is the fact that the document intended to regulate commercial space activities is elaborated under auspices of UNIDROIT, and not less is important that the unique experience of the UN COPUOS, in particular of its Legal Subcommittee, is used in this work. For the first time the Legal Subcommittee considered examination of the preliminary draft Protocol as item for discussion at its fortieth session in 2001[14]. Examination of the draft Protocol continues and the Subcommittee agreed that examination and review of the developments concerning the draft Protocol should remain on the agenda of forty-eighth session[15].

The Convention consists of Preamble, 14 chapters and 62 articles, the draft Protocol consists of Preamble, 6 chapters and 23 articles.

In the words of the Preamble of the draft Protocol, the Protocol "implements" the Convention in relation to space assets. The draft Protocol, in effect, provides special rules adapting the rules of the Convention to the specific characteristics of space assets. In accordance with Article II, p.1 of the draft Protocol "The Convention shall apply in relation of space assets as provided by the terms of this Protocol". According to p. 3 of the same Article "the Convention and this Protocol shall be known as the Convention on International Interests in Mobile Equipment as applied to space assets".

Pursuant to Article I, p. 2, subp. (f) of the draft Protocol *"space assets"* means:

"(i) any identifiable asset that is intended to be launched and placed in space or that is in space;

(ii) any identifiable asset assembled or manufactured in space;

(iii) any identifiable launch vehicle that is expendable or can be reused to transport persons or goods to and from space; and

(iv) any separately identifiable component forming a part of an asset referred to in the preceding sub-paragraphs or attached to or contained within such asset".

It is to be mentioned that the definition of "space assets" given in the draft Protocol is rather detailed but in some aspects less clear than the definition provided in Convention on International Liability for Damage Caused by Space Objects. (1972) and Convention on Registration of Objects Launched into Outer Space (1976)[16].

"International interest" under Article 2, p.2 of the Convention means interest granted by the chargor under a security agreement (a), or vested in a person who is conditional seller under a title reservation agreement (b) or a lessor under a leasing agreement (c). An interest is constituted as an international interest in case it is registered in the International Registry (see Chapter IV of the Convention "The international registration system"), is made in writing (a); relate to an object of which the chargor, conditional seller or lessor has power to dispose (b); enable the object to be identified in conformity with the Protocol (c); and in the case of a security agreement, enable the secured obligations to be determined, but without the need to state a sum or maximum sum secured (d) (Article 7 of the Convention).

The particular benefit of asset-based financing is that it enables the debtor to pay off his debt through the revenue that he manages to generate by his use of the asset, at the same time, enabling the creditor, in the event of default by the debtor, to go against the asset. The fact that the creditor has this guarantee improves his risk analysis on the transaction and thus simplifies the transaction and lowers the cost of the financing he can provide.

In view of the issue in consideration (formation of a new branch of law) important is the question of correlation between the Convention as applied to space assets and the founding documents of the contemporary international space law. The primary character of the established principles of space law, including those contained in the international space treaties under the auspices of the United Nations, is emphasized both in the Convention and in the draft Protocol. Moreover, in accordance with the Article XXIbis of the draft protocol "the Convention as applied to space assets does not effect State Party rights and obligations under the existing United Nations Treaties or instruments of the International Telecommunication Union".

As to the correlation of the provisions of the Convention and the draft Protocol, it is pointed out in Article 6, p.2 of the Convention that in case of any inconsistency between the Convention and the corresponding Protocol, the Protocol shall prevail.

At the forty-seventh session of the UN COPUOS Legal Subcommittee one of the delegations expressed the view that the draft Protocol is a good example of efforts being made to find a solution to the deficiencies in the existing United Nations treaties on outer space without compromising the interests safeguarded in those treaties. That delegation also noted that private and commercial space activities should be regulated[17].

Thus, the adoption of the Convention and the continuing work on the draft Protocol could be considered as one of the first efforts at the international level to modernize space law and to adapt it to the growing commercial space activity, on one hand, and as important preconditions for the formation of international space private law, on the other hand.

Both the Convention and the draft Protocol require more detailed and profound analysis but in this paper they are analyzed primarily as *potential first internationally applicable legal sources of international space private law.*

2. National Space Legislation Overview

In recent decades the tendencies of legal regulation of space activity have significantly changed. Individuals' and legal entities' space activities required national regulation. In accordance with the Article VI of the Outer Space Treaty, States Parties bear international responsibility for all national activities in outer space even if they are carried on by private entities. . Therefore, in practice the states adopt and will continue to adopt national space laws first of all regulating issues of licensing, insurance, export control in this field. Without any doubt national space legislation shall be in full conformity with international obligations of states.

In this paper *national space legislation* is considered as *one of the legal sources of the international space private law.*

The state that encourages the most commercial space activities is the *United States of America.* Legal regimen of national space activity regulation consists of:

- Communications Act (1934) that establishes licensing and operating requirements for satellites and associated ground stations;
- National Aeronautics and Space Exploration Act (1958);
- Communications Sattelite Act (1962);
- Commercial Space Launch Act (1984; as amended in 2004). This act envisages licensing of space Activity in the US.;
- Land Remote Sensing Policy Act (1984);
- Export Control Act (1998)
- Federal Aviation Administration (FAA) "Human Space Flight Requirements for Crew and Space Flight Participants" (2006).

Commercial Space Launch Act (CSLA) is administered by the Office of Commercial Space Transportation, FAA of the U. S. Department of Transportation. In accordance with CSLA provisions FAA license is required for launch/reentry in US; by US citizen outside US; by US citizen outside US and outside territory of foreign country, unless foreign country's government has an agreement with US on jurisdiction over the launch or operation; and by US citizen in foreign country if US has jurisdiction by agreement with government of foreign country. Thus, national activity includes both the activity of US citizens and activity that is carried on from US territory regardless of national identity of individuals or legal entities.

A licensee must obtain third party liability insurance or demonstrate financial ability to pay maximum probable loss arising from third party claims.

The US Minister of Transport supervises and coordinates operations on commercial space launches and reentries, is responsible for issuance and transfer of licenses authorizing such operations, for protection of national health and security, property safety and interests of national security and foreign policy of the US.

United Kingdom. The Outer Space Act was adopted on 18 July 1986. The

Act applies to United Kingdom nationals, Scottish firms, and bodies incorporated under the law of any part of the United Kingdom (p. 2(2) of the Act). The application of the Act is extended in particular to Isle of Man, Guernsey, Cayman Islands, and Bermuda (by the corresponding statutory instruments[18]).
The Secretary of State is empowered to license launching or procuring the launch of a space object, operating a space object or any activity in outer space. The Secretary of State exercises its powers through British National Space Center (BNSC).

In *Russian Federation* licensing (permission) procedure for the pursuit of space activity in scientific and national-economy purpose is established under Article 9 of the Federal Law of the Russian Federation № 5663-1 about Space Activities (18 July 1993; as in force on 12 December 2006). Subject to licensing shall be space activity of organizations and citizens of Russian Federation or space activity of foreign organizations and citizens under the jurisdiction of Russian Federation, if such activity includes tests, manufacture, storage, preparation for launching and launching of space objects, as well as control over space flights. Carrying out space activity by an organization or a citizen without a license or in willful violation of the terms of the license shall be punishable by virtue of the legislation of Russian Federation. Actions of state bodies to license space activity may be claimed in the court of law or in the arbitration tribunal.
In accordance with Article 6 of the aforementioned Law, the Russian Space Agency shall, within its competence, issue licenses for the types of space activity. This provision is confirmed in the List of Federal Licensing Bodies, adopted by Decree № 326 of the Government of the Russian Federation about Types of Activities Licensing (11 April 2000; as in force on 17 November 2000) and in the Regulation on Licensing of Space Activities, adopted by the Resolution № 403 of the Government of Russian Federation (30 June 2006).
Types, forms, and terms of licenses, the conditions and procedures for their issue, withholding, suspension or termination, as well as other questions of space activity licensing are governed by the Regulation on Licensing of Space Activities.

The growing role of national space legislation in commercial space activities regulation can not be denied, but nevertheless it shall have a "supportive" character since solely national legislation is insufficient for regulation of global activity in the interests of the whole humanity.

3. Space Law Cases
And Arbitration Practice

Space law cases and arbitration practice could also be considered as potential legal sources of international space private law for those states where they are applicable.

III.SUBJECTS OF THE INTERNATIONAL SPACE PRIVATE LAW.

Subjects of the international space private law are individuals, legal entities, states and international organizations. Of major interest are non-governmental entities.
In space law there is a specific correlation between public and private law aspects. It is challenged by certain character of the involved relations (e. g. only space law envisages absolute liability without indication of its limits[19]; states bear international

responsibility for all national activities in outer space (under Article VI of the Outer Space Treaty); cosmonautics effects interests of the whole global community etc.).

Interrelations of subjects of international space private law are predetermined by articles VI and VII of the Outer Space Treaty. In compliance with these articles states bear international responsibility also for space activities carried on by non-governmental entities (private companies). Therefore such activity shall be carried on under continuing supervision of the corresponding State Party to the Outer Space Treaty.

Subjects of international space private law in their turn are obliged to comply with international and national law provisions regulating space activities.

In view of the perspective of space tourism development actual also becomes the issue of legal status of space tourists.

PERSPECTIVES

Definition of international space private law as a body of legal norms comprising substantive law norms and rules of conflict of laws regulating connected with space activity property and personal non-property relations, complicated with "foreign element" is still relative.

However, international space private law can be already characterized by its specific subject of legal regulation and methods of legal regulation and this is necessary and sufficient for formation of a new branch of law.

In light of actual tendencies the Convention on International Interests on Mobile Equipment as Applied to Space Assets is of special importance. This document, consisting of typical for private international law substantive law norms and rules of conflict of laws, is one of the first potential internationally applicable legal sources of international space private law.

In view of increasingly growing space activities of non-governmental entities actual are issues of space activity licensing and insurance, responsibility of non-governmental entities. At present all this issues are more or less regulated in national space legislation. And moreover of special interest are issues of private property and intellectual property rights in space law.

The growing commercial space activity, various international space projects require and already get, to some extent, necessary legal regulation, but it is still much to be done[20].

The author thinks and hopes that international space private law could gradually fill in the existing "vacuum" in the regulation of space activities and provide adequate legal regulation for private space activities.

---

[1] А. В. Яковенко/ Прогрессивное развитие международного космического прав. Актуальные проблемы. – «Международные отношения» - Москва, 1999, Предисловие – Ю. М. Колосов -С. 8. (A. V. Yakovenko/ Progressive Development of the International Space Law. Actual Problems. – "International Relations"- Moscow, 1999. Foreword by Prof. Dr. Yuri M. Kolosov- P. 8)

[2] See the Article VI of the Outer Space Treaty.

[3] Международное космическое право/ Отв. редакторы проф. Жуков Г. П. и проф. Колосов Ю. М.. - «Международные отношения». – М.,1999 г., С.133. (International Space Law/Edited by Prof. Dr.

Gennady P. Zhukov and Prof. Dr. Yuri M. Kolosov. – "International Relations". – Moscow, 1999. - P. 133)

[4] Кунц О. Международное космическое право и международное частное право/Новое в космическом праве (на пути к международному частному космическому праву)// Отв. редактор проф. Верещетин В. С. – М., 1990. – С.13. (O. Kunc/International Space Law and Private International Law/Novelty in Space Law (Towards International Private Space Law)//Edited by Prof. V. S. Vereshetin – Moscow, 1990. –P.13).

[5] In this paper general theoretical issues of private international law are not covered. The author's aim is to raise the issue of formation of international space private law.

[6] International Space Law/Edited by Prof. Dr. Gennady P. Zhukov and Prof. Dr. Yuri M. Kolosov. – "International Relations". – Moscow, 1999. - P. 7.

[8]For the status of Convention please see: http://www.unidroit.org/english/implement/i-2001-convention.pdf

[9] For the Convention background please see: http://www.unidroit.org/english/workprogramme/study072/main.htm

[10] For the status of the Aircraft Protocol please see: http://www.unidroit.org/english/implement/i-2001-aircraftprotocol.pdf

[11] For the status of the Rail Protocol please see : http://www.unidroit.org/english/implement/i-2007-railprotocol.pdf

[12] The text of the preliminary draft Protocol used in this paper is that revised by the Committee of governmental experts at its first session (cf.C.G.E.SpacePr./1/Report/Appendix VI).

[13] In line with UNIDROIT practice, the Space Working Group at its fifth session, taking the view that the entry into force of the Convention as applied to space assets should be accomplished with the minimum number of ratifications/accessions possible, suggested that the appropriate number would be five.

[14] UN Document A/AC.105/C.2/L.225

[15] See UN Document A/AC.105/917, P. 18

[16] On this issue please see UN Document A/AC.105/C.2/2003/CRP.6. P.41,47.

[17] See UN Document A/AC.105/917, P. 18

[18] Statutory Instrument 1990 № 596 Outer Space Act 1986 (Isle of Man); Statutory Instrument 1990 № 248 Outer Space Act 1986 (Guernsey) Order 1990; Statutory Instrument 1998 № 2563 Outer Space Act 1986 (Cayman Islands) Order 1998; Statutory Instrument 2006 № 2959 Outer Space Act 1986 (Bermuda) Order 2006.

[19] International Space Law/Edited by Prof. Dr. Gennady P. Zhukov and Prof. Dr. Yuri M. Kolosov. – "International Relations". – Moscow, 1999. - P. 219.

[20] Actual and still "open", to some extent, also are the issues of private property and intellectual property rights in space law, legal status of space tourists etc.

ACHIEVING A LEVEL PLAYING FIELD IN PUBLIC-PRIVATE PARTNERSHIPS: CAN SOVEREIGN IMMUNITY UPSET THE BALANCE?

Diane Howard, J.D.
McGill University, L.L.M. Candidate 2008
Montreal, QC Canada H3A 2M1
ladydi814@gmail.com

ABSTRACT

This paper briefly explores the international and domestic legal responsibilities of both public and private parties in a space venture. It examines the doctrine of sovereign immunity in the United States and other countries in PPPs both outside and within the space industry context. Available waivers and exceptions from the doctrine are examined from the standpoint of maintaining a level playing field between all participants in a public private partnership.

I. INTRODUCTION

Increasingly important to the space industry are the hybrid entities known as public private partnerships, or P3s, that result along the spectrum between the public and private sectors. They are not new, nor peculiar to space ventures. P3s can be found in a wide range of applications, from public utilities, to infrastructure projects in developing countries, to social service delivery through the faith-based initiative in the United States.

Accountability in business is always a key concern, and certainly space business is no exception. If a government can avoid responsibility for its actions as a partner in space industry by invoking state immunity, the risks borne by the private side of the venture could be disproportionate to the possible upside potential. Such imbalance can create an uneven playing field and, perhaps, cripple commercial growth.

II. PUBLIC PRIVATE PARTNERSHIPS DEFINED

Methods of financing public services have undergone significant transformation since World War II. "[T]he international trend was to nationalize energy and other infrastructure assets and institute controls over private monopolies in order to limit abuses of market power."[1] Over time, the costs of public ownership and or subsidization, including the erosion of operational efficiency, became apparent, resulting in a restructuring trend.[2] Internationally, governments felt pressure to change the standard models of procurement, largely because of concerns for high levels of pubic debt, and moved toward a privatization model.[3]

Whereas privatization is on a downward spiral,[4] public-private partnerships are now hailed as "the new paradigm for economic development in the 21st century…increasingly being used as a policy tool to transform the role of national and local governments in public service delivery, infrastructure development, poverty alleviation, capital market improvement, and governance around the world."[5] This trend is global,[6] particularly in the European and Asian markets.[7]

Absent a universal legal definition of P3s, they are "generally recognized [to exist] wherever there is a contractual relationship between the public sector and a private sector company designed to deliver a project or service that traditionally is carried out by the public sector."[8]

P3s are creative arrangements. Usually, a governmental entity enters into contract with a private consortium which sets up a single purpose entity known as a special purpose vehicle (SPV). The private consortium is typically formed by a joint venture (JV) between a range of contractors, banks, investors, and suppliers willing to commit equity and/or resources to the project.[9]

Some underlying principles are indispensable to their success. Value for money (VFM) is crucial. It refers "to the best possible outcome after taking account of all benefits, costs and risks over the whole life of the procurement."[10] Risk is perceived from the public sector's perspective as "any event which jeopardizes the quality or quantity of service that they have contracted for: and from the private sector's perspective as any event which "causes the cash flow profile of the project to depart from the base case and jeopardize the debt servicing ability of the project or its ability to generate a dividend stream for shareholders."[11]

Internationally, examples of P3s abound. Commercial space mirrors this trend toward hybrid entities; examples can be found in a host of space applications encompassing remote sensing, international telecommunications, global navigation, proposed space solar power systems, and spaceports.

III. GOVERNMENT IMMUNITY

It is easy to imagine scenarios in which the public partner of a space-related P3 could attempt to evade a lawsuit. For instance, a contractual breach flowing from the P3 agreement itself could be avoided. Third party liability to private parties for accidents in a spaceport launch facility (such as the 2007 explosion at Scaled Composites) could be circumvented. A government partner could sidestep liability for any simple slip and fall in a spaceport or facility of a space P3. Responsibility for damage from the cessation or malfunction of a signal of a global emergency response system or navigation system could be dodged.

It is worth noting that the International Civil Aviation Organization (ICAO) listed sovereign immunity as an identified concern for GNSS in its Final Report on the Work of the Secretariat Study Group on Legal Aspects of CNS/ATM systems.[12] And, in litigation now before the US courts, an Israeli-owned and controlled remote sensing corporation, ImageSat, has claimed immunity in a shareholder's derivative action questioning business decisions.[13] An understanding of the background, purposes, and mechanics of government immunity can only help the space industry address these situations proactively.

A. <u>Background</u>

"Sovereign immunity encompasses immunity from both suit and liability."[14] A recognized state enjoys immunity from the jurisdiction of the courts of other states.[15] The doctrine can operate as a bar to actions between sovereigns, but is more often implicated in actions between private parties engaged in activities with governmental entities.

There are two schools of thought on the origin of the doctrine. One is based upon the theory that, as sovereign equals, one state cannot exercise authority over another,[16] while the other is based upon the common law tradition, originating with the Romans, that *rex non potest peccare*, or "the king can do no wrong."[17] Immunity can extend from the head of state to the government and its organs, the leader of the government (if a different person), ministers, officials, agents of the state for their official acts, some public corporations, and state owned property.[18]

Absolute immunity is just that – immunity for all government acts – "susceptible of no limitation not imposed by itself."[19] That view first appeared in US jurisprudence in an 1812 case, *The Schooner Exchange v. M'Faddon*,[20] and was fully embraced by 1926 in *Berizzi Brothers v. Steamship Pesaro*.[21] It is "the product of comity concerns, rather than a want of judicial power."[22]

However, the State Department and the United States Supreme Court were seemingly at odds in application of the immunity. The State Department was concerned that adjudications against foreign sovereigns could embarrass the executive arm of the government leading to the Supreme Court's reluctance to adjudicate at all.[23] By the mid-twentieth century, the Supreme Court recognized the commercial advantage that absolute immunity provided to foreign sovereigns over private businesses.[24] The State Department reconsidered its position, acknowledging the changes in the wind, and recognized that the extended immunities were based upon possibly outdated conceptions of sovereignty.[25]

The Acting Legal Advisor to the Department of State, Jack B. Tate wrote a letter to the Acting Attorney General in May 1952. While it expressed succinctly the shift from absolute to restrictive immunity, distinguishing between public and private acts of a state, it left unsettled the matter of who should determine whether immunity attached in a given situation – was it the courts or was it the State Department?[26]

For more than twenty years this ambiguity remained, allowing foreign states the alternative of seeking State Department approval for claims of immunity, which included the possibility that the Department would give in to political pressure. The European Convention on State Immunity, adopting restrictive state immunity, was signed by all members of the Council of Europe in 1972.[27] Eventually, the United States Congress in 1976 enacted the Foreign States Immunity Act, codifying the restrictive theory, thus reflecting the policy followed by a majority of States.

One of the most significant results of the Act was that it settled the question of where the determination of immunity would be made, placing the responsibility in the judicial system rather than in the State Department.[28]

The United Kingdom passed its State Immunity Act in 1978; Canada followed with its State Immunity Act in 1983. Both adopted restrictive immunity. The United Nations Convention on the Jurisdictional Immunities of States and Their Property[29] in 2004, adopted by the General Assembly in 2005,[30] The new instrument, too, reflects the restrictive theory of state immunity.

The United States Supreme Court officially espoused restrictive immunity for foreign states in *Alfred Dunhill of London, Inc. v. Republic of Cuba*; the Court found that the case essentially dealt with an issue of immunity, which it denied because the conduct was commercial in nature.[31]

B. FSIA

The text and structure of the FSIA "demonstrate Congress' intention that the FSIA be the sole basis for obtaining jurisdiction over a foreign state in [US] courts."[32]

As a starting point, then, "a foreign state is presumptively immune from suit unless a specific exception applies."[33] Courts employ a burden-shifting analysis; the defendant foreign state "must first establish a prima facie case that it is a sovereign state, creating a rebuttable presumption of immunity. Once the foreign sovereign makes that prima facie showing of immunity, the plaintiff has the burden of production to make an initial showing that an FSIA exception to foreign immunity applies."[34] The court must resolve whether foreign sovereign status applies – a question of law – as a threshold matter.

Though not yet directly applied, in *Dole Food Co. v. Patrickson,*[35] the United States Supreme Court affirmed as correct the application of the five-factor framework used by federal appellate courts to determine whether an entity is an organ or instrumentality of the state as defined in § 1604(b) of the FSIA. The factors are:

> (1) whether the foreign state created the entity for a national purpose; (2) whether the foreign state actively supervises the entity; (3) whether the foreign state requires the hiring of public employees and pays their salaries; (4) whether the entity holds some right in the [foreign] country; and (5) how the entity is treated under foreign state law.[36]

Some lower US courts have used "a 'characteristics' test, asking whether under the law of the foreign state where it was created, the entity can sue and be sued in its own name, contract in its own name, and hold property in its own name."[37] Other courts, mainly appellate, have adopted a "core functions test" limiting inquiry to "whether the defendant is an integral part of a foreign state's political structure and function is predominately commercial."[38]

Instrumentality status is determined by the facts at the time the action is filed.[39] In other words, if the foreign state's interest in a corporation was not a majority interest until after filing, then the claim of instrumentality would more than likely fail. Likewise, if a foreign-state owned entity, such as an airline, is privatized after a claim arose, the claim would not lose its status under the FSIA.[40] Furthermore, the US Supreme Court held that control could not be substituted for ownership interest, and that a subsidiary of an instrumentality could not, itself, be found an "instrumentality" as it was too far removed.[41]

Because the FSIA grants subject matter jurisdiction over the foreign state, objection to this jurisdiction may be raised at any time.[42] And, before a court may enter a default against a foreign state, the FSIA requires the plaintiff to establish his or her right to relief by evidence satisfactory to the court.[43] Once the defendant has made its case for immunity status, it is the plaintiff's burden to prove that one of the seven statutory exceptions applies, granting the court jurisdiction to hear the case. The commercial exception found in § 1605(a)(2) of the FSIA is the most significant of those enumerated.[44]

The courts have carved out definitions and tests to assure proper application of §§ 1603(d) and (e), and § 1605(a)(2) to the facts of a given case. Thus, in *Saudi Arabia v. Nelson*, the United States Supreme Court held that eligibility for the commercial exception required identification of the particular conduct upon which the claim was "based," or the gravamen of the complaint, as per the statutory language.[45] It is not enough that commercial activity have some loose connection to the basis of claim. The offending conduct must flow from genuine commercial activity. The statute expressly dictates that it is the nature of an act that determines its commercial character, not the purpose of that act.

The FSIA breaks the commercial activity exception into three alternative scenarios, each providing sufficient connection to the United States to afford a jurisdictional nexus. The commercial activity can be conducted in the United States by the foreign state; it can be an act performed in the United States but in connection with a commercial activity of the foreign state somewhere else; or, it can be predicated upon an act outside United States territory in connection with commercial activity of the foreign state elsewhere if that extra-territorial act caused a direct effect in the United States.[46]

In *Republic of Argentina v. Weltover*, the Supreme Court held that "an effect is direct if it follows as an immediate consequence of the defendant's ...activity."[47] This holding was expanded upon in *American Telecom Co. L.L.C. v. Republic of Lebanon*, a recent Sixth Circuit case, where that court applied the principle *de minimus non curat lex*, to ensure that jurisdiction was not based upon "purely trivial effects."[48] The court acknowledged the difficulty in applying *Weltover's* "immediate consequences" test and analyzed it in terms of another, the "legally significant act" test. The court found that second unnecessary, and held that

> the mere act of including an American company in or excluding an American company from the process of bidding on a contract, where both parties' performance is to occur entirely in a foreign locale, does not, standing alone, produce an immediate consequence in the United States, and therefore, does not "cause a direct effect in the United States" for the purposes of 28 U.S.C. § 1605(a)(2).[49]

Any discussion of adjudication over sovereign acts must include the act-of-state doctrine which "precludes [US] courts [] from inquiring into the validity of the public acts a recognized sovereign power committed within its own territory."[50] The act-of-state doctrine is important to consider when examining the types of actions for which a public partner in a P3 could be held accountable, even if a valid waiver is in place. In *Dunhill*, the United States Supreme Court was evenly divided on whether the commercial exception also limited the availability of an act-of-state defense.[51] *Weltover* revisited the issue when forced to rule on whether Argentina's issuance of commercial bonds to raise capital for its economy, and its unilateral extension of the time to pay on those bonds, was an act of a sovereign to fulfill its obligations when confronted with a national credit crisis, an act of state, or simply a commercial decision which any issuer of debt instruments could make.[52] The court used the nature v. purpose test to conclude that Argentina's actions were not an act of state but participation in the bond market in the manner of a private actor.[53]

C. Additional US law providing state immunity

Prior to enactment of the FSIA, the US Congress passed the International Organizations Immunities Act (IOIA). The statute granted international organizations "the same immunity from suit and every form of judicial process as is enjoyed by foreign governments, except to the extent that such organizations may expressly waive their immunity for the purpose of any proceedings or by the terms of any contract."[54] At the time the statute entered into force, the immunities extended to foreign governments by the United States were absolute. The FSIA, as described in great detail *supra*, restricted this immunity.

The United States Supreme Court has not, as yet, ruled on the scope of immunity offered to intergovernmental organizations (IGOs) under the IOIA, however, in a recent (2008) appellate case, *Inversora Murten, S.A. v. Energoprojekt-Niskograndnja Co.*, the District of Columbia Circuit Court held the immunity to still be absolute.[55] As a result, there are no exceptions for commercial activity, etc. The only available exception is achieved through the organization's own express waiver. International and national law governing immunity for international organizations requires that the language of such a waiver must not be broad on its face but narrowly construed, and must further the organization's objectives in entering the contract or agreement in which the waiver is found.[56]

Executive Order 9698 contains an extensive list of international organizations entitled to enjoy the absolute immunity of the IOIA.[57] A number of space-related organizations can be found on the list, to wit, the European Space Research Organization succeeded by the European Space Agency (ESA), the International Telecommunications Union (ITU), the International Telecommunications Satellite Organization, and the United Nations. Additionally, ICAO, the United International Bureau for the Protection of Intellectual Property and the World Intellectual Property Organization (WIPO) are listed. Great care should be taken to properly word express waivers of immunity in P3 agreements involving listed IGOs, taking into account that the organization must receive some benefit for the immunities it releases.

Several other laws in the United States deal with bringing suit against the United States in a US court. As this is very possible in light of the stated US policy to employ more entities along the public-private spectrum, they bear mention. First is the Tucker Act, in which the US government waives its immunity against suit for actions arising out of express or implied contracts with the government, or one of its agencies.[58] Claims may be for liquidated damages, or for constitutional violations, particularly for the taking of property. Tort claims are excluded under this Act. "The Tucker Act waives sovereign immunity and allows the Court of Federal Claims to hear certain suits against the government."[59] Claims for less than $10,000 US fall under the Little Tucker Act; concurrent jurisdiction is available for these in either federal district court or the Court of Federal Claims.[60]

As space business involves considerable interdependence between the military and the private sector, falling closer to the center of the public-private continuum, litigation involving contracts between the military and private contractors can be instructive. In *Dep't of the Army v. Blue Fox, Inc.*, a government sub-contractor on a construction project sued the Army and the Small Business Administration, seeking an equitable lien on property held jointly which had been distributed to the primary contractor who had failed to pay the sub-contractor.[61] The Army had not required a Miller Act bond,[62] which requires a contractor on a federal project to post two bonds, a performance bond and a labor and material payment bond.[63] The bonds cover first-tier claimants, or primary contractors, and their sub-contractors. However, claimants further down the chain are considered too remote and cannot assert a claim. And, as Blue Fox learned, with no bond, there was no possible recovery because sovereign immunity bars liens on government property.[64]

Lastly, the Federal Tort Claims Act (FTCA) permits private parties to bring an action against the United States in a federal court for most torts committed by persons acting on behalf of the United States.[65] Exceptions to the FTCA include the *Feres* doctrine, prohibiting suit by military personnel for injuries sustained incident to service;[66] the discretionary function exception, immunizing the United States for acts or omissions involving policy decisions;[67] and the intentional tort exception, unless the offending acts were committed by federal law enforcement or investigative personnel.[68] Pertinent to discussion of immunity in the context of space-related P3s is *Smith v. United States*,[69] as it dealt with a tort claim that arose in Antarctica, a region of indeterminate status in international law. Although the FTCA contemplated extra-territorial claims, the issue was whether a sovereign-less region constituted a foreign country for the purposes of applying the FTCA. The Court held that it did not, leading to the conclusion that a plaintiff injured from an event occurring in outer space caused by the negligence of the United States would not be able to sue a government partner under the FTCA.

D. Eleventh Amendment Immunity

The Eleventh Amendment of the United States Constitution removes a class of cases from Article III jurisdiction, establishing the judiciary and the federal court system.[70] This shield was extended t o arms, or alter-egos, of the state in *State Highway Comm'n v. Utah Constr. Co.*.[71] However, despite the breadth of the amendment's reach, some state-created and/or state-managed entities are not immune. The jurisprudence that has developed analyzes a number of structural factors in order to determine an entity's immunity or vulnerability to suit. In light of the many varied entities that have emerged in the trend away from the purely public end of the spectrum, courts have encountered difficulty in consistent application of these multi-factoral tests.

The arm-of-the-state doctrine appears in connection with three basic entities: 1) a political subdivision, like a city or county, or municipality; 2) an entity established by two or more states by congressionally-approved compact; and 3) a special purpose public corporation or agency established by or for the state.[72] The first is not entitled to Eleventh Amendment immunity; some forms of the other two are. Understanding the purpose of the immunity assists in the analysis required to determine whether an entity of the second or third type can avoid suit, and, ultimately, liability. "[I]t is not just the state's interest in its public treasury which is at stake in the assertion of Eleventh Amendment immunity. The state also has a 'dignity' interest as a sovereign in not being haled into federal court."[73]

The arm-of-the-state test generally includes a combination of five of the following eight factors: 1) whether a money judgment would be satisfied out of state funds or could be satisfied without direct participation of guarantees from the state; 2) the source of the entity's funding; 3) whether the entity performs central governmental functions or has a proprietary function; 4) whether the entity may sue or be sued and enter into contracts in its own name and right; 5) whether the entity has the power to take the property in its own name or only in the name of the state and whether the property is subject to state taxation; 6) whether the state exerts control over the agency and, if so, to what extent; 7) whether the state has immunized itself from responsibility for the agency's acts or omissions; and 8) the corporate status of the entity.[74]

These factors are considered in evaluation of the nexus between an entity and the state, to discriminate between governmental entities. Of these factors, the source of the entity's funding is considered to be the weightiest.[75] However, it is important to recognize that these factors are not intended to impute arm-of-the-state status to private entities.[76] This point was made recently by the US Court of Appeals for the Ninth Circuit, in a case where a private company claimed immunity from services it provided to the district attorney.[77] Contractors do not receive immunity.[78] The fact that a private company performs a central governmental function is not enough to grant state immunity, nor can immunity be extended to private companies as arms of the state.

While true that independent contractors cannot, per se, claim immunity from suit arising out of work done for the state, there are situations when a contractor can "acquire" immunity. This exception "provides that a contractor [that] performs its work with a governmental agency, and under the governmental agency's direct supervision, is not liable for damages resulting from its performance."[79] A contractor would not be protected for its negligence, only for acts done in accordance with its contract with the state.[80] Some states call this derivative sovereign immunity.[81]

Also important to the arm-of-the-state analysis is recognition of bi-state entities, creations of three sovereigns – two separate states and the federal government. These bodies are governed by compact agreements between the states (or between a state and an Indian tribe), with congressional consent.[82] Typically, these Compact Clause agencies fall outside the range of Eleventh Amendment immunity as that immunity is "only available to one of the United States."[83]

Another case reveals an organizational structure resembling a P3, in the context of a taking. The analysis of whether the organization was a part of the state is useful. In *Illinois Clean Energy Community Foundation v. Filan*, the legislatively created Foundation enjoined the state from enforcing a demand for its assets on the grounds that it would be a taking.[84] The state countered with the argument that the Foundation was, in essence, the state and therefore could not complain about the state taking its own property. Additionally, the state argued that it could amend the statute creating the Foundation to allow for the transfer of title. The court quickly disposed of that argument as the state cannot lawfully enlarge its regulatory power to allow taking someone's property through amendment of an existing statute but only through enactment of a new one. More telling was the court's holding that the "fact that the state's legislature authorized the creation of the plaintiff foundation does not make the foundation a state agency.[85]

Extension of immunity requires far more than pleading it. Axiomatic, too, is the fact that parties who successfully plead immunity avoid the litigation that could result in a published opinion.

IV. APPLYING THE TESTS TO P3s

Now it is time to apply the different doctrines that allow or deny immunity to the hybrid entities that are evolving in the space industry. "[E]ach PPP is sui generis, and consequently…no body of law or regulations…applies to all PPP contractual arrangements."[86] The decision of how to apply the various immunity tests to P3s becomes more difficult the closer to the middle of the continuum one finds the entity at issue. An activity carried out by a foreign state can be denied immunity if purely commercial. On the other hand, an independent contractor following his contract to the letter can enjoy immunity.

"[T]he most important factors to consider in deciding whether a hybrid entity is the state for purposes of sovereign immunity are the extent of state control and whether the entity was acting as the state's agent in conducting the activity that gave rise to the suit."[87] US courts are "extremely hesitant to extend this fundamental and carefully limited immunity to private parties whose only relationship to the sovereign is by contract."[88] This is not necessarily a bad thing. The commercial exception in restrictive foreign state immunity was adopted primarily because it leveled the playing field in a world where governments were behaving increasingly as ordinary trade partners. At this point in history, governments actually are trade partners.

What becomes apparent, looking at the state of the law, international and national, is that immunity does not extend in the face of clear, unambiguous waiver and obvious commercial activities. A P3 in the US, coalescing private interests with a government partner, requires use of the arm-of-the-state test. How is the deal structured? What is the source of funding? Who will have to pay a judgment – the entity or the government? Are the roles clearly defined, the risk spread equitably and transparently? Is the service provided governmental or proprietary? Already, telecommunications infrastructure has been widely recognized as a public service.[89] So, too, do earth observation, the internet, and military communication illustrate delivery of public services through private means. And, with a foreign partner, the nature v. purpose test will apply.

As noted earlier, it appears nearly impossible for a privatized concern to avoid suit under the auspices of the state, and only possible for a hybrid if the factors, considered together, render immunity inappropriate. In actions against states or their agencies/entities within the US, the inquiry is whether the entity is an arm of the state. Internationally, the inquiry will focus on 1) whether the entity is an instrumentality or organ of the state; 2) whether the P3's activity is commercial in nature, and, if so, where it occurred; and 3) was the activity from which the action arose an act of state?

The locus of the activity is significant. There is potential for a private partner with a foreign government in a P3 to be at a disadvantage in a US court. It is important to make sure that there is direct effect in the US if the P3 will be operating extraterritorially. A distant consequence, such as an adverse effect on stock price if a deal does not go through, would not qualify. Fraudulent activity in connection with a US bank account would. A claim based upon a bid, won or lost, for work to be performed outside the US does not create a direct enough effect for suit in the US.

V. RECOMMENDATIONS

P3s are based upon contractual agreement. In structuring the special purpose vehicle and drafting the joint venture agreements, care in drafting and, possibly, standardization of contracts are tools for keeping arrangements transparent. P3s are most successful when they survive long enough to realize the returns.

Six guiding principles have been identified for the sustainability of P3s in infrastructure contexts and they can easily be applied when creating space-related ventures.[90] They are: 1) design the project to deliver a balanced risk profile between the public and private partners; 2) win the commitment of critical stakeholders and operators; 3) develop a strong contract setting forth the rules of the game and clearly defining roles and responsibilities; 4) drive the bidding program allowing buy-in at all levels and stages of the process; 5) demonstrate improved service delivery; and 6) sustain change. Independent advisors have been recognized as useful in structuring P3 transactions to ensure the proper balance between public and private interests.[91] Transparency is a key issue.

Ultimately, the viability of P3s comes down to principles of equity and fair dealing, of fairness and natural justice, or due process, both substantive and procedural. The restrictive theory of immunity was adopted globally in recognition of the practical realities of business and government in the twentieth century, and in an effort to reduce legal maneuvering to avoid responsibility, even by sovereign states. These realities have solidified in the beginning of the twenty-first century.

Hence, it is safe to conclude that the restrictive theory of state immunity has achieved the status of customary international law, for it is followed by a majority of the international community.

It is important to address these realities in the early stages of a project. P3s make public services available to more users when done efficiently. The private sector has a better track record. Efficient delivery to more end-users is really an issue of freedom of access, found in Article I of the Outer Space Treaty,[92] and in customary international law – not only to space itself, but also to its benefits for all on Spaceship Earth.[93] If both sides of the spectrum proactively acknowledge the exposures and fairly apportion the risks between them, then the synergy created by P3s is an awesome resource available to all.

Clearly, the framers of the early space treaties contemplated space activities for both the public and private sectors.[94] It is a natural development that these activities have evolved to include formalized cooperation, not just between states, but between states and private entities. However, even with the lofty principles of the treaties to guide space actors, and fairness a cornerstone of the legal relationships that evolve, it is wise to address another reality – namely, that not all participants will play fair. For this reason, it is recommended t hat the claims arising from space-related P3s be resolved on the basis of private law, where decisions are binding and relief and enforcement possible. Government immunity does not have to upset the balance between partners in a P3. All that is required is clarity, transparency, and good planning.

---

[1] Robert Taylor, "Independent Regulation and Infrastructure Reform", online: ip3.org <http://www.ip3.org/pdf/2006_publicaation_012.pdf> (date accessed 20 April 2008).

[2] *Ibid.*

[3] Public-private partnerships, online: Wikipedia http://en.wikipedia.org/wiki/Public-private partnership (date accessed: 20 April 2008).

[4] Jerome Donovan, "Don't Want to Privatize? Then Corporatize (But Do it Right)" online: www.IP3.org

[5] Matthew L. Hensley, 'About 1P3: President's Welcome", online: www.ip3.org <http:..www.1p3.org/a_president.htm> date accessed 20 April 2008).

[6] Jamoke Jagun, Isabel Marques de Esa, "The Role and Importance of Independent Advisors in PPP Transactions: online: www.ip3.org <http://www.ip3.org/pdf/2006_publication_014.pdf> (date accessed 20 April 2008).

[7] Jacques Cook, "US PPP Market on the Upswing: Some Thoughts from Abroad: online: wwwIP3.org <http:///www.ip3.org/pdf/2007_publication_002.pdf> (date accessed 20 April 2008).

[8] Cook, *supra* note 7.

[9] A. Ng, Martin Loosemore, "Risk Allocation in the private provision of public infrastructure" (2007) 25 Int'l J of Project Management 66, 67.

[10] Xiao-Hua Jin and Hemanta Loloi, "Risk Allocation in Public-Private Partnership Projects – An Innovative Model with an Intelligent Approach", presented at The Construction and Building Research Conference of the Royal Institution of Chartered Surveyors, Georgia Tech, Atlanta USA, 6-7 September 2007, at 3.

[11] *Ibid.* at 3 – 4.

[12] ICAO Appendix to Final Report on the Work of the Secretariat Study Group on Legal Aspects of CNS/ATM Systems, C-WP/12197 3.4.1.

[13] *Wilson et al. v. ImageSat International N.V. et al*, Case No. 1:2007cv06176, New York Southern District Court; Jason A. Crook, "Corporate-Sovereign Symbiosis: *Wilson v. ImageSat Int'l*, Shareholders' Actions, and the Dualistic Nature of State-owned Corporations: (2007) 33 Journal of Space Law 2, 411.

[14] *Gulf Electroquip, Inc. v. University of Texas at Austin*, 2002 WL 480245 (Tex. App. Houston – 14th Dist. 2002); *see also GLF Construction Corp. v. LAN/STV*, 414 F.3d 553, 557 (5th Cir. 2005).

[15] Hugh M. Kindred, et. al *International Law: Chiefly as Interpreted and Applied in Canada*, 7th ed. (Toronto: Edmond Montgomery Publications Ltd., 2006) at 285.

[16] *Ibid.*

[17] Online: <http:..www.proz.com/kudos/latin_to_cnglish/law_patents/101818-rex_non_potest_peccare.html> (date accessed: 18 April 2008); M.Mofidi, "The Foreign Sovereign Immunities Act and the 'Commercial Activity' Exception: The Gulf between Theory and Practice" (1999) 5 J. Int'l Legal Stud. 95, 95.
[18] Kindred, *supra* note 15 at 290.
[19] *The Schooner Exchange v. M'Faddon*, 11 U.S. 116, 116 (1812).
[20] *Ibid.*
[21] 271 U.S. 562 (1926).
[22] *Permanent Mission of India to the United Nations v. City of New York*, 127 S.Ct. 2352, 2358 (2007) (Stevens, J. dissenting).
[23] Ex parte *Republic of Peru*, 318 U.S. 578, 588 (1943).
[24] Mofidi, *supra* note 17 at 99.
[25] Andreas F. Lowenfeld, "Claims Against Foreign States – A Proposal for Reform of Untied States Law" (1969) 44 N.Y.U. L. Rev. 901, 906.
[26] *Ibid.* citing The Tate Letter, 26 Dep't of State Bulletin 984 (1952).
[27] Council of Europe ETS no. 074, European Convention on State Immunity, Basle, 16 V.1972.
[28] 28 U.S.C. 1330(a).
[29] "Ad Hoc Committee on Jurisdictional Immunities of States and Their Property: online: <http://www.un.org/law/jurisdictionalimmunities/index.html> (date accessed 17 April 2008).
[30] The European Community is in discussion as to how best to denounce its prior Convention once it enters into force. 981 Meeting, 29 November 2006; 10.1 Committee of Legal Advisers on Public International Law, Appendix Three.
[31] *Ibid.*
[32] *Argentine Republic v. Amerada Hess Shipping Corp.*, 488 U.S. 428, 434 (1989).
[33] *Mission of India*, 127 S.Ct. at 2355.
[34] *Orient Mineral Co. v. Bank of China*, 506 F.3d 980, 991 (10th Cir. 2007) (citations omitted).
[35] 538 U.S. 468, 473 (2003) (affirming the Ninth Circuit holding that *Dole Foods* did not meet the definition of an instrumentality of a foreign state when the facts were applied to the five factors described herein. *See Patrickson v. Dole Food Co.*, 251 F.3d 795, 807 (9th Cir. 2001).
[36] *Board of Regents of the University of Texas System v. Nippon Telephone and Telegraph Corp.*, 478 F.3d 274, 279 (2007) (holding that regulation was not supervision, nor was research a government function or an exclusive obligation or right such that Japan's largest telecommunications company was immune from suit in a U.S. patent action as an instrumentality of the state); *see also Peninsula Asset Management (Cayman) Ltd. v. Hankook Tire Co.*, 476 F.3d 140 (2d Cir. 2007); *Filler v. Hanvit Bank*, 378 F.3d 213 (2d Cir. 2004).
[37] *Ministry of Defense and Support for the Armed Forces of the Islamic Republic Systems, Inc.*, 495 F.3d 1024, 1034 (9th Cir. 2007).
[38] *Ibid.*
[39] *Dole*, 538 U.S. at 478; *Olympia Express, Inc. v. Linee Aeree Italiane, S.P.A.*, 509 F.3d 347, 349 (7th Cir. 2007).
[40] *Leith v. Lufthansa German Airlines*, 897 F.Supp. 1115 (N.D. Ill 1995) (a US federal trial court decision).
[41] *Dole*, 538 U.S. at 474, 477.
[42] *Arbaugh v. Y & H Corp.*, 546 U.S. 500, 506 (2006).
[43] 28 U.S.C. 1608(e); *Cubic Defense*, 495 F.3d at 1028.
[44] *Weltover*, 504 U.S. at 611.
[45] 507 U.S. 349, 356 – 57 (1993).
[46] 28 U.S.C. 1605(a)(2).
[47] 504 U.S. at 618.
[48] 501 F.3d 534 (6th Cir. 2007).
[49] *Ibid.* at 540.
[50] *World Wide Minerals Ltd. v. Republic of Kazakhstan,* 296 F.3d 1154, 1164 -- 65 (2nd Cir. 2002) (citing *Banco Nacional de Cuba v. Sabbatino*, 376 U.S. 398, 401 (1964).
[51] *Weltover*, 504 U.S. at 613 – 14.
[52] *Ibid.* at 615.
[53] *Ibid.* at 616.
[54] 22 U.S.C. 288a(b).
[55] 2008 WL 441836 (D.C. Cir. February 14, 2008) at *2.
[56] *Ibid.* at *1.
[57] Executive Order 9698 of February 19, 1946, 11 F.R. 1809, [EO 9698 amended by EO 10083 of Oct. 10, 1949, 14 FR 6161, 3 CFR, 1949-1953 Comp., p. 284; EO 10533 of June 3, 1954, 19 FR 3289, 3 CFR, 1954-1958 Compl, p. 194].
[58] 28 U.S.C. 1491.
[59] *L-3 Communications Integrated Systems, L.P. v. United States*, 79 Fed. Cl. 453, 460 (2007). Interestingly, the Principal Deputy Secretary of the Air Force whose conviction for violating conflict of interest laws led to the suit against the United States was responsible for the largest P3 in US Air Force history, worth 10.1 billion USD.
[60] 28 U.S.C. 1346(a)(2).
[61] 525 U.S. 255, 256 (1999).

[62] 40 U.S.C. 3131 *et seq.*
[63] "The Miller Act: online: <http://www.sio.org/html/miller.html> (date accessed 17 April 2008).
[64] *Blue Fox*, 525 U.S. at 259 – 60.
[65] 28 U.S.C. 1346(b); 28 U.S.C. 2671 – 2680.
[66] *Feres v. United States*, 340 U.S. 135 (1950).
[67] 28 U.S.C. 2680(a).
[68] 28 U.S.C. 2680(h).
[69] 507 U.S. 197 (1993).
[70] U.S. CONST. Article III.
[71] 278 U.S. 194, 199 (1929).
[72] *Fresenius Medical Care Cardiovascular Resources, Inc. v. Puerto Rico and the Caribbean Cardiovascular Center Corp.*, 322 F.3d 56,61 (2003).
[73] *Ibid.* at 63.
[74] *Ali v. Daniel, Mann, Johnson & Mendenhall*, 355 F.3d 1140, 1147 (9th Cir. 2004); *Fresenius*, 322 F.3d at 62.
[75] *Barron v. Deloitte & Touche, L.L.P.*, 381 F.3d 438, 440 (5th Cir. 2004).
[76] *Del Campo v. Kennedy*, 517 F.3d 1070, 1077 (9th Cir. 2008).
[77] *Ibid.*
[78] Of note, the government contractor defense to the FTCA is a different thing entirely. There, a contractor of the federal government can claim immunity from torts committed if the contractor was following specifications given by the government, similar to the acquired immunity doctrine.
[79] *Lopez v. Mendez*, 432F..3d 829, 833 (8th Cir. 2005).
[80] *Ibid.*
[81] *McMahon v. Presidential Airways, Inc.*, 502 F.3d 1331 (11th Cir. 2007); *GLF Construction Corp. v. LAN/STV*, 414 F.3d 553 (5th Cir. 2005); *Butters v. Vance Intern., Inc.*, 225 F.3d 462 (4th Cir. 2000); *Hilbert v. Aeroquip, Inc.*, 486 F.Supp. 2d 135 (D. Mass. 2007); *Taylor v. Comsat Corp.*, Slip Copy, 2006 WL 3246508 (S.D. W. Va. 2006).
[82] "Interstate Compacts as a Means of Settling Disputes between States: (Jan. 1922) 35 Harvard L. Rev. 3, 322 – 26.
[83] *Hess v. Port Authority Trans-Hudson Corp.*, 513 U.S. 30, 43 (1994).
[84] 392 F. 3d 934, 935 (7th Cir. 2004).
[85] *Ibid.* at 936.
[86] Cook, *supra* note 7 at 1.
[87] *Takle v. University of Wisconsin Hosp. and Clinic Auth.*, 402 F.3d 768, 772 (7th Cir. 2005) (citing *Thiel v.State Bar of Wisconsin*, 94 F.3d 399 (7th Cir. 1996)).
[88] *Del Campo*, 517 F.3d at 1076.
[89] J.P.Singh, *Leapfrogging Development: The Political Economy of Telecommunications Restructuring* (Net Library 1999) at 19.
[90] Cledan Mandri-Perott, "Six Guiding Principles to Achieve Sustainable PPP Arrangements", online: www.ip3.org http://www.ip3.org/pub/2005_publication_002.htm (date accessed 18 April 2008).
[91] Jagun, *supra* note 5.
[92] *Treaty on Principles Governing the Activities of States in the Exploration and Use of Outer Space, including the Moon and Other Celestial Bodies*, 27 January 1967, 610 UNTS 205.
[93] Buckminster Fuller, *Operating Manual for Spaceship Earth*, (New York: E.P.Dutton & Co., 1963) section 8.
[94] *Outer Space Treaty*, *supra* note 92 at Art. VI.

IAC-08-E8.1.3

Capability and Fair Return in European and International Space Cooperation

Dipl.-Jur.-Ing. Cristian Bank
Astrium Space Transportation, Postfach 286156, D-28361 Bremen
cristian.bank@astrium.eads.net
Prof. Dr. Lesley Jane Smith
Weber-Steinhaus & Smith, Baumwollbörse, D-28195 Bremen
ljsmith@barkhof.uni-bremen.de

ABSTRACT

With the successful launch of COLUMBUS and ATV, Europe has become a partner in the operations of the International Space Station: COLUMBUS provides a working space not only for European scientists, but also for the international partners; the European astronaut corps participates in the operation and utilization of the ISS. Payloads, spare parts, consumables for the entire ISS are delivered by ATV. By doing so, Europe has demonstrated a political will to participate in the peaceful, cooperative endeavour of human spaceflight, following the Intergovernmental Agreement (IGA) between the USA, Russia, Japan, Canada and Europe

Although Europe has acquired a certain competence level in the area of human spaceflight through its own assets, one key asset is still lacking a solution: the launch and return of its astronauts. So far, European astronauts depend upon being launched by the US Shuttle and the Russian Soyuz.
This makes utilization of COLUMBUS difficult in the long run. Europe needs to decide if crew transport should be part of its human spaceflight activities. Such capabilities could be acquired through cooperation with other space-faring nations. In 2004, Russia has expressed such an invitation to Europe to jointly develop a new generation of spacecraft to follow the Soyuz family, at that stage under the name of Kliper. Cooperation of this kind would be governed by principles already applied for ISS, such as a "no exchange of funds" principle, leading to barter agreements rather than classical prime contractor – supplier contracts. Barter agreements are difficult to achieve, as the value of hardware development, services, engineering support, risk allocations etc. have to be agreed. Some lessons could be learned from previous projects such as ISS or Spacelab.

This presentation describes what are the challenges in elaborating a barter agreement for a crew transport vehicle from an industrial point of view – among them, the high frequency of deliveries, liability and risk issues, and most of all, the fair share in the utilization of the system. It also draws some conclusions from the ISS deals, again viewed from an industrial perspective.

1. INTRODUCTION

Through the combined efforts of the United States of America, Russia, Canada, Japan, and a number of European States, the International Space Station (ISS) will be assembled completely by late 2010. After more than 12 years of construction and utilization in parallel, the ISS will switch into a fully operational state.

These initial and subsequent steady state phase of operation and utilization were all agreed to by the signatories states to the legal framework of the ISS Intergovernmental Agreement (IGA), the bilateral Memoranda of Understanding (MoUs), and other side agreements such as the Astronaut's Code of Conduct (CoC) which were all negotiated and signed in the early 1990ies.

From a legal perspective, these international agreements have a dual nature: On the one hand, they clearly refer to the international space treaties of the 1960's and 70's on the other, they incorporate some elements of private law. They constitute a turning point in the concept of international space treaties.

Although the IGA and the MoUs are nowadays considered a contractual *sine qua non* for the formal conduct of the ISS programme and its control at highest political level, there are indisputable implications for industrial development of the programme. These could surface in the exploitation phase, resulting from e.g. delays, damages, warranty, and intellectual property issues. Given the complexity of the endeavour, not all of these effects can be anticipated, let alone avoided. Some of the observed effects are so substantial in terms of the financial, organizational, and technical implications that it is worth while identifying and analysing them to avoid similar results occurring in future cooperative space programmes.

One project requiring special attention and preparation in this respect is the Preparatory Phase for the Crew Space Transportation System (CSTS) to be performed jointly between Russia (represented by the Russian Space Agency RKA) and Europe (represented by the European Space Agency ESA)[1].

The main goal of the present paper is to highlight lessons learned in the contractual setup of the ISS which may serve for CSTS, concentrating in particular on the *industrial* implications of such regulations. Secondly, some emphasis is put on the identification of potential programmatic and organisational problems which may appear mainly in the development phase of CSTS.
These considerations may contribute to the preparation of the future agreements for the Development Phase of CSTS which is due for decision in November 2008.

2. The European - Russian CSTS Programme Initiative

CSTS today is the most prominent initiative of RKA and ESA that aims to identify and define a future cooperative space programme. It represents the first potential application of aspects and conclusions discussed within this paper.

2.1. Programme Initiative

The first initiative to cooperate in the development of a joint Euro-Russian crew transport vehicle was taken in 2003, when the Russian space systems company RSC Energia, Astrium Space Transportation, Bremen, Germany, and Thales Alenia Space Italy, Torino, conducted a joint study on a future space vehicle. RSC-E had developed a concept for an advanced space vehicle as successor for the Soyuz spacecraft, an idea dating back to the 1960s. The new spacecraft dubbed Kliper was intended to transport a crew of 6 astronauts to low Earth orbit (LEO), e.g. to the ISS, or to perform self-standing flights of some days with 2 professional pilots and 4 space tourists. The vehicle resembled Russian and US "lifting body" concepts of the 60s.

Based on this, a larger "Preparatory Programme for the Crew Space Transportation System" (CSTS) was initiated in 2006[2]. This programme encompassed both system study activities, as well as predevelopment of specific technologies. Some 14.7 M€ were invested by the governments of Belgium, France, Germany, Italy, the Netherlands, and Spain. Given their experience in Human Spaceflight (HSF), Thales-Alenia Space Italy and Astrium Space Transportation were selected to jointly lead the study.
Technical definitions for the vehicle aside, the elaboration of programmatic and organisational concepts for cooperation was a key issue to be tackled during the study which is to be concluded in January 2009.

2.2. Goals and Conditions of the Potential European-Russian Cooperation in CSTS

To identify open or potentially critical issues in the envisaged cooperation, it is important to understand the motivation of both sides in pursuing this programme. Geographical conditions impose constraints on the setup and conduct of the programme that in turn may give rise to implementation risks.

2.2.1. European Goals for Participating in CSTS

There are four main goals to be achieved for the European side in participating.
Firstly, the investment in COLUMBUS, ATV, the Nodes, Cupola, MPLM, ERA, payloads etc. have established a strong engineering and research community in industry and research institutes, as well as in ESA. With the end of the development phase of these programmes, the capabilities risk fading away if there is no follow-up activity in field of HSF. CSTS would be the natural successor programme for the European engineers.
Secondly, access for European astronauts to the ISS and in particular to COLUMBUS needs to be assured beyond 2013, when the current agreements on this come to an end. Europe assumes that, by participating in CSTS, it will obtain utilization rights – i.e. crew transport – in CSTS. Even if CSTS were only to become operative after 2016, the European hardware contributions to CSTS could be flight tested already

on Soyuz, and translate into crew transportation opportunities for European astronauts within that operational system before 2016.

Even though COLUMBUS, ATV and the other programmes have achieved recognized and well-established know-how in industry and research institutes, as well as in ESA, European competence in crew transport is lacking. CSTS should be the starting point for Europe for advancing the know-how in this area and building capability.

Last, but not least, although there is a well-established technology basis and engineering community in the area of HSF, the political support for HSF is not stable. As ISS related-activities have shown, an internationally anchored programme is much less vulnerable to being cancelled or de-scoped where political or public interest is dwindling.

2.2.2. Boundary Conditions for Inter-Agency CSTS Implementation

Among the conditions that influence the organisational and contractual backdrop to the CSTS programme, three aspects are the most decisive:

The participating agencies agree to a "no exchange of funds" approach, which leads to the need to barter or offset any contribution, either by a similar contribution from the other partner or through future utilization rights. This requires not only identifying comparable workshares, but also agreeing on a metrics to evaluate the contributions.

The successful launch of COLUMBUS and ATV, as well as the installation of Node 2 and the flawless operation of DMS-R, have led ESA to insist on a partnership among equals in the definition of the vehicle, its operation and utilization plans, and the sharing of system level engineering work. This has to be seen in the light of some 40 years of operation of Soyuz on the Russian side.

Finally one of the basic principles of how to install an industrial team in an ESA project has a strong impact on CSTS: the geo-return principle. This means that industrial contracts are placed in those countries which contribute financially to that programme and the value of the contracts corresponds to the financial share. The identification of the European workshare is not only determined by technical necessities, but also by georeturn constraints, which are even more difficult to achieve in a cooperative programme than in a purely European project.

3. Contractual Arrangements for European ISS Participation

In the past forty years, many practical solutions have been found by ESA to implement large space programmes within Europe taking into account the above or similar boundary conditions. The setup of an international programme necessitates to set up international agreements, which is not necessarily tailored to respond to inner-European considerations and agreements. One of the most prominent programmes which had to overcome such problems was the ISS, where Europe however only played a minor role. The European model for ISS participation needs to be assessed to see how far the solutions found for ISS are suitable as a model for CSTS.

The ISS is based both on general international space treaties as well as on ISS specific international and bilateral agreements.

3.1. General Legal Framework for the International Space Station

The ISS was developed with the major space treaties as the primary international instruments. As between the Registration and Liability Conventions and the Rescue Agreement, the REG has a particular role in relation to the registration of space objects. A "Space Object" encompasses components or parts of a space object, as well as its launch vehicle and parts thereof. REG was written at a time when only single satellites were launched into space. The ISS is, however, a space object assembled in space (not on Earth). The Russian elements are launched separately by individual launchers, and approach the ISS independently as individual spacecraft until they dock. At this stage they become part of the ISS. The US, Japanese, and European modules, as well as the other Canadian and US constituents (arm, truss etc.) are, in turn, launched by the US Space Shuttles and docked to the ISS only after the docking of the Shuttle to the ISS. As a result, these elements never become individual spacecraft, but are always either part of the US Shuttle or the ISS. This eliminates e.g. the risk of these elements leaving their foreseen trajectory and posing a risk to other spacecraft or the ISS.

It is evident that the content of these Conventions addresses risks given by the technical limitations and the spirit of the human spaceflight programmes of the late Sixties, such as Gemini, Soyuz and Apollo. For the ISS, conceived after introduction of the US Space Shuttle and following decades of successful operation of the Saljut, Almaz, Mir, and Skylab stations, these regulations were far from satisfactory. The new technical opportunities and the political goal to join forces in space following the end of the Cold War had to be reflected in new, specific treaties. Still, all ISS Agreements are connected to the main provisons of of these three Treaties.

3.2. Agreements Directly Applicable to European ISS Contribution

The specific legal framework of the International Space Station is based on two cornerstones in the form of international cooperation agreements. The first and key document on governmental level is the IGA from 1998. The second group of documents are the Agreements at agency level, which means a total of 4 Memoranda of Understanding (MoU) between NASA and each cooperating Agency (ESA, CSA, RKA, and JAXA).

3.2.1. ISS Intergovernmental Agreement (IGA)

As for a co-operative programme, the ISS IGA was signed by the Governments of Member States of the ESA, the Government of Japan, the Government of Canada, the Government of the Russian Federation, and the Government of the USA in January 1998. The IGA provides the framework for design, development, operation, and utilization of a permanently inhabited civil Space Station for peaceful purposes while maintaining a long-term and mutually beneficial relationship.

All partners join in their efforts under the lead role of the USA for overall management and coordination. Given their extensive experience in human space flight, the USA and Russia produce the main elements which serve as foundation of the ISS.[3] Canada is an essential part of the Station, because of its contributions, as well are the European Partners and Japan, who both enhance the Station's capability with their contributed elements.

Pursuant to this IGA, the ISS is to be developed, operated, and utilized in accordance with international law, including the Outer Space Treaty,[4] the Rescue Agreement[5], the Liability Convention[6], and the Registration Convention[7].
The following summary briefly reflects the most important points in the IGA:

Ownership of Elements and Equipment:
The Partners, acting through their Cooperating Agencies shall own the elements that they respectively provide. The rights and obligations of the Partners are not affected by transfer of ownership of elements or equipment, as long as it is not a transfer to a non-partner or private entity:, any equipment and elements are not to be owned by, or transferred to any non-Partner or private entity under the jurisdiction of a non-Partner, without the prior agreement of the other Partners.[8]

Utilization:
The rights of utilization are derived from the estimated value share of the total ISS value of each Partner's provision to the ISS, e.g. a user element, infrastructure elements, or both. Any Partner that provides ISS user elements shall retain use of those elements and obtains a similar share of the ISS resources, e.g. crew time, power, air, communication, etc. All Partners have the right to sell or barter any portion of their respective allocations, whereas all terms of conditions of any sale or barter determined on a case-by-case basis by transaction parties. (Barter is the exchange of goods or services for other products or services, crew transportation, resources etc.) In case one Partner is not able to provide a certain service (e.g. the launch of its element to the ISS), he can barter this service from another Partner in exchange of providing a to be determined product or service for that Partner. As an example, Europe did not pay for the launch of COLUMBUS, but provided elements to NASA.

Operation:
All Partners are responsible for their elements provided, as well as for sustaining the functional performance of their provided elements. All operation regulations are in accordance with all implementing arrangements and the MoU's.

Cross-Waiver of Liability:
The Cross-waiver of Liability (CWL) is to encourage participation in the exploration, exploitation and use of Outer Space through the ISS and it is important to establish it by the Partner States and the related entities. For any damage resulting from protected space operations, each

Partner State waives all claims against another Partner State, a related entity of another Partner State or the employees of any of the entities of another Partner State or related entity from this.
Furthermore each Partner is to extent that the CWL as mentioned above to its related entities by requiring them to waive all claims against another Partner State, a related entity of another Partner State or the employees of any of the entities mentioned before. Moreover each Partner require of its related entities that these waive all claims against another Partner State, against a related entity of another Partner State or the employees of any other Partner State or related entity of another Partner State. Both options are to be realized by contract or otherwise.
The CWL is not be applicable in claims between a Partner State and its related entity or between its own related entities, as well as claims made by a natural person, his or her estate, survivors or subrogates (except it's a Partner State) for bodily injury, or other impairment of health of, or death of such a natural person, plus claims for damage caused by willful misconduct, and claims for damage arising from a failure of a Partner State to extend the CWL to its related entities,.

Exchange of Data and Goods:
Each partner shall transfer all necessary technical data and goods to fulfill the responsibilities of their cooperating Agencies. At the same time, this shall not require any Partner State to transfer any technical data and goods in violation of its national laws and regulations. All partners shall provide the protection of technical data for the purpose of proprietary rights. In general, all Partner States and their Cooperating Agencies shall take all necessary steps to prevent unauthorized use, disclosure, retransfer of, or unauthorized access to, any kind of such technical data or goods.

Intellectual Property:
"Intellectual Property" shall include pursuant to Article 2 of the Convention establishing the World Intellectual Property Organization the rights relating to literary, artistic and scientific works, broadcasts, inventions in all fields of human endeavour, scientific discoveries, industrial design, trademarks, service marks, and commercial names and designations; protection against unfair competition; and all other rights resulting from intellectual activity in the industrial, scientific, literary or artistic field.[9] For any kind of activity occurring in or on a Space Station flight element, the purpose of intellectual property law shall be considered to have occurred only in the territory of a Partner State of that element's registry.[10]

3.2.2. MoU NASA - ESA

The MoU is a more detailed amendment to the higher ranking IGA, which provides interpretation of the responsibilities and the provided items of each participating agency and respectively their Governments. The purpose of this MoU is to establish arrangements between NASA and ESA implementing the provisions of the IGA, including the respective rights and obligations of each party. An MoU provides a good possibility to arrange some facts and regulations perfectly matched to each of the participating agencies.
Neither the MoU between NASA and ESA nor the MoU between NASA and RSA inform in a detailed way about handling cases of liability. Both MoU's refer again to the IGA and the Liability Convention which discusses damages caused by Space Objects.
Many other aspects of the above descibed IGA are defined in more detail in the bilateral MoU, the detailed discussion of which would however exceed the scope of this paper.

3.2.3. ISS Astronauts' CoC

The ISS Code of Conduct (CoC) defines the rules concerning the life and behaviour on the ISS. All ISS partners needed to find an appropriate balance between the military heritage of the USA and the Russian spaceflight programme and the civilian and multi-national character of the ISS. They approved the Astronauts' CoC on 15 September 2000 through the Multinational Coordination Board, the highest ISS coordination body.[11]
The CoC establishes a clear chain of command and relationship between ground and on-orbit management, responsibilities with respect to elements and equipment, standards of work, and disciplinary regulations, together with physical and information security guidelines.[12] Furthermore the Code has to define for the ISS the ISS Commander's authority and responsibility to enforce safety procedures, physical and information security and crew rescue procedures. The Agencies are interested to persuade the astronauts to abide by the rules outlined in the Code,

albeit on a voluntary basis, enabling them to pursue astronaut activities as employees of a Cooperating Agency.

Onboard the US Space Shuttle these matters are covered by regulations adopted under the authority of NASA. For crew members being launched or returned on a Russian space vehicle, specific Russian regulatory apply.

Legal requirements imposed on the ISS crew:
Primarily the provisions of the IGA, the MoU's and the ISS Flight Rules are directly applicable to astronauts' activities. The CoC applies to an ISS crew member from the time he or she is assigned to a specific ISS expedition until completion of post-flight activities.[13] It furthermore applies to visiting crew members, who stay just for a few days.[14]
All Crew members have the responsibility to protect and conserve all property to which they have access for ISS activities. Furthermore they don't have the permission of altering or removing such property for any other purposes than those necessary for the performance of the ISS duties.[15]
Moreover one of the main objectives of the Astronauts' CoC is to obtain the person's consent to be subject to the authority, direction and orders of the Commander, as well as to limit the disclosure of data which are protected, and to stop them from using their position or information obtained in the course of the mission for private or financial gain for himself or herself or other persons or entities.

Authority of the ISS Commander:
The ISS Commander can be any national of any Partner state, and he or she is the highest authority among the ISS crew members on-orbit. His task is to maintain a cohesive relationship among the ISS crew members which duly takes into account the international and multicultural nature of the crew. During on-orbit operations, the Commander is responsible for ensuring the safety of the ISS crewmembers and the protection of the ISS elements. Moreover he ensures the successful completion of the mission. He is directed by the Flight Director on ground. The ISS Commander is entitled to change the daily routine of the ISS crew members to address contingencies, perform urgent work associated with crew safety and the protection of the ISS elements, equipment or payload, or conduct critical flight operations.

Disciplinary Regulations:
All crewmembers are subject to a disciplinary policy, which was developed and revised by the MCOP (Multilateral Coordination Organisation Panel) and which is administrative in nature and intended to address violations of this CoC. However, this policy does not limit a Cooperating Agency's right to apply relevant laws, regulations, policies, and procedures to the ISS crewmembers it provided, consistent with the IGA and the MoUs. It shall be borne in mind that the IGA concerted that each State retains jurisdiction and control over personnel who are its nationals. Physical force or restraint may be used by the ISS Commander only when immediate safety is jeopardized for the crew and after exhaustion of other possibilities.

Proprietary and export controlled data generated in or on the ISS:
There is a need to protect data generated by activities conducted in or on the ISS when such data can be considered to be "proprietary" or "export-controlled". Under Article 19 IGA that protection of the corresponding data is linked to the fact that they are marked with an appropriate notice or otherwise identified. It is up to the Cooperating Agencies, the data owner, or provider to give instructions for the marking of data generated on board of the Station.

3.3. Implications for European ISS participation

Apart from the barter items such as Node 2 & 3, the Cupola, DMS-R, or MELFI, the main contributions of Europe to the ISS are COLUMBUS, ATV, and ERA. ERA is not be examined within this paper, as its operational role on the ISS still remains unclear. In contrast, COLUMBUS and ATV, as the two largest items under European design authority, are shortly to be introduced:

COLUMBUS was designed as a permanently docked general-purpose ISS laboratory to accommodate payloads in a pressurized environment.[16] It was constructed and developed by an industrial consortium under the lead of Astrium Space Transportation, and is the cornerstone of Europe's contribution to the ISS. Once COLUMBUS was attached to the ISS and checked, ESA became an active partner in the operations and utilization of the ISS. COLUMBUS is designed for ten years of operations, and will be controlled by the COLUMBUS Control Centre located near München. The laboratory fits into the cargo bay of a Space Shuttle and was pre-equipped with five internal racks during launch. It will be able to conduct experiments in life science, materials science, fluid physics, and a lot of other disciplines in orbit.[17]

The Barter Arrangements relating to COLUMBUS included a provision that NASA launch the COLUMBUS module and its initial payload on the Shuttle. This launch is barter in compensation for ESA's provisions to NASA of the fully integrated Node-2 and Node-3 ISS Modules, Cryogenic Freezer and Crew Refrigerator / Freezer equipment for ISS and spares and sustaining engineering for the Laboratory Support Equipment items.[18]
It also falls within ESA's provisions to NASA of hardware/support for software development and integration in the NASA ground software test and integration facilities for ISS.

Bartering for the COLUMBUS launch included at least two benefits for Europe: firstly, the risk of price uncertainties and cash payment to NASA was regulated through fixed conditions, secondly the creation of additional industrial work for Europe in high-technology domains.

ATV is the Automated Transfer Vehicle, which was developed in Europe for the supply of cargo to the ISS.[19] The ATV has acapability for carrying different kinds of freight to the stations in evacuated and/or pressurized containers, and is able to dock automatically to the ISS. The ATV provides delivery of dry and liquid cargo (e.g. experiments, foods, compressed air and water), refueling of the Station, which means the transport of propellant to the FGB, Reboost and attitude control during reboost of the whole station, i.e. orbit corrections using the ATV propulsion system to compensate for the continuous loss of altitude by the Station, and removal of waste from the Station followed by controlled destructive reentry of the ATV.[20] After completion of its mission, the vehicle will be brought on to a reentry trajectory which will cause the vehicle and its on-board disposable items to burn up in the Earth's atmosphere.

ATV is directly connected to the European COLUMBUS module because of ISS Common Operating Costs. The ATV provided the cargo which is used by the European ISS Partners as bartering for the utilization of the ISS with COLUMBUS. So far, the production of additional ATV's depends on the duration of use of COLUMBUS. The longer the ISS is operated as a whole, the longer COLUMBUS can be utilized by the Europeans. Under the agreement, Europeans must pay ISS Common Operation Costs to the United States for its use. To avoid price escalations, political changes and cash payment to the USA, the ESA pays with cargo transport through ATVs, so that this again is a barter deal. Until 2013, the ISS Operation Costs are bartered with a total of some 18t cargo transport. This is distributed to 5 ATVs. The capacity of cargo transport for one ATV is down to 9.5t, therefore each ATV has the capacity to transport some 5t for European purposes only.

Both programmes, COLUMBUS and ATV, after their successful development, launch and in-orbit operations, are now valuable contributions to the ISS. However, throughout the development phase and, to a certain extent even now, the legal arrangements contain some considerable risks for Europe, both for ESA and the industrial prime contractor. To mention a number of these issues:

Ownership of Hardware and Data

- COLUMBUS is registered as spacecraft by ESA, but first became part of the US Shuttle during launch, and later, after docking to Node 2, part of the ISS. It is doubtful whether, in case of a Shuttle launch failure and subsequent damage caused by COLUMBUS, there was a clear position in relation to ESA and NASA liability. The same consideration applies in case of damage by the ISS when reentering the atmosphere at the end of its lifetime (currently expected around 2020). It would be interesting to see whether, in such a case, the principle of "joint and several liability" does become applicable.
- The regulations about the "Exchange of Data and Goods" and the protection of IP between the industrial partners are anything but satisfactory for the European industry as against the US and Russian Primes. While industrial documents from the US and Russia are usually declared confidential and must not be disclosed to the international industrial partners by standard US and Rus-

sian domestic regulations, the Western European states have failed to install and enforce similar regulations. This is mainly due to the safety regulations which require all Partners to disclose their design documents to NASA and RKA for review, while NASA and RKA have no reason to disclose any of their design data to the Partners. The implementation of this regulation became even more stringent after Sept., 11, 2001. Practically, industry had to negotiate and install Non-Disclosure Agreements (NDA) on a very detailed level.

- Similarly, the fact that scientific and engineering data needs to be transmitted through the US data relay network[21] casts doubt as to confidentiality of the data, particularly when industry and research institutes consider using COLUBUS for non-public research. Even worse, in case of internal encryption, the key must be disclosed to NASA. The regulations in the IGA and the CoC are unfortunately weak – the data owner needs to be pro-active in declaring the data confidential, otherwise protection is difficult to maintain.

COLUMBUS Utilization

- One positive result of the regulations on utilization is the compatibility of all experiment installations in the Western modules (Destiny, COLUMBUS, Kibo) with the International Standard Payload Racks (ISPR), which makes them exchangeable.
- After experiencing the transfer of ownership of Spacelab to NASA after the first flight[22] in the 1980's, the fact that Europe keeps ownership of COLUMBUS and NASA gets 46.7% of the COLUMBUS research capacity may be considered a step forward. The limited research capacity and storage volume is theoretically a drawback for Europe. (It is questionable whether the contributions to the ISS by the international partners was – and could be - correctly evaluated at the beginning of the endeavour.) For the time being it appears that these constraints have been handled very flexible by the partners, but it remains to be seen if that still is true when the ISS starts to get crowded by stored spares and payloads.
- The fact that NASA payloads will be installed in COLUMBUS required the inclusion of the US Payload LAN and specific TM/TC avionics. This caused late definition of interface data, late changes and additional compatibility tests with schedule and cost impact on industry. As there was no direct contractual relation between the industrial consortia, these changes could only be introduced by a complex and painstaking process governed by the Agencies.
- Outside of the IGA and MoU, but part of the same discussion: A lot of effort was spent by ESA to obtain early access to the research capacity of the ISS: ESA provided experimental and system hardware to Russia and the US in exchange for ISS crew time and resources. Unfortunately, mainly due to delays in the ISS launch manifest and malfunctions, only few flight opportunities and crew time - which means research opportunities for European scientists & industry - was received in return or was provided very much later than originally foreseen.
- The most prominent "late delivery" of a bartered service was the launch of COLUMBUS: originally foreseen for 2002, it slipped considerably due to Shuttle launch manifest delays, which resulted in loss of research time or even "brain drain" within the European research community, additional costs for ESA and industry, late start of P/L development activities, while all along the bartering costs for Europe had been incurred much earlier. The legal construction did not foresee any offset for this kind of delay. On the other hand, payment in cash of the launch service would have been very costly for Europe and would have meant a loss of workload for the European industry.

ISS & COLUMBUS Operations

- The international agreements state that NASA & RKA operate the ISS, while ESA only contributes. This is a clear distinction in the roles of the International Partners. While this may have had

good reasons in the early 1990's, a similar distinction in the future would hardly be acceptable for Europe, given the experience meanwhile obtained in the meantime in Human Spaceflight. Such a distinction also has consequences for industry, as was shown in the COLUMBUS programme: not only that certain work packages would again not be accessible for the European industry, but also fundamental system operations principles could be taken into account only late in the design process, which would results in additional design and schedule risks.
- The Joint System Ops Panel with experts from each agency allowed the exchange of lessons learned and early coordination of P/Ls and experiments.

COLUMBUS Design
- While it makes sense that NASA provides the overall ISS specification, safety regulations, the launch manifest etc., there is an unavoidable impact on the assessment of the associated costs for the European industry by late changes or, even worse, late definition after contract signature within Europe, impact on milestone and review planning, etc.
- Even if safety regulations, test specifications etc. are defined, there is always room for interpretation. It is doubtful whether this interpretation was always the same for each of the International Partners.

ISS Code of Conduct
A definitely positive regulation was established with the CoC:
- the clearly defined hierarchy and roles within the ISS crew,
- the discussion of the use of force (even if the conditions remain somewhat vague), and
- the openness of any role to any nationality (onboard Soyuz / Shuttle there is always a Russian / US commander)

make the CoC certainly one of the most balanced of all ISS related agreements.

4. Identification of Potential Problems and Open Issues in CSTS

As discussed in the previous chapter, the international regulations governing the cooperation on the ISS have produced some innovative legal constructions, yet they have left some points open. The agreements address specifically those issues which could have been foreseen at the early 1990's as typical situations when operating an all-prototype, multi-national, permanently operated, revolutionary in-orbit research facility. CSTS would generate a completely different development and utilization scenario, as it is a series production type, evolutionary, bi-national vehicle which will see short missions and frequent launches. It is obvious that such a system requires an additional regulatory framework between Russia and Europe, the former has already achieved decades of successful manned spacecraft operation, while the latter is striving to obtain this capability.
The following paragraphs name but a few of these specific questions pertinent to CSTS that are already surfacing at the beginning of the development phase:

Relation of Contractual Situation between the industrial partners
- There must be a Technical Exchange Agreement put in place before the start of the next phase for a save transfer of information on technology employed as well as know-how, dealing with intellectual property rights, licensing of existing and protected IP if needed for CSTS as well as those IP generated in the programme. Also, clear directions concerning the transfer of Property and Possession respectively transfers of Ownership and Security Rights (see below CSTS vehicle) need to be formulated.
- Regarding the Exchange of Goods, Services, and Information encompassing the main products of the programme, it is understood that an up-front clarification of the general "payment" approach (e.g. bartering) should take place at agency level. However, industry and agencies need to agree on a "second level" approach and processing of direct B2B solutions for additional engineering support or maintenance, supply of attrition, spares or so-called "forgotten

items", commercial products, etc. at short notice.
- Furthermore, there should be an explicit position elaborated between both industrial partners related to liability (e.g. product liability, risk sharing…), in relation to the roles of ESA and RKA (defined conditions for share of risk, encompassing financial and liability risks). As NASA is doing more work itself than RKA, which involves more industry for performing review work etc., there is a higher risk for European industry when disclosing proprietary data to RKA.

Metrics to evaluate the values of each product and service
- It is necessary to clarify the value of each individual product, the integrated element, and any service. Therefore it is important to develop metrics or a method of evaluation. Each contribution to the jointly developed system should be evaluated and added up to determine its contribution to the overall value, and to deduce thereafter the corresponding utilization rights.
- This process needs to be consistent throughout the system's lifetime and allow for continuous update as a means of reflecting future evolutions, updates, add-ons, changes in workshare, etc.
- Any additional effort by one party resulting from requirements and interface changes with the other party during the development phase needs to be addressed, esp. liability and compensation (e.g. by utilization rights);
- As an operational transportation system, the launch & operations costs of CSTS are a significant part of the overall value and need to be assessed and fixed for a long time in advance, also cost escalation mechanisms need to be agreed.

Ownership of CSTS vehicle, elements, and equipment
- It is a major issue to align ownership of the CSTS vehicle and its elements when combining the elements to a full vehicle during the integration and test phase in order to allocate liability between the industrial partners. (note: Ownership is one key aspect which also governs all issues of registration and thereafter liability at international level., This aspect is, however, assumed to be regulated at agency level).
- Moreover, the basis of industrial claims relating to the following remain to be clarified:
    - Participation in intellectual property rights (IPR) & know-how obtained during the programme;
    - Handling of IPR relating to vehicle design;
    - Handover of IPR in case of withdrawal of one partner from the programme to allow the other partner to build and operate the system.

Identification of Settlement Procedures
- It is unlikely that, even with the utmost care, preparation and negotiation, all eventualities that may appear throughout the development, verification, and operation of the system can be fully clarified. It is therefore vital to set up bodies and procedures which allow for settlement of political issues. Any uncertainty may result in a considerable risk or even cost-impact on the partners. On the other hand, clear decision authority and responsibility for the product must be defined and allocated to industry.

Operation and Utilization
- A more detailed discussion about both the joint operation and utilization of the system as well as the related liability issues should be done between all governmental and industrial partners involved in the next phase of the programme. This is a critical point for industry, as it addresses e.g. the compatibility with more than one launcher, as the option to use the system for purely European missions requires a clear settlement of this issue.

5. Evaluation of the ISS Legal Framework with Respect to CSTS

In the light of the above discussion, there are indeed regulations developed for the "all-prototype" permanently operated orbital research facility ISS which could serve as blueprint for CSTS. Some questions specifically relating to the development, operation, and utilization of a quasi series production evolutionary crew transportation system require new answers.

Suitable regulations to be transferred from ISS to CSTS for adaptation and application could encompass

- the CoC with its definition of crew roles, tasks distribution, and conduct;
- the IGA/MoU regulations about bartering, safety, and operations coordination.

Regulations which seem to need review before being applied to CSTS are e.g.

- the IGA / MoU regulations about ownership of the system's constituents, launch, utilization rights, and data confidentiality.

The ISS Cross Waiver of Liability may have been suitable as a very special way of dealing with potential damages for ISS purposes, but whether all or part of it can be applied to a crew transportation system the purpose, operational mode and hence risk structure of which differs greatly from the ISS, requires thorough analysis.

Regulations to improve the direct interaction between the industrial partners, especially for interface definition and in critical phases of the programme such as specification, integration and test, are definitely missing from within the ISS framework.

All in all, the balance of bartering is positive, because it helps to substantiate cooperation at an early stage of programme definition. A barter arrangement avoids the risk of changing political and economic conditions, as well as the risk factor of scheduling problems. The allocation of additional work to the European industry was a huge support for the barter arrangements: this supported the spirit of partnership in the global environment of the ISS. Bartering has had an advantage, especially for ESA, in those cases where a barter arrangement avoided cash payments to non-Member States and instead permitted that money to be invested in contracts with European industry. On the other hand, some barter agreements causing considerable costs suffered from late or modified fulfillment.

6. SUMMARY

This paper has reviewed some of the industrial implications of the intergovernmental regulations governing the ISS. It also highlighted some open legal issues emerging at this early stage of discussion for the potential future cooperative project CSTS. It is evident that the regulations put in place in some instances for ISS resulted in implementation risks for industry. These are far from being suitable to serve as a blueprint for future cooperative space projects. This is particularly true when the cooperation encompasses not only operation and utilization, but the continuous production and launch of elements. Such shortcomings require significant work in the legal domain over the coming years, in particular with a view to the plans for Lunar exploration programmes. As the ISS experience has shown, it is highly recommended to perform this exercise in close cooperation with industry with a view to subsequent programme implementation. This is the best way of averting industry exposure to risks which are avoidable through proper preparation.

---

[1] ESA CSTS Programme Declaration, 2006
[2] ESA IPC Paper on CSTS, 2006
[3] ISS IGA, Article 1.2.
[4] Treaty on Principals Governing the Activities of States in the Exploration and Use of Outer Space, including the Moon and Other Celestial Bodies (3 Dec. 1968).
[5] Agreement on the Rescue of Astronauts, the Return of Astronauts, and the Return of Objects Launched into Outer Space (10 Oct. 1967).
[6] Convention on International Liability for Damage caused by Space Objects (1 Sept. 1972).
[7] Convention on Registration of Objects launched into Outer Space (15 Sept. 1976).
[8] ISS IGA, Article 6.3. / 6.4. and subsequent articles.
[9] http://www.wipo.int/treaties/en/convention/trtdocs_wo029.html#P50_1504 (Convention establishing the World Intellectual Property Organization, 1967, § 2.8.).
[10] Ibid. IGA, Article 21.2.
[11] http://portal.unesco.org/shs/en/files/8480/11224719601Farand.pdf/Farand.pdf, Seite 1/ Article 11 MoU.
[12] http://www.esa.int/esapub/bulletin/bullet105/bul105_6.pdf, pages 64 ff./ Article 11.6. MoU.
[13] Astronauts' CoC (hereinafter CoC), Article II.A.
[14]http://www.esa.int/esapub/bulletin/bullet105/bul105_6.pdf, p. 65.
[15] Ibid. CoC, Article II.B and subsequent articles.
[16] Messerschmidt, E./ Bertrand, R., Space Stations, 13.2.2, 480.
[17] http://www.esa.int/esaHS/ESAAYI0VMOC_iss_0.html
[18] http://www.nasa.gov/centers/johnson/news/releases/1996_1998/97-36.html
[19] Messerschmidt, E./ Bertrand, R., id. 12.1.3, 425.
[20] Messerschmidt, E./ Bertrand, R., Space Stations, id. 12.1.3. 426.
[21] Due to the deletion of the COLUMBUS Data Relay Terminal in the 1996 baseline development contract.
[22] The European Spacelab was considered an in kind contribution to the „US National Space Transportation System". The "utilization rights" were limited to one joint ESA-NASA mission.

SPACE TOURISM: private law implications

By

Dr. P.P.C. Haanappel[1]

Abstract

Space tourism, as envisaged at the time of writing in 2008, still rather experimental and potentially hazardous in nature, does not seem to call for new private international law instruments, in the form of treaties or otherwise. Existing freedom of contract, equality of bargaining power, contract law, tort law, the law of obligations, insurance law, private international law, and the law of conflicts will be able to solve liability and other private law issues arising out of space tourism activities. This stands in contrast with future, more elaborate space *transportation* systems where new private international law instruments might be required.

Text

Exactly thirty years ago, in the summer of 1978, Professor Hamilton DeSaussure and the author of this paper met on the shores of Lake Champlain in upstate New York to write an article, entitled "A Unified Multinational Approach to the Application of Tort and Contract Principles to Outer Space".[2] At that time, there was a lot of speculation about man's 'living and working in outer space'. In retrospect, a good deal of this proved to be science fiction like speculation, and certainly the presence of the 'ordinary' man, the non-professional astronaut in outer space, has been rather less extensive than foreseen. Much later, the scientific, technical and legal literature would move on to the subject of space transportation systems: ultra rapid transportation from one place on earth to another by a hybrid vehicle, with space shuttle capabilities, possessing characteristics of both aircraft and spacecraft, through air space and 'near' outer space, perhaps as successor to the now defunct civilian supersonic jet aircraft.[3] Today, 'space tourism' is a topical subject, but to date there are only two known applications: tourism to and from the International Space Station (ISS),[4] and Richard Branson's Space Ship Two which should be operational in 2009, operated by Virgin Galactic, initially expecting some 200 interested tourists to travel to a height up to 120 kilometers from the surface earth, in suborbital flight.[5]

In 1978, DeSaussure and Haanappel pleaded for a private international law approach, a conflicts of law approach to private law implications of torts and contracts (obligations) in outer space, possibly reinforced by a private international law convention or agreement unifying conflict rules.[6]

It is submitted that the two currently foreseen forms of space tourism are not extensive enough in scope to call for international private law conventions either unifying the substantive law applicable to such activities or the conflict rules surrounding their applicability.

Contracts for the transportation of private citizens to and from the ISS are so rare and expensive that complete freedom of contract between, presumably, parties of roughly equal bargaining power should govern them, including risk allocation for mishaps, insurance provisions (if insurance is at all possible), choice of applicable law and forum for dispute settlement.

Virgin Galactic contracts may not become commonplace, calling for international conventions or agreements governing them, but they are, it is submitted, consumer type contracts where the supplier of services dictates the contents of the contract to the consumer, the client, the tourist. This writer has not been able, on short notice, to register as a Space Ship Two client, let alone to obtain price and conditions of contract. May be the explosion that occurred during a Space Ship Two fuel flow test on 26 July 2007, killing three and injuring two persons, have set things back. At any rate, it begs the question whether Virgin Galactic / Space Ship Two is (still) a hazardous activity and whether it is insurable, both of which have a bearing on liability questions. This author would hope that EU Directive 90/3141 EEC on package travel would apply to trips on board Space Ship Two.

Not until we get to real space transportation systems will international conventions unifying substantive law or conflict rules become necessary. Perhaps the Warsaw 1929 / Montreal 1999 air transport liability system can server as a model or be declared applicable.[7]

References

1 Professor of Air and Space Law, Leiden University; Adjunct Professor of Law, McGill University; member IAA and IISL.

2 Published in (1979) Syracuse J. of International Law and Commerce 1-15.

3 See Haanappel, *The Law and Policy of Air Space and Outer Space, A Comparative Approach*, Kluwer Law International, The Hague / London / New York, 2003, at p. 161-163.

4 The first passenger / tourist being Mr. Dennis Tito who, in 2001, traveled from earth to the ISS and back on board the Russian Soyuz: *ibidem* at p. xiii and 161. To date, there have been five such passengers / tourists, paying large sums of money for the trip (reportedly some 20 million dollars). See also von der Dunk and Brus, eds., *The International Space Station, Commercial Utilisation from a European Legal Perspective*, Nijhoff, Leiden / Boston, 2006.

5 Thus, just at about what most consider being the boundary between airspace and outer space, and that at about one tenth of the price mentioned *supra* in footnote 4.

6 Cf. the website of the Hague Conference on Private International Law for treaties unifying conflict/ private international law rules, and also Haanappel, *Product Liability in Space Law*, in: *The Role of Private Enterprise in Outer Space: International Legal Implications*, (1979) 2 Houston J. of International Law 55-64.

7 On this subject, see Haanappel, *op.cit.* footnote 3, at p. 99-101.

THE NEED FOR A UNIFORM LAW SYSTEM PROTECTING INTELLECTUAL PROPERT RIGHTS IN OUTER SPACE

Dr. Hanneke van Traa-Engelman
The Netherlands
hlvantraa@box.nl

ABSTRACT

Effective intellectual property protection will become a key element in the regulatory environment for the commercial exploitation of outer space. Presently, intellectual property rights are protected by national or regional laws based on the very principle of territoriality. In order to protect intellectual property rights in outer space a solution has been found by extending the national law of the state of registry of a space object. Hence, as is the case for patents on earth, different national laws apply different regulations, providing conditions of uncertainty and conflict. This obstructs in particular private parties seeking to invest in space technology- and applications. A solution should be found in the establishment of an international framework for the protection of intellectual property rights in outer space, in particular in the field of patents. The chance of success of the creation of a Space Patent Treaty will depend on the willingness of states to reconcile their differences in patent law systems. Positive efforts will in turn accelerate the process of further harmonisation of terrestrial patent law. This might finally lead to a uniform and transparent patent law system for inventions made on earth as well as in outer space.

INTRODUCTION

Whilst advance in a wide range of space technologies and applications will create more potential for new space applications by the private sector, the protection of intellectual property rights will play an important role in attracting the participation of the private sector. Effective intellectual property protection will become a key element in the regulatory environment for the commercial exploitation of outer space, with specific emphasis on the most important intellectual property right, the right of invention. Furthermore, the World Intellectual Property Organisation (WIPO) noted that the effective acquisition and protection of intellectual property rights would have a positive effect on the participation of the private sector in the development of outer space activities and of the further development of space technology in general[1].

Moreover, as many space projects involve international co-operation there is a need for a simple, uniform and reliable international legal framework. Presently this is not the case.

THE CURRENT SITUATION

Intellectual property rights are established and protected by national or regional law of the countries concerned based on the principle of territoriality. Different national laws apply to different regulations. Although some harmonisation has taken place[2], there is still room for uncertainty and conflict. Moreover, inventions undertaken in outer space have brought another dimension to the legal environment in which these take place and where international space law plays an important role. First, the question arose whether international space law would permit the protection of intellectual property rights of activities in outer space and if so, on which basis those rights can be established and protected. The Outer Space Treaty (OST) determines outer space as a common area free for exploration and use by all states (Art. 1). Moreover it provides that outer space, including the moon and other celestial bodies, is not subject to national appropriation by claim of sovereignty, by means of use or occupation, or by any other means (Art.II OST). Some argued that the protection and enforcement of intellectual property rights in outer space may conflict with the aforementioned principles. However, the Outer Space Treaty also includes by its Article VIII that jurisdiction and control over an object launched into outer space and its personnel remains with the state on whose registry an object launched into outer space is carried. This would lead to the opinion that the territorial jurisdiction of the launching state, which registers the object launched into space object, is permitted to be extended to that object. Along these lines and in the absence of explicit international rules, under a number of international agreements[3], registered space objects are treated as quasi territory for the purpose of intellectual property. Apart from this notion of quasi-territorial jurisdiction, there are some who prefer the notion of extra-territorial jurisdiction for objects launched into outer space. They argue that, in analogy with the law of the Sea, the so-called "Flag-state[4] jurisdiction" allows states to assert extra-territorial jurisdiction over objects launched into outer space and its personnel in an area (outer space) not subject to territorial appropriation. In fact both viewpoints lead to the same consequence, that activities in outer space carried on board of space objects will be governed by the national law regime of the state of registry of the object launched in outer space, which can apply its national intellectual property system. However, this will not solve the problems generated by the application of different national law systems.

THE INTERNATIONAL SPACE STATION

The experience with the International Space Station (ISS) shows the approach taken so far. Confronted with the need to protect intellectual property rights for inventions on board the ISS, the U.S. amended its patent legislation by providing quasi-territorial effect on a space object or component thereof carried on the registry of the U.S., unless otherwise agreed by international agreement[5]. In fact, Title 35 of the US Code, section 105, Inventions in outer space[6] reads in paragraph a :

Any invention made, used, or sold in outer space on a space object or component thereof under the jurisdiction or control of the United States shall be considered to be made, used or sold within the United States for the purposes of this title, except with respect to any space object or component thereof that is specifically identified and otherwise provided for by an international agreement to which the United States is a party, or with respect to any space object or component thereof that is carried on the registry of a foreign state in accordance with the Convention with Registration of Objects launched into Outer Space.

We see that the U.S., in anticipation of the composition and development of the Space Station, consisting of different elements, which could be registered as different space objects in the registry by the relevant launching state, added the word *component* to the term space object. In accordance with this, Article 21 of the International Governmental Agreement governing cooperation on the International Space Station (IGA) lays down for purposes of intellectual property law, that an activity occurring in or on a Space Station flight element shall be deemed to have occurred only in the territory of that element's registry. (except that for ESA-registered elements any European Partner State may deem the activity to have occurred in within its territory). In practice, this system, again based on the principle of territoriality, will lead to the application of

different national intellectual property laws in so far as it concerns different elements of the Space Station. This means that one nation's patent laws ends and another's begins from one component/element of the ISS to another. Moreover, there is the problem that the majority of the Partner States of the IGA have not yet adapted their national laws by providing extraterritorial effect on an object or component thereof carried on their registry. This will lead to even more uncertainty.
Another point of discussion lies in the fact that the U.S. extended its extra-territoriality to space objects and components carried on its registry based on its jurisdiction or control. This is not in conformity with Article VIII of the Outer Space Treaty, which speaks of jurisdiction and control. In my opinion, the combination jurisdiction and control signifies an important requirement for a justified use of the application of quasi (or extended) national territoriality in outer space.

THE NEED FOR A NOVEL SYSTEM PROTECTING INTELLECTUAL PROPERTY RIGHTS IN OUTER SPACE

Due to the uncertainties lying in the present status of intellectual property rights of inventions in outer space, which cause hesitation by private parties wanting to invest in new technologies, there is a need for a new regime of patent protection in outer space to be agreed on by space faring nations. Such a regime should cover the whole outer space region including the moon and other celestial bodies. This is of particular importance since according to the Moon Agreement, the moon and other celestial bodies are, according to Article 11, par. 2 thereof, not subject to claims of national sovereignty, by means of use or occupation, or by any other means[7], which again might lead to a prohibition of territoriality as a basis for intellectual property rights. Such an opinion to prohibit the extension of state territoriality to the moon and other celestial bodies in light of the provisions of the Moon Agreement is even stronger than a similar opinion in relation to the Outer Space Treaty. This is because the Outer Space Treaty also includes, apart from the similar Article II (non-appropriation principle), Article VIII which does leave room for extending the terrestrial jurisdiction and control of the launching state who registers the object launched into outer space. Whilst by applying the Outer Space Treaty for space objects launched into outer space, state territoriality might also be applicable to those objects launched into outer space which have reached the moon or other celestial bodies, such applicability might be obstructed not only by referring to Article 11, par. 2 of the Moon Agreement but also by arguing that the Moon Agreement should be seen as a lex specialis in relation to the Outer Space Treaty, providing a specific legal regime for the moon and other celestial bodies. Also, the specific references in the Moon Agreement in relation to the exploitation of natural resources of the moon and the provision that States Parties should establish an international regime to exploit its natural resources[8] provide little or no room for claims based on territoriality. However, such an international regime could probably provide a basis for a novel kind of property rights including intellectual property rights based on an international agreement, which take due care of the interests of investors combined with the interests of mankind at large in the exploitation of the natural resources of the moon and other celestial bodies[9].

A SPACE PATENT TREATY

A new regime of patent rights and the protection thereof in the whole outer space region should emanate from a uniform patent law system[10]. Such a system has already been suggested in the form of a Space Patent Treaty[11]. This would create a single system to apply to outer space including the moon and other celestial bodies applicable to all States Parties. This would signify a system not based on the principle of territoriality, but rather on the environment of space as a region beyond the limits of states and their national competencies and laws. For instance, the earlier mentioned amendment of U.S. patent law provides for itself to be superseded by a novel international Space Patent Treaty[12]. Also, other states[13] could follow this example in order to facilitate a broadly based Space Patent Treaty which would establish a new regime of patent protection in outer space. This regime should, following the leading concept governing outer space as geographically separate from earthbound nations, be applicable on all inventions occurring in outer space. Hence through a single patent law system, space patents would be the only recognised patents in outer space. A Space Patent Office should be created to examine applications and issue patents in outer space[14]. Such a system would avoid many conflicts and uncertainties inherent of the current (quasi) territorial system.

SCOPE

As for the scope of the Space Patents Treaty it could follow the delimitation of outer space from airspace by applying to the geographical area above the limit of 100 km above the sea level of the earth.

A space patent would grant the owner being a citizen of a signatory state a period of protection on all use of the invention in outer space including the right to transfer, sell or license.

Apart from the extraterrestrial protection of the invention made in outer space it could be possible to provide the space patent holder with a separate right of protection to use its invention for application on earth. Filing a space patent could entail a general right of protection against all the contracting states, their nationals and residents.

At present within the framework of the IGA of the ISS an invention in outer space does not impact the ownership of the invention nor does it preclude the right to file for a patent in multiple countries. Only the country of the inventor ship will be determined by the ownership and registry of the Space Station's element in which the invention has taken place[15]. This approach could pave the way towards the general acceptance of a Space Patent Treaty providing for the filing for a space patent through a unique international filing procedure and prosecution phase by a Space Patent Office. This would stimulate incentives to create new inventions in outer space by saving costs for patent protection in multiple countries. The institution of a Space Patent Office would preferably be created under the auspices of the United Nations through the works of the Committee on the Peaceful Uses of Outer Space and its legal Sub-Committee in cooperation with the World Intellectual Property Organisation (WIPO).

PROBLEM AREAS

Areas where controversies might arise are first of all the choice between the first to file and the first to invent systems. Experience in the negotiations at the World Intellectual Property Organisation (WIPO) on a Substantive Patent Law Treaty (SPLT)[16] showed U.S. insistence to grant patents on the basis of invention date rather than filing date, although within the U.S. large corporations are generally prepared to change the system in exchange for the advantages that worldwide harmonisation would bring them. Another issue related to the first to invent system, the grace period during which information can be circulated between the invention and the filing date without invalidating the patent claim, will raise even more resistance to be relinquished by the U.S., as experienced during the negotiations on the SPLT. When the U.S. would eventually give up on the first-to-invent system, it would probably insist on keeping at least the grace period, which most other countries seem prepared to accept[17]. Another problem lies in the field of the invention secrecy laws of the existing different patent law systems in the world. However, regarding the aspect of secrecy laws it has been already agreed in the IGA of the ISS that it restricts the operation of Partner States' respective invention secrecy laws. This flows from the fact that an invention occurring on a specific element of the International Space Station does not effect its ownership. The inventor is still free to apply for a patent in a state of his/her choice. Hence restrictions based on invention secrecy laws may not be invoked to prevent a non-national to apply for a patent in a Partner State other than the State with territorial jurisdiction, as long as that Partner State has a patent protection regime[18]. However, a uniform patent law system should still have to solve the problems encountered by the existence of invention secrecy laws in many countries preventing to file a patent application in particular on national security grounds.

Apart from an international agreement to create a uniform patent law system, it is necessary to establish an international enforcement body, such as an international board of arbitration or an international court. According to Smith[19] this might be a gigantic undertaking, but this is surely more modest than attempting to harmonise the disparate national laws which exist.

Apart from the above mentioned problems there remain other issues to be dealt with. In the first place harmonisation is needed on the subject matter that is eligible for patent protection[20]. Also other requirements for patentability should be harmonised. Whilst the U.S. requires that an invention should be new, useful and unobvious, other nations require that the invention should be novel, manifesting an inventive step and involve industrial applicability.

In particular the novelty requirement can differ among the various countries of the world by applying a more or less stricter standard for prior art, an instrument that can be used to demonstrate what inventions have been invented before and that effects the novelty requirement. The difference between obviousness and inventive step, as a requirement for novelty, can also lead to a differentiation of patentability. Furthermore, the "best mode" requirement applied for instance by the U.S[21] is not been followed by most other countries. Another more general issue to be tackled, which also arose in relation to the ISS[22], is the situation that an invention is not entirely made in outer space (respectively on board the ISS) but partially on earth.

CONCLUSION

For more than a hundred years many efforts have been made in trying to harmonise patent laws worldwide. For eight years negotiations have taken place on a world patent system through a Substantive Patent Law Treaty with the aim to remove most of the remaining national flexibility in patent systems and pave the way for a future world patent granted directly by the World Property Organisation (WIPO), for inventions made on earth. In spite of the fact that this process of harmonisation appears to be slow and cumbersome, this should not prevent us from stressing the need for a uniform patent law system for invention made in outer space. Apart from the fact that such a system would advance the commercial uses of outer space, since it would stimulate the industry to invest in the development of space technology and products, it might also accelerate the process of further harmonisation of patent law for terrestrial inventions, finally leading to a uniform and transparent patent law system for inventions made by men on earth as well as in outer space.

---

[1] WIPO 2004, See Space 2030, Tackling Society's Challenges, p.181-182, OECD 2005,ISBN 92-64-00832-2

[2] i.a. The Patent Cooperation Treaty (PCT) and the WTO's Agreement in Trade Related Aspects of Intellectual Property Rights (TRIP's Agreement)

[3] i.a. the International Government Agreement on the International Space Station (IGA)

[4] The concept of Flag-State jurisdiction was developed in relation to the res communis of the High Seas, where the nationality of the ship "provides a necessary criterion …in locations …where the territorial criterion is inappropriate.

[5] See also below note 8

[6] Added Nov. 15,1990, U.S. Public Law 101-580, sec 1(a), Stat. 2863

[7] The 1979 Agreement Governing the Moon and Other celestial Bodies has been ratified by 13 countries. However none of which has the means to go to the moon.

[8] Article 11par 5

[9] See also H.L. van Traa-Engelman, Clearness regarding Property Rights on the Moon and Other Celestial bodies, Proceedings 39th IISL Colloqium, Beijing, AIAA 1996

[10] See also B.L.Smith, Société de Services en Propriété Industrielle, Paris, An industry Perspective on Space-Related Intellectual Property Rights.

[11] See Space Patents: Intellectual Property in Outer Space, presented at the 8th Mars Society Convention by Bryan E. Ericson and Gary.C. Fisher, August 12, 2005

[12] The U.S. amended its patent legislation by providing quasi-territorial effect on a space object or component thereof carried on the registry of the U.S., *unless otherwise agreed by international agreement*

[13] In particular those states parties to the IGA ISS

[14] See supra note 7

[15] Article 21 of the IGA of the ISS

[16] On June 1, 2000 in Geneva a Patent Law Treaty (PLT) was concluded. However the PLT is only concerned with patent formalities, providing speedier and less costly formalities. As of April 2008 the PTL has 18 contracting states, while 59 states and the European Patent Organisation have signed the Treaty.

[17] One global patent system? WIPO's Substantive Patent Law Treaty, October 2003-GRAIN

[18] Ibidem Article 21(3). See Also Sriram Swaminathan, University of Melbourne, The Applicability of Space Law Principles to basic space science: An update, Paper selected from workshops organized by the United nations Office for Outer Space Affairs, within the Progamme on Space Applications in 2004.

[19] Ibidem note 9

[20] For example, unlike the U.S., some countries do not allow patent protection for pharmaceutical inventions. See Michael D. Kaminski, Patent harmonization, International efforts are gradually unifying the world's patent laws.

[21]U.S. patent protection requires adequate disclosure to the public including a written description of the invention, an enabling disclosure, and the best mode of carrying out the invention. See 35U.S.Code par. 112

[22] Working Group on Space Stations, Project 2001, Legal Framework for the Commercial Use of Outer Space , International Institute of Air and Space Law of the University of Cologne.

IAC-08-E. 13

JURISDICTION AND APPLICABLE LAW IN CASES OF DAMAGE FROM SPACE IN EUROPE – THE ADVENT OF THE MOST SUITABLE CHOICE – ROME II

Prof. Dr. Lesley Jane Smith, LL.M.
Rector, Riga Graduate School of Law, Weber-Steinhaus & Smith, Bremen
Email: ljsmith@barkhof.uni-bremen.de
Telephone: 0049 421 63 93 60 Fax: 0049 421 63 93 622
Catherine Doldirina, LL.M. Eur.
PhD candidate Institute of Air and Space Law, McGill University
Email: catherine.doldirina@mail.mcgill.ca

ABSTRACT

Liability for space activities is a much discussed subject and the advent of commercial space operations has only added to its importance. Articles VI and VII Outer Space Treaty, together with Articles II and III Liability Convention, remain the main entry level for state liability for damage arising from the private space activities. Few space-faring nations have introduced national space statutes that include a down-flow of their international obligations. The European Union (EU) Regulation on the law applicable to non-contractual obligations could harbour developments for liability law in the context of damage resulting from space operations. Space activities were not the main focus of the Regulation but may well turn out to be an interesting spin-off. The Regulation prescribes general rules that will determine the applicable law in damage scenarios, where more than one legal system applies. It is important for trans-national tort cases and does not limit the systems of applicable law to those of the EU Member States only. This paper focuses on the common rules applicable in damage actions based on torts or other non-contractual obligations from the perspective of their applicability to damage caused by space activities. After assessment of the relevant international and national law norms the value of the EU Regulation will be addressed.

INTRODUCTION

Space activities can by nature lead to cross-border damage: this may not be confined to one location alone. Whilst every effort is employed in space operations to minimise risk, incidents leading to damage cannot be discounted. Cross-border damage can lead to competing jurisdictions and alternative rules of applicable law. Damage in orbit has its own particularities: it may manifest itself only after a considerable lapse of time. Space debris in LEO is a case in point here. Satellites can suffer damage long after an in-orbit collision or debris-producing incident.[1]

While the Liability Convention[2] contains clear rules of international liability for damage caused by space objects, either to the earth or to other space objects or persons in outer space, the current increase in satellite-related services could lead to an increase in third party claims arising at national level. Satellite navigation and

---

[1] "Anti-satellite test generates dangerous space debris" *New Scientist Space*, 20th January 2007, available at http://space.newscientist.com/article.ns?id=dn10999&print=true (last accessed 6th August 2008). *See* also K. U. Hoerl/ R. Jehn, C. Sarocco, LEO Constellations- Quo Vadis After End-of -Mission, in H. Saway-Lacoste (ed), *Proceedings of the 3rd European Conference on Space Debris*, Noordwijk, ESA Publications Division (2001).

[2] Convention on International Liability for Damage Caused by Space Objects (1971) 961 *U.N.T.S.* 187.

tracking systems are examples of satellite services where issues of manufacturer's or product liability could arise. While the international regime focuses on the role of the launching state,[3] national space legislation is generally designed to allow the state to recoup damages arising under its external liability for activities of its commercial space operators.[4] No particular liability model or regime has been elaborated for liability at a national level for commercial space-related activities.[5] National licensing rules, national space statutes, regulatory procedures, agency rules, and state practice all exist, but no systematic legislative treatment or prototype has yet been deployed as an interface between international state to state and private commercial liability for space related activities.

As a result, there is potential for third party space-related damage claims that can lead to substantive (and procedural) options within the law of tort /delict. More than one jurisdiction may be called upon by private parties to determine such disputes. Not only can this open the floodgates to potentially competing forums: there may also be questions about the law applicable to the issue of liability itself, particularly if there are factors linking the damage to another jurisdiction. Although the system of private international law provides a choice between rules that apply in cases involving conflicts between different legal systems, it can still add a dimension of unpredictability to the outcome of individual liability cases.

[3] Arts. II-III Liability Convention.
[4] E.g. in the provisions of the US, Canadian and Russian legislation. For the latest provisions of the French 2008 law, see below, at 3-4.
[5] *See* Michael Gerhard, Weltraumgesetzgebung (Köln, 2002) at 202, Annex: proposed model for national space law.

I. INTERNATIONAL LAW ON LIABILITY

While Article XI Liability Convention employs its own diplomatic regime for damage claims, coupled with a one year limitation from the date of damage or determination of the launching state, the enforcement of commercial liability claims at national level may lead to application of foreign law. Although the Hague Conference in Private International Law has gone a long way towards unifying private international law rules,[6] differences between legal systems regarding the classification and interpretation of legal obligations persist. As a discipline, private international law is criticised for being over-complex.[7] Courts must first identify or qualify the legal issue before them – as a tort, contract or other claim – and can only then establish the link between the type of claim (*sic*) and the rules of liability to be applied.

In cases involving space damage, concurrent issues of international and national liability cannot be excluded. Under international space law, the launching state remains liable towards the state that has suffered damage or loss to its property and/or nationals. In principle, state to state responsibility precludes any rights accruing to the private or individual party affected. The International Law Commission's Articles on Responsibility of States[8] reflect the grounds

[6] The Hague Conference on Private International Law that was first convened in 1893 and operates in the field of unification of private international law, *see* www.hcch.net/index_en.php (last accessed 08.08.2008).
[7] EC Commission, Proposal for a Regulation on the Law Applicable to Non-Contractual Obligations ("Rome II") of 22.7.2003, COM (2003) 427 final, Explanatory Memorandum, at 2 ff.
[8] The International Law Commission's Articles on the Responsibility of States for Internationally Wrongful Acts (hereinafter ILC Articles of Responsibility of States) were adopted by the UN GA in 2001 and constitute binding customary law on this issue. They are accessible at

for State liability for wrongful acts. Where private parties are involved, the basis of their claim will generally be independent of a state's international liability and rather founded on the national law regarding commercial relationships and legal duties between contractors and/or operators. This is also the case, for example, where manufacturer or product liability issues are raised, and also in cases of negligence. Equally, there are other situations of international state liability where claims by nationals are specifically excluded, thus opening up their access to courts at national level.[9]

Up until now, the development of space litigation in tort between private parties has been shielded through the practice of state licensing or regulation of commercial operations, compulsory launch insurance and a corresponding widespread use of cross-waivers of liability in launch service agreements that operate within the traditional concept of international liability of the launching state.[10] Some jurisdictions offer their commercial operators state immunity from suit, thereby precluding development of collateral space related litigation.[11] Next section describes why and how the provisions of the Liability Convention have to be incorporated into national legal systems.

II. <u>LIABILITY CONVENTION AND NATIONAL LAW</u>

National space laws may be formulated to allow state indemnification through compulsory commercial launch insurance that, in some cases, extends to bearing surplus risk beyond the compulsory insurance ceiling.[12] The international liability regime does not, however, appear to preclude private party litigation relating to commercial space activities at national level. Nor does it restrict those parties outside the ambit of the Liability Convention.[13] Public international law does recognise individuals as partial subjects of international law, in so far as they can derive rights and obligations from the state's international obligations.[14] Depending on the nature of the obligation, positive rights can be exercised in either of two ways in tort liability: in a claim against the state or against other private persons. On the one hand, states have obligations to respect international law, by accepting their international liability; on the other, they have obligations to ensure enforcement of such norms by transposing them into national legislation. The Liability Convention falls squarely into this category. Despite this view that states are under a duty to pass national space legislation imposing the same liability rules on the private sector

---

www.un.org/law/ilc/index.htm (last accessed 08.08.2008).

[9] Under Art. VII (a) and (b) Liability Convention, nationals and foreigners are excluded from the Convention's ambit, *see* I.H.Ph. Diedericks-Verschoor/ V. Kopal, *Introduction to Space Law*, 3rd ed. (the Hague, 2008) at 13.

[10] C. Kohlhase, P. Makiol in K.H. Böckstiegel (ed.) *Project 2001*, (Cologne, 2002), 78-79.

[11] For details of the Commercial Space Act of 1998 and the US Commercial Space Launch Amendments Act 2004, *see* P. Dempsey, United States Space law: Commercial Space Launches and Facilities, in: *IISL/IAF Proceedings of the 49th Colloquium on the Law of Outer Space* (2006), American Institute of Aeronautics and Astronautics; R. Hancock, Provisions of the Commercial Space Launch Act (CSLA), *Space Policy* 21 (2005) 227-229.

[12] Art. 13 French *Loi relative aux Operations Spatiales,* JO n 129 of 04/06/2008.

[13] Public international law contains positive and negative obligations in individual tort actions. The former requires effective support such as enabling provisions for reparation for the injured party, the latter extend to preventing the effective enforcement of such claims by reasons of jurisdictional immunity. Art 33 (2) of the Articles on State Responsibility expressly excludes individual tort claims from its ambit.

[14] This applies in particular to human rights law, *see* D. Shelton, *Remedies in Human Rights*, 2nd ed, (2005) 50 ff.

as those under the Liability Convention,[15] not all states have done so. Of those which have, several states formally refer to their own rules of civil law as regulating liability for space-related damage.[16] Beyond this however, and particularly in relation to commercial interests, the question arises as to which national rules of tort or delict become applicable in any one case that extends to more than one jurisdiction.[17]

National courts faced with commercial space liability issues may rely on any one of the following three possible sources of law to apply. In the *first* case, the national tort rules may be either the national civil law tort rules or, if there is a national space statute reflecting the liability provisions of the Liability Convention, then the provisions of that statute. In the *second* case, depending on the state's position under the monist or dualist theories, the provisions of the Liability Convention may be directly applicable or indirectly, via the transposition statute. In the *final* case, the rules of private international law may refer them to another country's rules of tort. Its substantive content may be independent of the provisions of absolute or fault-related liability under Article II and Article III Liability Convention.

Irrespective of the choice, the Liability Convention does not apply to the exclusion of other legal rules. The rules of civil law (e.g. manufacturer's liability), product liability, environmental conventions and other general rules of law may be applied, either by analogy or in conjunction with the Liability Convention itself. The Liability Convention merely ensures external liability of the state. Its transposition does no more than create a right of recourse against the private space operator.[18]

III. DIFFERENCES IN APPROACHES TO REGULATING RESPONSIBILITY AND LIABILITY FOR SPACE DAMAGE ON NATIONAL LEVEL

EU Regulation on the law applicable to non-contractual obligations[19] contains one important exception in the context of liability for space activities. Article 1(1) Regulation excludes liability for acts that fall within the exercise of state authority or *acta iure imperii.* In the pre-Cold War era, state participation in space activities was certainly viewed as exclusively falling within *acta iure imperii.* Fifteen years on, however, the increase of private participation in space activities has been given official international recognition.[20] States activities in space do not take place by reason of state interest alone. Their participation may be on a commercial level

[15] This is written into Art. VI Treaty on Principles governing the Activities of States in the Exploration and Use of Outer Space - Outer Space Treaty (1967) 610 *U.N.T.S.* 205.

[16] Art. 39 of *Law of Russian Federation on Space Activity* 1993, amended 2006; Art. 25 Ordinance of Supreme Soviet of Ukraine *on Space Activity* 1996. Art. 14 French *Loi relative aux Operations Spatiales*, n. 18 above, makes recourse dependent on extent of indemnification that has taken place through insurance or guarantee.

[17] This paper deals only with the law applicable to non-contractual obligations. Equal considerations apply to private international law aspects of contractual obligations. The EU has already passed equivalent conflicts legislation in Regulation 593/2008 of 17 June 2008 on the law applicable to contractual obligations (Rome I), *OJ L* 177/6 of 04.7.2008.

[18] On this point, *see* R.J. Lee, Liability Arising from Article VI of the Outer Space Treaty: States, Domestic Law and Private Operators, in: *IISL/IAF Proceedings of the 48th Colloquium on the Law of Outer Space* (2005) American Institute of Aeronautics and Astronautics.

[19] EU Regulation 864/ 2007/EC of 11.07.2007, *OJ L* 199/40 2007 of 31.7.2007. According to Art. 32 the Regulation enters into force on 11th January 2009. Also referred to as Rome II. Hereinafter Regulation.

[20] *See* UN GA Resolution 62/101 Recommendations on enhancing the practice of States and intergovernmental organisations in registering space objects, A/RES/62/101 of 10 January 2008; R. Skaar (ed.), *Commercialisation of Space and Its Evolution* (European Space Policy Institute, 2007), at 5.

and thus constitute *acta iure gestiones*.[21] Article 1 excludes application of the provisions of the Regulation where damage occurs in pursuit of *exclusively* state activities. The Liability Convention and, in particular, the claims procedure it establishes in Article IX would then make state to state liability rules applicable. The remaining question of internal (national) liability, or recourse by the state against the private commercial level, alongside the immediate issue of liability of providers of space services under private law, could arguably fall within *acta iure gestionis*. This distinction remains at least debatable, the dividing lines at best grey.

This is the crux of the demands that space law and liability law in particular, currently face. State liability, government liability and corporate (enterprise) liability are all based on a liability model which provides a limitation of liability (of the individual or legal person) in return for the assumption of liability by the external organ or actor, be it the state, the administrative authority or the company itself towards third parties. Among the most recent national space laws passed in Europe, particularly the French and Belgian laws, the individual commercial space operator is liable[22] on the same basis as those provided by the Liability Convention – i.e. absolute liability for damage on earth and air, but fault-based liability for damage occurring anywhere else. According to the French law, liability ceases when the conditions contained in the operations licence expire or, at the latest, one year thereafter.

Several points are of interest here. *Firstly*, that the state accepts its continued liability for commercial operators' liability on an international level after expiry of the one year limitation.[23] *Secondly*, Article 14 French Law foresees that the state can take recourse against an operator for any surplus over and above what has been indemnified under either insurance or guarantee. *Thirdly*, there will be no recourse against the private operator in the case of spatial operations carried out purely in interest of the state.[24]

It is debatable whether the imposition of commercial liability on space operators under French law leads to the exemption from the Regulation's ambit because of the *acta iure imperii* exception. Interpretation of its wording would not seem to support this line of argument. If the situation is compared to that in United States of America, where commercial space contracts and ventures are deemed by statute to fall under the broader umbrella of government contracts in order to benefit from government immunity, there could be a lack of consistency between states as to what constitutes – at least in space law terms – *acta iure imperii*.[25] Although this choice falls to each individual state on its own, there is much to be said for a coherent approach at EU level as to which conventions should be excluded from the ambit of the Regulation.[26]

IV. APPLICABLE LAW AND JURISDICTION: BRUSSELS I AND ROME II

Private international law prescribes two rules for determining the law applicable to

[21] This can lead to a lifting or restriction of state immunity. There are different patterns between jurisdictions as to how the doctrine is viewed and applied, so that no consistent pattern can be detected. There is, however, a recognisable trend towards restricting immunity, see I. Brownlie, *Principles of Public International Law*, 6th ed. (2003), at 323-332; in relation to Europe, see the European Convention on State Immunity, 11 *ILM* (1972) 470.

[22] cf. Article 13 French *Loi sur les Operations Spatiales*, n. 18 above.

[23] The period runs from the moment when the licensing conditions should have been met.

[24] Art. 14, *Loi relative aux Operations Spatiales*, id. last para.

[25] *See* e.g. US *Commercial Space Act*.

[26] Art 28(1) Regulation.

an action in damages: *lex loci commissi* denotes the law of the place where the act or event took place or was committed; *lex loci damni* denotes the place where the loss is sustained.[27] These notions may, but need not, point to the same result. There are variations in their interpretation and impact between States and this holds true for the EU Member States. In recent years, there has been a tendency for courts to pronounce in favour of the law of the country where the damage is sustained, with less attention being paid to the place where the harmful act occurred.[28] In such situations, there is a conflict between at least two differing systems of law. Defendants can generally be sued – at the claimant's choice – before courts of either jurisdiction.[29]

The Regulation was introduced as part of EU programme towards establishing an area of freedom, justice and security,[30] with a view to providing a base for harmonising and simplifying the private international conflict rules that apply in cases of non-contractual obligations arising out of a tort/delict within the EU.[31] Interestingly, the new EU rules have been hailed as the ultimate reform model that could and should re-inspire and re-structure complex tort jurisdictions with complicated conflicts rules, such as those in the USA.[32] The Regulation applies automatically in the event of tort or damage litigation within Europe. It is designed as legislation limiting the number of potentially applicable laws in any given conflict case "to improve the predictability of the outcome of litigation" in the Member States.[33] Its greatest success is in restricting and laying down the law that applies in cases of torts with a private international element. This bears well for all commercial and also space-related damage actions.

In order to limit the choices and options arising by virtue of the *lex loci delicit commissi* and *lex loci damni*, the Regulation has refined the complex rules of applicable law in tort or delict actions for European Member States. In so doing, it has clarified the rules applicable in cases relating to non-contractual damage within the EU.[34] A degree of complementarity has been built in between the Regulation and the original 1968 Brussels Convention (now Brussels I Regulation) on jurisdiction and recognition and enforcement of judgements in civil and commercial matters.[35] Brussels I Regulation provides various grounds of general and special jurisdiction, ranging from the defender's domicile under Article 2(1), to jurisdiction in matters of tort or delict under Article 5(3), to the place where the harmful event occurred or may occur. While Brussels I restricts the possibility of parallel actions before various jurisdictions through resort to defences such as *lis alibi pendens* and *forum non conveniens*, it radically reduces chances of forum shopping, at least in relation to

---

[27] P. North, J.J. Fawcett, *Cheshire and North's Private International Law*, 13th ed. (1999) ch. 20.

[28] Explanatory Memorandum to Proposal for Regulation, n. 7 above.

[29] ECJ, Case 21/76 Mines de Potasse d'Alsace [1976] *ECR* 1735; Case C-68/93 Fiona Sheville et al. v Press Alliance SA [1995] *ECR* I-415.

[30] Art. 65(b) EC Treaty.

[31] *See* Regulation; the Rome Convention on the Law Applicable to Contractual Obligations, 80/ 934/ EEC, *OJ L* 266 of 9.10.1980 has been reformulated as Regulation 593/ 2007 of June 17 2008 on law applicable to contractual obligations, *OJ L* 177/6 of 4.7.2008. This Regulation comes into force on 17th December 2009. The Convention, however, remains in force in relation to Denmark and the UK, although there is a provision allowing the UK to opt in.

[32] *S. Symeonidis*, Rome II and Tort Conflicts: A Missed Opportunity, 56 *Am. J. Comp. L.* 173 (2008); P.J.Kozyris, Rome II: Tort Conflicts on the Right Track! A Postscript to Symeonidis 'Missed Opportunity', 56 *Am. J. Comp. L.* 471 (2008).

[33] Recital 6 Regulation.

[34] With the exception of Denmark, *see* Art. 1(4) Regulation.

[35] Regulation 44/2001 of 22 December 2000, *OJ L* 12/1 of 16.1.2001. Hereinafter Brussels I Regulation.

jurisdiction of European courts in "intra-Community" situations and, to a certain extent, those involving an extra-community element.[36] After rationalising jurisdiction within Brussels I Regulation, the next logical step for the European legislator was to rationalise the rules applicable within the EU to claims arising out of contractual and thereafter non-contractual obligations, or tort.

Article 3 Regulation has universal application. It specifies that any law relevant by virtue of its provisions will be applied, whether or not it is a law of a Member State. This, too, reflects the international relevance of space-related damage. The main thrust of the Regulation is to be found in Article 4. This specifies that the law applicable to any tort/delict "shall be the law of the country in which the damage occurs, irrespective of the country in which the event giving rise to the damage occurred and irrespective of the country or countries in which the indirect consequences of that event occur."

With this rule, the EU has opted for a pragmatic solution or choice in favour of the *lex loci damni,* to the exclusion of *lex loci commissi.* This approach has been followed to allow predictability and foreseeability in the outcome for commercial operators within the EU. Under Article 4(2) the rule alters to allow the law of the country of claimant and defendant to apply where they both share the same country of residence. This, too, is a pragmatic restriction of the variety of options that exist within and around the *lex loci commissi.*

There is however, one important exception to the principle of *lex loci commissi.* In exceptional circumstances, Article 4(3) permits application of another system of law, where there is a manifestly closer connection to another country "based in particular on a pre-existing relationship between the parties, such as a contract" to allow this country's law to then be applied. Thereby the Regulation fulfils one of the main functions of private international law – linking a dispute to the rules of the best suited jurisdiction.

V. SPECIAL PROVISIONS FOR CERTAIN TORTS/DELICTS: PRODUCT LIABILITY AND ENVIRONMENTAL DAMAGE

Article 5 Regulation contains special provisions in relation to product liability.[37] In this case, the law of the place where the person damaged has his/her habitual residence applies,[38] with variations relating to the law in which the product was acquired[39] or to the law of the country in which the damage occurred if the product was put on sale there.[40] Here, too, there is a separate provision allowing for an alternative law to be applied if it is more manifestly connected with the country, other than that where the person had his/her residence.

In relation to environmental damage, Article 4 Regulation provides a general rule whereby the law of the country where the damage occurs and not where the damage arises is to apply. There is a second option allowing for the law of the country where the event leading to the damage took place to be applied.[41]

These provisions are all particularly relevant in the context of liability for damage arising from satellite services, given the potential

[36] Brussels I Regulation.

[37] Some EU Member States (Finland, France, Luxembourg, Netherlands, Spain, Slovenia) are already party to the 1973 Hague Convention on the law applicable to product liability, *see* Commission Explanatory Memorandum n.7 above, at 14.

[38] Article 5(1)(a) Regulation.

[39] Article 5(1)(b) Regulation.

[40] Article 5(1)(c) Regulation.

[41] Article 7 Regulation. *See also* ECJ Case 21/76 Mines de Potasse d'Alsace; Case C-68/93 Fiona Sheville et al. v Press Alliance SA, n. 24 above.

for a wide geographic range of damage resulting from loss of signal or disfunction.

VI. APPLICABLE LAW IN CONTEXT OF LIABILITY FOR GALILEO

The law applicable to actions in tort or delict has immediate implications for the development of the European navigation satellite system GALILEO. Despite the Commission's estimates that the risk factor attached to GALILEO is minimal, the risk of damage inherent in all space activities cannot be overlooked. Liability may arise from a variety of grounds, notably strict rules of product liability, services liability, manufacturer's liability, alongside institutional liability at EU level. Besides this, there is an inherent and permanent risk of debris damage to satellites.

Foreseeable damage ranges from delay in delivering services and economic loss, through to and including personal injury and loss of life. In view of the predicted geographical coverage of GALILEO and the complexities of cross-border claims in tort, the introduction of a special liability regime appears advisable, despite the existence of the Regulation, if complex conflicts of law are to be avoided. As with all product-driven damages claims, the decision to create a special liability regime has to be set against the complexities of multiple tort litigation on a world wide scale, alongside the variety of jurisdictions where damage could take place. States beyond the EU dealing with tort claims will still have their national rules of private international law choice between the *lex loci commissi* or *lex loci damni* approach, unless the Regulation applies through *renvoi.*[42]
The EU itself has made no formal provisions for a product or services liability regime to regulate GALIILEO related damage. On the institutional side, Article 17 Regulation 1321/2004 on the establishment of structures for the management of the European satellite radio-navigation programmes[43] contains within provisions parallel to the primary rules under Article 288(1) and (2) EC Treaty on the contractual and extra contractual liability of the EU institutions. These provisions are, however, targeted at damage resulting from acts and decisions of the GALILEO supervisor authority (GSA) within its role as an administrative or licensing body. Administrative liability implicates failure to maintain high standards by those taking administrative decisions.

Work is, however, underway on presenting a possible EU-wide liability regime.[44] UNIDROIT has presented a Draft Proposal for a Regulation on Civil Liability late in 2007 that could provide the impetus or basis for the introduction of a liability regime governing GALILEO. The current version of its Draft Regulation foresees a strict liability regime for Europe, limited to damage arising within the EU.[45] If agreement is reached on this text for GALILEO, its provisions could become the prototype for a liability regime, insofar as the issue falls to be decided as damage occurring within Europe under Article 4(1) Draft Regulation. For damage occurring in non-EU member states, its provisions would then become applicable in foreign jurisdictions by virtue of *renvoi,* failing which, the *lex loci damni* becomes applicable. (wording in this sentence is not very clear)

[42] See the thought-provoking exchange between S. Symeonidis and Kozyris on the respective advantages and disadvantages of modern EU and USA private international law approaches, n.36 above.

[43] Regulation of of 12 July 2004, *OJ L* 246/1 of 20.07.2004, at 1-9. Programmes are EGNOS and GALILEO.

[44] This was launched on the initiative of the Italian government.

[45] Unidroit 2007 – C.D. (8620, March 2007).

Outlook

The foregoing serves as a short appreciation of the impact of the Regulation. Its provisions come into effect as from January 11 2009. The list of international conventions excluded from its ambit was to be notified to the Commission by July 2008, but this list is not yet publicly available.[46] It should provide greater insight as to how a state regards conflicts between other liability conventions and modern commercial space activities and damage caused by them – whether as *acta iure imperii* or as *acta iure gestionis*. Clarification by states here is in itself important for continued development of space activities within the international legal regime.

Liability for damage from space activities is destined to attract greater attention with further development of commercial satellite applications. These will increase the risk interface and the potential for litigation, be it on issues of product liability, environmental law or beyond. It remains to be seen whether the Liability Convention, as *lex specialis,* can inspire continued development of a national private space liability regime that extends beyond the rights of recourse by the state against commercial operators. The Liability Convention does not contain *erga omnes* obligations so that effective enforcement of damage claims at national level is important.[47] The combination of tort-based rules of product, environmental and services liability could well give rise to interesting developments in the (future) law of liability for commercial space operations. A coherent response could optimise the operative conditions for the commercial sector and contribute to the increasing development of the law relating to private parties within what is now the third millennium of public international law.

[46] At the time of writing, August 2008.

[47] A. Fischer-Lescano, Subjektivierung völkerrechtlicher Sekundarregeln, *Archiv des Völkerrechts* 45 (2007) 299 ff.; D. Shelton, n. 20 above, *id.*.

IAC-08-E8.1.8

LIABILITY FROM THE LAUNCH OF A SPACE OBJECT AS A MATTER OF PRIVATE LAW

Souichirou Kozuka
Professor of Law
Sophia University, Japan
s-kozuka@sophia.ac.jp

Abstract

The liability for damages caused by the launch of a space object has been considered as an already settled issue under the Outer Space Treaty and the Liability Convention. However, the Treaty and Convention deals with the state responsibility only, the liability of relevant parties under the domestic law of each country being left unregulated. Faced with the progress of commercialisation of space activities, Japan is considering the enactment of a space law, which might be a good opportunity to introduce a special liability regime for space activities. However, it seems to be unlikely to take place, as no major problem is found with the current practice, especially because of the use of liability insurance.

I. Introduction

In the very early days of space activities, the liability for damages caused by space objects was found to be an important issue. As a result, a very unique scheme was incorporated in the Liability Convention: the strict liability against damages caused on the surface of the Earth or to aircraft in flight and the chanelling of liability to the launching state[1]. The liability mentioned here was the state liability under the international law: the civil liability regime among the entities involved was left to the national law of each country.

In most countries, the private law rules on liability remained untouched, notwithstanding the ever greater role carried by private entities, such as the launch service companies and satellite operators, in the space activities. However, when the legislator considers enacting a law on space activities, it may wish to examine whether any provision specific to the liability arising from space activities needs to be included. Some states, most recently the Republic of Korea, did introduce such rules on civil liability,

[1] Art. II, Liability Convention.

which can be understood as a special rules to the tort law. This article makes a brief analysis of the current liability rules under the Japanese law and examines the needs of having special rules on liability in Japan.

II. The Purpose of Tort Liability

When Focusing on tort liability under the private law rules, rather than the state liability, it may be significant to note that the purpose of tort law is understood to be not limited to the compensation for damages. If compensation is the only purpose to be achieved, the social security system that indemnifies any damage sustained by its nationals may suffice and there will be no need for the tort law. The reason for imposing the liability on the party causing damages may most plausibly be found in the effect of deterrence: the liability works as the incentive for the relevant party to employ necessary care to avoid the accident.[2]

This argument is valid with regard to not only the negligence liability but also strict liability. In fact, the strict liability is the simplest system for giving adequate incentives.[3] Even under the strict liability system, a party can avoid liability by using necessary care, because an accident will never take place if the party carefully behaves. If the accident is expected to take place however much care is used, the party can still avoid liability by choosing not to engage in the activity at all. The social cost of accidents are perfectly internalised and the level of activity and amount of care will be maintained at the optimal.

The negligence system differs from the strict liability in that the finding of liability lies in the hands of the court. As a result, there could be the gap between the level of care actually committed and that observed by the court. The gap will be even greater with regard to the level of activity, as it is harder for the court to declare that the party should have refrained from the activity, as compared with saying that the party should have been more careful in performing the activity.

Therefore, when designing the tort law regime about the space activities, the legislator must be mindful of its deterrence effect and try to achieve the level of activities and care considered to be optimal by the society.

III. Liability arising from space activities against the third party in Japan: the basic rule

We start with the simplest case: the launch fails due to the mistake in operation and the launcher is destroyed, causing the fragments fall down on the area beneath. A fishery boat may be in the area, disregarding or

---

[2] Cf. *Kenneth S. Abraham*, The Forms and Functions of Tort Law, Second edition 18-19 (Foundation Press, 2003).
[3] *Steven Shavell*, Economic Analysis of Accident Law 21-26 (Harvard U.P., 1987).

overlooking the warning issued before the launch, and as a result damages were susyained by the people and property on the boat. (Note that overlooking the warning itself does not infringe any law, though it may be considered as contributory negligence.)

If such an accident takes place in Japan, the relevant rule is the tort liability under the Civil Code. It is provided that any party that intentionally or negligently infringes any right or legally protected interest of another party shall be liable to compensate any resulting damages.[4] In other words, the rule is different from the principle under the Liability Convention in three respects: it is the negligence liability, not strict liability; there is no chanelling of liability; the scope of damages is not limited to personal or property damages[5] but covers pure economic loss as long as there is a causal link between the negligent act or omission and the damage.

Suppose, next, the case where the failure of launch is caused not by the mistake in operation but by a defect in the design of the launcher. Under the Japanese law, the liability of the launch service company is the same. The stricter liability for the defects in structures on land[6] is not applicable, since launcher or any other space object cannot be considered as a structure "on land." In this case, however, the manufacturer of the launcher may also be held liable for the damage caused by the defect under the Products Liability Law. The products liability is based on the finding of a defect, rather than negligence, in a way similar to the products liability under the European directive.

It is often argued that, if the launch is sufficiently insured, there is little practical concern with the liability of the launch service provider. The liability is covered by the insurance except for limited cases of exemption, no matter who is liable. Although a problem could arise when the amount of damages is so high as to exceed the insured amount, this is, it is argued, unlikely to be the case: contrary to the case of maritime accident, which often involves the spill of crude oil, serious environmental damages seldom take place from the accident of launch of space objects. Neither is it likely that a highly populated area is damaged,

[4] Article 709 of the Civil Code: A person who has intentionally or negligently infringed any right of others, or legally protected interest of others, shall be liable to compensate any damages resulting in consequence.

[5] Cf. Art. I (a), Liability Convention.

[6] Article 717 of the Civil Code: (1) If any defect in the installation or preservation of any structure on land causes damages to others, the possessor of such structure shall be liable to the victims to compensate for those damages; provided, however, that, if the possessor has used necessary care to prevent the damages arising, the owner must compensate for the damages.
(2) The provisions of the preceding paragraph shall apply mutatis mutandis to cases where there is any defect in the planting or support of bamboos and trees.
(3) In the cases of the preceding two paragraphs, if there is another person who is liable for the cause of the damages, the possessor or owner may exercise their right to obtain reimbursement against such person.

because the launch cites are cautiously located in a remote area in most space-faring countries, including Japan.

Thus, the request for a special rule providing for the strict liability may not become strong, as long as the launch is adequately insured and the financial soundness of the launching entity is monitored. However, such an argument may be looking at the compensation for damages only, not paying due regard to the deterrence effect of the tort law. The insurance is neutral to the deterrence effect only when the insurer can monitor the behaviour of the operator and give adequate incentive through requiring the premium (or refusing to undertake the risk, if necessary). Besides, it should be remembered that under the negligence system, the level of activities tend to be excessive, as the court is not in a good position to determine the optimal level of activities. In the era when the purely private space activity is not a fairy tale but an actual possibility, preventing an inappropriately planned space activity from being undertaken may be an important role of the tort law rules. Thus, there may be a good reason for the legislator to examine whether the strict liability, similar to the strict liability under the Outer Space Convention, should not be introduced into a domestic law as a special rule of tort law.

IV. Liability of the state under the domestic law

When the launch is carried out by the governmental or semi-governmental body, the applicable liability regime could be the law on the indemnification claim toward the state, rather than the basic rules of tort law. In Japan, such regime is founded by the Law on State Liability, which provides for two kinds of liabilities: the liability caused by the exercise of the public power and the liability deriving from the fault in establishment or management of a public architecture. As regards the first kind of liability, "the exercise of public power" has been interpreted very broadly so as to include any activity except for purely economic transactions.

It has been considered that the space activities by Japan Aerospace Exploration Agency (JAXA) based on the policy of the government falls upon "the exercise of the public power." Besides, a launcher of JAXA has been considered as "public architecture", because the latter term is understood to include not only real property but movables as well. Thus, in the case JAXA undertakes the launch, there is no doubt that any accident caused by it is governed by the State Liability Act and not the tort law rule of the Civil Code.

However, it has been decided that the space activities in Japan be "privatised" so that it is no longer JAXA but the manufacturer, Mitsubishi Heavy Industry, Inc. (MHI), that undertakes a launch from 2007 on. The role of JAXA is limited to the planning and supervising, besides offering the launch site for use by MHI. Under these changes in the circumstances, the applicable law seems to be less clear than before.

Even after privatisation, the launch by MHI is the "exercise of public power", in cases where the launch is based on the state policy. In this view, MHI is, after all, providing its launch service to state (JAXA). In other cases, the liability of state could be founded on the planning and supervising by JAXA. Such an argument is all the more persuasive, because the supervising activities of JAXA include the final decision of letting the launch proceed (decision of GO/NO GO). If this argument is accepted, the liability of the state under the State Liability Act arises in any way when an accident takes place.

The liability for the faulty establishment or management is also arguable, but with less conviction. It is doubtful that, after privatisation, a launcher itself qualifies "public architecture." If it were to be affirmed, an aircraft parked in a publicly managed airport could also be found to be a "public architecture." However, as long as the launch site remains the property of JAXA, it may be argued that the launch site is considered as the "public architecture" and the defective launcher parked in it may constitute the "fault in … management of public architecture".

The above argument, of course, does not exempt MHI from any tort liability under the Civil Code, if it or its servant and/or agent is found to have been negligent in the process.

V. The relationship of the state liability and the liability of the operator

If both the liability of the operator, as discussed under III above, and that of the state based on the engagement of JAXA, as discussed under IV above, are found to exist, they are joint and several liabilities. The victim can, therefore, raise either or both claims. The sharing of the liability between the state (JAXA) and the operator (MHI) depends on the extent to which each party contributed to the accident, unless otherwise agreed. In a usual case, an indemnification agreement will be concluded in advance.

In practice, these arguments are again irrelevant with regard to the compensation, as long as the launch is adequately insured. This is because the liability insurance for a launch covers the liability of any party concerned. As regards the deterrence aspect, the reasonableness of the indemnification agreement may depend on the possibility of monitoring between the parties of the agreement.

It is also ordinary that all the entities engaged in the launch of a space object enter into cross-waiver agreements and relieve each other from any liability. Although never tested in the case of space activities, it is doubtful that the Japanese courts enforce such a waiver in case of loss of life or bodily injury caused by an intentional act. Even in the case of a grossly negligent act, courts are very reluctant to enforce a waiver entered into in advance when it extends to the loss of life or bodily injury.

The policy behind such reluctance of courts about affirming the validity of waiver has not clearly stated. It could be that deterrence effect of liability is (implicitly) considered and the courts wish to control an intentional or grossly negligent behaviour. Whether such negative attitude of courts toward complete waiver should not be overruled by a statute is a policy question worth considering for the legislator.

VI. The Rule on the Conflict of Laws

When the failure of launch involves an international element, as in the case when the damages are caused to the territory outside of Japan or the manufacturer of the faulty component is a foreign corporation, the conflict of laws becomes an issue as well. The current law of Japan on the conflict of laws is the Act on the General Rules of Application of Laws, totally revised in 2007.

Under the Act on the General Rules of Application of Laws, the governing law in the case of a tort is the law of the place where the consequence of the infringing act takes place[7]. In a case of fragments from the failed launcher falling down, therefore, the law of the territory upon which the fragments fell applies. If the area is the High Sea, the law of the flag state of the damaged boat navigating beneath will apply. If, however, there is a place that is apparently more closely connected under the circumstances of the case, the law of such a place applies[8]. If, for example, the damaged party is on the contractual relationship with the damaging party (the launching entity), this super priority provision may be applicable.

These rules on the governing law do not seem to cause any particular difficulty, at least with regard to the case of damages on the surface caused by the failed launch. Therefore, here again, no need for enacting a special provision seems to be existent.

VII. Conclusion

The liability arising from space activities has been only partially addressed by the space law. The Outer Space Treaty, as well as the Liability Convention, addressed only the state liability and left the civil liability to domestic law. Faced by the progress of privatisation and commercilisation of space activities, a question arises whether the domestic law needs to be modified so as to provide special rules regarding the liability from space activities. However, as long as Japan is concerned, no pressing need for such special rules has been perceived. This is notwithstanding the fact that the applicable

[7] Art.17, the Act on the General Rules of Application of Laws.

[8] Art.20, the Act on the General Rules of Application of Laws.

rules of tort law differs from the principles of the Liability Convention, which consists of strict liability and the chanelling of liability.

The principal reason for such absence of interest in enacting liability rules is the availability of insurance that offers almost comprehensive cover. The insurance will indeed offer sufficient compensation, which is a part of the purposes of the tort law. However, another, rather more important, purpose of the tort law is the deterrence and the neutrality of insurance with regard to the deterrence aspect of tort liability is affirmed only when the insurer can monitor the activities of the insured and give adequate incentive to the latter. Based on these observations, it may be needed to reexamine whether there is a reason to have special tort rules with regard to the liability arising form space activities.

IAC-08.E8.1.10

THE FUTURE OF THE UNIDROIT DRAFT SPACE PROTOCOL: LEGAL PROPOSALS TO FAVOUR AND STIMULATE ITS SUCCESS[1]

Fabio Tronchetti
International Institute of Air and Space Law,
Leiden University, the Netherlands
f.tronchetti@leidenuniv.nl
fabio.tronchetti@tiscali.it

Abstract

The main purpose of the UNIDROIT Draft Protocol on matters specific to space assets is to encourage private financiers to invest in space activities. By establishing provisions dealing with financing space assets and securing the interests of creditors, the Protocol tries to set out a legal environment which makes financiers protected when loaning money for space ventures.

However the Draft Space Protocol fails to do so. Its provisions, indeed, not only are unable to define a clear legal framework governing security interests in space assets but also raise problems of compatibility and consistency of the Protocol with the existing *corpus iuris spatialis* and with national laws regulating the concession of licenses.

These limits of the provisions of the Draft Space Protocol have created vast uncertainty between both space operators and potential financiers of space activities by causing, thus, the refusal of States to accept and ratify the text of the Protocol.

Considering the fact that the negotiating process of the Draft Space Protocol has reached a deadlock, it is time to put forward some proposals aimed at providing the Draft Space Protocol with bigger chances of success.

Therefore, this paper will analyze the limits of the Draft Space Protocol and will propose several solutions aimed at making it a valuable and workable instrument for supporting the participation and involvement of financiers in current and future space activities.

Full Text

INTRODUCTION: THE NEED FOR AN INTERNATIONAL REGIME GOVERNING SECURITY INTERESTS IN SPACE ASSESTS

While States once dominated the exploration and use of outer space, since the early 1990s private operators have become increasingly involved in the carrying out of space activities.

This shift from a State/government dependent to a more and more privatized space industry, apart from generating questions related to the workability of the existing space law regime in the era of the commercialization of outer space, has created one major problem: the financing of private space activities. In order to start-up their space projects, indeed, private operators need an enormous amount of money.

Generally speaking, one way of attracting capital into industries is the existence of a reliable system for securing lending, thanks to its ability to diminish the cost of capital due to a reduced risk undertaken by creditors. While at domestic level, creditors may usually rely on secured lending, on the contrary, at international level, there are many difficulties associated with taking security in mobile assets that they may move from jurisdiction to jurisdiction. These problems are exacerbated with regards to financing of space assets, mainly due to the characteristics of the private operators carrying out space activities and to the nature of space assets.

On the one side, indeed, private operators usually have small or no credit history. This fact generates uncertainty among lenders who fear the risk of not being repaid, particularly in the case of bankruptcy and insolvency of the debtor. The effect of such uncertainty is the increasing of the costs related with space-assets based financing. On the other side, taking security in space assets is made particularly difficult by the same nature of space assets. For instance, a satellite in orbit is difficult to repossess. This means that a common remedy available to creditors in case of debtors 'default is not applicable, at least in its conventional meaning. Additionally, enforcement of remedies may be complicated by the fact that outer space does not fall under the jurisdiction of any State and that most of the facilities used in the course of space operations, such as ground stations for telemetry, tracking and control purposes, are usually located in a number of different jurisdiction. Finally, there are problems associated with the fact that more and more satellites carry on board transponders and that such transponders may be operated, leased or owned by several parties.

All these problems with taking security in space assets makes the legal regime applicable to financing of space assets very unclear and discourage financiers to invest money in private space projects. Clearly, this fact has a very negative impact on the space industry and has the capability for stopping the commercial development of outer space. Particularly, new space operators face financial difficulties since many of them do not have the financial standing to be attractive for lenders.

In order to solve these problems and to respond to the need for uniform and predictable rules applicable to high value mobile equipment, particularly with regard to conflict of law and remedies in case of default, the International Institute for the Unification of Private Law (UNIDROIT) [2] began working on a preliminary version of the Convention on International Interests in Mobile Equipment in the early1990's[3].

As the work on the Convention proceeded, it was decided to complement the base Convention with three asset-specific protocols, one for aircraft, one for railway rolling stock and one for space assets.

In 1997 a working group wit the task of preparing a preliminary draft Protocol on matters specific to space assets (hereafter, the Draft Space Protocol) was set up. Since the beginning of the operations of the working group, it was clear that its members agreed on the fact that creditors involved in space industry were not adequately protected by existing domestic secured transactions and that, consequently, opportunities for asset-based financing were reduced. Private international law or conflict of law rules were considered to be unsatisfactory for mainly two reasons. First, each domestic system has its own conflict of laws rules for transactions with trans-border elements. The problem is that solution provided in one domestic system may not be applicable in another. Second, conflict rules differ from jurisdiction to jurisdiction. This fact creates a risk of forum shopping which may finally weaken the effectiveness of a security right.

Relying upon these considerations, the members of the working group were strongly of the idea that an international regime governing security in space assets was an effective means of facilitating space commerce, expanding the private market for financing and reducing the costs of financing.

In 2001, a preliminary Draft Space Protocol was presented by the working group to UNIDROIT. After some revisions, a new version of the Protocol has been issued in 2003[4]. However, unlike the Convention on international interests in mobile equipment[5] (hereafter the Convention) and the two other protocols[6], whose texts have been adopted and opened for signature and ratification by States, the text of the Draft Space Protocol has not been accepted by States. Currently, the negotiating process of the Draft Space Protocol is facing a deadlock and its positive conclusion seems to be far from being reached. Why?

The basic point is that the Draft Space Protocol, as it stands today, is unable to create a safe and clear international framework to finance space assets. Both States and representative of manufactures, financiers, insurers and representatives of international organizations have expressed their doubts on the workability of the Protocol. The problem with the Draft Space Protocol is that its practical application raises several questions of compatibility with existing space law regime, for instance with regard to issues like jurisdiction and control of the launching States over its space objects, international liability of the launching State for damages caused by space objects and licensing procedures for space activities.

Additionally, some elements of the Draft Space Protocol, such as considering related rights, namely licenses, permits, etc. as "properties" which can be transferred and moved from one jurisdiction to another are rather questionable.

The present paper will address the major contradictions and problems of the Draft Space Protocol by putting forward solutions and proposal to solve them. In this way, this paper tries hopes to provide a contribution to the positive conclusion of the negotiating process of the Protocol. This paper, indeed, strongly holds the idea that an international instrument regulating financing of space assets is needed and that the benefits that may be generated as a result of its proper functioning are extremely relevant.

THE CONVENTION ON INTERNATIONAL INTERESTS IN MOBILE EQUIPMENT AND THE DRAFT SPACE PROTOCOL

Before going into the analysis of the specific problems of the provisions of the Draft Space Protocol, it is important to understand the relation between the Protocol and the Convention.

It has always to be kept in mind that the Draft Space Protocol is meant to complement the Convention. In accordance with Article 6 of the Convention, indeed, the Convention and the Protocol must be read and interpreted together, since they constitute one single instrument. In case of inconsistency, however, the Protocol prevails.

The Convention, read in conjunction with the Protocol, applies when at the time of the conclusion of the agreement, (for instance, security agreement, title reservation agreement, leasing agreement or sale) creating for "the international interest", the debtor is located in a contracting State [7]. The applicability of the Convention is not affected by the location of the creditor. This provision is based upon the theory that the location of the secured debtor[8] is the place from which the debtor mainly runs the business related to the collateral and where the third parties can verify whether or not the debtor is subject to the rules of the Convention and Protocol (for instance, by determining if the host State is a contracting State).

If the debtor defaults, the Convention gives to the creditors the following options: take control or possession or control of the assets, sell the assets or grant a lease, or collect income from the assets. Additionally, in case of default, the creditor may obtain court assistance to preserve the assets, gain possession, lease or manage them. Article IX, b (i) of the Draft Space Protocol states that: "*any remedy given by the Convention shall be exercised in a commercially reasonable manner*". Article IX, b (ii) specifies that a remedy is exercised in a commercially reasonable manner where: "*it is exercised in conformity with a provision of the agreement between the debtor and the creditor except where such provision is manifestly unreasonable*". The Convention also requires that remedies be exercised in conformity with the procedure of the law of the place where the remedy is to be exercised. The purpose of the provision is to preserve State sovereignty.

Another interesting feature of the Convention is that, in conjunction with the Draft Space Protocol, it foresees the establishment of an International Registry to secure the financial interests of investors in space system and other mobile assets. This Registry is to be created by a Supervisory Authority whose tasks range from appointing and terminating the Registrar, assuring continuity of the Registry in case the Registrar is dismissed, defines the operational rules of the Registry, and many others. There are several advantages connected with the creation of an International Registry. For example the increased publicity which not only may protect third parties form being mislead by the false perception of wealth projected by debtor's possession of an asset but also may allow creditors to check whether a debtor's asset is charged with security.

RELATION BETWEEN THE DRAFT SPACE PROTOCOL AND INTERNATIONAL SPACE LAW: A PROBLEM OF COMPATIBILITY

As mentioned in the introduction of this paper, one of the main reasons behind the refusal of States and private financiers to accept the Draft Space Protocol is the lack of compatibility between the Protocol and the existing international space law.

This incompatibility not only may prevent the implementation of the Protocol's provisions but also may undermine the ability of States to comply with the obligations they have accepted when ratifying the space law treaties[9].

It is certainly true that the Draft Space Protocol differs from the space treaties in terms of purpose and nature. The former is an instrument of private international law dealing with securing financial interests of investors who have provided funds for outer space activities. The latter are instruments of public international law aimed at regulating public and private activities in outer space.

However, the Draft Space Protocol deals with the uses of outer space and should not therefore conflict with space law.

The importance of avoiding conflicts and of ensuring compatibility between the provisions of the Draft Space Protocol and the space treaties was well spread among the drafters of the Protocol. Indeed, the Preamble of the Draft Space Protocol reads:

"*Mindful of the established principles of space law, including those contained in the international space treaties under the auspices of the United Nations*".

Nevertheless, in my opinion, the Draft Space Protocol conflicts with the space law treaties with regards to the following points:

1) terminology;
2) inclusion of related rights in the definition of space assets (related rights as "movable rights")
3) liability issues: relation between default remedies and obligations of the "launching State"
4) jurisdiction issues: relation between jurisdiction of a State over its space object and choice of law of the Parties to the Protocol
5) priority of competing rights/public service exemption from default remedies

<u>1) TERMINOLOGY</u>

Article I (b) of the Registration Convention as well as Article I (d) of the Liability Convention provide a definition of the term "space object" which reads: "*the term space object includes components parts of the space objects as well as its launch vehicle and parts thereof*".

The Draft Space Protocol uses the term "space asset" which, according to Article I (2) (g) of the Protocol, consists of: (i) *any separately identifiable asset that is in space or that is intended to be launched and placed in space or has been returned from space*; (ii) *any separately identifiable component forming part of an asset referred to in the preceding clause or attached to or contained within such asset*; (iii) *any separately identifiable asset or component assembled or manufactured in space*; and (iv) *any launch vehicle that is expendable or can be refused to transport persons or goods to and from space*".

The Protocol's definition of "space asset" is broader than the definition of "space object" provided in the Liability and Registration Conventions. This fact immediately generates an intuitive problem: some "space assets" may not fall within the scope and application of the space law treaties. This means that States may find themselves in a situation in which they apply different rules to the "space assets" belonging to one single space object. This situation clearly causes problems of understanding of the applicable legal regime relating to that space object.

The different extension of the terms "space object" and "space asset" raises some additional questions and problems. For instance would the Moon and other celestial bodies fall under the definition of "space assets" used in the Draft Space Protocol"? While, theoretically, celestial bodies, including the Moon, may be considered to be "space assets", however they are not "space property" since Article II of the Outer Space Treaty prohibits appropriation of and creation of property rights over outer space, including the Moon and other celestial bodies[10].

Similar problems may arise when referring to other "space assets" such as orbits, orbital positions, and the radio frequency spectrum (RFS) [11]. These "assets", indeed, cannot be appropriated. Using an orbital location does not confer any property rights on its user, due to the global commons nature of such location. The point is that while a "space object" may well be a "space asset", the contrary is not always true, especially when space property rights considerations are involved.

The present paper argues that when preparing the next version of the Space Protocol, its drafters must pay particular attention in analyzing and clearly defining the distinction between the terms "space object" and "space asset", in order to avoid contrasts and incompatibilities between them.

RELATED RIGHTS

According to Article I (2) (f) of the Draft Space Protocol related rights means:

"*Any permit, licence, authorization, concession or equivalent instrument that is granted or issued by, or pursuant to the authority of, a national or intergovernmental or other international body or authority to manufacture, launch, control, use or operate a space asset, relating to the use of orbits positions and the transmission, emission or reception of electromagnetic signals to and from a space asset*".

The problem with the interpretation of the term "related rights" provided by the Draft Space Protocol is that it seems to consider licenses as negotiable property, as commodities in which financial interests may be secured. This is clearly a very sensitive aspect from a financier's point of view, due to the fact that the value of a space object as collateral depends to a large extent on the transferability of licenses.

Licenses, however, are not negotiable properties, nor they are "rights". They are privileges and/or prerogatives, granted by an official governmental entity to a specific operator in accordance with national licensing procedures. Once the license has been issued, the licensee is allowed to perform a certain activity, whether on Earth or in outer space. It has to be kept in mind that the granting of a license is not an obligation of the official entity: it decides whether or not to grant a license in case the applicant meets certain requirements and criteria.

Considering licenses or any other government authorization as a transferable assets, as the Draft Space Protocol seems to do, could lead to a situation in which private operators take over a key government function, namely the right to provide licenses. This, of course, is a situation which States will never accept.

The crucial point is that in most jurisdictions, licenses cannot be transferred from the original applicant to another without the prior authorization of the entity which has issued the license in the first place. Therefore, if private operators were allowed to transfer licenses without official approval of the transfer, States will loose one of their primary prerogatives, namely the power to grant licenses. Thus, I think that the approach followed by the Draft Space Protocol, considering licenses and other authorizations as movable rights, has not real chances to be agreed upon by States. Additionally, other problems associated with considering licenses as movable rights may be provided. For instance, in case licenses to use an orbital location or the radio frequency spectrum are considered negotiable rights (rather than privileges granted by an official national entity), the outcome will be the appropriation by private parties of outer space resources. This result will represent a clear violation of the terms of Article II of the Outer Space Treaty prohibiting the appropriation of outer space and the resources contained thereof[12].

As a consequence of all the above mention problems related with deeming licenses and other government authorizations as transferable rights, this paper argues that any reference to related rights should be removed from the text of the Draft Space Protocol. As it has been analyzed, indeed, related rights are inadequate to be subject to security rights at international level.

This paper proposes to regulate licensing procedures at national level. A national space act could indicate conditions and procedures under which a license may be transferred. Such procedures should focus on the reliability and creditability of potential licensees. In case both conditions are given, it would be possible for a license to be transferred from a user to another.

LIABILITY ISSUES: RELATION BETWEEN DEFAULT REMEDIES AND OBLIGATIONS OF THE "LAUNCHING STATE"

According to Article VII of the Outer Space Treaty as well as Article II and III of the Liability Convention, the launching State is liable for damage caused by its space object.

While Article VI of the Outer Space Treaty sets out the principle that State are responsible for national activities in outer space and that such activities must be authorized and constantly supervision by the State, Article VIII of the Outer Space Treaty indicates the State that has registered a space object shall retain jurisdiction and control over it.

It may happen that as a result of default remedies provided by the Draft Space Protocol (Article VIII and X) a satellite which is subject to security agreements may be taken into possession and control, sold or leased to another operator from another country. In case an accident occurs through or after the changes in control, the launching State may find itself in a situation in which it can be held liable for damage caused by a foreign operator which is not under its jurisdiction.

As it has been rightly pointed out[13], it is true that a situation in which the launching State is unable to control its space object that has been transferred to a creditor is not unique to the Draft Space Protocol. There is, for instance, the case of an Indonesian satellite (Palapa B) which was repossessed by the insurer (Lloyds), and where the insurer took title to and control over the satellite, while assuming full liability for potential damage until the moment in which the satellite was retrieved by the Space Shuttle. This, however, remains as an exception and, at the moment, space law does not provide a clear set of rules applicable to this case[14].

Nevertheless, although the above situation is not unique, it does not mean that something must not be done. It is important, indeed, to clarify what are the right and duties of the parties involved. It has always to be kept in mind that the purpose of the Liability Convention is to protect third parties from damages caused by space objects. Therefore, it is extremely relevant to determine who, between the launching State and the operator/State which has assumed control of the satellites, has to pay compensation for the damage caused by the space object.

Let's take the case in which a creditor has made use of his default remedies and has taken control of the satellite on which he had security agreement. It may well happen that the creditor is not very familiar with space law neither with the concept of launching State. The creditor, indeed, relying upon normal principles applicable in case of insolvency, has taken possession of his secured satellite and he has decided to operate it. Therefore, at this stage of the process no agreement between the creditor and the launching State exists.

In case an accident occurs through or after the change in control, the damaged State will ask compensation from the creditor/operator. Following the advices of his legal advisors, the creditor will present to the launching State a claim for compensation. The launching State will refuse by arguing that when the satellite was transferred to the creditor, it had assumed that the creditor had accepted liability for damage which may occur after the transfer. The launching State will also rely on existing State practice to support its interpretation.

The result will be that they will not be able to decide who will have to pay for the damage and, thus, they will go to court to settle their dispute.

The major effect of this situation is that the primary purpose of the Liability Convention, namely to protect a third Party which have suffered damage caused by a space object and to guarantee them compensation for such damage, will be betrayed. Such a Party will have to wait for a long time before being compensated, always assuming that the court is able to settle the dispute.

The question, then, is: what can be done to protect third parties in case of damage caused by a space object?

The best way to deal with it is by means of a bilateral agreement signed between the launching State (or States) and the creditor when the change in control takes place. In this manner, the duties of the parties involved and the liability issues may be settled.

What can the Draft Space Protocol do in this respect? How can it help third parties?

This paper proposes to introduce in the text of the Protocol a provision which reads as follow: in case of damage caused by a space object to third parties following the implementation of default remedies, compensation to the damaged parties shall be guaranteed.

Such provision will not indicate which one of the parties involed will have to pay compensation for the damage. However, it will have the advantage to create an obligation for those parties to find an agreement regulating the issue of liability when the transfer of satellite from the launching State to the creditor takes place. This provision will, therefore, provide a higher level of protection for third parties damaged as a result of space activities.

JURISDICTION ISSUES: RELATION BETWEEN JURISDICTION OF A STATE OVER ITS SPACE OBJECT AND CHOICE OF LAW OF THE PARTIES TO THE PROTOCOL

According to Article VIII of the Outer Space Treaty, the State of registration retains jurisdiction and control over the space object it has registered. This means that not only the courts of the State of registration will be competent on matters related to the space object but also that the law of that State should be applicable to all transactions concerning that object.

Article 42 of the Convention gives parties the right to choose the forum for their transaction. This forum does not need to have any relation with the parties or the space object. In addition, Article 43 (1) of the Convention provides for another option for the relief pending final determination, which is explained in Article 13 of the Convention: "The courts if a contracting State chosen by the parties and the courts of the contracting State on the territory of which the object is situated have jurisdiction to grant relief under Article 13 (1) (a), (b), (c) and Article 13 84) in respect of that subject. The same principle is repeated in Article 20 of the Protocol. Regarding the relief pending, one may wonder how to interpret the term "the contracting State on the territory of which the object is situated", when the transaction concerns satellite. Should the State of Registration be considered the competent State?

Additionally, Article VIIII of the Protocol indicates that the parties may agree on the law governing their contractual obligations, unless the contracting State makes a declaration. Is this provision not inconsistent with the exclusive control over the space object of the State of registration?

Contractual rights and property law have a complex relationship. In a security agreement, parties may choose the law which will apply to their contractual relationship under the Draft Space Protocol. However, regarding property law, the law of the State of registration will apply. This means that the Draft Space Protocol does not offer a clear picture of the issue of the applicable law to the security agreement. Therefore, this paper recommends this issue to be further analyzed and clarified in the next version of the Draft Space Protocol.

PRIORITY OF COMPETING RIGHTS ASSIGNMENTS/PUBLIC SERVICES EXEMPTION FROM DEFAULT REMEDIES

Other problems related to the application of the Protocol concern the existence of priority among competing rights of States and creditors. As it is known, satellites carries a number if transponders through which different services may be offered to different users. What may happen in case a single user is in default and the creditor uses the remedies available under the Convention and the Protocol, and seizes control of that transponder or the entire satellites?

During the last meeting of the New Steering Committee on the Space Protocol, Germany has suggested that creditors should only exercise their default remedies if that exercise would not affect the use of other space assets physically or functionally linked to the secured space asset. This suggestion has been strongly opposed by the industry side. The sub-Committee will have to further analyze this issue.

Additionally, there are questions related to the possibility of creating an exemption from default remedies in case of public services. What will happen in case a creditor wants to seize a satellite which is offering public services, even of State interest? One suggestion would be that satellite services be continued irrespective of creditors' exercise of default remedies. On the other hand, creditors need protection of their investments. Therefore, a similar proposal is not likely to encounter their approval.

The relation between default remedies and public services needs to be further analyzed and clarified when drafting the new version of the Draft Space Protocol.

CONCLUSION

The current deadlock in the negotiation of the Draft Space Protocol requires an analysis of the causes and reasons of such situation. The present paper has revealed that there are some problems of incompatibility between the Protocol and the space treaties which are undermining the success of the Protocol. This paper has proposed a series of proposals, such as the deletion of any reference to related rights and the insertion in the text of the Protocol of a clause regulating the relation between the launching State and the creditor taking control and possess of a space object, which may contribute to the success of the Protocol.

---

[1] This particiaption in a conference has been sponsored by Leiden University Fund (LUF)/Van Beuningen.

[2] UNIDROIT is an independent intergovernmental organization whose purpose is to study needs and methods for modernizing, harmonizing and co-ordinating private and in particular commercial law as between States and groups of States. For more details see at: http://unidroit.org/dynasite.cfm?dsmid=84219

[3] See P.Larsen, "*Creditors' secured interests in satellites*", in Proceedings of the Thirty-Fourth Colloquium on the Law of Outer Space, (1991), p. 233.

[4] For the text of the Preliminary Draft Protocol on Matters Specific to Space Assets see at: http://www.unidroit.org/english/workprogramme/study072/spaceprotocol/main.htm.

[5] The text of the Convention on International Interests in Mobile Equipment has been adopted at Cape Town in 2001. For the text of the Convention see at: http://www.unidroit.org/english/conventions/mobile-equipment/main.htm.

[6] The Protocol to the Convention on International Interest in Mobile Equipment on Matters specific to Aircraft Equipment has been adopted at Cape Town in 2001. For the text of the Protocol see at: http://www.unidroit.org/english/conventions/mobile-equipment/main.htm#NR2. The text of Protocol to the Convention on International Interest in Mobile Equipment on Matters specific to Railway Rolling Stock has been approved in Luxembourg in 2007. For the text of the Protocol see at: http://www.unidroit.org/english/conventions/mobile-equipment/main.htm#NR3.

[7] As indicated in the main text, an international interest may be created by means of a security agreement, title reservation agreement or leasing agreement or sale. This characteristic of the Convention reflect reflects a functional approach to security which focuses on the effect of the transaction rather than on its form and which tries to avoid problems generated by different forms of security defined in different legal systems.

[8] The Convention's definition of the debtor's location is rather broad, including place of incorporation, registered office or statutory seat, centre of administration or its place of business (Art. 4 of the Convention)

[9] Currently, space law consists of five space law treaties, namely: the *Treaty on Principles Governing the Activities of States in the Exploration and Use of Outer Space, including the Moon and Other Celestial Bodies* (1967), usually referred as "The Outer Space Treaty"; the *Agreement on the Rescue of Astronauts, the Return of Astronauts and the Return of Objects Launched into Outer Space* (1968), usually referred as the "Rescue Agreement"; the *Convnetion on International Liability for Damage Caused by Space Objects* (1972), usually referred as the "Liability Convention; the

Convention on Registration of Objects Launched into Outer Space (1975), best know as the "Registration Convnetion"; the *Agreement Governing the Activities of States on the Moon and Other Celestial Bodies* (1979), usually referred as the the "Moon Agreement". See the text of the Treaties at: http://www.unoosa.org/oosa/en/SpaceLaw/treaties.html.

[10] See in this respect S. Ospina: "*The UNIDROIT registration of security interests and the Registration Convnetion: compatible-complementary, or contradictory?*, in Proceedings of the Forty-Sixth Colloquium on the Law of Outer Space, (2003), p. 464; P. Van Fenema, "*The UNIDROIT Space Protocol, the concept of launching State, space traffic management and the delimitation of outer space*", Air & Space Law, Vol. XVIII 4/5 (September 2003).

[11] *Ibidem* p. 469.

[12] See F. Tronchetti," *The non-appropriation principle under attack: using Article II of the Outer Space Treaty in its defence*", in Proceedings of the 50th Colloquium on the Law of Outer Space (2008), pp. 526; F. Tronchetti, "*The non-appropriation principle as a structural norm of international law: a new way of interpreting Article II of the Outer Space Treaty*", Air & Space Law, Issue 33, (June 2008), pp. 277-305.

[13] P. Larsen, "*UNIDROIT Space Protocol: Comments on the Relationship between the Protocol and Existing International Space Law*", in Proceedings of the Forty-Fourth Colloquium on the Law of Outer Space, (2001), p. 187; P. Larsen, "*Critical issues in the UNIDROIT Draft Space Protocol*", in Proceedings of the Forty-Fifth Colloquium on the Law of Outer Space, (2002), p. 2; P. Larsen, "*Space asset financing and trade issues*", in Proceedings of the Forty-Sixth Colloquium on the Law of Outer Space, (2003), p. 228; P. Larsen, "*Future Protocol on security interests in space assets*", in Journal of Air Law & Commerce (2002), p. 1072; O. Ribbelink, "*The Protocol on Matters specific to space assets*", in European Review of Private Law, (2004), p. 37.

[14] See, in this respect R. Lee, "*Effects of satellite ownership transfers on the liability of the launching States*", in Proceedings of the Forty-Third Colloquium on the Law of Outer Space, (2000), p. 48; M. Gerhard, "*Transfer of operation and control with respect to space objects – problems of responsibility and liability of States*", in ZLW 51, 4/2002.

THE UNIDROIT DRAFT OF A SPACE ASSETS PROTOCOL A CIVIL LAW INSTRUMENT UNDER A PUBLIC FRAMEWORK

Dr. Bernhard Schmidt-Tedd*
Deutsches Zentrum für Luft und Raumfahrt e.V. (DLR), Bonn, Germany
Bernhard.Schmidt-Tedd@dlr.de

Isabelle Arnold*
Deutsches Zentrum für Luft und Raumfahrt e.V. (DLR), Bonn, Germany
Isabelle.arnold@dlr.de

ABSTRACT

Further to the adoption of the UNIDROIT Convention on International Interests in Mobile Equipment, the UNIDROIT Space Assets Protocol has been under discussions for several years and addresses the question of the regulation of the financing of space activities. In this matter, international private law only brings limited solutions and the adoption of a uniform international regime might serve legal certainty and support private financing. Until now, consensus has not been achieved in all elements of the Protocol. Particular difficulties arise from the fact that the issue is at a cross road between civil law financing instruments and the international regulation of space activities. This paper addresses the legal issues resulting from this independency.

The Protocol addresses the specific difficulty arising from the location of the assets in space, implying limited access and ownership. This leads to the question of the identification of space assets, in particular as far as the registration of International Interests is concerned. This civil registration shall not impair the public register maintained by the Secretary of the United Nations, which is the baseline for an extension of the State's jurisdiction and control. A transfer of ownership in the case of the implementation of the remedies foreseen for the debtor's default under the Protocol might impact the rights related to the object and lead to new licence requirements.
A limitation of the remedies might also be examined, since their implementation might affect the launch while the Launching State remains liable. The purpose of the analysis is to provide solutions within the framework of the existing legal regime without narrowing more than it is necessary the scope of application of this civil law instrument.

I. INTRODUCTION

The Cape Town Convention on International Interests in Mobile Equipment (the "Convention"), was negotiated and opened to signature under the work of the UNIDROIT[1] Organisation in November 2001 and entered into force on 1. April 2004. The Convention sets out a general framework for the international financing of mobile equipment. For a case by case application, the Convention is to be completed by a series of three protocols, each of them dedicated to a particular international equipment. The first relevant Protocol is dedicated to Aircraft Equipment (the "Aircraft Protocol ") and was adopted at first, simultaneously with the Convention[2]. This was than followed by the adoption of the Luxembourg Protocol on matters specific to Railway Rolling Stock (the "Railway Protocol"), signed in Luxembourg on 23. February 2007[3]. The Space Assets Protocol to the Convention is the last of the three implementing instruments and is currently on the way to finalization[4]. It is not by chance that the Space Assets Protocol is the last one in this development. In this case, several elements unknown for the negotiation of the two first protocols come into question.

The first one lies in the fact that the assets the protocols deals with are located in outer space and are for this reason in most cases inaccessible to human contact. This is not without meaning for the Space Assets Protocol. Indeed, it aims at implementing an asset-based financing instrument, as opposed to a project-based one. The practical impact of this orientation is that international interests shall be taken on the space asset itself (*in rem*) and not on the economic value of the project as such, which means an aggregated value of contractual relationships under the law of obligations.

Further, the technology involved is still subject to evolution and it is impossible as of today to predict what space technologies will provide mankind in a couple of years from now. Yet, the protocol aims at providing a long lasting and flexible framework and its viability is dependant upon its adaptability to the technological framework for space activities. However, the amount of assets potentially covered worldwide by the field of application of the protocol is today still relatively limited, as the financing of private space assets is still almost exclusively made on a project-financed basis.

In this context, one of the most challenging issues for the finalization of the Space Assets Protocol is its interaction with the existing legal framework applicable to space activities. Indeed, an international financing tool for space assets is being negotiated for the first time. Until now, the international framework applicable to space activities is composed of five Treaties adopted at the beginning of the Space Age. Consequently, those concentrate on activities of States in Outer Space and have little consideration for private space activities. However, this international framework is relevant for the Space Assets Protocol, which has to find its place in this international context. The civil law instrument the Space Assets Protocol aims at creating will find application in a public law dominated area.

The present paper presents a critical overview on the relevant legal framework for private actors and presents the current negotiation status of the Space Assets Protocol. Finally, current issues of the negotiation process will be addressed.

II. INTERNATIONAL SPACE LAW: AN UNAVOIDABLE BACKGROUND FOR PRIVATE SPACE ACTORS

The first objective of the Convention is to facilitate the financing of high value mobile equipment, which by nature crosses borders more frequently than any other asset. Therefore, the necessity of having a unique instrument based on a unique international registration system is obvious. The Convention and the respective implementation protocols aim at facilitating the execution of remedies of creditors in case of non-payment. If this is achieved, creditors, and to a larger extent, financial institutions, might be drawn to new businesses in the space field. In the case of space assets, the Convention and the Space Assets Protocol could help to bring into the space field financing institutions, who for the moment rather concentrate on more terrestrial matters. The harmonised remedies could stand for more investment safety. For this reason, the Space Assets Protocol could support the creation of new financing solutions.

The Space Assets Protocol aims at implementing an efficient liquidation proceeding based on a new uniform legal framework and independent from local administrative customs. This logic of an international asset-based financing system assumes that the cross-border transfer of ownership under civil law does not fall under unexpected restrictions; in addition the applicable law should be easy to define. Those assumptions are partly in conflict with principles of public Space Law.

1. Applicable law

According to Art. VIII Outer Space Treaty (OST)[5] "a State Party to the Treaty on whose registry an object launched into outer space is carried shall retain jurisdiction and control over such object...". Art. VIII, 2nd sentence OST clarifies that "ownership of objects launched in outer space, including objects landed or constructed on a celestial body, and of their component parts, is not affected by their presence in outer space or on a celestial body or by their return to the Earth". Without Art. VIII OST, the private ownership would be isolated from a protecting legal regime; only Registration by a Launching State creates the link to a national legal system. This baseline for the applicable law in an environment free of national appropriation and sovereignty (Art. II OST) has as condition: the registration of such space asset, which is in practice effected only a couple of month after launch. Indeed, the final parameters for registration are determined only after the commissioning phase, meaning when the payload has reached its final orbit destination and is partially or fully exploitable. The registration is reserved exclusively to Launching States as defined in Art. VII OST, Art. I of the Liability Convention (LIAB)[6] and Art. I of the Registration Convention (REG)[7]. In the case there is more than one Launching State – which actually happens quite often – following the definition as provided above, according to Art. II (2) REG "they shall jointly determine which one of them shall register the object".

2. Responsibility for national activities

The international framework applicable to space activities created under the work of the United Nation does not accept any private space activity without a clear link and responsibility to an appropriate State. Art. VI OST states that "States Parties to the Treaty shall bear international responsibility for national activities in outer space, including the Moon and other celestial bodies, whether such activities are carried on by governmental agencies or by non-governmental entities, and for assuring that national activities are carried out in conformity with the provisions set forth in the present Treaty. The activities of non-governmental entities in outer space, including the Moon and other celestial bodies, shall require authorization and continuing supervision by the appropriate State Party to the Treaty." According to the reasons set out before, only Launching States concerned can be in a position to execute those authorization and control obligations. A Launching State remains responsible and liable (Art. VI, VII OST, LIAB) for that space object for an unlimited period of time in the frame of the field of application of the UN Treaties.

Under precise consideration, this presents several issues since a State, which is only involved in a limited way in a space activity (e.g. a State that is only involved in the launch operation because its territory was used for the launch, and in consequence has nothing to do with the further operation of the payload) can still be held liable for damages at an international level. This is also the case in the event of a transfer of ownership[8] under civil law: the transfer of ownership has no retroactive impact on the Launching State and it does not affect the initial public law constellation of Launching States / Registering State.

However, the State whose private entity has acquired as creditor a (foreign) space asset should have the opportunity to react to this new situation, since it is responsible for the activities of the creditor. The same is true for an authorizing State whose insolvent private entity lost the power of disposal to the creditor/private entity of a foreign State. Indeed, in this case also, the State is responsible for the activities of its nationals in space under the dispositions of the OST. This responsibility for authorisation and control is not linked with the quality of Launching State, but has another legal basis resulting from Art. VI OST. The French Act on space activities of 3 June 2008 [9] presents in this matter an adequate solution, since it imposes an authorization obligation upon the transfer of space objects launched. Under the authorization, the French State can ensure that the new space objects for which France is internationally responsible are conform to certain technical regulations and operated in an appropriate way. Those reflections are not only legal theory but have a concrete impact on the elaboration of the Space Assets Protocol.

Art. 3 (paragraph 1) of the French Act states: "transferring the control of a space object having been authorized under this law to a third party is subject to the prior authorization of the administrative authority." This authorization requirement applies according to Art. 3, paragraph 2 as well to the inverse case: "any French operator who intends to control a space object the launch or control of which has not been authorized under this law must obtain a prior authorization granted for that purpose." For financing and leasing institutions, which are the most concerned by the Convention, these boundary conditions might be quite unusual but it is a necessary condition to ensure that space activities, even conducted by private operators, are conducted under an internationally acceptable manner. Moreover, having in any case a State responsible for a space object ensures effective control over it and might help to avoid the development of the practice of flags of convenience, as seen in the maritime field. These issues of international law cannot remain ignored by the Space Assets Protocol. Should the Space Assets Protocol lead *de facto* to a transfer of ownership of a space asset because of bankruptcy or insolvency of an operator, this should be organised in a matter consistent with the international public law framework, even if this leads to some additional discussions.

3. <u>Restrictions of safety regulations</u>

Space assets can often have a dual use character, which means that they can serve for civilian as well as for military purposes. Since space assets also often rely on missile technology, they might also undergo transfer restrictions and export control dispositions. This is true for all kinds of space assets.

Art. XVI (2) of the preliminary draft of the Space Assets Protocol (version 2003) gives Contracting States the opportunity to restrict or attach conditions to the exercise of remedies in case "the exercise of such remedies would involve or require the transfer of controlled goods, technology, data or services".

As a consequence, Contracting States can place restrictions or conditions on the remedies available to creditors regarding controlled space assets or involving transfers or assignments of related rights[10].

Special attention must be given to the Wassenaar Agreement[11] and the Missile Technology Control Regime[12].

4. <u>Limitations for public interest reasons</u>

The issue of the limitation of remedies for reason of public interest is still under discussion. This reserve should avoid that the implementation of remedies under the Protocol affects a service run at a public level. In this matter, the negotiations about the Space Assets Protocol are confronted with the diverging conceptions that States have regarding the sole notion of public service. Options in this case could consist in leaving the definition of the notion of public service to the States concerned on a case by case basis or radically exclude the public service domain from the application field of the Space Assets Protocol.

III. <u>STATUS QUO OF THE SPACE ASSETS PROTOCOL: THE WAY FORWARD?</u>

The first reading of the preliminary draft Protocol on Matters specific to Space Assets took place in Rome from 15 to 19 December 2003 in the framework of a UNIDROIT committee of governmental experts. During the second session of governmental experts, held in Rome from 26 to 28 October 2004, a number of policy issues were raised, which are still under discussion today. Since this time inter-sessional work has been undertaken, especially those of the Space Working Group (SWG).

The Royal Bank of Scotland (RBS), a member of the SWG invited the governmental experts and the SWG to a Governmental-Industry Forum, which took place in London on 24 April 2006. A second government/industry meeting on invitation of Milbank, Tweed; Hadley & McCloy was held in New York on 19 and 20 June 2007. At its 61st session on 29 November 2007, the UNIDROIT General

Assembly endorsed the establishment of a steering committee with the goal to build consensus within the group of governmental and industry experts. The Steering Committee had its first meeting on invitation of the German Ministry of Justice, supported by the German Space Agency DLR in Berlin between the 7 and 9 May 2008.
The official reference text of the Draft Protocol on Space Assets is still the version of December 2003. However, the number of open points of discussion has drastically increased.

IV. OPEN POINTS OF DISCUSSION

Several fundamental issues remain unsolved and delay the finalization of the Space Assets Protocol. Those are not unessential and condition substantial elements of the Protocol. They concern its field of application, the question of related rights, the limitation of remedies and the question of the international register.

1. Field of application of the Space Assets Protocol

The first issue to be addressed in this matter is the time period of application, or in other words the question of the beginning of the field of application of the international legal instrument. Is it the beginning of the industrial production of the space asset on ground, the acceptance by the customer after delivery, the transfer to the launch pad or the launch itself? After long lasting discussions and options proposed, it was agreed that there should be no interruption within a financing concept of a space operation and the Space Assets Protocol should find application even on ground before launch.

The major problem regarding the scope of application is the definition of the space assets concerned. In the Aircraft Protocol and the Railway Protocol, an enumeration of the assets concerned by the respective Protocols (air frames, aircraft engines, helicopters and railway rolling stocks) is conclusive. In space, the situation is more complex and the technical developments are not settled yet. A limitation on today economically relevant objects, such as satellites would perhaps not meet the needs of the next generation and a revision for modernisation of such a legal instrument at a later stage is an incalculable effort. In general, the alternative is either to have a longer enumerative list of assets or a general clause.

Following the deliberations as of today, consensus has been reached on the relevance of at least satellites, space stations, space vehicles, launch vehicles, reusable space capsules in, or intended to be used as a launch vehicle. The main characteristic of this category of assets is that they are capable of being independently operated, used and commanded. The operative accessibility gives the practicable chance of recourse in the event of default.

Another layer of discussion concerns the possible inclusion of components of a space asset in the application field of the Space Assets Protocol. In contrast to the question of related rights, which will be further debated in this paper, components are clearly and obviously asset-related. They often have an enormous commercial value and utility, which makes them interesting for financers.

Components can be classified into two categories. Firstly, there are items which can be operated, used and commanded solely in connection with the space asset concerned (e.g. propulsion devices and solar cell panels). On the other hand, there are components which operation is not limited to the physical link with the space asset concerned and which can be operated, used and commanded independently (e.g. transponders and sensors). Only this last category should be included in the field of application of the Space Assets Protocol.

Therefore, the definition of the assets falling within the field of application of the Space Assets Protocol should be a combination of an enumeration of assets and a limited inclusion of uniquely identifiable items. Those components should be related to the enumerative list of assets and capable of being independently operated and commanded[13].

The question of independent use and operation is a general category for a well balanced system of default remedies. Thanks to this criterion, it will be possible to distinguish competing interests of creditors of different assets and or components, which are physically or by their function linked together. This can, for instance, be the case where several satellites are linked through an interposed orbital relay station and this entire constellation would no longer be able to function if an individual satellite were to be removed from the constellation. Here, too, recourse should only be possible to the extent that mutual impairment can be ruled out.

The consequence of this complex field of application is the necessity to balance conflicting interests and the risk of undue impairment of rights in a differentiated solution for the exercise of default remedies[14].

2. Associated and related rights

According to Art. I (2)(a) of the preliminary draft protocol (Version 2001), „associated rights" with respect to space assets means *inter alia* "(i) to the extent permissible and assignable under the national laws concerned, all permits, licences, approvals and authorisations granted or issued by a national or intergovernmental body or authority to control, use and operate the space assets, including orbital use authorisations and authorisations to transmit and receive radio signals to and from space assets".

In a footnote[15] of the draft protocol version 2003, the SWG recommendation[16] to introduce the new terms "debtor's rights" and "related rights" is mentioned; but further elaboration on this proposal is suggested by the Committee of Governmental Experts. The relevant definition of "related rights" as proposed by the SWG [17] is following: "any permit, licence, authorisation, concession or equivalent instrument that is granted or issued by, or pursuant to the authority of, a national or intergovernmental or other international body or authority to manufacture, launch, control, use or operate a space asset, relating to the use of orbits and the transmission, emission or reception of electromagnetic signals to and from a space asset".

From the point of view of the creditor, "there is a great significance in intangible rights and "control"... and contractual rights such as performance warranties". Associated rights are according to this opinion "inextricably linked to a physical satellite and are integral to the commercial value of a satellite"[18]. This argumentation jumps over a clear distinction between asset-related rights and contractual, project-related rights.

Two aspects have to be distinguished: a space asset in form of a satellite in orbit can only be controlled by indirect possession[19]. Possession as such is a crucial element for the transfer of ownership and the enforcement of remedies. Legal instruments to gain possession are therefore relevant. On the other side, operator-related permits and licences are only valid for a special (legal) person and are not transferable. Most of the time, they are granted *intuitu personae*. This is the same situation for an industrial plant on ground. The UN Treaties are insofar only an additional aspect to be observed with regard to a cross-border transfer of ownership. This necessity for the successor to apply for his own permits and licenses does not hinder on ground the transfer of ownership of industrial complexes. Why should it in orbit?

The critical aspects of the inclusion of related rights have been detailed in a German Working Paper during the Steering committee in Berlin in May 2008. The main arguments are following:

The intention to create an independent international interest in debtor's rights and related rights in addition to an interest in "space assets" is not in line with Art. 2 of the Convention, according to which the international interest can only be created in a uniquely identifiable object, as listed in Art. 2(3)(b) – here in the category "space assets" – and not in subjective rights.

Under Art. 2, an interest can only be created by virtue of a security agreement, a title or reservation agreement or a leasing agreement. By contrast, such "debtor's rights" and "related rights" would have to be effected in an assignment (by way of security).

Besides arguments of legal systematic, one should also take into consideration the practical aspect, that independent debtor's rights or related rights might be counter productive in the sense of weakening the position of the asset-related rights (in the event the debtor's rights are transferred to a different creditor). For systematic purposes, the question of access to debtor's rights in the event of default should be solved in the context of recourse, respectively exercising an interest, and not on the level of creation of an international interest.

From the aspect of national space legislation it is fundamental to avoid giving the impression that individual operator-oriented licences are transferable rights.

3. Limitations of remedies / launching phase

The launch is the most sensible and relevant phase for a space mission. A disturbance of a launch sequence could lead to extensive damage. Therefore, there is a need for a special temporary protection against enforcement measures by creditors. During the Steering Committee Meeting in May 2008 Germany introduced the following proposal:
"Article IX para. 7: The creditor shall not exercise default remedies according to Chapter III of the Convention during the launching phase. The launching phase begins on arrival at the final launch position; it ends on arrival at the first orbital position or on departure from the final launch position on account of termination of the launch." This is in line with similar exceptions for aircrafts with passengers on board.

4.

Interface UN -UNIDROIT - Register

The main provisions concerning the registration of space assets according to the UNIDROIT regime is contained in chapter IV of the Cape Town Convention and in chapter III of the Space Assets Protocol. The purpose of this registration is to ensure precedence in all Member States of the registered security interest in case of bankruptcy or insolvency.
Therefore only transferable rights should be admitted. Related rights in form of subjective licences are not the right category.
Under regular conditions, the UNIDROIT-registration takes place before the registration of the space object. Nevertheless, the latter criteria of the UN-Registration could be an additional identification criterion. On ground, a single number might be sufficient for the unique identification of an object. In orbit, a registration number is not really helpful. The identification has to be guaranteed by a combination of different indications.

V. CONCLUSIONS AND PERSPECTIVES

In order to fulfil the economic goal of this international financing instrument, the Space Assets Protocol must create an entirely new asset-based approach. It should not only improve the situation of established creditors, working today on a project-based financing regime but also open new perspectives for creative new-comers, potentially concentrated on smaller assets or components.

The Space Assets Protocol has to adhere to the existing international framework for space activities. Especially the transfer of ownership has to comply with the legitimate rights of launching states, responsible for jurisdiction and control as well as the related national space law regulations.

Adequate solutions are feasible but an intensive analysis of the specifics of the space sector is a pre-requisite.

*The opinions expressed in this paper are entirely those of the authors and do not in any way engage the organisations with which they are affiliated.

[1] www.unidroit.org

[2] Protocol to the Convention on International Interests in Mobile Equipment on Matters Specific to Aircraft Equipment.

[3] Luxembourg Protocol to the Convention on International Interests in Mobile Equipment on Matters Specific to Railway Rolling Stock.

[4] Goode, Sir Roy, Official Commentary on the Convention on International Interests in Mobile Equipment and the Protocol thereto on Matters specific to Aircraft Equipment, as approved for distribution by the UNIDROIT Governing Council pursuant to Resolution No. 5 of the Cape Town Diplomatic Conference, UNIDROIT 2002

[5] Treaty on Principles Governing the Activities of States in the Exploration and Use of Outer Space, including the Moon and Other Celestial Bodies, opened for signature on 27 January 1967, UN DOC. A/RES/2222/XXI, 610 UNTS 205

[6] Convention on International Liability for Damage caused by Space Objects, opened to signature for 29 March 1972, UN-DOC. A/RES/2777/XXVI, 961 UNTS 187

[7] Convention on Registration of Objects Launched into Outer Space, opened for signature 14 January 1975, UN DOC. 1023 UNTS 15

[8] Schmidt-Tedd, B.,/Gerhard, M., Registration of Space Objects: Which are the Advantages for States Resulting from Registration? In: Benkö, M./Schrogl, K.-U., Space Law: Current Problems and Perspectives for future regulation, Utrecht 2005, p. 121, 131 f

[9] Bill Nr. 2008-518 of June 3rd, 2008 relating to space operations, see Journal Officiel de la Rêpublique Francaise, 4 June 2008; for the English version see Schmidt-Tedd. B./Arnold, I., The French Act relating to space activities, From international law idealism to national industrial pragmatism, ESPI Perspectives 11, August 2008, http://www.espi.or.at

[10] See Larsen, Paul. B., Memorandum on National Restrictions on the Transfer and Operation of Space Assets, Space Working Group New York 19/20 June 2007, UNIDROIT Working Paper

[11] www.wassenaar.org/docs/IE96

[12] www.armscontrol.org/documents/icoc.asp

[13] A corresponding German proposal of May 2008 for a definition in Art. I (2)(g) Draft Protocol reads as follows: "Space Asset means (i) satellite, space station, space vehicles, launch vehicle, reusable space capsules in, or intended to be launched in or into space or used, or intended to be used as a launch vehicle and
(ii) any other uniquely identifiable item capable of being independently operated and commanded attached to, or intended to be attached onto the satellite, space station, space vehicles, launch vehicle, reusable space capsules"

[14] A corresponding German proposal of May 2008 for a definition in Art. IX (4)-(6) Draft Protocol reads as follows:
"(4) The creditor shall only exercise the default remedies according to Chapter III of the Convention insofar as this does not affect the use of, international interests in, and other rights relating to, other space assets physically linked to the secured space asset.
(5) Paragraph 4 shall apply with necessary modifications where space assets are not physically linked to each other but where the essential use of one such asset is not possible without the other asset.
(6) In the cases referred to in paragraphs 4 and 5 recourse shall be permitted where
a) the person impaired by recourse consents to the recourse or
b) the creditor offsets the impairment of the use of the international interest or of the other right in the space asset by taking equivalent technical measures. "

[15] Footnote 6 in the preliminary draft Protocol 2003

[16] Footnote 7 ibid. with reference to UNIDROIT C.G.E/Space Pr./1/W.P.13

[17] UNIDROIT 2004 C.G.E/Space Pr./2/W.P.4

[18] Memorandum on the Importance for Space property of the Inclusion of associated rights, 2nd Joint Session ICAO/UNIDROIT, 24 Aug.-3 Sept. 1999, ICAO Ref. LSC/ME/2-WP/7 and UNIDROIT C.G.E./Int. Int./2-WP/7 Appendix I Cif. II

[19] The situation for objects on a space station might be different

IAC-08-E8.1.12
CISG AND THE FINAL FRONTIER: CONTRACTING FOR THE INTERNATIONAL SALE OF GOODS THAT ORIGINATE FROM, ARE DELIVERED IN, OR TRANSIT THROUGH OUTER SPACE AND THE PASSAGE OF RISK OF LOSS

Michael C. Mineiro
Institute of Air and Space Law, McGill University- Montreal, Canada
Michael.mineiro@mail.mcgill.ca

Abstract

As the commercial potential of outer space is developed, goods will be demanded and so will contracts to govern the sale of those goods.

This paper examines the applicability of CISG to contracts for the international sale of goods that originate from, are delivered in, or transit through outer space. CISG default provisions are assessed for their adequacy to proportion risk of loss given the environmental and technical challenges of outer space.

Finally, based on the findings of these analyses, this paper concludes as to the sufficiency of CISG to govern the contracts for the international sale of goods that originate from, are delivered in, or transit through outer space.

I. Introduction

There is no agreement under international private law *specifically designed* to govern contracting for the international sale of goods that originate from, are delivered in, or transit through outer space.

The current void in international law raises the following questions: Does the United Nations Convention on Contracts for the International Sale of Goods (CISG)[1] apply to contracts for the international sale of goods that originate from, are delivered in, or transit through outer space? If applied, does CISG adequately take into account the unique characteristics of outer space to such an extent that CISG can be relied upon? Should default CISG provision be modified to better facilitate contracting for the international sale of goods in outer space?

This paper first assesses CISG's applicability to contracts for the international sale of goods that originate from, are delivered in, or transit through outer space. Thereafter, given the environmental and technical challenges of outer space, CISG's adequacy to proportion the risk of loss, passing of risk, and obligations to transfer goods in outer space is assessed.

The assessment of passage of risk provisions is undertaken via hypothetical scenarios that are designed to provide legal practitioners with a framework from which to begin analyzing the complexities of contracting for risk of loss in outer space. Proposals are made to modify the default CISG risk of loss provisions to better suit the unique characteristics of outer space. The need for outer space specific trade terms, i.e. Incoterms, is discussed. Cross-waiver of liability provisions are examined.

Based on the findings of these analyses, this paper concludes as to the sufficiency of CISG to govern the contracts for the international sale of goods that originate from, are delivered in, or transit through outer space.

II. What is CISG?

The United Nations Convention on Contracts for the International Sale of Goods (CISG) was adopted on April

[1] *United Nations Convention on Contracts for the International Sale of Goods* (CISG), 11 April 1980, 1489 U.N.T.S. 3 (entered into force 1 January 1988).

11th, 1980. It is an international treaty that provides a uniform text of law for the international sale of goods.[2] CISG does not apply to contracts for the sale of services.[3] Its subject matter is restricted to the formation of the contract and the rights and duties of the buyer arising from such a contract.[4] Seventy-one states are party to the convention.[5]

III. CISG and Outer Space: CISG can apply to contracts even if the goods originate from, are delivered in, or transit through outer space

While originally conceived as a convention covering contracts for the terrestrial international sale of goods, CISG does not prohibit its application to the international sale of goods that originate from, are delivered in, or transit through outer space.

Part I of CISG, *Sphere of Application and General Provisions*, sets the conditions necessary for CISG's applicability. If these conditions are fulfilled, CISG could apply, regardless of the geographic location of the goods.[6]

Article 1 provides the basic rules to determine if CISG is applicable.[7] Article 1(1) states: "This Convention applies to contracts of sale of goods between parties whose places of business are in different States."[8] In addition to this requirement, one of two other conditions must be fulfilled: either the parties' places of business are located in Contracting States or the rules of private international law lead to the application of the law of a Contracting State.[9]

These conditions can be fulfilled, even if the goods originate from, are delivered in, or transit through outer space. For example, S is a space-widget company, whose place of business is located in the United States.[10] S has 1,000 space-widgets in storage in an outer space cargo facility. B is a mining company whose place of business is located in China.[11] B wishes to purchase the 1,000 space-widgets for their mining operation on an asteroid in outer space. S agrees to this contract and delivers the goods (as contracted) to a location on B's mining asteroid.

The contracting parties, B and S, have their places of business in different States. Both of these States are Contracting States.[12] Article 1 conditions for CISG's application have been fulfilled, irregardless of the goods having originated in, transited through, and been delivered in outer space. So long as other relevant CISG provisions are met, CISG will apply to this contract.

---

[2] CISG merges two earlier conventions, the 1964 Hague Formation Convention and the 1964 Hague Sales Convention.

[3] CISG, Art. 3(2).

[4] CISG, Art. 4.

[5] As of July 8th, 2008, seventy-one States are party to the Convention. See UNCITRAL website <http://www.uncitral.org/uncitral/en/uncitral_texts/sale_goods/1980CISG_status.html>.

[6] During the initial demand for contracts involving goods in outer space, Sellers and Buyer will be located on earth. While the goods under contract may be located in outer space, business will still be conducted on earth. The cost of business operations, the convenience of contract formation, and practical reasons will make earth the location for places of business to be located.

[7] CISG, Article 1.

[8] Id.

[9] CISG, Art. 1(a) and 1(b).

[10] Both parties have places of businesses located in outer space (i.e. multiple places of business). In the contract they agree their respective business headquarters, which are located in China and the United States, are the relevant places of business for the purposes of the contract.

[11] Id.

[12] Both China and the United States are Contracting States.

IV. Other Articles Governing CISG's Scope of Application to Consider

Three key articles governing CISG's scope of application are Article 2(e), Article 6, and Article 90.

Article 2(e) excludes CISG application to the sale of "ships, vessels, hovercraft or aircraft."[13] This exclusion does not explicitly apply to spacecraft or other space objects. Some commentators have proffered that rockets, satellites, and space stations are not excluded under Article 2(e) and hence fall under the Convention.[14] This is mostly likely the case. However, interpreting the term vessel to include spacecraft that exhibit the functional characteristics of a transport vehicle is reasonable and if so interpreted such spacecraft would not be covered by CISG. To further complicate matters, it should be noted that in outer space a vessel may take on the dual-role of being both a means of transportation *and* a good.[15]

Article 6 grants parties the freedom to contract. Parties may exclude the application of CISG or derogate from or vary the effect of any of its provision.[16]

Article 90 regulates the relationship between CISG and other international agreements.[17] If an international agreement contains provisions concerning matters governed by CISG, the international agreement's provisions prevail over CISG, provided that the parties have their places of business in States parties to such agreement.[18]

V. General Considerations of Risk of Loss and Passage of Risk in Outer Space

On earth, goods may become damaged for a variety of reasons; flood, fire, shipwreck, carrier neglect, and theft, just to name a few. In outer space, the risk of loss to goods becomes even greater. Goods may be damaged by space-debris, radiation, loss of transportation vessel, and a variety of other foreseeable and unforeseeable events. When goods are damaged or loss, the party assuming the risk of loss must deal with costly and draining legal ramifications. If there is no insurance, the party bears the complete loss of the goods. If the party does have insurance, they must press a claim against the insurer, wait for a settlement, and may have the responsibility of salvaging damaged goods.[19] On earth, assignment of this risk is extremely important. Given the special environment conditions of outer space and the potential risks of loss, determining who bears that risk is paramount.

VI. CISG's Default Provisions on Risk of Loss and the Passage of Risk: Articles 66-70

CISG risk of loss and passage of risk provisions are default rules that

[13] CISG, Art. 2(d).

[14] Peter Schlechtriem & Ingeborg Schwenzer eds., *Commentary on the UN Convention on the International Sale of Goods* (CISG) (Oxford University Press, 2005), Commentary on Art.2 at 52 [para. 35] footnote 69.

[15] For example, B may contract for pre-fabricated space buildings. These buildings have the dual use of being space fairing vessels. The buildings can take off and land on different planets, although it is difficult and costly. Their primary use is to be space buildings. Do these buildings meet Article 2(e) vessel exclusion?

[16] CISG, Art. 6.

[17] Fritz Enderlein and Dietrich Maskow, International Sales Law: U. N. Convention on Contracts for the International Sale of Goods, (1992). Author commentary on CISG Article 90; reproduced with permission on http://www.cisg.law.pace.edu/cisg/biblio/enderlein-art90.html.

[18] CISG, Art. 90.

[19] Enderlein, *supra* note 17; Author commentary on CISG Article 67; reproduced with permission on http://www.cisg.law.pace.edu/cisg/text/secomm/secomm-67.html.

operate in the absence of the parties' contrary agreement. Depending on the desires of parties involved, the default CISG provisions may or may not be adopted. It is common practice for parties to derogate from CISG and draft their own risk of loss and passage of risk provisions. Nonetheless, these provisions are still important because they provide guidance to practitioners drafting alternative provisions.

CISG Chapter IV Articles 66-70 are the default provisions governing the passing of risk of loss. These articles allocate the risk of loss and passage of risk in four different situations. Article 67 allocates the risk of loss when the contract involves he carriage of goods. Article 68 allocates the risk of loss when the goods are sold in transit. Article 69(1) allocates risk when the buyer picks up the goods from the seller's place of business. Article 69(2) applies to all other transactions, such as "destination contracts, bailment contracts, and contracts in which the seller uses his own vehicle to deliver goods to the buyer."[20] Article 66 establishes that loss of or damage of goods after the risk has passed to the buyer does not discharge the buyer from his obligation to pay the price, unless the loss or damage is due to an act or omission of the seller.[21] Article 70 clarifies that buyer's remedies on account of the seller's fundamental breach of contract take priority over the risk rules.[22]

Article 67 applies when seller is required to ship the goods or is authorized to do so.[23] If the contract involves carriage of the goods and the seller is not bound to hand them over at a particular place, the risk passes to the buyer when the goods are handed over the first carrier for transmission to the buyer. Alternatively, if the seller is bound to hand the goods over to a carrier at a particular place, the risk does not pass to the buyer until the goods are handed over the carrier at that place.

Article 68 applies when the goods are in transit at the time the contract of sale is concluded.[24] The risk of loss transfers to the buyer at the time of contract conclusion, but the buyer's risk retroactively begins in certain circumstances when the goods were placed on the carrier.[25] The buyer is protected if the seller knew or ought to have known at the conclusion of the contract of sale that the goods had been lost or damaged and did not disclose this to buyer.[26]

Article 69 applies to cases not within articles 67 and 68. Article 69 "anticipates that the buyer will take possession of the goods and arrange for the necessary transport."[27] Risk passes to the buyer when he takes over the goods or, if he does not do so in due time, from the time when the goods are placed at his disposal and buyer commits breach for failing to

---

[20] Mitchell Stocks, *Risk of Loss Under the Uniform Commercial Code and the United Nations Convention on Contracts for the International Sale of Goods: A Comparative Analysis and Proposed Revision of UCC Sections 2-509 and 2-510,* 87 NW. U.L. Rev. 1415 at 1434.

[21] CISG, Art. 66.

[22] Schlechtriem, *supra* note 14; Commentary on Art.70 at 696 [para. 2].

[23] *Supra* note 17; Author commentary on CISG Article 67; reproduced with permission at http://www.cisg.law.pace.edu/cisg/text/secomm/secomm-67.html.

[24] CISG, Art. 68.

[25] Enderlein, *supra* note 17; Author commentary on CISG Article 68; reproduced with permission at http://www.cisg.law.pace.edu/cisg/text/secomm/secomm-68.html.

[26] CISG, Art. 68.

[27] Enderlein, *supra* note 17; Author commentary on CISG Article 69; reproduced with permission at http://www.cisg.law.pace.edu/cisg/text/secomm/secomm-69.html.

take delivery. In accordance with Article 69(2), if the buyer is bound to take over the goods or, if he does not do so in due time, from the time when the goods are placed at his disposal and buyer commits breach for failing to take delivery. If the buyer is bound to take over the goods at a place other than the place of business of the seller, the risk passes when delivery is due and the buyer is aware of the fact that the goods are placed at his disposal at that place.

CISG default passage of risk provisions needs to be read in conjunction the Convention as a whole, in particular with Article 36, Article 7(2), and Article 49(1) of CISG. Article 36 establishes the time at which goods must be in conformity with the contract. However with goods transiting through outer space, determining the time at which goods are in conformity may prove difficult. Article 7(2) fills the non-conformity timing gap with the general principle that a person who relies on a rule in his favor must prove that the preconditions for the application of the rule are satisfied.

Deciding who bears the risk of loss is crucial when negotiating and agreeing upon a contract. In practice, the default provisions listed *supra* may not allocate risk as desired among the contracting parties. It is likely, especially when outer space is involved, parties will want to alter or exclude the default CISG provisions on the passage of risk.

VII. CISG's Default Passage of Risk Provisions within the Context of Outer Space

Outer space poses new and sometimes riskier environmental conditions than earth. How goods are to be delivered, the risks involved in transport, the time needed to reasonably determine if goods have been damaged, the ramifications in case of breach; all of these contractual issues are given an extra dimension in outer space. To be sure, the lessons learned from contracting on earth can apply to outer space. However, practitioners unfamiliar with outer space need to take into special consideration how parties will want to contract in this new environment. The passage of risk provisions in CISG are an excellent example of contractual provisions that might need to be varied given the particulars of outer space.

In the following section, Articles 67, 68, and 69, are reviewed within the context of outer space. Each article's review begins with a hypothetical scenario. Each scenario applies CISG and the relevant passage of risk provisions when the goods are geographically located in outer space. Then, the provisions are examined and potential short-comings in default application of the provisions are discussed. Finally, alternatives to the default provisions are given. These alternatives attempt to harmonize CISG's passage of risk provisions with outer space.

1. Article 67: Shipping Contracts

Article 67 applies in two different scenarios involving the carriage of goods. In the first scenario, the risk of loss passes to the buyer when the goods are handed over to the first carrier for transmission in accordance with the contract of sale.[28] In the second, the Seller assumes the risk until goods are handed over to a particular carrier at a particular place.[29]

For example, under the first scenario, a typical contract may be as follows: Seller owns a factory on the Moon. Buyer contracts for the sale of

[28] CISG, Art. 67.
[29] Id.

goods. According to the terms of the contract, goods are to be picked up at Seller's factory and Seller does not need to transport the goods. Buyer independently hires a carrier to ship to the goods to earth. The carrier arrives at Seller's factory and the Seller hands over the goods to the carrier for transmission to the Buyer. When the goods are "handed over," the passage of risk has occurred.[30]

Allocating risk of loss under this provision, just as it would be on earth, works well in the given context of outer space. On earth, an overland carrier may pick up goods from the Seller factory and ship them to a second carrier for international transport. In outer space, a similar situation may arise if the Seller would lack a facility capable of accommodating the carrier or the carrier refused to go directly to Seller's factory. In either situation, the risk of loss passes when the goods are handed over to the first carrier for transmission.[31] This is a fairly clear passage of risk that provides predictability for the parties involved.

In the second scenario, the passage of risk occurs when the goods are handed over to a particular carrier. Multiple carriers are involved and the Seller assumes the risk of loss during part of the goods transport. For example, Seller owns a factory on the Moon. Buyer contracts for the sale of goods. According to the terms of the contract, Seller is to handover the goods at a space station in low earth orbit. Buyer has contracted for a carrier to ship the goods from the space station to Earth. The Seller assumes the risk until the goods are handed over to the Buyer's particular carrier.[32]

In the event that goods are lost or damaged while the goods were being handed over, an important question will be raised: At what point were the goods and the risk of loss handed over?

Outer space will require special methods and technology to pickup, handover, and transport goods. Depending on the method and technology used, it may be difficult for parties to determine exactly when the goods were handed over. There may be no precedent dealing with this method of transfer for the parties to rely on. The point in time when "goods are handed over" may be ambiguous or undefined. A good lawyer will draft an additional provision to Article 67, clearly defining the process of "handing over" and the procedure for determining when the goods were "handed over." Such a definition will need to be created for each contract and the special circumstances presented. Once defined, risks of loss will be more predictable. This is essential because it will allow both parties involved to consider the risks of the contract and take steps to mitigate that risk.[33]

Under Article 67 the passage of risk may become split in three cases, "namely if the seller (i) uses his own personnel to transport , (ii) is obliged to hand over the goods to the carrier at a particular place, or (iii) identifies the goods to the contract only after the transport has commenced."[34] If the place where damage occurred cannot be established in such cases "which party bears the risk will ultimately be decided by the question which party bears the burden of

[30] Id.
[31] Id
[32] Id.

[33] One of the actions a party can take to mitigate a risk is insurance. As discussed later in this paper, insurance and its successful adoption in outer space is directly related to the predictability of risks of loss.
[34] Schlechtriem, *supra* note 14; Commentary on Art.67 at 684 [para. 11].

proving the existence of conforming or non-conforming goods."[35] Ultimately, this is a matter of substantive law governed by the proper law of the contract.[36] When goods are transported through outer space, depending on the type of damage and container monitoring technology, it may be challenging to identify the place where damage occurred. In such a case, the issue of burden of proof will be of significance.

2. Article 68: Contracted while the goods are in transit

A scenario involving Article 68's application in outer space could be as follows: Seller has goods stored in an outer space warehouse orbiting earth. A carrier picks up the goods at the warehouse and begins transporting the goods to Mars. As the goods are being transported, buyer contracts with seller for sale of goods. Risk of loss passed to the buyer when the contract was concluded.[37]

Article 68 attempts to protect the buyer by placing the risk of loss on the seller if at the time of conclusion of the contract the seller "knew or ought to know" the goods had been damaged or lost and did not disclose.[38] In outer space, this protective provision may not be so effective and the buyer, absent a modification of this provision, may be assuming significant risks of loss.

For example, in outer space it is very likely the storage and transportation of goods will be unmanned, i.e. completely automated. Parties wishing to check goods for damage or loss will have to do so remotely, via robots, computers, and other equipment. Depending on the facilities the goods are stored in and the tools available to check for damage, the seller's ability to "know" or "ought to know" will be affected. If the seller has minimal means of determining the goods have been damaged or loss, the seller may limit his risk of loss without intentionally misleading the buyer.

Prior to contract conclusion the buyer will want to know the safe-guards seller has in place to determine if goods are damaged. Buyer may need to vary Article 68 if the seller has limited means of verifying goods. Failing to do so will place an additional risk of loss upon the buyer because the seller may not have known or not ought to have known that the goods were damaged or lost. As a result, at the time of contract conclusion the buyer will have assumed this risk of the unknown loss.

In this situation, the buyer can derogate from or vary Article 68 in at least two ways that will successfully mitigate this additional loss of risk.[39] First, the buyer can add verification standards for the goods to ensure the seller "ought to have known" the goods were damaged or lost at the time of contract conclusion. Such verifications standards will protect the buyer.

For example, special equipment can be used to test the goods in the outer space warehouse prior to handing over goods for transport. During the contracting process, the seller can provide the buyer with the test results, verifying the goods were not damaged.

A problem arises if the seller does not have the ability to provide the verification buyer is requesting. If the buyer cannot contract with a different seller, the buyer will be forced to accept the goods without the requested

---

[35] Id.
[36] Id.
[37] CISG, Art. 68.
[38] Id.

[39] CISG Article 6 allows parties to derogate or vary the effect of any of its provisions (subject to Article 12).

verifications.[40] In this case, the buyer may adopt a second solution to mitigate this additional risk of loss. The buyer may delay the passage of risk until the buyer has had a reasonable opportunity to examine the goods for damage or loss.[41]

This scenario raises an important issue: the need for standards that govern outer space storage, outer space transport, and the certification of outer space goods. Without such standards, buyers and sellers will have a high degree of inconsistency and unpredictability. A universal standard would avoid this issue, allowing parties to contract efficiently and with assurances. As the law governing outer space contracts for the sale of goods develops, such standards should be established.

3. Article 69(1): Picked up by the buyer at a place of business of the seller

Article 69(1) operates under the following scenario: Buyer contracts to buy goods from Seller. The goods are at a place of business of the Sellers, such as an outer space warehouse. The Buyer contracts to take over the goods at the outer space warehouse between July 1st and July 7th. The Buyer is responsible for arranging transportation himself.

Article 69(1), like Article 67, applies well in the context of outer space. Risk of loss is clearly allocated. The act of "taking over the goods" is the trigger for passing of risk.[42]

The term, taking over the goods, should be defined in the contract given the unique circumstances of outer space. Drafters should take special consideration of the method and technology of transfer used. Once again, standardization of method and technology would be useful and should be developed.

When the contract provides the buyer with a time-frame to pick up the contracted goods, seller assumes the risk of loss until the buyer "takes over the goods or does not do so in due time, from the time when the goods are placed at his disposal."[43] Depending on the situation, storing goods in outer space could be extremely risky. It is possible that seller will want buyer to assume some of the risk of loss before the buyer takes over the goods. For example, the buyer may assume a proportional risk of loss depending on the length of the time-frame granted for pickup.[44]

4. Article 69(2): Picked up by the buyer at a place other than a place of business of the seller

Article 69(2) applies to transaction not covered under Article 67, 68, or 69(1). These transactions include "destination contracts, bailment contracts, and contracts in which the seller uses his own vehicle to deliver goods to the buyer."[45] The scope of 69(2) makes it difficult to discuss every possible application in outer space. In this section, destination contracts are discussed because they

[40] This is a distinct possibility, especially at the beginning of space commerce. It is very likely only a few companies will initially be in outers-space providing goods for sale. As with all new commercial venture, it takes time before extensive competition develops. Once the sale of goods in outer space becomes proven a profitable venture and the risks are well known, extensive competition will develop.

[41] This represents a complete departure from Article 68's initial allocation of risk. This essential places the risk of loss on the buyer as if Article 69(2) was invoked.

[42] CISG, Art. 69(1).

[43] Id.

[44] The buyer may cover a portion of insurance costs during the storage of goods.

[45] Stocks, *supra* note 20 at 1422.

are likely to be a popular when contracting goods in outer space.[46]

In a destination contract, the seller is required to ship the goods to a location, in accordance with the contact, for tender of delivery to the buyer.[47] Seller bears the risk of loss during transport. Once the goods have arrived at the location, the risk of loss passes to the buyer when "delivery is due and the buyer is aware of the fact that the goods are placed at his disposal at that place."[48]

A typical destination contract for the sale of goods in outer space may be as follows: Seller and Buyer contract for the sale of goods. Seller is to deliver the goods to a location on the surface of Mars. Buyer, who has a scientific research facility on the Martian surface, is required to take control of the goods by the delivery due date. The delivery vehicle is a cargo container that has been launched from a spaceport. The cargo container is designed to withstand crash landing on the Martian surface. The contract specifies a geographic area the cargo container must land in. Once it has landed, the Seller transmits the location to the Buyer and Buyer has one week to take control of the goods.

A challenge when drafting destination contracts will be establishing a fixed due date when risk of loss passes. The environment of outer space and her celestial bodies may require the seller to grant buyer a non-fixed due date (i.e. a period of time) for buyer to organize a safe and efficient recovery of the cargo container and goods. For example, it's quite possible that destinations may be defined in general geographic areas, such as the "surface location of moon within 1km of lunar outpost X." One key factor will be whether infrastructure exists, and if so to what extent, to support cargo recovery.

VIII. INCOTERMS

"Incoterms" are rules of interpretation of trade terms published by the International Chamber of Commerce designed to clarify the distribution of functions, costs and risks relating to the transfer of goods from seller to buyer.[49]

CISG does not deal with the interpretation of trade terms and when CISG is applicable references to Incoterms do not exclude but merely supplement the Convention.[50] Parties contracting under CISG frequently rely on delivery clauses laid down in the Incoterms to govern obligations to deliver and the place of delivery.[51]

In theory, Incoterms 2000 could be applied to contracts for the sale of goods in outer space. However, current Incoterms are not designed and do not take into account the unique characteristics of functions, costs and

---

[46] Destination contracts are likely to be popular in outer space because delivery may be achieved with a disposable transport vessel. These disposal vessels, in certain situations, will be cheaper than a traditional freighter. For example, delivering goods to the surface of Mars would be very difficult if you wanted the transportation vessel to be reused. The vessel would have to survive the Martian environment and have the fuel to launch off Mars. It would essentially be a fully operational space ship that carries cargo. A cheaper alternative could be to deliver the goods in a disposable cargo vessel. This vessel could be launched from a location in outer space. Once it approached Mars, with minimal fuel it can navigate an approach and rely on Martian gravity to bring it to the surface. Once on the surface, buyer can go to delivery location and take control of the goods.

[47] Stokes, *supra* note 20 at 1424.

[48] CISG, Art.69(2).

[49] Jan Ramberg, *ICC Guide to Incoterms* 2000, (Paris: ICC Publishing, 1999) at 10.

[50] Schlechtriem, *Supra* note 14; Commentary on Art.6 at Pg.89 Section 12

[51] Schlechtriem, *Supra* note 14; Commentary on Art.30 Section 3 pg.338 commentary

risks relating to the transfer of goods from seller to buyer in outer space.

The ICC should consider adopting Incoterms tailored to transferring goods in outer space. These "Incospaceterms" would clarify concepts such as loading and unloading, outer space containerization, delivery and other important trade terms.

IX. Cross-Waivers

The principle of 'cross-waiver' of liability is an industry standard applied invariably by players in the space sector.[52] How would cross-waivers of liability effect contracts for the sale of goods in outer space?

To answer this question one needs to distinguish between the contracting parties to reach a conclusion. Clearly, the 'shipper' (i.e. a company operating a space cargo container vehicle analogous to a terrestrial seafaring cargo container ship), if contracting with either the seller or buyer, and with a cross-waiver of liability clause included, should receive the benefit of that clause so long as the clause is enforceable.

As between the seller and buyer of goods, the question will depend on how the contract is structured, who contracts with whom, and the scope of the cross-waiver.

For example, the U.S. Commercial Space Launch Act (CLSA) contains the following mandatory cross-waiver provision:

"A launch or reentry license issued or transferred under this chapter shall contain a provision requiring the licensee or transferee to make a reciprocal waiver of claims with its contractors, subcontractors, and customers, and contractors and subcontractors of the customers, involved in launch services or reentry services under which each party to the waiver agrees to be responsible for property damage or loss it sustains, or for personal injury to, death of, or property damage or loss sustained by its own employees resulting from an activity carried out under the applicable license."[53]

Based on the CLSA mandatory cross-waiver provision, one can conclude that sellers *or* buyer that contract with launch service providers are subject to cross-waiver provisions with regards to service they have contracted for and loss of goods sustained resulting from an activity carried out under the applicable license. This is a critical point because insurance providers for the seller *or* buyer who contract (and in turn assume some risk of loss) will adjust rates accordingly based on possible remedies available, including remedies from launch service providers.

However, it will be rare for both seller *and* buyer to be customers of the launch service provider. In cases where only one party is the customer, assumption of risk will pass either before or after 'launch' and hence the implications of cross-waiver provisions will only directly impact one party (the customer). As a result, the normal terrestrial contracting model should apply (as per the relationship between SELLER and BUYER) with the added caveat of contracting and insurance costs reflecting greater risk placed upon the customer of the launch service provider.

X. Conclusion

[52] Statement made by Professor Ram Jakhu in an email correspondence discussing the issue of cross-waivers within the context of contracting for the sale of goods in outer space (dated July 21st, 2008).

[53] *Commercial Space Launch Activities*, 49 U.S.C. §70112(b) (2008).

The assessment of CISG's risk of loss provisions illuminates the challenges of contracting for the sale of goods in outer space. While CISG default risk provisions do provide a useful framework, the unique environmental and technical challenges of outer space may require modification of default provisions to better serve the interests of contracting parties.

For the time being, CISG can suffice as the international legal regime governing contracts for the international sale of goods in outer space. CISG provides a useful framework on which to draft contracts and is progressive enough to allow contracting parties to modify default provisions to better facilitate the sale of goods in outer space. However, CISG should only be considered a stop-gap measure. Ultimately, an international agreement *specifically* governing contracts for the sale of goods in outer space will be necessary. The environment of outer space poses unique challenges to the transit, transfer, and storage of goods. While analogies and insights can be drawn from terrestrial activities (such as the containerization of goods and sea transport of containers), contracting parties, investors, and insurers will require agreements that fully integrate the unique characteristics of outer space.

LATEST DEVELOPMENTS IN RESPECT OF UNIDROIT'S PREPARATION OF A NEW INTERNATIONAL REGIMEN FOR COMMERCIAL SPACE FINANCING

by Martin STANFORD
(Unidroit)

I. *Background to, and status of implementation of Cape Town Convention in general*

The Convention on International Interests in Mobile Equipment, opened to signature on 16 November 2001, is, as most of you will already know, designed to make the cost-efficiency benefits of asset-based financing more widely available to those categories of high-value mobile equipment regularly moving either across or beyond national frontiers in the ordinary course of business, such as aircraft, railway rolling stock and satellites. The thinking is that the legal certainty that will flow from States adopting the new international regimen, in particular through operation of a fully computerised international registry recording financial interests taken in such highly mobile assets, can only encourage financial institutions to lend money against such assets, which otherwise would, in most cases, be seen as somewhat risky, in view of the uncertainty as regards the applicable law to any disputes that might arise.

Not surprisingly, these benefits are such that the Convention goes from strength to strength. One measure of this strength is to be seen in the one-third reduction by the Export-Import Bank of the United States of America of its exposure fee on the export financing of large commercial aircraft for buyers in Contracting States to the Convention and the Protocol thereto on Matters specific to Aircraft Equipment.

With the entry into force of the Aircraft Protocol, the Convention entered into force in respect of aircraft objects on 1 March 2006 - the Convention only enters into force as regards a category of assets to which a Protocol applies as from the time of the entry into force of that Protocol. As of 23 September 2008, the Convention as applied to aircraft objects is in force between 23 States, representing a good mix of developing and industrialised countries, and is set fair to enter into force among a considerable number of additional States, in view of the amended proposal submitted last month by the Commission of the European Communities for a Council decision on the conclusion by the European Community of the Convention and the Aircraft Protocol. The International Registry for Aircraft Objects has been in operation since the entry into force of the Convention as applied to aircraft objects. It is run by Aviareto Limited, an Irish company based in Dublin. In 2007 47,266 international interests were registered against 22,365 aircraft objects. It already covers over 50% of the world's commercial aircraft transactions. There can be no doubt that the International Registry for aircraft objects has got off to an excellent start and has earned wide acceptance in the user community.

A second Protocol to the Convention, this time on Matters specific to Railway Rolling Stock, was opened to signature in Luxembourg on 23 February 2007. Although the Protocol has not yet entered into force - to date it has only been signed by four States - it is attracting great interest, particularly among those developing countries anxious to develop their rail infrastructure.

II. *Preliminary draft Space Protocol*

(a) Specific features

As most of you, again, will know, the new international regimen which will apply to commercial space financing once the Convention is in force in respect of space assets is designed, first and foremost, to create a new international interest in space assets. This international interest amounts to what most of you would understand as an international mortgage. The international regimen sets out to endow the international interest with clear and commercially oriented international rules; these will provide, notably, for the prompt enforcement of remedies and guarantee that the priority rules and enforcement mechanisms of the international regimen will not be modified or qualified as a result of the insolvency of the debtor. As mentioned a moment ago, the key feature of the international regimen is the creation of an international registry and the fact that the new international interest will be enforceable against third parties in all States Parties provided it is registered in the International Registry to be set up under the future Protocol. It will be possible, through this Registry, for a financier considering advancing credit against a particular asset to evaluate instantaneously the exact status of the asset for financing purposes.

The general rules of this new regimen are carried in the Convention. The future Protocol, like the existing Protocols, is designed to carry those special rules necessary to implement the Convention in respect of space assets. First and foremost, this naturally means providing a clear definition of the categories of space asset covered and establishing the future International Registry for space assets. To the extent of any inconsistency between the Convention and a given Protocol, it is the latter that will prevail; in other words, the Protocol controls the Convention and not the other way round.

(b) Intergovernmental negotiations and referral of key issues to intersessional work

A preliminary draft Protocol, drawn up by a special working group representative of the different sectors of the international commercial space and financial communities, is currently before Governments. The Government of the Russian Federation has intimated that it is prepared to host the diplomatic Conference for the adoption of the draft Protocol that emerges from these negotiations, in the event of these negotiations being successful. A couple of sessions of intergovernmental negotiations have already been held. 44 Governments are participating in these negotiations. These Governments represent not only Unidroit member States but also members of the United Nations Committee for the Peaceful Uses of Outer Space. The degree of interest in this project by both Governments and industry is extremely encouraging.

However, a number of key issues were identified at the last session of governmental experts as requiring intersessional work. These issues are essentially three in number, first, examination of the issues arising out of extension of the Convention as applied to space assets to cover debtor's rights and related rights, secondly, examination of the treatment of public service under national law and in practice and consideration of possible solutions to the problem as to how best to balance the need of Governments to guarantee the continuation of a public service performed by a space asset where the debtor was in default with the rights of the creditor upon such default under the Convention as applied to space assets and, thirdly, examination of certain issues relating to the future international registration system for space assets, in particular the criteria for the identification of space assets for the purposes of their registration in the future International Registry for space assets.

Following important intersessional work on these issues spearheaded by Unidroit, with significant support from both the key space-faring nations and the international commercial space and financial communities, the General Assembly of Unidroit member Governments last November established a Steering Committee designed to build consensus, both within Government and industry, around the provisional conclusions reached by Government/industry meetings held over the previous 18 months, with a view to completing the project as timeously as possible. Both the Governments of the key space-faring nations and leading representatives of the international commercial space and financial communities having taken an active part in the intersessional work up to that point, the General Assembly decided that representatives of both the Governments and the international commercial space and financial communities having participated in that work should be invited to serve on the Steering Committee, on an equal footing, in particular with a view to ensuring that the end-product be commercially viable. The crucial message conveyed by the previous intersessional work was the need for the Steering Committee to take the necessary steps to simplify the preliminary draft Protocol so as to permit timeous completion of the project. In particular, the unanimous view of those participating in the intersessional work was that the sphere of application of the preliminary draft Protocol should be narrowed down to concentrate on those categories of space asset which were, realistically, the likely subject of asset-based financing transactions in the immediate future. The general view was, therefore, that the future Protocol should concentrate on the satellite, albeit viewed in its entirety, rather than aiming to cover also assets which might be the subject of asset-based financing one day, but not in the near future, such as assets manufactured in space. The thought was that the satellite, viewed in its entirety, accounts for some 80% of the potential asset-based financing market among those assets hitherto covered.

(c) Current status of project: work of Steering Committee

The Steering Committee got off to a great start, in no small measure thanks to the generous hospitality of the Government of the Federal Republic of Germany, which was quick to respond to the call issued by representatives of the international commercial space and financial communities during the intersessional work for Governments to give a lead on the prosecution of this project if the international commercial space and financial communities were to be convinced of the seriousness of the intention to complete the project timeously. The launch meeting of the Steering Committee was held in Berlin in May. A representative sample of Governments and the different sectors of the space industry and the financial communities attended. The Steering Committee, aware of the need to ensure a successful transition back to the Committee of governmental experts, elected Professor Sergio Marchisio (Italy), Chairman of the Committee of governmental experts, Chairman of the Steering Committee too.

The progress accomplished by the Steering Committee on all but one of the key outstanding issues was such as to warrant the decision, at the end of its launch meeting, that the co-chairmen of the Drafting Committee of the Committee of governmental experts (Canada and the United Kingdom), both of whom were present in Berlin, should essay an alternative version of those provisions of the preliminary draft Protocol affected by its decisions on these key outstanding issues. The item which the Steering Committee consciously deferred, to a sub-committee, was the issue of public service, it being agreed that the Sub-committee in question should seek to come up with various options capable of being put to the Committee of governmental experts with a good chance of commanding broad support.

What were the issues then on which the Steering Committee was able to reach consensus? First, it has to be borne in mind that the intersessional work accomplished on the criteria for the identification of space assets had thrown up two consequential ancillary issues, both related to the sphere of application of the preliminary draft Protocol. One of these was the categories of asset to be covered in general, notably in function of the extent to which they lent themselves to unique identification for the purposes of registration. The other was the issue of the coverage of components in particular, especially given that the Convention is, already in its preamble, described as applying to high-value assets and the question that must inevitably arise as to the compatibility of a high-value threshold and the average, common-garden category of components. The Steering Committee accordingly concluded as to the need to narrow the definition of space assets employed in the preliminary draft Protocol down to those assets which were independently identifiable, in particular the satellite as a whole, transponders and certain types of other component, referred to as "principal objects", as well as other assets capable of independent control, this last introduction being designed to permit the registration of future developments, such as space hotels. The advantage seen in this solution was its exclusion of those non-principal components, the value of which would not be consistent with the intention of the drafters of the Convention.

Secondly, there was agreement with the intersessional conclusions reached by Sir Roy Goode on the issue of the most appropriate manner in which to accommodate the extension of the Convention as applied, through the future Protocol, to space assets to those debtor's rights and related rights which are such important elements of a satellite's value as collateral in the event of the debtor's default. Most of you will not require further elucidation in this regard, I am sure, but, for those of you who may, I should mention that the typical sort of related rights are those permits, licenses or authorisations granted to control,. Use or operate a space asset, relating to the use of orbits positions and the transmission, emission or reception of electromagnetic signals to and from a space asset. Debtor's rights, on the other hand, is the term used to denote those contractual rights to performance or payment due to a debtor by any person with respect to a space asset. There was agreement within the Steering Committee as to the need for these rights also to be covered by the future international regimen, not, though, with the idea of their being registered separately but rather with the intention of their being noted in the future International Registry against the relevant space asset.

Thirdly, the Convention system of registration being posited on the basis of unique identification criteria being available for the different categories of asset to be registered, it has, ever since the outset of work on the preliminary draft Protocol, been recognised that, in order to be registrable, space assets will need to have some sort of unique identification criteria. The initial text laid before Governments in 2003 provided five specific criteria, such as the name and address of the debtor and the creditor, a general description of the asset and its intended location, the date and location of launch and, in the case of a component, a description of such component, the space asset of which it forms a part, to which it is attached or within which it is contained and each of the other identification criteria prescribed for such a space asset, as well as allowing for the regulations to be established by the Supervisory Authority under the future Protocol to specify additional criteria. This list of specific criteria was simply abandoned at the first session of governmental experts, the idea being to leave the question to be dealt with in the regulations. At the second session of governmental experts there was a feeling that the future Protocol needed to include at least some identification criteria, since otherwise the scope of the future Protocol would be left unclear. This matter was, therefore, referred to intersessional work. The solution decided upon by the Steering Committee was to take the system adopted in the Rail Protocol. In that context it had been agreed that there was no need for unique identification to be provided in the agreement between the contracting parties and that it would be sufficient for the asset to be capable of being identified as falling within the scope of the agreement. This would enable a class of assets or future assets to be covered by the same agreement and avoid the need for a separate agreement each time the debtor acquired a new asset. This said, it was agreed that identification criteria for registration purposes should not be left solely to the regulations but that certain basic identification criteria needed to be spelled out in the future Protocol, albeit with the idea that these basic criteria could then be supplemented by others to be laid down in the future regulations.

One issue on which it proved impossible to reach complete consensus concerned the question as to the enforcement regimen to apply in respect of a component functionally linked to another space asset in which another creditor has an interest. There was a widespread feeling among members of the Steering Committee that this was a matter customarily dealt with in inter-creditor agreements and that the future Protocol would, therefore, do better to stay silent on the issue, especially if the intention was not to put off financiers. A minority, nevertheless, remained wedded to the idea of the matter being covered in the future Protocol. The difference of opinion led to the decision to set up a sub-committee to find a solution agreeable to all. A questionnaire has already been drawn up on this issue and a number of responses received, notably from financial institutions and those advising them. The Sub-committee will be meeting in Berlin next month, at the invitation of Commerzbank.

III. Conclusions

As I hope you will agree, we are well on the way to sorting out the key outstanding issues and to bringing the results of the Steering Committee's work back to the Committee of governmental experts with a view to a speedy completion of the project at the diplomatic Conference which the Government of the Russian Federation has already indicated its preliminary willingness to host. Progress has been reached across a broad canvas of issues and we would anticipate later in the year being in a position to begin seeking to build consensus around the solutions advocated by the Steering Committee among those Governments and industry players not represented on the Steering Committee, with a view to preparing the ground for a successful resumption of the full intergovernmental consultation process. It has been most encouraging to see the way in which both the key space-faring nations and leading players from the international commercial space and financial communities have come together to work out solutions acceptable to both, for it is clear that only a Protocol that is going to work in practice and which may be expected, therefore, to enhance the chances of greater access to commercial finance for those operators and start-up companies in the less advantaged parts of the world is going to fly. I always find it egregious being drawn on precise time-tables but, as I have indicated, we are firmly committed to early completion of the project and would hope to be in a position to reconvene the Committee of governmental experts, with an accelerated programme for such completion, next summer or immediately thereafter.

51st Colloquium on the Law of Outer Space, Glasgow, Scotland

Session 2:
THE 40th Anniversary of the Rescue Agreement:
Looking Ahead

Chairmen:

Vladimir Kopal

Sergio Marchisio

Rapporteur:

Fabio Tronchetti

IAC-08-E8.2.2

THE RETURN OF OBJECTS LAUNCHED INTO OUTER SPACE: SOME LEGAL QUESTIONS

Juan Manuel de Faramiñán Gilbert*
Mª del Carmen Muñoz Rodríguez**

ABSTRACT[1]

The 1967 Agreement on the Rescue of Astronauts, the Return of Astronauts and the Return of Objects Launched into Outer Space, was drawn up under heavy pressure from the space powers and in the shadow of the first fatal accidents linked to space activities. It provided the opportunity to describe in greater detail the duties listed in article V of the 1967 Outer Space Treaty. The Agreement attempts to draw out the logical consequences of the principle whereby astronauts are considered the envoys of mankind and of the principle of cooperation, although, looking at this from a different angle, one might also think of the space powers' interest in preventing their space objects, or remains of those objects, from being recovered and analysed by countries other than the launching State". Paradoxically, in spite of this being a "precautionary" international treaty, in that the intention was to establish rules so that those involved would know how they could and/or should react in the face of dangerous situations or tragic outcomes, for astronauts or space objects, it has not to date been applied to any extent, as accidents have so far been dealt with by the launching States.

I. INTRODUCTION

The Agreement on the Rescue of Astronauts, the Return of Astronauts and the Return of Objects Launched into Outer Space[2], was drawn up under heavy pressure from the space powers and in the shadow of the first fatal accidents linked to space activities[3]. As can be seen, the text of this Treaty emphasises the dangerous, risky nature of space activities, given that during an actual activity or in the moments just before its completion, circumstances may fail to produce a positive outcome and unfortunately there have already been many accidents that have led to the loss of human life, involving either astronauts or staff on the ground, as well as of objects.

The Agreement provided the opportunity to describe in greater detail the duties listed in article V of the Outer Space Treaty[4], putting them in more concrete form and expressing human feeling. In this respect, Lacleta Muñoz points out that the Agreement "*only attempts to draw out the logical consequences of the principle whereby astronauts are considered the envoys of mankind and of the principle of cooperation, although,*

* Juan Manuel de Faramiñán Gilbert, Professor of Public International Law, University of Jaén (Spain) jmfarami@ujaen.es
** Mª del Carmen Muñoz Rodríguez, Assistant Lecturer of Public International Law, University of Jaén (Spain), mcrodri@ujaen.es

looking at this from a different angle, one might also think of the space powers' interest in preventing their space objects, or remains of those objects, from being recovered and analysed by countries other than the launching State"[5]. This was in line with the various statements by representatives of States at the session of the United Nations General Assembly that approved Resolution 2345(XXII)[6].

Paradoxically, in spite of this being a "precautionary" international treaty, in that the intention was to establish rules so that those involved would know how they could and/or should react in the face of dangerous situations or tragic outcomes, for astronauts or space objects, it has not to date been applied to any extent, as accidents have so far been dealt with by the launching States.

II. DRAFTING HISTORY

Although the USSR and the USA began their discussions in the recently created Legal Subcommittee, with their differences relating to the formal instrument to be used to regulate space cooperation[7], they finally agreed in 1963 to draw up a treaty that focused on the matter of assistance to and the return of astronauts and space objects[8].

The first substantive problem, posed at the 1962 and 1963 sessions of the Legal Subcommittee, and relating among other things to the general principles of space and liability, was whether to impose conditions on the return of space vehicles: the USSR proposed several conditions[9], but the USA, and some of its allies, rejected them. However, this impasse was, in a certain manner, resolved by General Assembly Resolution 1962 (XVIII)[10], which declares that in the exploration and use of outer space, States should be guided by several basic principles, among which are some references to the topics of assistance and return[11]. Subsequently, General Assembly Resolution 1963 (XVIII)[12] recommended that consideration be given to incorporating *in international agreement form*, the 1963 legal principles governing the activities of States in the exploration and use of outer space, and requested that UNCOPUOS continue to study and report on legal problems that might arise in the exploration and use of outer space. In particular, it was asked to arrange for the prompt preparation of draft international agreements on liability for damage caused by objects launched into outer space and on assistance to and return of astronauts and space vehicles.

Between that year and conclusion of the 1967 Outer Space Treaty, the Legal Subcommittee (UNCOPUOS)[13] made statements on the two traditional, clear positions regarding the urgent necessity of negotiating and concluding the Rescue Agreement. On the one hand, the USSR always attached major importance to the conclusion of a binding international treaty on assistance to and the return of astronauts and space objects - the USSR even thought priority should be given to this before anything else (such as the legal principles of outer space, liability, etc.). On the other hand, the USA, its allies and the non-space powers tried to deal with the complex issue of liability first[14]. Concerning space objects, many drafts were prepared and many debates took place during this period; however, the conflicting views of the USSR and the USA regarding making the return of space vehicles conditional or not constituted a serious obstacle to the work of the Legal Subcommittee[15].

The deadlock in the Legal Subcommittee was broken by the continuous steps forward in space exploration, including arrival near the Moon. The USA proposed to prepare a treaty on the legal status of celestial bodies, and the USSR accepted, provided that the text would be the general and legal framework for outer space and celestial bodies. In the end, the 1967 Outer Space Treaty reproduced paragraphs 7 and 9 of the 1963 Space Declaration in its articles V[16] and VIII[17]; thus, the final choice was not to make the return of space objects conditional.

At the Sixth Session of the Legal Subcommittee in 1967, the USSR tried to reach a definitive agreement on assistance to and the return of astronauts and space objects, and a brief, revised draft was proposed, without any references to the old discussion on whether to attach conditions. At last, and following the UN's usual method of working on the drawing up of space law, the final version of the text was prepared by the two superpowers, the USSR and the USA, as a "text negotiated backstage"[18], and was approved by the UN within less than a week[19].

III. THE LEGAL REGIME

The contracting States have several duties in regard to astronauts and space objects. Concerning astronauts, the Rescue Agreement establishes the following duties:

1. Notification of accidents: *"Each Contracting Party which receives information or discovers that the personnel of a spacecraft have suffered accident or are experiencing conditions of distress or have made an emergency or unintended landing in territory under its jurisdiction or on the high seas or in any other place not under the jurisdiction of any State shall immediately: (a) Notify the launching authority or, if it cannot identify and immediately communicate with the launching authority, immediately make a public announcement by all appropriate means of communication at its disposal; (b) Notify the Secretary-General of the United Nations, who should disseminate the information without delay by all appropriate means of communication at his disposal"*[20].

2. Search and rescue in a contracting State's territory: *"If, owing to accident, distress, emergency or unintended landing, the personnel of a spacecraft land in territory under the jurisdiction of a Contracting Party, it shall immediately take all possible steps to rescue them and render them all necessary assistance. It shall inform the launching authority and also the Secretary-General of the United Nations of the steps it is taking and of their progress. If assistance by the launching authority would help to effect a prompt rescue or would contribute substantially to the effectiveness of search and rescue*

operations, the launching authority shall cooperate with the Contracting Party with a view to the effective conduct of search and rescue operations. Such operations shall be subject to the direction and control of the Contracting Party, which shall act in close and continuing consultation with the launching authority"[21].

3. Landings in areas not under the jurisdiction of any State: "*If information is received or it is discovered that the personnel of a spacecraft have alighted on the high seas or in any other place not under the jurisdiction of any State, those Contracting Parties which are in a position to do so shall, if necessary, extend assistance in search and rescue operations for such personnel to assure their speedy rescue. They shall inform the launching authority and the Secretary-General of the United Nations of the steps they are taking and of their progress*"[22].

4. Return of astronauts: "*If, owing to accident, distress, emergency or unintended landing, the personnel of a spacecraft land in territory under the jurisdiction of a Contracting Party or have been found on the high seas or in any other place not under the jurisdiction of any State, they shall be safely and promptly returned to representatives of the launching authority"*[23].

With regard to space objects, the Rescue Agreement contains just one article[24], which provides a similar, but not equal, legal regime:

1. Notification of landing: "Each Contracting Party which receives information or discovers that a space object or its component parts has returned to Earth in territory under its jurisdiction or on the high seas or in any other place not under the jurisdiction of any State, shall notify the launching authority and the Secretary-General of the United Nations"[25]. In this case, the differences with the astronauts' regime are: first, that the obligation of notification is not to be immediate[26]; and, second, that this obligation applies only to space objects returning to Earth, so there is no mention in the text of return to celestial bodies[27]. Concerning the last point, it was necessary to wait until the Moon Agreement for clarification[28]; article 12.2 of this Agreement lays down explicitly that the rules of article 5 of the Rescue Agreement must be followed for the vehicles, installations and equipment or their component parts found in places other than their intended location on the Moon[29], and, in particular, that any State Party which learns of the crash landing, forced landing or other unintended landing on the Moon of a space object must promptly inform the launching State Party and the Secretary-General of the United Nations[30].

2. Recovery of foreign space objects in Contracting State's territory: "*Each Contracting Party having jurisdiction over the territory on which a space object or its component parts has been discovered shall, upon the request of the launching authority and with assistance from that authority if requested, take such steps as it finds practicable to recover the object or component parts"*[31]. In this second paragraph, the space object regime differs from that of the astronauts in two respects: first, a space object is recovered at the

request of the launching authority, while in the case of astronauts, all steps must be taken immediately to rescue them[32], with assistance from the launching authority if requested[33]; and, second, the State where the space object lands should take such steps as it finds practicable, while for astronauts the State must take all possible steps to rescue them and render them all necessary assistance. There is a difference in the words used but not in the meaning of the obligation, the USA having explained to delegations that the reference, for both astronauts and space objects, means "action within the limits of the facilities at the Contracting Parties' disposal"[34].

3. Return of space objects: *"Upon request of the launching authority, objects launched into outer space or their component parts found beyond the territorial limits of the launching authority shall be returned to or held at the disposal of representatives of the launching authority, which shall, upon request, furnish identifying data prior to their return"*[35]. This article essentially reproduces paragraph 7 of the 1963 Space Declaration[36] and article 8 of the Outer Space Treaty[37], that is, there is an obligation to return the space object[38], but also states that the space object is returned only on request received from the launching authority[39].

4. Elimination of possible danger of harm: *"Notwithstanding paragraphs 2 and 3 of this article, a Contracting Party which has reason to believe that a space object or its component parts discovered in territory under its jurisdiction, or recovered by it elsewhere, is of a hazardous or deleterious nature may so notify the launching authority, which shall immediately take effective steps, under the direction and control of the said Contracting Party, to eliminate possible danger of harm"*[40]. This is the launching authority obligation that is clearly in favour of the territorial State's interests; it could even have been considered "a method for the launching authority to discharge its obligation under international customary law not to cause damage to other States and to mitigate, and make reparation for, such damage as it may have caused"[41] - this possibility was removed with the 1972 Liability Convention[42].

5. Payment of expenses: *"Expenses incurred in fulfilling obligations to recover and return a space object or its component parts under paragraphs 2 and 3 of this article shall be borne by the launching authority"*[43]. Therefore, if the launching authority requests the recovery of the space object[44], it has to pay the expenses incurred in that action, and can be required to pay as work proceeds or even in advance[45].

Finally, after examining the background and legal regime of the Rescue Agreement, especially the return of space objects, it can be said that this has to be absolute and unconditional[46].

IV. CONCLUSION- FUTURE PROSPECTS

Regarding the the history of space accidents, whether 'controlled or uncontrolled', there is clearly a need to check, as far as possible, for calculation errors, as in the case of Challenger, Columbia and Mars

Climate Orbiter. It is also necessary to plan for obvious hazards to activities in outer space, such as space debris, and threats that could come from space, such as the close flyby of Earth by the asteroid Apophis in 2029. There is also the important question concerning the future of space exploration, which is whether spacecraft should be crewed or dangerous missions left to robots.

With respect to space debris and space contamination, there are currently three orbits in which these are to be found: the low Earth orbit (LEO), the graveyard orbit, and the geostationary orbit. The latter is the most worrying as it is occupied by many satellites. It is estimated to contain some 3000 fragments of varying sizes (between 15 cm and 1 metre) and all sorts of objects, ranging from a simple bottle to repair and assembly equipment, as well as the remains of satellites that have been damaged or involved in accidents. The Pegasus rocket, for example, placed in orbit in 1994, exploded two years later, generating hundreds of thousands of fragments. Even the smallest are extremely dangerous. The astronaut Edward White lost a glove measuring about 30 cm in space, which disintegrated in the atmosphere one month later, but in the meantime travelled at 28.000 km per hour and could have destroyed a spacecraft had there been one along its trajectory.

With respect to threats from space, the astronaut Pedro Duque stated[47] that there is absolute mathematical certainty that the Apophis asteroid will come very close to Earth in 2029, possibly bringing down commercial satellites, and seven years later, in 2036, attracted by the planet's magnetic field, could fall into the sea, causing a huge tsunami. Accordingly, he asserted that the European Space Agency (ESA) was looking into the possibility of deploying a mission involving the launch of two satellites, one, dubbed 'Don Quijote', to divert the asteroid, and the other, named 'Sancho', to provide coverage in orbit for the first satellite. Duque pointed out that this mission was in the process of being analysed and that its objective would include the possibility of detecting asteroids that might come close to Earth, adding that this operation was a priority for ESA.

Another major debate on the future of space exploration is to do with whether spacecraft should be crewed or whether dangerous missions should be left to robots. NASA has experienced some success with space robotics, as in the case of the Rover vehicle which, without some difficulties, was able to move around on the Martian surface, analysing rocks and sending back photos of the vehicle's surroundings. NASA is nevertheless continuing to place their faith in crewed flights, which imply greater control over the security of the astronauts and appropriate management of the investments required for the smooth running of the missions.

Recently, Discovery revealed the vulnerability of the space shuttle design when the spacecraft's cameras filmed a piece of insulation foam becoming detached during lift-off. It should be borne in mind that that problem is similar to the one which led to the Columbia disaster in 2003, so much so that Russia offered to help the USA if for some reason

Discovery's seven crew members had problems returning to Earth at the end of their mission onboard the International Space Station (ISS). After lift-off, NASA acknowledged that pieces of insulation had become detached from the fuel tank and that although they did not damage the spacecraft, they did present a serious hazard to the astronauts. Furthermore, photographs of the exterior of the shuttle were taken both by the astronauts and from Earth from every possible angle, which led to the detection of numerous dents caused by detached foam, although according to the technicians, they did not pose a risk to the spacecraft during its re-entry into the atmosphere. However, astronaut Steve Robinson had to perform a spacewalk to remove loose insulation material that could have been dangerous during Discovery's return to Earth. Subsequently, when the return date had been set, NASA detected more cracks in the insulation shield beneath one of the windows of the Shuttle's crew compartment, although decided not to repair it as it did not consider it to be dangerous. In the end, Discovery returned successfully, in spite of these technical problems being compounded by inclement weather at Cape Canaveral, which forced the shuttle to be diverted to California.

NASA was obliged to announce, during the Discovery mission, that it would suspend all shuttle flights until all problems relating to shedding of the main external tank insulation had been resolved. Despite that, as the Agency's Administrator, Michael Griffin, pointed out subsequently, preparations for the launch of Atlantis were maintained, to continue the ISS construction work. Bill Parsons, the new Director of the Kennedy Space Center, has said it would be difficult to carry out the five space shuttle missions NASA has scheduled in 2007 (March, June, September, October and December), owing to the countless logistical obstacles that the five missions are likely to encounter. The main problem is to do with NASA's ability to judge the optimum flight conditions of the three available shuttles after their return to Earth – bearing in mind that the objective of all the flights scheduled for 2007 is completion of the ISS, which will have to be achieved by 2010, when the shuttle programme ends

[1] This paper has been prepared in the framework of the Research Project "Convergences and divergences among the outer space law and particular sea spaces law", Spanish Ministry of Science and Education, SEJ2004-06116.

[2] The "Rescue Agreement", adopted by the General Assembly in its Resolution 2345 (XXII), 19 December 1967 (115 votes to none, with no abstentions) opened for signature on 22 April 1968, and entered into force on 3 December 1968 (89 ratifications, 24 signatures, and 1 acceptance of rights and obligations, as of 1 January 2007).

[3] Cheng, B.: *Studies in International Space Law,* Clarendon Press Oxford, 1997, p.417. See also below, part I.

[4] Treaty on Principles Governing the Activities of States in the Exploration and Use of Outer Space, including the Moon and Other Celestial Bodies, or "Outer Space Treaty", adopted by the General Assembly in its Resolution 2222 (XXI), opened for signature on 27 January 1967, entered into force on 10 October 1967, 98 ratifications and 27 signatures (as of 1 January 2007). Article 5: "States Parties to the Treaty shall regard astronauts as envoys of mankind in outer space and shall render to them all possible assistance in the event of accident,

distress, or emergency landing on the territory of another State Party or on the high seas. When astronauts make such a landing, they shall be safely and promptly returned to the State of registry of their space vehicle. In carrying on activities in outer space and on celestial bodies, the astronauts of one State Party shall render all possible assistance to the astronauts of other States Parties. States Parties to the Treaty shall immediately inform the other States Parties to the Treaty or the Secretary-General of the United Nations of any phenomena they discover in outer space, including the Moon and other celestial bodies, which could constitute a danger to the life or health of astronauts".

[5] According to Lacleta, this Agreement is moreover one whose application does not require *"the adoption of specific legal measures (perhaps in relation to customs and passports?)"*. Lacleta Muñoz, J.M.: "El Derecho en el espacio ultraterrestre", Real Instituto Elcano de Estudios Estratégicos Internacionales y Estratégicos, Documento de Trabajo 18/25, de 8.4.2005, p.8. http://www.realinstitutoelcano.org/documentos/187/Lacleta%20pdf.pdf

[6] See above. For instance, Mr Wyzner, the representative of Poland: "My colleagues will, no doubt, appreciate the significance, in humanitarian terms, of the Agreement for those brave men and gallant men who are, in the words of Article 5 of the Outer Space Treaty, the 'envoys of mankind in outer space', who are risking their lives, as recent tragic accidents have demonstrated, in endeavours which serve the interests of all... The Agreement on Rescue and Return is also a further collective step in the quest for peace since, among others, it eliminates possible sources of dispute and friction between States"; Mr Goldberg, the representative of the USA: "The treaty text represents agreement on implementing that famous phrase from the Outer Space Treaty that astronauts are 'envoys of mankind'... The United States considers that the Assistance and Return Agreement which we have adopted represents a just balancing of the interests of all Members of the United Nations, the space Powers, the near-space Powers, the cooperation space Powers and all who are interested in outer space"; and Mr Fedorenko, the representative of the USSR: "The Agreement on the Rescue of Astronauts will certainly be of great importance, ensuring the speedy rescue of astronauts in case of... The Agreement ... can be truly called a humanitarian act of international law on the part of the Member States of the United Nations towards the courageous explorers of those vast cosmic expanses, the men who are, in the words of the Outer Space Treaty, 'envoys of mankind' in space" (Doc. A/PV.1640, 22nd session of the General Assembly).

[7] While the USSR proposed concluding international treaties, the USA preferred to use just the UN General Assembly Resolutions.

[8] Doc. A/AC.105/C.2/SR.25 (30 April 1963).

[9] "(i) Foreign spacecraft would be returned without delay only if they had identification marks and their launchings had been officially announced, and (ii) spacecraft containing devices for the collection of intelligence information would not be returned" (Cheng, B.: *Studies in International Space Law,* see above, note 3, p.268).

[10] Resolution 1962 (XVIII) of 13 December 1963, Declaration of Legal Principles Governing the Activities of States in the Exploration and Use of Outer Space.

[11] "7. The State on whose registry an object launched into outer space is carried shall retain jurisdiction and control over such object, and any personnel thereon, while in outer space. Ownership of objects launched into outer space, and of their component parts, is not affected by their passage through outer space or by their return to the Earth. Such objects or component parts found beyond the limits of the State of registry shall be returned to that State, which shall furnish identifying data upon request prior to return. [...] 9. States shall regard astronauts as envoys of mankind in outer space, and shall render to them all possible assistance in the event of accident, distress, or emergency landing on the territory of a foreign State or on the high seas. Astronauts who make such a landing shall be safely and promptly returned to the State of registry of their space vehicle."

[12] Resolution 1963 (XVIII) of 13 December 1963, international co-operation in the peaceful uses of outer space.

[13]http://www.unoosa.org/oosa/COPUOS/Legal/index.html

[14] According to Cheng, the reasons are not difficult to identify: "whilst the United States [...] has a world-wide network of tracking and other facilities sited in its own territory and, thanks to numerous bilateral agreements, in a number of foreign countries, [...] the

Soviet Union does not have the same resources as the United States for recovering returning spacecraft and astronauts from the high seas in all parts of the world [...] The net result is that the chances of Soviet astronauts requiring assistance from third States are, relatively speaking, much higher [...] Furthermore, the imminence of man's landing on the moon and the possibility of large manned stations in space and other space probes in the near future unquestionably made the Soviet Union even more anxious than before to bring about an international agreement on assistance to and the return of astronauts" (Cheng, B.: *Studies in International Space Law,* see above, note 3, pp.266 and 267).

[15] "While no longer referring to intelligence-gathering devices, the revised Soviet draft at first made the return of foreign spacecraft conditional upon their having been launched 'for purposes of peaceful exploration and use of outer space' and the launching State having 'officially announced the launch of these objects and the purposes of launching'. Subsequently, the Soviet draft referred merely to 'foreign spaceships, satellites and capsules launched in accordance with' the 1963 Declaration [...] the United States and many States of the Western bloc voiced the fear that, especially in the absence of an arbitral clause, these conditions would give rise to endless difficulties in practice" (Cheng, B.: *Studies in International Space Law,* see above, note 3, p.270).

[16] See above note 4.

[17] "A State Party to the Treaty on whose registry an object launched into outer space is carried shall retain jurisdiction and control over such object, and over any personnel thereof, while in outer space or on a celestial body. Ownership of objects launched into outer space, including objects landed or constructed on a celestial body, and of their component parts, is not affected by their presence in outer space or on a celestial body or by their return to the Earth. Such objects or component parts found beyond the limits of the State Party to the Treaty on whose registry they are carried shall be returned to that State Party, which shall, upon request, furnish identifying data prior to their return".

[18] Phrase used by the Indian delegate, doc. A/AC.105/C.2/SR.86 (14 December 1967).

[19] The Legal Subcommittee received it on 14 December 1967, and in spite of complaints about lack of time, etc. from its members, the proposal was discussed, amended, approved and passed to the General Assembly (the First Committee was bypassed), which finally voted unanimously in favour of it, but with negative opinions voiced on the speed of the procedure and the fact that the liability problem was still unresolved.

[20] Article 1.

[21] Article 2.

[22] Article 3.

[23] Article 4.

[24] Article 5.

[25] Article 5.1.

[26] Canada pointed out this omission, which was understandable mainly because this immediacy requirement was included in previous drafts (doc. A/AC.105/C.2/SR.86, 14 December 1967, p.17).

[27] Except we consider that the Moon (or other celestial bodies) is implicitly included in the expression "any other place not under the jurisdiction of any State".

[28] Agreement Governing the Activities of States on the Moon and Other Celestial Bodies, the "Moon Agreement", adopted by the General Assembly in its Resolution 34/68, opened for signature on 18 December 1979, entered into force on 11 July 1984, 13 ratifications and 4 signatures (as of 1 January 2007).

[29] Article 12.2 of the Moon Agreement: "Vehicles, installations and equipment or their component parts found in places other than their intended location shall be dealt with in accordance with article 5 of the Agreement on the Rescue of Astronauts, the Return of Astronauts and the Return of Objects Launched into Outer Space".

[30] Article 13 of the Moon Agreement: "A State Party which learns of the crash landing, forced landing or other unintended landing on the Moon of a space object, or its component parts, that were not launched by it, shall promptly inform the launching State Party and the Secretary-General of the United Nations". These rules (articles 12.2 and 13) can be extended to the space objects in other celestial bodies, according to article 1 of the Moon Agreement: "The provisions of this Agreement relating to the Moon shall also apply to other celestial bodies within the solar system, other than the Earth, except insofar as specific legal norms enter into force with respect to any of these celestial bodies. 2. For the purposes of this Agreement reference to the Moon shall include orbits around or other

trajectories to or around it. 3. This Agreement does not apply to extraterrestrial materials which reach the surface of the Earth by natural means".

[31] Article 5.2 of the Rescue Agreement.

[32] Article 2 of the Rescue Agreement.

[33] However, this obligation of cooperation of the launching authority is the same independently of whether it is an astronaut or a space object, even if this obligation is better defined in article 2 than in article 5.2.

[34] Doc. A/AC.105/C.2/SR.86, 14 December 1967, p.21. Curiously, France proposed to use, for article 2, article 25 of the 1944 Chicago Convention ("to provide such measures of assistance ... as it finds practicable"); however, this suggestion was not taken up for article 2 (astronauts), though it was, to some extent, for article 5.2 (space objects), doc. A/AC.105/C.2/SR.86, 14 December 1967, p.13).

[35] Article 5.3.

[36] See above note 11.

[37] See above note 17.

[38] Without any other requirement, except the duty of the launching authority, upon request, to furnish identifying data prior to the return.

[39] Concerning the return of astronauts, there is no need for a prior request, the State just has to return them promptly (article 4). Article 5.3 also maintains the same duty to furnish information about the space object, as described in note 38. The French Delegation suggested referring to the representatives of the launching authority (for instance, an embassy), instead of the launching authority itself (doc. A/AC.105/C.2/SR.86, 14 December 1967, p.8), in order to make the return of both astronauts (article 4) and space objects (article 5.3) easier.

[40] Article 5.4.

[41] As was stated, for instance, in the Corfu Channel Case (*ICJ Report 1949*, p. 4). See above, note 3, Cheng, B.: *Studies in International Space Law,* pp.280 and 289.

[42] Convention on International Liability for Damage Caused by Space Objects (the "Liability Convention"), adopted by the General Assembly in its Resolution 2777 (XXVI), opened for signature on 29 March 1972, entered into force on 1 September 1972 (84 ratifications, 24 signatures, and 3 acceptances of rights and obligations, as of 1 January 2007).

[43] Article 5.5.

[44] Even if there is no explicit reference, it is absolutely clear that if there is no recovery request, the launching authority has no duty to pay for the cost of that operation.

[45] For that reason, France proposed changing the verb in the first draft "shall be reimbursed" to "shall be borne" (doc. A/AC.105/C.2/SR.86, 14 December 1967, p.13).

[46] However, it may be said that the return of the space object should be "subject to the receipt of compensation for any damage caused [...] under article VII of the Space Treaty" (see above, note 3, Cheng, B.: *Studies in International Space Law,* p.283), or, even, subject to the payment of the cost of the recovery (see above, note 45 and the related text).

[47]http://www.europapress.es/noticia.aspx?cod=20070313123909

Bifurcation of Spaceport Regulation Within the United States

Patricia M. Sterns*
Leslie I. Tennen**
Law Offices of Sterns and Tennen, Phoenix, Arizona U.S.A.

Abstract

The promise of commercial passenger space launch systems raises significant regulatory issues. Within the United States, the federal government has taken the primary responsibility for the formulation of the rules governing the spaceports from which these passenger systems will conduct their launch and landing activities. Several of the individual states have sought to seize upon the opportunity to take a leading role in this area, and have obtained spaceport licenses from the federal government. In addition, a number of states have enacted statutes to directly or indirectly regulate the operation of commercial spaceports. This article examines and compares these regulatory frameworks on both the federal and state levels.

Introduction

Over the past several years a small but growing number of states within the United States have explored the potential benefits of establishing an operational spaceport within their borders. In February, 2001, 14 states created the National Coalition of Spaceport States to advance their common interests.[1] Five currently active licenses have been issued by the U.S. Federal Aviation Administration of the Department of Transportation (DOT): two in California, and one each in Virginia, Alaska, and Oklahoma. Rocket launches have occurred from Florida, and several other states, most notably New Mexico, are in varying stages of spaceport development.

The federal government has the primary responsibility for authorizing the operation of spaceports, however, some states have enacted specific legislation which may impact this federal authorization. It is open to question whether any individual state statute may be subject to challenge on constitutional or other grounds. State statutes may be invalidated where they seek to regulate a matter which has been pre-empted by federal law.[2] Even where

---

1. George Mason University Policy Report 63 (2002)[hereinafter referred to as the "GMU Report"]. The initial membership consisted of Alabama, Alaska, California, Florida, Montana, Nevada, New Mexico, Oklahoma, South Dakota, Texas, Utah, Virginia, Washington, and Wisconsin. Nebraska later joined as an associate member.

2. *See* U.S. CONST. Art. VI, cl. 2; *Watters v. Wachovia Bank, N. A.*, 127 S. Ct. 1559; 167 L. Ed. 2d 389; 2007 U.S. LEXIS 4336; 75 U.S.L.W. 4176; 20 Fla. L. Weekly Fed. S 170 (2007); *Cipollone v. Liggett Group, Inc.*, 505

* Honorary Director IISL, Member IAA, COSPAR, AIAA
** Former Commissioner Arizona Space Commission, Member IISL, IAA
*** Member IAF

an area has not been pre-empted, state statutes may be unenforceable where they conflict with federal law.[3] Nevertheless, states have wide latitude to adopt statutes within these strictures that will withstand constitutional scrutiny. The requirements of the states for lawful adoption of statutes must, of course, be followed, as well as any requirements concerning specificity, subject matter, ambiguity, interpretation with other statutes, etc. that may exist under local law. This article will examine the licensing regime pursuant to federal statutes and regulations, as well as specific statutes adopted by individual states to govern the operation of spaceports.

Authorization by the Federal Government

The Commercial Space Launch Act (CSLA) authorizes the regulation of launch and reentry sites operated by US citizens or located within the United States.[4] This regulatory oversight is provided within the DOT by the FAA Associate Administrator for Commercial Space Transportation (designated AST), which licenses launch sites which are not operated by the Federal government or co-located with a Federal launch facility.[5] The CSLA seeks to promote and encourage the development of commercial launch activities by the private sector, consistent with the public health and safety, safety of property, and the national security and foreign policy interests of the United States.

The licensing process is outlined in the federal regulations, and begins with a mandatory pre-application consultation by the prospective applicant with the FAA.[6] The confidentiality of information contained in applications for licensing is protected,[7] and decisions on applications must be made within 180 days after the application is accepted by the FAA.[8] The contents of a license application are specified in the regulations. In addition to basic identification of the applicant and proposed launch site, are required disclosures concerning the layout of the launch site, including launch points; the types of launch vehicles to be accommodated at each launch point; foreign ownership of the applicant; and information relating to the environmental impacts which may be associated with the operation of the proposed launch site.[9] The applicant must demonstrate that the proposed launch operations can be conducted safely,[10] both at the launch facility as well as down range, and the regulations contain detailed criteria and specifications for the analysis of

---

U.S. 504, 516, 112 S. Ct. 2608, 120 L. Ed. 2d 407 (1992); *McCulloch v. Maryland*, 17 U.S. 316, 4 Wheat. 316, 4 L. Ed. 579 (1819); *Montalvo v. Spirit Airlines*, 2007 U.S. App. LEXIS 26146, (11-9-2007).

3. *See U.S. v. Locke*, 529 U.S. 89, 120 S. Ct. 1135, 146 L.Ed.2d 69 (2000); *SPGGC, LLC v. Blumenthal*, 505 F.3d. 183, 2007 U.S. App. LEXIS 24436 (2007); *Wachovia Bank, N.A. v. Burke*, 414 F.3d 305 (2nd Cir. 2005).

4. 49 U.S.C. §§ 70101 *et seq*; § 70105.

5. 14 CFR § 413.3(b)(2).

6. 14 CFR § 413.5.

7. 14 CFR §413.9.

8. 14 CFR § 413.15.

9. 14 CFR § 420.15.

10. A safe launch is one in which the risk level does not exceed "an expected average number of 0.00003 casualties (Ec) to the collective member of the public exposed to hazards from the flight ($Ec \leq 30 \times 10^{-6}$)." 14 CFR § 420.19.(a)(1).

risks.[11] Spaceports proposing to launch unproven vehicles are required to demonstrate safety by a clear and convincing standard.[12]

The FAA review of a license application includes an environmental assessment, and unless that assessment finds that the proposed spaceport facility will not have a significant environmental impact, the preparation of an environmental impact statement.[13] The application review process also includes a risk analysis of the launch site and proposed operations,[14] and a determination that the issuance of the license would not jeopardize the foreign policy or national security interests of the United States. An applicant, generally, also must enter into agreements with the Coast Guard and the FAA Air Traffic Control for the issuance of a Notice to Mariners and a Notice to Airmen, respectively, prior to a launch, and such other measures as deemed necessary for the protection of the public health and safety.[15]

A license to operate a launch site is valid for five years, and may be renewed.[16] The license authorizes the operator to offer to conduct launches of a specified vehicle from a specified launch site, however, the license does not authorize any specific launches, each of which is subject to separate authorization by the government.[17] Similarly, a launch vehicle reentry site is subject to separate licensing requirements.[18] A licensee is

---

11. 14 CFR Part 420, Appendices A through E.

12. 14 CFR § 420.29.

13. The environmental impact analysis is to be conducted in accordance with the National Environment Policy Act, 42 U.S.C. 4321 *et seq.* (NEPA), the Council on Environmental Quality Regulations for Implementing the Procedural Provisions of NEPA, 40 CFR Parts 1500–1508, and the FAA's Procedures for Considering Environmental Impacts, FAA Order 1050.1D. 14 CFR § 420.17(a)(2). For examples of environmental impact analyses regarding proposed spaceports, see FAA *Spaceport America Draft Environmental Impact Statement*, www.faa.gov/about/office_org/headquarters_offices/ast/licenses_permits/htm/sw_spaceport_eis/media/FAA_Spaceport_America_Draft_EIS.pdf.; Blue Origins Environmental Assessment 8-29-06 [4310-13] Department of Transportation Federal Aviation Administration Finding of No Significant Impact (FONSI); *Oklahoma Finding of No Impact* Federal Register / Vol. 71, No. 87 / Friday, May 5, 2006 / Notices 26593; *Final Environmental Assessment for the East Kern Airport District Launch Site Operator License for the Mojave Airport*, Prepared for the U.S. Department of Transportation Federal Aviation Administration Office of the Associate Administrator for Commercial Space Transportation; FAA Environmental Assessment of the Kodiak Launch Complex, www.faa.gov/about/office_org/headquarters_offices/ast/licenses_permits/media/faaexec.pdf.

14. *See, e.g.*, Insurance Requirement for Maximum Probable Loss With Respect to Launch of Alliant Techsystems Inc. ALV-X1 Suborbital Launch Vehicle at Wallops Flight, Federal Register: July 24, 2007 (Volume 72, Number 141) [Notices] [Page 40338].

15. 14 CFR § 420.31.

16. 14 CFR § 420.43.

17. 14 CFR § 420.41.

18. *See generally* 14 CFR part 433.

required, *inter alia*, to allow access by, and to cooperate with, federal officials authorized by the FAA to observe any activities of the licensee.[19] The licensee also is required to prevent unauthorized access to the launch site;[20] provide two days notice of all launches to surrounding land owners and local officials;[21] and develop and implement a launch site accident investigation plan.[22] A licensee is required to maintain records for three years.[23] A licensee also is required to obtain insurance or demonstrate financial responsibility,[24] and to implement a reciprocal waiver of claims with each of its contractors and subcontractors, each customer and each of the customer's contractors and subcontractors.[25]

The federal regulations expressly provide that "[i]ssuance of a license to operate a launch site does not relieve a licensee of its obligation to comply with any other laws or regulations; nor does it confer any proprietary, property, or exclusive right in the use of airspace or outer space."[26] Among the other laws and regulations which may be applicable to the licensing of launch sites are the Environmental Protection Act referenced above,[27] the Clean Water Act,[28] the Endangered Species Act,[29] and the National Historic Preservation Act.[30] The use of the airspace above the launch site, or course, also is subject to the regulation of the FAA.[31]

The first spaceport operator's license was issued to Spaceport Systems International, L.P., in 1996, to operate the California Spaceport from Vandenberg Air Force Base.[32] SSI is a limited partnership between ITT Federal Services Corporation and California Commercial Spaceport, Inc.[33] A second

---

19. 14 CFR § 420.49.

20. 14 CFR § 420.53.

21. 14 CFR § 420.57.

22. 14 CFR § 420.59.

23. 14 CFR § 420.61.

24. 14 CFR § 440.9.

25. 14 CFR § 420.17.

26. 14 CFR § 420.41(c).

27. See text & note 13, *supra*.

28. Section 404 of the Clean Water Act, 33 U.S.C. §§ 1251, *et seq*., requires a finding by the Army Corps of Engineers that a proposed launch site is located within a closed basin, and that it would not affect any jurisdictional waters.

29. Section 7 of the Endangered Species Act, 16 U.S.C. §§ 1531, *et seq*., requires a concurrence by the Fish and Wildlife Service in the FAA's finding that a proposed launch site "is not likely to jeopardize" any listed species.

30. 16 U.S.C. §§ 470, *et seq*.

31. 14 CFR part 71.

32. LSO 01-005, was renewed in 2006, and expires September 18, 2011.

33. 2004 Strategic Plan for Commercial Spaceport Development in Texas 31 (November, 2004) [hereinafter referred to as "Texas Governor Report"]; Maj. John W. Raymond, *Airports and Spaceports a Historical Comparison* 16-17 (1997) (presented to the Research Department Air Command and Staff College,

spaceport operators license in California was issued to the East Kern Airport District (EKAD) for the Mojave Airport, a former military base and current general aviation airport. This license authorizes the EKAD to launch suborbital, reusable launch vehicles, utilizing an air-drop design whereby the launch vehicle is dropped from an aircraft at a predetermined altitude.[34]

Virginia received a license for the operation of a spaceport at the Virginia Commercial Spaceflight Center at the Wallops Flight Center.[35] The spaceport is operated by DynSpace, LLC, a joint venture between the Commonwealth of Virginia and DynCorp. The VSFC can launch payloads of up to 8,500 pounds into orbit on solid, liquid, or hybrid fuel vehicles.[36]

Alaska was issued a license to operate a spaceport at the Kodiak Launch Complex on Kodiak Island as the first FAA-licensed, non-federally owned commercial spaceport in the U.S.[37] The spaceport is operated by the Alaska Aerospace Development Corporation (AADC), a public corporation established by the Alaska Legislature. The Kodiak Launch Complex can place up to 5000 pounds into orbit, and can launch up to 9 launches per year.[38] The KLC is located within a free trade zone, thereby permitting the importation and launch of foreign built space vehicles without paying customs duties.[39]

The fifth currently active spaceport operator's license was issued to the Oklahoma Space Industry Development Authority (OSIDA) for the operation of a commercial launch facility at the Clinton-Sherman Industrial Airpark (CSIA) near the town of Burns Flat.[40] The Oklahoma Spaceport specializes in horizontal take-off and landing reusable launch vehicles,[41] and has signed memoranda of understanding with more than a dozen companies seeking to develop that technology.[42]

The State of Florida, has created the Florida Space Authority (FSA), which operates a commercial spaceport from former federal facilities at the CCAFS.[43] The Florida Cape Canaveral Spaceport is the largest and has the most diverse capabilities of any spaceport, with

---

AU/ACSC/0368/97-03); GMU Report, *supra* note 1, at 59.

34. East Kern Airport District Final Environmental Assessment, *supra* note 13.

35. LSO 02-007 Virginia Commercial Space Flight Authority Wallops VA, expires December 18, 2012.

36. GMU Report, *supra* note 1, at 60; Texas Governor Report *supra* note 33, at 32.

37. LSO 03-008 Alaska Aerospace Development Corporation, Kodiak AK, expires September 24, 2008; Alaska Aerospace Development Corporation, www.akaerospace.com.

38. Kodiak Launch Complex Environmental Assessment, *supra* note 13, at 2.

39. Texas Governor Report, *supra* note 33, at 30-32.

40. LSO 06-010, expires June 11, 2011.

41. Oklahoma Space Industry Development Authority (OSIDA), www.okspaceport.state.ok.us.

42. Texas Governor Report, *supra* note 33, at 35.

43. *Id.* at 31.

the ability to launch more than 11 different types of rockets.[44]

The State of New Mexico has actively pursued the establishment of Spaceport America, in Sierra County approximately nine miles west of the White Sands Missile Range. The New Mexico Spaceport Authority (NMSA) proposes to conduct both horizontal and vertical launches of suborbital vehicles. Horizontal launch vehicles would land at the spaceport, while vertically launched vehicles could also land at White Sands. The spaceport seeks the capability to launch tourist flights.[45]

The seven states discussed above constitute less than half of the participants in the NCSS. The remaining participating states, among others, have expressed interest in establishing a spaceport within their borders, and many states have invested substantial resources in furthering that interest. For example, the Arizona Space Commission conducted a preliminary examination of several locations which could serve as a spaceport, including a former military base, and existing aerospace facilities.[46] Alabama's Commission on Aerospace Science & Industry has proposed a spaceport for suborbital reusable launch vehicles in Baldwin County.[47] Idaho began promoting the commercial development of a spaceport in the late 1990's, but the effort was abandoned after less than two years.[48]

Montana aggressively pursued plans to develop a commercial reusable launch vehicle site at Malstrom Air Force Base in Great Falls, and made $20 million of general obligation bonds available for aerospace research and spaceport infrastructure development. The Montana spaceport was geared for vehicles from Rotary Rocket, and Lockheed Martin's Venturestar. The state ceased its efforts after the Venturestar program was cancelled in 2001.[49]

Nevada also examined the prospect of establishing a spaceport designed for a specific vehicle, that is, the Kistler Aerospace K-1 reusable rocket. However, Kistler Aerospace relocated their planned test flights to Australia.[50] Ohio at one time considered the creation of a spaceport at a former military air base near Columbus.[51] Similarly, South Dakota identified a preliminary spaceport site near Ellsworth Air Force Base, but no formal

---

44. GMU Report, *supra* note 1, at 56. These capabilities, however, do not include the ability to reach polar orbit. Texas Governor Report, *supra* note 33, at 31.

45. FAA Spaceport America Draft Environmental Impact Statement, *supra* note 13, at ES-1.

46. *See* Sterns & Tennen, *Commercial Spaceports: Preliminary Site Selection and Regulatory Considerations*, 24 J. SPACE L. 29 (1996).

47. Texas Governor Report, *supra* note 33, at 34; GMU Report, *supra* note 1, at 63.

48. Idaho 2000 Economic Development Report, at 69.

49. Texas Governor Report, *supra* note 33, at 34-35; GMU Report, *supra* note 1, at 63.

50. Texas Governor Report, *supra* note 33, at 35.

51. Ohio considering spaceport deal, Space News, Dec. 1, 2006, http://www.msnbc.msn.com/id/15993507/.

effort has been made to obtain a spaceport license.[52]

The State of Texas has actively pursued the development of commercial spaceports within the state. The legislature authorized almost $1 million in block grants for the Gulf Coast Regional Spaceport Development Corporation (GCRSDC) in Brazoria County, the Pecos County/West Texas Spaceport Development Corporation, and the Willacy County Development Corporation for Spaceport Facilities (WCDCSF), in partnership with the South Texas Spaceport Consortium (STSC). These Spaceport Development Corporation's are independent agencies with taxing authority and the power of "imminent (sic) domain."[53] In addition, Blue Origins proposed the creation of a spaceport in Culberson County.[54] However, the Gulf Coast Regional Spaceport Development Corporation was terminated in 2007.[55]

The State of Utah created a Spaceport Advisory Board, which studied the possibility of launching RLV's. However, no action has been taken by the state subsequent to the cancellation of the X-33 and Venturestar programs.[56] Nevertheless, the Wah Wah Valley Interlocal Cooperation Entity has proposed a commercial spaceport near Milford to launch single-stage to orbit reusable launch vehicles.[57]

The State of Washington also examined the possibility of launching reusable vehicles such as the Venturestar. Spaceport Washington was proposed as a public/private partnership for the Grant County International Airport near Moses Lake in central Washington. However, similar to the efforts in other states, the proposal has not been actively pursued following the cancellation of the Venturestar project.[58] Finally, a launch facility has been established in Wisconsin, near the city of Sheboygan, on Lake Michigan, for the launch of sounding rockets.

Space Commissions and Spaceports

A number of states have established space commissions with authority that may extend to the regulation of spaceports. The Alaska Aerospace Development Corporation (AADC),[59] is authorized to issue bonds,[60] adopt regulations,[61] and to own and operate launch sites, payload and rocket facilities, and space business incubators.[62] However, the AADC is statutorily precluded from utilizing the power of eminent domain.[63] Nevertheless, the AADC is authorized to finance or develop space related projects with any public and

52. Texas Governor Report, *supra* note 33, at 35-36.

53. *Id.* at 4.

54. Blue Origins Environmental Assessment, *supra* note 13.

55. The Facts.com, February 28, 2007.

56. Texas Governor Report, *supra* note 33, at 36.

57. GMU Report, *supra* note 1, at 64.

58. Texas Governor Report, *supra* note 33, at 36.

59. Alaska Statutes §14.40.821 *et seq.*

60. *Id.* at § 14.40.866(a)(7).

61. *Id.* at § 14.40.871(a).

62. *Id.* at § 14.40.866(a)(10).

63. *Id.* at § 14.40.866(6).

private entities, up to $1,000,000, without additional legislative approval.[64] The creation of the Kodiak Launch Complex was subject to a review pursuant to the Alaska Coastal Management Program.[65]

The Space Florida Act[66] created Space Florida, as a special district and subdivision of the state.[67] The purposes of Space Florida include the promotion of space enterprise, and the preservation of the unique national role of Cape Canaveral. Space Florida seeks to accomplish these goals by reducing costs and improving regulatory flexibility for commercial sector launches, while also pursuing the development of complementary sites for commercial horizontal launches.[68]

Space Florida has a broad grant of authority to exercise all powers necessary or convenient to effect any or all of the purposes for which it is organized.[69] This includes owning launch vehicles and payloads;[70] issuing bonds, which are not the obligation of the state,[71] but which are subject to approval of the legislature or governor and cabinet;[72] and adopting regulations to prohibit or control pollution of air or water.[73] The statutes provide that Space Florida shall assist commercial launch operators to complete and submit the documentation required for federal approval to conduct launches from the state.[74]

The Wisconsin Aerospace Authority (WAA),[75] is charged with owning and operating spaceports, including the Sheboygan spaceport.[76] The WAA is empowered to issue bonds and fund any spaceport, facility or service of the authority with bond proceeds.[77] In addition, the WAA can exercise the power of eminent domain.[78] The WAA is charged with maintaining exclusive jurisdiction over spaceports, subject to the requirements of federal laws.[79] By statute, no spaceport "open to the general public" may be operated unless effective runway and landing strip lengths are properly reported, published and marked, as well as located such that "approaching and departing aircraft or spacecraft clear all public roads, highways, railroads, waterways or other traverse ways by a height which complies with applicable federal standards."[80]

The Wisconsin statutes contain a procedure for the issuance of a certificate of approval of spaceport locations by the secretary of transportation, based on compatibility with

---

64. *Id.* at § 14.40.886.

65. Alaska Administrative Code, Title Six, Chapter 80.

66. Florida S.A. §§ 331.301, *et seq.*

67. *Id.* at § 331.302.

68. *Id.* at § 331.3011(2).

69. *Id.* at § 331.305(7).

70. *Id.* at § 331.305(11).

71. *Id.* at § 331.347.

72. *Id.* at § 331.305(20).

73. *Id.* at § 331.320(1).

74. *Id.* at § 331.3051(7)(d).

75. Wisconsin S.A. §§ 114.60 *et seq.*

76. *Id.* at § 114.62(10)(a).

77. *Id.* at subsection (13).

78. *Id.* at subsection (17).

79. *Id.* at subsection (22).

80. *Id.* at § 114.134.

other existing or planned transportation facilities in the area.[81] Public hearings may be held prior to the approval, and shall be held on request of any applicant after the refusal to issue a certificate.[82] Aerial approaches to a spaceport may be regulated and taken for safety reasons, and an aggrieved property owner has a limited period of only six months in which to submit a claim for damages or such claim is barred.[83]

Spaceport Regulation Without a State Space Commission

Many states have statutory provisions concerning the regulation and operation of spaceports even though they do not otherwise have a form of space commission. The statutes of Montana, for example, authorize local governing bodies to create aerospace transportation districts.[84] California,[85] New Mexico,[86] and Texas,[87] on the other hand, have enacted more detailed statutory regimes for the operation of spaceports.

The California Space Enterprise Development Act[88] establishes a space enterprise development program within the Business, Housing and Transportation Agency.[89] The secretary of the agency is authorized to select a California non-profit corporation to serve as the California Spaceport Authority (CSA),[90] which shall designate spaceports for launch and re-entry sites, subject to appropriate licensing of the site operator by the federal government.[91] Intercity and county airport districts within the state are authorized to provide and maintain spaceports designated by the CSA.[92] These airport districts have the power to make rules,[93] and to issue bonds to purchase real property for spaceport purposes.[94] Further, the California Environmental Quality Act (CEQA) must be complied with in addition to the federal environmental protection regime.

New Mexico has enacted a comprehensive statutory framework for the creation, operation and funding of spaceports. The New Mexico Spaceport Authority has been created as a state agency, although it is administratively attached to the economic development department.[95] The NMSA is empowered to "initiate, develop, acquire, own, construct, maintain and lease space-related projects."[96]

---

81. *Id.* at § 114.134(3).

82. *Id.* at § 114.134(4).

83. *Id.* at § 114.135.

84. Montana C.A. § 7-15-4296.

85. California Government Code §§ 13999, *et seq.*

86. New Mexico S.A. §§ 58-31-1, *et seq.*

87. Texas V.T.C.A. Government Code, Title 83, Chapter 10, art. 5190.6, § 4E.

88. California Government Code § 13999.

89. *Id.* at § 13999.2(a).

90. *Id.* at §§ 13999.2(c);(d)(1).

91. *Id.* at § 13999.3(c).

92. *Id.* at § 22553.

93. *Id.* at § 22555.

94. *Id.* at § 22702.

95. New Mexico S.A. § 58-31-4(A).

96. *Id.* at subsection (A)(4). The term "project" defined to mean "any land, building or other improvements acquired as part of a spaceport or associated with a spaceport or to

The NMSA is authorized to issue revenue bonds,[97] which are exempt from state taxes.[98]

New Mexico has enacted the "Regional Spaceport District Act" by which local governments can establish and operate a spaceport.[99] Counties and municipalities may become members of a regional spaceport district pursuant to a procedure by which the governing body of the county or municipality must first adopt an ordinance imposing a "regional spaceport gross receipts tax," of not more than one-half percent.[100] The ordinance imposing the tax shall not go into effect until after it is approved by the voters in an election.[101] Tax revenues may be supplemented by bonds issued by the NMSA for the operation of the spaceport.[102] The voters of Dona Ana and Sierra Counties approved a regional spaceport gross receipts tax in March, 2007, and April, 2008, respectively,[103] for the establishment and operation of the New Mexico Spaceport America. It is interesting to note that the New Mexico statutes define "space" as the location above 60,000 feet above mean sea level.[104]

Texas has adopted statutes pursuant to which an "eligible entity" of a county or a combination of municipalities and counties may establish a corporation for the development of spaceport facilities.[105] This corporation has broad powers which include the power of eminent domain,[106] and which also include the authority to require the relocation or modification of a railroad, utility line, pipeline, or other facility that may interfere with a spaceport.[107] Furthermore, the corporation has the authority to issue bonds, which, however, are not an obligation or pledge of the faith and credit of the state.[108]

Finally, Hawaii's statutes include an express prohibition against the launch from within its borders of weapons of destruction or nuclear waste materials, or the use of radioactive materials as a power source in launch vehicles.[109]

---

aid commerce in connection with a spaceport and all real and personal property deemed necessary in connection with the spaceport." New Mexico S.A. § 58-31-3(B).

97. *Id.* at subsection (B)(7).

98. *Id.* at § 58-31-16.

99. *Id.* at §§ 5-16-1 *et seq.*

100. *Id.* at § 7-19D-15(A)(municipality); § 7-20E-25(A)(county).

101. *Id.* at § 7-19D-15(C)(municipality); § 7-20E-25(C)(county).

102. *Id.* at § 5-16-7. New Mexico Laws 2006, ch. 111, § 68, authorized $100,000,000 in severance tax bonds for a regional spaceport in Sierra County, to be expended in increments through 2008.

103. See http://spaceportamerica.com/news/press-releases/18-spaceport-press-articles/110-sierra-county-voters-app-tax.html.

104. New Mexico S.A.§ 7-9-54.2(E)(3).

105. Texas V.T.C.A. Government Code, Title 83, Chapter 10, art. 5190.6, § 4E.

106. *Id.* at § 4E(c)(2).

107. *Id.* at § 4E(c)(2)(B).

108. *Id.* at § 4E(k).

109. Hawaii R.S. § 201-75.

GLOBALISATION AND PRIVATISATION OF THE SPACE LAUNCH SECTOR AND ITS IMPLICATIONS ON THE RESCUE AGREEMENT

Ricky J. Lee*

Introduction

Since the beginning of the space age nearly sixty years ago, one of the major concerns of the spacefaring States has been the status in international law of astronauts in distress and returned space objects as well as the legal obligations that ought to be imposed on States as to their treatment and return.

It was in this context that, forty years ago in 1968, the General Assembly of the United Nations adopted the Agreement on the Rescue of Astronauts, the Return of Astronauts and the Return of Objects Launched into Outer Space (the **Rescue Agreement**).[1] Since the adoption of the Rescue Agreement, 91 States have ratified it and a further 23 States have signed but not yet ratified it, making it one of the more widely accepted of the five existing United Nations space treaties.[2]

Forty years later, the foundations on which the Rescue Agreement was built are no longer secure. The private space industry, particularly multinational firms, have been involved in the provision of commercial launch services for some time and are now forming a substantial segment of the commercial launch services market. Further, the Ansari X Prize, the advent of private manned space flight and the establishment of commercial space tourism ventures are eroding away what is left of the governmental monopoly on space activities.

This paper sketches the operative provisions of the 1968 Rescue Agreement and analyses the problems posed by the globalisation and privatisation of the launch sector and concludes by making some observations and suggestions for the adoption and reform of the relevant legal principles to adapt better to space activities in the next forty years.

Origins of the Rescue Agreement

As early as 1959, the report of the *ad hoc* Committee on the Peaceful Uses of Outer Space discussed the need for legal principles concerning astronauts in distress and returned space objects. Paragraph 21 of the report stated that:

> Problems of re-entry and landing of space vehicles will exist both with respect to unmanned space vehicles and later with respect to manned vehicles of exploration. Recognising that landing may occur through accident, mistake or distress, members of the committee called attention to the desirability of the conclusion of multilateral agreements concerning re-entry and landing. Among the subjects that might be covered by such agreements would be the return to the launching state of the vehicle itself and – in the case of a manned vehicle – provision for the speedy return of personnel.

Further and in particular, paragraph 74 of the report stated that:

> Where space vehicles re-enter the Earth's atmosphere either through design or misadventure and any equipment or instrumentation is recovered by countries other than the launching country, arrangements are needed for restoring such instrumentation and equipment to the launching country.

In 1963, the General Assembly of the United Nations adopted the legal principle that astronauts are to be regarded as envoys of all mankind and all States are thus obliged to render all reasonable assistance to astronauts in distress and are to return them to their State of national origin.[3] Specifically, Paragraph 9 of the Declaration of Legal Principles Governing the Activities of States in the Exploration and Use of Outer Space (the **Principles Declaration**) stated that:

> States shall regard astronauts as envoys of mankind in outer space, and shall render to them all possible assistance in the event of accident, distress, or emergency landing on the territory of a foreign State or on the high seas. Astronauts who make such a landing shall be safely and promptly returned to the State of registry of their space vehicle.[4]

This principle was later entrenched by the Legal Sub-Committee of the United Nations Committee on the Peaceful Uses of Outer Space, by then a permanent body, in the 1967 Treaty on Principles Governing the Activities of States in the Exploration and Use of Outer Space, including the Moon and other Celestial Bodies (the **Outer Space Treaty**).[5] Article V of the Outer Space Treaty already requires that:

> States Parties to the Treaty shall regard astronauts as envoys of mankind in outer space and shall render to them all possible assistance in the event of accident, distress, or emergency landing on the territory of another State Party or on the high seas. When astronauts make such a landing, they shall be safely and promptly returned to the State of registry of their space vehicle.
>
> In carrying on activities in outer space and on celestial bodies, the astronauts of one State Party shall render all possible assistance to the astronauts of other States Parties.
>
> ...

At the same time, there was agreement among the members of the Legal Sub-Committee that the legal principles concerning the rescue and return of astronauts and space objects should be contained in an international treaty, separate to the Outer Space Treaty. In 1964, the Legal Sub-Committee created a working group to consider the two draft instruments on this subject as submitted by the Soviet Union[6] and the United States.[7] Several other States also submitted amendments to the two existing drafts and a new proposal, based on those discussions and amendments, was later jointly submitted by Australia and Canada.[8] With two further drafts with some differences were submitted by Italy,[9] and Argentina,[10] the Secretariat of the United Nations circulated a consolidated working draft.[11] After debates and several revisions, the text was submitted to the Committee on the Peaceful Uses of Outer Space and, following its approval, the General Assembly adopted the Rescue Agreement on 19 December 1967.[12]

Important Provisions of the Rescue Agreement

The 1968 Rescue Agreement served to broaden the obligations of States in relation to the rescue, recovery and return of astronauts and to extend some of these obligations to the recovery and return of space objects. These obligations vary in specificity and scope but one clearly formulated and adopted in an environment in which the governmental agencies or military establishments of States are the exclusive actors in manned space exploration and launch activities.

The Rescue Agreement expands on these requirements and provides that:

(1) the Member States are required to notify the launching authority and the United Nations or, if the launching authority cannot be identified, announce publicly of any information or discovery of any accident, distress or emergency landing suffered by astronauts onboard a spacecraft;[13]

(2) if astronauts onboard a spacecraft have made an emergency landing within the territory of a State, that State is required to take all possible steps to rescue them, provide all necessary assistance and to return them safely and promptly to the launching authority;[14] and

(3) if astronauts onboard a spacecraft have alighted in the high seas, all States in a position to do so must extend assistance in search and rescue operations and, if they are discovered, to return them promptly to the launching authority.[15]

The term "launching authority" refers to the State that was responsible for the launch. In the case of an international organisation, "launching authority" refers to that organisation provided that it has accepted the rights and obligations of the Rescue Agreement and a majority of its members are party to the Outer Space Treaty.[16]

In addition to astronauts, the Rescue Agreement also imposes specific obligations on States in relation to returned foreign space objects in that:

(1) the Member States are required to notify the launching authority and the Secretary-General of the United Nations of any information or discovery of any returned space object or its component parts;[17]

(2) if a space object or one of its component parts has returned to the territory of a State, that State is required to recover the object or component part, if requested, and to return it to the launching authority;[18]

(3) if a space object or one of its component parts has been found by a State beyond the territorial sovereignty of the launching authority, that State is required to recover it and to return it to the launching authority;[19] and

(4) if the returned space object or one of its component parts is of a hazardous or deleterious nature, the State must notify the launching authority and to immediately take effective steps to eliminate possible dangers of harm.[20]

The only reference to the reimbursement of costs by the launching authority is found in Article 5(5) of the Rescue Agreement, which relates only to expenses incurred in recovering and returning the space object or its component parts. There is no specific provision in the Rescue Agreement that provides for the reimbursement of costs incurred in the rescue and repatriation of astronauts or the costs of any cleanup of hazardous materials or risks. It appears from the language of the provision that the reimbursement of the necessary costs is to be dealt with separate to the liability for damage caused by the space object.[21]

Similar to the provisions of the Outer Space Treaty, some commentators have asserted that some, if not all, of the provisions of the Rescue Agreement have crystallised into customary international law.[22] However, this view may be difficult to sustain considering the absence of a substantial amount of state practice relating to these principles.

Modern Controversies

Defining a Space Object

Article 5 of the Rescue Agreement refers to the return of a "space object or its component parts". Bin Cheng had suggested that "space objects" covers "any object launched by humans into outer space, as well as any component part thereof, together with its launch vehicle and parts thereof" and so objects launched into Earth orbit and beyond are *ipso facto* regarded as space objects.[23] A similar legal definition for "space object" has been proposed by Vladimir Kopal.[24]

In the modern context, this has particular relevance in the case of a space object launched by a rocket deployed from an aircraft in airspace. The definition as suggested by Cheng and Kopal would exclude the aircraft from the definition of "space object". However, Karl-Heinz Böckstiegel suggested that the aircraft may be considered the first stage of the launch vehicle.[25] Accordingly, the definition of "space object" under the Rescue Agreement would include the aircraft, which is respectfully submitted to be somewhat contrary to logic or practice.

The next issue is the definition of the "component parts" of a space object within the context of the Rescue Agreement. It is clear that pieces, fragments and other substances of an object would generally be regarded as "parts" of that object rather than its "component parts". As the term "component parts" can be considered to have a clear meaning, the argument may therefore be forcefully made that the drafters of the Rescue Agreement intended for such a distinction to be maintained in the case of the "component parts" vis-à-vis the "parts" of a space object. However, as Stephen Gorove suggested, such a technical distinction does not appear to be maintained by state practice and, in any event, there does not appear to be a sound policy justification for such a technical distinction.[26]

One practical consequence of not maintaining such a distinction between "component parts" and "parts" is that the obligations under Article 5 of the Rescue Agreement may be triggered even if the relevant part of the space object in question may amount to no more than fragments or debris. It is for this reason that William Wirin suggested that the use of the term "component parts" was to specifically exclude small pieces and fragments that are not capable of surviving a re-entry into the atmosphere of the Earth.[27] Accordingly, they would not be the subject of the Rescue Agreement, though this would produce the effect that anything that survived re-entry from outer space may be considered subject to the provisions of the Rescue Agrement. On the other hand, Cheng argued that "fragments of a space object that fall on the Earth are ... given the same status as the whole object ... [and] nothing suggests otherwise, or that shattered fuel tanks or flakes of paint from space objects in outer space should be treated any differently".[28] However, it is respectfully submitted that the drafters of the Rescue Agreement, in referring to the "component parts" of a space object, clearly intended to suggest an interpretation other than that proposed by Cheng, for then there would be no substantive difference between the meanings of "parts" and "component parts".

Returned to Earth

Article 5 of the Rescue Agreement applies to space objects or their component parts that have "returned to Earth". This provision applies only to space objects, as the other articles of the Rescue Agreement relating to the personnel of spacecrafts refer to them "landing" or "alighting". The fact that Article 5 refers to "Earth" rather than the "surface of the Earth" raises the question of whether the space object or its component parts must have reached outer space before it returned to Earth before the obligations arising from the Rescue Agreement can apply.

This technicality gives rise to certain practical difficulties. If the space object in question is a component part of the launch vehicle that was never intended to reach outer space, such as the first stage of a multi-stage expendable launch vehicle, then it is questionable whether the provisions of the Rescue Agreement can be invoked by the relevant launching authority. Similarly, in the event of a launch failure, it is also debatable whether the launching authority can invoke the Rescue Agreement for the return of the space object and the launch vehicle that never left the atmosphere. In the absence of a generally agreed delimitation between airspace and outer space, such a requirement may pose practical difficulties for launching authorities seeking to invoke the Rescue Agreement.

Article 5(3) of the Rescue Agreement would tend to suggest that the space object or component part in question must have reached outer space for the Rescue Agreement to be invoked. This is because Article 5(3) specifically refers to "objects launched into outer space". This would have the effect that, in the event of either a launch failure or the component parts of a launch vehicle that are not intended to reach outer space, the provisions of the Rescue Agreement relating to the recovery and return of space objects would not apply.

Implications of Globalisation and Privatisation of the Modern Commercial Space Industry

Launching Authority

One of the principal terms used in the Rescue Agreement is the concept of the "launching authority", a concept that is somewhat unique to the Rescue Agreement. Article V of the Outer Space Treaty refers to the return of astronauts to the State of registry of their space vehicle. In the Rescue Agreement, however, Article 6 defines the "launching authority" refers to the State responsible for launching the relevant space vehicle.

The "State of registry" for the "space vehicle" as referred to in Article V of the Outer Space Treaty presents a difficulty where the space vehicle is not registered voluntarily or under the mandatory provisions of the 1976 Convention on the Registration of Objects Launched into Outer Space (the **Registration Convention**).[29] Difficulty also arises where the astronauts have returned to the surface of the Earth as a result of a launch failure so that the space vehicle may not have been registered but would have been registered had the launch operation been successful. In such cases, it may be suggested that the obligations imposed under Article V of the Outer Space Treaty would not apply.

It may be suggested that, in the case of the Rescue Agreement, the use of the term "launching authority" as the "State responsible for the launching" of the space object or the personnel of the spacecraft would overcome this problem. This would have been sufficient for the intended purposes of the Rescue Agreement where only one single State is involved in the space activity that launched the spacecraft or object into outer space, as it had been the case for most of the past forty years. However, in the modern context of a globalised space industry, there may be more than one State involved in any particular launch and may thus be considered "launching authorities" for the purposes of the Rescue Agreement, giving rise to confusion when its provisions are applied.

In the context of the Convention on the Liability for Damage Caused by Space Objects (the **Liability Convention**) and the Registration Convention, it is possible for there to be more than one "launching State".[30] Indeed, both treaties specifically provide for such a possibility.[31] In the context of the Rescue Agreement, however, significant difficulties may arise where there is more than one State that may be regarded as the launching authority. This is because the rights and obligations under the Rescue Agreement may not be practically exercised by or imposed on more than one State. While this may not be of significant impact in relation to the personnel of a spacecraft as reference can be made to their State of nationality, the same cannot be said for space objects or their component parts. It is possible that, where the satellite of one State was launched by the launch vehicle of another State, the State whose satellite is being launched may seek to assert its rights as a launching authority and seek the recovery of the launch vehicle to obtain technology that otherwise would not be available to them. The reverse position can also occur, where the State whose launch vehicle it was may seek the recovery of the satellite with a view to attaining technological advancements that are otherwise unavailable to them.

Further, where there are more than one possible launching authorities and one is seeking the recovery of a returned space object, it may be open to the State that has recovered the space object to refuse the claim and instead proceed to return the space object to the other launching authority. This would have particular implications where the launch involved sensitive technology owned by one State that is sought by the other State.

In the context of a global and multinational space industry, such possibilities are not too remote or inconceivable. There are increasing instances of commercial launch services being provided by an industrialised State to a developing State (such as the launch of Indian satellites by Arianespace) and also vice versa (such as the launch of European and U.S. satellites by China). Without a resolution of the issue of defining a "launching authority", there is potential for future international disputes in the application of the provisions of the Rescue Agreement.

Flags of Convenience

One of the major implications arising from the globalisation of the space industry is the advent of possibility of utilising flags of convenience for the purposes of minimising its licensing and regulatory burden. While this is not a significant issue where there is licensing of private spacecraft in compliance with the authorisation and continuing supervision requirements of Article VI of the Outer Space Treaty, this would not be the case if the launch vehicle is not subject to licensing in the State that may be considered the launching authority.

In such a case, the State that is the launching authority may not have knowledge that its spacecraft personnel has landed or alighted within the jurisdiction of another State to invoke its rights under the Rescue Agreement. Further, it may also seek to refuse requests by the private concern to invoke the terms of the Rescue Agreement on its behalf as it may be seen to have no vested interest in expending its resources and efforts to do so.

Personnel of a Spacecraft

One of the central terms used in Paragraph 9 of the Principles Declaration and Article V of the Outer Space Treaty is the concept of an "astronaut", but there is no express definition for that term in either instrument. Similarly, the Rescue Agreement uses the term "personnel of a spacecraft" and there is no express definition for that term in the Rescue Agreement. There is also no express guidance in that or the other treaties to indicate whether there is a distinction between the terms, considering the adoption of the two treaties were separated in time by mere months in the late 1960s.

In the remainder of the Twentieth Century, when all manned space flight were conducted within the framework of governmental space programs conducted by State agencies, the absence of such definitions was not of much significance. This is because all individuals in outer space would be professional astronauts that have been launched into outer space within the framework of governmental space programs. However, as private individuals are launched into orbit as commercial space tourists by governmental space programs and the advent of private manned space flight, the implications of the lack of a definition of the term "personnel of a spacecraft" is becoming increasingly acute.

The subject has attracted some academic debate. On one view, a person must be conducting their activities for the benefit and in the interests of all States and regarded as an envoy of mankind in outer space to be considered an "astronaut" or "personnel of a spacecraft".[32] This means that there is discretion on the part of the States to decide whether a person is to be considered an astronaut or spacecraft personnel, depending on the nature of the functions and activities being conducted by the person in outer space, and thus be cloaked with the protection of the Rescue Agreement and Article V of the Outer Space Treaty. This view is supported by the distinction made by the partners of the International Space Station between a professional or governmental astronaut and a "space flight participant" or a space tourist.[33] In other words, a person may conceivably be excluded from the definition of an astronaut or spacecraft personnel either by official designation or by an objective determination by States based on considering the nature of the activities conducted by the person in outer space.

It is respectfully submitted that this restrictive view of the definitions of "astronaut" and "personnel of a spacecraft" is somewhat inconsistent with the language of the Outer Space Treaty and the Rescue Agreement. Article V of the Outer Space Treaty expressly requires States to regard astronauts as envoys of mankind. For the restrictive view to be correct, the language would need to be reversed so that envoys of mankind are to be regarded as astronauts. In other words, Article V of the Outer Space Treaty is to be read as imposing obligations on how astronauts are to be regarded and not as setting out the criteria as to who would constitute an astronaut for the purposes of the Outer Space Treaty. The same, of course, can be said in relation to Paragraph 9 of the Principles Declaration.

On the other hand, there is some academic support for a broader interpretation of "personnel of a spacecraft" so that it would include not only professional and governmental astronauts but also passengers and space tourists. In this context, the terms of the Rescue Agreement would apply in connection with all individuals in space, regardless of their designated status and activities. This would appear to be the more humanitarian expression of the obligations imposed under the Rescue Agreement. It may also be suggested that the use of the term "personnel of a spacecraft" in the Rescue Agreement in place of "astronaut" in the Outer Space Treaty is to remove the ambiguity that may exist in the use of the term "astronaut" so that all individuals on board a spacecraft would be considered within the scope of the Rescue Agreement.

Further, the paragraphs of the preamble of the Rescue Agreement make express reference to the provisions of the Outer Space Treaty in relation to "astronauts" and that the Rescue Agreement is to "develop and give further concrete expression to these duties". If the Rescue Agreement, taken as a whole, is an "expression" rather than an "extension" of Article V of the Outer Space Treaty, then "personnel of a spacecraft" must mean the same or less than "astronaut". Therefore, it would not follow that the term "astronaut" would have a narrower definition than the term "personnel of a spacecraft".

Concluding Observations

It is apparent that the definitions of "component parts" and "launching authority" and the potential for the use of flags of convenience have significant potential implications arising from the privatisation and globalisation of the contemporary space industry. To address these issues as well as other contemporary problems arising from the terms of the Rescue Agreement, some revision of the Rescue Agreement or an internationally agreed resolution of these definitional issues will be necessary.

[*] Associate, Legal, PricewaterhouseCoopers, Australia. Member of the Board of Directors of the IISL and the space law committees of the IBA and the ILA. Fellow, Commercial Law Association of Australia. The views and opinions expressed or implied in this paper are those of the author only and are not necessarily those of any organisation with which he is associated.

[1] Agreement on the Rescue of Astronauts, the Return of Astronauts and the Return of Objects Launched into Outer Space (1968) 672 U.N.T.S. 119; T.I.A.S. 6599; 19 U.S.T. 7570, opened for signature on 22 April 1968 (entered into force on 3 December 1968).

[2] United Nations Office of Outer Space Affairs, *Status of International Agreements Relating to International Activities in Outer Space* (2008), at <http://www.unoosa.org/oosa/en/SpaceLaw/treatystatus/index.html>, last accessed on 31 August 2008.

[3] Declaration of Legal Principles Governing the Activities of States in the Exploration and Use of Outer Space, General Assembly Resolution 1962 (XVIII), adopted without a vote on 13 December 1963.

[4] Declaration of Legal Principles Governing the Activities of States in the Exploration and Use of Outer Space, General Assembly Resolution 1962 (XVIII).

[5] Treaty on Principles Governing the Activities of States in the Exploration and Use of Outer Space, including the Moon and Other Celestial Bodies (1967) 610 U.N.T.S. 205; 18 U.S.T. 2410; T.I.A.S. 6347, opened for signature on 27 January 1967 (entered into force on 10 October 1967).

[6] U.N.Doc. A/AC.105/C.2/L.2.

[7] U.N.Doc. A/AC.105/C.2/L.9.

[8] U.N.Doc. A/AC.105/C.2/L.2.

[9] U.N.Doc. A/AC.105/C.2/L.21.

[10] U.N.Doc. A/AC.105/C.2/L.23.

[11] U.N.Doc. A/AC.105/C.2/L.28.

[12] General Assembly Resolution 2345 (XXII).

[13] Rescue Agreement, Article 1.

[14] *Ibid.*, Articles 2 and 4.

[15] *Ibid.*, Articles 3 and 4.

[16] *Ibid.*, Article 6. In the event of the launch being conducted by an international organisation that has not accepted the provisions of the Rescue Agreement, its Member States would presumably be regarded as the "launching authorities" for the purposes of the Rescue Agreement.

[17] *Ibid.*, Article 5(1).

[18] *Ibid.*, Article 5(2) and (3).

[19] *Ibid.*, Article 5(3).

[20] *Ibid.*, Article 5(4).

[21] The Principles Relevant to the Use of Nuclear Power Sources in Outer Space (the "**Nuclear Power Sources Principles**") establishes a contrary position; see the Nuclear Power Sources Principles, Principle 9(3).

[22] Vladlen S. Vereshchetin and Gennady M. Danilenko, *Custom as a Source of International Law of Outer Space* (1985) 13 J. SPACE L. 22; and Gennady M. Danilenko, *Outer Space and the Multilateral Treaty-Making Process* (1989) 4 BERKELEY TECH. L. J. 217.

[23] Bin Cheng, *"Space Objects", "Astronauts" and Related Expressions* (1991) 34 PROC. COLL. L. OUTER SP. 17.

[24] Vladimir Kopal, *Some Remarks on Issues Relating to Legal Definitions of "Space Object", "Space Debris" and "Astronaut"* (1994) 37 PROC. COLL. L. OUTER SP. 99 at 101.

[25] Karl-Heinz Böckstiegel, *The Terms "Appropriate State" and "Launching State" in the Space Treaties: Indications of State Responsibility and Liability for State and Private Space Activities* (1992) 35 PROC. COLL. L. OUTER SP. 15.

[26] Stephen Gorove, *Toward a Clarification of the Term "Space Object": An International Legal and Policy Imperative?* (1993) 21 J. SP. L. 11 at 13-14.

[27] William B. Wirin, *Space Debris and Space Objects* (1991) 34 PROC. COLL. L. OUTER SP. 45. This was a view supported in He, *Review of Definitial Issues in Space Law in the Light of Development of Space Activities* (1991) 34 PROC. COLL. L. OUTER SP. 32.

[28] Cheng, *supra* note 23, at 24.

[29] Convention on the Registration of Objects Launched into Outer Space (1976), opened for signature on 14 January 1975 (entered into force on 15 September 1976).

[30] Convention on International Liability for Damage Caused by Space Objects (1972), opened for signature on 29 March 1972 (entered into force on 1 September 1972), Article I.

[31] See Liability Convention, Article I; and Registration Convention, Article 1.

[32] See, for example, Stephen Gorove, *Interpreting Salient Provisions of the Agreement on the Rescue of Astronauts and Return of Objects Launched into Outer Space* (1968) 11 PROC. COLL. L. OUTER SP. 93; Yasuaki Hashimoto, *The Status of Astronauts Towards The Second Generation of Space Law* (1993) 36 PROC. COLL. L. OUTER SP. 93; and Patrick Collins and Koichi Yanemoto, *Legal and Regulatory Issues for Passenger Space Travel* (1998) 41 PROC. COLL. L. OUTER SP. 224.

[33] Marietta Benkö and Kai-Uwe Schrogl, SPACE LAW: CURRENT PROBLEMS AND PERSPECTIVES FOR FUTURE REGULATION (2002) at 202; and Andre Farand, *Commercialisation of International Space Station Utilisation: The European Partner's Viewpoint* (2003) 18 AIR & SP. L. 83.

IAC-08-E8.2.9

THE RESCUE AGREEMENT AND PRIVATE SPACE CARRIERS

Miss Zeldine Niamh O'Brien L.L.B.*
University of Dublin, Trinity College, Ireland
zeldineobrien@gmail.com

Abstract

The Rescue Agreement imposes several obligations on State parties, specifically in Article III, contracting Parties must extend assistance in search and rescue operations to assure the speedy rescue of spacecraft personnel where they have alighted in any other place not under the jurisdiction of any State if necessary and where they are in a position to do so. While the scope of those who may benefit from the application of the duties imposed by the Agreement is unclear, it is certain that only State parties and other contracting international intergovernmental bodies are bound by them. However, it may be envisaged that by the end of this decade, private commercial manned space crafts for the carriage of persons will also be active in low earth orbit and, in time, beyond that. This paper will address whether the duties in the Agreement should be extended to bind commercial entities engaged in space carriage, rather than the Agreement itself.

INTRODUCTION

"Danger invites rescue. The cry of distress is the summons to relief."[1]

The Rescue Agreement is an excellent example of the application of humanitarian principles in international law. The Agreement imposes both substantive and non-substantive duties. These include information-sharing obligations as well as the duty to conduct rescue operations. The duties imposed by the Agreement are on States vis à vis launching authorities but unlike maritime law, these duties do not currently exist between private parties. As pointed out by Jarvis, "a rescue operation involving private parties will not necessarily fall within the parameters of the treaties. Moreover the possibility exists that a rescue operation would involve both private and public parties."[2] The lack of specificity in both the Outer Space Treaty and the Rescue Agreement is problematic. Goedhuis suggested the imposition of an obligation to rescue as far back as 1963 when he observed that 'an obligation to render assistance to space personnel and spacecraft should not only be imposed on masters of ships and aircraft commanders but a mutual obligation to render assistance should also be imposed on space craft commanders."[3] This paper examines the duties imposed and the arguments in favour of imposing these duties on private space carriers as well as the methods of doing so.

THE RESCUE AGREEMENT[4]

The need for an international agreement governing the rescue and return of personnel was recognised by 1958.[5] The U.S. and the U.S.S.R. agreed separately on a working paper - this bilateral draft was pushed through into the multinational arena.[6] At this stage, the Americans had lost three astronauts and the Russians one. The Outer Space Committee was repeatedly requested to continue its work on formulating an agreement on the rescue

---

and return of astronauts in 1967[7] and, on the 19th December 1967, the agreement was unanimously adopted,[8] coming into force almost a year later. The Rescue Agreement is 'essentially an elaboration of Art.V of the Outer Space Treaty 1967'.[9]

DUTIES UNDER THE AGREEMENT

The Agreement provides for a number of duties on contracting party states. Some envisage the sharing of information and others the provision of more substantial assistance.

Information-Sharing Obligations

Article I envisages an obligation on the contracting state where it receives information or discovers that the personnel of a spacecraft have suffered accident or are experiencing conditions of distress or have made an emergency or unintended landing in territory under its jurisdiction or on the high seas or in any other place not under the jurisdiction of any State to immediately notify the launching authority or make an immediate public announcement where the authority cannot be identified. This information-sharing obligation is one of the least onerous and yet invaluable for those experiencing difficulty. The duty binds regardless of how the State learns of it. It may do so directly from its own emanations or it may learn from private entities. The Secretary General of the UN must also be informed. Under Art.II, the State must inform the launching authority and the Secretary General of the steps taken and its progress. This furthers inter-state co-operation and co-ordination on the rescue effort. Information-sharing obligations extend beyond personnel to cover space objects *simpliciter* under Art.V. Each State that receives information or discovers that a space object or its component parts has returned to Earth in territory under its jurisdiction or on the high seas or in any other place not under the jurisdiction of any State, must notify the launching authority and the Secretary-General.

'Consultation' Obligations

The Agreement envisages direct co-operation in certain circumstances under Art.II. If assistance by the launching authority would help to affect a prompt rescue or would contribute substantially to the effectiveness of search and rescue operations, the launching authority must co-operate with the State affecting the rescue. The operations are to be subject to the direction and control of that State although it must 'act in close and continuing consultation with the launching authority'. This is less onerous that the equivalent search and rescue provisions in air law which require States searching for missing aircraft to 'collaborate in co-ordinated measures'.[10]

Article V(5) of the Rescue Agreement also envisages interaction between contracting states and launching authorities, the launching authority must to take immediately effective steps to eliminate the dangers posed by hazardous or deleterious space objects or their component parts under the direction and control of the contracting state.

Substantive Obligations to Personnel[11]

Article II requires a contracting state to take all possible steps to rescue and render all necessary assistance to the personnel of a spacecraft where they have landed in its territory owing to accident, distress or unanticipated landing. Assistance in this regard may involve the 'launching authority, which would possess advanced knowledge and experience in locating space vehicles, and, perhaps, available aircraft or ships to join in a search' given that it might in certain circumstances "be crucial to saving the life of an astronaut."[12] However, where the State and the launching authority disagree, the former would have the final say. Article III provides for a more onerous duty by extending contracting states' obligations beyond their own jurisdiction. Where the personnel of a spacecraft have alighted on the high seas or any other place not under the jurisdiction of any State, contracting parties must, if necessary,[13] extend assistance in search and rescue operations for such personnel to assure their speedy rescue. This is subject to the proviso that this applies only to contracting states that are in a position to affect a rescue or aid in a search.[14] Therefore only states that have geographical proximity to the scene and technological capability to conduct a rescue are bound to act.[15] As Art.II applies only where the information concerning the astronauts is received or discovered, Art.IV fills the lacuna that would otherwise exist where a State finds those concerned without receiving or discovering any information. It provides that if, owing to accident, distress, emergency[16] or unintended landing, the personnel of a spacecraft land in territory under the jurisdiction of a Contracting Party or are found on the high seas or in any other place not under the jurisdiction of any State, then they are to be safely and promptly returned to the representatives of the launching authority. Some countries adopted Art.IV on condition that it would operate where there was no conflicting law, such as asylum.[17]

The proviso in Art.III is one of central limitations on the scope of the duties towards the personnel of the space craft. Other limitations also derive from the wording adopted by the Agreement. The most obvious limitation that gives rise to much debate is the use of the term 'personnel of the spacecraft' and 'astronaut'[18] as this gives rise to the question as to whether spaceflight participants/space tourists would be covered by the definition. One response to this *quare*, is to look to humanitarian law as a basis for imposing a duty to aid a space tourist in distress. Indeed, it would seem rather idiosyncratic if a private space launch vehicle for the purpose of space tourism was in distress but the duty to search for and rescue or otherwise aid extended to the pilot and crew but excluded all passengers. Gorove submits that while accompanying scientists are within the scope of the term, passengers and stowaways are not.[19] However, regardless of whether the Agreement in its current form extends to space tourists, the central argument of this paper is an examination of not whether the contracting parties owe duties to space tourists but rather whether the duties in themselves should bind private space carriers, especially those that engage in the carriage of persons given the potential risk to human life, albeit a risk taken with full knowledge and consent.

Another limitation derives from the use of the term 'alighted' which suggests some form of landing whether upon water or ground. Clearly, it applies to landings upon the Moon or other celestial bodies. Although, it is questionable if the personnel of a spacecraft are in distress while present in outer space,[20] whether they would come within a strict literalist reading of the provision. However, the spirit, if not the letter, of the provision would mandate contracting parties extend assistance in such circumstances, and it would appear to be covered by Art.V(2) of the Outer Space Treaty in any case.

Van Traa-Engelman observes that the Rescue Agreement, like Article V of the Outer Space Treaty, is primarily focused on distress and emergency situations arising from an unintentional landing.[21] Ogunbanwo notes that landings coming within the scope of the Agreement include 'those caused by a malfunction of the spacecraft, a collision between the spacecraft and another object, a physical disability suffered by the astronaut or by navigational error'.[22] This suggests another limitation on the scope of the duties envisaged by the Agreement as it:

> "... leads to the conclusion that intentional landings, with a specific amount of accent on landings on foreign territory, will probably fall outside its scope of application, in which case the principle of state sovereignty will prevail leaving it to the

discretion of the territorial state how it should act in such a situation."[23]

However, there is no limitation on the extent of the State's obligation to provide assistance to personnel where the spacecraft lands in territory under its jurisdiction. The obligation is to render all necessary assistance. The duty is not limited by practicability as is the case under Art.V. The extent of the remedial action is determined by the nature and scope of the emergency.

Substantive Obligations Regarding Spacecraft

Where a space object or its component parts has been discovered in the territory of a State, the State which has jurisdiction over that territory must take such steps as it finds practicable to recover the object or component parts when requested to do so by the launching authority and with its assistance if asked.[24] Those objects that are so recovered are to be returned to or held at the disposal of representatives of the launching authority upon its request. But where a State has reason to believe that a space object or its component parts discovered in territory under its jurisdiction, or recovered by it elsewhere, is or are of a hazardous or deleterious nature, it may notify the launching authority of this.[25]

Cost of Recovery

Diederiks-Verschoor notes that one of the most remarkable shortcomings of the agreement is the lack of any provision concerning expenditure for the rescue and return of the astronaut, due to their status as 'envoys of mankind'.[26] This is in contrast to the obligations under Art.V(2) and (3), where the expenses incurred are expressly stated to be borne by the launching authority.[27] Ordinary principles of restitution would fill this void. Thus assisting states may be able to recover the costs of aiding the astronaut and returning him/her.

Desirability of Duties Binding Private Parties

Should a commercial entity be held to have to respond to a distress call in outer space? It may be argued that given the difficulties inherent any space rescue operation, particularly in relation to the ability to in fact reach those in distress, the law should utilise the presence of privately manned vehicles in outer space to minimise any loss of life. If it is accepted that all aboard a space object will benefit from the duties as provided for under the Agreement, regardless of whether they are crew or not, it would not seem unreasonable to extend duties to the carriers whose crew and customers benefit from its provisions. In addition, international maritime custom, upon which the Agreement was in part modelled, would analogically favour the creation of such a duty. However, a key difficulty is the question of who must pay for the cost of the rescue. The Agreement itself requires the launching authority to bear the expenses involved in the recovery and return of the space object (Art.V(5)) but it does not extend this to the rescue of personnel. It is submitted that the burden should not fall on the rescuer in order to encourage acts of rescue. This would also acknowledge that while private carriers may be obliged to divert to respond to distress calls, they are not operating an emergency service but a business for profit.

Humanitarian Law

In 1984, Fawcett posited the question of whether *inter alia* humanitarian protection had become a general form of state responsibility in outer space.[28] Certainly, the ITU Convention prioritises absolutely all telecommunications concerning life at sea, on land in the air or in outer space.[29] Article V of the Outer Space Treaty and Article 10 of the

Moon Agreement appear to favour this conclusion as well as the Rescue Agreement,[30] although it does not expressly refer to outer space. The argument is that this humanitarian aspect is furthered by the imposition of responsibilities on private actors.

Another approach is to consider why the duty to rescue was adopted in the first place. Astronauts were first recognised as envoys of mankind in 1963 and thus States were obligated to render them all necessary assistance, a principle subsequently embodied in the Outer Space Treaty. As Matte notes, this 'attitude towards astronauts in particular reflects the spirit of international co-operation and mutual assistance governing space activities'.[31] The Rescue Agreement refers to the personnel of the spacecraft. It is questionable whether non-personnel, such as space tourists, should be considered envoys of mankind[32] but it is arguable that they should be protected under equivalent duties as arise under the Rescue Agreement for humanitarian reasons.[33] However, while this is uncertain, it is clear that the personnel of a private carrier's craft come within the terms of the Agreement and will have the benefit of the duties owed there under. This weakens any conceptual argument based on the distinction between public (in the sense of State-owned or run or owned and run by an emanation of the State) and private spacecraft personnel, as it is not borne out by the law. Given the current state of humanitarian law, it is also arguable that there should be no distinction between personnel and non-personnel. No distinction is made in maritime law between passengers and crew in terms of the obligation to render all necessary assistance or, indeed, to rescue. Once a spacecraft in distress has been reached, there would seem to be no reason to distinguish between personnel and non-personnel in the provision of aid from a humanitarian standpoint, although practical and pragmatic arguments may justify prioritising assistance to the crew before passengers (as some of the personnel may in fact possess information critical to bringing about a successful rescue). An explicit endorsement of an extension of the protections to space tourists would provide clarity and certainty and also strengthen fairness-based rationales for imposing a duty on private space carriers towards others.

However, the humanitarian argument does not stretch so far as to justify the recovery of a space object or a component part. In maritime law, the duty to render assistance or rescue does not extend to the recovery and return of crashed vessels; this is properly consigned to the law of salvage.[34] Therefore, it must be examined whether an alternate argument can be made to support the extension of this duty to private spacecraft commanders.

Fairness-Based Rationales

Currently, those personnel on privately-owned and manned space craft would be entitled to the protections under the Agreement, though the obligations are owed as and between the State(s) and relevant launching authority. However, as they are not emanations of the State, there is no binding obligation in international law on the commander of the spacecraft in such circumstances owed to any other spacecraft; this is even so in relation to non-substantive duties such as the sharing of information. A fairness-based argument can be made on a *quid pro quo* basis for extending such duties to the private craft operators. There is much value in collective action in relation to search and rescue operations. In this form, this argument for expanding the scope of duties may appear unduly simplistic in failing to take account resource issues – a state has far greater resources to fund a rescue operation than a private operator which may be crippled by the loss. It is possible that these difficulties could be resolved by applying strict restitution principles but this will still not answer the question of whether recovery should be permitted for pure economic loss consequent upon taking actions to rescue or recover others. While the response of 'what price for a human life?' does not address the underlying difficulty, it illuminates the core issue that States, in imposing such obligations on private space operators, are being asked to make a value judgment, a choice which they are not unfamiliar with making.

However, the human life counter-argument cannot be used in relation to the recovery and return of a space object. The resource arguments have therefore greater weight in denying existence of any duty in this regard. It is submitted that the duty to recover and return should not be extended to private operators and should be left to the law of salvage.

THE DUTY TO RESCUE

Unlike civil law systems which have Good Samaritan laws in place,[35] the majority of common law states (with the exception of Vermont[36] and Minnesota[37]) do not recognise a general duty to rescue[38] even where there is no risk to the potential rescuer.[39] While the common law will not generally require a person to engage in a rescue, rescuers may recover against rescuees in tort.[40] The most obvious parallel to draw is to the duty to aid ships in distress.[41]

The analogy with maritime law suffers from practical differences. It remains cheaper and easier to aid those on the high seas than to render assistance to those in outer space. While assistance in the form of expertise has a value, it may not always be welcome. Nonetheless, assistance could be rendered while on the High Seas as a part of maritime law. The main benefit of the analogy is however observing the existence of reciprocal duties to rescue and aid those in distress imposed on both public and private parties towards each other regardless of nationality.

OTHER RATIONALES

The notification requirement in Art.V(5) is justified by the need to protect and safeguard both earth and the space environment from hazardous and deleterious space objects and their component parts. This is particularly important in relation to space activities given the use of nuclear power sources. There is no reason why this rationale is less potent in relation to private operators where a notification requirement is envisaged.

METHODS OF IMPOSING DUTIES

While it is possible for custom to develop in this area, it is unlikely that a single instance of a private party aiding another when alighting on the high seas would be recognised at an international level to constitute what Cheng describes as instant customary international law.[42] Given these reservations, and the desirability of having a formal document, particularly where the obligations imposed by the document bind private parties that include those of varying size and economic strength is more pronounced, a formal multilateral agreement would be the preferable mode of imposing such obligations.[43] Article 98 of the UN Convention on the Law of the Sea provides an example of how this may be done. This article imposes the burden on States to ensure that the master of a ship flying its flag renders assistance to any person found at sea in danger of being lost or after a collision, to the ship, its crew and its passengers.

INFORMATION SHARING OBLIGATIONS

The least onerous of the obligations concerning the sharing of information could easily be adapted to apply to between private entities and their own launching authority. Once in the possession of the launching authority, the information could easily reach the relevant parties, indeed the authority as an emanation of the State may be bound by the obligations in Art.I of the Rescue Agreement. Under the Merchant Shipping Search and Rescue Manual and under domestic laws, information sharing obligations are imposed on the masters of ships. The relevant UK regulations require the masters of assisting vessels to inform the appropriate search and rescue services, if possible, that they are aiding the vessel in distress.[44] Applying this and borrowing from the terms of the Rescue Agreement, there should be an obligation imposed on all commanders of spacecraft to inform their own launching authority where they receive information or discover that the crew or passengers of a spacecraft have suffered an accident or are experiencing conditions of distress or have made an emergency or unintended landing or alighting immediately of that fact and any other pertinent information. If any substantive action is undertaken, the launching authority of the assisting vessel should be informed of the nature of the assistance being rendered or the rescue being undertaken.

In addition, there should be an obligation to inform their own launching authority where it has reason to believe that a space object or its component parts discovered or recovered by it is of a hazardous or deleterious nature.

Substantive Obligations

The basic substantive obligation would relate to the rendering of assistance to or rescue of another spacecraft, its crew or passengers. As the extension of the duty to private operators is in part justified on humanitarian grounds, a distinction in the duty owed to personnel and non-personnel or crew and spaceflight participants is unnecessary and undesirable. The duty, if modelled on Art.98, would extend to those found and in danger of being lost, those in distress who have requested assistance and to render assistance/rescue post-collision. These grounds have been moulded by time and are suited to the High Seas. However, given that the space environment poses *sui generis* problems, it may be preferable to develop the substantive obligations using the more familiar vocabulary of the Rescue Agreement.

The imposition of any substantive obligation would need to be limited by pragmatic considerations. The Rescue Agreement itself provides examples of the kind of limitations envisaged, phrased in terms of the practicability of rescues. Under Article 98, masters of ships are only required to render assistance to those found lost at sea or after a collision *in so far as they can do without serious danger to the ship, the crew or the passengers*. This pragmatic limitation also applies to the duty to proceed with all possible speed to the rescue of persons in distress, if informed of their need of assistance which is also further limited by the phrase 'in so far as such action may reasonably be expected of him.' These limitations are reflected in national instruments,[45] for example, Irish law imposes the duty unless the master is unable to comply, or in the special circumstances of the case considers it unreasonable or unnecessary.[46] The same law also releases the master from his duty in two other circumstances.[47] There is no duty to render assistance where the master is informed of the requisition of one or more ships other than his own and that the requisition is being complied with by the ship or ships requisitioned[48] nor if his ship has been requisitioned and if he is informed by the persons in distress, or by the master of any ship that has reached the persons in distress, that assistance is no longer required. Similar pragmatic constraints should be adopted in relation to spacecraft commanders.[49]

In addition, domestic laws which have provided for the duties in Article 98 have also made it an offence to send out distress signals for any purpose other than that prescribed.[50] Such a provision should be included when incorporating into national law in relation to spacecraft to prevent abuse. Some laws also make it an offence to ignore the duty to render assistance,[51] although the duty is generally unenforced against masters;[52] such a matter is better left to the discretion of the State.

Substantive obligations concerning the rescue of stricken spacecraft should be considered under the current admiralty rules on salvage.

THE FINANCIAL BURDEN

Given the high cost of launching and of space activities, as well as the physical constraints in launching such as fuel and oxygen, it may not always be possible to render assistance. However, in the event that these obstacles are overcome and a rescue is attempted, several issues in relation to costs arise. First, whether it should be possible to recover for the cost of a rescue including an attempted rescue. Second, whether it should be possible to recover for financial loss consequent upon engaging in a search and rescue operation, even where the loss is purely economic. Third, if financial recovery is possible, upon whom should the burden fall?

On the first issue, arguably a rescuer should be allowed to recover for a rescue. To hold otherwise would allow the commander of the craft and ultimately to its owners or charterers as the case may be, to suffer a pecuniary detriment where he/she acted to his disadvantage in order to preserve the life of others. The same argument can be applied to attempted rescues, provided they did not fail because of the gross negligence of the rescuer and there has been a sufficient proximate act to count as an attempt. This is however in direct contrast to the long-established rule in admiralty that prohibits recovery for 'pure life salvage'.[53] The reason for this is:

> "traditionally explained by reference to the costs of collecting an award where there is no property saved to constitute a fund out of which to pay it. An additional element is altruism…. Lives at sea are not only extremely valuable (as elsewhere) but usually can be saved at less cost than the ship itself or its cargo. Hence altruism may provide an efficient level of rescues in the pure life salvage situation notwithstanding the competitive constraints that limit altruism to professionals."[54]

However, by allowing recovery for property but not life, this creates a disincentive to rescuing people over property. In addition, while there may be some difference in cost between the rescue of persons and the rescue of property in outer space, the cost of launching where a rescue commences on the surface of the earth, which represents the bulk cost of such a rescue itself would exist regardless of the nature of that which is to be saved. The pure-life salvage rule has been heavily criticised[55] and it is not applied uniformly.[56] In addition, the rule does not prohibit an enhanced award where both lives and property has been salvaged.[57] Furthermore, it has been limited by case law so that it no longer operates between two seaworthy vessels where there is a transfer of an ailing seaman from one to another.[58] As to attempted rescues which ultimately fail, if reasonably undertaken with due skill and care, recovery should be permitted. However, aborted rescues, i.e. where there has been no proximate act of rescue before any rescue effort is abandoned, should not result in any recovery. Otherwise recovery is being permitted for nothing more than a diversion with no potential benefit or obviation of a disbenefit to the rescuee.

In relation to the second issue, while recovery for the actual costs of engaging in the rescue is unproblematic, the idea of permitting recovery for pure economic loss consequent upon fulfilling the duty to render assistance is. Different legal systems have different approaches to pure economic loss; divergent attitudes are apparent even among common law states.[59] Agreement on this issue may be difficult to achieve and the better approach may be to allow recovery for the actual cost and leave the issue of pure economic loss aside until there is a general consensus on how to treat the issue in law.

On the third issue, as noted above, under the Rescue Agreement, there is no provision for recovering costs for the rescue and return of an astronaut, however, the launching authority bears the cost of the recovery and return of the space object. This is, of course, a separate head from liability caused by the object. In relation to the rescue of persons by private operators, given the possible costs involved, that financial burden should be allocated by law. However, the costs of rescue are variable. In the case of a rescue on the High Seas or on the land, the cost of search and rescue of persons may be no greater than the cost arising from the equivalent operation in relation to an aircraft. The risk of requiring a rescue could be taken as a factor in assessing the insurance premium. It may rightly be imposed on the rescuee. In the case of State-owned and manned space objects, the recovery would be against the relevant state agency. For a privately-owned space object, recovery may be against the legal person that owns and manned the object. Although, such a position may discourage the parties at risk of harm from requesting assistance resulting in delays, leading to more complex rescues attendant with higher degrees of risk. In any case, it is untenable for a claim to be made against a spaceflight participant or other fee-paying passenger at the international level. Nonetheless, it is not likely that persons on a space object landing on earth or alighting on the High Seas would be aided by a private space operator before any other aid could reach them. Therefore the issue is more of form than substance.

But the possibility of rescue by a private space operator is much higher where the rescue occurs in outer space, such as in-orbit. The cost of a rescue in such cases will be significantly higher. It is by no means certain that the rescuer would be able to recover as against the rescuee even where the action is expressly permitted by law, for example, where the cost is high enough to drive the latter into liquidation. This could discourage potential rescuers from taking any substantive action. It is submitted that recovery should therefore be allowed directly against the launching authority of the rescuee in such circumstances. This would create a distinction regarding entities against which the rescuer may recover on the basis of the locus of the rescue. Such a distinction can be justified on the variability of the cost of rescue but is ultimately a decision of policy and not law. It may be preferable to permit uniform recovery against the launching authority, regardless of the locus of the rescue. By rendering the launching authority liable for costs, the law would also better harmonise with the existing space law.

If, however, the method of incorporating the obligation is phrased in terms of the private operator's spacecraft being requisitioned for a rescue attempt, then it is arguable that the appropriate entity to claim for recovery is the requisitioning state, which could be the state of registry or the launching state. This possibility depends entirely on domestic law and the terms used in the transposing instrument.

Conclusion

The Rescue Agreement was a great advancement for space law in imposing duties for the protection of astronauts. However, the lack of clarity endemic to international instruments has meant that the duties of private parties have not been clarified and certain situations involving private party rescuers may not come within the scope of the Agreement at all. For primarily humanitarian reasons, it is preferable to impose duties to rescue on the commanders of spacecraft akin to those on the masters of vessels with some modifications suitable to space law. The contentious pure life salvage rule should not be incorporated into space law given that the economic reasons for favouring the rule in admiralty are less potent in relation to space rescues. Information-sharing obligations, requiring commanders to inform their launching authority of any distress call received should be imposed in addition to more substantive duties to rescue the lives of those aboard. No distinction should be drawn between personnel and non-personnel in this regard and the burden for reimbursing for the rescue may fall either on the rescuee or the launching authority of the rescue depending on the circumstances, though the latter is preferable. The duty to render assistance or to rescue should be subject to requirements of practicability and no commander should be required to risk the safety of his own crew or vessel. Where another spacecraft has already responded or where the problem giving rise to the emergency has been resolved, the commander should be freed from his/her obligations. The recovery of another's craft should be left to the laws of admiralty. A new multilateral convention which requires States to ensure the obligation is placed on the commander of spacecraft is one method of imposing these duties and is the most desirable for the sake of certainty and clarity.

---

[*] My thanks to Dr. Gernot Biehler for his comments and advice on an earlier draft. All errors and omissions are my own.

[1] *Per* Cardozo J. in *Wagner v International Railroad Co.* (1921) 232 N.Y.S. 176; 133 N.E. 437.

[2] Jarvis, Robert M., "The Space Shuttle Challenger and the Future Law of Outer Space Rescues," (1986) 20(2) *International Law* 591 at p. 608.

[3] See "Conflicts of Law and Divergences in the Legal Regimes of Air Space and Outer Space" [1963 11 *Recueil des Cours* 332.

[4] See Agreement on the Rescue of Astronauts,the Return of Astronauts and the Return of Objects Launched into Outer Space [1969] 63(2) *American Journal of International Law* 382-385; [1968] 7 ILM 151.

[5] U.N.G.A. Resolution 1348 (XIII), 13 December 1958. See Matte, Nicholas M., *Space Activities and Emerging International Law*, CRASL, McGill University, 1984, pp.94-97; Doolittle, William, "Man in Space: The Rescue and Return of Downed Astronauts," (1967) 9 *U.S.A.F. J.A.G. L. Rev.* 4; Dembling and Arons, "The Treaty of the Rescue and Return of Astronauts and Space Objects," [1968] 9 *William and Mary L. Rev.* at 646 and Riccio, Charles, "Another Step for Mankind: The Agreement on the Rescue of Astronauts and the Return of Objects Launched into Outer Space," (1970) 12 *U.S.A.F. J.A.G. L. Rev.*142.

[6] See Houben, "A New Chapter of Space Law: The Agreement on the Rescue and Return of Astronauts," [1968] XV *Netherlands L. Rev* at p.121 quoted by Ogunbanwo, *infra*, at p.127.

[7] See Resolution 2222 (XXI) para. 4a and Resolution 2260 (XXII).

[8] See Resolution 2345 (XXII) passed 115-0 votes. See generally Cheng, Bin, "The 1968 Astronauts Agreement or How not to make a Treaty" [1969] 23 *Y.B.W.A.* 188 and Ogunbanwo, Ogunsola O., *International Law and Outer Space Activities: Proefschrift Ter Verkrijging Van de Graad Van Doctor in de Rechtsgeleerdheid Aan de Rijsuniversiteit Te Leiden, Op Gezag Van de Rector Magnificus Dr. A. E. Cohen, Hoogleraar in de Faculiteit Der Letteren*, Brill 1975, p.126 *et seq.*

[9] Diederiks-Verschoor, Isabella, *An Introduction to Space Law*, Kluwer Law International, Alphen ann den Rijn, 2008 p.31.

[10] Article 25 of the Convention on International Civil Aviation 1944 [the Chicago Convention].

[11] See Hall, R. Cargill, "Rescue and Return of Astronauts on Earth and in Outer Space," [1969] 63(2) *American Journal of International Law* 197-210.

[12] Reis, A/AC/105/C.2/SR.86, pp.7-8, quoted in full by Ogunbanwo, *supra*, p.132.

[13] See Riccio, *supra*, p.149.

[14] *Ibid*, at p.150.

[15] See Ogunbanwo, *supra*, p.133. Contrast with Art.25 of the Chicago Convention which is limited by the 'practicability' of a rescue.

[16] Doolittle defines 'accidents' as 'unintentional random event without deliberate design' and 'distress' as 'a state of danger or dire necessity,' *supra*, p.7.

[17] Such as France: A/AC/105/C.2/SR.86, p.14. See Dembling and Arons, *supra*, 653; Gorove, "Legal Problems of the Rescue and Return of Astronauts, (1969) 3 *Int'l Lawyer* at p.900 and Riccio, *supra*, p.147.

[18] See generally, Vereshchetin, V.S., "Elaborating the Legal Status of Astronauts," (1983-4) 7 *Hastings International & Comp. L. Rev.* 501.

[19] Gorove, *supra*, at p.898.

[20] Ogunbanwo submits that they do not come within the scope of the Agreement, p. 134.

[21] Van Traa-Engelman, Hanneke Louise, *Commercial Utilization of Outer Space: Law and Practice*, (Martinus Nijhoff Publishers, 1993), p.39.

[22] *Supra*, p.132, citing Dembling and Arons, *supra*, at 646.

[23] *Ibid.*

[24] Article V(2) of the Rescue Agreement.

[25] Article V(4) of the Rescue Agreement.

[26] *Supra*, 33.

[27] Article V(5) of the Rescue Agreement. The 'launching authority' for the purpose of the agreement refers 'to the State responsible for launching, or, where an international intergovernmental organization is responsible for launching, that organization, provided that that organization declares its acceptance of the rights and obligations provided for in this Agreement and a majority of the States members of that organization are Contracting Parties to this Agreement and to the Treaty on Principles Governing the Activities of States in the Exploration and Use of Outer Space, including the Moon and Other Celestial Bodies' under Article 6.

[28] Fawcett, J.E.S., *Outer Space*, Clarendon Press, Oxford, 1984, p.31.

[29] Article 39.

[30] Riccio, *supra*, p.146.

[31] Matte, *supra*, p.95.

[32] Nandakumar, S., "Legal Impasse – Commercialisation of Space Through Reusable Sub-Orbital Launchers" [2004] 47 *Proc of the Coll on the L of Outer Space* 452, at pp.458-9 and Collins, Patrick, "The Regulatory Reform Agenda for the Era of Passenger Space Transportation," *Proceedings of the 20th ISTS*, paper no 98-f-13, available at http://www.spacefuture.com/archive/the_regulatory_reform_agenda_for_the_era_of_passenger_space_transportation.shtml

[33] See Peyrefitte, Léopold and Confé, Patrick, *Droit de L'Espace*, (éditions Dalloz, Paris, 1993) pp.189-203.

[34] For more on this argument see Jarvis, Robert M., "The Space Shuttle Challenger and the Future Law of Outer Space Rescues," (1986) 20(2) *International Law* 591.

[35] See Ratcliffe, James, ed., *The Good Samaritan and the Law*, Anchor Books, 1966 and Long, Patrick J., "The Good Samaritan and the Admiralty: The Tale of a Statute Lost at Sea," (2000) 48 *Buff. L. Rev.* 591. See for example §323(c) of the Strafgesetzbuch.

[36] Duty to Aid the Endangered Act, Vt. Stat. Ann. Title 12 s.519.

[37] Good Samaritan Law, ch. 319, 1983 Minn Law Sess. 2329.

[38] See generally Ames, James, "Law and Morals" [1908] 22 *Harv. L. Rev.* 97, at 112 *et seq*, Lipkin, Robert Justin, "Beyond Good Samaritans and Moral Monsters: An Individualistic Justification of the General Legal Duty to Recue" [1983-4] 31 *UCLA L. Rev.* 252; Linden, "Rescuers and Good Samaritans" (1971) 34 *MLR* 241 and Rose, "Restitution for the Rescuer" (1989) 9 *Oxford J. of Legal Studies* 167.

[39] See Landes, William M., and Posner, Richard A., "Salvors, Finders and Good Samaritans, and Other Rescuers: An Economic Study of Law and Altruism" [1978] *Journal of Legal Studies* 83 at p.104 and Khalil, Elias L., "Are 'Good Samaritan' Laws Desirable?" (1992) 22(1) *Forum for Social Economics* 71.

[40] *Horsley v McLaren* [1972] SCR 441, (The Ogopogo Case); *Phillips v Durgan* [1991] Irish Law Reports Monthly 321. See generally, McMahon , B., and Binchy, W., *Law of Torts*, 3rd ed., Butterworths, Dublin 2000, paras 20.78-20.92.

[41] For the origins of the rule see Severance, Arthur, "The Duty to Render Assistance in the Satellite Age" (2005-2006) 36 *Cal. W. Int'l L.J.* 378. See Art.11 of the Convention for the Unification of Certain Rules of Law Respecting Assistance and Salvage at Sea 1910.; Art. 12(1) of the High Seas Convention 1958; Ch.2, para.2.1.10 of the International Convention on Search and Rescue 1979; Art. 98(1) UNCLOS. The IMO Treaty on the Safety of Life at Sea (SOLAS) as amended in 2006 following the *Tampa* case, also provides for a duty to assist those in distress regardless of their status and circumstances (see Regulation 10). These obligations are incorporated into national law, for examples see Regulation 5 of the Merchant Shipping (Distress Messages) Regulations 1998 (U.K.) (SI No. 1691 *of* 1998), 46 U.S.C. 2304 and s.37 Merchant Shipping (Safety Convention) Act 1952 (Ireland) (No.29 of 1952).

[42] Cheng, Bin, *Studies in International Space Law*, Oxford University Press, 1997, p.191 *et seq*.

[43] See Jarvis, *supra*, pp.618-619

[44] Merchant Shipping (Distress Messages) Regulations 1998 (U.K.) (SI No. 1691 *of* 1998) Regulation 5 (1).

[45] The U.S. Salvage Act provides that a "master or person in charge of a vessel shall, so far as he can do so without serious danger to his own vessel, crew, or passengers, render assistance to every person who is found at sea in danger of being lost" (46 U.S.C.§ 728).

[46] See section 37(1) of the Merchant Shipping (Safety Convention) Act 1952 (Ireland) (No.29 of 1952).

[47] Section 37(3) and (4) of the Merchant Shipping (Safety Convention) Act 1952 (Ireland) (No.29 of 1952).

[48] Ibid., section 37(3).

[49] Ibid., section 37(4).

[50] Ibid., section 36.

[51] Ibid., section 37(5).

[52] Severance, *supra*, p.387.

[53] Article 9 of the Convention for the Unification of Certain Rules of Law respecting Assistance and Salvage at Sea 1910; Art.16 of the IMO International Conventionon Salvage. See Force, Robert, Yiannopoulos, A.N. and Davies, Martin, *Admiralty and Maritime Law*, Vol.1, Beard Books, Washington D.C., 2005, pp.510-517 at p.515 and *Nourse v Liverpool Sailing Ship Owners' Mutual Protection and Indemnity Association* [1896] 2 QB 16.

[54] Landes, William M., and Posner, Richard A., "Salvors, Finders and Good Samaritans, and Other Rescuers: An Economic Study of Law and Altruism" [1978] 22(1) *Journal of Legal Studies* 83 at p.104 reprinted in Wahl, Jenny Bourne, (ed) *Economic Analysis of Tort and Products Liability Law*, Taylor & Francis Group, 1998, p.157 *et seq*, at p. 178

[55] See Jarett, L., "The Life Salvor Problem in Admiralty," (1954) 63 *Yale L.J.* 779; Severance, *supra*, p.388.

[56] The rule is not applied to lives saved in British territorial waters or to lives saved on British vessels: Jarett, *ibid*, pp.782-83.

[57] See Art.13 (1)(e) of the International Convention on Salvage 1989; Salvage Act; 46 U.S.C. § 729; *Grand Union Shipping Ltd. v. London SS Owners' Mutual Insurance Association Ltd. (The Bosworth No.3)* [1962] 1 Lloyd's Rep. 483 (Q.B.D.) and Hodges, Susan, *Law of Marine Insurance*, Cavendish Publishing, 1997, pp.427-428.

[58] *Peninsular & Oriental etc. v. Overseas Oil Carriers* 553 F.2d 830, 1977 AMC 283 (2nd Cir.), *cert. denied*, 434 U.S. 859, 98 S.Cr. 183, 54 L.Ed 2d 131 (1977), case note at p.512 of Force *et al*, *supra*.

[59] See McMahon and Binchy, *supra*, ch.10, p.*203 et seq*; *Hedley Byrne & Co. Ltd. v. Heller and Partners* [1964] AC 465 (HL).

51st Colloquium on the Law of Outer Space, Glasgow, Scotland

Session 3:
WEAPONISATION OF OUTER SPACE IN THE LIGHT OF THE OUTER SPACE TREATY

Chairmen:

Jonathan Galloway

José Monserrat Filho

Rapporteur:

Ricky Lee

IAC-08-E.2.14

THE JOINDER OF TELECOMMUNICATIONS AND GLOBALIZATION

By

Carl Q. Christol
Member, International Institute of Space Law
Distinguished Emeritus Professor of International Law and Political Science
University of Southern California
Los Angeles, Ca. 90089-0044
CarlQC@cox.net

Abstract

Many of the world's interests, values, wants, and needs are being served by the joinder of telecommunications and globalization. Dominant factors in the present era of the Space Age are security and commercialization, particularly telecommunications and remote sensing. It is not surprising that these two factors, which play a dominant role on the land, in the air, and in the sea should also prevail in outer space. Political and legal regimes, essential to the exploration, use, and exploitation of these areas, have been created for the necessary governance of outer space and its natural resources.

1. Introduction

The title to this essay contains the word "joinder." This raises the question of which of the two ideas, globalization[1] or telecommunications arrived on the scene first. In my view telecommunications and the driving forces of multilateralism were the immediate forces leading to the existence of globalization. Without the first presence and use of an advanced communication process there could not have been the evolvement of the multifaceted areas of discourse that contributed to the emergence of globalization. Without the gathering of raw data through remote sensing and the transmission via satellites of analyzed information there would not have been the dissemination of the specialized product of human endeavors.

In a sense the practical aspects of communications and globalization are opposite sides of a coin. The well-being of one requires the well-being of the other. Based on national self-interest a world communications regime has evolved in which interferences in space communication have been kept to a minimum. This has contributed in turn to the condition of globalization with attendant benefits to the members of the human race best able to obtain them.

2. Communications and National Security

With instant communications it is possible to provide assurances to foreign countries with factual statements on matters believed by them to impinge on or affecting their national security interests. Such advice can have a stabilizing influence and can contribute to the harmonious relationships from which globalization acquires its own beneficial qualities. One recent example was the very considerable effort made by the United States prior to and after February, 2008, to keep the world advised, and particularly Japan and China, of the means used in the destruction of a spent American satellite and its fuel tank in orbit above these two countries. This was important since there were fears that the fuel tank may have contained toxic substances and because the destruction was to be achieved through the firing of a missile from an American cruiser in the North Pacific. Monitoring allowed for the providing of assurances that the operation would not result in harms.

Another outer space security concern was reflected in 2006 when it was reported that China had engaged in targeting an American imaging satellite with a ground-based laser and that in 2007 China had conducted a kinetic anti-satellite weapons test on an obsolete Chinese weather satellite resulting in its destruction. Both American and Chinese exercises caused debate as to whether these countries were considering policies referred to as the "militarization" or the "weaponization" of the space environment.[2] Speculation was raised whether such exercises were a prelude to an arms race in space.

3. Terrorism

Terrorists approach globalization from several perspectives.[3] A basic outlook is that globalism constitutes a special threat to the culture and to the religious beliefs and practices of Muslims and that terrorism is an almost exclusive means to combat this perceived threat. Terrorists make full use of modern communication systems. At the same time they show no concern for environmental harms and the need for improvements in commercial relationships. A single-minded obsession with Western influences excuses reliance on unlimited force and violence. The openness and relaxed immigration and border policies of Western States have made them highly pervious to terrorist activities.

Moreover, most Western States place a high priority on the freedom of choice for individuals in such areas as the rights of women and children, family life, and religious and educational preferences. These have been coupled with social, cultural, political, economic, financial, and demographic values. These in turn are associated with Western preferences for the free enterprise system. This has resulted in a Western commercial hegemony. This has produced a comfortable and productive way of life in advanced countries requiring appropriate and balanced security systems

Terrorists perceiving the views and behavior of the West as being so threatening to their way of life have responded with acts of unspeakable horror and brutality directed not at individual persons because of individual animosities but against a non-preselected mass of individuals who just happened to be a convenient target of opportunity.

An additional perspective of terrorists is that scientific and technological advances and capabilities largely originating in the West, such as advanced methods of communication, should be employed by terrorists as they pursue their goals. Thus, a product usefully employed in world communications has been used by terrorists to advance their objectives.

In the existing battlefield the world views of the adversaries are perceived as being so irreconcilable that compromises and accommodations, never capable of being reached, should not even be attempted. Both sides have come to the view that basic differences are so irreconcilable that an interminable and never-ending conflict is permanently forecast.

Those in the West who have come to accept such an analysis also fear that anti-terrorist repressive measures may also have a detrimental affect on the cherished possession of civil and political rights and liberties. The resolution of such concerns will have to be entrusted to the American Constitutional process and to commitments to democracy.

4. United States Support for Globalism

On August 31, 2006 President George W. Bush announced a new American National Space Policy. It was supportive of the concept of globalization as being both in the short term and in the long term interests of the United States. In passing it should be noticed that globalism has become such an all pervasive concept relating to space activity that a mere pronouncement opposing it would have no or little value. The 2006 policy statement reads: "The United States is committed to the exploration and use of outer space by all nations for peaceful purposes, and for the benefit of all humanity. Consistent with this principle, 'peaceful purposes' allow U.S. defense and intelligence-related activities in pursuit of national interests." [4] In keeping with this pronouncement the United States stated that it would "seek to cooperate with other nations in the peaceful use of outer space to extend the benefits of space, enhance space exploration, and to protect and promote freedom around the world."[5]

5. The Role of International Organizations

The actions of many international organizations have materially influenced the availability to terrorists of effective communications. The UN Security Council in Resolution 1373 (2001) called for the creation of a Counter-Terrorism Committee to be administered by an Executive Directorate that has been able to call to the attention of participants subjects suited to cooperative efforts. On December 18, 2001 the Defense Ministers of NATO initiated measures seeking to compose a "Military Concept for Defense against Terrorism," allowing for case by case actions against terrorists. Its focus was on deterrence. The G-8, with its principal concerns for such subjects as energy sources, global warming, and the Middle-East conflict, has also voiced concerns respecting terrorist activities with a special reference to deterrence.

The United Nations is a living example of the benefits derived from globalization. Much of the formal activity is carried out by ECOSOC. It has a Department of Economic and Social Affairs which measures the rapid growth of economic interdependence. All of the specialized agencies of the UN are engaged in the progressive development of globalization. Although not directly concerned with deterring terrorism they have through their activities contributed to the beneficial conditions that have been opposed by terrorists. Through the ongoing activities of the specialized agencies over time the present social conditions of today's terrorists may be modified and they will come to entertain more modern perspectives.

6. Conclusion

Continuing advancements in science and technology will contribute to many new outer space activities. The transmission of new ideas will have a striking impact on human behavior with the enhancement of globalization being a significant factor. These developments will be employed to advance the common interests of mankind.

It may be supposed that scientific and technological progress may also be employed by terrorists as they pursue their stated goals. Terrorists cannot be prevented from seeking access to such advanced communication equipment and the "know-how" required to make use of it.

One response should be that terrorists and their supporters be brought under such detailed and ongoing scrutiny that they would be denied opportunities to make use of all of the hardware that is essential for their successful operations. If terrorists are to be taken at their word, and I deem it dangerous not to do so, their long-term plans and programs must be brought under the most careful surveillance. This will require highly coordinated efforts constructed on a globalized basis with the efficient and effective use of international communications.

Notes

[1]. DAVID J. BEDERMAN, GLOBALIZATION AND INTERNATIONAL LAW (2008).
[2]. Militarization of the space environment contemplates the presence of satellites designed to meet defensive purposes, such as gathering data relating to possible offensive activity on the part of another State. The satellite is considered to be "peaceful" since it is an information gatherer, with such information to aid it in determining the hostile intentions of an adversary. Weaponization refers to the emplacement in the space environment of a satellite designed to engage in hostile activity against another State. The intention of the launcher is critical. In the one instance there is no hostile intent. In the other there is an aggressive or hostile intent.
[3]. CARL Q. CHRISTOL, INTERNATIONAL LAW AND U. S. FOREIGN POLICY, Second, Revised Edition, Chapter 12 (2007).
[4]. U. S. National Space Policy, August 31, 2006, p. 1.
[5]. Ibid.

IAC-08-E8.2.10

CONTROLLING BALLISTIC MISSILE DEFENSE AND ANTI-SATELLITE MISSILES: SHOULD WE LOOK ELSEWHERE OR RATHER AMEND ART. IV OST?

Abhyudaya Agarwal
NUJS, India
abhyu5u@yahoo.com

Kaushika Vuppala
NUJS, India
vuppala.kaushika@gmail.com

Ramanuj Mukherjee
NUJS, India
ramanujwrites@gmail.com

ABSTRACT

Weaponisation and arms race in the outer space is a real threat facing humanity. While the cold war has been left behind, the threats once associated with it are still palpable. It is a grim fact that in absence of a legal regime countries are sometimes forced to join avoidable arms races to protect their strategic and regional interests. It is therefore natural that there has been strong international opinion in favour of further development of international law in this matter. However, the feasibility of the same can be questioned in face of consistent US objection to any further curbs on use of space. However, all the space exploring countries are obligated to respect their obligations under the Outer Space Treaty, the article IV of which places substantial restriction on weaponisation of outer space as signatory to the same.

There are two kinds of challenges that the existing regime of international space law is facing with respect to weaponisation of outer space, and these challenges are thrown up either by development of international politics or development of newer technology. This paper shows that reasonable interpretation of Art. IV adequately meets these challenges.

INTRODUCTION

Outer space holds unlimited possibilities for humankind. As our technology, population, imagination and attendant needs grow, it is not difficult to predict that resources available on earth itself would not be able to meet all the demands of human civilization. Once what is available on earth is harnessed optimally, or maybe even earlier, when it is economically and technologically feasible, human race will turn to the outer space – to explore, conquer and expand. Thus, outer space is a symbol of hope of prospects of the future, a dream that each of us nurture deep down.

However, outer space, which is so intricately connected to the survival and development of humanity in the coming centuries is continuously threatened today by the prospect of extensive weaponisation of the outer space that will render it a battleground where one must tread every moment with the fear of destruction of entire humanity, or even worse – a graveyard full of debris of weapons and destroyed space objects. This threat shatters the visions of outer space as a peaceful province of the mankind,[1] open to scientific and peaceful exploitation,[2] one the mankind today wants to leave behind for the future generations as a glorious heritage.

As strategic initiatives taken by different countries with capability of developing and deploying weapons in the space force

[1] Treaty On Principles Governing The Activities Of States In The Exploration And Use Of Outer Space, Including The Moon And Other Celestial Bodies, Art. I

[2] *Id.,* Preamble.

the international community to face the question of securing the outer space from military use that would harm long term interests of humanity, it seems the only long term solution lies in international cooperation and confidence building. This process of international cooperation must lead to development of an international legal regime governing outer space, which will be recognized through international instruments such as charters, treaties and declarations embodying a legal framework and the positive obligation of different countries in protecting outer space from weaponisation. Such development will be a judgment given by humankind in favour of peaceful enjoyment of the benefits the outer space offers as opposed to running a risk of 'assured destruction'. In the long run, international space law has to assume the prime role in limiting military activity in outer space, including possible conflicts *a la* Star Wars or an arms race.

As international space law is at present only in a nascent stage it is possible for a debate to arise over its application to specific issues which have already arisen or are likely to arise in the future, such as the use of space for *transit* of a weapon or military warship, engaging in conflict in outer space, etc. Nevertheless, it is already settled that the broad issue of weaponisation of space falls within the ambit of the subject matter of discussion in international space law.

This paper shall explore the importance of Art. IV of the Outer Space Treaty in international law and politics as it stands today. Has the understanding and context of Art. IV changed since its inception? What are the potentials of this provision today? Part I of the paper Ballistic Missile Defense ("BMD") systems and Anti-Satellite ("ASAT") Missile systems will be discussed to explore the ways and purposes, in and for which the outer space maybe weaponised. Part II briefly deals with existing law on armaments and emerging international opinion in relation to Ballistic Missile Defense and Anti-Satellite Missiles to appreciate the real value of Art.IV of OST in prevention of weaponisation of space. Part III discusses how Art.IV of the OST may successfully cover BMDs and ASATs.

PART I: CONNECTING BALLISTIC MISSILE DEFENSE WITH OUTER SPACE

For any weapon system to attract the application of Space law, it must have a connection with outer space. This part illustrates whether Ballistic Missile Defense ("BMD") and Anti-Satellite Missile ("ASAT") systems are amenable to international space law at all. If yes, whether space law applies to all BMD and ASAT systems or it excludes some of them is shown. This question is answered by reference to the definition of outer space and by resorting to several classifications based on existing weapons or those which may be used in future.

The definition of Outer Space

There are conflicting definitions on what would constitute outer space. In the 1980s, the former USSR had favoured an arbitrarily fixed distance of 110 km while the US refrained from a fixed delimitation. Some countries have suggested that the *function* of a particular object or activity should determine whether international space law *should apply* to that object or activity.[3] Another definition recognises that outer space begins where the atmosphere ends, and therefore an activity at exoatmospheric levels would be occurring in outer space. While there is no formal delimitation of outer space, it has been argued that state practice indicates an evolving rule of customary international law that earth orbit may be considered part of outer space.[4] For the purpose of this

---

[3] *See* UN Doc. A/AC.105/C.2/L.141/Add.6 (April 7, 1983).

[4] Burrus M. Camahan, "The Legality Of A High-Technology Missile Defense System: The ABM And Outer Space Treaties," 78 Am. J. Int'l L. 418 at 423. The geostationary orbit based classification has been emphasized by the United Report of the Committee on the Peaceful Uses of Outer Space, G.A.O.R. A/61/20, Para 206. *See also* Boleslaw Adam Boczek, International Law: A Dictionary, Scarecrow Press, 2005, *arguing* that state consensus to overflight by satellites and space

paper, a functional definition is adopted. Missile which are stationed or at any point of their trajectory or transit, pass through exoatmospheric levels are treated as being within the domain of space law as per the functional definition.

Understanding Ballistic Missile Defense and Anti-Ballistic Missiles

There is a difference in the terminology adopted in respect of missile defense in the Cold War era from that employed today. Whereas the former involved the usage of the term 'Anti-Ballistic Missile', the term 'Ballistic Missile Defense' is more frequently employed today. Therefore, the terminology requires clarification here. A Ballistic Missile Defense ("BMD") need not always employ missiles to destroy incoming intruder missiles. It may use advanced laser technology or particle-beam weapons, whether ground based, airborne, or space based. Even if such a system is missile based, an increase in the role of non-missile components in surveillance and detection of enemy missiles may have influenced the shift in terminology. Nevertheless, for convenience, a missile based BMD has been called an ABM in this paper.

The Classification of BMDs and ASATs: Connection with Outer Space

The term Anti-Ballistic Missile covers a bevy of weapons and it is far from a homogenous category. They are discernible by resorting to classifications based on features, characteristics, capacity and use. Such classification serves two purposes in this paper. Firstly, it helps to know which categories of ABMs may come within the discourse of space law. Secondly, it further enables one to accommodate certain kinds of weaponry which are not in existence at present, but may be developed in *future* in the discussion so that legal discourse is not flabbergasted by at least the foreseeable developments in technology. ASATs are functionally different from ABMs in the way that they are used against satellites as opposed to missiles. However, notwithstanding technical differences they will be amenable to most of the classification parameters such as base, range, speed, capacity and purpose. Further, the use of ASATs is *prima facie* comes within the scope of space law as they are deployed against satellites orbiting the earth and the definition of outer space is based on earth orbit. All satellites, on account of being placed in some or the other earth orbit, are within the space law.

The primary classification of Ballistic Missiles is based on the distance at which they can destroy a target. On this basis, Ballistic Missiles are classified into Intercontinental Ballistic Missiles ("ICBMs"), Intermediate Range Ballistic Missiles ("IRBMs") and Short Range Ballistic Missiles ("SRBMs"). Missiles and ABMs are also often classified into strategic, theatre and tactical depending on target range, speed and purpose. Strategic Ballistic Missiles are long-range and travel at the highest velocity.[5] Theater Ballistic missiles have a medium range and travel slower than strategic Ballistic Missiles.[6] They are aimed at destroying a limited localised area of the target territory, say, a particular defence installation. Tactical ballistic missiles are short-ranged and have the least velocity.[7]

ABMs, just like any other missile, may be further classified on the basis of the place they are stationed or launched from, into sea-based (submarine or otherwise), air-

vehicles of other states over their territory has created the rule that the path of transit of such objects constitutes outer space.

[5] The definition in the ABM Treaty classifies ABMs based on the nature of the Ballistic Missile they are intended to destroy. It restricts its application to *strategic* ABMs only, leaving out the other categories.

[6] For example, Japan has a Theater Ballistic Missile Defense System, which it has acquired with the co-operation of the US.

[7] It is to be noted that tactical ballistic missiles do not have an exoatmospheric phase as they are designed to hit targets within small distances. Therefore, they may not attract international space law.

based, space-based, or mobile land-based.[8] Space-based ABM systems and non-missile BMDs shall automatically come within the purview of space law.

BMDs may be classified based on the stage at which they strike their target, which may be at its boost, mid-course or the terminal stage. The mid-course phase of all ballistic missiles, except tactical ballistic missiles involves an *exoatmospheric* phase, implying a transit through outer space. The terminal phase is when it *re-enters* the atmosphere.

They may be classified based on the mode in which they destroy the target, that is, whether they rely on the kinetic energy generated by impact, or whether they carry payloads (which may be explosive, biological or chemical). Further, they may be distinguished on the basis of the *number of warheads* that they possess, so that they may destroy single or multiple targets.[9]

The above criteria of classification may be illustrated by examples. The Ground Based Mid-Course Defence of the US is an example which is an *exoatmospheric* missile, striking the target at its mid-course stage and relies on its *kinetic energy* to destroy its target by *impact only*.[10] The Airborne Laser is a non-missile BMD which is air-based (not space based) and strikes at a missile at its boost phase.[11]

From the above discussion, the deployment of ABMs may attract the application of Outer Space Treaty in one of several ways. Firstly, the ABM system itself may be space-based, in which case the ABM may be stationed and launched from outer space. Secondly, ABMs may involve the use of outer space in the course of transit. If they attack a ballistic missile in the boost phase, at least a portion of their trajectory would be through outer space. Thirdly, it may target an object in the outer space, as in the case of a strike during the mid-course phase of a ballistic missile. The kinetic impact or explosion, in such cases, may also result in space debris, which shall however, not be considered here.

Ballistic Missile Defense: Is It Only Defensive?

A preliminary argument that may pre-empt any further investigation as to the legality of BMDs is based on the notion that they are merely defensive in nature. In fact, this was the argument used to market BMDs. Arms control through BMD was equated with nuclear disarmament.[12] Thus BMDs were proclaimed as not in any way being antagonistic to the discourses on disarmament or peaceful uses of space. However, these statements belie the truth and have been condemned for 'stealing the language and cause' of the peace process.[13] A defense-based justification of a BMD system is fallacious. Firstly, an ABM system has been defined as, "[A] system to counter *strategic ballistic missiles* or their elements in *flight trajectory*... (emphasis added)"[14] This is clearly a purpose-based definition as there is nothing in the weapons system which makes it inherently defensive. Such justification is merely based on the *expected* use of the missile, and *not* on its capability. Depending on the necessity and will it can be very well used for offense. Unless there is an in-built check that prevents it from being so used, it is misleading to identify such a weapons system as purely defensive and ignore its

[8] *See* Treaty Between The United States Of America And The Union Of Soviet Socialist Republics On The Limitation Of Anti-Ballistic Missile Systems ("ABM Treaty") 1972, Art. V. The treaty is not in force now.

[9] These features are very important in determining WMD capability of such missiles. Refer to Part IV for more elaborate discussion.

[10] Missile Defense- Worldwide, 5th Edn., p. 27, available at http://www.mda.mil/mdalink/pdf/bmdsbook.pdf accessed 20/07/2008.

[11] *Id.*, at 19.

[12] *Id.*

[13] Gordon R. Mitchell, Japan-U.S. Missile Defense Collaboration: Rhetorically Delicious, Deceptively Dangerous 25-WTR Fletcher F. World Aff. 85 at 90.

[14] Treaty Between The United States Of America And The Union Of Soviet Socialist Republics On The Limitation Of Anti-Ballistic Missile Systems (hereinafter "ABM Treaty"), Art. II.1.

possible offensive use.[15] In fact, the same ABM technology may nevertheless be adapted for offensive purposes. Secondly, ABM systems, that is, missile based BMDs, are materially no different from other kinds of ballistic missile systems as they generally comprise of similar or the same components- interceptor missiles, launchers, radars and satellite sensors.[16]

PART II: LEGAL FRAMEWORK OUTSIDE THE OST: APPLICABILITY OF SPECIFIC ARMS CONTROL INSTRUMENTS

The ABM Treaty of 1972 between the US and the former USSR restricted the deployment, development and testing of ABMs in the Cold War era. Although the treaty was a bilateral measure, it was to a large extent sufficient in controlling the proliferation of ABMs and preventing an arms race at that time. It expressly prohibited the parties from testing, developing or testing *space-based* systems or components.[17] Following the break-up of the former USSR there had been argument as to whether the treaty still survived between the US and the successor states of the USSR. Such speculation as to the status of the ABM Treaty was laid to rest when the US formally withdrew in 2001, thereby indicating that it treated the treaty as being in force even after the break up.[18]

The Strategic Offensive Reductions Treaty, 2002 (also known as the Moscow Treaty) signed between the USSR and the US after the latter's withdrawal from the ABM Treaty seeks to control *strategic nuclear warheads* of the state parties.[19] Therefore, it brings within its ambit those ballistic missiles (and ABMs) which use nuclear warheads. Unfortunately, the Moscow treaty is also a bilateral instrument. Its application is more limited as it shall last only till 2012, unless extended by further agreement by either parties[20] and can be terminated by either party after giving a three month notice, without requirement of any specific reasons, whether or not they are related to a fundamental change in circumstances or otherwise.[21] The application of the Strategic Arms Reductions Treaty ("START I"), which extensively deals with ballistic missiles, including those deployed from space-launch facilities,[22] is another bilateral treaty between the US and Russia. A space-launch facility is defined as a specified facility from which objects are delivered into the upper atmosphere or space using ICBMs or SLBMs ("Submarine Launched Ballistic Missiles"),[23] and may not necessarily be space-based. The treaty shall remain in force until 2009 unless superseded by a *subsequent agreement* or *extended.*[24] By virtue of the Thirtieth Agreed Statement to the Treaty, the parties have *not foreclosed* the possibility of launching ICBMs and SLBMs *for delivering objects into the upper atmosphere or space* from waterborne vehicles other than submarines

---

[15] In fact, seemingly harmless civilian space-launch programs can be used to disguise potentially dangerous ballistic missile programs. *See* Paul Kerr, 'Code of Conduct Aims to Stop Ballistic Missile Proliferation', available at http://www.armscontrol.org/act/2003_01-02/icoc_janfeb03 accessed 01/08/08.

[16] Art. II.1 of the ABM Treaty identified only interceptor missiles, launchers and radars as components of an ABM system at that time. Now, such a system may include satellite-based sensors and guiding devices. E.g., Ballistic Missile Defense Space Sensors mounted on low earth orbit ("LEO") satellites. *See* Missile Defense-Worldwide, 5th Edn., p. 23, available at http://www.mda.mil/mdalink/pdf/bmdsbook.pdf accessed 20/07/2008.

[17] ABM Treaty, Art.V.

[18] In fact, a concern had been raised whether the US had actually withdrawn in terms of Art. XV of the ABM Treaty which permitted withdrawal in the event that extraordinary events related to the subject matter of this Treaty have jeopardized its supreme interests, thereby recognising the principle of *rebus sic stantibus* as embodied in Art.62.1 of the Vienna Convention on the Law of Treaties, 1969. *See* Frederic L. Kirgis, 'Proposed Missile Defenses and the ABM Treaty' available at http://www.asil.org/insights/insigh70.htm accessed 23/07/08.

[19] Strategic Offensive Reductions Treaty, 2002, Art.I.

[20] *Id.*, Art.IV.2.

[21] *Id.*, Art.IV.3.

[22] *See* The Treaty on Limitation of Strategic Offensive Arms, Arts. II, III, IV and V.

[23] START I Treaty, Definitions Annex, 104. (43) at http://www.fas.org/nuke/control/start1/text/defini.htm#104 accessed 20/07/08.

[24] The Treaty on Limitation of Strategic Offensive Arms, Art. XVII(2).

or from airplanes other than heavy bombers or former heavy bombers, although such use would be in after the party states reach an agreement.[25] ASATs are not regulated under the above regime, unless they are clothed with nuclear warheads or are designed to impact their targets only with the help of ICBMs and SLBMs from earth.

The Start I Treaty states that the parties shall not use ICBMs or SLBMs for delivering objects into the upper atmosphere or *space* for purposes inconsistent with existing international obligations undertaken by the Parties,[26] thereby referring to obligations of parties under international space law.

Beyond Law: the changed geo-political challenges

It is to be noted here that control of arms race and weaponisation of outer space through bilateral treaties were effective and justified in the cold war era, when the only countries with adequate techno logy or the capability of developing such capability were USSR and US. The end of cold war has markedly changed the face of geo-politics and diplomacy. Naturally the question that comes to one's mind is whether the same bilateral mechanism to contain arms race as well as weaponisation will work today.

Earlier, the dark threat of Mutually Assured Destruction was sufficient for both the sides to not engage in further weaponisation as new weapons would have given minimal strategic advantage if at all. United States tried to change the situation by inventing ABMs, but USSR responded with Multiple Independent Re-entry Vehicle and thus the threat of MAD prevailed. This mutual tension also forced both the countries to recognise the interest in not weaponising the outer space which would have augmented the threat under which they lived. This also influenced both US and USSR to bring about international control mechanisms such as the ABM treaty at that time. MAD was a dead end for arms race; when both the countries could destroy the other in retaliation to any attack, developing or deploying new weapons, be it in the outer space, didn't have much incentive compared to the augmentation of threat it would have brought along.

However, with the end of Cold War era, things took a different turn. Firstly, the relationship between USA and Russia is no more of the same tension as it was, and the threat of MAD has been left behind, having lost its relevance.[27] Even if any threat existed, both the countries have agreed to reduce their weaponry through the Moscow Treaty and other confidence building measures. The arms race is no more between Russia and USA.

However, this does not mean that arms race has become a non-existent threat in the 21st century. Rather now it is regional interests that propel arms race. Competing strategic powers such as India-Pakistan (both the countries having nuclear capability have been testing ballistic missiles and India is pursuing its own ABM development programme) have been engaged in a de facto arms race for the past decades.[28] So has been the case with

[25] *See* START I, Annex, Agreed Statements available at http://www.fas.org/nuke/control/start1/text/agreed.htm#30 accessed 22/07/08.

[26] START I, Art.V.15.

[27] One reason given by the US for withdrawing from the ABM Treaty was that it was no longer required as from 1991 its relationship with Russia had been co-operative rather than adversarial. *See* Richard Boucher, U.S. Department of State, Text of Diplomatic Notes Sent to Russia, Belarus, Kazakhstan, Ukraine (Dec. 14, 2001) at http://www.usembassy.org.uk/acda283.htm. Secondly, the US succeeded in quelling the anxiety of Russia and China at the time of its withdrawal from the ABM Treaty in 2001 by simply stating that its withdrawal and missile defense buildup was not directed at them. *See* Kerry M. Kartchner, 'Missile Defenses And New Approaches To Deterrence', US Foreign Policy Agenda, available at http://www.iranwatch.org/government/US/DOS/us-dos-foreignpolicyjournal-0702.pdf accessed 21/07/08.

[28] India plans to have a BMD system at the latest by the middle of the next decade. *See* 'India Developing Ballistic Missiles To Destroy IRBMs, ICBMs' available at http://www.india-defence.com/reports-3683 accessed 20/07/08. Another site states the date to be 2015. *See* Neha Kumar, 'India Ballistic Missile Defense Capabilities & Future Threats', at

countries such as Israel and the Arab nations and North Korea and Japan.[29] These countries having strategic ambitions fuelled by their regional interest as well as security concerns due to hostile neighbours have developed advanced weaponry including ballistic missiles and have proceeded to either develop or acquire BMD systems. ASAT capability also has expanded beyond the two superpowers with China testing an ASAT missile in 2007.[30] It is a matter of fact that similar strategic interests will propel and at times compel nations to engage in an arms race, in order to maintain a balance of power. The possibility that such an arms race may extend to outer space cannot be ruled out.

Therefore, it is at this juncture that the requirement of international law to ensure that there is no necessity of an arms race is felt. Binding obligations of nations to one another arising from an international law regime can successfully thwart the extension and amplification of such moves of one-upmanship to weaponise or strike targets in space.

Lastly, the risk of proliferation of weapon systems including that of missile technology and nuclear weapons is much higher today and many more countries hold keys to the technology.[31] This leads to a higher danger of misuse or irresponsible use of dangerous weapons or deploying of the same in the space.

A new legal framework: feasible?

The Herculean task facing the international community now is to come up with a legal framework to meet these challenges. International opinion to this effect is considered below.

The Prevailing Opinion of Nations

Since 1981, the General Assembly has consistently adopted every year, a resolution on the prevention of an arms race in outer space (PAROS).[32] The resolution reaffirms the belief of states that the legal regime applicable to outer space does not of itself guarantee the prevention of an arms race in that environment, and that there is a need to *consolidate*, *reinforce* and *enhance* the effectiveness of the existing regime.[33] It states that further measures are required to prevent an arms race in outer space[34] and calls upon states, especially to those with major space capabilities to proactively contribute to such an objective to refrain from actions against this objective.[35]

New Space Treaty proposed jointly by China and Russia

China and Russia had jointly drafted a new Treaty on the Prevention of the Deployment of Weapons in Outer Space, the Threat or Use of Force Against Outer

http://www.indiapost.com/article/perspective/1625/ accessed 20/07/08.

[29] Japan has built a Theater Ballistic Missile Defense system with the aid of the US. *See* Gordon R. Mitchell, Japan-U.S. Missile Defense Collaboration: Rhetorically Delicious, Deceptively Dangerous, 25 Fletcher F. World Aff. 85, 89 (2001). US also plans to provide a BMD to Poland. *See* 'US delegation to woo Czechs at missile shield seminar', at

http://www.spacewar.com/reports/US_delegation_to_woo_Czechs_at_missile_shield_seminar_999.html accessed 21/07/08.

[30] 'Chinese Anti-Satellite Capabilities' available at http://www.globalsecurity.org/space/world/china/asat.htm accessed 21/07/08.

[31] Although a legal framework to control proliferation of Ballistic Missile Technologies exists in the form of two instruments- the Missile Technology Control Regime ("MTCR") and the Hague Code of Conduct on Ballistic Missiles, it is incomplete in its coverage. The MTCR, in its attempt to restrict proliferation of missiles is limited in its application to export control measures only, and has only 33 members. On the other hand, the Hague Code of Conduct for Ballistic Missile Proliferation of 2002 restricts proliferation by requiring states to show *restraint* in their *own missile programs* and has a membership of over 120 states as per UNGA Res. A/RES/60/62. However, the Hague Code is only politically binding but does not provide for recourse to any legal mechanism in the event of violation of any of its requirements. *See generally* Paul Kerr, 'Code of Conduct Aims to Stop Ballistic Missile Proliferation', available at http://www.armscontrol.org/act/2003_01-02/icoc_janfeb03 accessed 01/08/08.

[32] Li Daoyu, Prevention of the 'Weaponization of and an Arms Race in Outer Space: An Urgent Task With No Time to Delay' available at http://www.fmprc.gov.cn/eng/wjb/zzjg/jks/kjfywj/t189569.htm accessed 03/08/08. *See* http://disarmament.un.org/vote.nsf for a list of the sessions where the PAROS resolution has been passed.

[33] 'Prevention of an Arms Race in Outer Space', Para 2, UN GA A/RES/62/20,

[34] *Id.*, Para 3

[35] *Id.*, Para 4.

Space Objects before the Conference Disarmament ("CD") in a Working Paper.[36] The same has been presented to the CD in February 2008.[37] They noted the absence of any legal impediments to the deployment of anti-satellite weapons.[38] The parties to the new treaty would be under the obligation not to place in orbit around the Earth any objects carrying *any kinds of weapons*, install such weapons on celestial bodies, or *station* such weapons in outer space *in any other manner.*[39] This is in contrast to the OST which merely prohibits placing in orbit *weapons of mass destruction* only.[40] Parties to the new treaty would not to resort to the threat or *use of force against outer space objects.* This provision is palpably aimed at regulating the use of ASATs. The draft treaty, unlike some other arms control treaties already discussed, is to be of permanent duration and shall permit withdrawal, like the ABM Treaty, only if extraordinary events, related to the subject matter of this Treaty, jeopardize the supreme national interests of a party state.[41]

The stance of the United States

The US withdrew from the ABM Treaty in 2001 and actively started pursuing a National Missile Defense program. The US National Space Policy of 2006 stated that the United States should "preserve its rights, capabilities, and *freedom of action* in space" and "*oppose* the development of *new legal regimes or other restrictions that seek to prohibit or limit U.S. access to or use of space.*"[42] The reason for rejecting any such regulation as given by US administration was that there was no need to do so in the absence of an arms race.[43] The US has also been consistently voting against the PAROS resolution in the General Assembly from 2005 to 2007.[44] The US has remarkably increased its expenditure on its National Missile Defense Program to $12 billion in the year 2008, which is three times its expenditure on ABM systems in any year during the Cold War,[45] justified by the US as being based on a perceived threat of a missile attack from Iran.[46]

In light of the above discussion, it is again emphasised that the legal and political mechanism as existing is not sufficient to deal with the global crisis that the threat of weaponisation of outer space poses. While a majority of countries including 2 major space superpowers, that is Russia and China would like to see a new legal regime for controlling military activity in space to come into being, the US is not agreeable to put any further restrictions on its potential military use of space rendering all efforts to develop any new international legal framework unviable and illusory. Unless the United States subscribes to such a treaty, the other space exploring nations would also not want to foreclose their options. Thus, at the moment it seems that further development of international law with respect to weaponisation towards prevention of weaponisation of outer space or arms race for that matter, has ended up in a blind alley. While the way forward is blocked, in such a situation, one must fortify the existing limitations on weaponisation and arms race. Many find

---

[36] 'Russia-China CD Working Paper on New Space Treaty', June 2002, available at http://www.acronym.org.uk/docs/0206/doc10.htm accessed 02/08/08.

[37] Prevention of an Arms Race in Outer Space, available at http://www.fmprc.gov.cn/eng/wjb/zzjg/jks/kjlc/wkdd/t410757.htm accessed 02/08/08.

[38] 'Russia China Working Paper', *supra.*

[39] Draft Treaty on the Prevention of the Deployment of Weapons in Outer Space, the Threat or Use of Force Against Outer Space Objects, Art.III.

[40] Outer Space Treaty, Art.IV.

[41] *Id.*

[42] U.S. Office of Sci. & Tech. Council, United States National Space Policy 1-2 (2006), available at http://www.ostp.gov/html/us%20National%20Space%20Policy.pdf.

[43] *Id.*

[44] Developing a new outer space treaty,available at http://www.reachingcriticalwill.org/legal/paros/ostreaty.html accessed 01/08/08.

[45] Joseph Cirincione, 'Incredibly Shrinking Missile Threat', Business & Economy, 26 June 2008 at p.70.

[46] *Id.*

the obvious ray of hope in Article IV of the Outer Space Treaty.

Part III: Implication of Art. IV of OST on BMD, ABMs, and ASAT: Legal Issues

Drafted by the United Nations Committee on Peaceful Uses of Outer Space (COPUOS), the *Treaty on Principles Governing the Activities of States in the Exploration and Use of Outer Space, Including the Moon and Other Celestial Bodies* (Outer Space Treaty) is the most comprehensive treaty governing use of the outer space.[47] It covers all aspects of exploration and use of outer space. The treaty has been signed by one hundred seven nations, including all of those with active space programs. It entered into effect on October 10, 1967. Art. IV of the Outer Space Treaty is a crucial legal instrument that envisages peaceful use of the outer space and aims at curbing international arms race in outer space. In fact it has been has been praised as the most important arms control document after CTBT.[48]

Analysis of the text of art. IV

Art. IV has two paragraphs. The first one is written in a negative language- it deals with nuclear weapon or *any other kind of weapons of mass destruction* and expressly prohibits

a) placing of such weapons on orbit around earth, b) installing such weapons on celestial bodies and c) stationing such weapons in any manner on the outer space. The 2nd paragraph of the article, a) confers positive obligations on the signatories to use the moon and other celestial bodies exclusively for peaceful purposes and b) forbids establishment of military bases, installations and fortifications, the testing of any type of weapons and the conduct of military manoeuvres on celestial bodies. The remainder of the paragraph excludes the use of military personnel for scientific research or for any other peaceful purposes and of any equipment or facility necessary for peaceful exploration of the Moon and other celestial bodies from the prohibition. It is only the first paragraph that is relevant to us for it places a bar on certain kind of weapons and this may, if at all, be applicable to BMD, ABMs, and ASATs.

Deploying BMD/ASAT in a space station, on the orbit around earth or other celestial bodies, or on a celestial body

Art. IV expressly prohibits such deployment *only if* BMD or ASAT carries nuclear weapons or if they will fall within the category of weapons of mass destruction. It is easy to identify BMD or ASAT involving nuclear weaponry as banned by this provision; however, the issue of what consists of weapon of mass destruction can be contentious and maybe interpreted in different ways as the definition is not provided in the document itself. If it is to be understood that conventional weapons are permissible, would all non-nuclear weapons be allowed on outer space? It is clear from the text of the provision that the intention of the parties is to ban more than just nuclear weapons. Therefore, one must look for a definition of weapons of mass destruction elsewhere.

According to a resolution adopted by the U.N. Commission for Conventional Armaments, weapons of mass destruction include lethal chemical and biological weapons having characteristics comparable in destructive effect to those of the atomic bomb.[49] Thus, it is clear that any missile that carries nuclear, biological or chemical warhead will come under the prohibition of Art. IV. However, it is not reasonable to restrict the definition to these three kinds of weapons only as other kinds of weapons which may cause mass destruction can be developed. The ideal of Art. IV was not to only prevent existing

---

[47] Jonathan N. Halpern, Anti-Satellite Weaponry: The High Road to Destruction, 3 B.U. Int'l L. J. 167, 180 (1985)

[48] Gorove, Arms Control Provisions in the Outer Space Treaty: A Scrutinizing Reappraisal, 3 Ga. J. Int'l & Comp. L. 114 (1973).

[49] Resolution Adopted by the Commission for Conventional Armaments, U.N. Doc. S/C.3/32 (1948)

kind of weapons from finding their way into space, but also preventing future weapons of mass destruction, and that is the reason it does not provide a list of weapons that are prohibited. For instance, if a nation deploys a weapon in the space that destroys thousands of lives by the use of radiological means or a weapon based on nanotechnology which causes similar levels of destruction, it would certainly defeat the purpose of Art. IV.

Several UN General Assembly Resolutions[50] and reports adopted by it[51] acknowledge that there can be new types of weapons of mass destruction other than these three types. Such weapons will be identified by the criterion that they would have similar effects in terms of destruction. Thus, the real parameter that decides whether a weapon is a weapon of mass destruction is the level of damage it causes. No one has defined the exact threshold of destruction that would qualify a weapon as WMD. However, some scholars take a inclusive approach towards defining WMD. One such scholar states: "While there is no indication in the Treaty as to how many people must be affected to constitute a weapon of mass destruction, a group of 20 to 30 people or less probably would not constitute such a mass. If on the other hand, bacteriological and chemical weapons were used, even against a small group, then these weapons would seem to fall under the category of weapons of mass destruction."[52]

This actually emphasises it is the *capability of destruction* of a weapon that determines whether it is WMD or not. Therefore, any BDM or ABM that can be used in such a way to cause wide spread damage can be argued to be covered under this article.

At the first impression, most BMD systems, ASAT or ABMs are to be directed towards a missile or satellite and not towards masses of people. However, in reality such missiles are quite capable of being used like conventional missiles if needed; and thus, many of them have capability of inflicting damage that would qualify them as WMD.

Payload based missiles that depends on an explosion caused at time of impact to destroy the target will more easily fit into the category of WMDs if the explosion is big enough to kill a large number of people. However, there are still other non-payload missiles that use the thrust of its impact[53] and weapons that depend on laser or particle beams to destroy missile or satellites.[54] Are such weapons permitted by Art. IV to be deployed in the space?

In reality, such weapons also can be used upon strategic targets such as nuclear reactors, gas stations, skyscrapers with hundreds of floors to cause death of multitudes of civilians. If used in such a way, all these weapons have WMD capability and thus Art. IV should be applicable to them prohibiting deployment of such weapons in outer space.

Another pertinent question is whether the moon is covered by the first paragraph of art. IV. One may argue that while in every other places in the article the moon has been separately mentioned along with other celestial bodies, only in this part moon is not mentioned. Is the intention to leave out moon from the application of this part? Such a proposition would not be consistent with the objectives and purpose of the article which is manifest from its general language and objectives of the

---

[50] *See,* e.g. General Assembly Resolution on Prohibition on the Development and Manufacture of New Weapons of Mass Destruction and New Systems of Such Weapons, UN Doc. A/RES/51/37.

[51] General Assembly resolution on Prohibition of the Development and Manufacture of New Types of Weapons of Mass Destruction and New Systems of such Weapons, A/C.1/45/L.27/Rev.1.

[52] Gorove, Arms Control Provisions in the Outer Space Treaty: A Scrutinizing Reappraisal, 3 Ga. J. Int'l & Comp. L. 114 (1973).

[53] *E.g.* The Kinetic Energy Interceptor of the Multiple Kill Vehicle of the United States. *See* Missile Defense Agency, Missile Defense- Worldwide, 5th Edn., p. 29, available at http://www.mda.mil/mdalink/pdf/bmdsbook.pdf accessed 20/07/2008.

[54] *E.g.* The Airborne Laser. *See* Missile Defense Agency, supra, p.19.

treaty itself. If one is allowed to deploy weapons no where in the space but only on the moon, which is mandated to be used in the second paragraph itself for exclusively peaceful purposes, then the whole purpose of the article fails.

Whether these weapons are allowed to travel through the space

While no Nuclear Weapons or WMD can be stationed in the outer space, there is no express bar on travelling of such weapons in outer space. Many ICBM and strategic ballistic missiles have an exo-atmospheric phase in their journey and at that time they can be said to travel through space. Apart from that, are nuclear weapons and WMDs allowed to travel through space to hit a target deep in the outer space?

It is to be noted that no nuclear weapon or WMD can be placed in an orbit around earth. This provision may potentially bar use of any weapon that has to be placed into a fractional orbit around the earth to hit the target as long as the usage of the word 'orbit' covers fractional orbits as well.

On the other hand, understanding the word 'stationed' may prove tricky. In reality, nothing in space is ordinarily fixed, any object in deep space is perpetually travelling. The phrase 'station in outer space in any manner' is wide[55] and is meant to cover all objects moving in a way or the other in outer space. In that case, the provision will naturally cover all the weapons in outer space even if they are travelling through the space.

Thus it is shown that reasonable interpretations of Art. IV will lead to a prohibition on sending any nuclear weapon or weapon of mass destruction, which includes ABMs carrying nuclear arsenal or BMDs and ASAT having WMD capability, into outer space even if they are travelling momentarily through it.

CONCLUSION

Ballistic Missile Defenses and Anti-Satellite Missiles in most cases attract the application of space law. Although there have been no legal instruments to control ASATs, BMDs have been controlled through specific arms control treaties. These treaties are bilateral and may have only sufficed in the Cold War era. Post the Cold War era, geopolitical realities are different and require multilateral efforts to prevent an arms race in space. Also, except for the ABM Treaty and the draft treaty proposed by China and Russia, the arms control treaties automatically terminate after a certain period of time and permit a nation to withdraw from its obligations *merely* by giving a notice, without any reference to a *fundamental* change in circumstances. While there is evidence of a desire of nations to create additional legislation to prevent weaponisation of space the vehement opposition of the US, a major space power, to any further regulation of arms control in space leaves one with no other resort except Art. IV of the OST. These observations and concerns hold true for BMDs and ASATs notwithstanding the terminology of defense or non-human target is used by the developers of these systems. It is time that international legal discourse on disarmament takes notice of the simple fact and acts upon it. The deployment of such kind of weapons in outer space is not justified by the argument that they are not violative of peace or aggressive. The question is of preserving sanctity of outer space as a peaceful province of humankind and any military use of the same which may in any way compromise future interests of humanity is totally unacceptable.

[55] The meaning of the phrase would have been narrower had a point of reference been mentioned in the provision. In absence of such a reference, it is reasonable to assume that the intention of the parties to the treaty was not to narrow down the provision in such a way.

(IAC-08-E8.2.11)

"SPACE TRAFFIC MANAGEMENT" FOR THE PREVENTION OF WEAPONIZATION OF OUTER SPACE

Setsuko AOKI

Faculty of Policy Management, Keio University, Japan

aosets@sfc.keio.ac.jp

Abstract

Since the setting-up of the CD, no proposals have been endorsed for the further discussions to adopt a treaty on the prevention of weaponization in outer space, and today, it seems difficult to even re-establish the ad hoc Committee on Prevention of Arms Race in Outer Space (PAROS). In contrast, entering into the 21st century, the notion of "space traffic management" (STM) has become prevalent in securing safety use of outer space. STM has been introduced as a broad concept including the space debris mitigation measures, frequencies and slot management, the pre-launch notification systems and concrete space traffic rules. Can the pursuit of the STM (space safety) be a proxy to the arms control measures (space security)? The conclusion would be positive if the following conditions are satisfied: (1) As a goal, emphasis is to be placed upon the safety itself, not security; (2) definition of space weapons, weaponization, or ASAT would be carefully avoided; (3) soft law frameworks would be selected; and (4) such soft law frameworks are adopted and carried out by the like-minded states outside the CD as a first step.

1. INTRODUCTION

It has been fourteen years since the re-establishment of the ad hoc Committee on the Prevention of Arms Race in Outer Space (PAROS) became impossible at the Conference on Disarmament (CD). It is true that various proposals have been repeatedly submitted to the plenary of the CD after 1995, but without any possibility of being seriously considered. The latest United States National Space Policy (2006), denying the possibility of entering into the negotiating the arms control agreements in outer space [1], is not encouraging either, to say the least. China's Anti-Satellite (ASAT) test in January 2007 also increased the concerns

[1] U.S. National Space Policy (2006), Point 6 in "2. Principles".

for the possible "weaponization" of outer space in the international community.

In the stalemate of the pursuing PAROS, alternative approaches have to be considered in order to maintain and improve the peaceful uses of outer space. Entering into the 21st century, the different approaches are gradually becoming prevalent. Such alternative approaches include pursuing a code of conduct on Transparency and Confidence Building Measures (TCBM) at the United Nations (UN) and promoting an international agreement with respect to the comprehensive space safety schemes in the name of "Space Traffic Management".

The purpose of this article is to study if the pursuit of safety measures for space activities could be a substitute of the agreement of the prevention of weaponization in outer space.

2 ACHIEVEMENTS AND LIMITATIONS OF THE CD

2.1 Military Use of Outer Space as a Stabilizing Effect.

Military use of outer space is one of the major motivations by the leading spacefaring states since the advent of the space development. Although it was not until 1978 when the US government officially acknowledged its use of photo-reconnaissance satellites for national security purposes, [2] it was internationally well recognized already in 1970's that the "national technical means of verification", [3] including satellite monitoring, carried out by the then both superpowers, made it possible to reach the bilateral arms control agreements including the SALT I (1972-1977), ABM Treaty (1972-2002), Threshold Test Ban Treaty (TTBT) (1974), Peaceful Nuclear Explosions Treaty (PNE) (1976) and SALT II (1979, not entered into force). Based on the hypothesis that the multilateral satellite monitoring systems would contribute to building confidence among international community, as the satellite-based monitoring by the superpowers increased the stability of the bilateral relationship by verifying the other Party's compliance with the treaty provisions, France proposed the setting-up of the International Satellite Monitoring Agency (ISMA) at the first Special Session of the General Assembly (SSOD I) held in 1978.[4] Military uses of

---

[2] See, e.g., S.A. Cohen, "SALT Verification: The Evolution of Soviet Views and their Meaning for the Future" (1980) 24 *Orbis* 657, pp. 661-662. The Soviet Union admitted using satellites to monitor SALT I accord in 1972. Ibid.

[3] See, e.g., Article V of the SALT I, Article XII of the ABM Treaty, Article II of the TTBT and Article IV of the PNE.

[4] A/S-10/AC.1/8 (1978). Also, see, paragraph 125 (d) of the Final Documents

outer space, at least some category of which, was not only been tolerated as the deplorable reality, but also being positively acknowledged as having stabilizing effects for the international community already around the end of the 1970's.

2.2 Three Categories of Proposals at the CD

The scope of the permissible military use has been the subject of the CD[5] since its establishment in 1979. However, no consensus has been reached on which kind of military uses shall be prohibited up until today irrespective of the various draft articles, institutional proposals and other schemes on the PAROS. While a variety of proposals have been made at the CD at the first look, the contents of major parts of the proposals are often similar. In other words, the similar proposals have been repeatedly submitted for almost three decades in order to pursue PAROS.[6] In order to find the measures to surmount the obstacles to reach consensus for the weaponization of outer space, it seems useful briefly studying the achievements and limitations of the PAROS proposals. Proposals submitted to the CD can be categorized into three:

(1) prohibition of the "weaponization" of outer space (comprehensive approach) ;

(2) prohibition of the ASAT weapons (individual approach); and

(3) TCBM as a first step.

Briefly, the present writer summarized the characteristics of the each category of proposals and the present situation, and concluded which category would have more possibility to be realized.

2.3 Prohibition of the "Weaponization" of Outer Space

2.3.1 General Trends

Among the various proposals found in this category, or prohibition of "weaponization" of outer space, some proposals define "space weapons" clearly, and others only provide for the type of behaviors or actions to be refrained from by the State Parties to a future Treaty. In case when "space weapons" are defined, differences exist in contents. Many proposals define space weapons based on their deployment positions and/or on their target positions. In those cases, some definitions include weapons to target to space and to the Earth from space and on

of the SSOD I, A/S-10/2 (1978).

5 Committee on Disarmament (CD) , established based on the Final Documents of the SSOD I was renamed as Conference on Disarmament (CD) in 1984. A/S-10/2, para. 120.

6 As a comprehensive guide on the various proposals, see, e.g., "Basic Documents of the Conference on Disarmament Related to the Prevention of Arms Race- Compilation Prepared by the Secretariat", CD/INF.50 (23 May 2006).

the Earth and others exclude weapons placed on the Earth to target the objects in space. [7] Some school of thoughts suggests that space weapons are weapons deployed in outer space regardless of their target position. The problem of this definition lies in the fact that currently outer space remains to be defined.

Also, pursuant to some definitions, space weapons are deemed as only devices which are specifically designed and developed to attack, destroy or damage other objects. According to other definitions, weapons capable of attacking, destroying or damaging other objects are judged as space weapons. In the latter case, e.g., manned space vehicles used for scientific purposes can be regarded as a potential space weapon. [8]

The most crucial difference is whether to take weapons deployed in areas other than outer space which target outer space object shall be defined as space weapons or not. Reaching a consensus on this specific point seems impossible since the view directly relates to the political position of a state if it would approve of the part of the ABM systems or Strategic Defense Initiatives (SDI), or since 1990's, missile defense (MD) systems to be deployed in space.

The following is the summarized process of the failure of the CD about reaching a consensus on weaponization of outer space. From the history, the conditions of possible measures for maintaining the peaceful uses of outer space could be found out.

2.3.2 Milestone Proposals at the CD

One of the earliest proposals on this subject was submitted by Italy in 1979, which requested that the wording of Article IV of the Outer Space Treaty (OST) be amended from the banning of the weapons of mass destruction in outer space to simply weapons of any other kinds in space. Italian proposal does not clearly define the deployment positions or target positions of prohibited space weapons. [9] Likewise, as mentioned above, similar provisions are found today, e.g., in the latest draft treaty proposal at the CD by China and Russia in 2008 (CD/1839).

The concept of "weaponization" was also established by the mid 1980's.[10] CD Report of 1986 session shows some examples of the definitions of "weaponization" of outer space, which

---

[7] Some proposals do not explicitly specify the deployment positions or target positions. See, e.g., CD/1487 (1988); CD/1549 (1999), p.1.

[8] See, e.g., CD/PV.170 (1982), pp. 14-15.

[9] CD/9 (1979), p.1. Italian proposal prohibits the development and use of any land- or space-based systems which are specifically designed to damage, destroy, or interfere with the operation of foreign satellites. Ibid.

[10] See, e.g., CD/641 (1985); CD/726 (1986).

remains the same even today.[11] It is identified that weaponization is the existence of "weapons systems in space or directed against targets in space".[12] Militarization of space, is, on the other hand, regarded as supporting of military systems and military operations on the Earth as an integral part of the whole weapons systems.[13] Such line of thought was reflected in, for instance, Chinese statement in 1985.

China stated in 1985 that the primary objective at that stage in the efforts to PAROS should be the "de-weaponization" of outer space, i.e., the banning the development, testing, production, deployment and use of any space weapons and the thorough destruction of all space weapons"[14] with the limited and prohibited use of military satellites of all types as the ultimate goal in mind.[15] For China, "space weapons should include all devices or installations either space-, land-, sea-, or atmosphere-based", which are designed to attack or damage space craft in outer space, or disrupt their normal functioning, or change their orbits; and all devices or installations based in space (including those based on the moon and other celestial bodies) which are designed to attack or damage objects in the atmosphere, or on land, or at sea, or disrupt their normal functioning. This is the most comprehensive definition of space weapons, repeatedly seen in the proposals in later years to date.[16]

Among the two of the Venezuelan proposals made in 1986 and 1988 respectively, the later one defines the space weapons in a comprehensive manner. It defines space weapons as "any offensive and defensive device, including its operational components, whatever the scientific principle on which its functioning is based: (a) capable of destroying or damaging from its place of deployment in outer space an object situated in outer space, in the air, in water or on land; (b) capable of destroying or damaging from its place of deployment in the air, in water or on land an object situated in outer space."[17] Venezuela went on to state that "[t]he following are also space weapons: any offensive or defensive device, including its operational components, and any system of such devices, whatever the scientific principle on which its functioning is based, that is capable of intercepting, from outer space or from land, water or the atmosphere,

---

[11] CD/726 (1986), pp.4-5.
[12] Ibid., p.4.
[13] Ibid.
[14] CD/679 (1985), para.5.
[15] Ibid., para.3.

[16] Parts of the ABM systems, SDI, and MD systems could be categorized as space weapons pursuant to this type of definition, depending on how "space object" should be defined. See, infra.
[17] CD/851, CD/S/WP.24 (1988), p.1.

ballistic projectiles during their flight."[18] While 1986 Venezuelan proposal did not clearly include weapons to damage or destroy from space to another space object in space,[19] 1988 proposal comprehensively includes weapons systems and components, from space to space, space to the Earth and the Earth to Space. By this comprehensive definition, ABM systems and then being developed Strategic Defense Initiative (SDI) could be comprised as a weapons to be forbidden. Bulgaria, Egypt, Mongol, Peru, Poland, and Zaire supported Venezuelan comprehensive definition of space weapons,[20] while majority of western countries regarded it too far-reaching to translate into a treaty provision in consideration of the ongoing ABM and SDI efforts.[21]

In the 21st century, Russia and China occasionally submit proposals on the PAROS. Joint proposal on a future treaty made in 2002 provides for the three basic obligations: first is not to place in orbit around the Earth any object carrying any kind of weapons, not to install such weapons on celestial bodies, or not to station such weapons in outer space in any other manner. Second obligation is not to resort to threat or use of force against outer space objects, and third is not to assist or encourage other States, groups of States, international organizations to participate in activities prohibited by this Treaty.[22] Space weapons are not defined in the proposal, but the working paper submitted to the CD four years later comprises the definitions of major technical terms. In that paper, an example of the definition of space weapons is cited as "any devices, installations or establishments based in outer space, including the Moon and other celestial bodies, which strike and damage objects in outer space, in the atmosphere, on the ground, in the sea or disrupt their normal functions, as well as any devices or installations based on the ground, in the sea or in the atmosphere, that strike and damage space objects, impair their normal functions or change their orbits."[23] It merits mentioning the model definition is not different from the Chinese statement 21 years before.

Latest proposal by China and Russia is the draft "treaty on prevention of the placement of weapons in outer space and of the threat or use of force against outer

[18] Ibid.
[19] CD/726 (1986), p.1. The narrower scope was the reason that some states did not support Venezuelan proposal including China, Sri Lanka and the USSR.
[20] CD/905, CD/OS/WP.28 (1989).
[21] CD/833 (1988), p.12.

[22] CD/1769 (2002). Note that the first obligation is the same as the Italian proposal in 1979.
[23] CD/1779 (2006), pp.2-3.

space objects (PPWT)" made on 12 February 2008.[24] Basic obligation of the PPWT is the same as specified in CD/1679, a 2002 China and Russia joint proposal (Article II of the PPWT).[25] Notable is that the definitions of the key terms are added and the consideration is found on the selection of the key technical terms to gather understanding of the usually skeptical former western nations. Instead of space weapons, "weapons in space" is defined as "any device placed in outer space, based on any physical principle, which has been specially produced or converted to destroy, damage or disrupt the normal functioning of objects in outer space, on the Earth or in the Earth's human existence or inflict damage on them" (Art. I (c)). For the purposes of the draft treaty, outer space is 100 km above sea level (Art. I (a)) and the meaning of "placed" includes the case a weapon orbiting the Earth at least once, or following a sections of such an object before leaving this orbit in addition to the case permanently located somewhere in outer space (Art. I (d)). Thus, a ballistic missile (BM), Fractional Orbital Bombardment System (FOBS) type of weapons, or even interceptor of the missile defense systems may be interpreted as "weapons in space". That is one of the reasons that Slovenia, on behalf of the EU, stated on 28 February that "it remains a difficult challenge to achieve consensus on the definitions needed as for a legally binding instrument."[26] Canada also expressed its preference on soft law frameworks than the draft treaty considering non consensus exists on the technical, legal and political issues in this field. [27]

2.3.3 Conclusion

Without the consensus of the definition of space weapons, yet, the greatest common divisor exits in the concept of the term "weaponization" of outer space in that the central concept of which includes the manner of the space use exceeding the scope of defensive or passive military use of space. The notion seems widely shared that the space use aiming at supporting terrestrial military activities are, in principle, permissible, while testing or stationing a device or an object which is capable of attacking or damaging other

[24] CD/1839 (2008). The document was issued on 29 February 2008.

[25] Article II provides that "[t]he State Parties undertake not to place in orbit around the Earth any objects carrying any kinds of weapons, not to install such weapons on celestial bodies and not to place such weapons in outer space in any other manner; not to resort to the threat or use of force against outer space objects; and not to assist or induce other States, groups of States or international organizations to participate in activities prohibited by this Treaty."

[26] EU Response to Russian Proposal of the draft PPWT, (28 February 2008) , p.2.

[27] CD/PV.1089 (2008), p.10.

objects in outer space or on the Earth from outer space should be forbidden.

The lesson from the great efforts to pursue the prevention of weaponization of outer space is that defining "space weapons" shall be avoided if the consensus building is seriously sought out. It is too difficult technically, legally and politically. Without the definition, international community acknowledges the core part of "weaponization" and no country is publicly opposed to the prevention of weaponization. Political commitments, or soft law frameworks may fit to stop in advance weaponization of outer space.

2.4 Prohibition of the ASAT weapons

Considerable efforts have also been made to reach a consensus to ban the most pressing category of the space weapons, or anti-satellite (ASAT) weapons. However, since ASAT weapons themselves entailing a variety of types of devices, methods and scientific principles, defining ASAT weapons turned out to be as difficult as that of space weapons. Although all the proposals submitted were designed carefully to avoid too wide a definition of the ASAT weapon systems, no one proposal was given a reasonably positive reaction. Proposals submitted to the CD include the prohibition of high-altitude ASAT weapons,[28] banning of ASAT weapons in combination with immunity for satellites,[29] the elimination of existing ASAT weapons along with the banning of the development of a new ASAT capability [30] and multilateral moratorium of any anti-satellite activities.[31] Although there was no consensus on the legally impermissible scope of ASAT weapons and/or tests, starting the latter half of 1980's, the US and the USSR entered into a tacit understanding of moratorium on physical ASAT tests. Let aside the USSR which made a proposal on partial ASAT prohibition at the CD, it has to be taken note of that the US, in principle, supported the idea of the banning of the ASAT, while it opposed to define ASAT weapons as an impossible undertaking.[32] It may be said that it is the definition factor that constitutes the insurmountable obstacle, not the concept of PAROS itself.

2.5 TCBM as a First Step

It became clear that concluding a treaty for that subject was most difficult by

---

[28] Proposal by France and the Netherlands.CD/905, CD/OS/WP.28 (1989), p.10 & p. 13.

[29] Proposal by France. CD/936, CD/OS/WP/35 (1989), p.6.

[30] Proposal by the USSR. CD/PV.486 (1989), p.17.

[31] Proposal by India. Ibid., pp.15-16.

[32] CD/905, CD/OS/WP.28 (1989), pp.15-16.

early 1990's at the latest. Then, focus was placed on the TCBM as a less unrealistic goal at CD.[33] Proposals were made often as a part of a soft law instrument such as "code of conduct", "rule of the road" or "rule of the behavior" because of the contents of commitments are not necessarily suitable for the legally-binding instruments and it is easier to be agreed upon. Major proposals for TCBM are as follows:

(a) Multilateral monitoring organizations to gather information in outer space

* PAXSAT A, space-to-space monitoring, proposed by Canada, first in 1987 and resurrected in 2006.[34]

* International Trajectory Centre (UNITRACE) proposed by France.[35]

(b) Pre-launch inspection

*International Space Inspectorate (ISI) by the UUSR.[36] The core of this USSR proposal is the multilateral on-site pre-launch inspection regime to verify that space objects to be launched would not possess space weapons.

(c) Strengthening the 1975 Registration Convention

A number of countries made proposals or supported the idea over the years. Registration Convention is not an arms control instruments, but a treaty to establish an international registry of space object for the purposes of providing practical effect to the 1972 Liability Convention. However, the transparency function of the Convention was taken note of at the CD.[37]

(d) The principle of non-interference among satellites

*Satellite immunity (France), keep-Out Zones and limitation on high velocity fly-bys (West Germany, France, etc.) [38]

*The establishment of minimum distances between space objects and speed limits imposed on space objects (West Germany).[39]

3. THE OUTLINE OF THE CONCEPT

[33] GARes45/55 (4 December 1990) for the first time requested that technical aspects of CBM be studied.

[34] CD/1785 (2006).

[35] CD/1092 CD/OS/WP.46 (1991) p. 5. Before UNITRACE proposal, see, e.g., CD/937 (1989) and CD/PV.570 (1989); ISMA, originally proposed by France in 1978 , was later proposed at various fora including the CD. A-S10/AC.1/7; A/C.1/33/PV.26 (1078); CD/641 (1985), para.4. ISMA is tasked with monitoring the earth not in outer space, but became a starting point of a series of French proposal such as Satellite Imaging Processing Agency (SIPA).

[36] CD/PV.385 (1987); CD/817 (1988). West Germany also maintained the necessity of the strict pre-launch notification with detailed parameters to notify. CD/905, CD/OS/WP.28 (1989) pp. 21-22.

[37] One of the first suggestions to the Convention for CBM measures is found at para.27 of CD/641 (1985), para. 27 of CD.786 (1987), para.24 of CD/833 (1988).

[38] CD/905,CD/OS/WP.28 (1989) pp.21-22; CD/1092, CD/OS/WP.46 (1991) p. 4.

[39] CD/786 (1987), para.27; CD/905/CD/OS/WP.28 (1989) p. 22.

OF SPACE TRAFFIC MANAGEMENT IN THE IAA REPORT

3.1 Brief History

The concept of Space Traffic Management (STM) was gradually evolved among academic circles, especially through the two important workshops organized by American Institute of Aeronautics and Astronautics (AIAA) in 1999 and 2001[40]. During the 2001 workshop, it was suggested that the International Academy of Astronautics (IAA) should conduct an in-depth study on STM. IAA accepted it in late 2001 and set up an interdisciplinary study group of approximately 20 experts.[41] IAA report titled "Cosmic Study on Space Traffic Management" was published in 2006.[42]

For the purposes of the report, STM is defined as "the set of technical and regulatory provisions for promoting safe access into outer space, operations in outer space and return from outer space to Earth free from physical or radio-frequency interference." [43] Space security is, presently, often defined as "the secure and sustainable access to and use of space, and freedom from space-based threats".[44] The overlapping area in the definition may be a good opportunity for the arms control advocate and at the same time the major obstacles for the STM.

3.2 Summary of The IAA Report [45]

The IAA report in 2006 aims at an outline of a space traffic regime in 2020. The concept of the STM would relate to the whole stages of space activities from pre-launch and launch phase, to the in-orbit operation phase, and finally to the re-entry phase. Also, the dimension of management of STM comprises scientific and technical area as well as regulatory and legal fields. Among the findings in the report, regulatory and legal requirements for an effective STM are summarized below. [46]

---

[40] AIAA, *Proceedings of the International Cooperation: Solving Global Problems* (1999) pp. 35-59; AIAA, *Proceedings of the International Cooperation: Addressing Challenges for the New Millennium* (2001), pp. 7-14.

[41] See, e.g., Corinne Contant, Petr Lála, Kai-Uwe Schrogl, "Status of the IAA Study Group on "Traffic Management Rules for Space Operations", IAA 03-5.5a06, Space Debris and Space Traffic Management Symposium (2003, IAA), pp. 576-577.

[42] IAA, Cosmic Study on Space Traffic Management (IAA, 2006).

[43] IAA, ibid.

[44] See, e.g., Canadian statement at the 44th session of the Science and Technical Subcommittee (STSC) of the Committee on the Peaceful Purposes of Outer Space (COPUOS) on 21 February 2007. Originally, it is a definition of the spacesecurity.org. See, Space Security series of spacesecurity.org.

[45] Ibid.

[46] IAA Report points out 21 findings, among which those related to space law are cited below. Supra, note 42, p. 64, p.75 & p.89.

3.2.1 The launch phase

(a) Safety certifications should be introduced.

(b) Clarification of the term "space object" is needed.

(c) The question of delimitation of air space and outer space should be revisited.

(d) The concept of "launching State" has to be clarified.

(e) A pre-launch notification system is necessary, although International Code of Conduct against Ballistic Missile Proliferation (widely known as Hague Code of Conduct: HCOC) includes non-regulatory binding provisions for such notifications of Space Launch Vehicle launches.

(f) Obligatory information in cases of damage is relevant.[47]

Among the six requirements, (a) and (e) relate to TCBM proposed at CD, especially ISI and the amendment of the Registration Convention. It is interesting that the IAA Report refers to the HCOC that imposes on the political commitment to prevent and curb the proliferation of ballistic missile systems capable of delivering weapons of mass destruction. Because of the similarity in functional principle between BM and space launch vehicles (SLV), subscribing states are obliged to provide launching information of BM and SLV in accordance with pre-launch notification mechanisms of the HCOC.[48] Drafted to supplement the task of Missile Technology Control Regime (MTCR), it was successfully adopted due to its non-binding nature of the instrument. While China and India are not subscribing states and the US postpones supplying pre-launch notifications, yet, it is remarkable that 130 states are now members to the TCBM mechanisms.[49] At least as a first step, a soft law framework has to be pursued to reach consensus in the fields of space security.

Points (b)-(d) and (f) are not unique for STM, but acknowledged as the pressing issues in general international space law[50].

3.2.2 The in-orbit operation phase

In this phase of activities, no specifically legal requirements are

[47] Ibid., p. 64.

[48] 4 iii) of the HCOC.

[49] When adopted in November 2002, the number of subscribing states to the HCOC was 93.

[50] Those issues are either currently being dealt with at the Legal Subcommittee (LSC) of the COPUOS or recently intensively discussed and solved to a certain degree, in most cases as UN General Assembly (GA) Resolutions. Examples would be the "application of the 'launching State' concept" and " recommendations on enhancing the practices of States and international organizations in registering space objects" adopted in 2004 and 2007 respectively in the part of the omnibus GA Resolutions under the title of "international cooperation in the peaceful uses of outer space". See, A/RES/59/115 (2004); A/RES/62/101 (2007).

mentioned. Instead, the efforts and limitation of the activities by the ITU are specified.[51]

3.2.3. The re-entry phase

(a) Space law and air law have to resolve the open issue of passage of space objects through airspace.

(b) The question arises, whether to introduce certain internationally recognized descent corridors and possibly even impact areas which are not frequently used by other traffic, and which could be dedicated to space traffic.[52]

In this phase, reusable vehicles and de-orbit of space debris mitigation measures are largely involved. Space debris mitigation guidelines, adopted at the Inter-Agency Space Debris Coordination Committee (IADC) in 2002 [53]and the UN Space Debris Mitigation Guidelines endorsed by the GA resolution in 2007 [54]can be regarded as one kind of STM instruments, and a proxy to the arms control measures.

3.2.4 STM model in 2020

The report provides how a comprehensive STM model in 2020 will look like (in chapter 4). According to the Report, a STM regime will comprise four elements: (1) the securing the information needs, (2) a notification system, (3) substantive space traffic rules and finally (4) mechanisms and organizations for implementation. Needless to say, the correct information on where space objects are flying should constitute a sound basis for any successful traffic control. In the report, the required areas include (a) definition of necessary data, (b) provision of data, (c) establishment of a database and distribution mechanisms for data and (d) establishment of an information service on space weather.[55] Presently, space situational awareness (SSA), the precision knowledge about the space traffic is only in the possession of the US and Russia.[56] Thus, it seems quite a challenge establishing technical basis as good as or better than the US Space Surveillance Network (SSN) information systems open and accessible to all actors, states, organizations and private entities.[57]

The second element is a notification system. Currently, it is the task of the

[51] Ibid., p. 75.

[52] Ibid., p. 89.

[53] It was supplemented in 2004 and amended in 2007.

[54] A/AC.105/890 (2007), para.99 & Annex IV; A/62/20, II.C.3, paras.116-128 & Annex 4; A/RES/62/217 (distributed in 2008).

[55] IAA, supra, note 42, p 14 & p.91.

[56] Russian Space Surveillance System (SSS) is evaluated slightly less sensitive than the US SSN. See, e.g., Nicholas L. Johnson, "Space Traffic Management: Concept and Practices", *Space Policy*, vol.20 (2004), p. 82.

[57] IAA, supra, note42, p. 14 & p. 91.

Registration Convention to which 51 states are parties as of April 2008, but the Convention has so far yielded the outcome less than satisfactory, partly because the obligatory notification provisions (Art. IV 1 (a)-(e)) insufficient and partly because the states parties are not duly abiding by the Convention. Required areas include (a) the adequate pre-launch notification of parameters, (b) the pre-notification of orbital manoeuvres and active de-orbiting and (c) the notification of the end of lifetime of objects, and re-entry. [58]

The third element is concrete traffic rules. Examples of space traffic rules, specified in the Report include (a) safety rules for launches for both unmanned vehicles and manned space flights, (b) rule on zoning (selection of orbits), (c) right-of-way rules for in-orbit phases, (d) prioritization with regard to maneuvers, (e) operational rules for GEO and LEO satellite constellations, (f) debris mitigation regulations, (g) safety rules for re-entry, and (h) environmental protection provisions in the atmosphere, troposphere, and on the earth.[59]

The fourth element is international control mechanisms. It is maintained that a legally-binding new international agreement designed specifically for the STM shall be adopted along with technical annexes modeled from the organizations such as International Civil Aviation Organization (ICAO) and International Maritime Organization (IMO).[60] After 2020, it is expected that "[t]he operative oversight- i.e. the task of space traffic management, could be taken up by an already existing forum or organization (such as UNCOPUOS/UNOOSA, or ICAO), which would evolve into a body shaped for that purpose. Looking 20 years later, it could also be handled by a non-governmental entity tasked by the State parties to an Outer Space Convention". ("Outer Space Convention" is a name for the future comprehensive treaty on space activities, which does not exist now.)

3.3 Lessons from the IAA Report

As drafters of the IAA Report admit that the STM regime would be realized "for decades now,"[61] and "the study does

[58] Ibid.

[59] Ibid. the numbering (a)-(h) is added by the present writer summarizing a number of possible rules specified in the Report; see, also, Kai-Uwe Schrogl, "Space Traffic Management", presentation at Space Policy Institute, The George Washington University, 24 January 2008, pp. 11-12.

[60] The Report foresees the possibility that both such a new international agreement and the present UN space treaties could be superseded by a comprehensive "Outer Space Convention" after 2020. Law of the Sea precedents might be envisioned. IAA, supra, note 42, p. 15 & p. 91.

[61] Kai-Uwe Schrogl, "Space Traffic Management: the New Comprehensive Approach for Regulating the Use of Outer Space", *ESPI Flash Report*, No.3,

not, however, advocate a rush to regulation", [62] the primary contribution of the IAA Report is to clearly identify pressing issues for the global society and indicate a model procedure to address them. Thus, it is not fair if the Report would be evaluated by the possibility of reaching an international agreement of space traffic rules in the near future. Noting that, a certain instruction can be deduced from the Report.

First, a soft law framework has to be chosen if an agreed instrument for space safety is sincerely sought out. Relatively successful STM, space debris mitigation guidelines both at the IADC and the UN are crystallized in the form of guidelines.

A good example would be Guideline 4 of the UN space debris guidelines. It functions not only for space safety measures as a planned goal, but also a certain arms control (space security) measures. It provides that "[r]ecognizing that an increased risk of collision could pose a threat to space operations, the intentional destruction of any on-orbit spacecraft and launch vehicle orbital stages or other harmful activities that generate long-lived debris should be avoided. When intentional break-ups are necessary, they should be conducted at sufficiently low altitude to limit the orbital lifetime or resulting fragments." Although the double limitations can be pointed out in this guideline in that (a) intentional destruction is not forbidden, and (b) the guideline can not pose legally-binding obligations, it is certain that guideline 4 was possible because it was adopted as a soft law rule.

Second, aversion and reluctance of spacefaring nations to any additional restraints in the freedom of activities must not be underestimated. Unless the merits of traffic rules clearly exceed, the giving up of the long-cherished freedom of space would not be possible. Thus, any future STM measures have to be compatible with the established scope of military use of outer space. To decide what the established practices are, three decades discussions at the CD would certainly help.

Third, it seems difficult to implement space traffic rules and enforce sanctions in case of non-compliance with the rules, even if a legally-binding instrument was adopted. Needless to say if merely political commitments exist. Then, the it seems more important to reach the pertinent gentlemen's agreement[63] among

(October 2007), p. 3.

[62] Petr Lála, "Traffic Management Rules for Space Operations" *Astropolitics*, vol.2 (2004), p.128.

[63] Export control regimes such as Nuclear Suppliers Group (NSG), Australia Group (AG), MTCR and Wassenaar Arrangements (WA) would be good models to follow.

like-minded countries, because through national legislation, the agreed rules are being implemented to a considerably reasonable degree. It cannot be too emphasized that harmonization of national legislation would play a critical role for the universal implementation.[64]

4 CODE OF CONDUCT ON SPACE TRAFFIC RULES

4.1 EU Proposal at the UNGA

On 18 September 2007, at the first Committee of GA, Portugal, on behalf of the EU, submitted a proposal on a comprehensive code of conduct on space objects and space activities for international space transparency and TCBM for the PAROS.[65] That proposal contains the concept of STM: "[t]he key activities to be covered under such a code of conduct could include, inter alia, the avoidance of collisions and deliberate explosions, the development of safer traffic-management practices, the provision of assurances through improved information exchanges, transparency and notification measures, and the adoption of more stringent space debris measures."[66] In the part of the implementation of the prospective code of conduct, enumerated best practices include avoidance of dangerous maneuvers for causing damage to space objects and for creating space debris, creating special areas of caution around satellites for the avoidance of collision, detailed information exchange on launching activities and enhanced registration system. The significance of using STM in a field of PAROS, or space security, cannot be too underlined. In the UNGA, STM is recommended as a possible alternative of arms control agreements. Considering overlapping recommended practices for the not necessarily the identical goal, a successful space traffic regime would have to be founded on the recognition that security use of outer space should be intact as long as such use is in accordance with the existing international space law and laws of arms control relevant to space activities.[67] In the name of STM, or a safety use of outer space, space activities would be restricted if it is not the strict legal obligations as the example of space debris mitigation measures. As time goes by, such restrictions might be crystallized into a customary international law rule. Otherwise, still, it is useful as a first step.

---

[64] Michael Gerhart, "Consequences of a Space Traffic Management Concept for National Space Legislation", IAA, ed., *Space Debris and Space Traffic Management Symposium 2004* (AAS, 2004), pp. 282-284.

[65] A/62/114/ Add.1 (2007), pp. 5-8.

[66] Ibid., p. 7.

[67] Ibid., p. 7, esp. 2.9.(c) .

4.2 Private Code of Conduct- The Henry L. Stimson Center

Nongovernmental organizations from Canada, France, Japan, Russia and the US, organized by the US Henry L. Stimson Center, released on 24 October 2007 a model Code of Conduct for Responsible Space-Faring Nations to promote the peaceful uses of outer space.[68] It provides both for the rights of space-faring states (5 items) and responsibilities of space-faring states (9 items). Among the 9 responsibilities, the following items are concerned with STM measures. Again, it would be the example that the space safety measures may be pursued easier than the space security provisions. Provisions concerned are as follows:

4. The responsibility to develop and abide by rules of safe space operation and traffic management;
5. The responsibility to share information related to safe space operations and traffic management and to enhance cooperation on space situational awareness;
6. The responsibility to mitigate and minimize space debris in accordance with the best practices established by the international community in such agreements as the Inter-Agency Debris Coordination Committee guidelines and guidelines of the Scientific and Technical Subcommittee of the United Nations Committee on the Peaceful Uses of Outer Space; and
7. The responsibility to refrain from harmful interference against space objects.

Some of the measures among items 4-7 are discussed at the CD as TCBM, and item 6, at IADC and STSC of the COPUOS. TCBM considered at CD may be intentionally used as space safety measures at least for the time being to reach consensus for the purposes of adopting international instruments.

5 CONCLUSION

Ever larger number of actors along with a wider variety of activities in outer space would make it imperative to reach a certain agreement on space traffic rules, taking a long-term view. Using that logic, space safety measures may be crystallized into soft law instruments at least as a first step as a substitute of legally-binding space security agreements. Without such legally-binding agreements, access to space and operations there in safer

---

[68] The present writer is one of the participating nongovernmental experts who helped draft the model Code of Conduct. See, e.g., Michael Krepon, "Model Text of a Code of Conduct for Responsible Space-Faring Nations" (released on 27 October 2007).

manner could be accomplished since the demarcation of customary international law and soft law has been blurring to some extent, and as the role of the national implementation is more important these days. It is possible that space safety measures represented by STM can be a proxy to arms control agreements with the conditions specified below would be satisfied:

(1) As a goal, emphasis is placed upon the safety, not security itself;

(2) Definition of "space weapons", "weaponization", "ASAT" and other key concepts are carefully avoided in the future international instruments;

(3) International instruments are made as soft law frameworks such as codes of conduct, guidelines, rules of the road, etc;

(4) Such soft law frameworks are, as a first step, adopted and implemented by the like-minded states outside the CD, since consensus-based adoption is almost impossible at the CD under the circumstances; and

(5) Such soft law frameworks are supported by the major space-faring nations.

VERIFICATION MECHANISM FOR PEACEFUL USES OF OUTER SPACE

Yuri Takaya-Umehara
Ph.D. Candidate, IDEST, Paris XI University, FRANCE
yuritakaya _ japan@hotmail.com

INTRODUCTION

The principle of peaceful uses of outer space has left the question how to define the term "peaceful uses" in the present context of international law and how to ensure "security" in the present legal regime. The legal regime for peaceful uses of outer space lacks its effectiveness in verifying the compliance. Although the Outer Space Treaty of 1967 (hereinafter: the 1967 OST)[1] has the provision of arms control as Article IV, no specific provision of verification is included. Some provisions are applicable to verification under international cooperation; however, they are not on mandatory basis.

With advances in space technologies, the concept of peaceful uses of outer space has shifted from "non-military" to "non-aggressive." Due to the invalidation of the ABM Treaty of 1972 in 2002, the legal implication on lawful military uses of outer space is being affirmed under international law in case of self-defense or collective security, which accelerates ABM or ASAT tests in outer space. In the present situation, it is urgent to take legal action to avoid space weaponization.

By learning lessons from other international areas, where embrace the same principle of peaceful uses and attempt to ensure security with or without verification measures, this article explores possibilities to figure out what elements are necessary for verification mechanism to ensure peaceful uses of outer space.

1. THE PRINCIPLE OF PEACEFUL USES IN INTERNATIONAL AREAS

The legal status of outer space has shift from *res communis* to an "international area" where States Parties have gradually extended their extra-territorial jurisdiction and substantial control. Such a shift is not exceptional to other international areas as Antarctica, high seas and the seabed and the ocean floors.

Although it is difficult to generalize the concept of peaceful uses due to its vague legal implication in the UN Charter, it is possible to understand how the principle of peaceful uses is applied to and function in legal regime governing international area. In this section, the concept of peaceful uses in each area is defined and the present legal regime governing there are studied.

1.1. Antarctica

Antarctica is the last continent discovered on the Earth by the UK, the then USSR and the US. Before International Geophysical Year (IGY) in 1957-58, there was no legal framework on Antarctica activities. In 1948, Chile, Argentina and the UK agreed not to send warships at 60 degrees of South latitude to avoid armed conflicts.[2] This is the first step for ensuring peaceful uses of Antarctica.

In the Antarctic Treaty of 1959[3], the principles related to peaceful uses of Antarctica are: the principle of peaceful uses of Antarctica,[4] the freedom of scientific invetigation and cooperation [5] and the principle of preservation of the environment.[6] The Treaty, first ratified by the twelve States who were active during the IGY,[7] consists of 14 provisions with effectiveness in demilitarizing Antarctica. It has gradually evolved in to a system governing potential activities there, which is known as the Antarctic Treaty System.[8] This development illustrates the consolidation of the Antarctic régime and its transformation from loose cooperation of States into a régime with a significant organizational structure.[9]

The concept of peaceful uses of Antarctica is outlined by provisions as: Antarctica should not be the battlfield (Preamble 1); the use of Antarctica should be in harmony with the UN Charter (Preamble 4); any measure of a military nature should be prohibited (Article I (1)); except for the use of military personnel or equipment for scientific purposes or for any other peaceful purpose (Article I (2)); Treaty meeting should be held for enhancing the transparecy among the Contracting Parties through exchange of information or communication (Article IX); and the dispute settelement should be considered by dialogue, directly or indirectly via the International Court of Justice (Article XI). Thus, those provisions function to ensure peaceful uses of Antractica aiming for the complete demilitarization of Antarctica with dispute settlement procedures.

Considering that enhancing transparency in Antarctica is prioritized, Article VII establishes a system of inspection. Article VII(1) provides that each Party whose representatives are entitled to participate in the meetings under Article IX has the right to designate observers to carry out any inspection provided for and that observers should be nationals of the Parties which designate them. The names of observers should be communicated to every other Party having the right to designate observers, and notice should be given of the termination of their appointment. Article VII(2) allows each observer designated under Article VII(1) to have complete freedom of access at any time to any or all areas of Antarctica. This provision dedicates to enhancing transparency in terms of ensuring the complete demilitarization of Antarctica. Article VII(3) opens all areas, including all stations, installations and equipment within those areas, and all ships and aircraft at points of discharging or embarking cargoes or personnel in Antarctica, for inspection by any observers designated under Article VII (1). Article VII(4) allows aerial observation at any time over any or all areas of Antarctica by any of the Parties having the right to designate observers. Article VII(5) requires information-sharing among the Parties as well as providing notice in advance of (a) all expeditions to and within Antarctica, on the part of its ships or nationals, and all expeditions to Antarctica organized in or proceeding from its territory; (b) all stations in Antarctica occupied by its nationals; and (c) any military personnel or equipment intended to be introduced by it into Antarctica subject to the conditions prescribed in paragraph 2 of Article I of the present Treaty. Thus, Article VII dedicates to ensuring peaceful uses of the Antarctica by opening all areas for inspection and by enhancing transparency though information-sharing.

1.2. High Seas

The principle of peaceful uses of the high seas is found in Article 88 of the UN Convention on the Law of the Sea of 1982[10], but not found in the Geneva Convention on the High Seas of 1958.[11] Article 88 provides that the high seas should be reserved for peaceful purposes. The principle is supported by Article 87 of the freedom of the high seas and by Article 89 of the principle of no sovereignty in high seas. Contrary to the aim of those provisions, there are other provisions allowing military uses of high seas in wartime. Article 95 provides that warships on the high seas have complete immunity from the jurisdiction of any State other than the flag State. In addition, Article 96 ensures that ships owned or operated by a State and used only on government non-commercial services should have complete immunity from the jurisdiction of any State other than the flag State. With the prevailing notion of the exclusive economic zone, the boarder line striding the plural EEZ or the high seas where the freedom of the high seas is applied might become provocative.

Despite the existing provision, Article 88, of peaceful uses of the high seas, there is no general or common understanding agreed on the definition among the Parties. While some States proposed to demilitarize the high seas completely, the US insisted on that the principle of peaceful uses does not prohibit military operations in general and any military operations for peaceful purposes are in accordance with general international law including the UN Charter.[12] Afterwards, ten developing States submitted a proposal that the general principle according to Article 2(4) of the UN Charter should be established, which prohibit the threat or use of force. Their initiatives were resulted into the inclusion of Article 301 in Part XVI. Article 301 provides that States Parties should refrain from any threat or use of force against the territorial integrity or political independence of any State, or in any other manner inconsistent with the principles of international law in the UN Charter in exercising their rights and performing their duties under the Convention of 1982. This provision still allows some kinds of military uses of the high seas for self-defense and collective security that are legal under some conditions.

Although the right of self-defense was denied by the Commission in the drafting Article 22 of the Convention on the High Seas "mainly because of the vagueness of terms like 'imminent danger' and 'hostile acts', which leaves them open to abuse."[13] Even in the case of disputes settlement, military uses of the high sea is not denied as Article 298(1)(b) provides that States Parties have the right not to accept any one or more of the procedures for disputes concerning military operations, including military activities by government vessels and aircraft engaged in non-commercial service, and disputes concerning law enforcement activities in regard to the exercise of sovereign rights or jurisdiction excluded from the jurisdiction of a court or tribunal under in case of dispute settlement under Article 297 (2) or (3).

Thus, without sharing any kind of common understanding of the concept of peaceful uses of the high seas, the legal regime of the high seas allows military uses of the high sea. The security in the high seas faces the same challenges as the security in outer space in questioning whether the existing principle of peaceful uses contributes to maintaining peace and security by preventing an arms race.

1.3. The Seabed and the Ocean Floors

The freedom of the high seas applies to the seabed and the ocean floor. The concept of peaceful uses of the seabed and the ocean floor is first stipulated in the Seabed Treaty of 1971[14] aiming for the denuclearization of the area, and second, the UN Convention on the Law of the Sea of 1982 aiming to reserve the area and its resources as the Common Heritage of Mankind (hereinafter: CHM). In applying the concept to their regimes, two different approaches are taken as: including the provisions of verification in the former, and establishing the International Seabed Authority (hereinafter: ISA) in the latter.

It is worth noting that the concept of peaceful uses of the area is linked with the CHM concept. The notion was advanced by developing States who concerned about the exceeding exploitation of marine mineral resources outside areas of national jurisdiction by developed States. The notion was first reflected to the 1967 OST and included in Article 11 of the Moon Agreement of 1979[15] however; it had already been a vague guiding principle. A similar approach was taken in UN resolution on the Committee on the Peaceful Uses of the Sea-Bed and the Ocean Floor beyond the Limits of National Jurisdiction of 21 December 1968.[16] In the end, the area and its resources were formally regarded as a common heritage by the Sea-Bed Declaration of 1970.[17] In the denuclearization of the area, the first paragraph in the Preamble of the Treaty of 1971 recognizes the common interest of mankind in the progress of the exploration and use of the seabed and the ocean floor for peaceful purposes. Furthermore, the Preamble indicates the measures to ensure the peaceful uses by excluding the area from an arms race, by taking a step on general and complete disarmament under strict and effective international control or continuous negotiations, and by furthering the purposes and principles of the UN Charter without the freedom of the seas.[18]

Apart from the Seabed Treaty of 1971, the concept of peaceful uses of the seabed and the ocean floor is included in the UN Convention on the Law of the Sea of 1982. Article 141 provides that the area should be open to all States exclusively for peaceful purposes, whether coastal or land-locked, without discrimination and without prejudice to the other provisions of the Part of the Seabed. Considering negotiations on the question of military uses of the seabed in the Committee on Disarmament in Geneva, Article 141 of the Convention of 1982 emphasizes the exclusive peaceful uses of the seabed by all States. [19] The concept of peaceful uses is also supported by other principles as: the CHM principle in Article 136; the principle of non-appropriation denying sovereignty and sovereign rights in Article 137; the principle of common interest and international cooperation in Article 138; and the principle of peaceful uses in Article 143 which repeating peaceful purpose in scientific research with an emphasis on the interest of humankind. Among those principles, the CHM concept has been an obstacle in administrating the ISA.

The core provision of verification in the Sea-bed Treaty of 1967 is Article III. The provision consists of six paragraphs as: Article III (1) on the affirmation of the right to observe activities in the area; Article III (2) on concrete verification measures as consultation, reporting, on-site inspection under certain conditions; Article III (3) on further measures to get rid of doubt of breaches; Article III (4) on the implication of possibility to appeal to the UN Security Council to take appropriate action in case of severe doubt of breaches; Article III (5) on the use of National Technical Means (hereinafter: NTM) for verification in accordance with the UN Charter; and Article III (6) on the requirement of such verification procedures to be conducted "with due regard for rights recognized under international law, including the freedom of the high seas and the rights of coastal states with respect to the exploration and exploitation of their continental shelves."

As a difference approach from the denuclearization of the area, the ISA is established on 16 November 1994 under the Convention of 1982 and the Agreement of 1994 Relating to the Implementation of Part XI of the Convention of 1982[20]. It aims to organize and control activities in the area, particularly to administrate the resources of the area. Due to applying the CHM concept to the control of the resources in the area, the ISA faces difficulties in involving key actors as the US. The CHM concept was elaborated in the Sea-Bed Decleration of 1970 [21] recognizing the lack of substantive rules for regulating the exploration of the area and the exploitation of its resources and defining those two as the CHM. Most principles in the Declaration are reflected to Article 133 to 149 of the Convention of 1982. In particular, Article 136 is the provision of the comon heritage of mankind

Although the legal regime under the ISA is accepted by States who prefer to reserving the area as the CHM, the industralized States were dissatisfied with the provision of Part XI.[22] The seabed regime of Part XI and the resolution of UNCLOS III on the Preparatory Commission for the ISA and for the International Tribunal for the Law of the Sea[23] failed in filling the gap between the developed and industrialized States. In order to avoid the development of conflicts, the UN Secretary-General initiated informal consultations among interested states from 1990 to 1994. The outcome of the consultation resuled into the Agreement of 1994 aiming to prevent the emerge of a dual regime as: the one regime fun by the ISA under the Convention of 1982 and the other independently by some industrialized States. [24] Section 4 of the Annex to the Agreement of 1994 ensures the CHM principle and equitable exploitation of the resources of the area for the benefit of all countries, especially the developing States in Article 155 (2) of the Convention of 1982.

Thus, the peaceful uses of the area and its resources requires the involvement of the industrialized States in the regime of the Convention of 1982 and the Agreement of

1994 under the CHM principle.

2. PRINCIPLE OF PEACEFUL USES OF OUTER SPACE

2.1. "Non-Aggressive" Definition

In the 1950s, the most prioritized issue in the international community was to control nuclear threat by establishing international regime on arms control and disarmament. For this purpose, the concept of peaceful uses of outer space was elaborated by the US with its proposal to establish the inspection system for launching activities.

2.1.1. The US Initiative in the 1950s

Even before the successful launch of the USSR's artificial satellite on 4 October 1957, the concept of peaceful uses of outer space was emphasized by the US. On 14 January 1957, US Ambassador Lodge to the UN presented a comprehensive proposal for nuclear conventional arms reduction, expressing that future developments in outer space should be devoted exclusively to peaceful and scientific purposes, by placing the testing of satellites and missiles under international inspection and participation.[25] In August 1957, Canada, France, the UK and the US submitted a working paper on proposals for partial measures of disarmament[26] to the Sub-Committee of the UN Disarmament Commission, calling for a study of an inspection system to assure that the launching of objects through outer space would be exclusively for peaceful and scientific purposes. On 12 January 1958, it was reemphasized by US President Eisenhower in his letter to USSR Premier Bulganin as "I propose that we agree that outer space should be used only for peaceful purposes."[27] The report was incorporated in the UN resolution 1148 (XII) of 12 December 1959 repeating the need of the inspection system to ensure that the sending of objects through outer space be used "exclusively" for peaceful and scientific purposes.[28]

The same US initiatives were also dedicated to bilateral negotiations with the then USSR during the period between the conclusion of the Antarctic Treaty of 1959 and the Limited Test Ban Treaty of 1963, revealing that the Parties attempted to include the verification mechanism in Article IV(1) of the 1967 OST. First, the concept of peaceful uses of outer space was derived from the US proposal by President Eisenhower submitted it to the General Assembly in 1960.[29] The US declared its position in ensuring peaceful use of outer space by recalling the Antarctic Treaty of 1959 as: "1. [W]e agree that celestial bodies are not subject to national appropriation by any claims of sovereignty; 2. [W]e agree that the nations of the world shall not engage in warlike activities on these bodies; 3. [W]e agree, subject to appropriate verification, that no nation will put into orbit or station in outer space weapons of mass destruction. All launchings of spacecraft should be verified in advance by the United Nations; and 4. [W]e press forward with a programme of international cooperation for constructive peaceful uses of outer space under the United Nations [...].[30]" The proposal was incorporated in the 1967 OST; however, due to the opposition of the then USSR, it lacks the monitoring scheme for disarmament of outer space in order to verify all launchings of spacecraft in advance[31]. Thus, the General Assembly resolution 1721 (XVI) on 20 December 1961 only called upon States to inform the UN when launching objects into orbit or beyond and asked the UN Secretary-General to establish a public register, later established as the UN Register, under the Registration Convention of 1975, to keep their records. Thus, the first concept of peaceful uses of outer space appeared in the UN resolution was "exclusively" peaceful, which is close to "non-military".

2.1.2. Creation of "Non-Aggressive" Definition

The US first defined "peaceful" as "non-military" in the National Aeronautics and Space Act of 1958.[32] In its section 101, sub-section (a) stipulated that "[T]he

Congress declares that it is the policy of the United States that activities in space should be devoted to peaceful purposes for the benefit of mankind." However, contrary to President Eisenhower's statement of 1960,[33] the US changed its definition in 1962 when Senator Gore, a UN delegate, stated before the UN First Committee that "[I]t is the view of the United States that outer space should be used only for peaceful - that is, non-aggressive and beneficial - purposes. The question of military activities in space cannot be divorced from the question of military activities on Earth. To banish these activities in both environments we must continue our efforts for general and complete disarmament with adequate safeguards. Until this is achieved, the test of any space activities must not be whether it is military or non-military, but whether or not it is consistent with the United Nations Charter and other obligation of law."[34] Thus, the US emphasized the lawfulness of defining the peaceful as "non-aggressive." In 1984, the US Senator Goldwater stated that "space is just another place where wars will be fought." Outer space has been regarded as "ultimate battlefield" contrary to increasing the dependence of human society on space-based infrastructure in orbit around the Earth, space objects still are fragile orbiting at high speeds and vulnerable to disturbance and destruction by simple impact.

2.2. No Specific Provision of Verification

"The principle of peaceful use of outer space laid down in the Outer Space Treaty of 1967 correlates with one of the leading principles of general international law – the principle of disarmament."[35] Article IV is the provision of arms control. Article IV (1) prohibits the placement of nuclear weapons or any kinds of weapons of mass destruction in orbit around the Earth[36] and Article IV (2) ensures the use of the Moon and other celestial bodies for "exclusively peaceful purposes."[37] However, no specific provision of verification is included in the legal regime for ensuring peaceful uses of outer space.

3. VERIFICATION MECHANISM FOR SPACE SECURITY

The introduction of verification provisions to modern international law is traceable to the creation of weapons of mass destruction as nuclear weapons, biological and chemical weapons. Those weapons were regarded as threat to international peace and security, therefore, the UN prioritized the issue on its agenda. Considering the existing arms control provision of Article IV, which deals with the non-proliferation of nuclear weapons into outer space, verification mechanism needs to be studied and designed to ensure its compliance.

3.1. The Need for Verification Mechanism in International Space Law

3.1.1. Verification of the Use of Space Weapons

There is no comprehensive definition of space weapons, however; it is important to understand how verification works to detect weapons being launched in outer space. The effective verification requires the consideration of the entire armament cycle from R&D to testing, production, deployment, stockpiling, possible use, transfer, reduction and destruction of weapons.[38] Space weapon is not exceptional in terms of requiring the phase of testing, deploying and use of weapons in outer space. Thus, the phases of verification require further identification of whole process.

It is pointed out that arms control agreements are most likely to focus on limitations just after completion of the research phase and before major funding for production is approved, or when the system is fully deployed.[39] In particular, considering that most of space technologies are dual-use, the target phase of verification must be well defined for ensuring space security. Such phases are roughly categorized as: R&D, production, deployment, storage, transfers, withdrawal or destruction. The importance of various "lifecycle" stages differ by each weapons to be verified. In case of the explosive testing of nuclear weapons, monitoring is more effective measure in the phase of R&D. Thus, a throughout examination of a weapon's life enable to permit certain critical points to be identified upon which verification efforts should productively focus on[40] considering that the result could be a more cost-effective verification system with a higher degree of reliability.[41]

3.1.2. The Right of Self-Defense in Outer Space

Under the UN Charter, the use of force is prohibited except for self-defense, collective security or peace keeping operation. In terms of accelerating space weaponization, the legal basis for pursuing military uses is the right of self-defense which leaves a question whether the exercise of the right is lawful in outer space.

The prohibition of the use of force is provided in Article 2 (4) of the UN Charter of and in customary international law constituting a rule of *jus cogens*, however; the paramount obligation not to resort to force has an exception stipulated in Article 51 of the Charter as "the right of self-defense".

The right of self-defense has been studied in the Committee for the Prevention of an Arms Race in Outer Space of the Conference on Disarmament. They concluded that the right of self-defense is not allowed to exercise in outer space, considering that the UN resolution [42] establishing the PAROS Committee itself is the consensus to prevent any arm race based on the right of self-defense.

Before defining conditions for exercising the right of self-defense in outer space, it is important to take other elements into consideration which differ from those on the Earth. First, no sovereignty exists in outer space but jurisdiction and control on space assets are claimed by States who registered under the Registration Convention of 1975. The interpretation of jurisdiction has been various as, for example, "territorial jurisdiction", "extra-territorial jurisdiction", "substantial jurisdiction", or "sovereign jurisdiction" in the legal regime of the high seas. To make a legal basis for the use of force in outer space, jurisdiction in outer space requires further consideration on its clarification. Second, environmental character of outer space requires more strict conditions to justify the use of force. Different from the Earth's environment, States need to refrain from causing space debris by using the force, which is against the freedom of outer space and against the Space Debris Mitigation Guideline of 2007[43], even though it is non-binding.

3.2. Verification Mechanism for Space Security

Verification is the process of gathering and analyzing information to make a judgement about Parties' compliance or non-compliance with an agreement, with the aim of building confidence between the Parties by assuring them that their agreement is being implemented effectively and fairly.[44]

Due to neither international nor comprehensive model/formula, the primary problem of verification is to define verification itself; however, the general process of verification are accepted which includes the following elements: (1) the existence of an obligation, the fulfillment and observance of which must be verified; (2) the gathering of information relating to the fulfillment of the obligation; (3) the analysis, interpretation and evaluation of the information from a technical, juridical and political viewpoint; and (4) the assessment concerning observance or non-observance of the obligation, which concludes the actual verification exercise. While the problem of appropriate reactions to the possible violation of an obligation appears to be a logical consequence of this exercise, it is not in itself an integral pat of verification.[45]

Major functions of verification may be distinguished as: fact-finding; political and legal qualification of facts; and enforcement to ensure the compliance of treaties.[46] Since the goals of disarmament and arms control are different, the goals of verification are also different. On the one hand, the primary objective of disarmament is to reduce the level and amount of armaments and relevant technologies, ultimately aiming for their complete removal or elimination. Its theoretical aspect is coherent with the aim in Article 2 (4) of the UN Charter which prohibits the use of force. On the other hand, the objective of arms control is to limit the level of quality and quantity of weapon systems while keeping military power balance. It imposes restrictions on the development, production, stockpiling, proliferation, and usage of weapons, in particular, weapons of mass destruction.

3.2.1. Possible Provisions in the 1967 OST

The 1967 OST only provides opportunities for the States Parties to verify on voluntary basis. Such provisions are Article IX, X, XI and XII. Although treaties aiming to prevent an arms race in outer space were elaborated by superpowers in the Cold War as the ABM Treaty of 1972[47] or the SALT II Agreement of 1979[48], the former was invalidated by the US withdrawal in 2002 and the latter was not ratified. Thus, the possible verification mechanism in international space law is on reciprocity basis without penalty but under international cooperation. The effectiveness of verification depends on the level of NTMs. Due to the limited number of space-faring States, Multilateral Technical Means (hereinafter: MTMs) are more effective in enhancing transparency in outer space.

3.2.2. NTMs vs. MTMs

Verification Measures are dividable into two categories as NTMs and MTM. The former relies on unilateral means of verification while the latter use multilateral means of verification.

NTMs are nationally owned and operated technologies and techniques used to monitor the treaty obligation of another States. In general, NTM include satellite, high-altitude and other aircraft and land-based remote detection systems, electronic signals intelligence (SIGINT) and electronic intelligence (ELINT) collection systems and systems for collecting open source information.[49] Due to its direct linkage with national security, States Parties were unwilling to define NTM.[50] On the other hand, there is an increasing tendency that States provide NTM-derived information for multilateral verification purposes. In order to involve non-space-faring States, verification mechanism using MTMs through international organizations is suitable for ensuring space security.

3.3. Verification by International Organization

Some verification mechanisms require the establishment of international organizations for monitoring or disarmament verification as the Agency for the Control of Armaments (ACA) of the Western European Union or the Agency for the Prohibition of Nuclear Weapons in Latin America (OPONAL) in Latin America. As to the present, the number of such organizations has increased and its role has varied. Such a tendency provides certain implication that verification mechanism and procedures have become more elaborate and cooperative, in short, multilatelized.

The proliferation of weapons has increased and the universal adherence to multilateral disarmament agreements leads to the need to operationalize the political principle of indiscriminate and equal access to verification capabilities.[51] Thus, internationalized verification measures enable all States to have the equal rights to participate in the process of the international verification of agreements to which they are parties.[52] Even in the process of establishing, elements for confidence-building measures are required which result into better compliance.[53] There is also another aim in establishing international organizations as the US proposal for an International Atomic Control Authority, known as the Baruch Plan, was aimed at denying the USSR access to atomic weapons; or as the Western Europe aim of establishing the Armament Control

Agency was at denying the right of Germany to produce atomic, biological, chemical and certain conventional weapons.[54]

Besides involving international organizations, the confidentiality of data acquired for verification is the issue to be considered or opposed among States Parties. The format of sharing information on advanced and sensitive technologies obtained by NTMs, which is directly related to intelligence gathering, is problematic.

3.3.1. Verification by UN Security Council

The United Nations Security Council has no statutory duty to ensure compliance with arms control treaties unless there is a threat to international peace and security[55] which is different from a test-ban treaty. The Council may initiate an investigation of alleged breach on request; however, there is no obligation to do so, even Member States of the Council may use the power of veto. Moreover, if none of Member States are parties to a treaty, a request for the investigation for the compliance may not to be considered in the Council. In any event, it is the organization of the parties that should perform the function of compliance evaluation, not an external international organization.[56] In order to enhance the effectiveness of treaties, their content should be clarified in case of breach, except the withdrawal, by inducing a defaulting State promptly to redress the situation.[57]

3.3.2. Verification by Specific Organization

There are three treaties which established international organizations for verifying the compliance as: (1) the Modified Brussels Treaty of 1954;[58] (2) the Treaty of Tlatelolco;[59] and (3) the Guadalajara Agreement between Argentina and Brazil. (1) established the Western European Union and the former Arms Control Agency which was already defunct; (2) the Agency for the Prohibition of Nuclear Weapons in Latin America (OPANAL) and (3) the Argentine-Brazilian Agency for Accounting and Control of Nuclear Materials (ABACC).

In 1978, France submitted a proposal to the UN to establish the International Satellite Monitoring Agency.[60]. And others followed, however, whole proposal failed in coordinating political will to share space-based information which is directly related to national security.

3.3.3. Challenges

All existing verification-dedicated organizations are treaty specific. The analysis of practices experienced by the three organizations above provides certain lessons. Complete controls of prohibited armaments and on-site challenging inspections are far acceptable to the suspected States. In the case of ACA, France did not agree with verification measures under the proposed Convention which allows on-site challenge inspection.[61]

CONCLUSION

The principle of peaceful uses applied to international area as, Antarctica, the high seas, the seabed and the ocean floor, has different definition and different legal regime for ensuring peace and security of those areas. Outer space is not exceptional in this respect.

Due to the military-oriented definition of peaceful uses of outer space, the legal regime in outer space fails in establishing an effective verification mechanism.

Considering the present tendency that the increasing number of States cooperate to MTMs-based verification through international organization by providing NTMs-based information, it is suitable for space security to establish such an organization responsible for verification with the aim of enhancing transparency in outer space.

---

[1] Treaty on Principles Governing the Activities of States in the Exploration and Use of Outer Space, including the Moon and Other Celestial Bodies, 27 January 1967, 610 UNTS 205.

[2] WOLFRUM R., KLEMM, U., Antarctica, in: R. Bernhardt (ed.), *Encyclopedia of Public*

International Law, Volume I, 1992, p. 174.
[3] The Antarctic Treaty, 1 December 1959, 402 UNTS 71.
[4] Article I of the Antarctic Treaty of 1959.
[5] Article II of the Antarctic Treaty of 1959.
[6] Article IX (1) (f) of the Antarctic Treaty of 1959.
[7] The twelve States were: Argentina, Australia, Belgium, Chile, France, Japan, New Zealand, Norway, South Africa, the UK, the US and the USSR.
[8] This term was first used by an Argentine scholar and diplomat, Roberto Guyer, in his 1973 lectures at The Hague Academy of International Law: see R. Guyer, "The Antarctic System", in Hague Recueil des Cours, vol. 139-II, 1973, pp. 149-226, at p. 156.
[9] Supre note. 2, p. 181.
[10] United Nations Convention on the Law of the Sea, 10 December 1982, 1833 UNTS 3.
[11] Convention of the High Seas, 29 April 1958, 450 UNTS 11.
[12] Official Records, p. 62.
[13] Year Book of International Law Commission, 1956, p. 284.
[14] Treaty on the Prohibition of the Emplacement of Nuclear Weapons or Other Weapons of Mass Destruction on the Sea-Bed and the Ocean Floor and in the Subsoil Thereof, Annex to UN General Assembly Resolution 2660 (XXV), 7 December 1970, 10 *International Legal Materials*, 1971, p. 145.
[15] Agreement Governing the Activities of States on the Moon and Other Celestial Bodies, 18 December 1979, 1363 UNTS 3.
[16] UN GA Res. 2467 A-D (XXIII).
[17] UN GA Res. 2749 (XXV), 17 December 1967,Declaration on Principles Governing the Sea-bed and the Ocean Floor, and the Subsoil thereof, Beyond the Limits of National Jurisdiction.
[18] Preamble of the Seabed Treaty of 1971.
[19] JAGOTA, S.P., "The Seabed Outside the Limits of National Jurisdiction", *International Law: Achievements and Prospects*, Martinus Nijhoff Publishers, UNESCO, Paris, 1991, p. 929.
[20] UN GA Res. 48/263, 28 July 1994, Agreement o 1994 Relating to the Implementation of Part XI of the Convention of 1982.
[21] UN GA resolution 2749 (XXV), 16 December 1970, "Declaration of Principles Governing the Sea-Bed and the Ocean Floor, and the Subsoil Thereof, beyond the Limits of National Jurisdiction.
[22] VITZTHUM, W.G., Sea-Bed and Subsoil, BERNHARDT, Rudolf L. (ed.), *Encyclopedia of Public International Law*, Volume I, 1992, p. 334.
[23] UN GA Res. A/RES/37/66, 3 December 1982.
[24] Agreement Concerning Interim Arrangements Relating to Polymetallic Nodules of the Deep Sea-Bed of 2 September 1982, ILM, Vol. 21, 1982, p. 950.
[25] UN GA (XI), 1st Com., A/C.1/SR 821, 14 January 1957, p. 41.
[26] DC/SC.1/66, Cmnd. 333, 1957, 96, p. 98.
[27] "President Eisenhower and Premier Bulganin Echange Correspondence on Proposals for Reducing International Tensions," *38 Dep't State Bull*, 1958, p. 126.
[28] *Documents on Disarmament: 1945-1959 (Vol.2)*, 1960, pp. 901-902. Also see, MATTE, N.M., BERNHARDT, Rudolf L. (ed.), *Encyclopedia of Public International Law*, Volume I, 1992, p. 552.
[29] A/AC.105/C.2/SR.65, 22 July 1966, pp. 9-10 (the US); ibid. /SR.66, 25 July 1966, pp. 6-7 (the then USSR).
[30] Official Records of the General Assembly, GA(XV) A/PV.868, 22 September 1960, p. 45.
[31] CHENG, B., "Military Use of Outer Space," *The Utilization of the World's Air Space and Free Outer Space in the 21st Century*, Kluwer Law International, the Hague, 1997, p. 315.
[32] P.L. 85-568; 72 Stat. 426.
[33] GAOR. GA(XV) A/PV.868 (22.9.60), p. 45, para. 48
[34] A/C.1/PV.1289, 3 December 1962, p. 13.
[35] DIEDERIKS-VERSCHOOR, I.H.PH, *An Introduction to Space Law (2nd ed.)*, Kluwer Law International, The Hague, 1999, p.29.
[36] Article IV (1) of the 1967 OST.
[37] Article IV (2) of the 1967 OST.
[38] SUR, S., "Introduction", *Verification of Disarmament or Limitation of Armaments: Instruments, Negotiations, Proposals*, UNIDIR/92/28, New York, United Nations, 1992, p. 5.
[39] KRUZEL, J., "From Rush-Bagot to START: the Lessons of Arms Control", *Orbis*, Spring 1986, p. 197.
[40] CRAWFORD, A., "The Armaments Lifestyle and Verification", *Verification of Disarmament or Limitation of Armaments: Instruments, Negotiations, Proposals*, UNIDIR/92/28, New York, United Nations, 1992, p. 6.
[41] KNOTH, A., "Implementing Treaty Requirements: Key Aspects of Verification", *International Defense Review*, No. 4, 1991, p. 340.
[42] A/RES/40/87, 12 December 1985,113th

plenary meeting, "Prevention of an arms race in outer space"

[43] UN GA Res.62/217, 21 February 2007, United Nation Guideline of Space Debris Mitigation.

[44] UNIDIR & VERTIC, *Coming to Terms with Security: A handbook on Verification and Compliance*, , United Nations, June 2003, p. 1.

[45] SUR, S., "Introduction", Verification of Disarmament or Limitation of Armaments: Instruments, Negotiations, Proposals, UNIDIR/92/28, New York, United Nations, 1992, p. 1.

[46] DE JONGE OUDRAAT, C., "Chapter 8 International Organizations and Verification", Verification of Disarmament or Limitation of Armaments: Instruments, Negotiations, Proposals, UNIDIR/92/28, New York, United Nations, 1992, p. 209.

[47] Anti-Ballistic Missile Treaty, 944 UNTS 13, entered into force on 3 October 1972.

[48] Treaty Between the United States of America and the Union of Soviet · Socialist Republics on the Limitation of Strategic Offensive Arms (ungratified).

[49] UNIDIR & VERTIC, *Coming to Terms with Security: A handbook on Verification and Compliance*, , United Nations, June 2003, p. 20.

[50] *Ibid.*

[51] Supra note, 46, p. 207.

[52] UN document A/S-15/3, 1988; it is the 10th principle of the 16 Principles of Verification elaborated by the Disarmament Commission in 1988.

[53] Supra note, 46, p. 207.

[54] WEU document, No.973, May 1984, March 1961, p. 11.

[55] GOLDBLAT, J., "Further Limitation and Prohibition of Nuclear Test Explosions: The Problem of Verification," *Verification of Disarmament or Limitation of Armaments: Instruments, Negotiations, Proposals*, UNIDIR/92/28, New York, United Nations, 1992, p. 35.

[56] *Ibid.*

[57] Supra note, 55, pp. 35-36.

[58] Protocol Modifying and Completing the Brussels Treaty of 1954.

[59] Treaty for the Prohibition of Nuclear Weapons in Latin America, 634 UNTS 326, 14 February 1967.

[60] UN GAOR, 10th Spec. Sess., UND Doc. A/S-10/AC.1/7, 1978.

[61] Convention concerning measures to be taken by Member States of Western European Union in order to enable the Agency for the Control of Armaments to carry out its control effectively, and making provisions for due process of law in accordance with Protocol No. IV of the Brussels Treaty as modified by the Protocol signed at Paris on 23 October 1954.

IAC-08-E8.2.13

ANTI SATELLITE MISSILE TESTING- A CHALLENGE TO ARTICLE IV OF THE OUTER SPACE TREATY

Author
Mr. Kiran Mohan
National University of Juridical Sciences
Kolkata, India
kiranmohaan@gmail.com

Co-author
Mr.Sandeepa Bhat
Faculty of Law
National University of Juridical Sciences
Kolkata, India
sandeep_nls@yahoo.com

Abstract

During the heights of Cold War, the research and development of anti satellite weapons (ASAT) were ardently pursued by the so-called superpowers, the USA and USSR. They were aware that a well developed ASAT in their arsenal would give them a definite edge over the rival, both militarily and psychologically. The disintegration of USSR seemed to slow down the arms race in the outer space in the twentieth century, which is reflected in the subsequent declarations made by the USA and Russia. But in 2007, China demonstrated its ASAT capabilities by destroying its own redundant weather satellite with a modified ballistic missile to become the third nation to possess such breed of weapons. This test was widely perceived as China's response to the doctrine of US space supremacy outlined in National Space Policy released in 2006 and as an attempt to bring US to the negotiation table. The recent US operation of disabling an out-of-control spy satellite is being criticized as a cover for testing its ASAT capabilities, and as a response to the Chinese act. These coupled with the fact that many states, in the recent past, announced their dedicated programmes for the space based military activities, is definitely indicative of resumption of arms race for supremacy in the outer space, which jeopardizes international peace and security.

Although Article IV of the Outer Space Treaty prohibits the deployment of nuclear weapons and any other kinds of weapons of mass destruction, owing to serious drafting faults and interpretative conundrums, ground-based Anti-Satellite Missiles are left out of its purview. The failure of the states to adopt an agreed definition of "peaceful uses", stipulated under Article IV, has left scope for attributing the colour of legitimacy to ASAT missile testing. This being the situation, the present paper ponders into the question of legality of testing and deployment of ASAT Missiles under the present legal framework. It highlights loopholes in Article IV of the OST, which aid states to transgress the barriers of international law. An exclusive treaty to control or prohibit anti satellite weapons is a far fetched dream given the non existence of such political will among concerned nations. This is evident from United States' rejection of the proposal of Russia and China for a new treaty regarding this. Hence the authors propose to suggest plausible solutions to this quandary from within existing international legal framework.

Introduction

ASAT (Anti Satellite) is a broad term which includes weapon systems with earth to space and space to earth capabilities.* ASAT Missiles are generally Intermediate Range Ballistic Missiles (IRBMs), ground-based or air-borne, modified for destroying artificial satellites. Currently, only the United States of America (USA), Russia (inherited from former USSR) and the People's Republic of China are known to have developed these weapons.

During the Cold War period, the superpowers, the USA and USSR relied heavily on their assets in space. Artificial satellites started playing a major role in national security of these nations. Satellites were increasingly used for photographic reconnaissance, intelligence gathering, navigation and defence communications. The development of anti satellite weapons were ardently pursued by both the countries since early sixties, as they were aware that a well developed ASAT in their arsenal would give them a definite edge over the rival, both militarily and psychologically. The United States started the bandwagon with its limited anti satellite programme in the early sixties as a part of its anti ballistic missile development. Soviet ASAT testing and development may have started as early as 1962. Though the SALT agreements deferred testing of these weapons temporarily, it was resumed in no time. By 1982, both the nations seemed to possess relatively developed ASAT systems. They appeared to be hedging their bets by pursuing anti-satellite arms control talks while pursuing anti-satellite technology, albeit at a low level.

However, military uses of space are no longer restricted to the superpowers. More and more, regional rivalries are being expressed in space with dedicated military or dual-use space systems. In 2007, People's Republic of China demonstrated its ASAT capabilities by destroying its own

* ASAT weapons basically include ballistic missiles (ground based and air launched) and interceptor satellites. Some satellites which are designed primarily to perform other functions are said to have ASAT capabilities.

redundant weather satellite with a modified ballistic missile to become the third nation to possess such breed of weapons. This test was widely perceived as China's response to the doctrine of US space supremacy outlined in National Space Policy released in 2006 and as an attempt to bring US to the negotiation table. . It is believed that much of China's accumulation of space power is also directed at Taiwan, which in turn is suspected of providing its military with images of China from its Formosat-2 research satellite.[†] The recent US operation of disabling an out-of-control spy satellite is being criticized as a cover for testing its ASAT capabilities, and as a reply to the Chinese act.

The Outer Space Treaty (OST) of 1967 provides the basic legal frame work for the governance of the outer space and has underpinned the expansion of one of the last great fields of exploration and accomplishment. . It was the first multilateral convention to enumerate "widely accepted guidelines designed to temper the intensity of potential disputes certain to raise in the future allocation of both the spatial and material resources of outer space".[‡] Entered into force in 10, October 1967, ninety eight countries have ratified the treaty so far while twenty seven remain signatories.[§]

Today, when the use of outer space is commonplace, the objectives of the OST have even greater relevance. In 1967 seven states had satellites in space. Today the number is 47.[**] Space use has expanded tremendously. All around the world people now depend on space capabilities for security, travel, communications, resource management and exploitation, early warning systems, search and rescue, medical services, and entertainment. These activities may not have inspired the imaginations of forty years ago, but they are central to our way of life in the contemporary space age. The Outer Space Treaty has been the foundation of this expanding use of outer space, but it is increasingly challenged by its own success.[††] Growing threats to the space environment, increasing rivalry between military space programs and the prospect of new technologies to threaten satellites and other assets in outer space are critical concerns which challenge the broad goals of the treaty. The primary objective of this paper is to analyse the OST, especially Article IV of the Treaty, its relevance and effectiveness in regulating anti satellite weapons.

[†] See http://www.atimes.com/atimes/China/IK09Ad02.html

[‡] Read, Lt. Colonel Walter, *The Outer Space Treaty: Freedoms -Prohibitions- Duties*, U.S.A.F Jag Law Review,1967, p. 26

[§] See http://www.unoosa.org/SpaceLaw/outerspt.html

[**] See http://www.stmary.ws/physics/97/AIQBAL.HTM (visited on 16th Dec, 2007)

[††] See Burton, Andrew, *Daggers in the Air: Anti Satellite Weapons and International Law*, Fletcher Forum, Winter 1988, p 144

Outer Space Treaty: A Brief Overview of its Objectives and Spirit

The Treaty on Principles Governing the Activities of States in the Exploration and Use of Outer Space, Including the Moon and Other Celestial Bodies (Outer Space Treaty is drafted by the United Nations Committee on Peaceful Uses of Outer Space (UNCOPUOS), the Outer Space Treaty governs all activities related to exploration and use of outer space. Commentators have touted the Outer Space Treaty as "the basic charter or constitution governing space activities,"‡‡ "a significant landmark in man's effort to control the use of atomic weapons and other weapons of mass destruction and to prevent military confrontations on celestial bodies,"§§ "the basic charter for the future conquest and utilization of outer space," and the establishment of "a general norm of peaceful uses of outer space."***

Rex Zedalis and Catherine L Wade, from their evaluation of the text of the agreement, derive three basic objectives the treaty attempts to conserve and promote.††† Those objectives are: (1) to guarantee that the outer space, including the moon and other celestial bodies remains the common heritage of mankind.(2) to promote co-operation among, and liability of, the parties exploring outer space, and (3) to prevent arm race from spreading to outer space. These objectives exemplify the widely shared values of security and well being. The following discussion is designed to treat the arm control provisions of the Outer Space Treaty in the context of anti satellite weapon problem.

The Preamble to the Treaty imbibes the spirit of the Treaty regarding demilitarization of the outer space. It reads:

"....taking account of United Nations General Assembly resolution 110 (II) of 3 November 1947, which condemned propaganda designed or likely to provoke or encourage any threat to the peace, breach of the peace or act of aggression, and considering that the aforementioned resolution is applicable to outer space...."

Provisions affecting Anti Satellite Weapons

Articles III, IV and IX seem to affect development, deployment and use of ASAT weapons. Article III specifically binds the treaty parties "to carry on activities in the exploration and the use of the outer space, including the moon and other celestial bodies, in accordance with international law, including the charter of United Nations". Among the provisions applicable to space activities are Articles 2(3) and Article 2(4). Article 2(3) of the U.N. Charter directs

---

‡‡ Hosenball, *The United Nations Committee on the Peaceful Uses of Outer Space: Past Accomplishments and Future Challenges*, Journal of Space Law, 1979, p.95.

§§ Vlasic, *The Space Treaty: A Preliminary Evaluation*, California Law Review,1967, p 50

*** Wulf, *Arms Control--Outer Space*, Journal of Space Law,1983, p 67

††† Zedalis Rex, Catherine Wade, Anti Satellite Weapons and the Outer Space Treaty of 1967, California Western International Law Journal, Vol.8, 1978, p 457, 458

nations to "settle their international disputes by peaceful means in such a manner that international peace and security, and justice, are not endangered." Article 2(4) requires that nations "refrain . . . from threat or use of force . . . in any . . . manner inconsistent with the purposes of the United Nations."‡‡‡

Article IV of the Outer Space Treaty establishes a clear prohibition against placing in orbit around Earth any objects carrying nuclear weapons or any other kinds of weapons of mass destruction. "The moon and other celestial bodies shall be used by the state parties to the treaty exclusively for peaceful purposes. The establishment of military bases, fortifications and installations, the testing of any type of weapons and the conduct of military maneuvers shall be forbidden." Any type of weapons naturally includes weapons with ASAT capabilities.

Article IX of the Outer Space Treaty directs nations to "undertake appropriate international consultations" before proceeding with any activity that might cause "potentially harmful interference with the activities of other states in the peaceful exploration and use of outer space." It is possible to argue that states developing ASATs should do so only after "appropriate international consultations. " Nonetheless, the vague wording of Article IX and the forced nature of such an interpretation reduce the Article's value as an arms control provision.

Taken together, the provisions of the Outer Space Treaty afford satellites some measure of legal protection against attack. The precise nature of this protection is unclear since the treaty was not drafted for the specific purpose of limiting deliberate hostile activities.

Ambiguities and Limitations of the Arm Control Provision of the OST and its Implications on ASAT

Specifically, Article IV provides that:

"States Parties to the Treaty undertake not to place in orbit around the earth any objects carrying nuclear weapons or any other kinds of weapons of mass destruction, install such weapons on celestial bodies, or station such weapons in outer space in any other manner.

The moon and other celestial bodies shall be used by all States Parties to the Treaty exclusively for peaceful purposes. The establishment of military bases, installations and fortifications, the testing of any type of weapons and the conduct of military maneuvers on celestial bodies shall be forbidden. The use of military personnel for scientific research or for any other peaceful purposes shall not be prohibited. The use of any equipment or facility necessary for peaceful exploration of the moon and other celestial bodies shall also not be prohibited."

‡‡‡ Article 51 of the Charter which entitles state to resort to self defence when an armed attack occurs is subjected to this provision.

This disarmament provision of the Treaty suffers from serious drafting faults which lead to various interpretative hassles. Also, it should be taken into account that the nature of the Treaty is general as opposed to a specific one. Article IV establishes a clear prohibition against placing "in orbit around Earth any objects carrying nuclear weapons or any other kinds of weapons of mass destruction." Orbiting weapons using nuclear power would presumably be included. But this provision does not limit ground-based ASATs or ASATs which use conventional explosives or other means to destroy a target. Neither does it ban nuclear armed "pop up" ASAT interceptors that ascend directly to their targets without entering into orbit. Markoff regards this provision as a clause for partial disarmament.[§§§] The following section discuss interpretative conundrums and drafting errors associated with Article IV which aids States to justify their programmes on ASAT development and testing.

Paragraph 2 of Article IV asserts that the moon and other celestial bodies shall be used exclusively for "peaceful purposes". It prohibits "the testing of any type of weapons" on "celestial bodies". Specifically and pointedly, the paragraph does not refer to outer space (or outer void space, as Bin Cheng calls it) as such; that is the space between the celestial bodies.[****] Does this mean, since "outer space" (in the narrower sense of the term) is excluded, the testing of weapons including ASAT missiles in the outer space is not proscribed by Article IV? This interpretation is undesirable as it goes contrary to the goals envisaged by the drafters which are well evident in the Preamble of the Treaty. This seems to be a classic case of drafting error since Article III, Article IX and Article XIII of the Treaty uses "outer space" in a much broader sense, including the Moon and other celestial bodies in its ambit.[††††] So considering the spirit of the Treaty and the intention of the drafters, it should be assumed that the outer space is not excluded from the purview of paragraph 2 of Article IV and it should be used exclusively for "peaceful purposes".[‡‡‡‡]

Unfortunately, the meaning of the term "peaceful purposes" used in the second paragraph of Article IV is another bone of contention. United States has consistently asserted that the term 'peaceful' means non-aggressive and therefore non-aggressive military activities are

§§§*See* Markoff, *Disarmament and "Peaceful Purposes" Provision in the 1967 Outer Space Treaty*, Journal Space Law ,Vol.4, 1976,p. 4

**** *See* Bin Cheng, Studies in International Space Law, Clarendon Press London, 1997, p. 518

†††† Some scholars suggest that the omission of the term "outer space" was not accidental. The rafters of the Treaty did not intend to enact a broad prohibition of military activity, and thus carefully constructed Article IV accordingly; *See* Dana James, *The Legality of Anti satellites*, Boston College International & Comparative Law Review, Vol.3, 1979-80, p. 467

‡‡‡‡ *See generally* Article 31 of Vienna Convention of the Law of Treaties.

permitted. The United States is of the view that, as the Charter of the United Nations permits the military activities necessary for individual and collective self-defense, non-aggressive military activities necessary to protect its space assets are within the ambit of peaceful uses. However, it should be noted that, though the view that defensive military actions are justified under the Charter, it does not say that defensive military actions are peaceful.

Article III of the OST, may be because of its generic nature, is often overlooked as an effective disarmament provision. It says:

States Parties to the Treaty shall carry on activities in the exploration and use of outer space, including the Moon and other celestial bodies, in accordance with international law, including the Charter of the United Nations, in the interest of maintaining international peace and security and promoting international cooperation and understanding.

The military activities in outer space counter many of the UN Charter provisions. The maintenance of international peace and security, being the most important agenda of the United Nations, the Preamble of the Charter of the UN imposes a duty on the States to practice tolerance and co-exist in peace. The purpose of the United Nations, as expressly stipulated in the Charter, include the removal and prevention of threat to peace[§§§§] and development of friendly relations among the nations[*****]. No doubt, the testing of ASAT weapon in outer space creates that element of suspicion required to disrupt the existing power equations among nations and spurs an arms race in space.

Vladlen Verschchetin, the Former Judge of the International Court of Justice says that "Semantic methods can not transform a military activity into peaceful activity and vice-versa; in any language, peaceful activity remains peaceful, and military, military."[†††††] Bin Cheng vehemently criticizes this approach when he says " For as long as the United States restricts its idiosyncratic interpretation of the word 'peaceful' to some non-existent limitation on the military use of the outer void space, perhaps no more harm is done than the emperor preening himself in his non existent clothes."[‡‡‡‡‡] But the Soviet Union and some other developing countries prefer to interpret the term to be non-military.[§§§§§] Their view suggests that when peaceful purposes clause is construed in conjunction with Article I paragraph, which calls for the use of celestial environment for the benefit and in the interest of all countries, peaceful must mean non-military.

§§§§ Article 1(1), Charter of the United Nations.

***** *Ibid*, Article 1(2)

††††† Quoted in Kamenetskaia, E.P, *International Legal Problems of Preventing an Arms race in Outer Space*, in W.E.Bulter (ed.), Perestroika and International Law, Martinus Nijhoff Publishers, 1990. p 149

‡‡‡‡ Cheng, Bin, *Studies in International Space Law*, Clarendon press Oxford, 1997, p. 521

§§§§§ Ibid.

Actually, the language of the Article IV, paragraph 2 is insufficient to suggest a conclusive definition. A clarification in this regard seems pertinent since testing of anti satellite weapons is a non aggressive activity but essentially a military activity.

As noted earlier, the second paragraph makes no reference to outer space (in the narrow sense of the term)[******]. Gorove wonders whether the failure to mention outer space mean that the outer space can be used for non peaceful purposes so long as the prohibitions of paragraph one and other relevant international law principles are not transgressed.[††††††]If this explanation is taken into face value, ASAT testing in the Outer Space is not at all prohibited by the Treaty.

Concluding Observations and Recommendations

Forty One years after the ratification of the Treaty, space is still free of weapons, the number of states accessing space continues to rise, and the benefits of space applications touch almost every aspect of human life. This accomplishment speaks to the continuing relevance of the Treaty as the cornerstone of outer space governance. Yet there are political, military, and technological challenges to this regime. In many ways these challenges are reminiscent of the concerns that initially drove the creation of the Treaty, both to prevent outer space from becoming a battleground, and to prevent colonial competition and damaging exploitation. But technologies, concepts, and geopolitics have developed and changed in the last four decades in ways that are interconnected and mutually reinforcing. Addressing these challenges and the changing security context demands significant international dialogue.

An exclusive treaty to control or prohibit ASAT weapons is a far fetched dream given the non existence of such political will among concerned nations. It is noteworthy in this context that the Moon Agreement is still not ratified by any space firing nation. The United States' outright rejection of the Chinese and Russian proposal for a comprehensive disarmament treaty in 2002 also serves an indicator of this trend. A plausible step towards controlling ASAT development and testing can be taken by the parties to the Outer Space treaty by adopting a position in this regard, after a review of state practices since 1967 and the negotiating history of the treaty. Unanimity among parties is not required for any formal interpretations, but a large majority of parties adopting a particular position would be persuasive. Similarly, the United Nations, acting through its First Committee and then through the

[******] Bin Cheng uses the term 'Outer void space' for the space between the celestial bodies; See Cheng Bin, *Studies in International Space Law*, Clarendon press Oxford, 1997, p. 517

[††††††] Gorove, Stephen, *Arm Control Provisions in the Outer Space Treaty: A Scrutinizing Reappraisal*, Georgia International & Comparative Law Review, 1973, p. 119

General Assembly (which recommended the Outer Space Treaty in the first place), could pass a resolution formally interpreting it. If there were significant dissent, pursuant to the UN Charter the General Assembly could request an advisory opinion from the International Court of Justice at The Hague confirming this interpretation.

It seems pertinent to note that the Outer Space Treaty does not include a formal process for international review. And although it contains provisions for international consultation if a planned event might cause harmful interference to the activities of another state, this provision has not been used. The Chinese did not hold international consultations prior to their anti-satellite test. While the details of US intelligence and actions regarding the event are not public, it would appear that the US neglected the possibilities of requesting consultations despite evidence of previous Chinese Anti satellite attempts. The Outer Space Treaty, while more or less observed, is not engaged, and risks growing stagnant. After 41 years it is time for a review of the letter, spirit, and application of the OST so that it can continue to guide the international community towards the type of security in outer space that can support the fulfillment of our imaginations.

Some Legal Problems on the Prevention of Proliferation of Weapons in Outer Space

Prof. Dr. Anatoly Kapustin

Peoples' Friendship University of Russia, Moscow

From the very beginning of space epoch military interests were one of determinatives of space activity of the States. In particular, in these purposes satellites of various assignment, including the early notification, communication, gathering of the information, check and navigation that helps to increase efficiency of ground armed forces are actively used. A similar sort satellites are not the weapon in the true sense this word: they do not create threat of a direct attack in space or from space.

Moreover, they promote maintenance of stability in the international relations. For this reason the intelligence satellites used for the control of performance of agreements on restriction of arms, use the international protection as national means of the control of observance of these agreements. Satellites of the early notification use similar protection. With the help of a space communication facility more reliable operative communication of Heads of States in the tense situations is organized. It also reduces probability of acceptance of wrong reciprocal decisions in critical military-political situations. Activity of such satellite systems in strong unit is connected to ground armed forces of the States. According to international obligations of the States, their national means of the control cannot repair handicapes, for example, to attack on them, to move or guzzle them, it is impossible to prevent their functioning by application of measures of masking on the Earth. Also it is impossible to create handicapes to systems of the early notification or to attack them. In process of restriction of arms on the Earth activity of such satellite systems will be reduced or transformed on their use extremely in the peace purposes.

For today of effort of the international community are directed on international legal prohibition of tests and expansions in a Space anti-satellites weapon, capable to damage, put out of action and destroy the satellite systems of the States serving them ground armed forces. Use of a similar sort of systems would be qualified from the point of view of international law as armed attack on the state with all tragical consequences following from it. Prohibition of anti-satellites weapon would be logic end of already achieved and working international agreements on prohibition of accommodation in a space of objects with the nuclear weapon or other kinds to the WMD, tests, creations and expansions of partial orbital system of bombing, expansion of systems or components of antimissile defense (AMD) space basings.

Whereas under effective international obligations of the States in a space it is forbidden to place separate kinds of the weapon from the point of view of international law it can be qualified as an establishment in this space of a mode of partial demilitarization. Only after an interdiction of anti-satellites weapon it will be possible to speak about an establishment in a space of an international legal regime of the full demilitarization excluding from this space in a peace time any kinds of the weapon.

Certainly, in connection with consideration of a problem of demilitarization of space inevitably there is a question on, whether it is necessary to deal with a problem of non-distribution of the weapon in space where it for the present is not present? Whether attracts it to a leaving from such important task as struggle against pollution of space by dangerous splinters or Debris? It is necessary to answer these questions positively as problems of a safety of space activity concern to number of the major in the international space law. Therefore the initiative on acceptance of effectual measures on prevention of race of arms in space is represented necessary and duly.

Really, necessity of acceptance of effectual measures on prevention of race of arms in space is determined by that it is necessary for maintenance of an international peace and security. If the international efforts appear insufficient for prevention of race of arms in space safety in the world begins much less. Under threat of there is a strategic stability which today is a key element of military-political balance in the world.

Timeliness of acceptance of the appropriate effectual measures on prevention of race of arms in space also is obvious, as recently scales and the importance of research and use of a space essentially grow. Interests of the further dynamical development of the international space cooperation urgently demand acceptance of measures with the purpose to not allow to transform a space in arena of an opposition to provide its preservation free from the weapon of any kind.

Thus it is marked, that *in the basis of the international space law* the idea of an anticipation of undesirable and dangerous consequences of space activity is incorporated. It induces the States to search of adequate answers to calls of modern development, including in sphere of military - space activity. So, according to item Y of the Treaty on principles governing the activities of States in the exploration and use of Outer Space, including the Moon and other Celestial Bodies of 1967 (further - the Outer Space Treaty) the State-participants immediately should inform other State-participants and the Secretary general of the United Nations on any phenomena established by them in a space, including the Moon and other celestial bodies which can present danger to life or health of cosmonauts. At the same time in an item IX of this Treaty the reference to its State-participants contains in case they have the bases to believe, that activity or the experiment planned by this State-participant or citizens of this State, will create potentially harmful handicapes of activity of other State-participants of the contract in business of peace research and use of a space, including the Moon and other celestial bodies, it should carry out the appropriate international consultations before to start such activity or experiment.

From its part, those State-participants of the contract which have the bases to believe, that activity or the experiment planned by other participant of the Treaty, will create potentially harmful handicapes of activity in business of peace research and use of space, they can request realizations of consultations concerning such activity or experiment. Though here nothing it is spoken about military activity or military experiments, evidence of application to them of the precautious approach does not cause doubts.

The history of development another not less an acute problem of the international space law - space Debris is indicative in this respect. So if effectual measures also pollution of space, in such degree as now, would not take place were in due time accepted.

In this respect it is impossible to ignore the initiative of two conducting Space Powers - Russia and China which February, 12, 2008 have presented at Conference on disarmament in Geneva the draft Treaty on prevention of the placement of weapons in Outer Space and of the threat or use of force against Outer Space objects. Russia has found conference by the most suitable forum for multilateral work under the project of the specified contract in view of its mandate, the agenda and high expert potential in the field of military space. Initiators of the project considered the offers made by the State-participants of conference during teamwork above elements of the treaty which were earlier offered by Russia and China together with group of co-authors and were fruitfully discussed here on an extent more than five years.

The Russian officials marked, that the draft treaty is brought with the so called research mandate. It is supported by the majority of the State-participants of Conference and does not create any complexities for search of the compromise in the program of its work. Russia and Chinese People's Republic expect that in the subsequent when for this purpose conditions will ripen, work can be transferred in a trunk-call channel with establishment appropriate Ad hoc Committee of the Conference on disarmament.

The modern international space law does not forbid placement in the Outer Space of the weapon which is not the weapon of mass destruction (WMD). At the same time, such weapon placed in the Outer Space, would have a global operative range, high readiness for application, an opportunity of reticent influences on outer space both ground objects and their removing out of operation. It would become - as against the weapon of mass destruction - not the tool of restraint, and the weapon of real application that would generate suspiciousness and intensity in relations between the States, would break a climate of trust and cooperation in an outer space exploration.

It is not necessary to forget also, that placement of the weapon in outer space by one State inevitably will cause chain reaction. And it, in turn, is fraught with a new coil of race of arms both in space, and on the Earth.

First of all, it is necessary to note, that the draft treaty provides an prohibition on the placement of the weapon of any kind in a outer space, application of force or threat force concerning space objects. The Treaty is called to remove existing blanks in the international space law and to create conditions for the further research and use of outer space, to strengthen general security and the control above arms.

As is known, the article IY of the Outer Space Treaty establishes a mode of demilitarization of the outer space, and also the Moon and other celestial bodies. In particular, the paragraph first of this article establishes a principle of partial demilitarization of the outer space as the State-participants of the treaty have undertaken to not put into orbit the Earth any objects with the nuclear weapon or any other kinds of the weapon of mass destruction to not establish the weapon on celestial bodies and to not place such weapon in the outer space any otherwise. In other wards, this condition of article IY contains an interdiction on place in orbit, an establishment and station only one kind of the weapon - the weapon of mass destruction.

The second paragraph of article IY of the Outer Space Treaty establishes full demilitarization of the Moon and other celestial bodies as the State-participants have agreed to use them extremely in the peace purposes. The article contains the provisional list of kinds of the activity forbidden on the Moon and other celestial bodies which was added and expanded in an item 3 of the Agreement on activity of the States on the Moon and other celestial bodies of 1979.

The draft treaty of Russia and Chinese People's Republic pursues three interconnected, but nevertheless various on the legal consequences "antimilitaristic" if so it is possible to be expressed, the purposes: to not allow placement in outer space of any kinds of the weapon; to not resort to application of force or threat force concerning outer space objects and to not render assistance and to not induce other States, groups of the States or the international organizations to participation in the activity forbidden to the draft treaty.

In article II of this project the obligation of the State-participants around of the Earth is provided to not put into orbit any objects with any kinds of the weapon to not establish such weapon on celestial bodies and to not place such weapon in outer space any otherwise. If to compare the given position to paragraph 1 of article IY of the Outer Space Treaty it is easy to be convinced that authors of the draft treaty practically have completely reproduced the formulation of the current Outer Space Treaty, having distributed its provisions on any kinds of the weapon, and not just on the weapon of mass destruction.

Whereas commentators of item IY of the Outer Space Treaty gave various interpretation to concept of "station" in the outer space of any weapon, composers of the draft treaty have decided to remove this blank. First, in point "á" of article I of the draft is defined "outer space" which means elevated space above 100 kms above a level of ocean. Such understanding of a high-altitude limit in definition of a outer space for the purposes of station of objects in Earth orbits today is shared by many lawyers. Second, in point "d" of the same article of the draft treaty it is established, that the weapon will be considered "placed" in outer space if it makes as a minimum one revolution (rotation) on an orbit around of the Earth or follows in connection with such orbit with the further leaving from it, or is on a constant basis somewhere in outer space. Unlike an article IY of the Outer Space Treaty, providing an interdiction on «placing into an orbit » the weapon, its establishment on celestial bodies and station in outer space any otherwise, the draft treaty contains the important specification that station includes as a minimum one revolution in a Earth orbit or following in connection with such orbit with the further leaving from it or a presence on a constant basis somewhere in a outer space. Though, as well as in the Outer Space Treaty, the term «station of the weapon» in the draft treaty is wider under the contents, including «placement into an orbit» and «an establishment on celestial bodies», the concept «placement into an orbit » has received more certain treatment.

Article I of the draft treaty defines one of key terms - «the weapon in the outer space» according to which it means any device placed in a outer space, based on any physical principle, specially created or converted for destruction, damage or infringement of normal functioning of objects in a outer space, on the Earth or in its air space, and also for destruction of the population, components of biosphere, important for existence of the person, or for drawing of damage by it. Thus, authors of the draft treaty recognize that under an interdiction the weapon should fall not only the weapon placed in a outer space and intended for destruction, damage or infringement of activity of any objects in a outer space, but also the weapon which can be used against the population of the States or components of biosphere, that is for defeat quite «terrestrial objects».

Probably, it is impossible to recognize the given formulation exhaustive as from it remains not clear why on the Earth as object of such illegal use of the weapon "population" acts only. And unless civil objects, and other objects forbidden by the international humanitarian law as the purposes of military attacks (cultural values, installations and the constructions containing dangerous forces), should not be a subject of regulation of given article? Probably, the expanded formulation of this preposition - «civilians and other objects the attack on which is forbidden by the international humanitarian law» would be more acceptable.

Separate question is on military objects against which the international humanitarian law does not forbid application of military force. How to be with them? Unless it is possible to place any objects with the weapon in a space with the purpose of destruction of military objects of other states, for example, in the order preventive self-defenses? Apparently with the approach recognizing legitimacy of accommodation of the weapon in outer space under a pretext of preventive self-defense to agree it is impossible, for obligations on prevention of accommodation of the weapon in outer space as a whole otherwise would be brought to nothing.

In general the question on self-defense in the draft law deserves the separate analysis. The draft treaty in article Y provides, that anything in the given document can not be interpreted as interfering realization by its State-participants of the right on self-defense according to article 51 of the Charter of the United Nations. As is known, according to article 51 of the Charter of the United Nations, each State has an inalienable right to individual and collective self-defense, but can take advantage of this right, only in case of armed attack and until the Security Council of the United Nations will not accept the measures necessary for maintenance of an international peace and security.

As the Earth outer space is not declared still excluded of sphere of military actions in case of a confrontation between the States according to effective standards of international law, including the Charter of the United Nations, it is necessary to recognize legitimacy of use of a outer space for drawing retaliation on an aggressor during realization of the right of lawful self-defense. At the same time it is necessary to declare fully completely, that such recognition has the logical limits. Really, in conditions of a possible nuclear rocket opposition of the countries having the nuclear weapon, especially dangerous are the statements of some military experts - supporters so-called against forcible strategy concerning legitimacy so-called preventive self-defenses, or, in other words, «anticipatory impact» from outer space or through it.

It is necessary to turn special attention to legitimacy of use of outer space in interests of national security. Outer Space activity of the States concerns to this under the control of observance international disarmament treaties and agreements (outer space systems of supervision), to the warning of a rocket attack (system of detection of starts of rockets), management of armed forces (system of communication), maintenance of their daily and fighting activity (system of definition of a site, meteorological, geodetic, cartographical maintenance etc.). Criterion of legitimacy of application of outer space means with a view of the international control are, first of all, the generally recognized principles and norms of international law. To control activity with the help of outer space means all general principles of international law, and, first of all, principles of non-interference to internal affairs of the States, a sovereign equality of the States, conscientious performance by the States of the international obligations, cooperation of the States and others are applicable.

Use by the States of outer space means for the control of observance of the international obligations should, except for it correspond to so called special principles of the international space law. As is known, the outer space is free and open for research and use by all states (an article 1 of the Outer Space Treaty).

However it does not mean, that for the States there are no restrictions of freedom at realization of outer space activity by them. Freedom of outer space is understood only as the allowable activity answering to interests of maintenance of an international peace, to interests of all mankind. Besides in article IX of the Outer Space Treaty necessity of the due account of the appropriate interests of all States by which they should be guided at realization of any outer space activity is fixed. To the full these requirements concern and to control activity with use of outer space means.

Certainly, it does not mean, that military outer space activity should conduct to transformation of a outer space into a potential battlefield.

As to the draft treaty the question on allowable and lawful military activity in outer space it is not affected. In the preamble of the draft treaty the positive role of agreements of the control on the arms and to the disarmament, concerning to outer space, including bilateral agreements, and also the existing legal regimes concerning use of outer space is only marked. Their inability effectively to prevent station of the weapon and race of arms in outer space in the same place is marked.

The second significant obligation of the States established in article II of the draft treaty is the obligation to not resort to use of force or threat of force is concerning outer space objects. In article 1 of the draft treaty it is specified, that «use of force» or «threat force» are understood as any hostile actions against outer space objects, including directed, in particular, on their destruction, damage, time or constant change of parameters of an orbit, or threat of fulfilment of such actions.

In same article it is stipulated, that the term «space object» means any device intended for functioning in a space, removed on an orbit around of any celestial body, or taking place on an orbit around of any celestial body, or on any celestial body, except for the Earth, or descending with an orbit around of any celestial body to this celestial body, or moving from any celestial body to other celestial body, or placed in outer space any otherwise. It is possible to approve, that in this definition which favourably differs from earlier attempts (the Rescue Agreement of 1968, the liability Convention of 1972 or Registration Convention of 1975), are stipulated all versions of technical devices and the constructions intended for use in a space, thus it corresponds to the established international legal practice, in particular to treaty practice of the States.

As to an ban of use of force or threat of force against outer space objects it logically follows from main principle of the general international law of non-use of force or threat of the force, stipulated by item 4 of the article 2 of the Charter of the United Nations. We already marked, that in outer space, as well as in any other sphere of human activity, the States should be guided by the basic treaty principles and the norms of international law fixed in the Charter of the United Nations. Thus, according to effective standards of the general international law, the States are obliged to abstain in the outer space activity from any hostile actions to resolve conflicts and disputes by extremely peace means. It, in particular, means inadmissibility of any violent intervention in lawful activity of space object of other country by its destruction or damage, and also capture or displacement from an orbit. The attention pays to itself, that as against earlier expressing offers on forbidding application of force both in outer space, and from outer space concerning the Earth, the draft treaty has more concrete object of regulation - non-use of force against outer space objects, that is forbids to use force extremely in outer space. At the same time, if to take into account the definition of outer space object given in article 1 of the draft treaty the use of force against the outer space objects located on other celestial bodies except for the Earth is forbidden. Alongside with prohibition of use of force or threat of force the draft treaty provides ways of settlement of disputes between its participants concerning application or interpretation of its provisions. The offer on creation of special conventional body is included in the list of means of peace settlement - the Executive organization of the treaty. This organization, alongside with assistance to settlement of disputes between participants of the treaty is allocated with the right of the control of observance of its provisions and acceptance of measures with a view of the termination of its infringements.

To a problem of an ban of the use of force and threat of force in outer space closely connects idea of full prohibition anti-satellites means which associates with the concept of immunity of artificial satellites of Earth (ASE). As is known, in initial variant this concept assumed prohibition of development, test and expansion anti-satellites means (ASM), and also liquidation of existing systems of ASM. It is necessary to note, that a direct prohibition on realization of tests ASM in agreements in force does not contain. The unilateral moratorium of the former USSR declared in 1983, has not affected aspiration of other countries to carry out test of similar systems. Nevertheless, Russia as a successor of the former USSR, continues to observe the specified moratorium. Then this idea has received development in the offer on maintenance of "immunity" of artificial satellites of Earth (ASE), or in their inviolability, that is an interdiction of attacks on ASE.

In the practical plan a number of bilateral Russian-American agreements (earlier the AMD, now - Agreements SOLT-1 and SOLT-2) provide inviolability of the satellites which are carrying out functions of the control of processes of restriction of arms and disarmament. Inviolability communication satellites of strategic purpose also use. In this connection the question on the right of other countries on similar inviolability repeatedly rose.

In the draft treaty as it is easy to be convinced of it, the problem of the prohibition of development or realization of tests of similar means is not put in general. Probably, it speaks that circumstance, that serious disagreements at sessions of Conference on disarmament concerning concepts of peace assignment of satellites, and also them and the military assignment which are not carrying destabilizing character etc., do not allow to hope for achievement of the fast arrangement on this question, therefore developers of the draft treaty have gone on a way of fastening of the least debatable provisions. Even the similar approach will promote, finally, to observance by the States in outer space of a principle of the international law forbidding threat by force or its use against outer space objects. It is necessary to notice, that aspiration to prevention of development or tests of any kinds of the weapon, is more characteristic for the international agreements forbidding use of various kinds of the weapons forbidden by the international humanitarian law (laser weapons, WMD etc.) that cannot be told about systems ASM.

It is necessary to notice, that absence of precise and distinct rules in this sphere creates ground for potential conflicts, including the armed character, between the States which are carrying out space activity or between those who fails it also outer space powers. In particular, in case of detection of any hostile actions against the outer space objects, the States have the right, as it was marked above to realize the right on self-defense, that in turn, can cause escalation of intensity and result in continuation of military actions on the Earth. Presence of anti-satellites weapon at the States does a similar opportunity quite probable, that once again emphasizes interrelation of possible conflicts on the Earth and in outer space. It also is additional argument for the benefit of the prompt end of work on the project of the given draft treaty.

At last, *the third object of the researched draft treaty is the obligation to not render assistance and to not induce other subjects of international law (the state or the international organizations) to participation in the activity forbidden by the draft treaty.* At first sight given rule looks superfluous as warns actions which can be defined as various forms of partnership in infringement of norms of the international law fixed in the given draft treaty. Illegality of similar activity follows from principles of the international responsibility on the general international law. However, inclusion of similar provision, in our opinion, was dictated by aspiration to not allow to occurrence of fears in its participants concerning easing national security, and, hence, to provide a high level of trust in relations between them. Not casually, in article YI of the draft treaty it is spoken about necessity of assistance of confidence for observance of provisions of the treaty and its maintenance of transparency and also consolidation of trust in outer space area to that realization of the coordinated measures of strengthening of trust on a voluntary basis if not will be arrangements on other should serve.

Summing up consideration of the draft treaty, it is possible to express opinion that its practical discussion and probable acceptance will promote strengthening of the international security, to the control above arms that is important for maintenance of strategic stability and for creation of conditions of the further research and use of a outer space in the peace purposes. By the way, Russia has shown the initiative to not place the first in a space the weapon of any kind which was maintained by its allies. And it is already good mark.

IAC-08-E8.3.2
51st INTERNATIONAL COLLOQUIUM ON THE LAW OF OUTER SPACE (IISL)
WEAPONISATION OF OUTER SPACE IN THE LIGHT OF THE OUTER SPACE TREATY, PART B
GLASGOW, 1 OCTOBER 2008

OUTER SPACE: ARENA FOR WAR OR PEACE

SA'ID MOSTESHAR
DIRECTOR, LONDON INSTITUTE OF SPACE POLICY AND LAW
LONDON ENGLAND
SA'ID.MOSTESHAR@SPACE-INSTITUTE.ORG

ABSTRACT

The Outer Space Treaty does not appear to prohibit weapons in space *per se*, but examination of the context in which it was agreed and the discussions at the time introduce a strong element of ambiguity. How the issues are resolved and whether space will be reserved for peaceful purposes can only be examined in the light of the political, military and international relations imperatives of each contender in the field. Although the United Nations has resolved to keep space free of weapons, the United States has been a notable abstainer, and more recently opponent, of the Resolution. In this paper it is argued that the pursuit of weapons in space weakens the international legal structure for peaceful uses of outer space and serves against the ultimate international relations objectives of those who follow such a course.

INTRODUCTION

In 1959 at Grünsbach in the Alsace, Dr Albert Schweizer was asked for his thoughts on man's attempt to reach the Moon. He replied "Poor Moon".

In this Paper the following issues will be examined.

1 Relevance of International Law to International Relations
2 Short History of Space Law
3 Relevant International Law
    (1) Peaceful Use Provision of the Outer Space Treaty
    (2) Anti-Ballistic Missile Treaty
    (3) Limited Test Ban Treaty
    (4) Strategic Arms Reduction Treaty
    (5) Conventional Weapons Convention
    (6) UN Resolutions
    (7) International Telecommunications Union
4 Changing US Policy
    (1) US Space Policy 1996
    (2) Policy Comparison 1996 – 2006
5 Impact of US Policy Shift on International Relations
    (1) National Missile Defence
    (2) Anti-Ballistic Missile Treaty
6 Peaceful Use of Outer Space
    (1) Alternative Interpretations
    (2) Pre-emptive Strike
    (3) Other Concerns
7 Prospects for the Future
    (1) International Containment Efforts
    (2) Need to Review US Policy
8 Conclusion

Views expressed here are the author's alone and do not reflect the views of the London Institute of Space Policy and Law, the IISL or of any other organisation.

RELEVANCE OF INTERNATIONAL LAW TO INTERNATIONAL RELATIONS

Public international law provides part of the framework for international relations between States. The policies pursued by each State will have a direct impact on the nature of those relations. US space policies invoke international laws of outer space, and have to be evaluated in the context of those laws. This paper is concerned with some of the consequences of US space policy on international relations.

The most important principle of international law is the sovereignty of States, recognising the right of each State to self-determination in its affairs.[1] Treaties and customary conduct define international law, which is further influenced by legal precedent and writings of academics and practitioners. Because it relates to the conduct of States and their relationships, it can only be invoked by a State. The United Nations (UN) provides the forum and administrative structure for the management of international law. States observe international law not because there are UN or other external enforcement provisions, but because it is in each State's best interest to do so. As of today, all but one of the sovereign States are members of the UN.[2]

It is fundamental to the UN that it has no direct enforcement powers or mechanism. Its primary charge in case of a dispute between States is to facilitate resolution by peaceful means. However, when there is a threat to international peace,[3] the UN Security Council can make certain determinations and mount collective international action to remedy a breach of international law.

SHORT HISTORY OF SPACE LAW

The first international treaty on space was concluded in 1967, two years after the launch of Early Bird[4] and six years after Yuri Gagarin became the first man in space on 12 April 1961. The Outer Space Treaty[5] (OST) is the primary instrument of international space law, and specifies permitted uses of space. The principles outlined in the OST have been elaborated in many subsequent treaties and conventions between various States. There are also numerous UN General Assembly Resolutions dealing with specific issues.

At the time the Outer Space Treaty was being negotiated, the Soviet Union and the United States were the two States with space capability. There was considerable discussion about permissible uses of outer space, including military uses. While every State was aware of the potential advantage that would be gained by controlling weapons in space, neither space power wanted the other to have military use of space. The declaration and provisions of the OST, providing that States may use outer space only for peaceful purposes, and in the interest of all States resolved the issue. Like the high seas and Antarctica, outer space is available for exploration and use by all States equally. Nor can any State exclude others from any part or use of outer space. Under the OST, sovereign rights cannot be exercised over any part of outer space.

RELEVANT INTERNATIONAL LAW

A substantial body of international law has been developed that is relevant to consideration of national space policies. A 1999 UN list of relevant documents runs to 55 pages.[6] The following summarises only a few of these.

(1) Peaceful Use Provision of the Outer Space Treaty

As has been stated, the OST precludes an individual State's exclusive right to space, and its unilateral use. More specifically, the peaceful use provision of the OST[7] can be interpreted to prohibit space weapons entirely. Interpretation of the phrase "peaceful purposes" is discussed later in this paper.

(2) Anti-Ballistic Missile Treaty

The Anti-Ballistic Missile Treaty (ABM Treaty or ABMT) was signed in 1972 by the US and the USSR, and remained in force until 2002. The Treaty limited ABM systems used in defending against missile-delivered nuclear weapons.[8] Placement of anti-ballistic missile systems elsewhere than those specified was forbidden under the treaty.

(3) Limited Test Ban Treaty

Any attempt to place nuclear weapons in space will inevitably require some level of testing. The Limited Test Ban Treaty explicitly prohibits such tests.[9]

(4) Strategic Arms Reduction Treaty

Signed between the Soviet Union and the US in 1991, START, later re-named START I,[10] prohibits the placing of nuclear and other weapons of mass destruction in space.[11] Five months after its signature the Soviet Union dissolved, and the rights and obligations under the Treaty devolved to the constituent parts of the Union: Russia, Belarus, Kazakhstan and Ukraine, the latter three of which no longer have nuclear weapons. (For START II, see note 61.)

(5) Conventional Weapons Convention

Protocol IV to the Conventional Weapons Convention[12] was adopted in 1995.[13] The Protocol goes some way towards limiting the use of laser weapons in space. Although most of the Protocol focuses on technology designed to blind individuals, it also requires parties employing laser systems to "take all feasible precautions to avoid the incidence of permanent blindness to unenhanced vision."[14] To comply with this provision, any use of laser weapons against land-based targets must be "legitimate."[15]

(6) UN Resolutions[16]

The UN General Assembly has passed resolutions each year for the past 25 years calling for the continued peaceful use of space and the prevention of an arms race in space. The resolution asks all States to refrain from actions contrary to the peaceful use of outer space and calls for negotiation in the Conference on Disarmament on a multilateral agreement to prevent an arms race in outer space. Most of these resolutions have been unanimous and without opposition, although the United States and a few other governments have abstained. In the 2006 version, adopted by the First Committee of the General Assembly in October 2006, there were 166 votes for the resolution with only the US opposing. Israel and the Ivory Coast abstained, and 23 permanent representatives were absent from the First Committee. The resolution was again adopted in 2007.[17] The US opposed and Israel abstained from the resolution.

These repeated, nearly unanimous resolutions - against which the United States had not until 2005 found it expedient to vote[18] - demonstrate the existence of a norm against the weaponisation of space. They also indicate a widespread desire to expand existing multilateral agreements to include an explicit prohibition against all weapons in space.

Beyond this, there are five relevant General Assembly resolutions. They are: the Declaration of Legal Principles Governing the Activities of States in the Exploration and Uses of Outer Space (1963), which preceded the Outer Space Treaty and laid out most of its content; the Declaration on International Cooperation in the Exploration and Use of Outer Space for the Use and Benefit and in the Interest of All States (1996); and resolutions on Direct Television Broadcasting, Remote Sensing of the Earth from Outer Space (which seeks to ensure affordable access by developing countries to non-military satellite imaging), and the Use of Nuclear Power in Outer Space (which deals with limiting exposure in the crash landing of nuclear-powered satellites and the liability for such accidents).

(7) International Telecommunications Union

Other international instruments are pertinent to space. The International Telecommunication Union (ITU) allocates radio frequencies used by satellites. It would be difficult for any one country to operate satellites without coordinating its efforts through the ITU. This regime encourages state cooperation and also provides a locus of influence should any State pursue behaviours, such as the deployment of space weapons, that are dangerous for other states.

CHANGING US POLICY

(1) US Space Policy 1996

The US interest in space-based weapons found its most visible expression in the Strategic Defence Initiative (SDI), announced by President Reagan on 23 March 1983. The SDI programme had as its stated aim the development of non-nuclear missile defences.[19] Early plans for the SDI also included the use of space-based technologies such as lasers and particle beam weapons. The space-based elements of the system proved impractical and were abandoned.

Despite these developments, the US government's policy on space remained largely in line with the OST, as manifested by its policy statement of 1996.[20] Militarization of space was not an explicit aim.

US policy underwent a shift after the September 11, 2001 attacks in the United States, when military-based policies and spending increased. Policies relating to space-based weapons also changed. Commenting on these policies, the Washington DC-based Centre for Defence Information (CDI) observes:

> Unlike in Star Trek, the 'final frontier' has yet to become a battlefield. But if the current trends continue, that will change — not in the distant future of science fiction, but within the next several decades. Emerging US plans and policies are clearly aimed at making the United States the first nation to deploy space-based weapons. There are several drivers behind this goal, including the perceived vulnerability of space assets that are increasingly important to how the US military operates, and the administration's decision to pursue missile defence.[21]

The CDI report also points out "The Administration's views were directly reflected in the 2001 Quadrennial Defence Review (QDR), released October 1, 2001." This states: "A key objective ... is not only to ensure US ability to exploit space for military purposes, but also to deny an adversary's ability to do so." In this context, then, by 2002 the US no longer saw space as a resource available for all of humanity, but as another arena from which to fight geopolitical and economic battles. This view became official US policy by 2006.[22]

(2) Policy Comparison 1996 - 2006

The 1996 US Space Policy recognises the equal rights of all States over space[23] including the acquisition of data from space. There have been many developments before and after the publication of this document, which moved in a new direction; some driven by Military and some by Government.

However, it was not until the release of the 2006 Policy that there is an official assertion of a specific US right to operate freely in space.[24] It is also clear that there has been an important change in the U.S. view of acceptable activities in space.

The list of guiding principles of US space policy in 2006 includes the right to assert US interests over all others,[25] and to oppose any attempt to restrict total freedom of action by the US.[26] The latter declared policy guide, together with the first[27] and second[28] stated 2006 Policy[29] goals, are as clear a statement by the US as is likely to be forthcoming of its intention to place weapons in space.

To understand the current policy, and how it differs from previous policy, it is helpful to examine the two policy statements made by the US Government, ten years apart.

(a) Changing Context

A superficial consideration of the policies outlined in 2006 may suggest little change in the US policy from that of 1996. Although both 1996 and 2006 Policy documents discuss the same aspects of military space operations, including denying freedom of action in space to other countries, the 1996 Policy places it in the context of international legal obligations,[30] and recognizes that diplomatic and legal means are important tools in solving space security conflicts.

In contrast, the 2006 Policy[31] asserts US rights of freedom to act and to exclude others, without recognition of legal constraints. This aspect of the 2006 Policy has been said to be indicative of US disregard for international law and its institutions.[32] The 2006 Policy can be interpreted to assert a US right to place weapons in space.[33]

The changes in context include a stated interest by the US administration in new military uses of space and increased funding for military space programmes,[34] and a much greater reliance on space assets in the conduct of U.S. military operations.[35]

While US authorities characterize their policies as defensive (as with Missile Defence and Star Wars), the US military explicitly says it wants to "control" space to protect its economic interests and establish superiority over the world.

(b) Shifting Emphasis

Evidence of a shift in policy is found in the changes in emphasis, all of which must be carefully interpreted within the context of surrounding circumstances.[36] This is particularly so because, like the 1996 document, the language of the unclassified version of the 2006 Policy is vague. Since the current context of US policy differs significantly from that in 1996, ambiguities in the wording are likely to be interpreted in a new light.[37]

In its emphasis, it is immediately noticeable that the 2006 Policy priorities have been rearranged, with national security uses of space brought to the forefront and civil space and exploration given lower priority. The top two goals listed are related to national security.[38]

There is also an express willingness to undertake pre-emptive[39] and military action unauthorized by the UN.[40]

(c) Vision for 2020

US plans are revealed in several documents. They are explicitly incorporated in the US 2006 Policy on space. They also appear in *Vision for 2020*, a 1996 report of the US Space Command, a body which was set up in 1985 to "help institutionalize the use of space" and that "co-ordinates the use of Army, Navy, and Air Force space forces."[41]

The cover of *Vision for 2020* shows a weapon shooting a laser beam from space at a target below. The report opens with the following: "US Space Command— dominating the space dimension of military operations to protect US interests and investment. Integrating Space Forces into war-fighting capabilities across the full spectrum of conflict." A century ago, "Nations built navies to protect and enhance their commercial interests" by ruling the seas, the report notes. Now it is time to rule space.[42]

IMPACT OF US POLICY SHIFT ON INTERNATIONAL RELATIONS

There is international concern about US motives for pursuing such policies. The changes in US policy have already had an impact on international relations, particularly over four issues: the ABMT; the US National Missile Defence system (NMD); the doctrine of peaceful uses of outer space; and the willingness to use pre-emptive strike. The case of the ABMT and the related developments in NMD provide an interesting illustration of the link between US space policy changes and consequences for international relations.

(1) National Missile Defence

Other States fear that the creation of a national missile defence by the US will allow it to pursue its own national interests even more aggressively, through globalization and other policies, and that it would be a precursor to space-based military developments.[43] Even prior to 2006, concern was being voiced internationally.

(a) International Reaction

French President Jacques Chirac, in an interview with The New York Times on December 17, 1999 observed:[44]

> If you look at world history, ever since men began waging war, you will see that there's a permanent race between sword and shield. The sword always wins. The more improvements that are made to the shield, the more improvements are made to the sword. We think that with these systems [missile defence], we are just going to spur sword-makers to intensify their efforts. China, which was already working harder than we realized on both nuclear weapons and delivery vehicles

> for them, would of course be encouraged to intensify those efforts, and it has the resources to do so. India would be encouraged to do the same thing, and it, too, has the resources. And it would also increase tensions within NATO, which would be too bad.

At the beginning of 2001, the UK Conservative opposition party leader announced that he would support the US NMD scheme and urged the UK government to do so as well. This caused concern within UK political circles and across Europe.[45] If the UK government were to support the US policy, it would be at odds with the rest of Europe. In addition, US proposals to install parts of missile defence systems in the UK and other European countries raised concerns that these countries could consequently become targets of terrorist attacks.[46]

On 11 May 2001, Dr Theodore Postol,[47] sent a letter to the White House in which he asserted a cover-up of fundamental flaws in the missile defence system.[48] Eight days later, the Los Angeles Times published an interview with a high-level US intelligence official who flatly contradicted the US Administration's contention that China has nothing to fear from a limited US NMD system. The official also noted that the North Korean and Iranian missile threats have not been moving along as rapidly as expected, and he asserted that the concept of the "rogue state" was in itself an impediment to objective analysis of the missile threat.[49]

An analysis of potential missile defence threat to US and world security can be found in a report issued in 2000 by the National Intelligence Council. That report suggested that deployment of such a system would be likely to provoke "an unsettling series of political and military ripple effects ... that would include a sharp build-up of strategic and medium-range nuclear missiles by China, India and Pakistan and the further spread of military technology in the Middle East."[50]

The restructuring within the US government, creation of the Missile Defence Agency,[51] and the creation of the National Missile Defence system[52] further strained international relations. Both China and Russia required assurances that the NMD system was not aimed at either. In 2002 the signing of the Strategic Offensive Reductions Treaty (SORT) allowed Russia to reduce its spending on missiles, without decreasing its comparative strength.[53] Given the cash shortage in the Russian armed forces, it is interesting to note that it is cheaper to maintain a strategic missile than to dismantle it.[54] China has already reacted to the new US policy by destroying one of is own satellites with a missile.[55]

The position of Russia has recently become further entrenched with the deal between Poland and the US to locate part of the NMD system in Poland, and by increasing tension between Russia and the US over Georgia.[56]

(b) <u>Proliferation and Arms Race</u>

Instead of reducing security concerns, the missile defence policy has heightened them. The consequence of heightened security fears in other States would be the consideration of further arms procurement. Ironically, this could be the reason that the US could eventually be threatened. The US move equates not to a reaction to arms proliferation, but a cause of it.

This observation is worth noting:[57]

> Once testing [of space weapons] begins, the "need" for destructive capabilities in orbit induces a mindset opposed to rational restraint. The mindset becomes unassailable if testing is completed, for then the system "must" be deployed since, if we have developed the capability, others will want to follow suit and rapidly will do so.

Thus, as more nations have become capable of accessing outer space there is a consequent risk of triggering an arms race in space. India, Pakistan and North Korea, for example, have confirmed their nuclear capabilities and India, together with China and Russia have expressed their concerns at US aggression, and frequent violations of international law. Deployment of a

US national missile defence system therefore risks drawing South Asia into an arms race.

Washington's own intelligence community has warned that NMD deployment is likely to set off a new arms race that could ripple across the Eurasian continent, provoking first China, then India, and then Pakistan to either build up their own missile forces or trying to deploy anti-missile defences of their own.[58]

(c) Associated Risks

There are of course pitfalls inherent in this situation. To win a space arms race, the US or any other State must rely on the uncertain superiority of its technology, intellectual capital and financial resources.

In addition, in order to legitimize a policy of militarizing space, the militarizing State will either have to withdraw from or breach relevant space treaties. Consequently, there can be no reliance on any legal constraints on competitors, since the relevant international legal framework will have been undermined. Other States will inevitably be less likely to join the international space treaties and conventions.

(2) Anti-Ballistic Missile Treaty

At the beginning of May 2001, the US President announced the intention to withdraw from the ABM Treaty, which he considered an outdated Cold War relic. Supporters of the withdrawal argued that it was necessary in order to test and build a National Missile Defence system to protect the United States from nuclear blackmail by a rogue state.

In June 2001 Russian President Vladimir Putin warned that US violation of the ABM Treaty would force Russia to augment its nuclear capability by mounting multiple warheads on its missiles. At the same time, he suggested that both the START I,[59] which prohibits even partial use of the Earth orbit for delivery of nuclear or other weapons of mass destruction, and the START II treaties would be negated if the US abrogated the ABM Treaty. Nevertheless, the US did so in December 2001. The termination of these treaties would eliminate verification and inspection requirements and allow Russia to hide its nuclcar capabilities.[60]
The international community, particularly Russia, was concerned about the US withdrawal from the ABMT. When the decision was announced, President Vladimir Putin commented that it was a "mistake", but not a threat to his country. However, On June 14, 2002, one day after the U.S. withdrew from the Anti-Ballistic Missile Treaty, Russia withdrew from START II.[61] That treaty, which banned the use of a particular class of warheads,[62] would have compelled it to destroy its most powerful intercontinental nuclear missiles. Russia instead announced the overhaul more than 100 of these.[63] Ironically the US policy change, which had been explained by a desire to reduce nuclear threat, led instead to an increased nuclear risk not only to the US, but to the world.

The US withdrawal in 2002[64] can be said to have removed one of the legal obstacles to placing weapons in space.[65]

PEACEFUL USE OF OUTER SPACE

Adherence to the principles of the OST serves the interest of all States, as it seeks to ensure that outer space is not used for military activities such as launching attacks from space. The US policy of pursuing the military use of space[66] has the potential to undermine the principle of peaceful use of outer space, and therefore, the international order in space.

(1) Alternative Interpretations

It has been argued, notably by some seeking to justify US placement of weapons in space, that "peaceful use" in the context of outer space law means "non-aggressive" as opposed to "non-military".[67] Some have argued that the OST restricts *only* nuclear weapons and weapons of mass destruction.[68]

There is an obvious risk inherent in the acceptance of these standards. As has been stated, any military use of outer space will lead to a weakening of international law by diminishing incentives for non-party nations to accede to or ratify the Outer Space Treaty and other space-related international conventions.

It is arguable that any intention to place in space weapons capable of attacking other space objects is not a defensive posture. Since there is no threat to the US space objects by any other country, *it poses at least a threat of use of force (cannot be defensive),* contrary to UN Charter Article 2(4)[69] and therefore falls outside the self-defence provisions of the UN Charter.[70]

In addition, other international documents[71] support the view of many legal scholars that "peaceful" means "non-military", rather than "non-aggressive."[72]

Indeed, the historical context of the OST is indicative of the importance placed on peaceful use. In 1958, U.S. President Eisenhower and Soviet Premier Khrushchev each asked the United Nations to consider the legal issues associated with space activity. The U.N. created the Committee on the Peaceful Uses of Outer Space ("COPUOS"). Its Legal Subcommittee[73] has been the primary forum for discussion and negotiation of international agreements relating to outer space, including the Outer Space Treaty.

More recently, the principle of "peaceful use" and the ban on nuclear weapons in space was reaffirmed by the unanimous UN Resolution on International Co-operation in the Peaceful Uses of Outer Space,[74] by which States have committed to peaceful uses of outer space, and prevention of a space arms race.[75]

(2) Pre-emptive Strike

A further area of potential concern is the US stated policy of pre-emptive strike against potential aggressors. Aside from any legal objection, there is obviously a risk that other States would adopt similar policies.[76]

If such attacks included the use of any space-based weapons, legal questions would arise under the UN Charter. The first-use of military power in outer space is per-se illegal, if undertaken without justification as outlined in the UN Charter[77] (self-defence), or unless authorised by the Security Council,[78] as are space deployment of nuclear and other "weapons of mass destruction."[79]

(3) Other Concerns

It is unarguable that the domination of space would enable the US or any other State to maintain, expand and enforce those policies that will serve the national interest. There is a perception that the US is seeking to create a dominant position in space that would allow it to be even more powerful and influential terrestrially.[80]

PROSPECTS FOR THE FUTURE

(1) International Containment Efforts

There have been many proposals to fill the gap in the Outer Space Treaty's prohibition of weapons. Canada and many NGOs have made proposals. In addition, the PAROS Resolutions of the UN General Assembly[81] address this issue.

On 27 June, 2002, Russia and China presented a working paper to the UN Conference on Disarmament (CD) containing possible elements of an international legal agreement on prohibiting the deployment of any weapons in outer space. It would also prohibit the threat or use of force against space objects, a concept that would ban anti-satellite weapons, either mounted on aircraft or ground-based.

On 12 February 2008, Russia and China submitted a draft treaty[82] for a ban on weapons in outer space to the Conference on Disarmament, based on the elements outlined in their 2002 working paper. The US administration dismissed the proposal out-of-hand, characterizing the offer to preserve space for peaceful uses "a diplomatic ploy by the two nations to gain a military advantage."[83]

At present, there is no prospect that this treaty outline will make progress at the CD, owing to the conference rule of consensus decisions and the outright opposition of the United States. The United States has said it is willing to discuss this issue at the CD, but not to negotiate a treaty on it. China had long insisted that, in addition to discussion, the possibility of negotiation must be mentioned in the agenda, however in August 2003, China signalled it was prepared to compromise on this point. There is even some agitation to change the consensus rules of the Conference on Disarmament. In the meantime, the Russian-Chinese draft can be refined further and developed into a usable treaty text, with help from other governments and NGOs.

There are other, more peaceful ways to promote better security and cooperation between nations, as Michael Wallace hints:

> [T]he poorer and weaker nations and peoples of the world regard the entire BMD controversy with a mixture of disbelief and disgust. For the world's richest nation [USA] to spend such enormous sums on unproven and provocative technologies while failing to pay the full amount of their dues to the UN, refusing to agree to total debt relief for the poorest nations, and denying full access to American markets for such key Third World products as textiles and sugar, seems utterly incomprehensible. To put this in specific perspective: it was estimated by a Greenpeace activist from the Cook Islands that the $100 million wasted on the failed July 7, 2000 test could have built and run a hospital and provided free university education for the entire population of the Cook Islands for many decades. Surely, American security would be better served by spending money on such worthy projects than by a futile attempt to create an unattainable Fortress America.[84]

The parties to the Outer Space Treaty could decide whether to formally interpret the peaceful use provision to preclude weapons in space.[85] This could be based on a review of state practices since 1967 and the negotiating history of the treaty.[86] Similarly, the United Nations, acting through its First Committee and then through the General Assembly (which recommended the Outer Space Treaty in the first place), could pass a resolution formally interpreting it. If there were significant dissent, pursuant to the UN Charter the General Assembly could request an advisory opinion from the International Court of Justice at The Hague confirming this interpretation.[87]

Repercussions from violating space law, such as lawsuits and international legal actions, should also be included in the calculation of gains and losses from weaponisation.

(2) Need to Review US Policy

On 21 February 2008, the US military shot down a failed satellite with a Standard Missile-3, whose primary vocation is interceptor for the US Navy's missile defense system.[88] It has also evoked criticism from many space security experts,[89] who have voiced two primary concerns.

One concern is the debris. While Marine Corps Gen. James E. Cartwright, vice chairman of the Joint Chiefs of Staff, said most of the debris will come down within two orbits, Jeffrey Lewis of the New America Foundation said, "modeling of debris creation isn't an exact science," arguing, "the debris that the light-weight interceptor will kick into higher orbits when it hits the massive (bus-sized) satellite" will remain in orbit, posing a risk to the International Space Station.

The second concern is the political implications of conducting what amounts to an anti-satellite test. The US administration has argued the test is not the same as the Chinese anti-satellite test in January 2007, which it condemned. The US government says the Chinese test was "designed specifically" to test the ability to destroy satellites, and argues that its own plan is only aimed at protecting civilians on the ground. However, the Russian Defence Ministry asserted the US plan is "in many ways close" to China's test, arguing, "The impression arises that the United States is trying to use the accident with its satellite to test its national anti-missile defence system as a means of destroying satellites." Bruce Gagnon, Coordinator of the Global Network Against Weapons and Nuclear

Power in Space, likewise argued, "The [US] administration is magnifying the risk to justify the testing of new dangerous and provocative offensive space warfare technologies."

There has been little public thinking about the potential for far-reaching military, political and economic ramifications of a US move to break the taboo[90] against weaponising space. Doing so could actually undermine, rather than enhance, the national security of the United States, as well as global stability. There is nothing to be gained, and potentially much to be lost, by rushing such a momentous change in US space policy.[91]

Thus it behoves the Administration, as well as Congress, to undertake an in-depth and public policy review of the pros and cons of weaponising space.

Such a review would look seriously at short-term and long-term threat, as well as measures to prevent, deter or counter any future threat using all the tools in the US policy toolbox: diplomatic, including arms control treaties; economic; and military, including defensive measures short of offensive weapons.

CONCLUSION

The militarization of space by the United States will weaken the international laws of outer space, and affect the interpretation of international and US domestic laws governing space activities. If the US persists in militarizing space it will do so despite the unanimous UN Resolution on International Co-operation in the Peaceful Uses of Outer Space[92] by which the US has joined other countries committing to peaceful uses of outer space, and prevention of a space arms race.[93]

In itself a retrograde step, US actions and policy will also have a chilling effect on the UN efforts to promote ratification of space treaties among countries not already parties to them.

Poor Moon and poor Man.

---

1 Sovereignty is an anachronistic concept originating in bygone times when society consisted of rulers and subjects, not citizens. It became the cornerstone of international relations with the Treaty of Westphalia in 1648. Today, though not all nation-states are democratically accountable to their citizens, the principle of sovereignty stands in the way of outside intervention in the internal affairs of nation-states. True sovereignty belongs to the people, who in turn delegate it to their Government [PCIJ, Advisory Opinion, Nationality Decrees Issued in Tunis and Morocco, Series B, N° 4, p. 24.] Sovereignty is not, and has never been, an unlimited power to do all that is not expressly forbidden by international law [See however PCIJ, Judgment, *Lotus Case*, Series A, N° 10, p. 18.] It can only be defined as the very criterion of States, by virtue of which such an entity "possesses the totality of international rights and duties recognized by international law" as long as it has not limited them in particular terms by concluding a treaty. [ICJ, Advisory Opinion, Reparation for Injuries Suffered in the Service of the United Nations, ICJ Rep. 1949, p. 180.]

2 Switzerland and East Timor joined in 2001. The Holy See is not a member of the UN.

3 Articles 41 and 42 of the U.N. Charter declare that no member state has the right to enforce any resolution with armed force unless the Security Council decides there has been a material breach of its resolution, and determines that all non-military means of enforcement have been exhausted. Then, the Council must specifically authorize the use of military force, as it did in November 1990 with Resolution 678, in response to Iraq's occupation of Kuwait in violation of Security Council resolutions passed the previous August. See also, *Invading Iraq Would Violate US and International Law*, Professor Marjorie Cohn, Thomas Jefferson School of Law, JURIST Contributing Editor. http://www.lawyersagainstthewar.org/legalarticles/cohn2.html.

4 Early Bird was the world's first commercial communication synchronous orbit satellite, launched by COMSAT in 1965.

5 Treaty on Principles Governing the Activities of States in the Exploration and Use of Outer Space, including the Moon and Other Celestial Bodies; Signed 1967; 98 Parties (including the United States); 27 Signatories.

6 International Agreements and Other Available Legal Documents Relevant To Space-Related Activities, United Nations, Vienna 1999.

7 Article IV; reference to "peaceful purposes" also occurs in the 4th paragraph of the preamble.

8 By 1972 agreement had been reached to limiting strategic offensive weapons and strategic defensive systems. Each country was allowed two sites at which it could base a defensive system, one for the capital

and one for ICBM silos (Art. III). The treaty was signed in Moscow on 26 May, 1972 and ratified by the US Senate on 3 August, 1972. The 1974 Protocol reduced the number of sites to one per party, largely because neither country had developed a second site. The sites were Moscow for the USSR and Grand Forks Air Force Base, North Dakota, since its Safeguard facility was already under construction, for the US. The treaty was undisturbed until Ronald Reagan announced his Strategic Defence Initiative (SDI) on 23 March, 1983. After the dissolution of the Soviet Union in December 1991 the status of the treaty became unclear, and in 2002 George W Bush announced that the US was withdrawing.

9 Treaty Banning Nuclear Weapon Tests in the Atmosphere, in Outer Space and Under Water; Signed at Moscow August 5, 1963; in force 10 October 1963. Article I, Para. 1(a) provides: "Each of the Parties to this Treaty undertakes to prohibit, to prevent, and not to carry out any nuclear weapon test explosion, or any other nuclear explosion, at any place under its jurisdiction or control: (a) in the atmosphere; beyond its limits, including outer space; ..."

10 Treaty Between the United States of America and the Union of Soviet Socialist Republic on the Reduction and Limitation of Strategic Offensive Arms; Done at Moscow on 31 July 1991.

11 Article V, Para. 18(c): Each Party undertakes not to produce, test, or deploy ... systems, including missiles, for placing nuclear weapons or any other kinds of weapons of mass destruction into Earth orbit or a fraction of an Earth orbit.

12 The Convention on Prohibitions or Restrictions on the Use of Certain Conventional Weapons Which May Be Deemed to Be Excessively Injurious or to Have Indiscriminate Effects; concluded Geneva, 10 October, 1980, entered force December 1983, and is an annexe to the Geneva Conventions of August 12, 1949. It consists of five protocols, including Protocol IV on Blinding Laser Weapons.

13 To date, the United States has ratified Protocols I (fragmentation weapons) and II (landmines). It has assumed certain obligations under Protocol III (incendiary weapons) within stated limits, but has not ratified it, and is obligated under international law not to take actions that defeat the object and purpose of Protocol IV. The United States did not join the consensus approval of the text of Protocol V (clearance of explosive remnants of war, adopted 2003). Accordingly, the view of the United States is that the provisions of Protocol V will not apply until ratified by the U.S. Senate.1 Twenty signatories must ratify Protocol V for it to become legally binding. Currently thirteen states parties have ratified Protocol V.

http://www.ntip.navy.mil/certain_conventional_weapons.shtml

14 Article 2.

15 Article 3 of the Protocol provides: Blinding as an incidental or collateral effect of the legitimate military employment of laser systems, including laser systems used against optical equipment, is not covered by the prohibition of this Protocol.

16 Union of Concerned Scientists, Citizens and Scientists for Environmental Solutions, International Legal Agreements Relevant to Space Weapons, February 2004, http://www.ucsusa.org.

17 Sixty–first General Assembly, First Committee, 20th Meeting, GA/DIS/3334. Sixty-second General Assembly, First Committee, A/62/251 96 Prevention of an arms race in outer space, A/RES/62/20, adopted 10 January 2008. See http://www.reachingcriticalwill.org/political/1com/1com07/ga/Res20.pdf; and http://www.reachingcriticalwill.org/legal/paros/parosindex.html.

18 The US voted against the PAROS Resolution of 2005, breaking with its previous practice to abstain. The US vote was seen by some commentators as an indication of the US drive to weaponise space, and by others as consistent with Ambassador John Bolton's hard line; The Ploughshares Monitor, Winter 2005, Vol 26, no. 4.

19 "SDI consolidated missile defence programs that were scattered among several government offices and moulded them into a coherent program guided by a clear strategic vision-produce non-nuclear defences. The technologies and systems developed under SDI fed into variants of the original SDI program that were managed first by the Ballistic Missile Defence Organization and then by today's Missile Defence Agency." MDA History, from the Missile Defence Agency MDALink, http://www.mda.mil/mdalink/html/history.html

20 President Clinton's Presidential Decision Directive/NSTC-8', National Space Policy, dated 19 September 1996, and released 19 September 1996.

21 Theresa Hitchens, *Weapons in Space: Silver Bullet or Russian Roulette? The Policy Implications of U.S. Pursuit of Space-Based Weapons*, Center for Defence Information, 18 April 2002.

22 National Security Presidential Directive, (NSPD-49) dated 31 August 2006. Only an unclassified summary was released on 6 October 2006.

23 "The United States rejects any claims to sovereignty by any nation over outer space or celestial bodies, or any portion thereof, and rejects any limitations on the fundamental right of *sovereign nations* to acquire data from space." (Emphasis added)

24 "The United States rejects any claims to sovereignty by any nation over outer space or celestial bodies, or any portion thereof, and rejects any limitations on the fundamental right of the *United States to operate in* and acquire data from space." (Emphasis added)

25 "The United States considers space capabilities -- including the ground and space segments and supporting links -- vital to its national interests. Consistent with this policy, the United States will: preserve its rights, capabilities, and freedom of action in space; dissuade or deter others from either impeding those rights or developing capabilities

intended to do so; take those actions necessary to protect its space capabilities; respond to interference; and deny, if necessary, adversaries the use of space capabilities hostile to U.S. national interests."

26 "The United States will oppose the development of new legal regimes or other restrictions that seek to prohibit or limit U.S. access to or use of space. Proposed arms control agreements or restrictions must not impair the rights of the United States to conduct research, development, testing, and operations or other activities in space for U.S. national interests."

27 "Strengthen the nation's space leadership and ensure that space capabilities are available in time to further U.S. national security, homeland security, and foreign policy objectives."

28 "Enable unhindered U.S. operations in and through space to defend our interests there."

29 Presidential Decision Directive/NSC-49/NSTC-8, National Space Policy, dated 14 September 1996.

30 "DoD shall maintain the capability to execute the mission areas of space support, force enhancement, space control, and force application. ...Consistent with treaty obligations, the United States will develop, operate and maintain space control capabilities to ensure freedom of action in space and, if directed, deny such freedom of action to adversaries. These capabilities may also be enhanced by diplomatic, legal or military measures to preclude an adversary's hostile use of space systems and services."

31 "Maintain the capabilities to execute the space support, force enhancement, space control, and force application missions; ... Develop capabilities, plans, and options to ensure freedom of action in space, and, if directed, deny such freedom of action to adversaries."

32 Global Policy Forum, *US, UN and International Law*, http://www.globalpolicy.org/empire/un/unindex.htm

33 See also Jon Kyl (Republican Senator, Arizona; Senate Committees: Finance, Judiciary, and Sub-Committee on Terrorism, Technology & Homeland Security), *China's Anti-Satellite Weapon and American Security*, 29 January 2007, Heritage Lecture No. 990.

34 Military Space Operations: Planning, Funding, and Acquisition Challenges Facing Efforts to Strengthen Space Control; GAO Report to the Secretary of Defense, September 2002.

35 National Security Presidential Directive, NSPD-49. Cited Federation of American Scientists website, http://www.fas.org/irp/offdocs/nspd/space.html

36 Laura Grego and David Wright , *U.S. Space Weapons Policy*, The Bush Administration's National Space Policy, Union of Concerned Scientists, 13 October 2006.

37 Jon Kyl, Republican Senator for Arizona, *China's Anti-Satellite Weapon and American Security*, 29 January 2007, Heritage Lecture No. 990.

38 See Notes 29 and 30.

39 "The United States has long maintained the option of pre-emptive actions to counter a sufficient threat to our national security. The greater the threat, the greater is the risk of inaction—and the more compelling the case for taking anticipatory action to defend ourselves, even if uncertainty remains as to the time and place of the enemy's attack. To forestall or prevent such hostile acts by our adversaries, the United States will, if necessary, act pre-emptively." National Security Strategy, 17 September 2002, http://www.whitehouse.gov/nsc/print/nssall.html; quoted in Taylor, Rachel S., *International Law - War in Iraq - United Nations - Iraq*. World Press Review Online, www.worldpress.org/Mideast/2230.cfm.

40 O'Connell, Mary Ellen, *UN Resolution 1441: Compelling Saddam, Restraining Bush.* Jurist, 21 November 2002. http://jurist.law.pitt.edu/forum/forumnew73.php#6.

41 Report by United States Space Command; www.fas.org/spp/military/docops/usspac/visbook.pdf.

42 Karl Grossman, *Master of Space*, Progressive Magazine, January 2000.

43 *NMD Again 'Hogging Headlines' As World 'Faces Up to U.S. Missile Plan'*, 2 February 2001, http://www.fas.org/news/usa/2001/usa-010202.htm.

44 Craig R Whitney, *With a 'Don't Be Vexed' Air, Chirac Assesses U.S.*, New York Times, 17 December 1999.

45 *Hague defends 'Star Wars' stance*, BBC News, 12 January 2001, http://news.bbc.co.uk/1/hi/uk_politics/1113018.stm.

46 See James Fergusson, *Nato, Europe and Theatre Missile Defence*, Canadian Military Journal, Spring 2002, p 45.

47 Dr Theodore Postol, Professor of Science, Technology and National Security Policy, Massachusetts Institute of Technology.

48 In the letter dated 11 May 2000, to John Podesta, Whitehouse Chief of Staff, Dr. Postol said Pentagon sensor data he had obtained from the first antimissile test flight in June 1997 showed that the ground-based interceptor was inherently unable to make the distinction and that the Pentagon and its contractors had tried to hide this failure. The cover-up, he said, was "like rolling a pair of dice and throwing away all outcomes that did not give snake eyes." An inability to tell cheap decoys from costly warheads in theory could force a defender to fire interceptors at every threatening object, which as a practical matter could make the system useless.

49 William Hartung and Michelle Ciarrocca, *Star Wars II; Here We Go Again*, The Nation Magazine, June 19, 2000.

50 See *Missile Shield Analysis Warns of Arms Buildup*, Los Angeles Times, 19 May 2000.

51 The Missile Defence Agency (MDA) is the section of the United States government's Department of Defence responsible for developing a layered defence against ballistic missiles.

52 The NMD system consists primarily of radar and ground based interceptor missiles that are intended to intercept incoming warheads in space.

53 SORT, signed on 24 May 2002, mandated the deepest ever cut in deployed strategic nuclear warheads without actually mandating cuts to total stockpiled warheads.

54 Yevgeny Miasnikov of the Centre for Arms Control, Energy and Environmental Studies in Moscow, *Moscow extends life of 144 cold war ballistic missiles*, Nick Paton Walsh, The Guardian, Tuesday 20 August 2002.

55 On 23 January 2007, the Chinese government confirmed that it had conducted a successful test of a new anti-satellite weapon, but said it had no intention of participating in a "space race." Liu Jianchao, the foreign ministry spokesman, issued the first official comment on the matter, stating, "This test was not directed at any country and does not constitute a threat to any country. . . . What needs to be stressed is that China has always advocated the peaceful use of space, opposes the weaponisation of space and an arms race in space. . . . China has never participated and will never participate in any arms race in outer space." On 25 January 2007, Russian President Vladimir Putin criticized US plans for space-based weapons, saying they were the reason behind the Chinese anti-satellite weapons test.

56 On 14 August 2008, The United States of America and Poland announced a deal to implement the missile defense system in Polish territory, with a tracking system placed in the Czech Republic. The Russians responded by saying such action "cannot go unpunished." [*Russia Lashes Out on Missile Deal*, New York Times, 15 August 2008.] "The fact that this was signed in a period of very difficult crisis in the relations between Russia and the United States over the situation in Georgia shows that, of course, the missile defense system will be deployed not against Iran but against the strategic potential of Russia," Dmitry Rogozin, Russia's NATO envoy, said. [*Russia Angry Over US Missile Shield*, Al Jazeera, 15 August 2008.]

57 *Space Wars*, by Colonel Daniel Smith, USA (Ret.) Chief of Research, Centre for Defence Information, February 2001.

58 Jim Lobe, *Bush's New Era of Missile Defence Heightens Tensions*, Inter Press Service, 2 May 2001.

59 Strategic Arms Reduction Treaty (START) I – The bilateral treaty on the Reduction and Limitation of Strategic Offensive Arms; Signed by the United States and Soviet Union, 1991. Article V.18 commits both parties "not to produce, test, or deploy...systems, including missiles, for placing nuclear weapons or any other kinds of weapons of mass destruction into Earth orbit or a fraction of Earth orbit".

60 Michelle Ciarrocca and William Hartung, *Star Wars Revisited*, Foreign Policy In Focus, Volume 6, Number 25, June 2001.

61 Strategic Arms Reduction Treaty, START II, was a weapons treaty signed by George Bush Sr and Boris Yeltsin in 1993.

62 The weapons banned were multiple independently targetable re-entry vehicle (MIRVs) when carried on a single intercontinental ballistic missile (ICBM). Using a MIRV warhead, a single launched missile can strike several targets, or fewer targets redundantly. MIRVed land-based ICBMs were considered destabilizing because they tended to put a premium on striking first.

63 A total of 144 of the missiles, which weigh 200 tonnes and can each carry 10 warheads to the US from silos behind the Ural mountains, were due to be dismantled by 2007 under the Treaty.

64 On 13 December 2001, George W. Bush gave Russia notice of the United States' withdrawal from the treaty, in accordance with the clause that requires six months notice before terminating the pact. This was the first time in recent history the United States has withdrawn from a major international arms treaty. Nick Paton Walsh, *Moscow Extends Life of 144 Cold War Ballistic Missiles*, The Guardian, 20 August 2002.
http://www.guardian.co.uk/russia/article/0,2763,7773 79,00.html

65 Corah Ong, *Withdrawal from the ABM Treaty Threatens US and International Security*, Nuclear Age Peace Foundation, December 2001. www.wagingpeace.org/2001/12/00_ong_wihdrawal.h tm

66 Space-based communications, remote sensing and GPS systems, while not in themselves necessarily "military", are increasingly used to provide an advantage in warfare.

67 Michael N. Schmitt, *Bellum Americanum: The U.S. View of Twenty-First Century War and Its Possible Implications for the Law of Armed Conflict*, 19 MICH. J. INT'L L. 1051, 1087 (1998). Cited in Christopher M. Petras, Military use of the International Space Station and the concept of "peaceful purposes", Air Force Law Review, 22/3/2002.

68 "Under this...interpretation, none of the exotic future weapons systems currently being proposed or researched by the United States would violate this provision of the Outer Space Treaty. ...violations would only occur if any of the weapon systems included a nuclear explosion to propel them or as a means of destroying a target." Major Douglas S. Anderson, *A Military Look into Space Law: The Ultimate High Ground*, Nov. ARMY LAW 19, 22 (1995). 24–25. Cited in Jackson N. Maogoto, The *Military Ascent into Space: From Playground to Battleground–The New Uncertain Game in the Heavens*, Berkeley electronic press Legal Series. 2006, 1347, 22.

69 "All Members shall refrain in their international relations from the threat or use of force against the territorial integrity or political independence of any State or in any other manner inconsistent with the purposes of the United Nations."

70 UN Charter Article 51 provides: Nothing in the present Charter shall impair the inherent right of

individual or collective self-defence if an armed attack occurs against a Member of the United Nations, until the Security Council has taken measures necessary to maintain international peace and security. Measures taken by Members in the exercise of this right of self-defence shall be immediately reported to the Security Council and shall not in any way affect the authority and responsibility of the Security Council under the present Charter to take at any time such action as it deems necessary in order to maintain or restore international peace and security.

71 The thirteenth session of the General Assembly, held in 1958, provided a forum for the debate on 'Questions of the Peaceful Use of Outer Space'. During this session the term 'peaceful' was used as an antonym to 'military'. Sweden appealed to fellow Member States to 'safeguard outer space against any military use whatsoever' and the Soviet Union put forward a proposal to ban the use of outer space for military purposes. The General Assembly adopted resolution 1348 (XIII), which recognized the 'common aim' of humankind that outer space 'should be used for peaceful purposes only.' [Quoted in M.S. McDougal, H.S. Lasswell and I.A. Vlasic, 1963, *Law and Public Order in Space,* New Haven, Yale University Press, p. 395. See also http://www.oosa.unvienna.org/SpaceLaw/gares/html/gares_13_1348.html.

72 But see Bin Cheng, Studies In International Space Law150 (1997), Note 83; "[T]he outer void space as such can be used for any military activity that is compatible with general international law and the Charter of the United Nations, so long as no nuclear weapons or any other kind of weapons of mass destruction are stationed there."

73 The other subcommittee is the Scientific and Technical Subcommittee.

74 Resolution adopted by the UN General Assembly 1982 (and readopted each year, last in 2006) against an arms race in outer space, which "Called upon all States, in particular those with major space capabilities, to contribute actively to the objective of the peaceful use of outer space and of the prevention of an arms race in outer space (PAROS) and to refrain from actions contrary to that objective and to the relevant existing treaties in the interest of maintaining international peace and security and promoting international cooperation." In 2003, the US, Israel, and US supporters Marshall Islands and Micronesia abstained from this vote in the General Assembly. There were 174 'yes' votes and no negative votes. See also under UN Resolutions above. Resolution adopted by the General Assembly; A/RES/61/58; 18 October 2006, PAROS.

75 Para. 34: Urges all States, in particular those with major space capabilities, to contribute actively to the goal of preventing an arms race in outer space as an essential condition for the promotion of international cooperation in the exploration and use of outer space for peaceful purposes. The US, Israel and one other country abstained.

76 Other countries are already adopting similar policies. For example see statement by President Putin of Russia; *The Washington Times*, 13 September 2004, http://www.washtimes.com/upi-breaking/20040913-104239-9091r.htm.

77 UN Charter, Article 51.

78 UN Charter, Chapter VII.

79 Article IV informed states to "undertake not to place in orbit around the Earth any objects carrying nuclear weapons or any other kinds of weapons of mass destruction, install such weapons on celestial bodies, or station such weapons in outer space in any other manner." Weapons of mass destruction (WMD) are weapons designed to kill large numbers of people, typically targeting civilians and military personnel alike. Coined in 1937 to describe aerial bombardment, today they are often referred to as NBC weapons or ABC weapons, comprising: nuclear weapons (including radiological weapons); biological weapon; and chemical weapon.

80 *Academy Paper Examines Russian and Chinese Views of U.S. Plans for Space Weapons*, American Academy of Arts, 21 February 2008; http://www.amacad.org/news/russia_china.aspx.

81 See UN Resolutions above.

82 Draft *Treaty on the Prevention of the Placement of Weapons in Outer Space, the Threat or Use of Force Against Outer Space Objects*, (PPW Treaty), http://www.reachingcriticalwill.org/political/cd/papers08/1session/Feb12%20Draft%20PPWT.pdf.

83 See http://www.reachingcriticalwill.org/political/cd/speeches08/reports.html#feb12.

84 *Ballistic Missile Defence: The View from the Cheap Seats*, WagingPeace.org (Nuclear Age Peace Foundation)

85 "The prohibition on orbiting of weapons of mass destruction, including nuclear weapons, strongly suggests the distinction between those weapons, and conventional weapons of lesser destructive power, including those directed at satellites. Though Article IV (1) could easily be modified to effect the de-weaponisation of space, conventional weapons are not proscribed." Major Robert A. Ramsey, *Armed Conflict on the Final Frontier: The Law of War in Space* 48 A.F.L. REV. 1, 84. (2000). Cited in Jackson N. Maogoto, *The Military Ascent into Space: From Playground to Battleground–The New Uncertain Game in the Heavens*, Berkeley electronic press Legal Series. 2006, 1347, 29.

86 Unanimity among parties is not required for any formal interpretations, but a large majority of parties adopting a particular position would be persuasive.

87 George Bunn & John B Rheinlander, *Outer Space Treaty May Ban Strike Weapons*, June 2002 Letter to the Editor, Arms Control Association, www.armscontrol.org/act/2002_06/letterjune02.asp.

88 This incident has been said to highlight one of the deficiencies of the draft treaty proposed by Russia and China, which does not address attacks from ground- or sea-based interceptors such as the SM-3. But see Draft *PPW Treaty* at note 79.

[89] Security experts have criticised the US on the basis that it has considered the worst case for the tank of hydrazine coming down into a populated area from which people cannot evacuate. This is an unlikely scenario. See http://www.reachingcriticalwill.org/political/cd/speeches08/reports.html#19feb.

[90] It is contended that such action is also against international law.

[91] Theresa Hitchens, *Weapons in Space: Silver Bullet or Russian Roulette? The Policy Implications of US Pursuit of Space-Based Weapons*, Centre for Defence Information, 18 April 2002.

[92] Resolution adopted by the General Assembly; A/RES/58/89; 17 December 2003.

[93] Para. 34: *Urges* all States, in particular those with major space capabilities, to contribute actively to the goal of preventing an arms race in outer space as an essential condition for the promotion of international cooperation in the exploration and use of outer space for peaceful purposes.

IAC-08-E8

Updating the Outer Space Treaty – a U.S. perspective

Stephen M. D. Day
IVA LTD., U.S.A.
ssday1@aol.com

Abstract

In a nutshell, the U.S. administration (and much of Congress) perceives that the U. S. has a technological advantage in space-based weapons - so why give up that advantage in such uncertain times? A variation on this theme, from the U.S. Administration perspective, is that there is no space weapons race.

The World relies on a treaty banning certain space-based weapons that is over forty years old, when ASAT weapons, lasers, particle beam technologies, and other dual-use technologies were in their infancy. As a reminder, the 1967 Outer Space Treaty, eventually signed by 189 countries, 1) contains an undertaking not to place in orbit around the Earth, install on the moon or any other celestial body, or otherwise station in outer space, nuclear or any other weapons of mass destruction, and 2) limits the use of the Moon and other celestial bodies exclusively to peaceful purposes and expressly prohibits their use for establishing military bases, installation, or fortifications; testing weapons of any kind; or conducting military maneuvers.

These are eminently sensible rules regarding the non-proliferation of weapons of mass (or even moderate) destruction in outer Space; however, times have changed and the devil is once again in the details. The 1967 treaty was principally aimed at nuclear non-proliferation. At the time non-nuclear space-based weapons, such as high powered lasers, were not practical. However, these technologies are now becoming increasingly so - particularly in the U.S.

The time has come for us in the U. S. to take a leadership position, and support a new international treaty to ban weapons used in space – and nip in the bud a space-based arms race that has already begun.

Introduction

This is the perspective of a private citizen and non Space lawyer, who has been involved in the U.S. and international satellite and Space industries for over twenty-five years.

This year, at a U.N. sponsored forum, February 2008, Sergei Lavrov, Russia's foreign minister, repeated his country's plea to ban space weapons,. Both Russia and China are on the same page on this issue. The U.S. is currently not.

In a nutshell, the U.S. administration (and much of Congress) perceives that the U. S. has an advantage in space-based weapons technology - so why give up that advantage in such uncertain times? A variation on this theme, from the U.S. Administration's perspective, is that "there is no space weapons race."

The World continues to rely on a treaty, banning certain space-based weapons, that is over forty years old - when anti-satellite weapons (ASATs), lasers, particle beam technologies, and other dual-use technologies were in their infancy. Outer Space however, is still generally agreed to begin 100 kilometers from Earth. As a reminder, the 1967 Outer Space Treaty (OST), eventually signed by 189 countries:

1) Contains an undertaking not to place in orbit around the Earth, install on the moon or any other celestial body, or otherwise station in outer space, nuclear

or any other weapons of mass destruction, and

2) Limits the use of the moon and other celestial bodies exclusively to peaceful purposes and expressly prohibits their use for establishing military bases, installation, or fortifications; testing weapons of any kind; or conducting military maneuvers.

3) Does not prohibit the placement of weapons in orbit, except for weapons of mass destruction and nuclear weapons.

These were sensible rules regarding the non-proliferation of weapons of mass (and perhaps moderate) destruction in outer Space back in the nineteen sixties; however times have changed and the devil is once again in the details. The 1967 treaty was stimulated by nuclear weapons proliferation in a Cold War context, and was designed to ensure that Space would remain the common property of mankind - and would be used for peaceful purposes. All sorts of military related satellites such as observation (spy) and specialized communication satellites were tacitly allowed because any reference to them was omitted from the treaty. Also, at the time, earth-based ASATs, space-based weapons such as high powered lasers, and particle beams, were not practical. However, these technologies are becoming increasingly so now - particularly in the U.S.

Perhaps the time has come for the U. S. public to demand a different leadership position on this issue, and support a new international treaty to ban weapons in space – to nip in the bud a space-based arms race that in reality has already begun. This paper explores an enlightened, but hopefully realistic, approach the next U.S. Administration might take to update the Outer Space Treaty - for the benefit of humankind.

The current situation

Two recent anti-satellite (ASAT) tests shook up the world's complacency, regarding the creeping "weaponization" of Space.

In January 2007, China destroyed one of its aging weather satellites with an Earth based anti-satellite weapon creating approximately 2300 observable pieces of Space debris. According to General Kevin Chilton, the U.S. general in charge of military operations in space at the time, the Chinese were not transparent about the reasons for their test.

Following the Chinese test, the United States decided to strike a dying spy satellite (USA 193) with missiles launched from a Navy cruiser on 20 February 2008, with the ostensible intention of protecting the population from space debris, including the toxic fuel hydrazine. According to many space professionals in the U.S. and abroad (including this one), the U.S. response was also not entirely credible or transparent about the reasons for this action. However, others felt the action was justifiable, in that there was a non-zero probability that debris from the failed satellite could have caused damage to populated areas on Earth.

In June of this year, India's army chief stated India needs a military space program to defend its satellites from threats such as China's ability to shoot down targets in Space. Specifically, General Deepak Kapoor said India urgently needs to "optimize space applications for military purposes." Another general, H.S. Lidder, stated "With time we will get sucked into a military race to protect our space assets and inevitably there will be a military contest in space.

So there you have it, with the U.S., Russia, and China already having tested ASAT weapons, the space arms race is under way.

It's fair to conclude that the current U.S. Administration, from the President down, is in favor of protecting the perceived technological advantage of the U.S. in the military aspects of space. Hence the unwillingness to acknowledge that *there is a space weapons race,* in the first place. Nowhere, was this clearer than in the U.S.'s last National Space Policy doctrine announced October 2006, which effectively stated that the Outer Space Treaty (OST) is sufficient, and that the U.S "will oppose the development of new legal regimes or other restrictions that seek to prohibit or limit U.S. access to or use of space," and that "The United States will preserve its rights, capabilities, and freedom of action in space . . . and deny, if necessary, adversaries the use of space capabilities hostile to U.S. national interests." Taken at face value, this was and is an extraordinarily overreaching and pugilistic statement.

However, since all nations are seldom completely altruistic in their intentions, it is also fair to conclude that Russia and China perceive they are at a current strategic technological disadvantage regarding the "weaponization" of space, hence their willingness to push for a new or expanded treaty.

It is not clear how many countries currently have the capability to "shoot down" satellites. In the fourth annual report by the Space Security Index (Space Security 2007), co-author Thomas Graham estimates that thirty nations now have such capability. This sounds high, but whatever the number; it's probably greater than seven. The other co-author, Ray Williamson with Secure World Foundation, stated "There is growing tension between the U.S. and China over the security of outer space, largely driven by mistrust and suspicions over weapons programs."

In 2007, Mike Moore, a U.S. journalist, wrote an insightful book *Twilight War: The folly of U.S. space dominance* (published by The Independent Institute, 2007). Moore provides a solid, fact-based argument that the U.S. is wittingly or unwittingly provoking an arms race in space by its current posture and actions.

In a May 9, 2008 commentary from the Independent Institute, the current debate on the future of U.S. military space policy continues to evolve. Speaking recently in Colorado Springs, where the U.S. Air Force Space Command is headquartered, Republican Senator Wayne Allard and Democratic Representative Mark Udall agreed that the next (U.S.) president, to quote Allard, "will have to chose which direction to take."

The commentary goes on to state that the options are both stark and clear. Allard is among those who believe the United States needs missile defense weapons in space – weapons that could also be turned against other nations' satellites. Udall, chairman of the congressional subcommittee that overseas NASA, opposes space weapons. "My vision would be that all nations of the world share the high ground of space," he said, and not engage in a new arms race "that results in the weaponization of space."

Where do we go from here?

It is tempting to offer solutions to these two disparate U. S. visions or *beliefs* of national security in Space. Either way, the efficacy of the forty-one-year-old UN Outer Space Treaty will have to be debated. To add a further layer of complexity, there is an intellectual debate bubbling concerning the basic relevance and verification of multinational treaties. Examples are the failed Treaty of the Moon (1984), the non- ratification by the U.S. Congress of the 1982 Law of the Sea, the Kyoto protocol (2005), the Comprehensive Test Ban Treaty (CTBT), plus the current U.S. Administration's general disdain for certain international treaties and agreements (see http://fti.neep.wisc.edu/neep533/FALL2001/lecture40.pdf for chronology of international space agreements).

In order to shed a little light on this critical issue relating to world peace over the next twelve or so years, a more useful approach might be to examine a few credible scenarios, rather than offering specific solutions. Scenarios, of course, are not predictions of the future but, rather, are possible future situations. The determination of which scenario prevails, when, where, and how, is unpredictable and will be determined by events, and actions/reactions of key global participants. As a reminder, *Scenario Analysis* was developed and successfully used by the Shell Oil Company to anticipate the 1973 and subsequent oil "shocks," by examining the critical forces (especially *independent forces*) affecting supply and demand. It is now an accepted tool of management consultants and planning groups.

Scenario I: *Status Quo/Creeping arms race*

- The U.S. is slow to respond to the Russian and Chinese request to update the 1967 Outer Space Treaty. Reasons are varied, from not wanting to dismantle or weaken the (only) existing treaty; to difficulties of verification; mutual mistrust; beliefs that perceived U.S. strategic technological advantage in (military) space could/should be maintained; further ASAT tests by other countries such as India; other military and domestic priorities taking precedent; and no *independent* force emerges to galvanize world bodies to take action to stop the gradual increased (U.S.) weaponization of Space over next twelve years.
- In addition, since 1981, the UN General Assembly has voted year after year to ask the Conference on Disarmament in Geneva to begin negotiations of a treaty that would prevent an arms race in Space. The U.S. has historically abstained from voting on the UN General Assembly resolution, though in recent years it has voted against the resolution. The U.S. has also used its veto power over the work agenda.
- By 2020, a fairly wide array of international ASAT capabilities have been tested and implemented. However, caution still exists about overtly placing offensive weapons in Space, in contrast to the capability to disable satellites in Space from Earth (ASATs).

Scenario II: *Arms Race accelerates in Space*

- For any number of reasons, the 1967 Outer Space Treaty remains effectively the only multinational agreement limiting weapons in Space. Selective interpretations of the Treaty allow certain nation states, including the U.S., to continue non-nuclear ASAT testing - with deployment. Space rapidly becomes the theatre for weapons both defensive and offensive, and a run-away arms race in Space is underway. Each nation justifies its actions by the dual "*independent forces*" of perceived 1) self- defense, and 2) reactions to provocative actions taken by competing nations in an environment of mistrust.

- Numerous efforts at the UN fail to slow the proliferation of weapons involving Space.

Scenario III: *Effective ban of weapons used in Space is agreed & implemented by ~2020*

- Two *independent* forces, 1) the perceived growing uncontrolled space arms race and 2) the continuing threat from failed States and "terrorist organizations," cause the major powers to agree that the use of weapons in space must be controlled, and that verification will be monitored by an independent agency such as an International Space Weapons Agency (ISWA), analogous to the International Atomic Energy Agency (IAEA) formed in 1957.
- The next U.S. Administration (2009) proposes this approach by supporting a revised or new Treaty that bans all weapons in Space by ~2020 including ASATs, and the creation of an organization to verify that weapons are not being deployed or used in space. It is understood that, ultimately, every treaty requires the good will of the signatories to abide by treaties they sign. Earnest negotiations with Space powers begin in 2009 under the new U.S. Administration, with a revised agreement or new treaty signed before 2020. The U.S initiates a significant R&D effort, with strong multinational support, on space technology so that international compliance can be effectively monitored and verified.

The audience for this paper will make their own judgment concerning the probability of each scenario. If the next U.S. Administration and Congress (hopefully) decide that the growing arms race in space requires nipping in the bud; and that the U.S. people and humanity are best served by avoiding a rampant uncontrolled space arms race, Scenario III could prevail. If, however, mutual distrust, suspicion, international deception, the quest for dominance, preemptive control, phony ideologies and hubris prevail, then Scenario I and II are more probable. Optimistically, Scenario 1 would be the middle ground, and perhaps most likely. Pessimistically, Scenario II could prevail. The choice is ours.

<u>If Scenario III (banning weapons in Space) prevails, how could/would it be implemented?</u>

From a practical consideration, the critical questions are: should the 1967 OST be left in place, revised, replaced, or scrapped? Cogent arguments exist for all four possible actions. Some space lawyers and arms control negotiators argue that opening up and revising the OST would risk weakening its impact. Others argue that the treaty is an anachronism from the Cold War and requires replacing with a new treaty that recognizes and anticipates new space weapons technology, including more advanced ASATs, lasers, and particle beam weapons. There are also arguments decrying the *a priori* usefulness of treaties in the first place, and this one in particular, because of verification and enforcement difficulties.

There is no consensus on this issue in the U.S. However, there is the potential that the next Administration will lean towards a new treaty – one aimed expressly at the weapons issue. One mechanism to achieve this could be via a resurrected National Aeronautics and Space Council to coordinate all civilian, commercial and military space programs. This organization was disbanded in 1992.

For Scenario III to prevail, a different leadership from the last U.S. Administration and Congress is a prerequisite. Something must change on the political front in Washington**.** The U.S. is critical because, of the approximately 800 active satellites in orbit, over fifty percent are U.S. satellites.

There are, of course, multiple routes to a changed position by the U.S.:

Dr. Detlev Wolter, Political Councilor German Mission to the UN, 2003-5, proposed a "Legal foundations and essential elements for a system of Common Security in Outer Space (CSO-Treaty)," in May 2007. He argued "There is an urgent need for a comprehensive space security order that starts with a space arms control regime and also encompasses positive elements of cooperative space security like confidence-building measures, rules of the road, international verification, as well as institutional structures. The need for such a preventative arms control regime cannot be overemphasized." He went on to quote U.S. Ambassador Jonathan Dean, who works on issues related to national and European security, arms control, and international peacekeeping, ". . . humanity is on the verge of an irreversible shift to active, destructive, military use of outer space, a global revolution in human security which will almost certainly surpass in significance the introduction of nuclear weapons.*"*

In addition to Dr. Wolter's timely remarks, his paper published by the Global Security Institute includes a detailed bibliography that readers can access (www.gsinstitute.org).

Whether a new treaty is required to replace OST, or an evolving set of related treaties would be more realistic to implement, has to be determined. The complex and ambiguous evolution of the nuclear test ban treaties SALT 1, SALT II, START, and the Comprehensive Test Ban Treaty (CTBT) and Nuclear Non Proliferation Treaty (neither of which the U.S. has ratified to date), provide a cautionary example of the international complexity, national and political self-interest and beliefs, and inherent distrust concerning intent and verification. Nowhere, was the latter more apparent than Reagan's Star Wars initiative and the prevailing philosophy of "Trust - but verify."

Spin, Bubbles, and the Space Arms Race

The Washington journalist Steven Pearlstein noted recently, during the political campaign of 2008, that the analogy of an "arms race" or "race to the bottom" describes Washington's incumbent political bubble culture. "It's the kind of competition in which players – acting (seemingly) rationally to maximize gain, or to defend, attack or push a policy agenda – wind up producing an irrational outcome that leaves everyone worse off." This has been evident in the culture of both the Clinton and Bush II Administrations.

For a new or revised comprehensive treaty banning weapons in space, to supersede the existing 1967 OST, there has to be a change in the 'spin and bubble" culture of Washington by the next Administration – as well as an informed domestic (and international) public that demands it.

Come January 20th 2009, there will be a window of opportunity for the next U.S. president and Congress to join other major Space powers, especially Russia, China, and India to propose a ban of all weapons (especially including ASATs) in Space, with verification through a new independent international organization similar to the venerable IAEA**.** The U.S could propose initiating the necessary funding research for verification, and seek similar funding initiatives from other space-faring nations. This would be a startling *volte face* for the U.S., but could be part of a very different foreign policy (especially from the Bush II Administration and Congress), that would once again seek international consensus and accommodation, respect, and a leadership position for developing, implementing, and abiding by international rules of law. It's time for the U. S. Congress to debate the future of the OST, and to support a (new) treaty for the Non Proliferation of Weapons in Space.

To quote a great American wit, W. C. Fields: "There comes a time in the affairs of man when he must take the bull by the tail and face the situation."

Bibliography: See www.gsinstitute.org

IAC-08-E8.3.4

LEGAL ASPECTS RELATING TO DISARMAMENT, SPACE CONTROL, SPACE SITUATIONAL AWARENESS AND INTERNATIONAL SPACE TRAFFIC MANAGEMENT

Johanna Catena LLB Hons, LLM

Space Generation Advisory Council (SGAC)τ, UK
johanna.catena@dsl.pipex.com
+44 7986423450

ABSTRACT*

A major Commission, chaired by the former Secretary for Defense Donald Rumsfeld to assess security space management in 2000 to 2001, warned of future space pearl harbour. In 2003 the US Air Force published the Transformational Flight Plan detailing future capabilities such as the Hypervelocity Rod Bundles, nicknamed "Rods from God", Ballistic Missile Defense (BMD) systems, space based defense systems and directed energy weapons.

This paper will take a two-fold approach highlighting disarmament treaties, Articles I, IV, and IX contrasting Articles VI, VIII and IX of the Outer Space Treaty (OST) 1967, Liability Convention and the Conference on Disarmament to assess traditional methods of prevention. The second is a pragmatic approach, highlighting space situational awareness, recommending a collective method for alternative measures to space control. For example introducing an appropriate international system to regulate space traffic in order to safeguard space assets and maximise the use of outer space without interference from other space objects.

*τ

The Commission to Assess United States National Security Space Management and Organization was established pursuant to Public Law 106-65, the National Defense Authorization Act for Fiscal Year 2000, Section 1622, commonly referred as the Space Commission.

The Rumsfeld Space Commission of 2001 was mandated to exploit military assets in particular to analyse the relationship between intelligence and non-intelligence national security spaces. Four sectors of space were used as markers: - defense, intelligence, civil and commercial space. Each sector's capabilities were assessed and the Commission made recommendations with respect to national security. The Commission stated that by vigorously pursuing the National Space Policy and in order to deter threats and defend against any possible attack to US interests this would leave no other option but to deploy weapons in space. Also the US must have capabilities to defend its space assets against hostile acts to negate the hostile use of space against US interests.

The Commission recognized the dependency of US national security on space assets by

τ The views expressed in this paper are those of the author and do not represent the views of SGAC unless otherwise stated.

stressing that an event that would "disrupt, disable or destroy" a satellite would have potentially devastating consequences for both defense and intelligence systems. The Commission was concerned that this could make 'US an attractive candidate for a "Space Pearl Harbor"'. [‡]

Therefore, this doctrine justifies the technological developments based on a deterrence strategy that would both prevent any infringement of US sovereign rights and, should attacks happen, provide the means to negate threats temporarily or permanently. Thus according to the Commission applying cost effective technologies would assure access to space and on-orbit operations as well as defence in space with active and passive protection of satellite systems.

The Space Commission's recommendations[§] and concepts were further developed in the Transformational flight Plan of 2003 and 2004 both of which drew upon the six core competencies of Joint Vision 2020 aerospace, information superiority, global attack, precision engagement, rapid global mobility and agile combat support enabling *"global vigilance to anticipate and deter threats, strategic reach to curb crises and overwhelming power to prevail in conflicts and win America's wars."* [**]

The Transformational Flight Plan of 2003 detailed active and passive measures envisioned for space defense in particular listing future weapons that could facilitate such measures. Protection of space assets, denial of an adversary's access to space services and detection of ballistic missiles and subsequent destruction during flight phase are among 16 listed transformational goals as well as six supporting attributes. One of which was lethality, which involved the use of kinetic and non-kinetic means. The sixteen goals are organised under the competencies of Joint Vision 2020 with the addition of air and space superiority.

The 2003 plan also listed future technologies and weapons that could facilitate such measures. Under informational superiority suggested space systems included Advanced Extremely High Frequency system (AEHF) in the near term and mid term space-based radar. To protect space assets and deny adversary's access to space services suggested technologies include counter satellite communication system, counter surveillance and reconnaissance system, rapid attack identification detection and reporting system, space- based infra-red and surveillance system. Other future systems cited in the mid-term include the common aero vehicle a hypersonic glide vehicle dispenses conventional weapons, sensors, payloads from and through space, orbital deep space imager and Space Tracking and Surveillance System (STSS). In the long term suggested weapon systems include air launched anti-satellite missile, ground based laser, space based radio frequency energy weapon, Evolutionary Aerospace Global Laser Engagement (EAGLE) and hypervelocity rod bundles, commonly referred to as the "rods from gods". Other systems include the space manoeuvres vehicle and space operations vehicle to assist with space operations with regard to offensive and defensive counterpace.

In its 2004 edition, the US Air Force focused the flight plan on passive measures, omitting any proposed or future weapon systems. Some of the main changes include replacing

---

[‡] see chapter 2 of the Rumsfeld Space Commission.

[§] Historically the Space Commission's recommendations with regard to providing active and passive protection of space systems can also be found in pg v & vii 04/11/1996 Joint Doctrine 3.01-1 Aerospace Defense of North America

[**] America's Air Force Vision 2020 pg2

'denial' with negating an adversary's space system, minimizing space debris and using active space defense as a last rcsort instead pursing temporary and reversible means. However denying or destroying adversary's space systems is still being considered.[tt]

Negating an enemy's space system involves five elements: deception, disruption, denial, degradation and destruction.[:t:t] Deception involves misleading the enemy into acting unfavourable to their interests. Disruption and denial are temporary measures to the utility of the space system whereas degradation and destruction are permanent involving physical damage or elimination. Examples of space defense systems that are in development include[*]:- AEHF, a jam resistant satellite communications system independent from ground relay stations; counterspace systems, which includes counter communication system designed for temporary disruption; STSS; Space Based Infrared System High detection system for intelligence; missile warning and defence; Space Radar surveillance and reconnaissance capabilities; Space Test Program (e.g. XSS-11); mid-course missile defense plus space based interceptor test bed; Starfire Optical Range funded under PE 0603605 advanced weapons technology; directed energy research e.g. relay mirror (previously known as EAGLE); Multi Kill Vehicle and Autonomous Nanosatellite Guardian for Evaluating Local Space.

In light of the Commission's report and subsequent military doctrines if one analyses Article IV of The Treaty on Principles Governing the Activities of States in the Exploration and Use of Outer Space, including the Moon and Other Celestial Bodies 1969, commonly referred to as the Outer Space Treaty, this prohibits placing "in orbit around the Earth any objects carrying nuclear weapons or any other kinds of weapons of mass destruction, install such weapons on celestial bodies, or station such weapons in outer space in any other manner." In the second part of Article IV the moon and other celestial bodies must be used exclusively for peaceful purposes. Also military bases, installations and fortifications, testing of any weapon or any military manoeuvres on celestial bodies is prohibited. If military personnel are used for peaceful purposes such as scientific research this not prohibited including any equipment or facility for the peaceful exploration of the moon and other celestial bodies.

Weapons of mass destruction are one of the key terms in Article IV. This was defined by United Nations Commission for Conventional Armaments Resolution dated 12 August 1948 as '*atomic explosive weapons, radio-active material weapons, lethal chemical and biological weapons and any other weapons developed in the future which have the characteristics comparable in destructive effect to those of the atomic bomb or other weapons mentioned above."*

Equally, the USA defines weapons of mass destruction in Title 18 Part 1 Chapter 113B §2332a (c) (2) of the United States Code. The term means A) any destructive device which is defined in §9 of Title 18 and includes "any explosive, incendiary or poison gas" for example bomb, grenade, rocket with propellant more than four ounces or a device similar to any of devices

---

[tt] see D-21/D22 with respect to space control and also destroying adversary's missile launch capabilities see for example D-15/16 Transformational Flight Plan 2004

[:t:t] These elements can also be found in the CounterSpace Operations Air Force Doctrine 2-2.1 2 August 2004.

[*] www.mda.mil / wpafb.af.mil / kirtland.af.mil/afrl_vs/ Space Weapons Spending in the FY 2008 Defense Budget Feb. 21, 2007 Theresa Hitchens, Victoria Samson, and Sam Black Space Weapons Spending in the FY 2007 Defense Budget March 6, 2006 Updated March 8, 2006 Theresa Hitchens, Michael Katz-Hyman, and Victoria Samson www.cdi.org

described. The list is not exhaustive. Also B) any weapon that is designed or intended to cause death or serious bodily injury through the release, dissemination or impact of toxic or poisonous chemicals or their precursors or C) any weapon involving a biological agent, toxin or vector or D) any weapon that is designed to release radiation or radioactivity at a level dangerous to human life. Nevertheless, these clarifications still leave important room for interpretation. For example: are the provisions mutually exclusive or is it equating a destructive device with a weapon causing death or serious injury releases a biological agent, toxin or vector or radiation? If they are exclusive terms could an anti-satellite, directed or kinetic energy weapon also be defined as a destructive device?

WMD is also defined in US Code Title 50 Chapter 40 §2302 as a function of the quantity of human casualties. There, it is equated to technologies causing death or serious bodily injury to a significant number of people through the release dissemination or impact of toxic substances, disease and radiation. A definition can also be found Joint Publication 1-02 in the Department of Defense Dictionary of Military and Associated Terms as amended 30 May 2008 describing WMD *"weapons that are capable of a high order of destruction and/or being used in such a manner to destroy large number of people"* and are *"high-yield explosives or nuclear, biological, chemical or radiological weapons."* Moreover, it is worth noting that there is no official definition in international law for WMD. One of the relevant text related to the space environment, is the Limited Test Ban Treaty which prevents any nuclear testing / explosions in space.

The question is whether one could constitute space weapons that have directed energy or kinetic energy weapon capabilities as an explosive device within the meaning of WMD in Title 18 Part 1 Chapter 113B §2332a (c) (2) of the United States Code. Regardless of the above a vacuum would still remain for example other such weapons that are designed to jam would not fall within the meaning of destructive device. Alternatively the action of the state that instigated the attack causing a "high order of destruction" through the loss of one or more state's space asset (s) whilst in-orbit, (which also contributed to the death or serious bodily injury or loss of property). Following the Liability Convention 1972 Article III the state/commercial operator that suffered the damage would have to prove that the launching state was at fault. Also supposing a fragment from the destruction accidently disabled a functional satellite owned by another state which state would be at fault?

A Launching State under Article 1 of the Liability Convention means a state that launches or procures the launching of a space object and a State from whose territory of facility a space object is launched. There can be more than one launching state. Attempted launching is also included. However what if the state no longer has operational control of the object, for example the satellite is sold or even leased, for example what if a satellite was sold by state/company A to State B and State B used the satellite to jam signals who would be the launching state that would have international liability and responsibility under Articles VI and VII of the Outer Space Treaty? Would State A and B be jointly liable under Article IV of the Liability Convention, which covers third party liability in the event of two or more launching states where the first two shall be jointly and severable liable with the burden of compensation apportioned between the first two states to the extent of their fault. Transferring registry may not simply be enough to mitigate any international responsibility and state registries are not immediately filed which can be problematic.

A further issue to consider is whether "procuring" and "space object" include for example the state, (or the private and/or juridical persons) who owns the launch facility; manufactured and installed the launch facility; supplied or transported some of the components to the launch facility, owns the land in which the facility is on; owns the satellite to be launched; manufactured and or supplied components to the satellite; transported the satellite; arranged for the launch; supplied the launch vehicle or financed the project etc. A further clarification on procuring may be called for in the future.

Article IX of the Outer Space Treaty where space exploration should be conducted in such a way that avoids harmful contamination and also adverse changes to the environment of the Earth resulting from the introduction of extra-terrestrial matter, and if necessary measures should be adopted. The provision does not state what those measures should be. However if we apply Article IX to the failing USA-193 satellite which, if not intercepted, according to the US DoD would have deposited 453kg hydrazine fuel potentially causing an adverse change to the environment of the earth. Also, if a state has reason to believe that the activity or experiment would cause potential interference with the activities of another state in the peaceful use of outer space, appropriate consultations should be made.

A statement was made by Ambassador Christina Rocca, Permanent Representative of the United States to the Conference on Disarmament 15th February 2008 notifying plans to intercept the satellite by targeting the fuel tank. It was anticipated that debris would re-enter posing no threat to other satellites and peaceful uses of outer space, as well as choosing an engagement point that would minimise any impact to populated areas. Ambassador Rocca also stated that, *"our transparency in notifying foreign governments and the broader international community is consistent with our commitment to safe and responsible space operations,"* as well as reassuring that *"this engagement is not part of an anti-satellite development and testing program, and we do not intend to retain the technical capability resulting from the modifications required to carry out the engagement."* The Aegis BMD deployed aboard the USS Lake Erie, a sea-based to midcourse BMD system, successfully intercepted using a Standard Missile-3 the failing US 193 satellite on 21st February 2008, which Ambassador Rocca informed the Conference of Disarmament. In contrast an estimated 2,200 trackable pieces of debris plus another estimated 33, 000 objects that are too small to track from the Chinese anti-satellite weapon test, which destroyed the Fengyun-1C weather satellite on 11th January 2007. A survey was held among Space Generation members where 58% of those who took the survey had no objections to the ASAT tests.

Space weaponisation has been a cause for concern since the 1970's, although discussions did not begin until the next decade in the Conference of Disarmament with regard to preventing an arms race in outer space. A draft Treaty was submitted by Soviet Union in 1981 entitled "Treaty on the Prohibition of Stationing of Weapons of Any Kind in Outer Space " (Document AI 36/192 of 20 August 1981) citing a prohibition of weapons of any kind and not to "destroy, damage or disturb the normal functioning of space objects of other state parties." Although any missile defense or ground to space systems appear not to fall under the draft treaty. In 1983 Soviet Union submitted another draft treaty, "Conclusion of a treaty on the prohibition of the use of force in outer space from outer space against the earth" which included broader prohibition including testing and creating anti-satellite weapons. Although a caveat for BMD was still in place. Coming full circle in 2002 Russia again introduced a draft treaty this time in collaboration with China, which has been subject to a few developments over the last years. The new draft version, entitled "Treaty on Prevention of Placement Of Weapons in Outer Space and the Threat or Use of Force Against Outer Space Objects (PPWT) was introduced 29th February 2008 to Conference of Disarmament meeting and Committee on the Peaceful Uses of Outer Space 51st session. Article I of the PPWT defines outer space as "in excess of 100km above sea level". The second part defines an outer space object, although it is very clumsy in the wording for example;

"any device designed to function in outer space which is launched into an orbit around any celestial body, or located in orbit around any celestial body, or on any celestial body, except Earth, or leaving orbit around any celestial body towards this celestial body, or moving from any celestial body towards another celestial body, or placed in outer space by any means."

The one sentence provision is misleading and could be interpreted to mean that Earth is literally the exception for any device to function in space, which is launched into orbit around any celestial etc. It is probably meant to read as "or on any celestial body except earth" (minus the comma). Confusingly this celestial body suggests a particular body, but does not say which one. Is it referring to earth? The first part could read 1) any device designed to function in outer space, which is launched into orbit or is already located in orbit around any celestial body; 2) any device located on any celestial body except earth; 3) any device that deorbits and moves between celestial bodies. An alternative space object definition is offered later in the paper in the context of space situational awareness and space traffic management.

Article 1) c) defines weapons in outer space which is "any device placed in outer space, based on any physical principle which has been specially produced or converted to destroy, damage or disrupt the normal functioning of objects in outer space, on the Earth or in the Earth's atmosphere, or to eliminate a population or components of the biosphere which are important to human existence or inflict damage on them." The treaty is silent on any ground to space or air to space systems thus is it really required to include a boundary definition in this respect? Placed in outer space is defined in Article 1 (d) although it is not really needed. Also perhaps substituting "placed" for "deployed" would be more effective with respect to the terminology and meaning of launching weapons in space. The difficulty with defining a space weapon is distinguishing between offensive and defensive systems.

In a statement with regard to treaty prevention measures by George Pataki, who was nominated by President George W. Bush to serve as Public Delegate to the 62nd UNGA, at the Thematic Debate on "Outer Space (Disarmament Aspects)" in the First Committee of the General Assembly October 22, 2007, he said *"any object orbiting or transiting through outer space can be a weapon if that object is placed intentionally on a collision course with another space object."* Determining the intent of the operator was the only way to distinguish between co-orbital satellite interceptor from non-threatening autonomous servicing vehicle and that could be achieved through transparency of each other space policies and strategies.

The last sub clause (e) defines the use of force or threat of force towards space systems which includes any hostile actions against outer space objects which will destroy, damage, temporarily or permanently disrupt their normal functioning as well as deliberately change their orbits. Article 1(c) and (e) go to the very heart of military activities in space and what constitutes peaceful and non-peaceful purposes. One school of thought is to make a distinction between military space and space weapons, given the dual nature of space systems. This must be read in light of Article I of the OST which states that outer space, moon and other celestial bodies is a province of mankind and shall be free for exploration by all states. Additionally Article III stipulates that exploration must be in accordance with international law and the Charter of the United Nations (UN). With regard to space weapons and peaceful uses Article 2 of the Charter prohibits the use of force or threat of the use of force thus 1 (e) is really elaborating on what is a space weapon. However Article 51 of the UN Charter gives "the inherent right of individual or collective self-defense if an armed attack occurs," and this has been interpreted to include space defense. If we follow the US Air Force Space Operations Doctrine 2-2 27 pg 27 November 2006, it states the traditional view that peaceful purposes do not prohibit military activities and this has continued without any significant protest. Thus activities must be non-aggressive following the UN charter except with the right of self-defense.

In a separate address Ambassador Rocca to the CD 13th February 2007 stated that they reserve the right to protect their space systems and that by maintaining the right to self-defence this does not mean that US will claim space or weaponise it. The statement also described how defining space weapon has come to no avail and efforts should be focused on deterring and dissuading the misuse of space.

However one could argue that developing and deploying offensive weapons potentially capable of an aggressive act could fall short within the meaning of the Outer Space Treaty and United Nations Charter. Additionally would it also not fall within the meaning of the Environmental Modification Convention 1980, which prohibits all hostile actions that might cause long-lasting, severe or widespread environmental effects in space, where any resultant debris could pose a significant threat to other space systems?

According to the 2001 Rumsfeld Space Commission "there is no blanket prohibition in international law on placing or using weapons in space, applying force from space to earth or conducting military operations in and through space." Anticipatory self-defence is included within Article 51 of the UN Charter. The danger is misinterpretation given the nature of the space environment how would one know an attack from an accidental collision and/or any damage that has resulted from space debris? After the Rumsfeld report the US subsequently pulled out of the Anti-Ballistic Missile (ABM) Treaty 1972 to enable the development of National Missile Defense System in December 2001.

In order to arrive at an appropriate definition of a space weapon one could consider the following justifications; 1) a space object has the potential to act as a space weapon;
2) Article 51 of the UN Charter the right to self-defense and 3) the operator's intent for

example whether it is offensive and/or aggressive.

If we then divide space weapon into four categories; potential weapon, modified in-orbit weapon, launched weapon and any other weapon. A potential weapon is an in-orbit space object that is deliberately manoeuvred into the pathway of another state's object with the intent to permanently degrade or destroy that object. A modified in-orbit weapon is any autonomous in-orbit system/object or any ground operations that modifies any space object to deny, deceit, disrupt, degrade and/or cause the destruction to another space object. A launched space weapon includes any object launched from earth or from any autonomous vehicle and/or satellite in space that is designed to deny, deceit, disrupt, degrade and/or cause the destruction to another space object. Lastly any other weapon includes any system that is designed to deny, deceit, disrupt, degrade and/or cause the destruction to another space object from any ground systems or air-launched systems. However any autonomous in-orbit system/object that is used for peaceful purposes is not a space weapon, where the autonomous in-orbit system/object services the normal functioning of a satellite and or tracks and collects space debris.

Following Article 3 of the OST one could additionally add that nothing would preclude the states right to self-defence under Article 51 of the UN Charter. However this would have to be from a direct attack as opposed to a pre-emptive or anticipatory defense action, which too problematic and could have a high probability of making costly mistake and adversely affecting international peace and security. Even if states did exercise their right under Article 51 of the UN Charter how would the state be able to distinguish between an accidental interference and accidental collision with either functional space object or space debris? Thus it could be said in order for space arms control regime to work the treaty may have to call upon states/commercial organizations to include obligatory sensors that would be able to detect and distinguish the difference. Thus on the premise that the difference can be made after a direct first strike this would then not preclude the state's right to temporarily deny or disrupt the space weapon instigating the damage in order to prevent any further or additional attacks. Also if a first strike was to occur the state would be obligated to report it to the appropriate bodies e.g. CD, Legal Sub-Committee of UN Committee of the Peaceful Uses of Outer Space (COPUOS) and the Security Council as well as any Space Traffic Management system in place. With regard to liability issues in using space weapons where collateral damage from the targeted satellite could cause the loss of another state's space object and/or the target that was hit which was not part of the initial strike, a financial burden should be placed on the state that initiated the attack for any damages caused to space objects. This is also problematic under current space law (see below).

SPACE SITUATIONAL AWARENESS

The Transformational Flight plan 2004 defines *"Space Situational Awareness (SSA) combines command, control, intelligence, surveillance, and the environment to understand space operations, threats to operations and impacts".* Deleted from the 2003 edition was that "*SSA enables.... offensive and defensive operations to gain and maintain space superiority."*

For example the Space Surveillance Network (SSN) is part of SSA and consists of various phased array sensors and conventional radars that are distributed worldwide as well as including ground based electro-optic deep space surveillance system. The SSN tracks operational satellites and space debris with a diameter of 10cm or more or 1 meter in geostationary orbit. All the data is transmitted to the Air Force Space Command's Joint Space Operations Center Mountain (JSpOC-Mtn) located at Cheyenne Mountain Air Station.

The Outer Space Treaty (OST) 1967 provides under Article VIII states shall retain jurisdiction and control over objects launched into outer space that are registered to that state and their presence in space or return to Earth remains unaffected. Following the definition of a space object under Article 1(d) of the Liability Convention "which includes component parts of a space object as well as its launch vehicle and parts thereof" states would retain ownership and control of any pieces of satellite that should for example fragment. The application of space debris with particular regard to establishing fault to Article III is problematic. To reiterate the state/commercial operator that suffered the damage would have to prove the other state/commercial operator was at fault, but there is no system to provide for identification of component parts and it would be difficult to prove jurisdiction and control under Article VIII of the OST.

The Space Debris Mitigation Guidelines issued by Inter-Agency Space Debris Coordination Committee (IADC) covers the environment impact of space missions by recommending the (1) Limitation of debris released during normal operations, (2) Minimization of the potential for on-orbit break-ups, (3) Post-mission disposal and (4) Prevention of on-orbit collisions. According to the guidelines space debris[:t] are all man-made objects including fragments and elements thereof, in Earth's orbit or re-entering the earth's atmosphere that are non functional. Utilizing this definition a space object could be defined as *"any "object" launched from a spaceport whilst ascending above or re-entering the transitional "airspaceway" is a space object, which includes all component parts as well as the launch or human operated vehicle and parts thereof."* Thus *"any launch pad or runway that has vertical or horizontal take-off that is designed to place one or more objects in space or coordinate the return of such objects is a "spaceport."* Transitional airspaceway is discussed further under space traffic management.

"A space object is also any object that has de-orbited from its original orbital position and/or any component parts that have separated or fragmented and/or located in a graveyard orbit."

Another difficulty is that the line between the technologies used to develop space weapons could be blurred with the technology that could used to mitigate space debris. Listed in the 2003 Transformational Flight plan is the compact environmental anomaly sensor II ACTD which will help to reduce any satellite down-time through the sensor as well as providing warnings as to the space environment and assist with information as to any malfunctions that may occur, particularly as to whether it is hostile. This type of technology could be useful in a space weapon control regime and also for space debris mitigation.

The telescope at the Okno facility located in Nurek, Tajikistan and the Krona complex in

[:t] Most of the debris is located in LEO and picks up a electrostatic charge perhaps taking advantage of this and using a debris cloud to actuate (e.g. high temperature superconductor) a satellite in order to collect (smaller) debris using aerogel grids. Two satellites would work in a given sector. One of the satellites would, using a Polyvinylidene Fluoride (PvDF) dust sensor, first track the position and measure velocity of smaller debris within the cloud. From the initial sweep by the dust detector one should be able to calculate the direction and speed for a successful collection. The aerogel grids would be disposed of by re-entry.

Zelenchukskaya Northern Caucasus, plus nine radar stations forms Russia's own space surveillance network. The information or data generated by the network is private and not generally shared outside the Russia Government. But there could be cause for some optimism with regard to space surveillance. For example the European Parliament has adopted a resolution 10 July 2008 (2008/2030(INI)) with regard to space and security, approving the "creation of a European space surveillance system leading to space situational awareness (including, for example, GRAVES and TIRA) to monitor the space infrastructure, space debris and, possibly, other threats." But is this enough?

Under Public Law Chapter 135, Title 10 of the United States Code §2274 permits the Department of Defense (DoD) to provide satellite-tracking support to entities outside the US Government under a pilot program subject to an agreement and where the Secretary of Defense determines the service to be in the national security interests of United States. The entity agrees to pay a fee, which will be charged by the Secretary, in order to reimburse the Department for the costs of providing space surveillance data support. The entity must also agree, "not to transfer any data or technical information received under the agreement, including the analysis of tracking data, to any other entity without the express approval of the Secretary."

The Air Force Space Command initiated the CFE Pilot Program through the space-track.org website. The full Congressional pilot program has yet to implemented. The pilot study was extended to September 30 2009 through Public Law 109-364 in 2006, which amended Section 2274(i). The full program is envisioned to supply shared Space Situational Awareness to Commercial, Allied, Public and Foreign Interests.

Services under the fee program include launch support, conjunction assessment, end-of-life/re-entry support, and anomaly resolution support.

However in the future could managing space traffic be equivalent to managing air transport? Could the line between air and space traffic join as one continuous flow of traffic? How can we overcome the risks the risks posed by space debris and any potential hostile interference? In the first instance working together to expand the current radar range makes sense. Secondarily introducing a robust international communication and tracking network could be a first-step solution to facilitating a space traffic management system.

SPACE TRAFFIC MANAGEMENT

Using rules enshrined within civil aviation a roadmap can be formed for a basic framework, which could have the potential to operate in the near-term. Two other major studies have been conducted on Space Traffic Management one by the International Space University (ISS) and the other by International Academy of Astronautics (IAA) providing even more solid grounding in taking the responsibility for safeguarding both the space and earth environment. There is not enough space to make a descriptive analysis of these studies however this paper draws upon the ISS and IAA reports to propose a model framework.

During shuttle missions NASA is informed of any space objects that come within theoretical safety box that measures 40 by 40 by 10 kilometres orientated along the flight path of the Joint Space Operations Center (JSOC) at Cheyenne Mountain Air Station. JSOC also co-ordinate any rendezvous manoeuvres, for example docking with ISS. Thirty-six hours to seventy hours ahead a flight plan is submitted to JSOC, which is analysed for any possible close approaches to the spaceflight mission. Co-ordination of the Shuttle and the ISS are the only STM services currently provided by SSN. Infact there are no formal international procedures or system for designating, identifying, routing and re-routing space traffic or international organisation to implement such measures.

Establishing rules for managing the space highway could help safeguard both the space and earth environment.

A new model framework for a "Space Traffic Management System" was introduced by the author on behalf of the Space Generation Council, at the United Nations Committee on the Peaceful Uses of Outer Space 51st session. The proposed model draws upon principles established in air traffic management and International Civil Aviation Organization annexes as well as a range of ideas and concerns from young students and professionals regarding the space environment.

What would be the organisational structure to set up a STMS?

The framework calls for the creation of an international (regional) communication ground Telemetry, Tracking and Control (TT&C) network, consisting of all Launch Mission Control Centres (MCC) and Satellite Control Centres (SCC). This would be organized under the nine regionally designated Mission-Telemetry, Tracking and Control (M-TTC) Centres' located in the following regions

Africa- Indian Ocean (AFI)
Asia (ASIA)
Caribbean (CAR)
European (EUR)
Middle East (MID)
North America (NAM)
North Atlantic (NAT)
Pacific (PAC)
South America (SAM)

What is Space Traffic?

Space traffic includes space objects ascending and descending (including re-entry) between Initial Sub-Orbital/Orbital Launch Phase and Spacecraft Sub-Orbital/Orbital Approach Phase.
The term space object is defined in the Convention on International Liability for Damage Caused by Space Objects 1972 Article 1(d), which includes component parts of a space object as well as its launch vehicle and parts thereof. Alternatively the definitions proposed on pg 9 of this paper.

How would traffic be identified and designated under a STMS?

Following the nine air navigation regions set out by ICAO an identical network is envisioned separating 'space equivalent' operational traffic and general air traffic. The framework provides for non-civil traffic, which would be identified under "Military Operational Space Traffic" and "General Space Traffic" could encompass any Civil (CIVST) or Commercial Space Traffic (COST). The owner/operator would be responsible for designating their space objects. Designating an appropriate regional Centre would be based on the location of launch site/provider.

The Initial Sub-Orbital/Orbital Launch Phase incorporates the term 'launching', which includes attempted launching. Launching also incorporates all flight phases from stage separation to coasting. The Framework introduces a method to achieve a continuous STM system across border by establishing an international 'transitional airspaceway', which could be located within the

Mesosphere layer of the atmosphere from 50km to 90/100km above the Earth's surface.

The Spacecraft Sub-Orbital/Orbital Approach Phase includes payload separation and orbit injection/insertion. Also transfer from Spacecraft Approach Phase to Final Graveyard Orbit. Under this segment the GST orbital parameters and status including any planned orbital manoeuvres in particular re-orbiting would be made available to the M-TTC Centres.

How would the STMS work?

The MTT&C would facilitate a communication network whereby all State Mission Control Centre's, (who co-ordinate all space launches) and Satellite Control Centre's, (who after the launch take over in-orbit operations) would report to their appropriate regional centre. The MCC would co-ordinate any space launches through their regional Centre. Taking into account that the MCC who provided the launch service may be of different geographical location to that of the owner/operator then the regional centre that provided mission control analysis and launch clearance would transfer TT&C services to the appropriate regional centre, which is again determined by the location of the owner/operator's Satellite Control Centre (SCC).

Prior to launch the framework calls for international spaceflight plan to be filed, which would be made available to all regional M-TT&C Centres. For example the information provided on a spaceflight plan could consist of the; launch date and time; launch provider / spaceport; space launch vehicle / spacecraft; operator; payloads; the designated regional M-TT&C Centre; pre-launch Keplerian Elements (if applicable and is not MOST); MOST Identification Code; descending/re-entry information and any astronauts and/or civilian passengers onboard if it is a human spaceflight mission. The M-TT&C should be immediately informed if the launch date and time should change or is postponed by the state MCC. A single automated tracking software and graphical user interface must be utilised and networked between all Centres.

Regional M-TT&C Centre under Mission Control services will be responsible for launch and re-entry clearance. A pre-launch safety assessment will be performed ensuring that the flight/mission path is clear after the spaceflight plan has been filed. With regard to re-entry clearance all Regional M-TT&C Centres' will adhere to Air Defence Identification Zones, providing information and identification of any space object re-entering the Earth's atmosphere.

Tracking Telemetry and Control services include providing orbit screening, which will also form part of the launch clearance. Other services include maintaining minimum distances for the safe separation of space objects and guiding any planned orbital manoeuvres. Advanced notification of any planned manoeuvres, satellite fragmentation and de-orbit information should be provided by the owner/operator to their regional centre. The information would be treated with confidentiality amongst the nine regional centres.

How would space traffic be co-ordinated with air traffic?

The flexible use of airspace concept is used is civil aviation where the airspace is viewed as one continuum providing for flexible usage on a daily basis to all users, i.e. both civil and military users. The same principle according to the framework could be applied to any space launch from any launch pad or commercial spaceport. Thus after the spaceflight plan is filed the airspace is reserved for the duration of the launch and/or re-entry. Regional M-TT&C Centres' would co-ordinate with Air Traffic Control Services to ensure notification and re-direction of traffic, including maritime traffic.

How would a STMS take into account national security aspects to space traffic?

The military agency responsible for the space object would direct their own traffic, which would be co-ordinated in co-operation with GST M-TT&C as well as with civil and commercial operators. An identification system for MOST traffic without revealing orbital parameters (due to national security requirements) should be provided by the responsible agency in order to facilitate co-ordination with GST.

Under air traffic services co-ordination between civil and military operators is provided for under Annex 11 of ICAO. Following similar terminology for the framework whereby arrangements can be made: -

"To permit information relevant to the safe expeditious conduct of GST to be promptly exchanged between regional M-TT&C Centres and appropriate military units." Thus "Regional M-TT&C Centres shall, either routinely or on request provide appropriate military units with pertinent spaceflight plan and other data concerning GST."

Providing a spaceflight plan and data concerning GST to appropriate military units will help to preserve national security by eliminating the need to provide information from military to military units regarding potential collisions with respect to MOST and with GST.

The STM framework outlined is a first step of many that is intended to naturally develop that will enable future generations to safely continue the exploration and use of outer space, including the moon and other celestial bodies, following the Article I of the Outer Space Treaty 1967 as new purposes and endeavours are utilized by both Governmental and commercial operators.

CONCLUSION

This paper has identified three parameters based on international and US legal and strategic frameworks, with regard to initiating a space weapon treaty:- for any space object operating potentially as a space weapon, the right to self-defence under Article 51 of the UN Charter as well as distinguishing the operator's intent. Each of which must be considered in order to initiate renewed discussions within the Conference of Disarmament. This study also identified remaining problematic areas, in particular liability and collateral damage that could harm other space objects due to the difficulty to identify the origin of the object causing the damage.

The four proposed categories of space weapons: potential weapon, modified in-orbit weapon, launched weapon and any other weapon could be a method for initiating further discussions with regard to a treaty. However if one considers traditional treaty banning methods within the Conference of Disarmament, one must recognize that they have been fraught with difficulties. For this reason, alternative measures and methods need to be considered. Space Traffic Management Systems, for example, would enable deterring and dissuading the misuse of space by working together through the proposed regional centres to ensure the safe passage of traffic.

The model framework facilitates a communication and cooperation network that is designed to protect space assets. It is a passive defense measure that will help to foster further international peace and security as well as promote international co-operation and strengthen international relations.

Specific examples of future goals of the STM framework include; developing debris liability and arbitration system which is a protocol to the Liability Convention; equivalent flexible use of airspace policy with regard to MOST and GST; transition between ATC and STM with an ultimate end goal; a single flexible use of the airspaceway.

Ultimately it is method of safeguarding the resources of outer space, and human explorations for future generations, as well as ensuring a safe space and earth environment for all users whether that is civil, commercial or military space.

BIBILIOGRAPHY

1. The Commission to Assess United States National Security Space Management and Organization January 11 2001 - http://www.dod.mil/pubs/space20010111.html
2. U.S. Air Force Transformation Flight Plan November 2003 & 2004 www.af.mil/library/posture/AF_TRANS_FLIGHT_PLAN-2003.pdf www.oft.osd.mil/library/library_files/document_385_2004_USAF_Transformation_Flight_Plan.pdf
3. CounterSpace Operations Air Force Doctrine 2-2.1 2 August 2004.
4. Joint Vision 2020 and America's Air Force Joint Vision 2020 www.dtic.mil/jointvision/jvpub2.htm www.army.mil/thewayahead/afvision.pdf
5. Space Weapons Spending in the FY 2008 Defense Budget Feb. 21, 2007 Theresa Hitchens, Victoria Samson, and Sam Black
6. Space Weapons Spending in the FY 2007 Defense Budget Updated March 8, 2006 Theresa Hitchens, Michael Katz-Hyman, and Victoria Samson http://www.cdi.org
7. UN Commission for Conventional Armaments Resolution dated 12 August 1948
8. The Office of the Law Revision Counsel http://uscode.house.gov
9. United States Mission to the Conference on Disarmament http://geneva.usmission.gov/CD/statements.html
10. International Space Law & the UN- Nandasiri Jasentuliyana –Kluwer Law International 1999
11. The Space Debris Mitigation Guidelines - Inter-Agency Space Debris Coordination Committee
12. European Parliament http://www.europarl.europa.eu/sides/getDoc.do?pubRef=-//EP//TEXT+TA+P6-TA-2008-0365+0+DOC+XML+V0//EN
13. 2006 International Academy of Astronautics COSMIC Study on STM http://iaaweb.org/iaa/Studies/spacetraffic.pdf
14. STM International Space University Summer Session Program 2007

IAC-08-E8.3.5

LIMITS ON SPACE WEAPONS: INCOPORATING THE LAW OF WAR INTO THE *CORPUS JURIS SPATIALIS*

P.J. Blount
National Center for Remote Sensing, Air, and Space Law
University of Mississippi School of Law
Oxford, Mississippi, U.S.A
pjblount@olemiss.edu

Abstract

Article III of the Outer Space Treaty applies the principles of international law to the territory of outer space. This paper investigates international law sources for limits on "space weaponization" found in the laws of war. Through analysis of both customary and treaty made limitations on space weapons, the paper presents a more unified theory of the *corpus juris spatialis* as called for by Article III. Furthermore, definition of these limits allows for a more focused debate on "space weaponisation."

I. Introduction

In the past two years the world has seen China successfully test an ASAT weapon on a defunct weather satellite, Iran successfully launch a rocket that achieved Earth orbit, and the United States successfully intercept and destroy a failed military reconnaissance satellite. These events and others have prompted renewed interest in an old debate, that of space security. The current tenor of this debate focuses on space weapons, which has been highlighted by a new draft treaty on space weapons that was submitted to the 2008 Conference on Disarmament by Russia and China.[1]

While a push for new governing principles may be valuable, it is vital to understand the current base line restrictions in place that govern space weapons. In order to do so one must examine not only international space law but also international law on the use of force (*jus ad* bellum) and international humanitarian law (*jus in bello*). It is important to note that Article III of the Outer Space Treaty incorporates the application of international law and specifically the Charter of the United Nations in outer space, making it a vital part of the *corpus juris spatialis*.[2] While

[1] *See* Sergey Lavrov, Prepared Statement, Statement by H.E. Mr. Sergey Lavrov at the Plenary Meeting of the Conference on Disarmament (Feb. 12, 2008) *and* Yang Jiechi, Prepared Statement, Message from Foreign Minister Yang Jiechi of the People's Republic of China to the Conference on Disarmament (Feb. 12, 2008). The treaty can be found in CONFERENCE ON DISARMAMENT, LETTER DATED 12 FEBRUARY 2008 FROM THE PERMANENT REPRESENTATIVE OF THE RUSSIAN FEDERATION AND THE PERMANENT REPRESENTATIVE OF CHINA TO THE CONFERENCE ON DISARMAMENT ADDRESSED TO THE SECRETARY-GENERAL OF THE CONFERENCE TRANSMITTING THE RUSSIAN AND CHINESE TEXTS OF THE DRAFT "TREATY ON PREVENTION OF THE PLACEMENT OF WEAPONS IN OUTER SPACE AND THE THREAT OR USE OF FORCE AGAINST OUTER SPACE OBJECTS (PPWT)" INTRODUCED BY THE RUSSIAN FEDERATION AND CHINA, U.N. Doc. CD/1839 (2008).

[2] Treaty on the Principles Governing the Activities of States in the Exploration and Use of Outer Space, including the Moon and Other

the principles that flow from the Outer Space Treaty are important guiding legal principles, it is crucial to acknowledge the incorporation of international law made by the treaty. By applying the body of international law in space it is possible to achieve a holistic view of the law applicable in outer space.

This paper will briefly address the areas of *jus ad bellum* and *jus in bello* in order to analyze the current restrictions on space weaponry that can be found in the law of war. The principles and rules found in this body of law were developed to apply to the traditional areas of warfare, mainly land, sea, and air. Not all rules are directly translatable into the space environment, but many can be applied through analogy. More importantly the underlying humanitarian principles that form the foundation of these rules are translatable and will serve as the basis for the extension of the law of war into space.

II. Peaceful Purposes and *Jus ad Bellum*

The principle of the peaceful uses of outer space can be found throughout the literature on space law; however, the Outer Space Treaty only uses the term "peaceful purposes" to refer to outer space in the preamble of the treaty.[3] It is used in the body of the treaty to refer to the Moon and other celestial bodies but not to outer space in general.[4] There is, however, strong support for the term applying to outer space via customary international law from the term's use in the preambles to both the Declaration of Legal Principles Governing the Activities of States in the Exploration and Use of Outer Space[5] and in the Outer Space Treaty[6] to its use in laws, policies, and official statements of numerous States dealing with their respective space programs.[7]

While the principle of the use of outer space for peaceful purposes has most likely entered customary international law, the content of that term is somewhat nebulous, because different nations apply it in different ways. However, a base line meaning can be attributed to the term through an

---

Celestial Bodies art. III, Jan. 27, 1967, 610 U.N.T.S. 205 [hereinafter Outer Space Treaty].

[3] Outer Space Treaty, *supra* note 2, preamble,

[4] *Id.*, art. IV.

[5] Declaration of Legal Principles Governing the Activities of States in the Exploration and Use of Outer Space, preamble, UN GA res. XVIII, preamble, U.N. Doc. A/RES/18/1962 (1962). *See* Bin Cheng, *United Nations Resolutions on Outer Space: "Instant" Customary International Law*, 5 INDIAN J. INT'L L. 23 (1965).

[6] Outer Space Treaty, *supra* note 2, preamble.

[7] For example: United States: U.S. National Space Policy, N.S.P.D. 49 (2006) ("The United States is committed to the exploration and use of outer space by all nations for peaceful purposes, and for the benefit of all humanity."); Russian Federation: Government of Russian Federation Resolution of May 15, 1995 N 468 Moscow, *available at* http://www.unoosa.org/oosa/SpaceLaw/national/russian_federation/resolution_468_1995E.html ("The Russian Space Agency (RSA) is a federal body of executive power which ensures implementation of the state policy in the field of research and use of outer space for peaceful purposes. . ."); China: Statement by Hu Xiaodi to the U.N. First Committee, October 15, 2002, in General Assembly Records, 57th Session, 1st Committee, 12th Meeting, UN Doc. A/C.1/57/PV.12 (Oct. 15, 2002) ("Using outer space for peaceful purposes reflects the common will and fundamental interests of the international community."); Iran: *Iran rocket launch non-military – ambassador*, RIA NOVISTI, February 8, 2008, http://en.rian.ru/world/20080208/98732321.html ("The recent launch of an Iranian research rocket was strictly for peaceful purposes, and was designed to obtain meteorological data, the Islamic republic's ambassador to Moscow said on Friday.").

analysis of international law on the use of force.[8] The Outer Space Treaty, in art. III, states that international law, and specifically the U.N. Charter shall be applicable in outer space "in the interest of maintaining international peace and security." The adoption of the U.N. Charter is a watershed moment in the law concerning the use of force as it marks a change from *jus ad bellum* to *jus contra bellum*.[9] Art. 2(4) of the Charter adopts the rule that States "shall refrain from the threat of or use of force against the territorial integrity or political independence of any state."[10] In fact Art. 2(3) of the convention uses the phrase "international peace and security" (later echoed in the Outer Space Treaty) when it requires states to use "peaceful means" to settle disputes.[11] The U.N. Charter does not however place a unilateral ban on the use of force in all situations. Self defense[12] and force used by direction of the Security Council are still allowed.[13] Essentially, the U.N Charter outlaws the aggressive use of force,[14] and the aggressive use of force has become an international crime.[15] The repetition of the words "international peace and security" in the Outer Space Treaty is evidence that if nothing else peaceful purposes refers back to norms on force enunciated in the U.N. Charter.

[8] This is not to say that this is what "peaceful purposes" means. The analysis is meant to take into account what peaceful purposes means at a minimum.

[9] Yoram Dinstein, *Notes on War*, 27 HARV. J. L. & PUB. POL'Y 877, 880 (2003-2004).

[10] U.N. Charter, art. 2(4).

[11] *Id.* at art. 2(3).

[12] *Id.* at art. 51

[13] *Id.* at art. 42.

[14] *Id.* at art. 1(1).

[15] *See generally* ANTONIO CASSESSE, INTERNATIONAL CRIMINAL LAW 110-125 (2003).

The exact meaning of aggression is often debated, but guidance can be found in the Definition of Aggression adopted by the U.N. General Assembly in 1974.[16] This resolution defines aggression as "the use of force by a State against the sovereignty, territorial integrity or political independence of another State, or in any other manner inconsistent with the Charter of the United Nations."[17] Since one of the UN Charter's purposes is to maintain international peace and security, States may use force in a way that disturbs international peace and security. Such actions would include the use of force from space or in space when not consistent with the exceptions found within the UN Charter. Furthermore, in light of the assertion by the United States that "purposeful interference" with its space systems "will be viewed as an infringement on our sovereign rights," any attack on one of these spacecraft would be considered an attack on the sovereignty of the U.S. and possibly on its territorial integrity.[18]

Since peaceful purposes can be construed to, at least, mean nonaggressive, then any use of a weapon in space would have to conform to the exceptions to the ban on the use of force found in the U.N. Charter. This first is the use of force when authorized by the Security Council. Under this exception the Security Council may authorize the use of force in order to maintain international peace and security. The second, and more limiting, exception is force used in self defense. The U.N.

[16] The Definition of Aggression, U.N.G.A. Res. 3314 (1974).

[17] *Id.* at art 1.

[18] U.S. DEPARTMENT OF DEFENSE SPACE POLICY, DEPARTMENT OF DEFENSE DIRECTIVE 3100.10, at 4.2.1, (July 9, 1999).

Charter states that States have an "inherent right" to self-defense.[19] States may use force to defend themselves or to defend others; however, there are accepted limitations to this exception: necessity, proportionality, and immediacy.[20] Therefore weapons used in self defense must be able to conform to this paradigm. This is especially important with the idea of proportionality, which disallows States from responding in self-defense in a disproportionate manner. The requirement of immediacy could also be relevant in space due to physics which could cause delay in mobilizing weapons. The necessity requirement allows States to respond with force only when a peaceful settlement of the dispute cannot be negotiated.[21] While necessity is a very important limitation on the recourse to force, it does not add any limitations to what weapons can be used to respond.

Since the U.N. Charter (via art. III of the Outer Space Treaty) applies to the actions of nations in outer space, then the principle of non-aggression also applies. This places an affirmative duty on States not to station weapons of an aggressive nature in outer space. It does not prohibit States from placing weapons of a defensive nature in space (unless some further meaning can be attributed to the term "peaceful purposes") or from placing weapons required by order of the Security Council in order to maintain international peace and security. Of course the difference between an aggressive weapon and a defensive weapon can almost always be found in its use. The proportionality rule is really the only factor under *jus ad bellum* in determining the legality of a weapon, and even this limitation is lacking as more often than not it will be an issue of the use of a weapon rather than the weapon itself. If a weapon can never respond in a proportional manner, then any response from it in self defense would violate international law and such a weapon would be illegal whether placed in space or terrestrially.[22] The factor of immediacy would not necessarily make a weapon illegal. If a weapon cannot be mobilized in time to make an initial and immediate response to an act of aggression, then it could still be legally used during a protracted conflict.

IV. Weapons in Space and Jus in Bello

International humanitarian law seeks to keep civilians from being the target of attack and to reduce the suffering of combatants.[23] This is accomplished through a variety of rules that States agree to apply in times of international armed conflict. Those that are most important to the space weapons debate are those that govern the potential weapons and their use and those banning specific types of weapons. It should also be noted that space assets can enhance a states ability to fulfill its humanitarian responsibilities under the law of international armed conflict.

A. Weapons banned by their Nature

While some weapons are banned specifically, weapons are generally restricted in *jus in bello* due to their nature. The ICJ has recognized two

---

[19] UN Charter, art. 51.

[20] YORAM DINSTEIN, WAR AGGRESSION AND SELF-DEFENCE 237-243 (2005).

[21] *Id.* at 237.

[22] *But see* Legality of the Threat or Use of Nuclear Weapons, Advisory Opinion, 1996 I.C.J. 266 (July 8) *and infra* at 6.

[23] *Id.* at 257.

cardinal principles that "constitut[e] the fabric of humanitarian law."[24] The first is that civilians must not be the object of attacks, so States "must consequently never use weapons that are incapable of distinguishing between civilian and military targets."[25] The second principle is that "it is prohibited to cause unnecessary suffering to combatants."[26] Both of these principles place substantial limitations on the potential space weapons, the first more so than the second.

The first principle states that the weapon must be able to be effectively targeted so that it will only damage military targets and not civilian targets. This could be a crucial limitation on both space based weapons and on weapons used against space based assets. In the first instance a weapon launched to the Earth from space must be reasonably precise; it must be able to re-enter the Earth's atmosphere and hit a legal target.[27] The technical problems of re-entry are amplified if a weapon has limited maneuverability. For instance a Soyuz capsule recently reentered and landed 420 km from its target site.[28] If a technical problem such as this were to be recurring on a weaponized reentry vehicle, it could affect the legality of a space based weapon. Also, a weapon designed to take advantage of the physics of space in order to attack large swaths of the Earth's surface would be illegal due to the fact that it would be "incapable of distinguishing between civilian and military targets."[29] A weapon that cannot be effectively targeted is illegal.

In the second case of a weapon used against a space based asset, the weapon must also be able to be targeted with discrimination. Assuming that the weapon is able to be targeted correctly, the creation of debris could be considered an indiscriminate effect. If a weapon destroys an object on orbit, it is feasible that the resulting debris could cause damage to civilian assets in space. It could be reasonably argued that the State has perpetrated an indiscriminate attack, which can be defined as an attack "which employs a method or means of combat the effects of which cannot be limited a required" under Additional Protocol I.[30] The protocol specifically places limits on attacks on civilian objects[31] and attacks that cause "widespread, long-term and severe damage" to the environment.[32] It should be noted that Additional Protocol I does limit these restrictions' applicability to land, sea, and air combat.[33] However, these limitations are echoed in other treaties and in customary international law. For instance, the International Committee of the Red Cross states that it is customary international law that attacks must be limited as required by

[24] *Id.*
[25] *Id.*
[26] *Id.*
[27] For a discussion of legitimate military targets see YORAM DINSTEIN, CONDUCT OF HOSTILITIES UNDER THE LAW OF INTERNATIONAL ARMED CONFLICT 82-140 (2004).
[28] *Soyuz Bumpy Re-Entry Caused By Technical Glitch*, SPACE DAILY, May 22, 2008, http://www.spacedaily.com/reports/Soyuz_Bumpy_Re_Entry_Caused_By_Technical_Glitch_999.html.
[29] Legality of the Threat or Use of Nuclear Weapons, supra note 22, at 257.
[30] Protocol Additional to the Geneva Conventions of 12 August 1949, and relating to the Protection of Victims of International Armed Conflicts (Protocol I), art. 51(4) 1125 U.N.T.S. 3 [hereinafter Additional Protocol I].
[31] *Id* at art. 52.
[32] *Id.* at art. 55.
[33] *Id.* at art. 49(3).

international law.[34] Liability for such attacks would be covered under the liability convention, such actions taken during a time of international armed conflict could trigger other forms of state responsibility[35] and the violation might even rise to the level of a war crime with individual criminal responsibility. For instance the International Criminal Court has jurisdiction over war crimes, including:

> Intentionally launching an attack in the knowledge that such attack will cause . . . damage to civilian objects or widespread, long-term and severe damage to the natural environment which would clearly be excessive in relation to the concrete and direct military advantage anticipated.[36]

It could be argued that an ASAT weapon that created space debris that damaged civilian objects in orbit during wartime could be a war crime if it met the excessiveness threshold. It could also be argued that the damage to the natural environment during an armed conflict could be a war crime, as it would most certainly be long-term. The question would be whether it was widespread and severe.

The I.C.J.'s second cardinal principle on humanitarian law is that an attack cannot cause unnecessary suffering. This principle can also be found in the Additional Protocol I to the Geneva Conventions which prohibits State parties from "employ[ing] weapons, projectiles and material and methods of warfare of a nature to cause superfluous injury or unnecessary suffering."[37] The ICJ interpreted this to be a ban on weapons that cause "harm greater than that unavoidable to achieve legitimate military objectives."[38] Any such weapon would be illegal in outer space.

Of course, these restrictions are often difficult to apply since many weapons have the capability of both being used in a legal manner and being used in an illegal manner. Thus a weapon that *could* be used in a nondiscriminatory manner or in a way that would cause unnecessary suffering is not banned under *jus in bello* if it can *also* be used in a discriminatory manner and cause limited suffering. In such a case it is the illicit use of the weapon that is outlawed and not the weapon itself.

B. Specific Weapons

i. *Conventional Weapons*

In the Outer Space Treaty there is no ban on the stationing or use of conventional weapons in outer space.[39] This does not, however, give States free license to use or station weapons in outer space.[40] Any such weapons must still be

---

[34] INTERNATIONAL COMMITTEE OF THE RED CROSS, CUSTOMARY INTERNATIONAL HUMANITARIAN LAW 40 (2005).

[35] DRAFT ARTICLES ON RESPONSIBILITY OF STATES FOR INTERNATIONALLY WRONGFUL ACTS, WITH COMMENTARIES, art. 1, *available at* http://untreaty.un.org/ilc/texts/instruments/english/commentaries/9_6_2001.pdf.

[36] Rome Statute of the International Criminal Court, July 1, 2002, art. 82(b)(iv), 2187 U.N.T.S. 90.

[37] Additional Protocol I, *supra* note 30, at art. 35(2).

[38] Legality of the Threat or Use of Nuclear Weapons, *supra* note 22, at 257.

[39] However Article IV of the Outer Space Treaty does prohibit the stationing of conventional weapons on the moon or other celestial bodies. Outer Space Treaty, *supra* note 2, art. IV.

[40] Hague Regulations Respecting the Laws and Customs of War on Land (Annex to Hague Convention IV, 1907), Art. 22, *available at*

permissible under international law. There are several restrictions on the uses of specific conventional weapons in outer space. Poison,[41] weapons that cause non-detectable fragments in their victims,[42] incendiaries when used in the vicinity of civilians,[43] and weapons whose sole purpose are blind combatants[44] are all prohibited to some extent under international law. While it would not be technologically feasible to station some of these weapons in space, if the technology were available they would none the less be prohibited.[45]

Of note are restrictions placed on unsecured naval mines. These are contact mines that are not secured by a mooring or anchor and have the ability to be swept away in a current. According to the Hague Convention VIII these mines must be disabled within an hour of release, due to the way in which they might move and destroy non military objectives.[46] While the ban is not directly translatable to space due to physics, the principle behind this ban is. By analogy, it is probably impermissible to put a weapon in space that would target at random and might destroy a civilian satellite, which may not be a legitimate military objective. The principle could be extended by an analogy to torpedoes which must be disabled if they miss their targets.[47] A weapon in space that misses its target and still poses a threat due to its capabilities might also be illegal (e.g. a warhead being used as an ASAT that misses its mark). However, there probably would not be a ban on a purely kinetic kill weapon that entered orbit as a piece of space debris, as that debris would be no different from the other debris around it (whereas a warhead would still have the potential to explode). While these rules are extended by analogy, it is the underlying principle that States must discriminate amongst targets that governs.

ii. Nuclear Weapons and Weapons of Mass Destruction

It is well known that the Outer Space Treaty creates an affirmative ban on the stationing of nuclear weapons and weapons of mass destruction in space.[48] The term weapons of mass destruction generally encompass nuclear, chemical, and biological weapons.[49] Additionally, the Nuclear Test Ban treaty prohibits State parties from causing nuclear

---

http://www.icrc.org/ihl.nsf/0/1d1726425f6955aec125641e0038bfd6 [hereinafter Hague Regulations]; Additional Protocol I, *supra* note 30, art. 35(1).

[41] Hague Regulations Respecting the Laws and Customs of War on Land (Annex to Hague Convention IV, 1907), *supra* note 40 Art. 23(a). The ban has been recognized most recently in article 3(a) of the Statute of the International Tribunal for the Former Yugoslavia *available at* http://www1.umn.edu/humanrts/icty/statute.html.

[42] Protocol on Non-detectable Fragments (Protocol I), Oct. 10, 1980, 1342 U.N.T.S. 168.

[43] Protocol on Prohibitions or Restrictions on the Use of Incendiary Weapons (Protocol III), Oct. 10 1980, 1342 U.N.T.S. 171.

[44] Protocol on Blinding Laser Weapons (Protocol), Oct. 13 1995, 35 I.L.M. 1213.

[45] *See generally* DAVID WRIGHT, LAURA GREGO, & LISBETH GRONLUND, THE PHYSICS OF SPACE SECURITY A REFERENCE MANUAL (2005).

[46] Convention (VIII) relative to the Laying of Automatic Submarine Contact Mines, Oct. 18 1907, art. 1(1) *available at* http://www.yale.edu/lawweb/avalon/lawofwar/hague08.htm. *See also* SAN REMO MANUAL ON INTERNATIONAL LAW APPLICABLE TO ARMED CONFLICTS AT SEA art. 80-81 (1994) *available at* http://www.icrc.org/ihl.nsf/FULL/560?OpenDocument [hereinafter SAN REMO MANUAL].

[47] Convention (VIII) relative to the Laying of Automatic Submarine Contact Mines, *supra* note 46, at art. 1(1) *and* SAN REMO MANUAL, *supra* note 46, at art. 79.

[48] Outer Space Treaty, *supra* note 2, at art. IV.

[49] DINSTEIN, *supra* note 27, at 73-80.

explosions in outer space.[50] So it seems that the use of nuclear weapons in space (aside from transit of a nuclear warhead which is an accepted and legal practice) is completely foreclosed. However, the International Court of Justice's (ICJ) *Advisory Opinion on Legality of the Threat or Use of Nuclear Weapons* might have created an exception to this rule. The ICJ ruled that in general the use of nuclear weapons would be "contrary to the rules of international law applicable in armed conflict, and in particular the principles and rules of humanitarian law."[51] However, the court states that a State may use a nuclear weapon when the "very survival of a State would be at stake."[52] Since the court treats this as a moment of necessity in which both customary and treaty law can be suspended it is feasible that the Outer Space Treaty and the Limited Test Ban Treaty could also be suspended and that a State may, during "an extreme circumstance of self-defence" use a nuclear weapon in space.[53]

C. Space assets furthering humanitarian responsibilities

It should be noted that the regimes of *jus ad bellum* and *jus in bello* are related but separate bodies of law. While a State has the responsibility not to engage in the threat of or use of force, if it does so the principles of *jus in bello* still apply to both sides of the conflict. This is due to the different goals of the two regimes. *Jus ad bellum* seeks to protect international peace and security while *jus in bello* seeks to protect combatants and civilians. This is important because the question of whether it is a violation of international space law for a State to use a remote sensing satellite to image its opponent's territory while waging an aggressive war could be asked. When analyzing this issue, it is vital to remember that a State has a duty to attack with discrimination. Thus while a State is violating international law by engaging in an illegal use of force, it must still uphold its duties under international humanitarian law.[54] If a state can use a reconnaissance satellite to reduce the loss of life to a civilian population when targeting its weapons (whether space based or not), it is required to do so due to the high priority that international law places on the protection of human life. It is incumbent on a State to do "everything feasible to verify that the objectives to be attacked are military objectives."[55] The objective of this rule most certainly authorizes and requires the use of satellite imagery, satellite navigation systems, satellite communication systems, and even meteorological data gathered by satellite.

D. The Duty to Distinguish

Another general principle in *jus ad bello* is that combatants should distinguish themselves from civilians: "Parties to the conflict shall at all times distinguish between the civilian population and combatants and between civilian objects and military objectives."[56] In order to do this belligerents must behave in such a manner that military objectives are not

[50] Treaty Banning Nuclear Weapon Tests in the Atmosphere, in Outer Space and Under Water, Oct. 10, 1963, Art. 1, 480 U.N.T.S. 43.
[51] Legality of the Threat or Use of Nuclear Weapons, *supra* note 22, at 266.
[52] *Id.*
[53] *Id.*

[54] DINSTEIN, *supra* note 27, at 4-5.
[55] Additional Protocol I, *supra* note 30, at art. 57(2)(a)(i).
[56] *Id.* at art. 48

confused with civilian objects. For instance, combatants must carry their arms openly and wear "fixed distinctive emblem" so that they will not be confused with civilians.[57] Military ships and aircraft must also bear markings that distinguish them from civilian ships and aircraft.[58] This general principle may apply in space as well, but it is unclear how.[59] With current technology military emblems and markings on a spacecraft would not *de facto* serve to distinguish military spacecraft from civilian spacecraft. This objective could be achieved by the use of signals sent on a certain frequency, but States would be very resistant to the idea of broadcasting the location of a target to their enemies. At present, it seems best that states rely on the U.N. Registry of Spacecraft in order to distinguish military from civilian space crafts.

However, it is unclear what the effect of a State registering a military spacecraft as a civilian spacecraft would be. While camouflage and legitimate ruses of war are allowed, it is generally not acceptable for a State to disguise a military objective as a civilian object. An analogy can be drawn from the rules of naval warfare. Warships are allowed to fly false neutral flags when at war as a disguise.[60] However, they may not fly false flags during an attack or false flags that identify them as certain classes of ships such as hospital vessels, Red Cross vessels, or civilian passenger vessels.[61] In light of the underlying principle on these rules, it is unlikely that a state would be allowed to register a military spacecraft as a civilian spacecraft as a ruse of war. However, it might be acceptable if a state were to make declarations about its military space assets at the onset of an armed conflict.

V. Conclusion

Outside of those weapons expressly forbidden, much of the body of law is concerned with of the legality of use of a weapon more than the legality of the weapon itself, since a weapon that *can* be used illegally is not illegal unless it *cannot* be used legally. Unfortunately, not all of the principles of the laws of war can be applied directly to the space environment due to the significantly different physical attributes of outer space as compared to land, sea, and air. However, the laws of war as applicable in the traditional arenas of warfare can serve as guiding principles for limitations on space warfare. Space is a feasible arena for war, and despite the wisdom of such actions, it is important to be able to identify the law applicable to belligerents in such operations. This holds true not just for the issue of space weapons, but for the full gamut of military space operations.

---

[57] Hague Regulations, supra note 40, art. 1.

[58] Rules concerning the Control of Wireless Telegraphy in Time of War and Air Warfare. Drafted by a Commission of Jurists at the Hague, December 1922 - February 1923, pt. II art. 3, *available at* http://www.icrc.org/ihl.nsf/FULL/275?OpenDocument; SAN REMO MANUAL, *supra* note 46, art. 13(g).

[59] It is important to note that rules affecting the duty to distinguish would affect all military space craft and not just space weapons, since all military craft would be considered a military objective.

[60] SAN REMO MANUAL, *supra* note 46, art. 110.

[61] *Id.*

IAC-08-E8.3-6

Distinction Between Militarisation and Weaponisation of Outer Space, A Misleading Concept

A.A.Golroo
Faculty Member, Aerospace Research Institute
Ministry of Science Research and Technology
Tehran- Iran
ali@ari.ac.ir

Prof. M. Bahrami
Director, Aerospace Research Institute, Iran
Professor, Amir Kabir University of Technology, Iran
mbahrami@ari.ac.ir

Abstract

The Article IV of the Outer Space Treaty (OST) Prohibits to place in orbit around the earth any objects carrying nuclear weapons or any other kinds of weapons of mass destruction, install weapons on celestial bodies or stations such weapons in Outer Space in any other manner and clearly indicate that the Moon and other Celestial bodies shall be used exclusively for peaceful purpose also the establishment of military bases, installations and fortifications, the testing of any type of weapons and the conduct of military maneuvers on celestial bodies shall be forbidden.

After September 11, 2001, Terrorist attacks in United States. Security became the most important issue in space faring states specially United States and they are using space technology to protect their national interest. Use of space based technology for military purposes seem to be inevitable. In order to overcome the prohibition in Art. IV of the OST many of these states giving a different interpretation for Militarisation and Weaponisation in Outer Space and try to distinguish between these two subjects.

When Commercialization and Privatization of activities in Outer Space became inevitable, the OST and other treaties related to activities in Outer Space were interpreted in such a manner which these treaties permit such activities. It seems the same rational and methodology are used to interpretate the Militarisation and Weaponisation of outer space, despite the fact that the first interpretation is inline with the merit of the Treaties which is the concept of "Peaceful use of Outer Space " ,but the second interpretation for sure is not in the same line.

This paper tries to shed some lights on these issues

Introduction

Assembly of Western European Union, the Interparliamentary European Security and Defence Assembly in its 53^{ed} session adopted Recommendation 804 on weapons in space which among other things defined the militarisation of space as the use of space systems in support of ground-based military operations and the weaponisation of space as the placing in orbit of weapons[1]. Almost the same position is exist among other space faring nations such as USA.

It seems that is a short sighted definition and the main purpose is to overcome the legal problems which prohibit the military use of outer space. Infact this type of interpretation give ground to legitimize any type of operations by satellites including spying, missile control and guidance, first hand information about natural and strategic resources,... and at the same time any reaction to stop such operation by any state is prohibited. As this type of interpretation is in favor of those advanced countries which by using space technology dominate their supremacy in air and make states with no such ability defenceless, therefore states to defend their national interest have no other choice but to react with all their ways and means, and that is why we saw China anti-satellite lasers gun and no doubt that many others states will go the same way, however we should not forget that two super powers went the same path during the cold war.

Definitional Problems:

Although the Assembly of Western European Union, the Interparliamentary European Security and Defence Assembly tried to distinguish between the militarisation and the weaponisation of outer space with the definition of the militarisation of space as the use of space systems in support of ground-based military operations and of weaponisation of space as the placing in orbit of weapons, but it seems, there are many serious problems with this definition.

First of all it seems that distinction is purely based on practical approach, infact they believe that according to international space law using the outer space for military purposes are permissible but placing the weapons in space is prohibited, and therefore space systems has been considered as part of militarisation. According to this definition, the supporting systems must be in outer space, the destroyers and the targets must be in airspace or on the ground which has no legal base.

Secondly a weapon is a system which all parts of it work together, what Assembly has defined is not definition for militarisation and weaponisation but infact give different definitions for different parts of a weapon, control and guidance has been considered as supporting system therefore under "militarisation" concept, destroyers under "weaponisation" concept, and targets must be in air space or on the ground. Obviously this is a legal cover up for what militaries are practicing today and if they change the games, the Assembly has to change the definition to cover up the legality of the new operation.

Commercialisation and Privatetisation Approach:

Where the space based activities be came so important to the day to day life of human being and space based technology be came inevitable part of global economy and space based activities made commercial sense. Therefore private sectors wanted to have their share in space based activities which obviously there is noting wrong to that end, infect private sectors opened new horizon to the space.

However existing international legal instruments such as Outer Space Treaty (OST) or other legal instruments have no direct reference to the commercialisation or privatisation of outer space activities, infect existing instruments, refer to state activities and state responsibilities [2]. The lack of direct reference to the commercialisation and privatisation were not a big obstacle for private sector to inter in commercial based activities in outer space.

Two main reasons exist to justify the legality of commercial activities in outer space by private sectors. Although there is no direct reference in international legal instruments for private or commercial use of outer space, however there is no direct reference in opposite direction to prohibit such activities. The more important reason is the merit of existing international legal instrument, which among other issues emphasis the exploration and exploitation of outer space for the benefit of mankind, there shall be freedom of scientific investigation in outer space including the Moon and others celestial bodies. The importance of international cooperation in the field of activities in the peaceful exploration and use of outer space [3]. It is common believe that commercial activities in outer space by the private sector is in the line of human Endeavour for exploration and use of outer space for peaceful purposes, but for sure militarisation of outer space is not in the same line.

Policy Mistakes:

Although politicians in order to gain public support in international conflicts try to minimize the number of personnel casualties by fighting from distance, for example using long rang missile or Unmanned Aerial Vehicle (UAV), which in order to be precise and successful. They need to be controlled and guided through space based technology, as are the case in Iraq and Afghanistan. The consequence of this policy will be the shifting of the conflicts on the earth to the outer space, and therefore increasing of militarisation of outer space.

As the use of the space based technology for military purposes in actual battlefield is limited against those countries or group which have no capability to stop such usage or to react accordingly, therefore the danger of this policy has not been magnified yet. But in case of achievement of space based technology by terrorist group or real conflict between major power then there will be real disaster in outer space.

The core fear is that any conflict in space would cause the most injury to those states which using it the most. Damaged planes crash to the ground and destroyed ships sink to the bottom of the sea. But the weightlessness of space means that debris keeps spinning around the Earth for years, if not centuries. Each destruction of a satellite creates, in effect, thousands of missiles zipping round randomly, each subsequent impact provides yet more high-speed debris. At some point, given enough litter, there would be a chain reaction of impacts that would render parts of low- Earth orbit- the location of about half the active satellites- Unusable [4].

Self-defence:

Article 2(4) of the Charter provides that states are to refrain 'from the threat or use of force against the territorial integrity or political independence of any state, or in any other manner inconsistent with the Purposes of the United Nations' [5]. As discussed above, this principle has been found by the International Court of Justice to be binding on all states not only as an international customary norm but also as a norm of jus cogens [6].

One of the two exceptions to this principle is the use of force as authorized by the Security Council under Article 42 of the Charter 'to maintain or restore international peace and security' if there is a 'threat to the peace, breach of the peace, or act of aggression' for which economic and trade sanctions would be inadequate [7].

The other exception is the collective right to individual or collective self-defence as recognized in Article 51 of the Charter 'until the Security Council has taken measures necessary to maintain international peace and security'. In any event, the right of states to individual and collective self-defence is also well established in customary international law [8].

UN's charter does state that outer space and celestial bodies are free for exploration and use by all states in conformity with international law, and they are not subject to national appropriation. But the charter's Article 51 also provides the right to self-defence. It states that "nothing in the present Charter shall impair the inherent rights of individual or collective self-defence if an armed attack occurs against a UN member, until the Security Council has taken measures for maintaining peace and security". Though this provision was enunciated in 1945, much before space explorations began, it also extends to outer space.

The Vienna Convention on the Law of Treaties provides that a later treaty, such as the Outer Space Treaty, prevails over an earlier treaty, such as the Charter of the United Nations, in the event of any inconsistency, subject only to Article 103 of the Charter. In this context, the prohibitions contained in Article IV of the Outer Space Treaty would arguably prevail in all circumstances except where the Security Council decided expressly or impliedly that military action, including the deployment and the use of force in contravention of Article IV of the Outer Space Treaty, was sanctioned under Article 42 of the Charter of the United Nations. On the other hand, while this position would be correct in the context of the effects of Article IV of the Outer Space Treaty on Article 51 of the Charter of the United Nations, such a discussion must also take into account that the right to individual and collective self-defence has an existence as a jus cogens norm of customary international law external to the terms of Article 51.This can be seen from the actual wording of Article 51 of the Charter, which provides for the recognition of the 'inherent right' to self-defence rather than providing for the right to self-defence within its own terms.

Communications satellites that transfer civilian communications for civilian purposes also can transfer military communications in times of armed conflict, as in the Gulf War. Does the intent to aid a military purpose render the activity as aggressive and contrary to the peaceful purpose language of the Treaty?

Aid for military purpose is not aggressive in an unlawful sense under the United Nations Charter if it is pursuant to a United Nations Security Council Resolution or done in self-defence. If an activity does not violate the United Nations Charter, then arguably it does not violate the peaceful purpose of the Outer Space Treaty. The same dilemma arises with the use of satellites for mapping, weather, navigation, early warning and reconnaissance when the activity aids a military conflict.

Each state has an inherent right to self-defence, and Article III of the Outer Space Treaty, references Article 51 of the United Nations Charter, expressly preserves the right to use space in self-defence. However, Article 51 of the United Nations Charter authorizes self-defence only in circumstances of an armed attack. Some narrowly interpret this to mean only those situations "resulting from an instant overwhelming necessity leaving no choice of means and no moment for deliberation." This view requires an armed attack before self-defence measures can be invoked. In light of the rapid and massive destructive capabilities of modern weaponry, this view may leave insufficient time to effectively exercise the self-defence option from space. The more realistic approach is to recognize the need of nations to anticipate the threat of armed attack and react defensively to the threat without waiting for the actual attack [9].

Under the Outer Space Treaty, while the principle of self-defence remains intact, the method of that defence is limited. Even for self-defence purposes, the Outer Space Treaty prohibits the use of nuclear, chemical, biological, or other weapons of mass destruction. Thus, the Outer Space Treaty limits the self-defence principle. With this precept in mind, a wide range of military activity can still fit under the self-defence umbrella.

The United States is currently developing means to equip satellites with warning or impact sensors to signal when a satellite is being approached or has been attacked. Satellites designed with weapons, other than weapons of mass destruction, to sense and preemptively destroy other "killer satellites" seeking to attack are lawful under the self-defence exception to the peaceful purpose of the Outer Space Treaty.

Space control measures to preemptively deny other nations from gaining space superiority in a future armed conflict poses a more difficult problem. In this scenario, judge advocates must consider the type of weapon to be used and the nature of the underlying armed conflict. If the space-based system is used in support of an unlawful conflict of aggression, in violation of the United Nations Charter or other recognized international law, then the space control measure is likewise unlawful. However, if the space control measure serves a United Nations sanctioned defensive response to aggression, as in the Gulf War, and no weapons of mass destruction are used, then it likely would fit within accepted activity under Article III of the Outer Space Treaty.

The principles of public international law, this right of self-defence remains subject to express legal limitations the requirements of necessity and proportionality [10].

In its Advisory Opinion in the Legality of the Threat or Use of Nuclear Weapons, the International Court of Justice observed: "The submission of the exercise of the right of self-defence to the conditions of necessity and proportionality is a rule of customary international law [11] ".

Moreover, even where the right of self-defence is lawfully exercised, the State acting in self-defence remains subject to the jus in bello principles.The sentiments encapsulated in the United Nations Charter were strengthened further by the restrictions imposed in relation to nuclear weapons and weapons of mass destruction by Article IV of the Outer Space

Treaty, although, as has been well documented by leading commentators, this provision in and of itself does not represent a complete restriction on the placement of weapons in outer space. Indeed, there have been, from time to time, proposals put forward to amend Article IV in order to enhance these restrictions, but this has not (yet) eventuated.

Conclusion:

Distinction between militarisation and weaponisation of outer space not only do not promot the international peace and security, rather transfer the international conflicts from the Earth to the outer space. As the states under UN Charter have the legitimate right for self defence, they can destroy any satellite which are supporting the military activities of their enemy during a conflict.

That means those who are the main user of outer space loss the most. Each destruction of a satellite creats, in effect, thousands of missiles zipping round randomly, and makes outer space specially the near Earth orbit unusable.

References:

1. Assembly of Western European Union, the Interparliamentary European Security and Defence Assembly, 53th Session 6 June 2007, Dec. A/1966

2. Treaty on Principles Governing the Activities of State in the Exploration and Use of Outer Space Including the Moon and other Celestial Bodies (The Outer Space Treaty of 1967) Art. III and VI

3. Ibid Art.I

4. Disharmony in the spheres. http:// www. Economist. Com/science/Printer Friendly. Cfm? Story id=10533205 Jan. 17th 2008

5. United Nations Charter Art. 2(4)

6.Military and Paramilitary Activities in and against Nicaragua (Merits) [1986] ICJ Rep 14

7. UN Charter, Arts 39 and 42

8.Ochoa-Ruiz and Salamanca-Aguado, 'Exploring the Limits of International Law Relating to the Use of Force in Self-defence', 16 EJIL, 2005, p499

9. Initial efforts by the United States to develop a ballistic missile defence system were made in 1956 with the United States Army's Nike-Zeus program.

10. referring to an additional requirement of immediacy. However, this was not mentioned in the recent decision of the International Court of Justice concerning Oil Platforms (Iran v. U.S., 1996 I.C.J. 803 (Dec. 12).

11. Legality of the Threat or Use of Nuclear Weapons, Advisory Opinion, 1996 I.C.J. 259 (July 8).

IAC 2008 - E8.3

Weaponization of Outer Space and Impact on Peaceful Uses[▲]

Mr. V. Gopalakrishnan
Indian Space Research Organisation, Bangalore, India, gopal@isro.gov.in

Mr. KRS. Murthi
Antrix Corporation, Bangalore, India, krs@antrix.gov.in

Dr. MYS Prasad
Satish Dhawan Space Centre, Sriharikota, India, mys@shar.gov.in

The advent of access to outer space and successful placement and operation of man-made objects therein promised many potential applications to the humankind. The extension of cold war rivalry beyond the terrestrial limits to outer space cautioned the then super powers, of the potential security related implications and brought to an understanding on non-militarization of space, which culminated into the inclusion of Article-IV in the Outer Space Treaty. The carefully chosen language of this article gave sufficient scope for interpretation of the term 'peaceful purposes' as 'non-aggressive', thus permitting certain types of military uses of outer space. However, over a period of time, several new dimensions for using space for defence related purposes and for testing certain technologies had emerged, rising concerns on escalating an arms race in outer space. This paper examines the initial phase concerns, approaches and understanding, the transitional phase alterations and the present scenario of the principle of peaceful purposes vis-à-vis weaponization of space.

Under the above backdrop of international development, the paper also attempts to elaborate the trends of Indian Space Programme and its space policy that focuses on developmental applications of space. Potential impacts of any arms race in outer space on the peaceful uses of outer space are analysed and need for additional legal measures beyond Article-IV are discussed.

[▲] *All the opinions expressed herein belong to the authors and do not represent those of their employer or their country. Any errors contained herein remain entirely the authors'.*

1. Introduction

Space has become very vital for global economy and strategic planning. The space based applications have stretched their wings from advanced exploratory activities to every day activities of a common man. Because of the military potential / advantages, Space is becoming the fourth physical military environment, after land, maritime and air. Placement of weapons under the pretext of national security, are actively considered by some space faring nations. The protection of outer space from weaponization thereby eliminating an arms race in outer space is the minimum desire of the global community beyond the space faring nations. The security of the space systems would eventually depend on the harmonious ambience in outer space, which obviously would necessitate a legal instrument supported by the State practice in letter and spirit.

In this paper it is attempted to review the initial phase developments in the space law formulation and subsequent developments in UN forums, analysis on arms control provisions of Outer Space Treaty (OST), a review on the policy changes towards space militarization and preparations for space security and technology demonstrations of few space faring nations and an analysis on the ongoing activities in various international forums towards reaching a legal instrument, in the first part. Later, a summary of the India's space programme, its objectives and approaches and a brief on the application programmes for peaceful purposes are dwelt upon. Finally, an analysis on the need for a legal instrument to proscribe the weaponization of outer space and the *modus operandi* of realizing this objective is presented.

2. Initial phase developments in the international arena

In this section it is attempted to give a broad look on the initial phase developments in the UN forums to prevent arms race in outer space, particularly pursed by the then super powers with a basic desire to prevent the other side from achieving a military advantage through the use of outer space. As we know, the advent of access to outer space and successful placement and operation of man-made objects therein ushered in a new era of high technology applications to the benefits of human kind along with certain undesirable side effects as well. It is worth noting that even before the launch of Sputnik, the United States of America proposed to UN General Assembly in its first memorandum devoted to arms control in outer space, pleading for the establishment of multilateral control system with international inspection and participation.[1]

Starting from the first outer space achievement of launching of Sputnik on 4th October 1957 by the then USSR, the rivalry between the super powers continued throughout the Cold War era in exhibiting their technical supremacies. This scientific achievement taken up as a programme of International Geo Physical Year triggered a military revolution with extraordinary strategic consequence [2]. Subsequently, the public statements, proclamations and letter correspondences between Mr. D. Eisenhover, the President of USA at that time and the then Soviet Premier, during the period October 1957 to April 1958 revealed their true intentions based on the military potentials of the outer space. The statement of President Eisenhower in his letter dated January 12, 1958, to Soviet Premier (Mr. Bulganin), which reads as - *"I propose that we agree that outer space should be used only for peaceful purposes. We face a decisive moment in history in relation to this matter. Both the Soviet Union and the United States are now using outer space for the testing of missiles designed for military purposes. The time is to stop now"*[3], is a testimony of the severity of the situation that prevailed. The international efforts in controlling the perceived threat of arms race in outer space culminated in addressing this issue for the first time in the UN General Assembly Resolution [No. 1148 (XII)] on 14th November 1957 under Disarmament matters handled by the First Committee. This resolution urged States to reach a disarmament agreement which *inter-alia* provide for '*the Joint study of an inspection system designed to ensure that the sending objects through outer space shall be exclusively for*

---

1 *US memorandum submitted to First committee of UN on 12th January 1957, UN Doc. A/C.1/738, as quoted in UNDIR/2005/29 – 'Common Security in outer space and international law', pp-9*

2 *'Background paper: peaceful and military uses of outer space: law and policy', February, 2005- Institute of Air & space Law, Mc Gill University*

3 *Bess CM Reinjnen, 'The United Nations Space Treaties Analysed'- Frontiers, 1992- p.p 41*

peaceful and scientific purpose". [4] In the subsequent year, multi-national debate on the question of peaceful uses of outer space helped to reach a major milestone of setting up an eighteen nation ad-hoc Committee on the peaceful uses of outer space and requested it to submit a report on the activities of UN bodies relating to peaceful uses of outer space, international cooperation in promoting the outer space activities for peaceful uses and any associated legal problems that might arise[5]. Further boost was given by making this ad-hoc body a permanent United Nations Committee On Peaceful Uses of Oouter Space (UNCOPUOS) with 24 member states through General Assembly Resolution under the title 'International Cooperation in the Peaceful Uses of Outer Space'[6]. The major objectives that were brought under the scope of UNCOPUOS include the promotion of international cooperation for giving effect to programmes in the peaceful uses of outer space under the auspices of UN and developing an adequate legal framework for the use of outer space. *[In the subsequent years and even till now it has been a customary to pass resolutions to this effect with special mentions on the current issues].*

In the year 1961, UNCOPUOS prepared yet another resolution 1721, which was adopted by the General Assembly on December 20, 1961. It included two leading principles, namely application of international law and in particular the UN Charter to outer space activities and the prohibition of national appropriation of parts of outer space and its resources[7]. Another directive in relation to the furnishing of information on the space objects launched by the States to UN was also included, obviously to keep a watch on the space activities of the States. *[It may be noted that this is the forerunner provision with respect registration of space objects, which is being followed even now by some space faring nations which are not parties to Registration Convention].*

The year 1963 witnessed major developments relating to disarmament in general and demilitarization of outer space in space in particular. The treaty banning Nuclear weapon tests in the Atmosphere, **Outer Space** and Under Water, which is popularly known as Partial Test Ban (PTB) Treaty came into effect at the initiatives of USA, USSR and UK. This is considered to be the first legal instrument that addressed the issue of weapons in outer space. Taking a cue from this, the First Committee on UN dealing with disarmament matters passed a resolution[8] which called upon the States to refrain from placing nuclear weapons or any kind of Weapons of Mass Destruction (WMD) in outer space. Interestingly this resolution, which acknowledged the basic initiatives of USA and USSR in this regard, is construed to be the precursor for the subsequent development of Article - IV of Outer Space Treaty (OST) by the scholars[9]. The other resolution [10] that was taken up by the UNCOPUOS was on the 'Declaration of legal principles governing the activities of States in the exploration and use of outer space'. This 'Principles Declaration', is considered to be the precursor of the formulation of the Outer space Treaty (OST) in 1967, as the nine basic principles enunciated therein were subsequently developed as the major articles of the OST. It is unfortunate that this major resolution did not include any provision prohibiting the stationing of weapons in orbit, though it was specifically addressed by the Fourth Committee, two months ago. The lack of such provision was even regretted by India and Austria[11].

The 'Principles Declaration' of 1963 was further debated in the UNCOPUOS in order to enhance its binding obligations through a Treaty mode. Thus born the OST, popularly known as Magna Carta or Constitution of Outer Space, through a UNGA resolution[12], which came into force on October 10, 1967[13].

---

4 *Section 1 (f) of UNGA Resolution 1148 (XII) Dt.14th Nov 1957*

5 *UNGA Resolution 1348 (XIII) dt 13th December 1958*

6 *UNGA Resolution 1472 (XIV) dt 12h December 1959*

7 *UNDIR/2005/29 - 'Common Security in outer space and international law'*

8 *UNGA Resolution No. 1884 (XVIII) dt.October 17, 1963 on 'Question of general and complete disarmament'.*

9 *Ogunsoloa O. Ogunbanwo, 'International law and outer space activities', Martinus Nijhoff,, 1975*

10 *UNGA 1962 (XVIII) of December 13, 1963.*

11 *C. Wilfred Jenks, 'Space Law', Stevens & Sons' 1965 – pp.303*

12 *UNGA Resolution. No. 2222 (XXI) dt. December 19, 1966.*

13 *Some of the articles of OST were further elaborated as further treaties - Agreement on Rescue of Astronauts, Liability conventions, Registration Convention and Moon Treaty*

3. **Mid term developments - Multilateral negotiations at the United Nations on the preventions of arms race in Outer Space - UNCOPUOS and CD:**

The United Nations General Assembly at it 10th Special Session in 1978, sensing the potential danger of an arms race in outer space, called upon the States to initiate multi lateral negotiations on Prevention of Arms Race in Outer Space (PAROS) in accordance with OST[14]. The deliberations culminated in the formation of a single multilateral disarmament negotiating forum called Conference on Disarmament (CD). An agenda item on PAROS has been regularly debated in the UN CD since 1981 to evolve a treaty on PAROS [15]. While some states like Italy proposed for additional protocols to OST for prohibitions of military or hostile uses of outer space to be taken up by UNCOPUOS, supported by other states too, USA rejected any discussion on disarmament of outer space in UNCOPUOS and preferred to be taken up in CD. Subsequently, an ad-hoc committee was set up in CD 1985 to work on PAROS. While a majority of the States were favouring the formation of a new treaty on PAROS, USA maintained the stand the existing UN space treaties, bilateral and multilateral arms control provisions, customary international law and national laws, which are complementary to each other are adequate to tackle outer space weaponization issues. Without any progress, the extensions of Ad-hoc committee of CD on PAROS after 10 years became difficult and hence got suspended. Renewed efforts to re-establish the Working Group in 1998 too did not fructify followed by a complete standstill since 1998.

4. **Other International Treaties, Legal instruments relating regulation of Military Activities in outer space:**

In addition to the treaties discussed in the previous sections such as OST, PTBT, and UNGA Resolution on PAROS, few other legal instruments that are directly or indirectly related to regulation of the military activities or advocacy for activities in outer space for 'peaceful purposes' are listed below for the sake of reference.

- Article 2(4) of UN Charter which obligates the Members to refrain from the threat or use of force in international relations.
- Article -51, the right of self-defence in case of an armed attack, which balanced with certain restrictions.
- Anti-Ballistic Missile (ABM) Treaty (1972) and Protocol (1976), between the USA and USSR, which was intended to limit the deployment, testing and use of missile systems designed to intercept incoming strategic ballistic missiles. *(Later in 2002, USA unilaterally withdrew from this Treaty thereby increasing the speculations on its preparedness for waging an arms race in outer space).*
- Chemical Weapons Convention of 1993, imposing wide-ranging bans on chemical weapons anywhere under the jurisdiction or control of a state party and verifiable. *(Article VIII of OST provides for retaining the right of jurisdiction and control over a space object by the State of Registry).*
- Convention on the Prohibition of Military or Any Other Hostile Use of Environmental Modification Techniques, 1978 - Article II says that "environmental modification techniques" refers to any technique for changing -- through the deliberate manipulation of natural processes -- the dynamics, composition or structure of the Earth, including its biota, lithosphere, hydrosphere and atmosphere, or of outer space.

5. **Arms Control and Outer Space Treaty:**

Having seen the background of the developments leading to the formation of Outer Space Treaty and other related developments in multilateral forums of UN to prohibit the arms race in outer space, it is pertinent to analyse the ambiguities in the provisions of OST with respect to peaceful uses and prohibition on stationing of weapons in outer space, in particular the Article-IV, that lead to various interpretations and state practices, though such analyses have been already made by various scholars.

---

[14] *UNDIR/2005/29 – 'Common Security in outer space and international law' pp.55-56*

[15] *Supra note at 2*

Article IV of OST reads as –

[Paragraph -1] States Parties to the Treaty undertake not to place in obit around the Earth any objects carrying nuclear weapons or any other kinds of weapons of mass destruction, install such weapons on celestial bodies or station such weapons in outer space in any other manner.

[Paragraph -2] The moon and other celestial bodies shall be used by all the States Parties to the Treaty exclusively for peaceful purposes. The establishment of military bases, installation and fortifications, the testing of any type of weapons and the conduct of military manoeuvres on celestial bodies shall be forbidden. The use of military personnel for scientific research and for any other peaceful purposes shall not be prohibited. The use of any equipment or facility necessary for peaceful exploration of the Moon and other celestial bodies shall also not be prohibited.

It is apparent that a clear distinction is made between the outer space and the celestial bodies. The 1st operative paragraph, which is referred to as 'no bomb in orbit[16]' provision, bans the placement of nuclear weapons or any kinds of WMDs in the orbit around Earth. It further continues and bans the installation on celestial bodies or stationing such weapons in outer space. If questions are raised whether it allows the placement of conventional weapons in Earth orbit or does it allow the installation of conventional weapons on celestial bodies, the interpretation could be affirmative. It can further be extrapolated that the placement of weapons in the airspace which falls under the territorial jurisdiction of a State is also allowed, as there is no legal definition or demarcation of airspace and outer space[17].

The first operative part of second paragraph emphasizes on 'peaceful uses' and prohibits the establishment military bases etc., only on the celestial bodies. Does it not prohibit the non-peaceful or military activities such as establishment of military bases, installations and fortifications, the testing of any type of weapons etc., in 'outer void space' [18] ? Logically the answer should be affirmative. Observing these lacunae, the Indian delegate cautioned that in such a situation Article IV is meaningless and the Space Treaty might become nugatory[19]. Thus it is very clearly seen that the Article-IV intends for a limited demilitarization of outer space and total demilitarization of celestial bodies. An inquisitive approach would clarify this differential treatment. It was intended to accommodate the ICBMs with nuclear warheads, which were being tested and just introduced to arsenals by the USA and the then USSR. It is obvious that the major trajectory of ballistic missiles would be in outer space even though they exclude Earth's orbit.

Coming to the major point of contention, the phrase 'peaceful purposes', which is quite often used in the space treaties, leads to controversial interpretations as it has not been defined anywhere in such treaties. Equally the term 'peaceful uses' was not defined when the UNCOPUOS was formed, the title of which was centred on the same term. The phrase 'peaceful purposes' adopted from the Antarctic Treaty was ambiguous and led to two schools of thought namely, one interpreted as 'non-military use' and the other as 'non-aggressive use'. Interestingly, the super powers took the opposite stands; USSR argued that all military uses (even passive uses) are non-peaceful and hence illegal; whereas USA defended that all military activities in outer space are 'peaceful' so long as they are not 'aggressive'. It was further urged that the term 'peaceful' should include 'peaceful military activities' too[20]. The dual use nature of a space system further complicates the bifurcation. For example, use of satellite communication network for military purposes like any other user segment or remote sensing satellite data which is used for surveillance by defence personnel or satellite based navigation systems for peace time uses are taken as granted. Here yet another class of grouping comes into picture,

---

16 *Ogunsoloa O. Ogunbanwo, 'International law and outer space activities', Martinus Nijhoff, 1975-pp92-93*

17 *Ge`radine Meishan. Goh, 'Keeping the peace in outer space: a legal framework for the prohibition of the use of force'- Space policy 20 (2004) 259-278*

18 *A terminology used by Prof. Bin Cheng - Keith Wilson- 'Why are the missiles (and the missile defence) called Peace Keepers? - Corroding the concept of Peaceful Use' - 14, Leiden Journal of International Law 789-828 (2001)*

19 *Supra note at 16- pp31- referred to UN Doc. A/C.I/PV.1493*

20 *Supra note at 16 - p.p29.*

as 'force support ' and 'force application[21]'. The former is considered to be 'peaceful purpose' or 'non-aggressive' and the latter as 'non-peaceful' or 'aggressive'. There are also other interpretations which are categorized as minimalist and maximalist representing 'non-aggressive' and non-military' uses respectively. However, the maximalist interpretation, which was more prevalent in the beginning of the space age, has now given way to the minimalist concept as demonstrated by state practice, which has not been contradicted in a forceful manner, by any state formally protesting military utilisation of space[22].

A scholarly definition on 'peaceful purposes' offered by Galloway is quite pragmatic as it does not *per se* categorize the space systems or projects to any category based on their nature but attributed to their use. It reads as - '*Peaceful purposes are those incorporated into plans and designs for space projects and whose uses are beneficial and / or not estimated to be harmful*[23]'.

Besides Article -IV, to secure the outer space arms control provisions Article 1 (para 2) and Article XII are also quoted. Article 1 Para 2 provides for free access to the States to all areas of celestial bodies, thereby establishment of military bases or testing of weapons could be put under check. Further, Article XII provides for the visit of any Member State representative on the basis of reciprocity and advance notice. However this is viewed by Ivan A. Vlacic that the conditions of reciprocity might offer an unwilling party to nullify the right of inspection, in contrast to Antartica Treaty provisions[24].

6. Recent developments towards Space Militarization - Policy aspects and demonstrations

Having seen the scope of 'peaceful purposes' and the regulatory provisions as professed by the OST, it is attempted to present a bird's eye view on the recent developments with respect to policy aspects or changes thereupon, demonstration of activities that cause concern to international community on possible arms race in outer space.

As regards to military applications of outer space, the USA is categorical in declaring its intentions through its space policy, which is proclaimed from time to time. Interestingly, the Rumsfeld Commission, which was chartered to assess the organization and management of space activities that support US national security interests, in its report submitted in January 2001, concluded that USA while continuing to promote the peaceful uses of space, should use the nation's potential in space to support its domestic, economic, diplomatic and national security objectives and further *develop and deploy the means to deter and defend against hostile acts directed at U.S. space assets and against the uses of space hostile to U.S. interests*[25]. The US Space Command's Vision 2020 called for 'full spectrum dominance' arguing that 'the medium of space is the fourth medium of warfare'[26].

Subsequently, the new space policy of USA declared on August 31, 2006 under the authorization of the President George Bush, included certain principles which are indicative of the US position on military use of space in the context of national security. A few principles relevant to the present context of discussion are reproduced here below;

- The United States is committed to the exploration and use of outer space by all nations for peaceful purposes, and for the benefit of all humanity. Consistent with this principle, "peaceful purposes" allow U.S. defense and intelligence-related activities in pursuit of national interests;
- The United States considers space capabilities -- including the ground and space segments and supporting links -- vital to its national interests. Consistent with this policy, the United States will: preserve its rights, capabilities, and freedom of action in space; dissuade or deter others from either impeding those rights or developing capabilities intended

[21] *Supra note at 2*

[22] *UN Document. CD/1165 of 12 August 1992 and Dr. Rajeev Lochan, ISRO, 'Some reflections on collective security in space' - Collective security in space- Asian Perspectives',Space Policy Institute, January 2008,*

[23] *Supra Note at 17; Galloway, E, International Institutions to ensure peaceful uses of Outer Space, (1984) IX AASL 303at 323.*

[24] *Ivan A. Vlacic, 'The Space Treaty: A preliminary Evaluation', California Law Review- Vol 55:507, 1967*

[25] *Report of the Commission to Assess United States National Security Space Management and organization Pursuant to Public Law 106-65, January 11, 2001 (Emphasis added)*

[26] *Supra note at 2*

to do so; take those actions necessary to protect its space capabilities; respond to interference; and deny, if necessary, adversaries the use of space capabilities hostile to U.S. national interests;

- The United States will oppose the development of new legal regimes or other restrictions that seek to prohibit or limit U.S. access to or use of space. Proposed arms control agreements or restrictions must not impair the rights of the United States to conduct research, development, testing, and operations or other activities in space for U.S. national interests.

As a sequel to the space policy, the Secretary of Defense, USA was asked to submit to the congressional defence committees a report setting forth a plan, including eight required elements, for the acquisition by the Department of Defense (DoD) of capabilities for operationally responsive space (ORS)[27] to support military users and military operations[28]. The DoD's plan was to establish and stand-up an initial ORS Office by May 2007 with full implementation by October 2007 which include the Tacsat demonstrations, deployment of operational satellites including the planning and procuring small launch vehicles.

Yet another strategic programme taken up by the US Air Force (USAF) is Space Situational Awareness (SSA), which was quoted in the Rumsfeld Commission's Report. The SSA programmes of USA is intended to monitor the threat from human-made objects such as other satellites and space vehicles, anti satellite (ASAT) weapons as well as space debris. It is believed that SSA can serve military objectives such as preparing for defensive and offensive operations and it is expected to constitute a basis for polices of transparency and confidence in space[29].

With this background on the preparedness of USA towards pursuing its national interests in the outer space, which is proclaimed as the province of all mankind, in the above mentioned manner, the recent shooting down of a defunct spy satellite through SM-3 interceptor on February 2008, on the basis of the hazardous potential of the onboard fuel of the satellite, has been viewed as a demonstration of its space capability to counter any harmful interference in space.

The China's aggressive progress in all spheres of space activities such as human space flight demonstrations, lunar mission and the planning of the manned missions to Moon and Mars have been over shadowed by its ASAT test in January 2007[30]. However, this demonstration exhibited by China, which was very actively pursuing along with Russia in Conference on Disarmament for the establishment of a treaty to effect the PAROS, was viewed by the strategic analysts as a caution to USA[31].

More recently the Japanese Government has passed a bill in order to pursue the military support activities in outer space to the extent of using remote sensing spy satellites for the cause of national security[32]. This policy change of Japan, which has demonstrated as a peace lover to the core, is again a noteworthy development. Though *per se* it does not propose installation of any weapons, it is inferred that this is a mandatory reaction to the trajectory of the North Korean missile, Taepodong over Japanese territory in August 1998, in the interest of national security[33].

The French Government too followed the suit by declaring its military space programmes in its White Book on Defence Promises, in July 2008. It promises for the establishment of a Joint Space Command,

---

[27] *Operationally Responsive Space (ORS) has been defined broadly in DoD as assured space power focused on timely satisfaction of Joint Force Commanders' needs. This definition considers ORS as a subset of space activities designed to satisfy Joint Force Commanders' (JFCs') needs, while also maintaining the ability to address other users' needs, for improving the responsiveness of space capabilities to meet national security requirements.*

[28] *Under the Section 913 (c) of the John Warner National Defense Authorization Act (NDAA) for Fiscal Year (FY) 2007 - The letter from Deputy Secretary of Defense to the Chairman of the Committee on Armed Services, Dt. April 20, 2007, available at <www. .. > accessed on*

[29] *Lawrence nardon, 'Space Situational Awareness and International Policy', Document de travail 14, October 2007; Available at <www. ... > dt of acces: ...2008*

[30] *Smashing of the Feng Yun -1C satellite by the KT-1 rocket by kinetic energy killing method.*

[31] *Aviation Week on Space Technology (Jan29, 2007)*

[32] *Space War, May 21, 2008 Dt. Of access May 23, 2008*

[33] *Andrew L Oros, 'Explaining Japan's tortured course to surveillance satellite', Review of Policy Research-Vol24, No. (2007)© Policy Studies Organization*

development of reconnaissance satellites and Early Warning Systems etc[34].

7. International efforts on legal measures towards non-weaponization of Space

As seen in the initial part of this paper, international efforts were pursued through various forums like UNCOPUOS or UN Conference on Disarmament to reach a viable solution in terms of an agreement or a treaty, beyond Article -IV of OST. Though these attempts have not really achieved desired results, it must be admitted that they have served the purpose to the extent of restraining the arms race in outer space. Beyond this, it has been proved that space and space systems have become very vital for an effective warfare as seen from the Operation Desert Storm in 1990 for the first time and successive warfares or military operations undertaken by USA[35]. Because of military role of space a vulnerable situation has arisen for the security of space assets, as they are also becoming potential targets of attack in order to suppress the military capabilities of an opponent. The recent ASAT tests have further aggravated the situation. Thus the issue of space security is becoming more prominent for space faring nations as well as space users and again the international effort on reaching Common Space Security has become a need. The nations where space systems are utilized more effectively for military applications are preparing to face the eventualities and started implementing various military models involving space systems under the context of national security. Integrated space command has become a component of the military force of advanced space faring nations.

With these developments, the topic of space weaponization or arms race in outer space is becoming more serious amongst various international forums. But the consoling factor is, though certain advanced nations have fully prepared for engaging themselves in space warfare, the reports indicate that so far the deployment of any space weapon has not taken place[36]. An approach of restraint is essential as the result of any aggressive actions would be catastrophic. It is also obvious that loss of space assets would be strategically more disadvantageous to a nation which engages the space systems to the maximum extent either for military or civil applications.

With this background it is attempted to review the various legal measures proposed by various multilateral forums and other institutions. Though the proceedings on PAROS have been stalled in CD in 1998, an interesting development of proposing a treaty based prohibition for the deployment of weapons in outer space, jointly by People's Republic of China and the Russian Federation has taken place in June 2002, which could be more significantly related the abrogation of anti Ballistic Missile Treaty by USA in early 2002. A Working Paper entitled 'Possible elements for a future international legal agreement on the prevention of the deployment of weapons in outer space, the threat or use of force against outer space objects[37]' was placed before the CD for discussion. This working paper proposed for certain basic obligations on the States towards the objective of not placing 'any kind of weapons' in orbit or on celestial bodies and not station such weapons in outer space, in order to fulfil the ambiguities seen in the Article -IV of OST. It included some measure for confidence building measures and proposed for setting up an Executive Organization for implementation of the Treaty. USA has been opposing to this move as it is not in favour of any treaty or agreement which would impair its rights to conduct research, development, testing, and operations or other activities in space for U.S. national interests.

Interestingly, in US Congress a bill 'To preserve the cooperative, peaceful uses of space for the benefit of all humankind by prohibiting the basing of weapons in space and the use of weapons to destroy or damage objects in space that are in orbit and for other purposes' has been placed in the successive

[34] *Space Review - ' French Military Space Policy; More of the same', Monday, July 7, 2008.*

[35] *Kosovo in 1999 and Iraq in 2003 - 'Background paper: peaceful and military uses of outer space: law and policy', February, 2005- Institute of Air & Space Law, Mc Gill University*

[36] *Keith Wilson, 'Why are the missiles (and the missile defence) called Peace Keepers? - Corroding the concept of Peaceful Use' - 14, Leiden Journal of International Law page798 (2001)*

[37] *CD/1679 Dt. June 28, 2002*

years since 2001(H.R.2977), subsequently in 2002 (H.R.3616)[38] and lastly in 2005 (H.R.2420 Dt May, 18, 2005) by Rep. Kucinich et.al. This bill called 'Space Preservation Act of 2005'[39], provides for ban on basing weapons in space and use of weapons in space as its detailed title implies. Furher, it suggests for formulating an international treaty in this regard. Section 4 reads as -"*The President shall direct the United States representatives to the United Nations and other international organizations to immediately work toward negotiating, adopting, and implementing an international treaty banning space-based weapons and the use of weapons to destroy or damage objects in space that are in orbit*[40]." This bill was referred to Committee on Science, Committes on Armed services and International Relations and it is indicated that this bill would never become law[41].

Subsequently, in the forum of UN Conference on Disarmament, as the initiative of PRC and Russian Federation taken up in 2002 was not successful. With a renewed interest, they put up a set of working papers in May 2006 on the definitional issues [42] relating to the proposed arms control treaty, on the Transparency and Confidence Building Measures (TCBMs)[43] and on the Verifications Measures[44] in order to provide clarifications to the apprehensions and to gain confidence of the States, towards reaching the objective. However, in the UN First Committee on 25th October, 2006, USA voted against the annual resolution on PAROS for the first time and subsequently proclaimed its new space policy in 2006, which vehemently opposed any move to curb its rights to access to space and non-aggressive military research activities in outer space. However to overcome this impasse, based on the past year discussions on the working papers, the PRC and Russian Federation presented a Draft Treaty the Prevention of Weapons in Outer Space, the Threat or use of force against outer space objects, in February 2008. This includes the definitions on outer space, outer space object, weapons in outer space, placement of weapons etc. Interestingly the definition on placement covers the sub-orbital flights too, which do not get into orbit. Besides covering the weapons of any kind, it assures the right of self defence of a state as enunciated in UN Charter and includes certain confidence building and verification measures.

It is interesting to note that some initiatives are volunteered by some research institutions and scholars who articulate their proposals on non-weaponization of space. The Henry L. Stimson Center, Washington has proposed a Model Code of Conduct for Responsible Space-Faring Nations in 2007, with a central objective as - *To preserve and advance the peaceful exploration and use of outer space.* It offers certain rights such as access to space, self defence etc. and obligates certain responsibilities such as refraining from harmful interferences against space objects, sharing of information relating to space operations, mitigate & minimize space debris etc, on space faring nations. The international response to this code of conduct is to be watched.

Having seen the developments on the international efforts to prohibit the weaponization of space and before going to discussion on the need for an internationally acceptable legal measure and its realization plan, it is felt necessary to provide a glimpse on the space activities of India, which is centered around the theme of peaceful and developmental uses of outer space.

8. Indian Space Programme: *the approach and accomplishments*

As seen through, while the space activities were vigorously pursued by the super powers for technology demonstrations and thereby achieving technical supremacy and gaining military advantages, a unique approach was adopted by India in the early sixties with a strong conviction that high technology applications could be used to serve the common man and society. Probably the Indian socio-economic conditions under the state of a developing country that got its freedom a decade ago, with a large geographical sprawl filled with a massive population of diverse cultures has prompted the scientist cum industrialist Dr. Vikram

---

38 *Jhon Carter McKnight, 'Let's weponize space' - Space Daily, Dt. Jan 30, 2003; Dt of access: Aug 28, 2008*

39 *Bill- H.R.2420-109th Congress, 1st Session- Thomas Library of Congress- Dt of access: Aug 28, 2008*

40 *Space News, Dt. 30 July 2002.*

41 *http://www.govtrack.us/congress/bill=h109-2420 (dt. of access:Aug27, 2008)*

42 *CD/1779 Dt. May 22, 2006*

43 *CD/ 1778 Dt May 22, 2006*

44 *CD/ 1781 Dt.May 22, 2006*

Sarabhai[45] to think differently to conceive a long term vision of pursuing high technology programmes and reaping the benefits there from to serve the society. Thus Indian space programme was born out of necessity and pursued with a passion.

The conceptualization of the programme with a strong backing of the Government support, regardless of the resources crunches that prevailed, gave birth to two major application programmes in the areas of satellite communication & meteorology and space based remote sensing, besides the formulation of an overall programme to achieve end-to-end capability of sustaining a space system. Foreseeing the threat of denial of critical technologies from the advanced nations, which would obviously disrupt or stall the operational services that would be ushered in through space systems, India judicially chose the path of self reliance to develop the critical space technologies. Self reliance was targeted as a strategy for sustainable development since large scale benefits can accrue to a large country like India only when it has its own space segment specifically tailored to meet its requirements[46].

The approach of India in space activities has undoubtedly paid rich dividends in the building of the nation. The establishment of country wide telecommunication networks, application of broadcasting services (far beyond routine entertainment services), for tele-education, tele-medicine and operationalization of remote sensing applications and services for a variety of national needs in the areas of agriculture, forestry, fisheries, rural development and urban planning etc., using state-of-the art space systems are a reality to-day. It is hardly to be emphasized that the Indian space programme has set a model through this approach.

While military applications provided a strong rationale for several programes across the globe, India practically demonstrated that peaceful uses of outer space can be a strong contributor to development and can justify investments in space. For example, India's Tele-education programme with more than 34,200 class rooms through 61 networks, Tele Medicine programme connecting 261 remote, rural and district hospitals with 45 super speciality hospitals and 417 Village Resource Centres in 21 States are enabled by space based systems. These services and applications are being provided with eleven satellites in geostationary orbit and nine satellites in low earth orbits.

The space infrastructure built over four decades and the operational services implemented through space systems, demand a high level of security for ensuring uninterrupted services to the nation. However the concern about the space security is not unique to India as every developing and developed country, which depends on space services, would experience the same. But the approach of India towards the space security is again somewhat unique. The Indian space security architecture include five crucial elements[47] of a firm foundation namely – *i) Self reliance to evade the perils of technology denials and export control threats , ii) Sustained unflinching support-base from society including the policy maker, law maker, media, industry, iii) International Cooperation– a crucial apparatus for cooperative activities and enhancing national, regional and global security, iv) The Rule of Law- compliance to national and international legal regimes, and v) Security Commitments- peaceful purposes and non-proliferation.*

9. Need for additional legal measures and *modus operandi*

It has been seen that the need for preventing and /or controlling the militarization of outer space has been pursued right from the initial phase of space explorations and operations. There existed a consensus amongst the states that the trends of space activities during and even after the Cold War era could lead to an arms race in outer space which should be controlled. The threat perceptions on the space assets and the

[45] *Dr. Vikram Sarabhai, fondly remembered as the Father of the Indian space programme, served as Vice President and Scientific Chairman of the UNISPACE-I held in Vienna in 1968*

[46] *Rao, U. R. Remote Sensing for National Development, Indian Space Research Organisation, ISRO-SP-56-91. October 1999.*

[47] *Dr. Rajeev Lochan, ISRO, 'Some reflections on collective security in space' - Collective security in space- Asian Perspectives', Space Policy Institute, January 2008,*

resulting catastrophic effects to human kind in general and the concerned states in particular added strength to the argument.

Further, the threat perceptions posed on to the space systems of any country, which engages such space systems for very vital services like national development or national security, would cause severe repercussions. Thus the need for security of space assets arises, which is only one dimension of common space security. In order to avoid the loss of space systems, some states are planning redundant space systems. Also, the space systems are shielded in order to protect against minute debris or laser dazzling from ground or space. But the irony of this situation is the manufacturing and operational cost of the space systems would be increased exponentially. Whether a developing or less developed country could afford to sustain this? In the case of India, space systems are effectively used for peaceful purposes for national development. If the systems meant for such applications are to be proofed against perceived threats, cost would make such application programmes unaffordable or unsustainable. Yet another dimension to this complex issue can be shown. The countries have to depend on the select states only for getting critical technologies required for such proofing, which either would be denied or exorbitantly priced. Thus in order to clear up the threat perceptions and to ensure the common space security in the interest of all nations, there is a strong need for appropriate legal measures.

Coming to the legal measures being considered in various international forums, it is proposed to analyse the pros and cons of treaty based compliance system. As is seen in the CD proceedings, the draft treaty on prevention of placement of weapons could not see progress though it is acceptable to most of the nations. Another possibility is to amend the OST. Unfortunately UNCOPUOS is not willing to amend Article-IV as it considers that the topic of 'military uses' is beyond the scope of 'peaceful uses'. Further decision making processes in such UN forums are extremely slow. It may be recalled that the proposal for formulating a comprehensive space treaty in place of the existing treaties on various topics which call for many amendments, is in the preliminary stages of discussions/ exchange of views and publication of papers.

It is also to be remembered that a treaty does not offer an everlasting binding on a State after the ratification, as it also provides for withdrawal with sufficient notice. The withdrawal of USA from the ABM treaty in order to proceed on the outer space militarization process is a classic example.

At this juncture, the proposal[48] articulated by Ge`radine Meishan. Goh[49], a scholar of Leiden University for a protocol to the existing OST is interesting. The protocol includes provisions on comprehensive prohibition on the use of force in outer space, right to self defence, pacific resolution of disputes etc., and proposes for three bodies for three different functions namely - enforcement, monitoring and interpreting. This approach could be a via-media instead of cumbersome amendment process of the existing treaty through UNCOPOUS or formulation of a new treaty through CD.

As it is seen, over a period time, the scope of military use of outer space has been growing and this trend if unchecked will be an impediment to enhance the peaceful uses and also affect the transfer of knowledge and technologies thereby reducing overall economic, social and cultural benefits of space to humanity as a whole. Peaceful use of outer space can not be compromised under the pretext of national interests. Appropriate forums should continue the debate and evolve certain 'code of conduct' or 'rule of road' for outer space activities for sustained use of outer space for peaceful purposes.

10. Conclusion:

The space activities encompassing technology and applications, since inception and subsequent evolution have proved to be beneficial to human kind and have also enabled military applications due to their inherent dual use nature. The efforts of international community in different UN forums to curb the possible arms race in outer space have not received adequate response, as the use of space systems for military support services have been practiced by many nations.

---

[48] *Ge`radine Meishan. Goh, 'Keeping the place in outer space: a legal framework for the prohibition of the use of force', Space policy 20 (2004) 259-278.*

[49] *International Institute of Air & Space Law, Leiden University, The Netherlands (2004)*

Convincing interpretation of the phrase 'peaceful purposes,' is yet to be resolved as it is used for the convenience of the state practices. The enactment of a new legal instrument of international law to prohibit the weaponization of outer space is taking its own course. In the absence of placement of any weapons in outer space by any nation so far, the preparations of a few advanced nations in this direction with national security concerns, are posing some threat perceptions on space systems that are engaged even for peaceful purposes. Common space security architecture should be evolved and practised by all the space faring nations in order to maintain a weapon-free outer space.

IAC-08-E8.3.12

The 2008 Russia / China Proposal for a Treaty to Ban Weapons in Space: A Missed Opportunity or an Opening Gambit?

Steven Freeland[1]

Abstract

Events during the early months of 2008 have highlighted further the dangers of an arms race in space. At an important disarmament conference in Geneva, Russia and China jointly submitted their latest proposal for a Treaty banning the placement of weapons in outer space. This came only days before the United States deliberately destroyed one of its own satellites, ostensibly to minimize the risk of highly dangerous pollutant falling on the earth. This action was criticized by both Russia and China, who alleged that it was simply a 'test' by the United States of its missile defence shield capability. A few days later, China announced that it proposed to increase its military spending by almost 20% in 2008, this following a report from the United States voicing concerns about China's advances in space weapons technology.

It is in this context that this paper seeks to examine the latest Russia / China draft Treaty, and analyze its provisions within the broader framework of the major Space Treaties, the United Nations General Assembly Resolutions aimed at the 'Prevention of an Arms Race in Outer Space' and the strong desire of the broader international community for increased measures designed to promote greater transparency and confidence-building in relation to the use and exploration of outer space for peaceful purposes. This paper concludes that, although the draft Treaty has significant weaknesses, it is imperative that all space faring nations, as well as the wider community, engage as a matter of urgency in good faith negotiations and discussions aimed at finalizing a more comprehensive legal regime banning the use of any weapons in outer space. In this regard, the draft Treaty represents another step in the negotiation process and is thus an important instrument for reflection.

Introduction

In February 2008, the Minister of Foreign Affairs of the Russian Federation, Sergey Lavrov, presented a draft document headed *Treaty on the Prevention of the Placement of Weapons in Outer Space, the Threat or Use of Force Against Outer Space Objects* (*PPWT*) (see annexure) to the 65 members attending the Plenary Meeting of the United Nations Conference on Disarmament (CD) in Geneva.[2] The *PPWT* had been developed by Russia and China, two of the major space superpowers in the world. An earlier draft had been informally circulated the previous June, resulting in comments from a number of other countries.[3]

The formal submission of the *PPWT* to the CD followed several years of diplomatic discussion, ostensibly aimed at agreeing the terms of legally binding rules addressing the dangers of an 'arms race in space'. In presenting the *PPWT*, Minister Lavrov noted that the terms of the document were supported by a majority of the Member States of the CD. He warned that:[4]

> [w]eapons deployment in space by one state will inevitably result in a chain reaction. And this in turn, is fraught with a new spiral in the arms race in space and on the earth.

In his supporting comments, the Foreign Minister of the People's Republic of China, Yang Jiechi, added that:[5]

> [a] peaceful and tranquil outer space free from weaponization and [an] arms race serves the common interests of all countries.

> It is therefore necessary for the international community to formulate new legal instruments to strengthen the current legal regime on outer space.

In responding to this proposal, the United States Administration has reiterated that it opposes any treaty that seeks 'to prohibit or limit access to or use of space', adding that, in any event, such a treaty would be impossible to enforce.[6] Instead, the United States has indicated that it prefers 'discussions aimed at promoting transparency and confidence building measures'.[7]

Within the constraints that apply as to its length, this paper provides a brief overview of the principle terms of the *PPWT* within the broader framework of the major United Nations Space Treaties, the United Nations General Assembly (UNGA) Resolutions aimed at the 'Prevention of an Arms Race in Outer Space' and the desire of the international community for increased measures to promote greater transparency and confidence-building in the use and exploration of outer space for peaceful purposes. There is undoubtedly a need for significant further research and analysis regarding this proposal, and the broader issue of the 'weaponization' of space, which probably represents the most pressing concern facing humankind in relation to the exploration and use of outer space.

Weapons in Space – Current Legal Restrictions under International Space Law

Article IV of the *Treaty on Principles Governing the Activities of States in the Exploration and Use of Outer Space, including the Moon and Other Celestial Bodies* (*Outer Space Treaty*)[8] imposes various restrictions on weapons in outer space, specifically 'nuclear weapons or any other kinds of weapons of mass destruction'.[9] However, as has been well documented, this provision in and of itself does not represent a complete restriction on the placement of weapons in outer space.[10] Indeed, there have, from time to time, been proposals to amend Article IV in order to enhance its scope, but these efforts have not been successful.[11]

Moreover, the 'peaceful purposes' provision set out in Article IV – that '[t]he moon and other celestial bodies shall be used by all States Parties ... exclusively for peaceful purposes' - has been the subject of much discussion. While there is general agreement – but not complete unanimity – among space law commentators that this is directed against 'non-military' rather than merely 'non-aggressive' activities, the reality is different. In addition to many commercial and scientific uses, outer space has increasingly been used for an expanding array of military activities. As this trend continues, there is a danger that the 'weaponization' of space, as well as its evolution into a distinct theatre of military operations, may become a reality,[12] particularly given the extent to which the major powers rely on space capability as part of their national security infrastructure.[13]

In this context, as this author has previously suggested,[14] if one were to adopt a hard-line pragmatic (and perhaps non-legal) view of the current situation, one could suggest that the 'non-military v. non-aggressive' debate has ceased to have practical relevance, even though it represents an extremely important issue of interpretation of the principles set out in the *Outer Space Treaty*. Instead, the focus of the discussion has now shifted to the weaponization of space and its implications for international relations, particularly between the major powers.

Indeed, there have been a series of UNGA Resolutions on the specific issue of preventing an arms race in outer space, the latest adopted in December 2007.[15] In addition, the UNGA has repeated its invitation to Member States to:[16]

> continue to submit ... concrete proposals on international outer space transparency and confidence-building measures in the interest of maintaining international peace and security and promoting international

> cooperation and the prevention of an arms race in outer space

These measures focus the attention of the broader international community on the need to respond to various military initiatives taken by major space powers in their use of outer space. Moreover, the UNGA has, at the same time, also emphasized the importance of international cooperation in the peaceful uses of outer space, an important element of which is that:[17]

> all States, in particular those with major space capabilities ... contribute actively to the goal of preventing an arms race in outer space

Recent Worrying Developments

Despite these multilateral efforts, several recent developments have given rise to heightened concerns. On December 14, 2001, in an effort to consolidate its policy of 'space control,' the United States announced its withdrawal from the 1972 *Anti-Ballistic Missile Treaty*,[18] invoking Article 15 of that instrument.[19] The key reason given by the United States for the decision to withdraw was that the *ABM Treaty* was an outdated relic of the Cold War, negotiated in a 'vastly different world'.[20] However, in practical terms, withdrawal from the *ABM Treaty* had the effect of removing significant conventional restrictions on the United States developing what would otherwise have been prohibited weapon systems. These include, in particular, space-based devices, which now form an integral part of its national security infrastructure.[21] As a result, since its withdrawal, the United States has actively pursued its (so-called) 'defensive' military space capabilities, particularly its National Missile Defence system (NMD).

The dangers associated with such a strategy are clear – once a sophisticated defensive capability has been developed (either in reality or, at least, in the perception of military and political leaders), this opens the way for 'first-strike' capability on the space assets of other States. Moreover, plans by the United States to deploy parts of its NMD infrastructure - including radar systems and missile bases – in the former communist-bloc States of Poland, the Czech Republic and Lithuania, have, not surprisingly, drawn a fierce response from Russia. In turn, Russia has, among other things, threatened to aim nuclear weapons at Europe.[22] There is much speculation that Russia is, at the same time, also progressing with plans to develop a similar system to the NMD, as well as other space-related weaponry, a process that was accelerated almost immediately following the withdrawal by the United States from the *ABM Treaty*.[23] Indeed, over the eight years since Vladimir Putin first came into power, the country's military spending budget has increased by over 800% (albeit from a relatively low base).[24]

As worrying as these developments are, tensions were ratcheted up even further in January 2007, when the Chinese military launched a KT-1 anti-satellite rocket to successfully destroy a redundant Chinese Feng Yun 1-C weather satellite at an altitude of several hundred kilometres. By that time, China had already proven itself to be a space superpower, being only the third country to have sent a human into space, but this latest action was perceived as its 'way of saying that it will cede control of space to no one'.[25]

More was to follow when, in February 2008, barely a week after Russia and China submitted the *PPWT* to the CD, the United States fired an SM-3 missile from USS Lake Erie that destroyed a failed satellite approximately 150 kilometres above the Pacific Ocean. Although the United States argued that this action was necessary to prevent the fuel tank of the satellite – containing hydrazine – from breaking up and polluting the atmosphere, others have suggested that this was simply a 'test' by the United States of its anti-satellite capability. At the time, the Chinese Communist Party newspaper, *The People's Daily*, reported that:[26]

> [t]he United States, the world's top space power, has often accused other countries of vigorously developing military space technology … But faced with the Chinese-Russian proposal to restrict space armaments, it runs in fear from what it claimed to love.

The Principle Terms of the *PPWT*

It is within this context of claim and counter-claim that it is useful to consider the main provisions of the *PPWT*. What the foregoing discussion indicates is that talk of a 'spiral' in space weapons technology is not misplaced, with each of the three major space powers – and perhaps other countries as well – devoting increasing resources to the development of space-related weapons systems that, ultimately, may also involve the placement of weapons in outer space.

It is therefore important to examine any rational proposal, such as the *PPWT*, that may restrict – or ideally prohibit – such an eventuality. The principle terms of the *PPWT* are as follows:

A. Primary Obligations

In general terms, the *PPWT* focuses on three primary obligations of States Parties, each of which are specified in Article II:

(a) not to place in orbit around the Earth, install on celestial bodies, or station in outer space in any other manner 'any objects carrying any kind of weapons';

(b) not to 'resort to the threat or use of force against any outer space objects'; and

(c) not to encourage another State(s) or Intergovernmental Organization to 'participate in activities prohibited' by the *PPWT*.

As well as applying to a broader range of weapons, the prohibitions in Article II, specifically Article II(b), rectify another major shortcoming of Article IV of the *Outer Space Treaty*, in that they appear to cover the *use* of such weapons, as well as their placement.

B. Definitions

The scope of these obligations is dependent upon the definitions of those key concepts set out in Article I. It is here that the *PPWT* makes some interesting and positive steps. Of course, the definitions only relate specifically to the *PPWT* itself; however, it can be argued that the *PPWT* represents an important example of State practice, developed by at least two of the most important space faring nations and supported by other States. In this regard, these definitions may be relevant to the future amendment / expansion of the formal international space law documents.

Probably the most striking of these is the definition of 'outer space' as 'space beyond the elevation of approximately 100 km above ocean level of the Earth' (Article I(a)). Apart from the curious use of the word 'approximately'- in what circumstances would it *not* be 100 km? – this represents a rather revolutionary suggestion by two major superpowers, which, along with the United States, have tended to stifle previous attempts to designate a formal demarcation, primarily for strategic and political reasons. Indeed, it was only a few years ago that a Chinese Foreign Ministry spokesperson referred to outer space as the 'Fourth Territory'.

As is well known, ever since Sputnik 1 was launched in October 1957, there has been intense discussion regarding both the need for, and the criteria of a clearly defined point of where outer space begins.[27] At present there is still no widely accepted legal demarcation between air space and outer space – despite their very different legal bases – although there has been some other relevant State practice,[28] which perhaps may, along with the formulation of this definition in the *PPWT*, contribute towards the eventual formation of a customary international law principle.

Another important definition in the *PPWT* is that of an 'outer space object' (Article I(b)). This expands considerably on the traditional (and generally unsatisfactory) definition of 'space object' in Article I(d) of the *Convention on International Liability For Damage Caused by Space Objects* (*Liability Convention*)[29] and Article I (b) of the *Convention on Registration of Objects Launched into Outer Space* (*Registration Agreement*),[30] both of which simply refer to 'component parts of a space object as well as its launch vehicle and parts thereof'. It remains to be seen whether this expanded view of a (outer) space object might be useful in addressing other space-related issues, such as space debris.

One of the most important definitions in the *PPWT* is that of 'weapons in outer space' (Article I(c)), since that clarifies the scope of the treaty. Whilst it is a relatively broad description – including 'any device' – it still leaves some room for doubt, particularly as to assets that may initially be 'peaceful' but are subsequently utilized to 'damage or disrupt normal functions of objects in outer space', such as through the generation of various electromagnetic pulses. Moreover, if an object is deliberately allowed to become debris, and then affects the space assets of other States, query whether this falls within the requirement of 'produced or converted'?

From a broader public international law perspective, the definitions of 'use of force' and 'threat of force' (Article I(d)) are of interest. Of course, the concept of 'force' is a fundamental principle of international law, with the prohibition of the 'threat or use of force', specified in Article 2(4) of the *Charter of the United Nations* (*UN Charter*),[31] as well as by way of a customary norm,[32] underpinning the conduct of international relations. Under traditional international law principles, 'force' is regarded as an act of 'violence', so that, for example, economic sanctions were not to be regarded as such, despite arguments to the contrary raised by developing countries at the time that the *UN Charter* was being negotiated.

The definition in the *PPWT* appears to be considerably broader than these traditional views of what constitutes force, and was presumably drafted in this way to encompass (non-violent) actions such as 'jamming' and the use of electromagnetic interference, as long as they constituted a 'hostile' act. Should this new approach to force become more widely accepted, it may also raise interesting questions about the legal nature of actions such as 'cyber-attacks'.

C. Necessary Measures

Article III of the *PPWT* requires State Parties to take 'all necessary measures' to prevent violations of the Treaty on its territory or 'in any other *place* under its jurisdiction or control' (emphasis added). This does not appear to be sufficiently broad to cover all potentially relevant activities occurring in space, or carried out by nationals (either legal or juridical) of a State that are conducted outside of its territory. This represents a significant potential 'loophole' that should be rectified, although, of course, this is not to ignore the difficulties this raises in terms of enforcement. Widening the obligation in this way would require any relevant national implementing legislation to have an extra-territorial effect, at least in relation to the acts by nationals of that State. However, this is an increasingly common characteristic of national law – for example, the Australian Space Activities Act requires an Australian national to obtain an overseas launch certificate to 'launch a space object … from a launch facility located outside of Australia'[33] – and would not pose any particular legal problems for the State passing such laws.

D. Peaceful Purposes

Article IV of the *PPWT* refers to the exploration and use of outer space for peaceful purposes. As noted above, this principle has in practical terms had a 'difficult' history and, notwithstanding its interpretation by (most) space lawyers, State practice has indicated that military uses of outer space are 'acceptable'. Interestingly, unlike in the *Outer Space Treaty*, the term is qualified by the expression 'in accordance with international law', although this does not conclusively resolve the 'non-military' / 'non-aggressive' debate. However, given that the *PPWT* relates to weapons in space, it impliedly assumes that military uses of space are already taking place (which they obviously are) and seeks simply to restrict those uses.

It would have been preferable had the *PPWT* provided a clear definition of 'peaceful purposes', so as to provide further teeth to the document. However, without wishing to sound too cynical, that would probably have been asking too much of any of the major space powers.

E. Self-Defence

As one would expect, Article V of the *PPWT* expressly reserves to States the 'inherent right' of self-defence 'if an armed attack occurs', pursuant to Article 51 of the *UN Charter*. In this regard, however, it should be noted that, under the principles of public international law, this right of self-defence remains subject to express legal limitations – the requirements of necessity and proportionality.[34] In its Advisory Opinion in the *Legality of the Threat or Use of Nuclear Weapons*, the International Court of Justice observed: 'The submission of the exercise of the right of self-defence to the conditions of necessity and proportionality is a rule of customary international law'.[35] Moreover, even where the right of self-defence is lawfully exercised, the State acting in self-defence remains subject to the *jus in bello* principles.[36]

On the other hand, recent events, as well as assertions by a number of States of the existence of a 'pre-emptive strike' doctrine of force in response to the (perceived) threats posed by weapons of mass destruction, are seen by many as challenging the traditional international law limitations to this right. In 2003 the United Nations Secretary-General created a High-level Panel on Threats, Challenges and Change to, in part, consider the 'relevance' of these principles in light of current and future challenges to collective security.[37] It is not yet apparent whether, and how far these challenges will impact upon the (legal) scope of Article 51 of the *UN Charter* – suffice to say that reference to this provision highlights the political nature of the *PPWT* and clearly leaves a degree of flexibility for those States that may seek to ignore its underlying principles.

F. Voluntary Compliance

As noted above, the UNGA has resolved on a number of occasions that States should take appropriate steps to enhance transparency and confidence-building measures in relation to their activities in outer space. As is well known, such resolutions are *per se* not binding, but can form part of the evidence of State practice and / or *opinio juris* that may lead to the creation of a binding principle of customary international law.[38] It is unfortunate, therefore, that the drafters of the *PPWT* still found it necessary to emphasize again the voluntary nature of such measures, since this appears to 'weaken' the impact of the instrument. It would have been preferable had the *PPWT* instead strengthened those measures by making them compulsory for all State Parties.

As noted above, verification measures in relation to the obligations of State Parties under the *PPWT* may prove to be difficult and complex to implement. It therefore remains to be seen what the terms of an additional protocol on this question may provide (Article VI).

G. Executive Organization

The proposed establishment of an Executive Organization (Article VIII) is a sensible suggestion; however, the effectiveness of this body will very much depend upon its make-up and procedures (ie which States / IGOs are represented? what are their terms? what is its decision-making mechanism? what are its powers? what criteria must it take into account when making decisions?) These matters are also to be addressed in an additional protocol.

H. Resolution of Disputes

One of the shortcomings of the current international space law regime – indeed, of many other areas of public international law – is the lack of compulsory and / or binding dispute resolution mechanisms. With the exception of the *Liability Convention*, the space law treaties generally provide for the resolution of disputes by peaceful means and through a process of consultation and negotiation. Even the more formalized Claims Commission process under the *Liability Convention* has some significant weaknesses.[39]

In this regard, the reference of a dispute concerning the application / interpretation of the *PPWT* to the Executive Organization (in the event that consultation does not resolve the matter first) may represent a positive development. Once again, however, this will depend upon the powers and procedures of the Executive Organization, which are yet to be specified.

In addition, it would have been preferable had the *PPWT* provided that, in the event that the Executive Organization process is also not able to resolve the dispute, then the relevant State Parties would refer the matter the International Court of Justice. This type of provision is found in a number of other significant international instruments.[40]

Concluding Comments

The formal submission of the *PPWT* by two of the world's space superpowers represents more than a symbolic (and otherwise pointless) gesture designed to gain political mileage but little else. Despite its shortcomings, the document is worthy of detailed consideration, both in relation to its substantive terms (some of which are briefly considered in this paper), but also because it raises issues of crucial importance to the future use and exploration of outer space; indeed to the very *nature* of space activities. It is therefore unfortunate that the document has so quickly been rejected out of hand by the United States. However, it is to be hoped that the *PPWT* will generate further momentum in relation to moves to address the impending perils associated with the possible weaponization of space. These are questions that will, ultimately, affect all of humanity.

Annexure

Treaty on the Prevention of the Placement of Weapons in Outer Space, the Threat or Use of Force Against Outer Space Objects

The States Parties to this Treaty,

Reaffirming that outer space plays an ever-increasing role in the future development of humankind,

Emphasizing the rights to explore and use outer space freely for peaceful purposes,

Interested in keeping outer space from turning into an arena for military confrontation, in assuring security in outer space and safe functioning of space objects,

Recognizing that prevention of the placement of weapons and of an arms race in outer space would avert a grave danger for international peace and security,

Desiring to keep outer space as a sphere where no weapon of any kind is placed,

Noting that the existing agreements on arms control and disarmament relevant to outer space, including the bilateral ones, and the existing legal regimes concerning the use of outer space play a positive role in exploration of outer space and in regulating outer space activities, and should be strictly complied with; although they are unable to effectively prevent the placement of weapons and an arms race in outer space,

Recalling the resolution of the General Assembly of the United Nations "Prevention of an arms race in outer space", where, inter alia, a conviction was expressed in the need for examination of further measures in the search for effective and verifiable bilateral and multilateral agreements in order to prevent an arms race in outer space,

Have agreed on the following:

Article I

For the purpose of this Treaty:

a) the term "outer space" means space beyond the elevation of approximately 100 km above ocean level of the Earth;

b) the term "outer space object" means any device, designed for functioning in outer space, being launched into an orbit around any celestial body, or being in the orbit around any celestial body, or on any celestial body except the Earth, or leaving the orbit around any celestial body towards this celestial body, or moving from any celestial body towards another celestial body, or placed in outer space by any other means;

c) the term "weapons in outer space" means any device placed in outer space, based on any physical principle, specially produced or converted to eliminate, damage or disrupt normal function of objects in outer space, on the Earth or in its air, as well as to eliminate population, components of biosphere critical to human existence or inflict damage to them;

d) a weapon will be considered as "placed" in outer space if it orbits the Earth at least once, or follows a section of such an orbit before leaving this orbit, or is stationed on a permanent basis somewhere in outer space;

e) the "use of force" or "threat of force" mean any hostile actions against outer space objects including, inter alia, those aimed at their destruction, damage, temporarily or permanently injuring normal functioning, deliberate alteration of the parameters of their orbit, or the threat of these actions.

Article II

States Parties undertake not to place in orbit around the Earth any objects carrying any kind of weapons, not to install such weapons on celestial bodies, and not to station such weapons in outer space in any other manner; not to resort to the threat or use of force against outer space objects; not to assist or encourage other states, groups of states or international organizations to participate in activities prohibited by the Treaty.

Article III

Each State Party shall take all necessary measures to prevent any activity prohibited by the Treaty on its territory or in any other place under its jurisdiction or control.

Article IV

Nothing in this Treaty can be interpreted as impeding the rights of the States Parties to explore and use outer space for peaceful purposes in accordance with international law, which include but are not limited to the Charter of the United Nations and the Outer Space Treaty.

Article V

Nothing in this Treaty can be construed as impeding the realization by the States Parties of the sovereign right for self-defense in accordance with Article 51 of the Charter of the United Nations.

Article VI

With a view to facilitate assurance of compliance with the Treaty provisions and to promote transparency and confidence-building in outer space activities the States Parties shall practice on a voluntary basis, unless agreed otherwise, agreed confidence-building measures.

Measures of verification of compliance with the Treaty may be the subject of an additional protocol.

Article VII

When a dispute arises between States Parties concerning the application or the interpretation of the provisions of this Treaty,

the parties concerned shall first consult together with a view to settling the dispute by negotiation and cooperation.
When the parties concerned do not come to an agreement after consultation, the disputed situation that has arisen may be referred to the Executive organization of the Treaty along with provision of the relevant argumentation.

Each State Party shall undertake to cooperate in the settlement of the disputed situation that has arisen with the Executive organization of the Treaty.

Article VIII
To promote the implementation of the objectives and the provisions of the Treaty, States Parties shall establish the Executive organization of the Treaty which shall:
a) receive for consideration inquiries by any State Party or a group of States Parties related to the grounds that have arisen to believe that the violation of the Treaty by any State Party is taking place;
b) consider matters concerning the compliance with the obligations taken by States Parties;
c) organize and conduct consultations with the State Parties with the view to settle down the situation that has arisen in connection with the violation of a State Party of the Treaty;
d) take measures to put an end to the violation of the Treaty by any State Party.
The title, status, specific functions and forms of work of the Executive organization of the Treaty shall be the subject of an additional protocol to the Treaty.

Article IX
International intergovernmental organizations may take part in the Treaty. Provisions defining variants and modalities of their participation in the Treaty shall be the subject of an additional protocol to the Treaty.

Article X
Any State Party may propose amendments to the Treaty. The text of any proposed amendment shall be submitted to the Depository who shall promptly circulate it to all States Parties. Upon the request of at least one third of the States Parties, the Depository Governments shall convene a conference to which all States Parties shall be invited to consider the proposed amendment.

Any amendment to the Treaty shall be approved by a majority of the votes of the States Parties. The amendment shall enter into force for all the States Parties in accordance with the procedures of the entry into force of the Treaty.

Article XI
The Treaty shall be of unlimited duration.

Each State Party shall in exercising its national sovereignty have the right to withdraw from the Treaty if it decides that extraordinary events, related to the subject matter of this Treaty, have jeopardized the supreme interests of its country. It shall notify the Depository in written form of the decision taken six months in advance of the withdrawal from the Treaty.

Article XII
The Treaty shall be opened for signature by all States at the United Nations headquarters in New York. Any State which did not sign the Treaty before its entry into force may accede to it at any time.

The Treaty shall be subject to ratification by signatory States in accordance with their constitutional norms. Instruments of ratification or accession shall be deposited with the Secretary-General of the United Nations, who is hereby designated the Depository of the Treaty.

Article XIII
The Treaty shall enter into force upon the deposit of instruments of ratification by twenty States, including all Permanent Member States of the United Nations Security Council.
For States whose instruments of ratification or accession are deposited after the entry into force of the Treaty, it shall enter into force on

the date of the deposit of their instruments of ratification or accession.

Article XIV

The Treaty, of which the Arabic, Chinese, English, French, Russian and Spanish texts are equally authentic, shall be deposited with the Secretary-General of the United Nations, who shall send duly certified copies thereof to all signatory and acceding States.

---

[1] Associate Professor of International Law, University of Western Sydney; Visiting Professor of International Law, University of Copenhagen; Visiting Professional at the International Criminal Court, The Hague. Email s.freeland@uws.edu.au.

[2] The United Nations Conference on Disarmament was established in 1979 as the single multilateral disarmament negotiating forum of the international community, following the first Special Session on Disarmament (SSOD I) of the United Nations General Assembly held in 1978. It succeeded other Geneva-based negotiating fora, which include the 10 Nation Committee on Disarmament (1960), the 18 Nation Committee on Disarmament (1962-68), and the Conference of the Committee on Disarmament (1969-78): The United Nations Office at Geneva (UNOG),'Disarmament', <http://www.unog.ch/80256EE600585943/(httpHomepages)/6A03113D1857348E80256F04006755F6?OpenDocument> at 29 July 2008. The last agreement that it produced was the 1993 *Convention on the Prohibition of the Development, Production, Stockpiling and Use of Chemical Weapons*, 32 ILM 800.

[3] For example, Canada submitted 'detailed comments' to Russia in relation to an earlier draft: Comments by Ambassador Grinius of Canada, Geneva, 12 February 2008, page 1.

[4] Statement by H.E. Sergey Lavrov, Minister of Foreign Affairs of the Russian Federation at the Plenary Meeting of the Conference of Disarmament (unofficial translation), Geneva, 12 February 2008, page 6.

[5] Message from Foreign Minister Yang Jiechi of The People's Republic of China to the Conference of Disarmament, Geneva, 12 February 2008, page 2.

[6] Nick Cumming-Bruce, 'U.N. Weighs a Ban on Weapons in Space, but U.S. Still Objects', *The New York Times*, 13 February 2008 <http://www.nytimes.com/2008/02/13/world/europe/13arms.html?partner=rssnyt&emc=rss> at 31 July 2008.

[7] Ibid.

[8] 610 UNTS 205.

[9] *Outer Space Treaty*, article IV.

[10] See, for example, Gyula Gál, '"Threat or Use of Force" - Observations to Article 2 of the U.N. Charter and Article III of the Outer Space Treaty' (1989) 17:1 *Journal of Space Law* 54, 57.

[11] See, for example, Vladimir Bogomolov, 'Prevention of an Arms Race in Outer Space: The Deliberations in the Conference on Disarmament in 1993' (1993) 21:2 *Journal of Space Law* 141, 141, where the author refers to a failed Venezuelan proposal to amend Article IV.

[12] Jackson Nyamuya Maogoto and Steven Freeland, 'From Star Wars to Space Wars – The Next Strategic Frontier: Paradigms to Anchor Space Security' (2008) XXXIII:1 *Air and Space Law* 10, 12.

[13] At a recent conference in Australia, a senior official in Australia's military stated that up to 90% of that country's defence acquisition is now reliant in some way upon space capability: Group Captain Dennis Davison, Australian Defence Space Coordinating Office, 'National Security', 10th Australian Space Development Conference, Adelaide, 21-23 July 2008.

[14] See Steven Freeland, 'The Applicability of the *Jus in Bello* Rules of International Humanitarian Law to The Use of Outer Space' (2006) 49 *Proceedings of the Colloquium on the Law of Outer Space* 338.

[15] See United Nations General Assembly Resolution 62/20, 22 December 2007 on the 'Prevention of an arms race in outer space' (UNGA Resolution 62/20).

[16] See United Nations General Assembly Resolution 62/43, 5 December 2007 on 'Transparency and confidence-building measures in outer space activities', article 2.

[17] See United Nations General Assembly Resolution 62/217, 22 December 2007 on 'International cooperation in the peaceful uses of outer space', article 29.

[18] *Treaty on the Limitation of Anti-Ballistic Missile Systems* (*ABM Treaty*), 26 May 1972, 23 U.S.T. 3435. Article V(1) of the *ABM Treaty* provided that '[e]ach Party undertakes not to develop, test or deploy ABM systems or components which are sea-based, air-based, space-based, or mobile land-based'.

[19] Article XV(2) of the *ABM Treaty* provided *inter alia* that '[e]ach Party shall, in exercising its national sovereignty, have the right to withdraw

from this Treaty if it decides that extraordinary events related to the subject matter of this Treaty have jeopardized its supreme interests. It shall give notice of its decision to the other Party six months prior to withdrawal from the Treaty.'

[20] See Terence Neilan, 'Bush Pulls Out of ABM Treaty; Putin Calls Move a Mistake, *The New York Times*, 13 December 2001, <http://www.nytimes.com/2001/12/13/international/13CNDBUSH.html?ex=1217563200&en=d6306da3e8b8a4c1&ei=5070> at 31 July 2008.

[21] Apart from the development of its NMD capability, there appear to be at least three types of space weapon being developed by the United States; 1) Multiple Kill Vehicles; 2) Microsats; and 3) Laser Weapons: Paul Marks, 'Space weapons could make orbit a no-fly zone' *New Scientist*, 15 April 2006, page 30. See also Jackson Maogoto and Steven Freeland, 'The Final Frontier: The Laws of Armed Conflict and Space Warfare' (2007) 23(1) *Connecticut Journal of International Law* 165.

[22] See Mark Franchetti, 'Russia may aim nukes at Europe', *The Australian*, 14 July 2008, page 11.

[23] See 'Russia test launches intercontinental missile', *SpaceDaily*, 10 December 2002, <http://www.spacedaily.com/2002/021210064449.qc6szaa8.html> at 31 July 2008.

[24] 'Tanks to roll again in Red Square', *The International Herald Tribune*, 6 May 2008, page 3.

[25] See 'A New Arms Race in Space?', *The Economist*, 25 January 2007, page 5. It should also be noted that China has increased its military spending by over 10% each year over the past 20 years, with its military spending for 2008 to increase by nearly 18% over the previous year: 'China to raise military spending', *BBC News*, 4 March 2008 <http://news.bbc.co.uk/2/hi/asia pacific/7276277.stm> at 31 July 2008.

[26] As quoted in James Randerson and Mark Tran, 'China accuses US of double standards over satellite strike' *The Guardian*, 21 February 2008 <http://www.guardian.co.uk/science/2008/feb/21/spaceexploration.usa> at 31 July 2008.

[27] See, for example, I.H.Ph. Diederiks-Verschoor, *An Introduction to Space Law*, (2nd ed) (1999), pages 17-21 and the references therein.

[28] For example, for the purposes of the Australian Space Activities Act, the definitions of a 'launch', a 'launch vehicle', a 'return', and a 'space object' all make reference to 'the distance of 100 [kilometres] above mean sea level': for details of the Space Activities Act, see Steven Freeland, 'Difficulties of Implementing National Space Legislation Exemplified by the Australian Approach', in Stephan Hobe, Bernhard Schmidt-Tedd and Kai-Uwe Schrogl (eds), '*Project 2001 Plus' - Global and European Challenges for Air and Space Law at the Edge of the 21st Century*, (2006), page 65.

[29] 961 UNTS 187.

[30] 1023 UNTS 15.

[31] 1 UNTS 16.

[32] See *Military and Paramilitary Activities in and against Nicaragua (Nicaragua v. United States of America)* (Merits) (Judgment) [1986] ICJ Rep 14.

[33] Space Activities Act 1998 (Cth), Section 12(a).

[34] See *The Caroline Case* 29 BFSP 1137-1138; 30 BFSP 195-196, which also referred to a requirement of immediacy, although this was not mentioned in the decision of the International Court of Justice in *Oil Platforms* (*Merits*) (*Iran v. United States*) [2003] ICJ Rep. Judgment of 6 November 2003.

[35] [1996] 1 ICJ Rep. 245, para 41.

[36] See Steven Freeland, 'The Applicability of the *Jus in Bello* Rules of International Humanitarian Law to The Use of Outer Space' (2006) 49 *Proceedings of the Colloquium on the Law of Outer Space* 338.

[37] See Report of the High-level Panel on Threats, Challenges and Change 'A More Secure World: Our Shared Responsibility', attached to Note by the Secretary-General, UN Doc A/59/565, 2 December 2004, paras 188-192.

[38] See *North Sea Continental Shelf Cases (Federal Republic of Germany v. Denmark and Federal Republic of Germany v. The Netherlands)* (Judgment) [1969] ICJ Rep 3.

[39] See Steven Freeland, 'There's a Satellite in my Backyard! - Mir and the Convention on International Liability For Damage Caused by Space Objects' (2001) 24 *University of New South Wales Law Journal* 462.

[40] See, for example, *Convention on the Prevention and Punishment of the Crime of Genocide*, 78 UNTS 277, Article 9, which provides that: '[d]isputes between the Contracting Parties relating to the interpretation, application or fulfilment of the present Convention … shall be submitted to the International Court of Justice at the request of any of the parties to the dispute'.

51st Colloquium on the Law of Outer Space, Glasgow, Scotland

Session 4:
LEGAL ASPECTS OF NATURAL NEAR EARTH OBJECTS (NEOs)

Chairmen:

Richard Tremayne-Smith

Maureen Williams

Rapporteur:

Yuri Takaya-Umehara

IAC-08-E8.4.1

International Responsibility vis-à-vis natural near-Earth objects (NEOs) and their possible implications

Professor Dr. Maureen Williams
Professor of International Law, University of BuenosAires/Conicet
Chair of the ILA Space Law Committee (HQ London)
maureenw777@yahoo.co.uk and swilliams@derecho.uba.ar

Introduction

The Scientific and Technical Subcommittee of Copuos has been addressing this topic for some time now and the moment appears ripe to embark in the analysis of NEOs from a legal angle. The subject, currently in a somewhat stormy infancy, calls for a permanent interdisciplinary approach.

This paper is a first step in the legal approach to some of the major aspects involved in the presence of natural near-earth objects, their characteristics, implications and threats, in light of the Outer Space Treaties, the UN Principles and UNGA Resolutions and Declarations on Outer Space, and other rules of international law which might be applicable. Emphasis will be given to issues of international responsibility and liability, having in mind the dangers the presence of these objects may entail to mankind. The duties of countries having the necessary means to deflect or destroy NEOs which pose a serious risk to planet Earth and its resources will be the object of special treatment.

On the private level this question was recently brought up on the occasion of the 73rd Conference of the International Law Association held in Rio de Janeiro, (17-21 August 2008[1]) as a possible new topic on its agenda.

Stating the problem : man-made objects and natural space objects

(a) man-made objects

Different reactions are perceived from the legal standpoint in these initial stages. Part of the doctrine view the question as a subchapter of space debris, except that, on general lines, the latter term is applied to man-made objects and collisions in outer space between active satellites and otherwise inactive or abandoned satellites which originate small particles -referred to as 'second generation debris'. These particles travel at enormous speeds and their impact on an active satellite may have untold consequences. Figures are alarming: there are tens of thousands of these particles currently orbiting the Earth.

The question is aggravated by the existence of minute particles which, in the state-of-the-art, cannot be detected from Earth. Indeed this is a technical limitation which, as technology advances, should gradually be overcome.

In this field an important landmark are the 'United Nations Guidelines on Space Debris Mitigation' elaborated in the framework of the Scientific and Technical Subcommittee of Copuos (hereinafter the 'STSC') and adopted by a UN General Assembly Resolution in December 2007[2].

The seven guidelines considered -and subsequently adopted- by the STSC for the launch, mission and disposal phases of spacecraft and launch vehicle orbital stages are as follows:

1. to limit debris released during normal operations;
2. to minimise the potential for break-ups during operational phases;
3. to limit the probability of accidental collision in orbit;
4. to avoid intentional destruction and other harmful activities;
5. to minimise potential for post-mission break-ups resulting from stored energy;
6. to limit the long-term presence of spacecraft and launch vehicle orbital stages in the low-Earth orbit (LEO) region after the end of their mission;
7. to limit the long term interference of spacecraft and launch vehicle orbital stages with the geosynchronous Earth orbit (GEO) region after the end of their mission.

The fact that the Guidelines have reached the status of 'UN Guidelines on Space Debris Mitigation', plus the response given by a number of states concerning domestic measures taken in accordance with those Guidelines, was a powerful indication that the topic would be finally included on the agenda of the Legal Subcommittee of Copuos (hereinafter the 'LSC'). In fact, the latest development is that a single item for discussion entitled 'General exchange of information on national mechanisms relating to space debris mitigation measures' was included in the Report of the LSC for consideration at its 48th Session in 2009 [3].

In the private field, one of the first instruments of the kind is possibly the *ILA Instrument on the Protection of the Environment from Damage caused by Space Debris (hereinafter the 'ILA Instrument on Space Debris' or the 'Buenos Aires International Instrument on Space Debris')*, adopted in 1994 at the 66th Conference of this institution and which is kept under permanent review by its Space Law Committee[4].

(b) natural near-earth space objects (NEOs)

From the above considerations it follows that, from a legal viewpoint, NEOs are objects of an entirely different nature to space debris. They come under the general category of celestial bodies. In accordance with Article II of the 1967 Outer Space Treaty (hereinafter the '1967 OST') they are not subject to national appropriation by claim of sovereignty, by means of use or occupation or by any other means.

This provision should be read together with Article I of the OST establishing that the exploration and use of outer space, including the moon *and other celestial bodies*[5] shall be carried out for the benefit and in the interests of all countries, irrespective of their degree of economic or scientific development, and *shall be the province of mankind*[6].

The Article of reference further adds that outer space, the moon and *other celestial bodies*[7], shall be free for exploration and use by all states without discrimination of any kind, on a basis of equality and in accordance with international law, and there shall be *free access to all areas of celestial bodies*. Finally this Article states that there shall be freedom of scientific investigation in outer space, including the moon and *other celestial bodies*[8], and that states shall facilitate and encourage international cooperation in such investigation.

In this light, and without getting involved in the corresponding provisions of the 1979 Moon Agreement -many of which are not part of customary international law today- added to the fact that very few states have ratified this Agreement, it is clear -pursuant to the afore-mentioned Articles- that all states have a right to explore and carry out research concerning NEOs with the objective of anticipating risks and threats of collision with planet Earth.

By way of example, mention should be made of the announcement by NASA earlier this year concerning a 250-metre-length asteroid travelling at 33,000 km an hour which, according to NASA's latest radar information, would be coming very close to the Earth on 29 January 2008. As oberved by NASA, this is the closest ever an asteroid has been to date. And no asteroid or object of the kind will get closer to Earth before, at least, the end of next century[9].

Even though, in this particular instance, the possibility of collision was discarded, this may not always be the case. However so, the exploration and further research carried out by NASA in this field to enable this institution to make the above-mentioned announcement was undoubtedly lawful under the 1967 OST and consistent with international law. NASA, it is clear, met the requirements of the OST laid down in Articles I, II and III[10]. Moreover, it may be fairly assumed that it observed the provisions of Article IX of this Treaty as well, inasmuch as it had no reason to believe that the activity in question would cause harmful contamination or adverse changes in the environment of the Earth.

Yet it may be wondered whether, had NASA predicted -as a result of its exploration and research activities- a possibility of collision, could this governmental body, or any other state, group of states or international intergovernmental agency be held responsible for not avoiding an imminent catastrophe?

Let us assume further that NASA and/or other governmental bodies involved had the necessary technology to deflect the threatening object and, in so doing, had caused some damage to planet Earth. What would the legal consequences be? Indeed there are no specific rules on this issue.

Hence the conclusion that questions of responsibility and liability and their many intricacies should be carefully sorted out when addressing NEOs from the legal angle.

In addition to the duty of international cooperation underlying the UN Space

Treaties, Principles and UNGA Resolutions relating to Outer Space, and the ILA International Instrument embodying rules which may be seen as part of customary international law, this situation, viewed from a legal standpoint, calls for further precision. A slight step forward may be perceived in the 'Declaration'[11].

As Castillo Argañarás holds[12], point 4 of the Declaration is a sound basis for the duty to provide technology to deflect NEOs when stating that *international cooperation should be conducted in the modes that are considered most effective and appropriate by the countries concerned, including, inter alia, governmental and non-governmental; commercial and non-commercial; global, multilateral, regional or bilateral; and international cooperation among countries in all levels of development.*

At first sight, the situation is remindful of the 1968 Astronauts Agreement - hereinafter also referred to as the 'Rescue and Return Agreement'- where the obligations enshrined are essentially humanitarian. In fact, in its Preamble this Agreement calls for the rendering of all possible assistance to astronauts in the event of accident, distress or emergency landing, and the prompt and safe return of astronauts, and the return of objects launched into outer space. Later on in this text a reference to 'sentiments of humanity' may be found.

In like manner, the need to deflect NEOs which constitute a serious risk of collision with planet Earth may be viewed as based on humanitarian reasons. However, unlike the case of the Rescue and Return Agreement, it is the whole of mankind which would be here at risk.

Unlike the 1972 Liability Convention, it must borne in mind that we are dealing with natural -and not man-made- objects which, consequently, are not subject to registration nor do they come under the provisions of this Convention. This makes matters somewhat more complex.

As indicated earlier, NEOs cannot be subjected to national appropriation by claim of sovereignty, by means of use or occupation, or by any other means, as Article II of the OST clearly provides. NEOs are celestial boodies and -doubtless- the province of mankind. Their presence in the vicinity of planet Earth is a threat to the whole of mankind. And, as in the case of space debris, there are currently tens of thousands of NEOs from which a small number may imply a considerable threat of collision with the Earth in the present century[13]. Such the situation today.

Interesting, for its implications, is the Safeguard Survey Programme established by the US Congress and put into practice by NASA in recent years. The figures released are telling. As Schweickart indicates, in the past eight years 4,600 NEOs have been discovered of which 713 are greater than 1 km diameter. The new directive from the US Congress to NASA anticipates the discovery of 400,000 NEOs of which approximately 40,000 will go beyond the 140 metre diameter goal[14].

The question has clear international dimensions and calls for a strict interdisciplinary approach. It is well-known that the precise data provided by science and technology becomes essential where the building up of realistic legal proposals is concerned. In this context the legal implications of NEOs is no exception.

In the search for viable suggestions and interdisciplinary approaches to NEOs a look at the recent work of the STSC on the matter becomes pertinent. In the first place, this Subcommittee has agreed on a definition -which may perhaps be seen as a non-exhaustive description- of NEOs which appears extremely useful as a working tool.

NEOs, as noted by that body, are asteroids and comets with orbits that could cross the orbit of planet Earth. The STSC also considered that the interest in asteroids was largely due to their scientific value as remnant debris from the inner solar system formation process, the possibility of their collision with the Earth and potentially devastating consequences, and their possesion of a wide range of natural resources[15].

The STSC noted that early detection and precision tracking were the most effective mechanisms for the management of threats posed by near-Earth objects. In that regard, the STSC noted with satisfaction that a number of international teams in various countries were currently searching for, investigating and cataloguing NEOs[16].

UNGA Resolution 62/217 (22 December 2007) reconvened its Working Group on NEOs which started work on 18 February 2008 with Richard Croucher (UK) as its Chairman[17]. This Working Group considered that the task accomplished in the intersessional period had resulted in important contributions to international cooperation in that area.

One of the main objectives of this Group is to review policies and procedures related to the handling of the NEO threat at the international level and consider drafting international procedures for handling this threat[18]. At the end of the 45th Session of the STSC, and on the basis of the multi-year plan[19] reviewed by the Working Group, it was recommended that the STSC continued to consider NEOs in acordance with a new multi-year plan (2009-2011)[20].

This workplan should be followed closely when elaborating general guidelines or codes of conduct addressing the legal aspects of NEOs so that the ensuing proposals are consistent with the international context of our time.

Finally it may be observed that contributions from the doctrine to the legal aspects of NEOs have been sparse so far. Nevertheless the subject seems to be gaining momentum in recent months as an increasing number of institutions are becoming gradually aware of the dimensions of this challenge and listing the topic on their agendas for discussion.

Perceptions, suggestions and conclusions on the legal aspects of NEOs

1. The general opinion within the STSC seems to be moving towards the drafting of some guidelines on the subject and, as perceived by the present writer, in a manner similar to that followed when addressing space debris mitigation.

2. Should that pattern be continued it would not be unreasonable to anticipate the possibility of adoption of guidelines on NEOs by the STSC in a not too distant future to be followed -as recent precedents are indicating- by the adoption of a set 'UN Guidelines on NEOs' leading, in

turn, to the inclusion of the topic on the agenda of the LSC for discussion of its legal aspects.

3. Any possible guidelines proposed on the legal sides of NEOs should be general enough to survive the times and follow a strict interdisciplinary approach. In other words, law should keep pace with scientific and technological developments.

4. A number of provisions enshrined in the UN Space Treaties and related instruments are clearly applicable to NEOs, as observed at the outset. Customary international law is slowly growing as well. International cooperation -seen as a 'general obligation to cooperate' and a *conditio sine qua non* for an activity in outer space to be lawful- is at the very root of the protection of mankind from risks implied by NEOs. On these bases, and in light of the threat posed by NEOs to planet Earth, some new law should also be developed.

5. In this context, pride of place should be given to issues of international responsibility of states and international organisations for failing to take quick action in case of serious threats. Similarly, rules on liability for damage caused in the course of deflecting operations should be considered. To this end it is important to have in mind the humanitarian sides of the questions involved and the need to strike an appropriate balance between all converging interests with a view to protecting mankind from a nowadays rather unmanageable menace.

6. Effective dispute settlement mechanisms are, in this field, essential. They should be sufficiently agile to operate promptly when faced with an imminent risk of collision.

7. Some thought should be given to the possibility of setting up a Fund to meet the costs of deflection operations and compensation for damage arising therefrom. Therefore, the duties incumbent on states and international organisations having the proper technology for deflection or mitigation ought to be carefully analysed.

8. Suggestions point at the need to draw up a Protocol to supplement and give a more precise meaning to the duties stemming from the UN Space Treaties, Principles and UNGA Resolutions, as well as from customary international law applicable in this field.

9. In these initial stages, it would be helpful to think in terms of an international instrument the nature of which -binding or non-binding- could be decided later. The ILA International Instrument on Space Debris appears useful for this purpose as certain duties of states embodied in that Instrument would be applicable to NEOs as well.

10. Many of the duties relating to space debris are part of customary international law today, and should be included in any future Guidelines for NEOs, *inter alia,* the obligations to prevent, to inform and exchange information, to consult and hold consultations, and to negotiate in good faith[21] concerning problems involving the presence of NEOs entailing serious threats of collision with planet Earth.

11. Much further research is required in view of the uncertainty surrounding the exact position and orbit of NEOs, which may take several years to be tracked. In this context one of the most crucial points is the decision-making process to establish at what point to go ahead -if at all- with deflection operations and the most appropriate moment therefor.

12. The above-mentioned question remains outstanding. It is a major problem and a real obstacle for advancing on the preparation of any international instrument or guidelines on the matter. It will be looked at by the STSC during its 2009 session in February 2009 on the basis of studies presently underway by an interdisciplinary group of experts.

---

[1] The *ILA Report of the Rio Conference,* in book format, should be expected for early 2009. In the meantime, the 2008 *Report of the Space Law Committee* may be found on the ILA website www.ila-hq, clicking on "Committees" and then "Space Law".

[2]UNGA Resolution 62/217 of 21 December 2007.

[3] Report of the LSC on its 47th Session, Doc. A/Ac.105/917, p.24, paragraph 151, item 10.

[4] See *Report of the 66th Conference of the International Law Association* (Buenos Aires), Space Law Committee, pp.305-324. At the ILA 73rd Conference (Rio de Janeiro 17-21 August 2008) it was agreed to continue reviewing this topic in light of the adoption of the *'UN Guidelines on Space Debris Mitigation'* and its inclusion on the agenda of the Legal Subcommittee of Copuos for discussion in 2009.

[5] Italics of the author.

[6] Italics of the author.

[7] Italics of the author.

[8] Italics of the author.

[9] See La Nación, Buenos Aires, 29 January 2008, p.10. The asteroid would be coming as close as 538 000 Km, i.e one third further away from the distance between the Earth and the Moon.

[10] Article III of the 1967 OST provides that states parties to the Treaty shall carry on activities in the exploration and use of outer space, including the moon *and other celestial bodies*, in accordance with international law, including the Charter of the United Nations, in the interest of maintaining international peace and security and promoting international cooperation and understanding (italics of the author).

[11] UNGA Declaration on International Cooperation in the Exploration and Use of Outer Space for the Benefit and in the Interest of All States, Taking into Particular Account the Needs of Developing Countries (1996), Official Records of the General Assembly, Fifty-first Session, Supplement N° 20 (A/51/20).

[12] Castillo Argañarás, L.F., *'Natural Near-Earth Objects and the International Law of Outer Space'*, paper submitted to the 51st Colloquium of the IISL (Glasgow 2008). Manuscript by courtesy of the author.

[13] Russell L. Schweickart, *Eilene M. Galloway Symposium on Critical Issues in Space Law,* 49th Colloquium on the Law of Outer Space, Valencia 2-6 October 2006, pp. 574-579.

[14] Schweickart, Russell L., *op.cit.loc.cit supra,* pp.574-575.

[15] Doc. A/AC.105/911, at p. 27. Paragraph 159.

[16] Ibid., paragraph 160.

[17] It should be noted that the British National Space Centre (BNSC) participates actively in research on NEOs and is a recognised leader in the field. As observed by Castillo Argañarás, Richard Tremayne-Smith (UK) was the initial chair of the UN Working Group on NEOs. Various UK universities such as Queen's (Belfast), Southampton, Glasgow and the Open University are currently conducting important research programmes on different aspects of the matter

with encouraging results. See Castillo Argañarás, L.F., *'Natural Near-Earth Objects and the International Law of Outer Space'*, op.cit. in note 12 *supra.*

[18] Doc. A/AC.1105/911, pp.42-43, Annex III *on 'Report of the Working Group on Near-Earth Objects'.*

[19] Doc. A/AC.105/C.1/2008/CRP.12.

[20] Doc. A/AC.105/911, p.43, paragraph 11, Annex III.

[21] See the *ILA Instrument on Space Debris* (66th Conference, Buenos Aires 1994), articles III, IV, VI and VII. See also op.cit. note 4 *supra.*

IAC. – 08 – E8. 4. 2. NATURAL NEAR EARTH OBJECTS AND THE INTERNATIONAL LAW OF OUTER SPACE

Dr. Luis F. Castillo Argañarás
National Council of Scientific and Technical Research of Argentina (CONICET) and
Universidad Argentina de la Empresa – Instituto de Ciencias Jurídicas y Sociales (UADE)
lcastillo@uade.edu.ar

ABSTRACT

Natural Near Earth objects (NEOs) are celestial bodies such as asteroids and meteorites, which may go past the Earth orbit. They imply an impact risk for our planet. Such threat, though improbable, cannot be ignored. The damage produced would not affect only one country but it may have cross – border effects without discrimination. It would affect the ecosystem and human population. On September 2007, the Spaceguard Association of Japan had observed a total of 212 asteroids and 90 comets near the Earth.[i]

The Scientific and Technical Subcommittee of COPUOS stated that "the NEO hazard should be recognized as a global issue that can be effectively addressed only through international cooperation and coordination"[ii]. Therefore, the following questions arise: a) Which activities are the States performing in this field? b) Is there a legal framework that regulates this activity or is it necessary to draft a new one? This paper intends to answer those questions according the following structure: 1) Introduction; 2) Activities carried out; 3) Applicable legal frame; 4) Conclusions
Special thanks to Daniela Costa, attorney at law and legal translator, for her collaboration in the English version of this paper. dcosta@thelinguacorp.com

I. INTRODUCTION

Natural near-Earth objects, such as asteroids and meteorites, are celestial bodies that go across the terrestrial orbit. These objects may impact against our planet and, though improbable, such threat cannot be ignored. The damage caused would not only affect a single State but it would rather have trans-boundary and indiscriminate effects. The ecosystem and the human population would be affected. By September 2007, The Japanese Spaceguard association observed a total of 212 asteroids and 90 comets near Earth.
The COPUOS Scientific and Technical Subcommittee stated that "the NEO hazard should be recognized as a global issue that can be effectively addressed only through international cooperation and coordination."[iii] In that sense, and in compliance with the United Nations General Assembly Resolution 61/111 of 14 December, 2006, a working group on near-Earth objects was established in the framework of the Subcommittee. The

working group was presided by Mr. Richard Tremayne – Smith. A multi-year plan for years 2008-2010 was recommended[iv], which can be summarized as follows:[v]

2008: Continue inter-sessional work and consider the reports submitted in response to the annual request for information on near-Earth object activities. The presentations would focus on national, regional and international collaborative activity for observation and analysis of near-Earth objects. Drafting of the interim report.
2009: Continue annual reporting on near-Earth object activities, which will include an update on near-Earth object missions and submission of draft procedures related to threat handling at the international level. Review and update the interim report.
2010: Continue analysis on international procedures for threat handling of near-Earth objects. Review progress with international cooperation and collaboration on observations. Review and update the interim report.
In view of the foregoing, some questions arise. First, as regards the cooperation between the States and the work performed by the Working Group created in the framework of the Scientific and Technical Subcommittee, the question is: a) Which activities are the States performing in this field? Since the works are performed in the context of the Scientific and Technical Subcommittee, this issue is not addressed by the Legal Subcommittee. However, no adequate set of rules in this field can be prepared without being aware of the scientific and technical issues. So, the question is: b) Is there a legal framework that regulates this activity or is it necessary to draft a new one?

As regards the methodology applied in this paper, factual issues will be addressed first, making reference to the activities of those involved in this field, to later deal with the legal issues and finally draw conclusions. Due to space limitations, the analysis will be focused in only four States: USA, UK, Germany and the Russian Federation.

II. ACTIVITIES PERFORMED BY THE STATES

Pursuant to the interim Report of the Working Group on near-Earth objects, the first step in addressing the risk posed by a near-Earth object is to detect its presence and infer its size from its trajectory and observed brightness.[vi] Due to space limitations, we will only concentrate on the activities of some States, considering the relevance of the activities performed by those States.

1. UNITED STATES OF AMERICA

This country has created the "*Near Earth Object Program*" of the National Aeronautics and Space Administration (NASA). This program provides funding to five near-Earth objects search teams to operate nine separate 1-metre class survey telescopes across the south-western United States and one in Australia.[vii] The United States also operates two planetary radars capable of observing NEOs. The Goldstone radar is located in southern California, in the Mojave Desert. The second radar, located at Arecibo, Puerto Rico, is owned and managed by the *National Science Foundation* and operated by Cornell University under a cooperative agreement with the Foundation.[viii]
A partnership of astronomers in the United Kingdom, from Durham

University, Queen's University Belfast and the University of Edinburgh, has joined a group of German and American institutions in using an advanced new telescope, which is equipped with the world's largest digital camera and is located in Hawaii, on the island of Maui, to observe and determine the characteristics of NEOs. [ix]

As part of its NEO observation programme, NASA has established the Near-Earth Object Program Office at its Jet Propulsion Laboratory (JPL). Through the JPL Sentry System, risk analyses are automatically performed on objects that have a potential for Earth impact. Such an analysis usually occurs for an object that has recently been discovered and for which there is not yet a data interval sufficiently long to make its orbit secure.[x]

The Sentry System automatically updates the orbits of approximately 40 near-Earth objects per day, and close approach tables are generated and posted on the Internet: http://neo.jpl.nasa.gov/ca [xi]

Approximately five risk analysis cases are performed each day, with each analysis providing 10,000 multiple solutions. That process is also performed in parallel in Pisa, Italy, and by JPL. Since the inception of the Sentry System in 2002, approximately 400 objects have appeared on the Sentry risk page: http://neo.jpl.nasa.gov/risk. For recently discovered objects of unusual interest, the Minor Planet Centre, JPL, and the centre at Pisa will often alert observers that additional future or pre-discovery data are needed.[xii]

The NASA with support from the University of Santa Cruz study the threat posed by impact-induced tsunamis. The University of Arizona has created an easy-to-use, interactive website for estimating the environmental consequences of an impact on Earth. By providing inputs for the distance from ground zero and the projectile's diameter, density, velocity and impact angle, the program will estimate the ejecta distribution, ground shaking, atmospheric blast wave, the thermal effects of an impact and the size of the crater produced. The website is found at http://www.lpl.arizona.edu./impaceffects [xiii]

2. UNITED KINGDOM OF GREAT BRITAIN AND NORTHERN IRELAND

The British National Space Centre (BNSC) maintains an active role in this field. This leadership role is demonstrated by the United Kingdom's chairmanship of the Action Team on Near-Earth Objects and the Working Group on Near-Earth Objects.[xiv]

Astronomers at Queen's University of Belfast, in cooperation with colleagues in Germany, the Czech Republic, Slovakia and the United States, successfully measured the Yarkovsky-O'Keefe-Radzievskii-Paddack (YORP) effect for the first time, by studying NEO (54509) 2000 PH5 over several years. That theoretical effect causes NEOs to spin up or spin down through the action of being heated by the Sun. The YORP effect is probably the primary method of the creation of binary NEOs and acts in unison with the Yarkovsky effect in delivering asteroids to Near-Earth space.[xv] Astronomers at Queen's University of Belfast continued to obtain astrometric data on NEOs with an identified small risk of hitting the Earth in the next 100 years, with the aim of improving their orbits. As of year 2007, they have measured positions and improved the orbits of over 200 NEOs.[xvi]

At the Open University, a number of experimental programmes aimed at understanding the formation of smaller bodies in the solar system are under way. In addition, studies to mitigate and deviate NEOs are conducted.[xvii]

The Astronautics Research Group at the University of Southampton is conducting a research into the effects of NEO impacts on the Earth. This research is focused in the world threat posed by NEOs of a diameter under 1 km. A software tool was developed in the period 2004-2005 to assess how impact-generated effects will affect the human population, and the software was exploited in 2006 to analyse particular impact case studies.[xviii]

The first of these was an evaluation of the casualty numbers resulting from ground and sea impacts in the regional neighbourhood of the United Kingdom, and the second was a study of the effects on the human population resulting from the potential impact of the asteroid 99942 Apophis in 2036. This work is being further advanced with the development of a software tool called the "NEO impactor", which will be used for global studies of the effects of NEO impacts on both the population and on infrastructure. Completion of the research programme was expected in 2007.[xix]

The objective of the work conducted by the University of Glasgow is to develop fundamental optimal control theory and apply it to the interception of hazardous NEOs. Different parameters (time, mass, orbital corrections and maximum deviation) will be optimized.[xx]

The website of the Near Earth Objects Information Centre, http://www.nearearthobjects.co.uk, provides a wide range of information about this issue.

3. RUSSIAN FEDERATION

The Russian Federation and the countries of the Commonwealth of Independent States have accumulated significant scientific and technological capabilities. These could be used to develop a planetary defence system to protect the Earth from the threat posed by asteroids and comets.[xxi]

A number of Russian and Ukrainian organizations founded the *Planetary Defence Centre* as a nonprofit-making partnership, with a view to combining the efforts of organizations and experts working in various fields towards the establishment of a planetary defence system.[xxii]

The main activities of the Planetary Defence Centre are as follows:[xxiii]

a) Design of a planetary defence system to protect the Earth from the threat posed by asteroids and comets;
b) Elaboration of possible space threat scenarios and methods and means of countering such threats;
c) Participation in the preparation and conduct of simulation and demonstration experiments to test the components of the planetary defence system;
d) Conduct of public awareness campaigns and other activities.

The activities of the Planetary Defence Centre are based on the conceptual design for the *Citadel* planetary defence system. The Citadel system will consist of an Earth- and space-based service for rapid interception. The interception service will make use of space rockets and nuclear and other resources from the Russian Federation, the United States of America and European and other countries. These resources will include reconnaissance and interceptor spacecraft.[xxiv]

The Space Patrol Project provides for a spacecraft to be built and launched towards asteroids passing near the Earth and, in particular, objects moving in meteor showers. The following missions could also be carried out: Prolet (fly-by), Udar (impact), Vnedrenie (penetration) and Perekhvat (interception).[xxv]

4. GERMANY

Scientists at the Institute of *Planetary Research of the German Aerospace Centre* in Berlin-Adlershof have been engaged in international research on near-Earth objects for many years. The work includes observation campaigns for physical characterization of NEOs using major ground-based and space-borne astronomical telescopes. Data reduction and analysis, theoretical investigations and the publication of results in major refereed journals are also major activities of the group in this field. This work is carried out mainly by the *Asteroids and Comets Department* of the Institute by scientists and research students.[xxvi]

An observation programme of NEOs has been implemented in collaboration with Scandinavian institutes (the University of Helsinki, the University of Uppsala, the University of Oslo and the University of Copenhagen). The programme uses the Nordic Optic Telescope on the island of La Palma, Spain, to carry out photometric light curve observations of NEOs and perform astrometric follow-up of NEOs as from 2007.[xxvii]

A strong future participation in the planning of the Don Quijote mission is foreseen. Don Quijote is a mitigation precursor mission of European Space Agency currently under feasibility study by a consortium of European industrial and academic partners. The Institute of Planetary Research is a member of the consortium, which has successfully completed a phase A study of the mission.[xxviii]

An online database of physical properties of all known NEOs is maintained and available at http://earn.dlr.de. The annual reports of the research activities conducted in Germany are available at http://solarsystem.dlr.de/KK and http://elib.dlr.de/perl/search.

5. THE SPACEGUARD FOUNDATION

The purpose of this Foundation is the protection of the Earth environment against the bombardment of objects of the solar system (comets and asteroids). It was officially created on 26 March, 1996 in Rome. Apart from Italy, it is also hosted in different countries such as Germany, Japan, the United Kingdom, etc. This Foundation constitutes a liaison between the general public and the scientific community. It provides the general public and the government agencies with scientific information in easily understandable terms.[xxix]

III. APPLICABLE LEGAL FRAMEWORK

A. THE OBLIGATION TO PREVENT AND INFORM

Maureen Williams understands that the obligation to prevent damage arises from international customary law.[xxx] Generally, two levels of prevention are recognized: a) unilateral measures to be adopted by the source State and b) adoption of rules.[xxxi] Source State means a contaminating State. In my opinion, according to the *prater – legem* equitable principle, this concept involves a State with sufficient technology available to monitor NEOs and prevent any potential

damage. If such State fails to do so, it is held internationally liable.
The obligation to inform upon contamination is found in Article IX of the Outer Space Treaty. Therefore, based on the arguments presented above, we can conclude that if a State has technology available, is aware of the information and fails to inform the potentially affected State, such behaviour would result in the consequences outlined by Williams, *i.e.* "both States may be involved in a serious and never ending controversy about the limits of the exclusive jurisdiction of the affected State to investigate the activity in question."[xxxii]

B. THE CONSULTATION OBLIGATION ARISING FROM ARTICLE IX OF THE OUTER SPACE TREATY

Aldo Armando Cocca[xxxiii] explains that the drafting of this article was based on the experiences of the US Project West Ford. The project consisted in launching into orbit a ring of tiny dipole antennas (needles). In the first attempt in October 1961, the needles failed to disperse. The second attempt was conducted on 12 May, 1963, achieved the expected results. The mere announcement of the project raised comments and protests, among others, from the International Telecommunication Union. In September 1961, the International Council for Science (ICSU) entrusted the Committee on Space Research (COSPAR) with the analysis of experimental projects and other space related activities which effects were not appropriate for scientific activities and observations. Based on its report, and although the above mentioned experiences caused no noticeably harmful interferences, the COSPAR adopted several resolutions whereby its members were requested to provide reports on these type of projects beforehand, and recommended measures to avoid celestial bodies contamination. According to Bin Cheng[xxxiv], afterwards, the consultation procedure through the COSPAR was satisfactorily implemented, which would explain why two of the great superpowers agreed with the consultation procedure set out in Article IX.

In that regard, Merrills[xxxv] explains that when a government expects that a proposed decision or course of action causes potential damage to another State, the discussions with the affected party may lead to a dispute settlement, thus creating opportunities to settle the controversy. Consultation is particularly valuable because it provides useful information in due time, that is, before any actions are implemented. Therefore, we understand that the consultation obligation is totally applicable to this field, given that the threat of an impact entails a potential damage greater than contamination itself.

C. INTERNATIONAL COOPERATION

Article 1 of the 1967 Outer Space Treaty states the following:

"...There shall be freedom of scientific investigation in outer space, including the Moon and other celestial bodies, and States shall facilitate and encourage international cooperation in such investigation"

Furthermore, Article 3 provides that:

"States Parties to the Treaty shall carry on activities in the exploration and use of outer space, including the Moon and other celestial bodies, in accordance with international law, including the Charter of the United Nations, in the interest of maintaining international peace and security and promoting international cooperation and understanding."

Article 9 of the Outer Space Treaty regulates on the following issue:

"In the exploration and use of outer space, including the Moon and other celestial bodies, States Parties to the Treaty shall be guided by the principle of cooperation and mutual assistance and shall conduct all their activities in outer space, including the Moon and other celestial bodies, with due regard to the corresponding interests of all other States Parties to the Treaty..."

According to Manuel Augusto Ferrer (h) "international cooperation is a statutory obligation rather than a mere aim";[xxxvi] and he further states that "international cooperation is an aim that all the international community expects to achieve in all respects, except for cooperation as regards space-related activities, which is already binding upon the States."[xxxvii] Maureen Williams claims that "The Argentine thesis that imposes the international cooperation obligation as of the entry into force of the 1967 Outer Space Treaty, and which considers such obligation a previous requirement to develop space-related activities, is widely accepted by most of the scholars worldwide."[xxxviii]

D. UNITED NATIONS GENERAL ASSEMBLY RESOLUTION 51/122 OF 1996: "DECLARATION ON INTERNATIONAL COOPERATION IN THE EXPLORATION AND USE OF OUTER SPACE FOR THE BENEFIT AND IN THE INTEREST OF ALL STATES, TAKING INTO PARTICULAR ACCOUNT THE NEEDS OF DEVELOPING COUNTRIES"

In 1996, the United Nations General Assembly adopted without vote resolution 51/122, entitled: Declaration on International Cooperation in the Exploration and Use of Outer Space for the Benefit and in the Interest of All States, Taking into Particular Account the Needs of Developing Countries"
Article 1 thereof provides that "International cooperation in the exploration and use of outer space for peaceful purposes shall be conducted in accordance with the provisions of international law, including the Charter of the United Nations and the Treaty on Principles Governing the Activities of States in the Exploration and Use of Outer Space, including the Moon and Other Celestial Bodies."

Article 3 states the following:

"All States, particularly those with relevant space capabilities and with programmes for the exploration and use of outer space, should contribute to promoting and fostering international cooperation on an equitable and mutually acceptable basis. In this context, particular attention should be given to the benefit and the interests of developing countries and countries with incipient space programmes stemming from such international cooperation conducted with countries with more advanced space capabilities"

Article 4 provides that:

"International cooperation should be

conducted in the modes that are considered most effective and appropriate by the countries concerned, including, inter alia, governmental and non-governmental; commercial and non-commercial; global, multilateral, regional or bilateral; and international cooperation among countries in all levels of development"

D .1. LEGAL VALUE OF RESOLUTION 51/122 OF 1996

In 1965, Bin Cheng published a paper posing the following question: "United Nations Resolutions on Outer Space: Instant International Customary Law?"[xxxix] The answer is to deny the meaning of the usual practice and the time factor in the formation of international customary law and base it on the *opinio juris,* as expressed in non-binding resolutions and declarations, as an integral aspect of international customary law. Bin Cheng's position can be supported by the North Sea Continental Shelf case, in which it was stated that a rule of customary international law may arises even in a short period of time.[xl]

Peter Malanczuk considers that the "instantaneous international customary law rule" is associated to certain declarations and resolutions issued by international organizations and to the "soft-law"[xli] phenomenon. According to this jurist, the soft law discussion is mainly related to relevant General Assembly Resolutions, such as the first ones that set out the basic principles of space law, or the 1982 resolution on "Direct Satellite Television Broadcasting Principles", the 1986 resolution on principles related to satellite teledetection or the 1992 resolution on use of nuclear power sources in outer space.[xlii] In my opinion, Resolution 51/122 can be added to the aforementioned list. Malanczuk concludes that certain principles and regulations arise as new rules in the drafting process: even though they are not legally accepted as binding rules, these have a limited early effect in court or arbitration decisions as valid arguments in the interpretation of the law.[xliii]

E. INTERNATIONAL RESPONSIBILITY

In the case "Chorzow Factory" (jurisdiction), the Permanent Court stated that: "it is a principle of international law that the breach of an engagement involves an obligation to make reparation in an adequate form."[xliv] Article 1 of the International Law Commission's Articles on State Responsibility provides that: "Every internationally wrongful act of a State entails the international responsibility of that State"; and Article 2 states that: "There is an internationally wrongful act of a State when conduct consisting of an action or omission: a) is attributable to the State under international law; and b) constitutes a breach of an international obligation of State". Therefore, the States shall be internationally responsible upon breach of the prevention, information, consultation and cooperation obligations.

GENERAL CONCLUSIONS

The activities performed by states such as the United Kingdom of Great Britain, the United States of America, Germany and the Russian Federation have been addressed in this paper without limitation and only as relevant examples due to space reasons.

It is worth emphasizing the **international cooperation** of States with space technology available, such as the German observation programme of NEOs in collaboration with Scandinavian institutes (the University of Helsinki, the University of Uppsala, the University of Oslo and the University of Copenhagen). This programme uses the Nordic Optic Telescope located on the island of La Palma, Spain, to carry out photometric light curve observations of NEOs and perform astrometric follow-up of NEOs as of 2007.

Furthermore, the creation of the Spaceguard Foundation entails a quite remarkable mission: informing the population in simple and clear terms to enable them to understand this issue.

From a legal approach, it is worth mentioning that a set of legal obligations arising from the treaties currently in force, and the United Nations General Assembly resolutions, are now comprised by the **international customary law**.

So far, the prevention and information, international cooperation and consultation obligations and the United Nations General Assembly resolution 51/122 of 1996 provide a suitable legal framework to this international issue and, therefore, no new legal instrument need be drafted in this respect. The existing rules are validly applicable to the issue in question, considering the potential risks posed by near-Earth objects. Upon breach of the legal obligations mentioned above, the international responsibility of the States shall undoubtedly arise.

---

[i] A/Ac.105/863/Add.2. (5 de Diciembre de 2006). Page. 5

[ii] A/AC.105/C.1/L.290, Page 9

[iii] United Nations. General Assembly. A/AC. 105/C. 1/L. 290, (12 December 2006) Page 8.

[iv] United Nations. General Assembly. A/Ac.105/890. Annex III (6 March 2007). Page. 40.

[v] Ibídem. Page 40.

[vi] United Nations. General Assembly. A/Ac. 105/C.1/L.290 (12 December 2006). Page 2.

[vii] Ibidem

[viii] Ibidem. Page 3

[ix] Ibidem

[x] Ibidem. Page 5

[xi] Ibidem

[xii] Ibidem

[xiii] Ibidem

[xiv] United Nations. General Assembly. A/AC. 105/896 (4 December 2007). Page 5.

[xv] Ibidem, Page 5

[xvi] Ibidem

[xvii] Ibidem

[xviii] Ibídem

[xix] Ibidem. Page 6

[xx] Ibidem. Page 7

[xxi] United Nations. General Assembly. A/AC.105/863/Add.1 (28 March 2006). Page 2

[xxii] Ibidem

[xxiii] Ibidem

[xxiv] Ibidem.

[xxv] Ibidem

[xxvi] United Nations. General Assembly. A/Ac.105/863/Add.2. (5 December 2006). Page 2

[xxvii] Ibidem, Page 3

[xxviii] Ibidem

[xxix] See: http://spaceguard.rm.iasf.cnr.it/SGF/INDEX.html and http://spaceguard.rm.iasf.cnr.it/index.html

[xxx] Williams, Silvia Maureen: *Derecho Internacional Contemporáneo*. Abeledo Perrot. Buenos Aires. 1990. Page 116

[xxxi] Ibídem. Page 117

[xxxii] Ibídem

[xxxiii] Cocca, Aldo Armando: *Consolidación del Derecho Espacial*. Editorial Astrea. 1971. Page 41 and following pages.

[xxxiv] Cheng, Bin: *Studies in International Space law*. Oxford University Press. New York. 1997. Page 257.

[xxxv] Merrils,J. G.: *International Dispute Settlement*. Third Edition. Cambridge University Press. UK. 1998. Page 3.

[xxxvi] Ferrer, Manuel Augusto (h): "Contenidos Éticos y Jurídicos de la Transferencia de Tecnología Espacial" Pág. 223 in Cocca. Aldo Armando, Serie Estudios Internacionales Avanzados: *Ética, Derecho, Ciencia, Tecnología y Cooperación Internacional*. Consejo de Estudios Internacionales Avanzados. Argentina. 1985.

[xxxvii] Ibídem, Page 224.

[xxxviii] Williams, Silvia Maureen. 1990. Page 94.

[xxxix] As published in Indian Journal of International Law. 1965. Pages 23 and following pages.
[xl] North Sea Continental Shelf Case, ICJ Rep. 1969. P. 4
[xli]Malanczuk, Peter: "Space Law as a Branch of International Law" in Netherlands Yearbook of International Law. Volume XXV. 1994. Martin Nijhoff Publishers. 1994. Page 161
[xlii] Ibidem
[xliii] Ibidem. Page 162
[xliv] PCIJ, Ser. A, Nª 17, p29 (1927).

IAC-08.E8.4.3

DEFINING SUBJECT MATTER UNDER SPACE LAW: NEAR EARTH OBJECTS *VERSUS* SPACE OBJECTS

Frans G. von der Dunk
University of Nebraska, College of Law, Space and Telecommunications Law Program
Fvonderdunk2@unl.edu

Abstract

It may seem to be an obvious, instinctive distinction, the one between (natural) near earth objects and (man-made) space objects. However, the very recent proposal tabled by the Russian Federation and the People's Republic of China for a treaty on the de-weaponisation of space apparently makes reference in this context to a prohibition to use or threaten the use of force against "outer space objects". Such varying use of terminology may raise appropriate concerns about the applicability of any such agreement, or indeed other present or future rules of space law, to the specific case of NEO's and any possible future actions to protect the earth against their potentially devastating impact. The present paper represents an effort to clearly outline the definitional issues involved, in the hope of precluding any potentially stifling confusion about the applicability or non-applicability of relevant rules of international space law.

Thus, the issue of the definition of 'space object' will be revisited and discussed in juxtaposition with such definitions as those of 'celestial bodies' and 'near earth objects', with reference to the applicability of relevant rules of international space law.

1. The definitional issue in perspective

Incoming asteroids that threaten human civilisation have been subjects not only of Hollywood movies (*Deep Impact*, *Armageddon*) but also of serious scientific research. As a consequence, it is now generally assumed that the extinction of the dinosaurs some 65 million years ago was at least partially caused by a giant asteroid (presumed magnitude: in the range of 10 kilometres across) principally hitting in the area of what is now the Yucatan peninsula.[1] More recently an originally mysterious blast over the Siberian Tunguska area in 1908 turned out to have been caused by an asteroid of perhaps 40 meters across, exploding in the air at a few kilometres altitude.[2]

Barring fantasy, however, little attention was paid to possibilities to actually *do* something about incoming asteroids. It is only since fairly recently, that developments in space science and space technology have made it possible *both* to predict with some accuracy (and sometimes decades in advance) whether an asteroid will come too close for comfort, *and* to undertake serious efforts to minimise the chance of actual collision with the earth.

In this context, the Association of Space Explorers (ASE) has taken the initiative, by means principally of establishment of a NEO Committee

(Near-Earth Objects, or NEO's, being the generic label used for both asteroids and comets) to develop a, preferably international, framework for dealing with such issues.[3]

Or, as the Open Letter of the ASE of 14 October 2005 phrased it:[4]

> Due to advances in both the discovery of these objects and in space technology (especially advanced space power and propulsion) we are aware of the unique fact that these infrequent cosmic collisions are, using advanced space technology, both predictable and preventable. This distinctive and providential characteristic of NEO impacts allows the prevention of these largest of natural disasters, if, and only if, national governments and relevant international institutions understand these inevitable events and act together to prevent their occurrence.
>
> In our opinion responsible action consists of three components; the extension of the current discovery and tracking program (Spaceguard Survey) to include the more populous smaller but still highly dangerous NEOs, the continued development of the essential space power and propulsion systems necessary for deflection of any NEO found to be on a collision course, and the cooperative development of international legal and operational policies to facilitate timely and equitable disaster prevention decision-making.

The ASE NEO initiative thus *inter alia* points at the possibility that relevant actors need to take deflective action *vis-à-vis* a NEO of which it has been calculated that the risk of collision with earth is not negligible enough to sit idle, which action might well lead to the partial or complete destruction of the NEO.[5]

Whilst this raises several international institutional and legal ramifications, this paper focuses on the definitional issue only – what, exactly, *is* a NEO in legal terms? Once legal texts would come to underpin the ASE NEO initiative, they should be clear on what triggers relevant action in order to prevent unnecessary complications – including geo-political ones.

This is also where the Russo-Chinese proposal for a treaty on 'space deweaponisation' comes in.[6] As it proposes to prohibit any use of force against "outer space objects", a literal interpretation of that proposal would or might include action against Near-Earth Objects – even if for the purpose of keeping them out of the earth's way.

2. The Outer Space Treaty

To start with, the term 'Near-Earth Object' as such does not appear in any of the five 'classical' space treaties – and neither do the terms 'asteroid' or 'comet'. This raises the question whether other terms in the treaties may still implicitly include any of these concepts. Analysis, obviously, starts with the first and most fundamental of them, the Outer Space Treaty[7].

The Outer Space Treaty repeatedly uses the words "object" and "objects", for example when it provides for relevant prohibitions regarding "objects carrying nuclear weapons or any other kinds of weapons of mass destruction".[8] Also, the Treaty makes reference to an (apparently related) concept of "space object", when providing for a duty to seriously consider requests "to observe the flight" thereof.[9]

Though these concepts have not been defined closer by the Treaty, and the lack of precision in terminology may be somewhat confusing, following the standards for treaty interpretation set by the 1969 Vienna Convention on the Law of Treaties[10] a fairly precise idea can be obtained of what they were supposed to mean.

Firstly, the intention, indeed the whole – 'teleological' – context of the Outer Space Treaty was to deal with activities of mankind and "man's entry into outer space"[11], and to provide a legal context and framework for those. The 'objects' considered at the time were probes, satellites, manned space vehicles and in a further future manned space stations or stations on celestial bodies used in the context of that entry.[12]

Secondly, the concepts were always used in close conjunction with that of "launching",[13] clearly an activity irrelevant in the context of asteroids or comets. Thus, liability for damage caused by space activities was framed as damage caused *by an object launched* into outer space.[14] Similarly, the Treaty provided for jurisdiction over an "object launched into outer space", further addressing "ownership" thereof, and even of "component parts" thereof, more or less suggesting a process of 'composition' by human hands.[15] Also the reference to the observation of "the flight of space objects" (as "launched by those States") makes clear that the Treaty addressed hardware manufactured by humans and then launched into outer space.[16] In other words, the Treaty used such concepts as 'objects' and 'space objects' to deal with *human artefacts* happening to traverse outer space.[17]

Actually, the Outer Space Treaty uses a quite different term for the category of things NEO's *would* easily fit in: that of 'celestial bodies'. That term is used usually in conjunction with the moon and adorned for the purpose with the adjective 'other', so as to make clear that, at any rate, the moon is one specimen of that category.[18]

A summary analysis of the contexts in which the term is used, makes already clear that the concept of 'celestial bodies' refers to natural objects of a tangible and visible nature, to pieces of more or less solid substance traversing outer space.

Thus, celestial bodies should be explored and used freely as long as "for the benefit and in the interests of all countries".[19] They are not to be subjected to any territorial sovereignty, whether "by means of use or occupation, or by any other means".[20] More forcefully still, the Treaty prohibits the installation of weapons of mass destruction on celestial bodies as well as the establishment of military bases, installations and fortifications thereon: such activities logically require a solid natural body of some size upon which to occur.[21] A similar conclusion arises from Article V, which refers to "activities (…) *on* celestial bodies".[22]

Though the focus was clearly on natural objects of some size, on the other hand there is no principled exclusion of any smaller-size objects. Indeed, for example the "use" of such small-size celestial bodies would still be feasible, and regulating it therefore would still make sense.[23]

As to the Outer Space Treaty, the conclusion thus arises that NEO's, asteroids and comets are included in the concept of 'celestial bodies', as natural objects of a tangible and visible nature – and as opposed to 'objects (in outer space)' or 'space objects', which are man-made objects launched into outer space.

Next, it has to be analysed in how far this conclusion is reconfirmed and/or augmented by the other space treaties, firstly the three later treaties which are more or less globally applicable and secondly the Moon Agreement which requires special treatment for a number of reasons.

3. The three later space treaties of global application

3.1. The Rescue Agreement

The first space treaty to follow upon the heels of the Outer Space Treaty, the 1968 Rescue Agreement[24], deals with the issue of 'objects' very prominently – its very title in the comprehensive version already speaks of "the Return of Objects Launched into Outer Space". Again, though it was not considered necessary to define the concept of 'object', the reference to its being 'launched into outer space' makes clear that, as in the Outer Space Treaty, this concerns man-made objects, not natural objects such as NEO's.

Interestingly, the operative text of the Rescue Agreement hardly refers to 'objects launched into outer space' beyond the title and the Preamble; rather, it refers to "spacecraft" where such objects are (to be) launched into outer space with humans on board,[25] and to "space objects" whenever unmanned objects are also concerned.[26]

With a view to the general focus of the Agreement, this also means that (space) objects are man-made and launched, as opposed to being of natural origin and just happening to pass by.

Once more, also, the inconsistent and somewhat confusing alternation of 'objects (launched into outer space)' and 'space objects' should be blamed on the absence at the time of any need to further define those terms – it was clearly not foreseen that the question addressed in the current paper might come up. There is little doubt that the two terms are, for all practical purposes, identical – as a matter of fact the very same Article 5 uses "space object" in its paragraphs 1, 2, 4 and 5; "object" without further ado in its paragraph 2; and "objects launched into outer space" in its paragraph 3.[27]

In any event, the same conclusion applies here as with the Outer Space Treaty: the Rescue Agreement deals with man-made objects whenever referring to 'objects' and 'space objects'. As a matter of fact, the Rescue Agreement does not at all deal with the other concept of 'celestial bodies'[28] – which of course logically follows from the aims and objectives of the Agreement, once the mutual exclusivity of the two concepts is accepted.

3.2. The Liability Convention

The 1972 Liability Convention[29] no more refers to 'objects' or 'objects launched into outer space', but to 'space objects' throughout its provisions – which of course deal with the liability for damage caused by such space objects; this already indicates the unlikelihood of encompassing NEO's, asteroids or comets in that context.

This time, the Convention does also define the concept, albeit not very precise – and partly in circular fashion: "The term "space object" includes component parts of a space object as well as its launch vehicle and parts thereof."[30]

Further analysis of the text of the Convention clarifies that, wherever relevant, the term 'space object' comprises manned spacecraft, which obviously makes sense only for man-made objects.[31]

Finally, in line with earlier analysis, the close conjunction of the concepts of '(space) object' and 'launching', explicitly reconfirmed here and the reference to 'component parts' leave room for no other conclusion than that also the Liability Convention is not concerned at all with the natural objects the ASE NEO initiative is concerned with – and does not refer to 'celestial bodies' anywhere, either.

3.3. The Registration Convention

Since it was drafted in close conjunction with the 1972 Liability Convention, the 1975 Registration Convention[32] follows the same approach to the subject matter.

Using a (semi-)definition identical to that of the Launching Convention, the Registration Convention provides for obligations related to the registration of 'space objects', in principle without exception.[33] Such obligations pertain to "a space object (…) launched into Earth orbit or beyond", and rest upon the "launching State" or states.[34]

'Registration', with a further goal of allowing relevant states to maintain jurisdiction over registered space objects,[35] does indeed only make sense with regard to man-made artefacts, not NEO's. This is finally confirmed by the requirements which Article IV of the Registration Convention imposes when it comes to providing the UN Secretary General with relevant registration information, which include the "name of launching State or States", an "appropriate designator of the space object or its registration number", the "date and territory or location of launch" and the "general function of the space object".[36] Needless to say, the concept of 'celestial bodies' does not figure anywhere in the Convention.

4. The Moon Agreement

The 1979 Moon Agreement[37], the last of the five 'traditional' space treaties, requires special attention from the current perspective. On the one hand, its relatively poor ratification, with to date only thirteen states being parties,[38] means that its provisions find only limited direct legal application. This the more so as in particular the only states currently capable of the monitoring and response activities required with regard to NEO's – principally the United States, to a lesser extent possibly Russia, China and the major European states jointly through the European Space Agency (ESA)[39] – are not amongst those parties.

On the other hand, the Moon Agreement focuses precisely on that category of 'objects' in outer space which is to be contrasted, being effectively mutually exclusive, with that of 'objects (launched into space)' or 'space objects' and hence, presumably, encompasses NEO's: the celestial bodies.

Unfortunately from our perspective, the term 'celestial bodies' is not defined by the Agreement, although the combination in its comprehensive title of the words "the Moon" with the words "other celestial bodies" already gives away a first clue: the moon is a prime example of a celestial body, however that might be defined.[40]

Whereas the Moon Agreement furthermore refers prominently to the 'exploration' of the moon and other celestial bodies,[41] this term has never been applied in any of the space treaties to 'objects (launched into outer space)' or 'space objects'. Apparently, space objects are not a object of exploration – which is only logical, if 'space objects' indeed would be concerned exclusively with man-made objects.

Many other provisions refer to issues, scenarios, events or situations that do not make sense *vis-à-vis* man-made space objects as well, whilst making a lot of sense *vis-à-vis* natural bodies. Information is due on activities "in the same area of (…) the Moon" where another State is active[42]; information is also due in case of a "crash landing, forced landing or other unintended landing" on a celestial body[43]; the collection and removal of "samples of its mineral and other substances" is allowed[44]; whereas the "natural resources" of celestial bodies should be considered "the common heritage of mankind"[45].

In addition, several Articles more generally address activities that simply presume an 'object' of a size man-made space objects for the time being are not likely to be, such as "the establishment of military bases, installations and fortifications"[46], "expeditions to or installations on" the moon and other celestial bodies[47]; activities "on or below (...)[the] surface"[48]; "the placement of personnel, space vehicles, equipment, facilities, stations and installations on or below the surface"[49]; a right of access to those[50]; operations of "manned and unmanned stations"[51]; or even the need to protect the environment[52].

Most interestingly, however, the Moon Agreement "does not apply to extraterrestrial materials which reach the surface of the Earth by natural means",[53] which at the very least suggests that in the absence of such a clause extraterrestrial material would, or might well, have fallen within the scope of this agreement regulating human behaviour *vis-à-vis* the moon and other celestial bodies.

In short: the Moon Agreement throughout its provisions underscores that 'celestial bodies' on the one hand, and 'objects (launched into outer space)' / 'space objects' on the other hand, are indeed mutually exclusive categories – and that NEO's unequivocally belong to the former category.

Of course, the poor rate of ratification of the Moon Agreement, in particular with the space powers that primarily matter from the NEO perspective in that they can actually respond if perceived necessary, may question the global legal validity and applicability of this conclusion. It should be noted, however, that it was above all the clauses regarding the 'common heritage of mankind' and the prospective regulation of (commercial) exploitation of the moon (and their further expected elaboration) that caused the large majority of states to refrain from signing and ratifying the Agreement.[54]

In other words, this does not take away the general acceptability for purposes of legal analysis of relevant phrases – including "natural resources"[55] or "natural resources in place"[56], logically applicable only to 'celestial bodies'.

5. Back to the Russo-Chinese proposal

The next question would then be: what does the Russo-Chinese proposal for an international treaty on 'space deweaponisation' mean for the development of a NEO- threat response initiative, taking into consideration that the former is as of yet nothing more than a *proposal* for a treaty, rather than an established part of international law on the issue?

The primary issue – of potential conceptual confusion – is defused directly by the definition of "outer space object" in the draft: it means "any *device*, *designed for functioning* in outer space, *being launched* into an orbit around any celestial body, or being in the orbit around any celestial body, or *on* any celestial body except the Earth, or leaving the orbit around any celestial body towards this celestial body, or moving from any celestial body towards another celestial body, or *placed* in outer space by any other means".[57]

Thus, whatever the other merits or problems of the proposal, the prohibition that it most fundamentally tries to establish (of using or threatening the use of force against such outer space object[58]) applies only to man-made artefacts – the same objects that the space treaties refer to as "space objects" or "objects (launched into outer space)".

That the proposal is not intended to prohibit the use of any force in outer space comprehensively is also confirmed by a fundamental, almost standard escape clause, where the use of force in outer space for purposes of self-defence is carved out from the general prohibition.[59] Of course, the reference to 'self-defence' is to the classical concept as this refers to defence against 'an armed attack' against a sovereign state – by another state or (arguably) other congregations of humans such as non-official armed bands, terrorists or guerrillas.[60] Yet, if that were to mean *a contrario* that a form of 'defence' against a 'non-human' attack would not be permissible, that could only be qualified as "a result [of interpretation] which is manifestly absurd or unreasonable"[61] – in other words: would be an incorrect interpretation.

Actually, the Russo-Chinese proposal – *if* it were to become a treaty ratified by a relevant number and category of states, and thus would become part of international (space) law – would only come to affect actions against NEO's in a different manner, namely where any response to a perceived NEO threat would involve the use of something that could be considered a 'weapon'.

The draft firstly defines a 'weapon in outer space' as "any device placed in outer space, based on any physical principle, *specially produced or converted* to eliminate, damage or disrupt normal function of objects in outer space, on the Earth or in its air, as well as to eliminate population, components of biosphere critical to human existence or inflict damage to them".[62] As this definition focuses on 'production or conversion' of a 'device', not on its (intended) usage, it automatically applies to any device produced or converted for the purpose of disruption of operation of man-made artefacts in outer space, whether it actually is *used* for that purpose or not. Unless devices to be used for response to NEO threats would by contrast be produced or converted "specially" for dealing with such threats, they could well be considered a 'weapon in outer space'.

Secondly then, the Draft prohibits not only the use of force or threat of such use involving weapons in outer space, but also "to place in orbit around the Earth any objects carrying any kind of weapons, not to install such weapons on celestial bodies, and not to station such weapons in outer space in any other manner".[63]

Thus, if the Russo-Chinese draft were to become part of international law, any state bound by it would be restrained from placing into orbit or otherwise in outer space any anti-NEO device falling within the definition of 'weapon' as given above, unless *either* the clause of Article V on self-defence should be interpreted as encompassing 'defence' against a threatening NEO *or* an amendment under Article X would be incorporated in the draft treaty so as to carve out the relevant exception here. In view of the dual-use nature of any anti-NEO device that would fall within the definition of a 'weapon in outer space' it is not feasible here to refer to the clause of the Vienna Convention on the Law of Treaties mentioned before, on 'manifest absurdity or unreasonableness': the draft treaty for obvious reasons and quite consciously does focus on the character and potential of the devices, not on the actual usage.

6. Concluding remarks

Whilst thus, from the perspective of the NEO-response discussions, any entry into force of the Russo-Chinese draft treaty without further ado would indeed generate some problems, this is *not* the consequence of any definitional confusion. In spite of several inconsistencies in the space treaties as discussed, where terms such as 'objects', 'objects in outer space' and 'space objects' are frequently used in interchangeable fashion, it is sufficiently clear that all of those concepts refer to man-made artefacts and not to NEO's of any sort. The same conclusion also applies regarding the term 'outer space object' introduced by the Russo-Chinese draft.

If anything, the extended analysis of these concepts has thus resulted in further support for the conclusion that, by contrast, NEO's form part of that larger concept of 'celestial bodies' (which even recurs in the Russo-Chinese draft[64]) and that those two concepts – of '((outer) space) objects (launched into outer space)' respectively 'celestial bodies' are mutually exclusive. Though the confusion might be to some extent unavoidable, recognising that the word 'object' is part and parcel of the concept of 'NEO', one is thus left to wonder whether it would not be a feasible alternative to speak no longer of 'defence against Near-Earth Objects' but rather of 'response to threatening near-earth celestial bodies'...

Endnotes

[1]. With an estimated explosive power of millions of nuclear bombs; see e.g. http://en.wikipedia.org/wiki/Cretaceous–Tertiary_extinction_event#cite_note-Alvarez-62; http://www.enchantedlearning.com/subjects/dinosaurs/glossary/K-T.shtml.

[2]. With an estimated explosive power of 10-15 Megatons, which is still some 1,000 times that of the bomb dropped in 1945 on Hiroshima; see e.g. http://en.wikipedia.org/wiki/Tunguska_event.

[3]. For more information on the ASE NEO initiative, see http://www.space-explorers.org/committees/NEO/neo.html.

[4]. Open Letter of the Association of Space Explorers, 19th Annual Congress, Salt Lake City, UT, 14 October 2005; see http://www.space-explorers.org/committees/NEO/docs/Open_Letter.pdf.

[5]. Note that on the other hand not every deflection of a potentially threatening NEO requires physical destruction; technologies have been developed (of which some are essentially operational) that amount to slowing down alternatively speeding up the NEO in its orbit, thus decreasing its chances of intersecting the earth's orbit at the moment the earth is actually passing such intersection.

[6]. Draft PPWT Treaty, or Draft Treaty on the Prevention of the Placement of Weapons in Outer Space, the Threat or Use of Force Against Outer Space Objects; presented 12 February 2008 to the Conference on Disarmament; e.g. http://www.reachingcriticalwill.org/legal/paros/parosindex.html (click on the treaty's comprehensive name).

[7]. Outer Space Treaty, or Treaty on Principles Governing the Activities of States in the Exploration and Use of Outer Space, including the Moon and Other Celestial Bodies, London/ Moscow/Washington, done 27 January 1967, entered into force 10 October 1967; 610 UNTS 205; TIAS 6347; 18 UST 2410; UKTS 1968 No. 10; Cmnd. 3198; ATS 1967 No. 24; 6 ILM 386 (1967).

[8]. Art. IV, Outer Space Treaty.

[9]. Art. X, Outer Space Treaty.

[10]. Cf. Art. 31, Vienna Convention on the Law of Treaties, Vienna, done 23 May 1969, entered into force 27 January 1980; 1155 UNTS 331; UKTS 1980 No. 58; Cmnd. 4818; ATS 1974 No. 2; 8 ILM 679 (1969).

[11]. Preamble, 1st para., Outer Space Treaty.

[12]. See e.g. already G. Gal, *Space Law* (1969), 207-9; also C. Zanghi, Aerospace Object, in *Outlook on Space Law over the Next 30 Years*, Eds. G. Lafferranderie & G. Crowther (1997), 115-23.

[13]. See also e.g. V. Kopal, Some Remarks on Issues Relating to Legal Definitions of "Space Object", "Space Debris" and "Astronaut", in *Proceedings of the Thirty-Seventh Colloquium on the Law of Outer Space* (1995), 100-3.

[14]. See Art. VII, Outer Space Treaty, which provides in particular: "Each State Party to the Treaty that *launches or procures the launching of an object into outer space*, including the Moon and other celestial bodies, and each State Party from whose territory or facility *an object is launched*, is internationally liable for damage to another State Party to the Treaty or to its natural or juridical persons *by such object or its component parts* on the Earth, in air space or in outer space, including the Moon and other celestial bodies." Emphasis added.

[15]. Art. VIII, Outer Space Treaty. Also Art. VII already referred to "component parts"; see *supra*, n. 14.

[16]. Art. X, Outer Space Treaty.

[17]. Cf. further in detail e.g. Zanghi.

[18]. See further already the full title of the Outer Space Treaty.

[19]. Art. I, Outer Space Treaty; further Art. III.

[20]. Art. II, Outer Space Treaty.

[21]. See Art. IV(2), Outer Space Treaty.

[22]. Emphasis added.

[23]. Cf. e.g. Artt. I, II, Outer Space Treaty.

[24]. Rescue Agreement, or Agreement on the Rescue of Astronauts, the Return of Astronauts and the Return of Objects Launched into Outer Space, London/Moscow/Washington, done 22 April 1968, entered into force 3 December 1968; 672 UNTS 119; TIAS 6599; 19 UST 7570; UKTS 1969 No. 56; Cmnd. 3786; ATS 1986 No. 8; 7 ILM 151 (1968).

[25]. Artt. 1-4, Rescue Agreement.

[26]. Art. 5, Rescue Agreement.

[27]. Furthermore, also in the Rescue Agreement, reference is made to "component parts" as suggesting some human activity of 'composition' to have been involved; Art. 5(1), (2), (3), (4) & (5), Rescue Agreement.

[28]. The Rescue Agreement does not refer to 'celestial bodies' anywhere; at best it could be presumed included in "any other place not under the jurisdiction" of any state (see e.g. Artt. 3, 4, (1), cf. also 5(3), Rescue Agreement); but obviously this concerns cases where astronauts are in

distress *on* such celestial bodies, not any obligation to 'return' one.

[29]. Liability Convention, or Convention on International Liability for Damage Caused by Space Objects, London/Moscow/Washington, done 29 March 1972, entered into force 1 September 1972; 961 UNTS 187; TIAS 7762; 24 UST 2389; UKTS 1974 No. 16; Cmnd. 5068; ATS 1975 No. 5; 10 ILM 965 (1971).

[30]. Art. I(d), Liability Convention.

[31]. Cf. e.g. Artt. III, IV(1), Liability Convention (referring to "persons (...) on board such a space object").

[32]. Registration Convention, or Convention on Registration of Objects Launched into Outer Space, New York, done 14 January 1975, entered into force 15 September 1976; 1023 UNTS 15; TIAS 8480; 28 UST 695; UKTS 1978 No. 70; Cmnd. 6256; ATS 1986 No. 5; 14 ILM 43 (1975).

[33]. See, respectively, Art. II, esp. (1) & (3), and Artt. II, III, IV, Registration Convention. That *de facto* registration is often lacking with regard to specific space objects, is more a consequence of abusing several 'escape' clauses qualifying the obligation to register, for example the one stating registration should take place "as soon as practicable"; see also Y. Lee, Registration of space objects: ESA member states' practice, 22 *Space Policy* (2006), 42 ff., esp. 44, 50.

[34]. Art. II(1), Registration Convention.

[35]. See Art. VIII, Outer Space Treaty, in conjunction with Art. II(2), Registration Convention.

[36]. See Art. IV(1), Registration Convention, resp. sub (a), (b), (c) and (e).

[37]. Moon Agreement, or Agreement Governing the Activities of States on the Moon and Other Celestial Bodies, New York, done 18 December 1979, entered into force 11 July 1984; 1363 UNTS 3; ATS 1986 No. 14; 18 ILM 1434 (1979).

[38]. Status as of 1 January 2008; see http://www.unoosa.org/oosa/en/SpaceLaw/treaties.html.

[39]. ESA was established by means of the Convention for the Establishment of a European Space Agency (hereafter ESA Convention), Paris, done 30 May 1975, entered into force 30 October 1980; 14 ILM 864 (1975); Space Law – Basic Legal Documents, C.I.1. Currently, it comprises eighteen member states: Austria, Belgium, the Czech Republic, Denmark, Finland, France, Germany, Greece, Ireland, Italy, Luxembourg, the Netherlands, Norway, Portugal Spain, Sweden, Switzerland and the United Kingdom.

[40]. Cf. further e.g. Gal, 186-8; also the discussion at C.Q. Christol, *The Modern International Law of Outer Space* (1982), 20-2.

[41]. See e.g. Preamble, Artt. 2, 4, 5, 7, 8, Moon Agreement.

[42]. Art. 5(2), Moon Agreement.

[43]. Art. 13, Moon Agreement.

[44]. Art. 6(2), Moon Agreement.

[45]. Art. 11(1), Moon Agreement.

[46]. Art. 3(4), Moon Agreement.

[47]. Art. 6(3), Moon Agreement.

[48]. Art. 8(1), Moon Agreement.

[49]. Art. 11(3), Moon Agreement. See also e.g. Art. 12(1), providing for jurisdiction and control over such items.

[50]. See Art. 15, Moon Agreement.

[51]. Art. 9, Moon Agreement; cf. further e.g. Art. 10(2).

[52]. Art. 7(1), Moon Agreement.

[53]. Art. 1(3), Moon Agreement.

[54]. See on this argument further e.g. the author's The Moon Agreement and the Prospect of the Commercial Exploitation of Lunar Resources, 32 *Annals of Air and Space Law* (2007), 98-109.

[55]. E.g. Art. 11(1), (5), (7) & (8), Moon Agreement.

[56]. Art. 11(3), Moon Agreement.

[57]. Art. I(b), Draft PPWT Treaty; emphasis added.

[58]. See Art. II, Draft PPWT Treaty; emphasis added.

[59]. See Art. V, Draft PPWT Treaty, which runs as follows: "Nothing in this Treaty can be construed as impeding the realization by the States Parties of the sovereign right for self-defense in accordance with Article 51 of the Charter of the United Nations." This clause furthermore echoes the general provision of Art. III, Outer Space Treaty, that "States Parties to the Treaty shall carry on activities in the exploration and use of outer space, including the Moon and other celestial bodies, in accordance with international law, including the Charter of the United Nations, in the interest of maintaining international peace and security and promoting international cooperation and understanding."

[60]. See Art. 51, Charter of the United Nations, San Francisco, done 26 June 1945, entered into force 24 October 1945; USTS 993; 24 UST 2225; 59 Stat. 1031; 145 UKTS 805; UKTS 1946 No. 67; Cmd. 6666 & 6711; CTS 1945 No. 7; ATS 1945 No. 1. This Article specifically refers to "an armed attack (...) against a Member of the United Nations", but without going any deeper into the discussion regarding the extent to which a customary right of self-defence exists broader than and beyond the UN Charter, for purposes of the present analysis it can be safely equated to 'an armed attack against a state'.

[61]. Art. 32(b), Vienna Convention on the Law of Treaties, as providing for some "Supplementary means of interpretation" of treaties.

[62]. Art. I(c), Draft PPWT Treaty; emphasis added. A weapon "will be considered as 'placed' in outer space if it orbits the Earth at least once, or follows a section of such an orbit before leaving this orbit, or is stationed on a permanent basis somewhere in outer space"; Art. I(d).

[63]. Art. II, Draft PPWT Treaty.

[64]. See Artt. I(b), II, Draft PPWT Treaty.

PELLA VILYA: NEAR EARTH OBJECTS – PLANETARY DEFENCE THROUGH THE REGULATION OF RESOURCE UTILISATION

Dr. Gérardine Meishan Goh*
Deutsches Zentrum für Luft- und Raumfahrt e.V. (DLR), Bonn, Germany /
Institute of Air and Space Law, University of Cologne, Germany
Gerardine.Goh@dlr.de

ABSTRACT

Reactions to Near Earth Objects (NEOs) in the past decade have run the gamut from expectations of Armageddon-type scenarios to Eureka moments of revolutionary scientific ideas. Concerns over the potentially devastating effects of an unmitigated collision jostle with forecasts of untold economic returns from the utilisation of NEO resources. Drawing from recent analogies and examples from the field of international environmental law, this paper proposes the development of a legal framework for the regulation of NEO resource utilisation. The proposed legal framework also includes a mechanism to ensure the political will and economic investment necessary for technological advances in planetary defence. By twinning the threats and opportunities presented by NEOs, this paper also analyses the position of theme-specific space law development in the overall legal framework of space exploration and traffic management.

I. INTRODUCTION

Immense fascination has been generated by Near Earth Objects (NEOs) with public, commercial and scientific interest vacillating between the two extremes of Armageddon- and El Dorado-type scenarios. A catastrophic collision between the Earth and an NEO is a clear "low probability, high consequence event"[1] – while there has been no record of fatalities caused by such a collision, it is now undisputed that various NEO collisions in the past has led to mass localised and global destruction.[2] It must be noted that NEO collisions with the Earth is not a thing of the distant past. There have been several instances of large NEOs colliding with the Earth or exploding in the Earth's atmosphere in the last century, with one of the most significant collisions by a 60-meter asteroid in Tunguska, Siberia on 30 June 1908.[3] As of 31 August 2008, 208 NEO impact risks are listed by the website of the Near Earth Object Program of the United States' National Aeronautics and Space Administration (NASA).[4] Of these, only one asteroid 2007 VK_{184} is listed with Torino Scale 0,[5] with the other 207 listed as being equal or smaller than 50 meters in diameter, and therefore of little or no threat to the general public.[6] However, without prompt preventive action, a large NEO is likely to collide with the Earth at some point, with catastrophic consequences.[7]

On the other hand, it has been demonstrated that NEOs are comprised of minerals that are extremely valuable for various applications.[8] With burgeoning costs of Earth-based resource acquisition and rapidly advancing space-oriented technology, NEO resource extraction and utilisation is fast becoming a reality. Aside from supporting space exploration, the large-scale acquisition and utilisation of resources from NEOs for Earth-based activities may soon become economically and technically feasible.

Presently however, there is a lacuna in the international legal framework with regard to NEOs. Aside from a blanket prohibition on appropriation of outer space,[9] including celestial bodies, there is no mention in any legal texts of NEOs. This is a dangerous void, as without a clear and consistent regulatory framework, there cannot be an effective development of technologies, scientific exploration, and economic policies with regard to NEOs. Coupled with the potential hazards of NEOs and the fact that capital investment into both the deflection and utilisation of NEOs will only be viable with a clear regulatory framework, the need for an enunciation of the international legal standards in this field is increasingly pressing.

A possible legal framework is proposed in this paper that twins the economic utilisation of NEO-extracted resources with the scientific research and development necessary for planetary defence against a possible collision. It is mooted that public funding into NEO deflection and planetary defence is unviable for a long-term, sustained effort. On the other hand, NEO resource utilisation for economic benefit is presently prohibited by law. This paper makes the case for a comprehensive, doubly-dependent regulatory system, which combines the motivation from economic exploitation of NEO-extracted resources with the public good provided by deflection technologies. Economic policy instruments are proposed, based on an analogy from international environmental law, for a harmonised, integrated approach to NEO utilisation and planetary defence. The historical context relating to the international legal framework and the scientific and economic background of NEOs is first discussed.

II. CONTEXT – NEAR EARTH OBJECTS: THE GOOD, THE BAD AND THE UGLY

Issues raised by NEOs typify the classical "the good, the bad and the ugly" debates. The "good" comprise the possibility of economically sound resource extraction and utilisation – providing a potential infinite source of presently-scarce resources for both Earth-based applications and space-based exploration. The "bad" relates to the threat caused by an uncontrolled NEO collision with the Earth, and the catastrophic consequences of impact. The "ugly" fact is that the international legal regime is not ready to deal either with the "good" or the "bad" scenarios. This section looks in more detail at these three contextual issues underlying present and future approaches to NEOs.

1. The Good: NEO Resource Extraction and Utilisation

The Earth's resource base is complex, differentiated, and limited. Resource extraction from the Earth has taken place from the cradle of civilisation.[10] Although supplies of fossil fuels and mineral resources are not likely to be fully depleted for decades, an important junction has been reached in the history of Humanity where the limits of the Earth's resources can be identified.[11] As the most easily-accessible resources are being depleted, lower-grade substances obtained through more costly, damaging and dangerous methods are being used. In the light of these developments, the extraction and utilisation of space-based resources, especially those of near-Earth space, is becoming more viable and attractive.[12]

In particular, NEOs may be the most viable source of valuable minerals and compounds.[13] This is mainly due to two reasons: the mineral composition of NEOs, and their low surface gravity, which decreases the mission energy requirements for orbital manoeuvre.[14] Spectroscopic studies and ground truthing have shown that NEOs comprise stony silicates with semiconductor- and platinum-group metals,[15] bituminous and carbonaceous compounds,[16] ice and clay,[17] non-metals,[18] and large amounts of nickel-iron alloys.[19] Of particular interest are the platinum-group metals, consisting of platinum, palladium, rhodium, ruthenium, iridium and osmium; as well as the occurrence of gold.[20] These metal ores are of especial economic and manufacturing value. Further, some of the volatiles believed to be in NEOs include water ice, carbon monoxide, carbon dioxide, and a conglomerate of carbon, hydrogen, oxygen and nitrogen (CHON), which may be harnessed to produce volatile propellants.[21]

Once engineering-wise possible, NEO mining is expected to transform the economic and technical dynamics of resource production. Robotic NEO mining has been argued to have several advantages over terrestrial mining. These include a higher success rate in prospecting, higher grade ore for processing and less environmental damage to the Earth.[22] This ability to cost-effectively meet existing market demands will allow the minimisation of capital injections as well as the time to return on investment.[23]

2. The Bad: The NEO Threat

NEOs are estimated to impact the Earth approximately twice every million years, causing global ecological disaster.[24] International efforts to create a comprehensive catalogue of NEOs has been undertaken by several governmental, scientific, non-governmental and professional associations, including Spacewatch,[25] NASA's Near Earth Object Program,[26] the Lincoln Laboratory's Near-Earth Asteroid Research,[27] the Japanese Hayabusa mission[28] and the European Space Agency's proposed Don Quijote mission.[29] As a result, the number of known NEOs and their orbits has exponentially increased. Figure 1 below shows the discoveries of Near-Earth Asteroids at half-year intervals in the period 1995 – 2008.[30] At present, more than 400 Near-Earth Asteroids are catalogued.

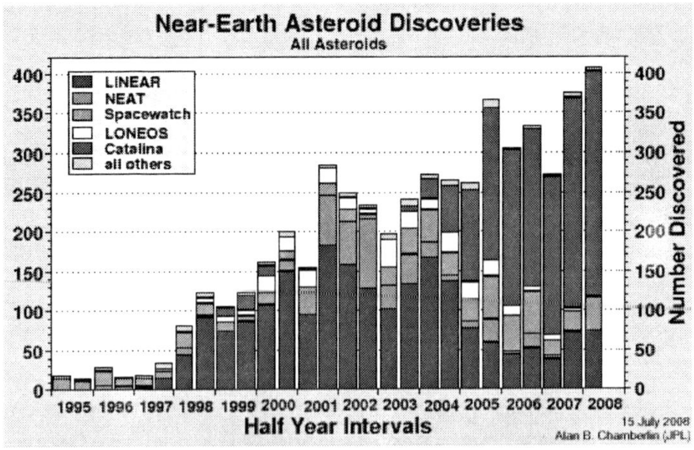

Figure 1: Near-Earth Asteroid Discoveries (1995 – 2008)

Immediate consequences[31] of a major NEO impact include the radiation from the explosion, firestorms, acid rain, pyrotoxin production, volcanism and seismic activity, significant environmental disaster (including tsunamis, the possible loss of the ozone layer and a nuclear winter), and massive fatalities, casualties and property damage.[32]

In the case of a potential major NEO impact scenario, scientific studies have found that a deflection of the NEO resulting in a change in orbit is the best method by which to avoid impact.[33] Deflection would ensure that the NEO does not continue on a trajectory directly for the Earth, but would also avoid splintering the NEO body, thereby creating the scenario that instead of one big NEO impact, the Earth would face a series of smaller NEOs, all of which would in any case have great consequences. It has also been pointed out that a nuclear powered deflection is the only means at present that is of sufficient energy to adequately deflect an approaching NEO.[34] It must be noted however, that the deflection of a threatening object is still in a "conceptual phase".[35] Serious technical issues are still under consideration at the international level.

3. The Ugly: International Space Law is Not Ready

Given the high-risk potential consequences of an NEO impact, as well as the immense economic and scientific benefits of NEO resource extraction, it may come as a surprise that the international legal community has not yet jumped on the bandwagon. There is no legal framework that either ensures the development of technologies to protect the Earth in case of an impact, or enables the effective extraction and utilisation of resources from NEOs. In fact, the present international legal framework may actually hinder both.

Article II of the Outer Space Treaty provides that

> Outer space, including the moon and other celestial bodies, is not subject to national appropriation by claim of sovereignty, by means of use or occupation, or by any other means.[36]

Ostensibly this was to ensure the preservation of outer space and its resources for the benefit of all Mankind as elucidated in Article I. Read together with Article VI of the Outer Space Treaty, which stipulates the obligation of States to bear international responsibility for national activities in outer space, whether such activities are carried on by governmental agencies or non-governmental entities, it is clear that the international legal framework does not allow the commercial exploitation of resources from celestial bodies. In light of the fact that Article VI also requires States to authorise and continually supervise national activities in outer space, this meant that States were under the obligation to ensure, through a licensing system, that commercial entities registered in their territories did not violate Article II.[37] Without being able to "appropriate", by "use or occupation", outer space and its celestial bodies, there is arguably no real motivation either for commercial entities to invest in NEO mining. There is no further provision in any of the international legal texts relating to space activities for the extraction and use of resources from outer space.

With regard to the deflection of NEOs, the international legal regime appears to run counter to convention scientific thought. As discussed above, it has been agreed that a nuclear-powered deflection would be one of the only feasible methods by which to deflect a threatening NEO. However, given the prohibition on the placement of nuclear weapons in orbit according to Article IV of the Outer Space Treaty, as well as the general prohibition on the use of nuclear weapons as envisaged by the Comprehensive Test Ban Treaty,[38] it seems that the international legal community is set against the use of nuclear weapons in outer space. It can of course be mooted that, in the case of an impending impact, the international legal community will not stand in the way of the use of a nuclear warhead for the non-aggressive deflection of the NEO. However, the prohibition on the placement of nuclear weapons in orbit (as well as on the Moon)[39] can be seen as a *prima facie* opposition to the use of nuclear weapons in outer space.

The foregoing discussion evinces a clear lack of specific legal provision for the equitable and effective commercial mining of NEO resources, as well as an equivocal stance on the use of nuclear weapons in outer space. It is submitted that this status of the law cannot be left as it is. Firstly, with progressive developments in space technology and the crippling prices of Earth-based resources, it is likely that the commercial use of space resources, such as those on NEOs, will become quickly a reality. In this case it would be desirable to have a clear and consistent legal framework in place so as to safeguard the principles of the Outer Space Treaty, while remaining relevant to the ambient developments in the field. Secondly, the legal framework has often served as a possible means of motivation by which to achieve certain aims. In this case, the international legal framework would be a mechanism by which to stimulate research and development into the requisite technology to deflect a threatening NEO. It cannot be overstated that time of is the essence. Given the slowdown in available public funding for such research, and the fact that an NEO collision, while not soon, is in any case likely, it is submitted that commercial and private funding would be the most feasible means of sustaining a long-term perspective into such research. Further, it is submitted that the international legal framework should evolve a mechanism by which to stimulate such commercial and private funding through the use of legal and economic policy instruments.

Before elaborating on the proposed framework, the next section will take a look at recent developments in international environmental law. In that field, legal and economic policy instruments have been employed to quite some success in motivating research and development, as well as behaviour and action, in the furtherance of environmental protection. It is submitted that the example of such legal and economic policy instruments can be adapted to deal with the issue of NEOs in international space law.

III. THE INTERNATIONAL ENVIRONMENTAL LAW ANALOGY

Environmental protection is an overarching goal that has direct implications on a plethora of stakeholders on various levels. States, international organisations, non-governmental organisations, private corporations and individuals all count amongst the potential beneficiaries of positive environmental policy; as well as the potential victims of any environmental disaster. The global reach of the environment, however, has not in itself correspondingly produced global responses in terms of research, development and pro-active initiatives for its protection.

1. Motivation from Economic Policy Instruments

The lack of tangible results in achieving environmentally cleaner technology and products through environmental regulations led in recent years to a re-thinking of motivational instruments. Amongst the carrots suggested is the use of economic policy instruments to protect the environment.[40] With the belief that traditional legal and policy mechanisms have failed to provide adequate incentives to ensure the achievement of environmental objectives, the use of economic instruments is based on the logic that market incentives can guide State, business and organisational behaviour.[41]

A growing number of States have supported the use of economic policy instruments at the international level for global environmental protection.[42] However, the use of economic policy instruments at the international level complementary to legal regulation is a novel development.[43] However, once the idea was mooted at the international level, it quickly found its way into various significant environmental fora, including the 1992 Climate Change Convention,[44] the 1992 Biodiversity Convention,[45] and the 1992 Rio Declaration.[46]

Various forms of economic instruments are recommended for promoting environmental protection, including charges and taxes, marketable permits, deposit-refund systems, financial assistance, enforcement incentives, administrative charges, liability and compensation for damage, trade measures and consumer information incentives, non-compliance fees and performance bonds, and subsidies. For the purposes of this paper, emphasis will be put on joint implementation and tradable permits, as well as investment incentives, in the framework of integrated pollution control.

2. Joint Implementation and Tradable Permits

The original idea of using tradable permits in the implementation of environmental initiatives was drawn from the 1990 amendment to the United States of America's Clean Air Act.[47] The notion of tradable permits involved granting to certain regions or utilities a number of pollution rights. If these regions or utilities used less than the allocated amount, the excess rights may be sold to another region or utility. The first international framework employing this format was adopted by the European Commission (EC) in 2002, granting parties the right to jointly implement programmes without specifying criteria or specific pollution limits.[48] As such, there was no intention by the EC drafters for inter-State trading in the allocated permits. Tradable permits were first enunciated in Article 2(7) of the 1987 Montreal Protocol,[49] by which member States agreed to "jointly fulfil their obligations respecting consumption" of certain ozone-depleting substances so long as their total consumption level did not exceed the levels stipulated by the Protocol.

A break-through step was taken by the 1992 Climate Change Convention. In that framework Convention, developed member States, together with other Annex 1 parties, were allowed to implement policies required under Article 4(2)(a) and 4(2)(b) "jointly with other parties", subject to decisions made by the conference of parties "regarding criteria for joint implementation".[50] Details with regard to the specific implementation of this framework Convention were answered by the 1997 Kyoto Protocol,[51] which enunciates clear provisions in respect of the joint implementation of the obligations in the Climate Change Convention.[52] A system of tradable permits, called "emission reduction units", emerged from the Kyoto Protocol, as elaborated by the States Parties at their first meeting.[53]

The next positive step was undertaken by the EC Council in March 2003. The Council issued a Directive establishing the scheme for greenhouse gas emission allowance trading, aimed at the reduction of greenhouse gas emissions in a "cost-effective and economically efficient manner".[54] The EC Directive also indicated that such reduction projects be linked with the mechanisms envisaged by the Kyoto Protocol, including joint implementation and the Clean Development Mechanism. The Directive thus established an international trading scheme premised on allocated allowances to certain installation operators for the reduction of greenhouse gases as a result of their activities.[55] Further, harmonised joint implementation of the trading scheme is ensured by the EC Directive's requirement that all member States ensure as of 01 January 2005 that these activities and the trading scheme provide be authorised by a competent authority.[56] What is significant is that each member State is required to adopt a national allocation plan consistent with their obligations under the Kyoto Protocol and EC law, with allowances for the period 01 January 2005 – 01 January 2008 to be allocated free of charge. Moreover, 90% of allocations made for the period 01 January 2008 – 01 January 2013 are also to be allocated free of charge. These allowances will be transferable between persons within the EU and between persons within the EU and third States as listed in Annex B to the Kyoto Protocol. The EC has also adopted guidelines on monitoring and reporting, requiring member States to ensure the due monitoring of emissions and the verification of reports submitted by operators.[57] Each member State is also obliged to establish a registry for the accurate holding, transfer, accounting and cancellation of allowances. These registries are to work in close cooperation with the EC's central independent transaction log to allow for automated checks.[58]

3. Investment Incentives

Mechanisms exploring the use of investment incentives for environmental protection have recently been adopted. In particular, such investment incentives are targeting investment into clean technologies for developing countries and countries with economies in transition. An excellent example of a framework put up for investment incentives is the Clean Development Mechanism established by the Kyoto Protocol. The Clean Development Mechanism ensures the provision of credits to States whose private companies invest in greenhouse gas reduction activities in other States.[59] There has also been a recent trend at the international level to provide financial and technological assistance to developing countries that invest in clean technologies pursuant to international ozone-related and environmental agreements.[60]

4. Integrated Pollution Control

There is a growing awareness of the cross-cutting nature of pollution and environmental degradation. The traditional method of addressing specific activities or substances belies the fact that each environmental medium has effects on a range of media as they accumulate in the environment. Therefore, dealing with environmental protection on a medium-by-medium basis is certainly not the most efficient or effective method.

In 1991, the Organisation for Economic Co-operation and Development (OECD) Council issued a Recommendation for "integrated pollution prevention (or control)".[61] The harmonised, integrated system is aimed at a more holistic perspective on environmental protection, including a "cradle-to-grave" approach to entire "streams" of environmental degradation.[62] This initiative has led to the adoption of the first international rules on integrated pollution control by the European Union in 1996. The OECD Recommendation acknowledged that certain policies were significant in ensuring an integrated approach, including a change in institutional arrangements, management techniques and technical know-how. Co-ordination mechanisms and cooperative arrangements within different levels of governments amongst various States would be vital to the integrated approach. Such mechanisms would take the form of integrated inspection and enforcement mechanisms, economic policy instruments, life cycle analysis and the more effective monitoring of inventories and registries.

It must be noted that to date, there is no international legal framework that establishes the principles of integrated pollution control in environmental law. Despite the efforts of the International Union for Conservation of Nature's (IUCN) Commission on Environmental Law, it is not likely that such a legal framework will be adopted in the near future. The lack of such a central framework document has resulted in a fragmentary corpus of legal regulations, together with all the well-known problems and drawbacks of a piecemeal development of the law. In this regard, it has been suggested that three issues must be addressed urgently: the establishment of improved mechanisms for the identification of critical priorities for law-making; the guarantee that all stakeholders will be able to effectively participate in the law-making process at the international and transnational level; and the improvement of coordination between international organisations, States, non-governmental organisations and other relevant entities. Any legal rules established must also be accompanied by a range of effective economic policy instruments to motivate compliance. The framework initiated by the 1992 Climate Change Convention, and detailed by the 1997 Kyoto Protocol, as well as subsequent developments thereto, will be significant in the evaluation of the efficacy of economic policy instruments as compliance motivators.

IV. WHERE OPPORTUNITY FITS THE CHALLENGE: PROPOSED FRAMEWORK FOR NEOs

A similar framework of economic policy instruments encased in a legal treaty system is proposed with regard to NEOs. This potentially thorny and complex proposal involves the creation of an international agreement, as a Protocol to the Outer Space Treaty, which allocates tradable NEO mining rights to public or private entities in return for scientific and technical developments in NEO threat deflection. This builds upon the experience gained from embedding economic policy instruments as compliance motivators in the international environmental law framework. By twinning the urgent need for technological advancement in NEO deflection together with the economic incentive of commercial mining, it is submitted that adequate motivation will be provided to ensure private long-term funding of research into NEOs.

1. Proposal: International Legal Framework for NEOs

To adequately address issues relating to NEOs, it is proposed that an international framework treaty be established under the auspices of the United Nations. This proposed United Nations Framework Convention on Near-Earth Objects (FCNEO) should enunciate, in line with the 1967 Outer Space Treaty, the international legal standards applicable to issues relating to NEOs. In compliance with Article VI Outer Space Treaty, the FCNEO should be open to States Parties, which will also represent the interests of their non-governmental and commercial entities. Further, a harmonised integrated approach to NEOs should be coordinated by a United Nations NEO Taskforce, which reports directly to the UN Committee on the Peaceful Uses of Outer

Space (COPUOS).

In essence, the proposed FCNEO will allocate a currency of rights to mine and commercially exploit resources from NEOs in exchange for the publication of certified advancements in NEO threat deflection and planetary defence. These mining and exploitation rights can be allocated to both public and private entities, and are tradable on a newly-created exchange in return for technology transfer, or economic benefits. These rights will function as exceptions to the blanket prohibition on the commercial appropriation and exploitation of natural resources from outer space. The tradability of these rights means that public entities that invest capital into NEO deflection research may also have rights accredited to them, which can be traded with private companies intending to mine NEOs for more research capital injections. Upon the accumulation of a threshold number of these rights to mine NEOs, the entity in question will be allowed to extract and commercially utilise NEO resources, with the caveat that a certain percentage of profits must be put into a UN trust. The framework for this UN trust will be based upon that used for the deep seabed mining regime.[63] Each member State to the FCNEO will be required to establish a registry for the accurate holding, transfer, accounting and cancellation of these rights. These State-maintained registries are to work in close cooperation with the UN's independent registry to allow for automated checks.

The proposed FCNEO structure should have criteria for the joint implementation of initiatives relating to NEOs. A Commission of nominated and highly qualified experts, from the scientific, technical, economic, engineering, legal and policy fields should be set up to provide guidance as to the categories, amount and types of tradable rights to be allocated. This Commission should make annual reports to the COPUOS sessions as to the progressive developments in the field, and should supervise the allocation of such mining rights.

The proposed FCNEO mechanism will broadly work as follows:

a) Negotiation and establishment by the expert Commission of the system of tradable NEO mining rights in exchange for publication of NEO deflection technology, including weighting currency (i.e. what each right credit entails, and what the credit is worth) and exchange mechanisms;
b) Each member State to establish registry of allocated NEO mining rights;
c) Entities with possible technologies for NEO threat deflection to submit these to the expert Commission for verification and weighting;
    i. If the technology is not accepted for credit allocation, the intellectual property remains with the submitting entity. No publication will be made.
    ii. If the technology is accepted for credit allocation, the Commission will indicate the amount of credits to be allocated. If the entity does not agree with the amount, no publication, and no allocation, will be made. In this situation, the intellectual property remains with the submitting entity. If the entity agrees to the allocated credits, the technology must be published in the public domain and the intellectual property over the technology in question is released into the public domain. The Commission allocates the indicated number of mining credits to the entity.
    iii. Entities must register the allocated mining credits with their State.
d) Where there is a dispute or disagreement between the Commission and the entity in question, it should first be settled by negotiation or mediation. If a resolution cannot be brought about by such means, the dispute should be submitted to a three-person arbitral tribunal in accordance with the procedures of the Permanent Court of Arbitration.
e) Entities may trade and transfer the allocated mining credits. Each trade or transfer must be approved by the UN FCNEO Commission.
f) Entities with a threshold number of mining credits, as stipulated by the FCNEO, may undertake mining on NEOs for commercial exploitation of mined resources. A certain X% of profits, as stipulated by the FCNEO, must be deposited in the UN Trust Fund.

2. <u>The Legality of the Proposed Economic Policy Incentive Scheme</u>

Prima facie the legality of the above proposal may be challenged by the argument that it apparently runs counter to the principles as established in the 1967 Outer Space Treaty, especially that in Article II. However, it is submitted that the proposed framework is legal, for three reasons:

a) Article II Outer Space Treaty must be read and applied in light of the principle of the common benefit of all Mankind.
   Article II Outer Space Treaty is to be read subject to Article I, and in light of the Preamble, which stipulates that the exploration and use of outer space is to be for the benefit of all Mankind. Clearly, the progressive development of technology for the successful deflection of threatening NEOs is in the interest of the global community. As such, it is submitted that where the allocation of such mining rights brings benefits for the entire global community, Article II Outer Space Treaty is not violated. Further, the stipulation that a certain percentage of profits from NEO mining should be deposited into a trust similar to that for deep seabed mining also ensures that NEO mining is carried out for the benefit of all Humanity.

b) Progressive developments in the field of international space law must learn, through necessity, from analogous developments in other fields of international law.
   Events in international environmental law have shown that economic policy instruments, when embedded into an international legal framework, may provide the necessary impetus and motivation to achieve goals that are beneficial to the global community. In the case of international environmental law, economic policy instruments in the form of tradable emission credits and investment incentives have kickstarted the momentum in favour of environmental protection and action against climate change. It is submitted that the experience from international environmental law may be commuted to space activities, and that the economic policy instruments through tradable credits for NEO mining will also attract private investment into research and development for technologies that can deflect threatening NEOs. Given that the technology to effectively mine NEOs is not so far from technology to deflect NEOs, it is submitted that this economic policy instrument will ensure continued and sustained funding from the private sector. Due to the impossibility of long term public financing for research into NEO deflection, this is a necessary step.

c) The potentially infinite amounts of resources available in outer space will offset the amounts allocated in the proposed framework; furthermore, the technologies gained for the deflection of threatening NEOs outweigh the one-off allocation of mining rights.
   Many highly qualified technicians and scientists have argued that the amount of resources available in outer space is potentially infinite.[64] As such, it is submitted that the good gained from allocating mining credits, in the form of publicly available and published technologies for the deflection of threatening NEOs and planetary defence, vastly outweighs the risks inherent in this proposed framework.

3. <u>Perspectives: Steps towards Establishing a Workable Framework</u>

This proposed framework is admittedly rather bold, and flies in the face of conventional thinking in international space law. However, it is submitted that this proposed framework is workable, and indeed contributes to ensuring the relevance of international space law through progressive development in step with the needs and evolution in the field. To establish this framework, the following steps are further proposed:

a) The issue of NEO mining, and the deflection of threatening NEOs, should be undertaken seriously both at the international level (within the UN COPUOS), as well as amongst the experts in the field, for example, within the Committee on Space Research (COSPAR).
b) A Working Group comprising of technical and legal experts should be established to negotiate and draft the proposed Framework Convention, including detailed specifications on the type of deflection technology acceptable for credit allocation, the type and weight of mining credits to be allocated, the trading system, the dispute settlement mechanism, the rights and obligations bestowed by the allocation of such credits, and the functioning of the proposed UN Trust Fund. A Draft Framework Convention should be produced by the Working Group at the end of no more than three years from its

establishment.

c) The Draft Framework Convention should be passed through the UN COPUOS, and then through the UN General Assembly, before being finalised. The finalised Framework Convention should then be opened for signature by interested States. It is important to note that States that are not members of the Convention must abide by the customary international law that outer space, including celestial bodies, cannot be appropriated. Therefore, only the exception as stipulated in this Convention allows commercial mining and resource exploitation of NEOs. This should provide the necessary incentive for States to engage in the proposed framework.

V. CONCLUSION: THEME-SPECIFIC SPACE LAW DEVELOPMENT

It is a rare coincidence that one issue should provide both the problem and the potential solution – and especially so when the solution comes with an incentive that costs nothing and is potentially infinite. Thus, it is the proposal of this paper that the international legal community should seize the opportunity presented by NEOs and enunciated a legal framework that is both creative and necessary. Planetary defence through the regulation of NEO resource utilisation is an elegant solution that should be considered by the international legal community.

The juxtaposition of the threats and opportunities presented by the NEO issue brings to light an interesting point in theme-specific space law development. Just like the horizontal, cross-cutting impact of environmental degradation and protection, it is submitted that such theme-specific space law development adds considerable substance to the overall space traffic management framework. In the case of NEOs, a comprehensive regulatory framework that not only deals with the hazards posed, but also addresses the contemporary issues raised by commercialisation, ensures a thorough handling of all pressing issues related to NEOs. It is submitted that such detailed, compartmentalised handling of a specific theme is a good method by which to ensure that the final, comprehensive legal framework for space traffic management deals with these issues in depth.

Ultimately, such creative legal development methods ensure that when dealing with activities beyond the Earth's airspace, the law can also reach above and beyond the stifling atmosphere of mundane tradition and become truly *pella vilya* – *"beyond air"*.[65]

---

[*] Ph.D. (Leiden), LL.M. (London), LL.B. (Hons.) (Singapore). Project Manager and Legal Advisor, Deutsches Zentrum für Luft- und Raumfahrt e.V. (German Aerospace Center) (DLR); Lecturer, Institute of Air and Space Law, University of Cologne. Member, IISL, ECSL, ILA; Fellow, IAASS; Associate, COSPAR. The opinions expressed in this paper are entirely those of the author and do not in any way engage the organisations with which she is affiliated.

[1] Gerrard, M.B. and Barber, A.W., "Asteroids and Comets: U.S. and International Law and the Lowest-Probability, Highest Consequence Risk", (1997) 6 New York University Environmental Law Journal 4.

[2] The impact of a 10-kilometer-wide NEO at Chicxulub, Mexico, for example, left a 180-kilometer diameter crater. It has been argued that the impact of this NEO was one of the factors that caused the mass extinction at the end of the Cretaceous Period (approximately 65 million years ago). See Bottke, W.F., Vokrouhlicky, D. and Nesvorny, D., "An asteroid breakup 160 Myr ago as the probable source of the K/T impactor", (September 2007) 449 Nature 23; for an opposing view see Keller, G. et al, "Chicxulub impact predates K-T boundary: New evidence from Brazos, Texas", (2007) 255 Earth and Planetary Science Letters 1.

[3] Other significant collisions include the airborne explosion at Curuca River, Brazil (1930), at Sikhote-Alin, Siberia (1947), over Revelstoke, Canada (1965), over Ontario, Canada (1966), over Alaska, United States of America (1969) and in Honduras (1996). The 1908 Tunguska event incinerated hundreds of acres of forest in Siberia, and had it impacted on a populated area (over St. Petersburg, for example, had it been approximately four hours later), would have caused catastrophic fatalities and damage. The 100 million kilogramme asteroid entered the Earth's atmosphere at about 54,000 kilometres per hour, and exploded at a height of approximately 8.5 kilometres above the ground. The explosion released energy equivalent to 185 times that released by the atomic bomb at Hiroshima. A century on in 2008, the requisite technology required to deflect an NEO such as the Tunguska asteroid is still lacking. See Space Daily, "Tunguska Event Still a Mystery 100 Years On", (30 June 2008), online at http://www.spacedaily.com/reports/Tunguska_Event_Still_A_Mystery_100_Years_OnTunguska_Event_Still_A_Mystery_100_Years_On_999.html/, accessed 31 August 2008.

[4] Online at http://neo.jpl.nasa.gov/, accessed 31 August 2008.

[5] The Torino Scale provides an assessment of asteroid and comet impact hazard predictions based on a 0 – 10 point scale. See Morrison D. et al, "Impacts and the Public: Communicating the Nature of the Impact Hazard" in Belton, M.J.S. et al (eds.), *Mitigation of Hazardous Comets and Asteroids*, (2004). NASA's Near Earth Object Program provides a graphical illustration of the Torino Scale online at http://neo.jpl.nasa.gov/images/torino_scale.jpg, accessed 31 August 2008.

[6] NASA Near Earth Object Program, "Sentry Risk Table", online at http://neo.jpi.nasa.gov/risk/, accessed 31 August 2008.
[7] Tate, J.R., "Near Earth Objects – a threat and an opportunity", (2003) 38(3) Physics Education 218.
[8] See on the topic of the utilisation of NEO and near space resources, Wingo, D., *Moonrush: Improving Life on Earth with the Moon's Resource*, (2004); Binzel, R.P. et al, "Physical Properties of Near-Earth Objects", in Bottke, W.F. et al (eds.), *Asteroids III* (2002) 264; Sonter, M.J., "Near Earth Objects as Resources for Space Industrialization", (2001) 1(1) Solar System Development Journal 1; Lewis, J.S., Matthews, M.S. and Guerrieri, M., "Resources of Near-Earth Space", (1993); Hartmann, W.K., "The Resource Base in Our Solar System", in Finney, B.R. and Jones, E.M., *Interstellar Migration and the Human Experience*, (1985) 26.
[9] Article II, Treaty on Principles Governing the Activities of States in the Exploration and Use of Outer Space Including the Moon and Other Celestial Bodies, (1967) 610 UNTS 205, hereinafter "Outer Space Treaty". As of 01 January 2008, 98 States have ratified the Treaty, and an additional 27 States have signed it.
[10] See for example Lynch, M., *Mining in World History (Globalities)*, (2004) at p. 3.
[11] See on the topic, Guthery, F.S., *A Primer on Natural Resource Science*, (2008); Pirages D. and Cousins K., *From Resource Scarcity to Ecological Security: Exploring New Limits to Growth*, (2005); Rogers, J.J.W. and Feiss, P.G., *People and the Earth: Basic Issues in the Sustainability of Resources and Environment* (1998); and Constanza R. et al, *A New Century for Natural Resources Management* (1995).
[12] See generally Schrunk D. et al, *The Moon: Resources, Future Development and Settlement* (2007); and Lewis, J.S., *Mining the Sky: Untold Riches from the Asteroids, Comets and Planets* (1997).
[13] An excellent overview of contemporary issues relating to NEOs can be found in Milani, A., Valsecchi, G.B. and Vokrouhlicky, D., *Near Earth Objects, our Celestial Neighbours (IAU S236): Opportunity and Risk (Proceedings of the International Astronomical Union Symposia and Colloquia)*, (2007).
[14] This means, in practical terms that less scalar effort is required to carry the payload from one orbit into another – that it would be easier to lift the load off the surface of the NEO, and also easier to transfer it into Earth orbit.
[15] Kuck, D.L., "Near-Earth Extraterrestrial Resources", (1979) Space Manufacturing Facilities 3. See also Gertsch R., Gertsch L. and Remo J.L, (eds.), "Mining near-Earth resources in near-Earth objects, the United Nations International Conference", (1997) 822 Annals of the New York Academy of Science 468.
[16] Zuppero, A.C., "Discovery of Abundant Accessible Hydrocarbons Nearly Everywhere in the Solar System", (1996) Proceedings of the 5th International Conference on Engineering, Construction, and Operations in Space, American Society of Civil Engineers 96.
[17] Lewis, J.S., "Resources of the Asteroids", (1997) 50 Journal of the British Interplanetary Society 51.
[18] These include olivine, pyroxene, plagioclase feldspar, arsenic, selenium, germanium, phosphorous, carbon and sulphur. See O'Leary, B. et al, *Retrieval of Asteroidal Materials, Space Resources and Settlements*, (1979) NASA SP-428 pp. 142 – 155.
[19] Ostro S.D. et al, "Asteroid 1986 DA: Radar evidence for a metallic composition", (1991) 252 Science 1399.
[20] Horan, M.F., Walker, R.J. and Morgan, J.W., "High Precision Measurements of Pt and Os in Chrondrites", (2001) Lunar and Planetary Science XXX.
[21] Smitherman D. and Fikes, J., "Space Resource Requirements for Future Propellant Depots", (2001) Space Resource Utilisation Roundtable III; also Zuppero, A.C. and Jacox, M.G., "Near Earth Object Fuels (neo-fuels): Discovery, Prospecting and Use", (1992) IAA-92-0159, paper at the World Space Conference of the American Institute for Astronautics and Aeronautics, Washington DC, (30 August – 02 September 1992)
[22] See Sonter, M.J., "Near Earth Objects as Resources for Space Industrialization", see *supra* note 8.
[23] Gerlach, C.L., "Profitably Exploiting Near-Earth Object Resources", paper given at the International Space Development Conference, National Space Society, Washington DC, 19 – 22 May 2005 at p. 11.
[24] Morrison, D. et al, "Dealing with the Impact Hazard", in Bottke W.F. et al (eds.), *Asteroids III*, (2002) pp. 739 ff.
[25] The Spacewatch Project, Lunar and Planetary Laboratory, University of Arizona, website at http://spacewatch.lpl.arizona.edu/, accessed 31 August 2008.
[26] Near Earth Object Program, NASA, see *supra* note 4.
[27] Lincoln Laboratory, Near-Earth Asteroid Research (LINEAR), Massachusetts Institute of Technology, website at http://www.ll.mit.edu/mission/space/linear/, accessed 31 August 2008.
[28] Hayabusa mission, Japan Aerospace Exploration Agency, website at http://www.muses-c.isas.ac.jp/e/index.html, accessed 31 August 2008.
[29] European Space Agency, NEO Space Mission Preparation, Don Quijote mission, information online at http://www.esa.int/SPECIALS/NEO/SEMZRZNVGJE_0.html, accessed 31 August 2008.
[30] Figure 1 is from Chamberlin, A.B., Jet Propulsion Laboratory, online at http://neo.jpl.nasa.gov/stats/, accessed 31 August 2008.
[31] Chapman, C.R., Harris, A.W. and Binzel, R., "Physical properties of near-Earth asteroids: implications for the hazard issue", in Gehrels, T., (ed.) *Hazards Due to Asteroids and Comets* (1994) 537.
[32] See *supra* note 7 at p. 218.
[33] NASA, *Near-Earth Object Survey and Deflection Analysis of Alternatives*, Report to Congress, (March 2007), online at http://www.nasa.gov/pdf/171331main_NEO_report_march07.pdf, accessed 31 August 2008.
[34] See generally Brown, P. et al, "The flux of small near-Earth objects colliding with the Earth", (2002) 420 Nature 294; and Remo, J.L., "Energy requirements and payload masses for near-Earth object hazard mitigation", (2000) 47(1) Acta Astronautica 35.
[35] George Washington University, *Summary and Recommendations from the 2007 Planetary Defense Conference*, 05 – 08 March 2007, (25 April 2007), at p. 3.
[36] See *supra* note 9.
[37] See on the topic of non-appropriation and Article II of the Outer Space Treaty generally Li, S., "Thoughts on relation between individual and the principle of inappropriation of outer space", in Papers Collection: International Forum on Air and Space Law (2007) 390 (in Chinese); von der Dunk, F.G., Back-Impallomeni, E., Hobe, S. and Ramirez de Arellano,

R.M., "Surreal estate: addressing the issue of 'Immovable Property Rights on the Moon'", (2004) 20 Space Policy 149; Doyle, S.E., "Issues of Sovereignty and Private Property", in Benkö, M. and Kröll, W., *Luft- und Weltraumrecht im 21. Jahrhundert (Air and Space Law in the 21st Century: Liber Americorum - Karl-Heinz Böckstiegel)*, (2001) 313; and White, W.N., Jr., "Interpreting Article II of the Outer Space Treaty", (2003) 46 Proc. Coll. Law of Outer Space 171.

[38] Comprehensive Nuclear Test Ban Treaty, (1996) 35 ILM 1439.

[39] Article 3(3), Agreement Governing the Activities of States on the Moon and Other Celestial Bodies, (1979) 18 ILM 1434, hereinafter "Moon Agreement". As of 01 January 2008, 13 States have ratified the Agreement, and 4 States have signed it.

[40] Sands, P., *Principles of International Environmental Law*, (2nd ed., 2003) at p. 158 ff.

[41] For an excellent overview of the use of economic instruments in international environmental protection, see generally Stewart, R., "The Importance of Law and Economics for European Environmental Law", (2002) 2 Yearbook of European Environmental Law 856; Bosselmann K. and Richardson B., *Environmental Justice and Market Mechanisms* (1999) and Galizzi, P., "Economic Instruments as Tools for the Protection of the International Environment", (1997) 6 European Environmental Law Review 155.

[42] One of the earliest initiatives can be found in "Report of the Working Group of Experts from the EC Member States on the Use of Economic and Fiscal Instruments in EC Environmental Policy", (1991) 14 BCICLR 447.

[43] It was only in April 1990, for example, that the European Commission (EC) Environmental Council "acknowledged the value of supplementing existing regulatory instruments...by the use of economic and fiscal instruments", see "Report of the Working Group of Experts from EC Member States on the use of Economic and Fiscal Instruments in EC Environmental Policy", *ibidem* at p. 448. The first global declaration in support of economic instruments in this regard was in November 1990, with the Ministerial Declaration of the Second World Climate Conference.

[44] Developed States Parties to the Convention are obliged to coordinate their relevant economic instruments in regard of climate change issues, see Article 4(2)(e), United Nations Framework Convention on Climate Change, (1992) 31 ILM 851, hereinafter "Climate Change Convention".

[45] States Parties are called upon to adopt economically and socially sound measures that act as incentives for the conservation and sustainable use of components of biological diversity in Article 11, Convention on Biological Diversity, (1991) 31 ILM 818.

[46] Rio Declaration on Environment and Development (1992) 31 ILM 874, hereinafter "Rio Declaration".

[47] USC §§ 7401 – 7671 (1988), and amendments there to in Supplement III to USC (1991). See for an overview on the topic Nash, J. and Revesz, R., "Markets and Geography: Designing Marketable Permit Schemes to Control Local and Regional Pollutants", (2001) 28 Ecology Law Quarterly 569.

[48] Article 4(2), Convention for the Prevention of Marine Pollution from Land-Based Sources (Paris), (1974) 13 ILM 352, as amended by the Protocol of 1986, (1988) 27 ILM 625.

[49] Protocol on Substances that Deplete the Ozone Layer (Montreal), (1987) 26 ILM 1550.

[50] An excellent discussion of the topic can be found in Sands, P., *Principles of International Environmental Law, supra* note 40 at pp. 357 – 368.

[51] Article 208, Protocol to the Framework Convention on Climate Change (Kyoto), (1997) 36 ILM 1436, hereinafter "Kyoto Protocol".

[52] Gosseries, A., "The Legal Architecture of Joint Implementation", (1999) 7 New York University Environmental Law Journal 49.

[53] Stewart, R.B., Connaughton, J.L. and Foxhall L.C., "Designing an International Greenhouse Gas Emissions Trading System", (2001) 15 Natural Resources and Environment 160.

[54] See Articles 1 and 26(3) of the European Council Directive, online at http://europa.eu.int/comm/environment/climat/030318commonposition_en.pdf, accessed 31 August 2008.

[55] Article 2, European Council Directive, *ibidem*. The target operations are energy, production and processing of ferrous metals, the mineral industry and other activities such as the production of pulp and paper (Annex I). Gases targeted are carbon dioxide, methane, nitrous oxide, hydrofluorocarbons, perfluorocarbons and sulphur hexafluoride (Annex II).

[56] Articles 4 – 7, European Council Directive, *supra* note 54.

[57] Articles 14 and 15; and Annexes IV and V of the European Council Directive, *supra* note 54.

[58] Articles 19 and 20, European Council Directive, *supra* note 54.

[59] See for an overview Sands, P., *Principles of International Environmental Law, supra* note 40 at p. 373 ff.

[60] See for an overview Sands, P., *Principles of International Environmental Law, supra* note 40 at p. 1021 ff.

[61] OECD Council Recommendation on Integrated Pollution Prevention and Control, (1991) C(90)164/FINAL, paragraph I(a).

[62] See for example the draft Directive on Integrated Pollution Prevention and Control: COM (93) 423, (14 September 1993).

[63] On the topic of the legal regime for deep seabed mining, see Mwenda, K., "Deep Sea-bed Mining under Customary International Law", (2000) 7(2) Murdoch University Electronic Journal of Law, online at http://www.murdoch.edu.au/elaw/issues/v7n2/mwenda72.html, accessed 31 August 2008; also see generally Hauser, W., *The Legal Regime for Deep Seabed Mining under the Law of the Sea Convention*, (1983) and Schmidt, M.G. and Richardson, E.L., *Common Heritage or Common Burden? The United States Position on the Development of a Regime for Deep Sea-bed Mining in the Law of the Sea Convention*, (1990).

[64] See *supra* note 15.

[65] Tolkien, J.R.R., *The Lord of the Rings: The Two Towers*, (1966).

IAC-08-E3.5.XX

THE ONGOING IAA STUDY ON PROTECTING THE ENVIRONMENT OF CELESTIAL BODIES (PECB):

THE CONCEPTUAL BACKGROUND AND INTERMEDIATE RESULTS

Mahulena Hofmann
University Giessen, Germany, mahulena.hofmann@recht.uni-giessen.de
Petra Rettberg
DLR, Germany, petra.rettberg@dlr.de
Marc Haese
ESA, The Netherlands, marc.haese@esa.int

ABSTRACT

Intensive activities on the Moon and an increase of man-made objects to be sent to Mars can be expected over the next decade and will result in a growing risk of damages to the natural environment of the celestial bodies concerned. The ongoing IAA Study on "Protecting the Environment of Celestial Bodies" (PECB) has been initiated in order to address this development and to respond to the most pressing questions related to it. It was kicked off in 2007 and is intended to end in 2010. The Study uses a multidisciplinary approach to analyze related issues and to identifying possible solutions which would be acceptable to the international community. The study team developed the first draft chapter texts by July 2008. It can be noted that the present planetary protection (PP) approach is highly developed and performed by several space faring countries, but, however, focused on scientific exploration activities. Space faring nations seem not interested in an appropriate binding norm of international law. The next steps of the Study will focus on the analysis of potential multidisciplinary measures and, possibly, the development of a set of recommendations for a comprising approach to the protection of the environment of celestial bodies.

1. MOTIVES AND SCOPE OF THE STUDY

Space activities bring undoubtedly a wide range of benefits to humankind –in science, technology, economy and many other areas. Like all human activities, they leave *per se* more or less measurable traces in the space environment. Some special forms of this phenomenon - the issue of space debris or the problematics of biological contamination - have been analysed in the space community already. Individual studies have been devoted to other questions such as the issue of abandoned objects in planetary orbits or the environmental impact of planned planetary mining or human settlements, however, the space community as a whole has not considered them of particular importance until now. It can be stated that the issue of protection of planetary environment has not met any systematic, international and multidisciplinary approach yet.

The example of Earth-bound environmental problems and their currently hardly manageable range shows how difficult it is to cope with the detrimental consequences of human activities in comparison with preventive measures to avoid them. Even if the precautionary steps might be expensive and time consuming, economy teaches that at a certain point in time they prove less demanding than the cleaning measures later.

The IAA Study on Protecting the Environment of Celestial Bodies aims at providing an overview of existing methods of planetary protection and their efficiency from the perspective of current and future needs – biological, technical, legal, economical and other methods. By doing so, it goes deliberately beyond the interpretation of "Planetary Protection" by COSPAR (Committee on Space Research), which is generally used as a set of methods used for protecting the planets from *biological* contamination avoiding compromising future exobiological research. Because of the limits given by the planned volume of the study, it concentrates mainly to the Moon and Mars environment.

Its general goal is not necessarily the draft of a new, formal international document, because the required knowledge and international consensus is missing at this stage. The ambition of the study team primarily is to initiate an international discussion and to raise awareness of the whole spectrum of the issues connected with the protection of the environment of celestial bodies, and in deliberating ideas how to organize this protection more efficiently on international scale. Depending on the findings during the study, a proposal of a code of conduct "how to avoid future pollution of environment of celestial bodies" might be developed at the final stage of the project.

2. THE ORGANIZATION OF THE WORK

The interdisciplinary character of the subject of the study requires a research done by a **multidisciplinary team**. The chairs and the secretary of the study challenged many prominent authors active in the field of Planetary Protection. Until now, nine authors have delivered their contributions to the study: Ivan Almar, Hungary (What Could Cospar do for the Protection of the Planetary and Space Environment?), Cassie Conley, USA (Current planetary protection activities), Petra Rettberg, Germany (Currently applied methods for bioburden measurements of spacecraft – are they sufficient?), Gernot Groemer, Austria (Mars Astrobiology: Introductory Aspects and Context), Mahulena Hofmann, Czech Republic and Germany (Is there any Legal Regime for the Protection of the Moon's Environment?), Francis Lyall, UK (Environmental Regulations of and in Outer Space), Annelie Schoenmaker, The Netherlands (The Potential of Education and Media as Instruments for Planetary Protection), Kazuto Suzuki, Japan (Potential of Politics for Protecting the Environment of Celestial Bodies), Mark Williamson, UK (Protection of the Space Environment: the Advantage of an interdisciplinary Approach), and Vasilis Zervos, Greece (Instruments of the Protection of the Environment of Celestial Bodies: The Potential of Economy).

The study team is an open group and is currently working on the initial draft text. Therefore, experts that have interest in participating are invited to contact the co-chairs or the secretary.

The study has been performed in several **steps**: It has been accepted by the Commission V of the International Academy of Astronautics (IAA) and was approved by its Scientific Activity Committee (SAC) in spring 2007. The first step has been the definition of key questions related to the protection of celestial bodies which took part mostly during the Kick-off Meeting at the occasion of the 58th International Astronautical Congress in Hyderabad, India. The present study team has been formed in the first half of 2008. After the distribution of the relevant themes among the potential authors, first drafting work for each chapter has started in spring 2008. The 2008 Glasgow IAF Congress will serve as a platform for presenting and discussing the very first results from the work begun and the steps to follow. The first draft of the study will be presented in the IAA and at the IISL session connected with legal issues of the environmental protection. Until March 2009, the texts of the chapters will be finalized, in order to enable to draft the part "Proposals and Recommendations" on their basis until March 2009. The team plans to meet again at the IAC 2009 in Daejon, South Korea and submit the final version to the IAA afterwards. The final Cosmic Study is intended to be published in 2010.

3. PRELIMINARY RESULTS

3.1. The Challenges for the Environment of Celestial Bodies

In their contributions, the members of the study team have identified several challenges for the environment of celestial bodies:

A significant amount of space debris has already contaminated the surface of Moon, Venus and Mars. The spacecraft populations in orbital environments of planetary bodies, notably the Moon and Mars, are already on the rise. Commercial space tourism is increasing and so is its environmental impact. Industrial activity, mining in particular, may destroy the original environment of smaller celestial bodies. The commercial endeavors can evolve very quickly. Each research, also in-situ research, is always producing a certain amount of pollution. More planetary protection issues and risks will arise with the addition of the human component to the planetary missions: neither space suits nor space habitats will be closed systems; cross contamination will be reduced, but not fully avoided – at least at the landing and habitat site. Any permanent base on a celestial body can be a source of pollutants which can destroy or degrade in-situ research. Developing space based energy sources, including in-situ fuels for use in space or transfers to Earth, can affect the planetary environment detrimentally. Colonization and terraforming mean a large scale transformation of the environment – the reforming of the environment of a planet to accommodate human life. Polluting planetary space can have harmful effects also to astronomy. Space based weapons would be damaging the space environment.

3.2. Instruments of the Protection of the Environment of Celestial Bodies

Following instruments of protection of environment of planetary bodies have been identified by the members of the study team:

For a long time, the tendency to concentrate on biological protection of potential life forms on planetary bodies against contamination form spacecraft has dominated this field; other aspects of protection have been sidelined. The wide range of the activities which can have detrimental effects on the environment of celestial bodies requires a **consolidated, interdisciplinary approach** to this issue.

Historically, the protection against **biological contamination** is the most experienced form of protection of planetary bodies. The activities concentrated so far on the avoidance of contamination with viable organisms, but also contamination with organic / biochemical compounds is already addressed. The aim of these planetary protection activities is exclusively the enabling of future scientific investigations of possible extraterrestrial life forms, precursors, and remnants without any environmental pollution resulting in the misleading interpretation of scientific results. These activities are organized the **COSPAR** since the sixties when it was charged to work on the planetary quarantine prescriptions. Based on its survey, a list of the most important/ interesting planetary environments of a classification scheme for territories with gradually decreasing interest for science should be established (space wilderness areas). In October 2002, COSPAR has formulated the present **Planetary Protection Policy** which embodies a set of **guidelines and recommendations** based on Article IX of the Outer Space Treaty; the last amendment to the policy was adopted in July 2008. The intensity of measures the states have to take during their space activities varies according to the category of the target body (e.g. Mars, Venus, and Moon) and mission type (e.g. flyby, orbiter, lander, and rover). The Policy understands the environment of the Moon as a part of the Earth-Moon system; in general, it must be protected from back contamination to fulfill planetary protection requirements on Earth-Moon travel. However, in 2008 the Moon was reclassified into another category (PP category II) which requires simple documentation to outline intended or potential impact targets, brief pre- and post-launch analyses and an organic inventory.

The COSPAR members are recommended to provide information to COSPAR within a period not exceeding six months after any launch about the procedures used for planetary protection for each flight and again within one year after the end of a solar-system exploration mission about the areas of the target(s) which may have been subject to contamination. COSPAR will make a repository of these reports; make them available to the public, and annually delivers a record of thee reports to the UN Secretary General.

The acceptance of the COSPAR rules is high in the solar system exploration community. In the European sphere, they have been implemented formally by the definition of the **ESA planetary protection policy** together with related requirements and procedures in 2007. The European Cooperation for Space Standardization (**ECSS**) is a normative system jointly elaborated by European space agencies and companies. The goal of standards defined by ECSS working groups is to guarantee that the entire European space community works under the same rules and same procedures – not only ESA member states but also organizations from non-member states involved in ESA missions or launched from launching pads located in the territory of an ESA member state. A set of standards aimed at the application of appropriate methods and procedures for biological and chemical contamination control is under development.

In addition to international and European standards, the COSPAR recommendations have been implemented by national systems, such as **national standards** issued by the CNES or NASA.

The specific feature of the COSPAR rules is that they are focused strongly on the issue of scientific investigation of possible extraterrestrial life forms ("The conduct of scientific investigations of possible extraterrestrial life forms, precursors, and remnants must not be jeopardized"). Materially, they do not concern the question of a simple "use" or "exploitation" of outer space. The measures in stake should be reported only after the completion of a mission. These rules are of a recommendatory character, the compliance with their content is therefore only voluntarily. Their significance as "best practices" rules broadly accepted in the scientific community cannot be underestimated.

Other instruments of planetary protection are or can be focused on the protection of specific areas of outer space, such as **Mars**. It shows that contamination vectors during a crewed surface sojourn are unknown due to uncertainties in the exploration procedures, tools to be used and areas to be explored. Currently, there is a paradigm in place not to allow humans to access the astrobiologically most interesting regions in observance of the planetary protection measures set forth by COSPAR. At the same time, it is generally recognized that limiting subsurface sampling activities only to robotic drilling and analysis systems put severe constraints on the reachable soil depth and sample procurement. Thus, proper sample procurement protocols have to be elaborated to reduce cross contamination on the basis of further simulation missions.

In addition to the methods of biological contamination control driven by scientific interests, the scale of the influence of space activities on the environment of celestial bodies can be influenced by the instruments of **economy**: Cost is an obvious automatic mechanism for the protection of much of the environment both on the Earth and celestial bodies. The potential discovery of minerals on celestial bodies might lead not only to rethinking the concept of "common benefits" and allocation of resources but also to new approaches to the sharing the environmental burden. The even more meaningful way how human activities can affect the environment of celestial bodies – their habitation - allows several scenarios: The variant that the pollutant continues to pollute and compensation is too ineffective and risky. A decisive moment in this model is the dynamic element of colonial expansion: from a certain number of inhabitants, it is most costly to first pollute and then be in need to clean, rather than utilizing the clean technology from the beginning.

The **law** attempts to prevent the detriment of space environment since the 1960s. Until now, there are three specific provisions dealing with this issue: Article IX of the Outer Space Treaty obliges the states to explore and use the Outer Space with due regard to corresponding interests of other states (due diligence principle). Further, states shall pursue studies of Outer Space, including the Moon and other celestial bodies, so as to avoid their harmful contamination and, if necessary, adopt appropriate measures for this purpose. The next special provision is Article 7 of the Moon Agreement prohibiting to explore and use the Moon and other celestial bodies in a manner disrupting the existing balance of their environment; the states parties are obliged to take measures preventing such impact of their activities.

The well known problem of these provisions is the fact that they are too vague or too narrow in case of the internationally reputable Outer Space Treaty or binding only a minimal number of states in case of the Moon Agreement. There does not exist any specific, generally binding and enforceable provision prohibiting deteriorating the environment of celestial bodies by both their exploration and exploitation. This rule can be, however, deduced from general international law: A general international duty towards the preservation and conservation of the environment, both within and outside areas of national jurisdiction is developing, primarily through the International Court of Justice (ICJ) case law. This developing international Law on the environment is not confined to the post hoc action: The Precautionary Principle which originates in municipal law is making its way in international law. In relation to space activities, harm done to "generations unborn" as the ICJ put it, could well include the degradation of the space environment, both near and far. However, further deliberation has to be devoted to the question as to who is entitled to bring an action to enforce a duty owed to many or to the population as a whole.

Politics as a process of people seeking and using power to realize their objectives helps to understand that if one seeks to limit the behavior of a state or individual to contaminate celestial bodies, there should be a person or an entity who has enough power to set up a rule or law for protection the environment of celestial bodies.

However, there are a very limited number of states capable of exploring celestial bodies and having power to establish international standards. Additionally, there is no current consensus or normative understanding among these states with the respective capacities. The lack of "power" for enforcing environmental criteria is multiplied by the fact that there is no effective governing body of space activities; there are bodies such as UNCOPUOS and OOSA (UN Office for Outer Space Affairs) which may act as governing body, but there is no power in these institutions for enforcing the rules for rule-breakers. Constructivist arguments, which explain that actors behave according to what they believe is good or appropriate, lead to the presumption that it is necessary to construct viable ideas and discourse which would convince all states to protect the environment of celestial bodies.

The awareness of the necessity to prevent the deterioration of the planetary environment can be raised by several means, which also include **education** and **media**. Education is meant to inform the public on this sensible issue and to create public opinion. Public opinion can make a difference, for example for political pressure, especially in space exploration where passion and fear easily appear. In this area, a number of initiatives have already started – e.g. regular joint courses by ESA and NASA on planetary protection policies and practices to familiarize current and future practitioners with the COSPAR planetary protection policy and guidelines. The presentation of the respective policies in media targets at journalists and the general public: all media tools – such as specialized journals, general publications, or the internet can be used in order to spread information and awareness on planetary protection issues. The way forward would be to elaborate a communication plan aimed at promoting the results of the IAA study and the topic in general to make education and media coverage happen.

4. CONCLUSIONS

At the present stage of the IAA study, a lot of important information on present approaches to planetary protection, its gaps and future concepts has been collected and analyzed. This information enables to draw the conclusion that the present planetary protection approach is highly developed by several space faring countries, internationally coordinated and duly performed.

On the other hand, it cannot be overlooked that this system bases on recommendatory and posteriori rules, is focused on scientific exploration activities and does not handle the possibility of deterioration of environment of celestial bodies by their exploitation or human settlement. However, in the latest version the COSPAR Planetary Protection Policy principles and guidelines for human missions were already outlined by the statement that a comprehensive planetary protection protocol for human missions should be developed. It should encompass both forward and backward contamination concerns, and address the combined human and robotic aspects of a mission, including subsurface exploration, sample handling, and the return of the samples and crew to Earth.

If there are more comprehensive national concepts of prevention of planetary environment by space activities, they have not been included into any general international network until now.

The present knowledge of the situation allows also a conclusion that the space faring countries do not show any apparent interest to be bound by an appropriate binding norm of international law which would make them internationally responsible for environmental failures. The elaboration of a new, recommendatory document (UN resolution) encompassing such a rule does not seem to be of any immediate use for them.

The next steps of the IAA study should, therefore, focus on raising awareness in the space community about the fact that a common approach in avoiding the deterioration of space environment would be of general interest for pursuing sustainable long-term science and exploration programs (also involving e.g. countries with only potential space programs); it should be evaluated, whether the planetary protection authorities of space faring countries would be willing to assist in the creation of an international network for exchange information about protection issues. Additionally, the content and potential of present international general law obliging states to ensure that activities within their control respect the environment of areas beyond their national control should be further elaborated in context with space activities. At the final stage of the project, a set of recommendations aimed at the protection of the environment of celestial bodies might be developed, depending on the findings during the preceding steps of the study.

DECISION PROGRAM ON ASTEROID THREAT MITIGATION

Russell L. Schweickart
Chairman, Committee on Near-Earth Objects
Association of Space Explorers
Sonoma, California
rs@well.com

Abstract

The Association of Space Explorers Committee on Near-Earth Objects (NEOs) and its Panel on Asteroid Threat Mitigation have prepared a decision program to aid the international community in organizing a coordinated response to asteroid impact threats. The program is described in the ASE's report, *Asteroid Threats: A Call for Global Response*, which will be considered by the United Nations Committee on the Peaceful Uses of Outer Space in its 2009 sessions. The findings and recommendations of this report are presented here as well as some of the major implications of the complex decision-making involved in developing a coordinated international response to the challenge of protecting the Earth from NEO impacts.

BACKGROUND

In its 2009 sessions the United Nations Committee on the Peaceful Uses of Outer Space (UN-COPUOS) will be presented with a decision program on asteroid threat mitigation developed over a two year period by the Association of Space Explorers[i] (ASE), the international organization of astronauts and cosmonauts from 34 nations. The program was developed by the ASE Committee on Near-Earth Objects (ASE-NEO) and its Panel on Asteroid Threat Mitigation[ii] (Panel), a distinguished international group of experts in science, law, diplomacy, and disaster management.

The ASE effort was initiated during its 2005 Congress when the members took note of the series of international disasters which had occurred that year (especially the Indian Ocean tsunami, hurricanes Katrina and Rita, and the Pakistani earthquake) and the recognition of the critical role of preparation and warning in saving lives. Being also aware of the devastation caused by NEO impacts with Earth, the accelerating discovery rate of NEOs and the emerging technical capability (with adequate early warning) to divert such NEOs from impacting Earth, the Association realized the need for systematic preparation for this eventuality by the international community.

Recognizing the significance of this need the ASE issued an open letter[iii] to world institutions and leaders calling on them to "acknowledge this challenge and accept the responsibility for prevention of these most devastating of all natural disasters." To support such efforts ASE created an ASE-NEO committee and charged it with supporting "national and international responses by providing relevant information, organizing meetings or workshops, and providing expert witnesses." In 2006 the ASE-NEO committee, utilizing the ASE's Observer status in UN/COPUOS, assumed membership on Action Team-14 (NEO) of COPUOS and initiated an effort to develop a decision program on asteroid deflection for consideration by the international community. To support this effort ASE-NEO formed its international Panel on Asteroid Threat Mitigation and initiated a series of four workshops to develop this program. The result of this effort is the report, *Asteroid Threats: A Call for Global Response*, currently in process by Action Team-14 for introduction to COPUOS in its 2009 sessions.

SUMMARY FINDINGS

Our highly interconnected society is vulnerable to the destructive power of impact events ranging from the 1908 Tunguska event in which the impact of an estimated 45 meter diameter object destroyed 2000 square kilometers of Siberian forest to the 12 kilometer diameter object responsible for the Chicxulub impact 65 million years ago which is thought to have caused the extinction of the dinosaurs and 70% of all species alive at the time. Such cosmic collisions occur infrequently juxtaposed with a human lifetime, and yet when they do happen they dwarf other natural disasters more common in human experience.

Yet surprisingly in the instance of this most devastating of natural disasters we are far from helpless. With our telescopic and spaceflight capabilities we can detect and predict potential impacts, and with adequate early warning we can deploy space systems capable of altering the orbit of threatening NEOs sufficient to cause them to pass harmlessly by the Earth thereby avoiding an impact. In the event of a discovery where insufficient time is available to successfully divert a threatening NEO we will nevertheless, if prepared, be able to mitigate the effects of an impact by evacuation and other disaster preparedness measures.

What is needed to match the technical capability for responding to the NEO impact challenge is an in-place international system of preparation, planning and timely decision-making. The need for attention to this issue now by the international community is driven by the rapid expansion of the number of NEOs which will be discovered and tracked in the next 10-15 years, and the inherent geographic variability associated with impact prediction and deflection operations.

New telescopic resources coming into service within the next decade will dramatically increase the number of NEOs discovered and tracked. The US Congress has charged NASA with discovering and tracking 90% of all NEOs larger than 140 meters in diameter by 2020. While meeting this goal poses a considerable challenge, it is clear that with new telescopes coming online (e.g. Pan-STARRS[iv] and LSST[v]) this goal will be approached in the 2020-2025 timeframe. In the process of achieving the 140 meter goal many smaller but still dangerous NEOs will be discovered with the number of such objects likely to exceed 300,000. Based on current empirical experience the number of potentially damaging NEOs with a non-zero probability of impact within the next 100 years is likely to exceed 10,000 by this time. Of these NEOs with at least a small probability of impact over the next 100 years many are likely to appear threatening enough to necessitate a decision of whether action should be taken to prevent an impact.

The need for international coordination in making such a decision is determined by the natural uncertainty regarding which specific populations are at risk in predicting an impact and the inherent shifting of risk in the process of deflection. All measurements have an associated uncertainty and in the instance of NEO observations these measurement uncertainties, projected forward in time, manifest as a risk corridor across the face of the Earth within which, if it impacts, the NEO will hit. While in the end an impact would occur at a specific point, at the time a decision must be made to deflect a threatening object the impact zone will extend for some distance along the risk corridor and, in fact, in many instances may well extend beyond the Earth's limbs for many Earth diameters in both directions. Hence, at the time a deflection decision must be made (to provide adequate time to conduct the operation and for the deflection to take effect) it is likely that the people of many nations will be at risk. Furthermore in the process of deflection per se, there will be a temporary shifting of risk between populations as the NEO impact point is itself shifted from a point on the Earth's surface to a safe distance along the risk corridor either ahead of or behind the Earth.

Because NEO impacts can occur anywhere on our planet and affect the entire international community, a collaborative, global response is required. Furthermore it is highly desirable that a decision process, with agreed criteria, policies and procedures be established prior to the development of a specific threat in order to assure that minimization of risk to life and property prevail over competing national self interests.

PRIMARY RECOMMENDATIONS

A global, coordinated response by the United Nations to the NEO impact hazard should ensure that three logical, necessary functions are performed:

Information Gathering, Analysis, and Warning

An Information, Analysis, and Warning Network should be established. This Network would operate a global system of ground- and/or space-based telescopes to detect and track potentially hazardous NEOs. The Network, using existing or new research institutions, should analyze NEO orbits to identify potential impacts. The Network should establish criteria for issuing NEO impact warnings.

Mission Planning and Operations

A Mission Planning and Operations "Group," drawing on the expertise of the space-faring nations, should be established and mandated to outline the most likely options for NEO deflection missions. This group should assess the current, global capacity to deflect a hazardous NEO by gathering necessary

NEO information, identifying required technologies, and surveying the NEO-related capabilities of interested space agencies. In response to a specific warning, the group should use these mission plans to prepare for a deflection campaign to prevent the threatened impact.

NEO Threat Oversight and Recommendation for Action

The United Nations should exercise oversight of the above functions through an intergovernmental Mission Authorization and Oversight "Group." This group would develop the policies and guidelines that represent the international will to respond to the global impact hazard. The Mission Authorization and Oversight Group should establish impact risk thresholds and criteria to determine when to execute a NEO deflection campaign. The Mission Authorization and Oversight Group would submit recommendations to the Security Council for appropriate action.

THE NATURE OF THE CHALLENGE

The NEO environment has remained virtually constant for the past three billion years. The question therefore logically arises regarding the basis for needing a decision process to address NEO impact threats now. The simple answer to this fundamental question is that our telescopic observations of the NEO population are rapidly converting what has been a statistical threat into direct knowledge of specific impact threats which necessitate response.

What is the nature of these threats? And what responses are possible? How often might actual threats materialize? These and a host of additional questions arise when it is realized that effective action can be taken in response to an impact threat thereby either materially reducing the loss to life or preventing an impact occurrence entirely.

A Coming Wave of Discovery

The current NEO search program (Spaceguard Survey), initiated in 1998, has resulted in the discovery and tracking of over 5600 NEOs in the past 10 years. These NEOs, of all sizes, have been discovered in the process of achieving the goal of discovering 90% of all NEOs greater than 1 kilometer in diameter by the end of 2008. Approximately 80% of the statistical population of these large NEOs has been discovered, and the search continues.

In 2005 the US Congress established a new goal for the Spaceguard Survey; to discover 90% of all NEOs greater than 140 meters in diameter over the next 15 years. This revised goal reflected the understanding that these smaller NEOs will also cause terrible destruction on impact and should therefore be discovered and tracked in order to enable responsive action.

While a specific revised search program has not yet been developed there are large new telescopes currently in development and testing that will enter operation within the next 5-10 years. (i.e. Pan-STARRS and LSST[vi]) As these powerful telescopes begin their observations the rate of NEO discoveries will take a dramatic turn upward. Based on expected performance and current Spaceguard Survey experience it is estimated that within the next 15 years over 500,000 NEOs will be discovered and enter the tracking database, and that 200-400,000 of these will be of a size capable of doing substantial damage at the Earth's surface on impact.

Based on tracking experience to date the vast majority of these NEOs will have zero probability of impacting Earth within the next 100 years. Nevertheless on the order of 3% of these new NEO discoveries will likely have some, generally small, probability of impact in that timeframe. Within these thousands of potential Earth impactors there will likely be dozens which will appear threatening enough that they will require proactive decisions regarding mitigation or deflection.

Mitigating an Impact

When a threatening NEO is discovered too late to permit a deflection the only response option, if an impact is confirmed, is to mitigate the potential damage through conventional disaster response mechanisms, primarily an evacuation.

A similar case arises when, at the time a deflection campaign could have been mounted, it was not, due either to an insufficient probability of impact to justify such action or the inability of the international community to reach a decision. Should the probability of impact subsequently increase to 100% the remaining option is again mitigation.

In either of the above instances, as the time of the potential impact is approached, we will know in which direction to point our telescopic resources to determine whether or not an impact is imminent.

If a potential impact is destined for the nighttime hemisphere of Earth optical telescopes will be able to confirm its approach by looking up the "final approach" path and provide a month or more of warning.

If, on the other hand, the potential impactor is approaching toward the daylight hemisphere optical telescopes cannot be used due to the solar glare and early warning will be possible only by using radar telescopes. Since the range of active radar telescopes is considerably less than their optical counterparts the warning time may be as low as a couple of days.

It should be emphasized, however, that in both of the above situations the probability of impact will likely be extremely low and that the almost certain outcome of looking up the "final approach" path will be the issuance of an "all-clear." Nevertheless it is the possibility of that one chance of an impact materializing, and the saving of many lives, that justifies the required telescopic observations.

Preventing an Impact

Whenever a sufficiently threatening NEO is discovered early enough to mount a deflection campaign an impact can be averted. For such a scenario to be realized three essential elements have to be in place; a capable early warning system, a deflection capability, and an institutional process capable of making timely decisions.

In order to provide the time required for preparation and deployment of a deflection campaign, and adequate time following the deflection action per se for it to take effect, the early warning system will have to provide the information required to make a deflection decision at least 10-15 years ahead of impact.

Confidence in the time required to prepare and accomplish a deflection should be provided by designing and flight testing deflection systems well in advance of their use. Such preparation will not only provide technical confidence in the planning of an actual deflection operation, but also develop confidence in the general public regarding their personal safety.

Time Criticality

Most critical, however, is the requirement that the international community be prepared to authorize a deflection campaign in a timely manner. Failing to provide a decision-making framework before a threatening NEO situation is discovered will risk lengthy argument, political delays, and collective paralysis. Such avoidable inaction will preclude a deflection and force the world to absorb a damaging – and preventable – impact. With the lead time for an authorization decision typically needed 10-15 years ahead of a potential impact, we should begin now to forge that vital decision process.

DEFLECTION AND ITS IMPLICATONS

Deflection Means

NEO impact deflection concepts have been discussed now for several years and while there is not yet a consensus on the most appropriate techniques there is a growing understanding of the nature of the task. Furthermore there is considerable agreement that there currently exist deflection concepts adequate to divert the vast majority of the potential NEO impact threats.

The diversion of a NEO on a path toward an Earth impact requires, depending on the specific circumstances, either a precise, but modest orbit change or the combination of a robust orbit change followed by a precise orbit trim. A successful deflection campaign is one in which both an immediate impact and all near-term return impacts are prevented. These conditions correlate with the robust and precision deflection requirements respectively.

Robust orbit change, i.e. orbit changes requiring substantial total impulse[vii] applied to the NEO, can be provided by either a kinetic impact (KI) or a nuclear stand-off explosion. Both are available technologies and KI was demonstrated conceptually during the 2005 Deep Impact mission, albeit that impact was designed for a different purpose. A nuclear stand-off explosion has not been demonstrated but the technology is arguably available for use. Both of these techniques, while capable of transferring substantial momentum change to a NEO cannot do so with adequate precision to assure a fully successful deflection.

A precise NEO orbit adjustment is conceptually available via a number of techniques. However the most simple and readily available concept is that of the gravitational tractor (GT)[viii] which provides precise adjustment to the NEO orbit by "hovering" in close proximity to the NEO thereby using mutual gravity to change the NEO's velocity.

While the non-nuclear combination of a kinetic impact and gravity tractor will suffice for approximately 98% of the statistical impact threat, the use of nuclear means cannot be ruled out absent further technology development. The frequency of NEO impacts which would require the use of nuclear means is approximately 1 in 100,000 years based on the current best estimate of NEO size-frequency distribution[ix].

Impacts and Impact Precursors

A NEO impact will occur when the orbits of the NEO and the Earth intersect in space and both bodies reach that intersection at the same time.

Such intersections are not static over significant periods of time due to the occasional gravitational perturbation of NEO orbits due to close passes by planets or other large NEOs. Most frequently a NEO threatening an impact with Earth has experienced prior close passes by the Earth which have, in fact, set up the subsequent impact. Such close gravitational encounters with the Earth can substantially change the orbit of a NEO, and on occasion, cause a precise change which brings the NEO back several years later for a direct impact.

The small region near the Earth through which a NEO must pass for such a resonant impact to occur is referred to as a gravitational keyhole. Whenever a NEO passes nearby, whether passing in front or behind the Earth, it passes through a field of dozens of such keyholes. Since the size of keyholes is very small compared with the distance between them near-misses seldom result in a resonant return and subsequent impact. Conversely virtually all impacts, projected back in time, are the result of a precursor close gravitational encounter, most often with the Earth.

Consequently a NEO deflection is most easily accomplished when knowledge of its orbit is gained sufficiently in advance of a possible impact that it can be diverted from a precursor keyhole passage rather than diverting it subsequently from a direct impact. A deflection operation to avoid a target of a kilometer or so in size (i.e. a keyhole) is far less daunting than a deflection to avoid a target of several tens of thousands of kilometers (i.e. the Earth).

Missing a collision with the Earth (or a keyhole) is a matter of slightly increasing or decreasing the size of the NEO orbit, far enough in advance of impact, such that the NEO arrives either slightly late or early at the intersection. While such a deflection will have avoided a direct impact with Earth the NEO will now be passing through the nearby field of resonant keyholes and risks a return impact unless its passage between them is assured. Should a deflected NEO in fact be headed for a return keyhole following a primary deflection it is the precise capability of the gravity tractor (or other precision adjustment means) that will assure a successful deflection.

Deflection Implications

The deflection process, regardless of the technology used, can be understood as an operation which, by slightly adjusting the velocity of a NEO, will cause it to arrive at the impact point slightly earlier or later than it otherwise would have. With a sufficient change in the NEO velocity the change in the arrival time is sufficient that the impact point has moved completely off the Earth.

A deflection can thus be seen as a process whereby an impact point is shifted from its original ("act of God") location to a point ultimately off the Earth's leading or trailing limb. Should, for any reason, the deflection be terminated or only partially completed, the impact point will now be displaced along the risk corridor in the direction of the intended goal. Whether the deflection is a continuous gravity tractor process or an impulsive kinetic impact or nuclear explosion, a partial completion will result in a new impact point displaced along the risk corridor from the original undisturbed impact point.

Hence people not originally near the impact point will experience a temporary increase in risk in the process of reducing the risk of an impact to zero for everyone. Risk shifting is an inseparable element of risk elimination in NEO deflection.

Agreeing to deflect a NEO and deciding on which direction the impact point should be shifted are therefore clearly decisions which must be coordinated among the international community.

FUNCTIONAL NEEDS

In fact the above example is just one of many challenges which will necessitate coordination between nations in order to protect life on Earth from future NEO impacts. In preparing its recommended decision program, *Asteroid Threats: A Call for Global Response,* for the United Nations the ASE and its Panel on Asteroid Threat Mitigation recognized the vital role the UN must play in providing the forum for affecting such international coordination.

While the final shape of an international regime for responding to the NEO challenge will appropriately be determined only following substantive discussion among and between nations, the ASE and its Panel have identified three functional areas which would need to be addressed within any such agreement.

Information Gathering, Analysis, and Warning

An Information, Analysis, and Warning Network should be established. This Network would operate a global system of ground- and/or space-based telescopes to detect and track potentially hazardous NEOs. The Network, using existing or new research institutions, should analyze NEO orbits to identify potential impacts. The Network should establish criteria for issuing NEO impact warnings.

Among the responsibilities that might be assigned to this Information, Analysis, and Warning Network are the following:

a) To serve as the official source of information on the NEO environment.
b) To designate and maintain the official clearinghouse for all NEO observations and impact analysis results.
c) To review the existing NEO information set provided by JPL/Sentry and NEODyS and possibly recommend modifications to them.
d) To recommend policies to the Mission Authorization and Oversight Group regarding criteria for warning and, with MOAG approval, issue NEO warnings and "all-clear" notices.
e) To consider and recommend to the Mission Authorization and Oversight Group a NEO threat public information policy, and explore what threshold should trigger release of information like the risk corridor, potential tsunami simulations, and other potential impact information.
f) To identify in cooperation with United Nations Member States a focal point to engage designated national/international disaster response entities.
g) To assist in mitigation response planning.
h) In cooperation with the Mission Planning and Operations Group, to recommend to the Mission Authorization and Oversight Group the criteria for initiating the planning of a deflection campaign.
i) To develop and recommend to the Mission Authorization and Oversight Group the threshold NEO characteristics that warrant international community attention.
j) To develop and recommend to the Mission Authorization and Oversight Group a NEO impact public information plan.

Mission Planning and Operations

A Mission Planning and Operations "Group," drawing on the expertise of the space-faring nations, should be established and mandated to outline the most likely options for NEO deflection missions. This group should assess the current, global capacity to deflect a hazardous NEO by gathering necessary NEO information, identifying required technologies, and surveying the NEO-related capabilities of interested space agencies. In response to a specific warning, the group should use these mission plans to prepare for a deflection campaign to prevent the threatened impact.

Examples of responsibilities which might be assigned to a Mission Planning and Operations Group are:

a) To determine specific decision and event timelines for all NEOs selected for preliminary deflection campaign analysis
b) To develop and recommend to the Mission Authorization and Oversight Group a process for assigning operational responsibility for a deflection campaign.
c) To evaluate and recommend to the Mission Authorization and Oversight Group alternative deflection concepts proposed by space-faring nations (SFNs).
d) To develop the necessary information requirements for mission planning, and transmit them to the Information, Analysis, and Warning Network.
e) To develop cost models for each approved deflection campaign concept, including each planning and mission operations event.

Mission Authorization and Executive Oversight

The United Nations should exercise oversight of the above functions through an intergovernmental Mission Authorization and Oversight "Group." This group would develop the policies and guidelines that represent the international will to respond to the global impact hazard. The Mission Authorization and Oversight Group should establish impact risk thresholds and criteria to determine when to execute a NEO deflection campaign. The Mission Authorization and Oversight Group would submit recommendations to the Security Council for appropriate action.

This institutional element, which must represent the international community as a whole, should have responsibility, inter alia;

a) To develop a policy to fund those United Nations Member States who conduct authorized NEO activities on behalf of the international community. Submit final recommendations on such a funding policy to the United Nations Security Council for adoption and implementation.
b) To consider and propose for adoption, by the appropriate United Nations organs, threshold criteria submitted by the Information, Analysis, and Warning Network concerning NEO alerts, warnings and actions.
c) To consider and decide those general policy questions presented and/or recommended by the Mission Planning and Operations Group
d) To sit ex-officio on all Information, Analysis, and Warning Network and Mission Planning and Operations Group sessions

These, and other necessary actions that will emerge in the substantial discussion on this challenge to follow, must be systematically integrated into a coordinated decision system by the international community.

IN CONCLUSION

An adequate global action program must include deflection criteria and campaign plans which, when conditions warrant, can be implemented rapidly and with little debate by the international community. In the absence of an agreed-upon decision-making process, we may lose the opportunity to act against a NEO in time, leaving evacuation and disaster management as our only response to a pending impact. A single such missed opportunity will add painful fault-finding to the devastating physical effects of an impact. The international community should begin work now on forging its warning, technology, and decision-making capacities into an effective shield against a future collision.

Now that humankind has the scientific, technical and operational capabilities both to predict whether an asteroid will come too close for comfort, and to launch operational missions to divert a potential impact, it is time for the international community to identify the decision-making institutions and begin the development of a coordinated decision-making process. This decision-making program proposed by the International Panel is only the first step in that direction.

The Association of Space Explorers and its Panel on Asteroid Threat Mitigation are confident that with a program for concerted action in place, the international community can prevent most future impacts. The Association of Space Explorers and its Panel are firmly convinced that if the international community fails to adopt an effective, internationally mandated program, society will likely suffer the effects of some future cosmic disaster—intensified by the knowledge that loss of life, economic devastation, and long-lasting societal disruption could have been prevented. Scientific knowledge and existing international institutions, if harnessed today, offer society the means to avoid such a catastrophe. We cannot afford to shirk that responsibility.

---

[i] Association of Space Explorers (ASE); http://www.space-explorers.org
[ii] Panel on Asteroid Threat Mitigation (PATM), convened by ASE-NEO Committee in its first international workshop on the NEO threat, Strasbourg, France, May 9-12, 2007 (http://www.space-explorers.org/committees/NEO/workshop1.html)
[iii] Open Letter of the Association of Space Explorers, 19th Annual Congress, Salt Lake City, 14 October 2005. (http://www.space-explorers.org/committees/NEO/docs/Open_Letter.pdf)
[iv] Panoramic Survey Telescope & Rapid Response System, (http://pan-starrs.ifa.hawaii.edu/public/)
[v] Large Synoptic Survey Telescope (http://www.lsst.org/lsst_home.shtml)
[vi] ibid
[vii] Total impulse equals the amount of momentum change imparted to an object
[viii] *Gravitational tractor for towing asteroids,* Ed Lu and Stan Love. Nature magazine, Vol. 438, 10 November 2005, and also JPL report on gravity tractor performance, #18, http://www.B612Foundation.org/press/press.html
[ix] *What Spaceguard Did,* Alan Harris, Nature Magazine, Vol. 453, June 26, 2008, pgs 1178, 1179

51st Colloquium on the Law of Outer Space, Glasgow, Scotland

Session 5:
OTHER LEGAL MATTERS

Chairmen:

Joanne Gabrynowicz

Joanne Wheeler

Rapporteur:

Sylvia Ospina

IAC-08-E8.4.11

THE MOON TREATY IN THE 21ST CENTURY

Antonella Bini
United Nations Office for Outer Space Affairs (UNOOSA), Vienna, Austria
Antonella.Bini@unoosa.org

ABSTRACT

Forty years after the Promethean achievement of Apollo 11, the major space-faring nations are showing a renewed interest in missions to the Moon. While in the 1960s, in the context of the Cold War, the decision to land humans on the Moon was perceived as a strategic political instrument between the two blocs, now in the 21st century, a considerably increased number of space actors complicates the scene with multiple sectorial and national interests. Security still predominates, although economic considerations also play an increasing role. The commercialisation and privatisation of space activities leads to new challenges which raise the issue of the role of the international legal framework.
In this context, the Agreement Governing the Activities of the States on the Moon and Other Celestial Bodies of 1979 (known as the Moon Agreement) should be able to play a relevant role. There are several political and legal questions that arise for discussion. At its forty-sixth session, the Legal Subcommittee of the United Nations Committee on the Peaceful Uses of Outer Space (UNCOPUOS) decided to include a discussion on the activities which are being or to be carried out on the Moon and other celestial bodies in the forty-seventh session of the Subcommittee in 2008. The result is a joint statement by the State Parties to the Agreement. This paper addresses the questions raised during these discussions.

5 DECEMBER 1979: THE ADOPTION OF THE MOON AGREEMENT

As far as space exploration is concerned, the United Nations, through the efforts of the Committee on the Peaceful Uses of Outer Space (COPUOS) and its Legal Subcommittee (LSC), have taken the lead role in the development and codification of a body of law to govern the activities of space in this field. Its work began in 1961 with a declaration of the legal principles for governing space activities, and continued with the adoption of the five space treaties.[1]

At the end of the 1960s, as soon as the perspective of mankind's landing on the Moon became a reality, the interest of the international community was heightened with respect to short or long-term peaceful or military gains to be made from the possible colonisation of the Moon and other celestial bodies.

In a letter dated 9 May 1966 addressed to the Chairman of the Committee on the Peaceful Uses of Outer Space,[2] the Permanent Representative of the United States of America requested an early

[1] Treaty on Principles Governing the Activities of States in the Exploration and Use of Outer Space, Including the Moon and Other Celestial Bodies (Outer Space Treaty, 1967), Agreement on the Rescue of Astronauts, the Return of Astronauts and the Return of Objects Launched into Outer Space (Rescue Agreement, 1968), Convention on International Liability for Damage Caused by Space Objects (Liability Convention, 1972), Convention on Registration of Objects Launched into Outer Space (Registration Convention, 1975), Agreement Governing the Activities of States on the Moon and Other Celestial Bodies (Moon Agreement, 1979). For further information, see: http://www.unoosa.org/oosa/en/SpaceLaw/treaties.html.
[2] A/AC.105/32.

convening of the Outer Space Legal Subcommittee to prepare a "*draft treaty governing the exploration of the moon and other celestial bodies*" for submission to the General Assembly at its next session. Shortly after that request, the United States consulted with members of COPUOS, presenting them an outline of points for inclusion in the draft treaty.

At the same time, the Soviet Union Government submitted a "*draft treaty on principles governing the activities of space in the exploration and use of outer space, the moon and other celestial bodies*"[3] for consideration by the General Assembly.

The Legal Subcommittee reconvened in its fifth session in July and then September of the same year[4] and examined the two drafts as well as all proposals presented by the other Member States with respect to particular articles. Although an agreement was reached on a series of articles,[5] there was no consensus on several others. The discussion was then adjourned and not reopened until June 1969 when, upon the proposal of Argentina, France and Poland, the Legal Subcommittee decided to include in its agenda an item dealing with new activities on the Moon and other celestial bodies, including a legal regime for governing their natural resources.[6]

Less than a year after Neil Armstrong's first step on the Moon (20 July 1969), Argentina again submitted for discussion a proposal for a draft agreement.[7] This included the provision that the Moon and other celestial bodies were to be declared a "*common heritage of mankind*".

[3] A/AC.105/35 Annex 1, pp. 10 and 11.
[4] A/AC.105/35, 16 September 1966.
[5] A/AC.105/C.2/L.16.
[6] Two proposals were made by Poland (A/AC.105/C.2/L.53) and Argentina (A/AC.105/C.2/L.54) on the same date - 13June 1969. They were later amalgamated (A/AC.105/C.2/L.66), and still later a joint proposal was made by Argentina, France and Poland (A/AC.105/C.2/L.69) which was adopted by the LSC.
[7] A/AC.105/C.2/L.71 and Corr.1.

Wider questions, including military uses, were introduced into the discussions by a new draft treaty proposal of the USSR in 1971[8] in which, among other things, the principle of space as an *international area of common use* was reiterated. The United Nations General Assembly placed the subject as a separate item on the agenda of the Legal Subcommittee for priority consideration at its 1972 session.[9]

While informal and behind-the-scenes discussions took place, the work of the Legal Subcommittee on producing a draft treaty continued. There were in total 16 proposals from the United States and nine from other countries in addition to the Argentine and Soviet ones. Following the finalisation of extensive negotiations on the agenda item in the COPUOS and its Legal Subcommittee, the draft agreement sponsored by 38 delegations was adopted by the Special Political Committee of the General Assembly on 2 November 1979[10] and by the General Assembly itself on 5 December 1979 in Resolution 34/68,[11] in both cases by consensus without a vote. The agreement was opened for signature and ratification by the Secretary General of the United Nations on 18 December 1979. On that date, the Agreement was signed by 6 nations (Chile, France, Romania, the Philippines, Austria and Morocco) at the United Nations Headquarters in New York.[12]

THE INSPIRATIONAL PRINCIPLES OF THE MOON AGREEMENT

The 21 principles of the Moon Agreement adopted by consensus by the UN Member States (after 7 years of negotiation) presents an effort on the part of the 152 nations involved to establish a basic legal

[8] A/83/91.
[9] GA Res. 2779(XXVI), para. 2, of 29 November 1971.
[10] See A/SPC/34/SR.20 (2.11.79), paras. 9 and 10; Report of the Special Political Committee, A/34/664, paras. 8 and 9.
[11] A/34/PV.89 (Prov.), pp. 7-10.
[12] See UN Press Release L/T/3439, 23 March 1983.

framework for the exploration and later exploitation of the Moon and other celestial bodies and reflects the United Nations' general commitment to the ideals of peace and development through international cooperation.

The Agreement provides broad and general guidelines since at that time, it was recognised that more particular provisions would become possible only once the reality of space was better known.

Several considerations guided the drafting of the Agreement:

1. One major consideration was that the prepared treaty should assure that the Moon and other celestial bodies were used exclusively for peaceful purposes. In line with the basic principles of the Charter of the United Nations as well as of the Treaty on Principles Governing the Activities of States in the Exploration and Use of Outer Space, Including the Moon and Other Celestial Bodies (known as the Outer Space Treaty), the Moon Agreement was clearly designed to ban any non-peaceful use of space, and it explicitly prohibits any threat or use of force or any other hostile act or threat of a hostile act. At that time, the provisions were considered as providing a basic assurance that outer space would be used for peaceful purposes only.

2. Another major consideration was that the Moon Treaty should protect the right of all people to a share of the resources of outer space, and to secure for developed and developing countries alike the opportunity to benefit from space activities. Along these lines, the treaty established that those resources should be regarded as the common heritage of mankind. It has been pointed out that this concept replaced the vague expression "province of all mankind" of the Outer Space Treaty. The expression "common heritage of mankind" implies, according to its authors, the element of a "beneficial domain" which includes the enjoyment of, profit from, and partaking in space benefits which had been excluded by the provision of the Outer Space Treaty.

3. The third major consideration that guided the drafting is reflected in the provision for the freedom of exploration and use of the Moon and other celestial bodies and the encouragement of scientific investigations thereof. The Moon Treaty is designed to permit governmental as well as private entities to explore and use the Moon and other celestial bodies in an orderly manner. It specifically establishes the right to collect and remove samples from the Moon and other celestial bodies to support scientific missions.

Since its adoption, vigorous national debates have taken place in many countries. In attempting to elaborate an international agreement of such magnitude, every nation participating in that process had to make compromises in the hope of creating a more effective global policy, with the result that not all specific interests could be satisfied by the final document. This is certainly true of the Moon Agreement. However, while criticising it, one should be reminded of its larger significance and of the general spirit and intent with which it was elaborated.

THE MOON AGREEMENT IN THE 21ST CENTURY

Forty years after the unprecedented achievement of Apollo 11, the major space-faring nations are again showing interest in missions to the Moon. At the beginning of the 21st century, various countries press ahead with plans for human space exploration: the United States, China, India and Japan are developing lunar exploration programmes, and Canada, Germany, the Russian Federation, the United Kingdom, the European Space Agency and others have revealed their future plans for the human and robotic mission exploration of both the Moon and Mars.

We are at the threshold of a new era of space exploration in which mankind's use of outer space will expand and diversify rapidly with a considerably increased number of space actors pressured by sectorial and national interests. Security issues still predominate, although economic considerations now play an increasing role. The commercialisation and privatisation of space activities leads to new challenges and their vast potential raises the need for being able to plan and act with a more specific and rational direction, as well as the issue of the role of the international legal framework.

Moreover, space exploration and its practical applications are, by their very nature, of global concern and will require to an ever-increasing degree the cooperation of all nations. This is certainly the ultimate aim of the heads of 14 space agencies who, on 31 May 2007, met to coordinate their exploration planning. The result of the discussion is a document entitled the "*Global Exploration Strategy*". Some words of this strategy are particularly elevating: "*this new era of space exploration will strengthen international partnerships through the sharing of challenging and peaceful goals*".[13]

In this context, the Agreement Governing the Activities of the States on the Moon and Other Celestial Bodies could be able to play a relevant role. The Moon Agreement represents an effort on the part of the international community to establish conditions promoting the peaceful uses of the Moon and other celestial bodies. But there are several political and legal questions that arise for discussion. While more specific laws or regulations governing outer space need to be added as the realities of space warrant such provisions, some of the core principles of the Agreement continue to be the object of debate. As of today, only 13 states have ratified the Moon Agreement and an additional four have signed but not ratified it.

[13] See http://esamultimedia.esa.int/docs/GES_Framework_final.pdf.

LEGAL IMPLICATIONS

As pointed out above, the most important concern at the time of the drafting of the Moon Agreement was to ensure the peaceful and equitable exploration of outer space – the first essential phase which would serve as the foundation of later efforts. The Moon Agreement is therefore designed to encourage resource development, and the only major constraint imposed by the treaty is the requirement that the Moon and other celestial bodies and their resources be used for peaceful purposes only.

The Agreement also specifies that the State Parties must take measures to prevent the disruption of the existing environmental balance in space (art. 7, para. 1); to allow the designation of international scientific preserves for which special protective arrangements may be made (art. 7, para. 3); to ensure concrete and positive forms of cooperation concerning assistance to persons on the Moon (art. 10); to establish conditions facilitating activities of states on or below the surface of the Moon (art. 8) such as scientific investigations (art. 6) or the installation of personnel, stations and other objects (art. 8 and 15, para. 1); to define their international responsibilities (art. 14); and to establish peaceful means for resolving disputes, including provisions for consultations in enforcing treaty obligations (art. 15, paras. 2-3).

While some of these provisions are similar to those of the Outer Space Treaty, several changes and innovations were made. Some of the new provisions have helped to clarify issues which were left unresolved by the Outer Space Treaty. Article 1, for example, sets out the cosmographical scope of the Agreement, stating that "*the provision [...] relating to the Moon shall also apply to other celestial bodies within the solar system,* [and] *shall include orbits around or the trajectories to or around it" (art. 1, paras. 1-2).*

Other new provisions are still the object of debate in relevant fora and constitute the reason why several countries which actively participated in the drafting of the Moon Treaty did eventually not sign it. Cases in point are articles 4 and 11. Article 4 repeats the provisions of the first article of the Outer Space Treaty (OST), stating that the exploration and use of the Moon shall be the *province of all mankind* and shall be carried out for the benefit and in the interest of all countries. This principle is related to the principle of the *non-appropriation* of the Moon and other celestial bodies enshrined in article 11, which contains the most novel and at the same time controversial provision of the Moon Agreement.

Article 11 stipulates that the Moon and its natural resources are the *"common heritage of mankind"*. This term does not appear in the OST, although the underlining ideas are certainly there. The term has appeared for the first time in the Moon Agreement and was initially suggested by Argentina, but proposed formally by the United States in 1972. Since the exploitation of the natural resources of celestial bodies other than the Earth is about to become feasible, Article 11 aims at ensuring that all nations enjoy equal rights in exploiting and benefiting from space resources. Moreover, it makes clear that the parties to the treaty intend to enter into negotiations for establishing a mutually acceptable international regime governing the exploitation of the mineral and other substantive resources which may be found on the surface or subsurface of a celestial body (paragraph 5, article 11 and reiterated in article 18). This regime will serve to provide a framework for various nations and non-governmental enterprises to pool their creative energies and cooperate in seeking the best means of exploration and development. The establishment of the international regime would be the task of a conference convened by the UN at the request of one third of the State Parties to the Moon Treaty.

This last article of the Moon Treaty has received special attention from several countries. The main criticisms are related to the principle of the "common heritage of mankind", and several discussions - held since the adoption of the Agreement - have centred on the structure and power of the above-mentioned international regime. According to article 11, the regime will be developed by all nations which are parties to the Moon Agreement, and the treaty itself is open to all states for signature.[14] The purpose of the future international regime recognises the equitable sharing of the benefits derived from any resources in space, and special consideration is given to the efforts of the countries which contributed either directly or indirectly to the exploration of these resources, as well as to the interests and needs of developing countries.[15]

THE ROLE OF COPUOS AND ITS LEGAL SUBCOMMITTEE

As set out by article 18 of the Agreement, ten years after its entry into force, COPUOS considered the question of a first review of the Agreement and the prospective of the establishment of an international regime at its thirty-seventh session in 1994. After the discussion, the Committee recommended to the General Assembly at its forty-ninth session that the Assembly should take no further action at the time.[16]

Today, the intensive discussion of a possible revision of the Moon Agreement and in particular of article 11 is becoming topical, as numerous new realities of international and national space activities, including growing commercialisation and the appearance of new actors, have had an

[14] Article 11, para. 5, art. 18, and article 19, para. 1.
[15] Article 11, para. 7.
[16] General Assembly Resolution 49/34.

impact on the interpretation of the concept of the "common heritage of mankind".

The ongoing discussion in the United Nations Committee on the Peaceful Uses of Outer Space and its Legal Subcommittee reflects these needs. Already in 2004, at the forty-third session of the Legal Subcommittee, the International Institute of Space Law (IISL) and the European Centre for Space Law (ECSL) were invited by Member States to hold a symposium on space law entitled "New developments and the legal framework covering the exploitation of the resources of the Moon".[17] In 2007, the High-Level Panel on Space Exploration held a meeting to discuss the ongoing and planned exploration programmes including the projects aimed at returning to the Moon.[18]

At the forty-sixth session of the Legal Subcommittee, some delegations expressed the view that consideration should be given to the reasons behind the low participation of states in the Agreement Governing the Activities of the Moon and Other Celestial Bodies, and that efforts should be made to resolve any identified obstacles to participation.[19]

In order to do so, the Working Group on the Status of Application of the five United Nations Treaties on Outer Space agreed to: a) address the activities currently being carried out or to be carried out on the Moon and other celestial bodies in the near future; b) identify the international rules governing the activities on the Moon and other celestial bodies; and c) assess whether the existing international rules adequately address the activities on the Moon and other celestial bodies, also incorporating information from states already parties to the Moon Agreement about the benefits of adherence to that Agreement.

[17] A/AC.105/C.2/2004/CRP.11.
[18] See http://www.unoosa.org/oosa/COPUOS/2007/panel.html.
[19] A/AC.105/891, Annex I.

THE RESPONSE OF THE STATE PARTIES TO THE MOON AGREEMENT

In response to this call for action, some of the State Parties to the Moon Agreement (Austria, Belgium, Chile, Mexico, the Netherlands, Pakistan and the Philippines), at the forty-seventh session of the Legal Subcommittee held in March 2008, presented the "*Joint statement on the benefits of the adherence to the Agreement Governing the Activities of States on the Moon and Other Celestial Bodies by States parties to the Agreement*".[20]

This joint statement is based on the experience of the State Parties to the Agreement and emphasises a number of aspects and considerations concerning the benefits of Member States from becoming parties to the Agreement, with the purpose of providing "*the UN Committee on the Peaceful Uses of Outer Space with elements for reflection in the framework of its activities aimed at the development and wider application of outer space law*". It takes into consideration the main issues related to the Moon Agreement such as the fact that "*some States regularly question whether the Agreement is part of international law or should be considered to be on the same level as the other four United Nations treaties on outer space*" or the poor participation of States.

In answer to the first issue, the statement recalls that the Agreement was registered according to article 102 of the UN Charter, entered into force on 11 July 1984, and therefore considers the Moon Agreement as part of international law. The statement also comments that it offers a dedicated international legal framework commented by the UN General Assembly and accepted by the international community.

The statement provides a commentary on the Agreement, emphasising that some

[20] A/AC.105/C.2/L.272.

provisions unique to the Agreement are of "*particular interest for the implementation of projects, activities and missions either because they provide a better understanding of or a complement to principles, procedures or notions enshrined in the other outer space treaties that are applicable to the Moon and other celestial bodies*[21] *or because they facilitate international scientific cooperation*".[22]

Concerning the principle of the "common heritage of mankind" contained in article 11, the State Parties of the Moon Agreement point out that this is the only provision within the UN outer space treaties which foresees the possibility of exploiting resources in outer space. In so doing, the provision offers an obvious legal solution. Moreover, it is underlined that the Moon Agreement, following a constructive approach, leaves the responsibility to define, set up and implement an international regime responding to the status of the common heritage of mankind and to other principles of outer space law to the involved states for which the exploitation of the Moon's and other celestial bodies' natural resources becomes feasible. In particular, it is highlighted that the "*Agreement does not pre-exclude any modality of exploitation, by public and/or private entities, nor forbids commercial treatment, as long as such exploitation is compatible with the requirements of the Common Heritage of Mankind regime*".

The statement encourages states that have signed but not yet ratified the Agreement as well as other states to become parties to it, stressing that the Agreement represents a "*mutual commitment to seeking a multilateral solution*", in particular "*considering their possible involvement in forthcoming missions and projects aimed at exploring celestial bodies*".

[21] See article 1, paras. 1 and 2; article 3, para 4; article 7, paras 1-2; and articles 10, 12, 13, 14 and 15 of the Moon Agreement.
[22] See article 5, paras 1-3; article 6, paras. 2-3; and article 7, para. 3 of the Moon Agreement.

WHAT'S IN THE NEAR FUTURE?

The initiative taken by the State Parties to the Moon Agreement was welcomed by some delegations of COPUOS during its fifty-first session held in June 2007.[23] Those delegations expressed the view that "*the consideration of the joint statement by the Legal Subcommittee at its forty-eighth session might assist in finding a mutually acceptable approach on legal issues relating to the exploration of and the use of the resources of the Moon and other celestial bodies*".

The new perspectives of the use of the Moon for the needs of mankind have been a central issue in the discussions in COPUOS this year, and they will continue to be debated in the forthcoming session of its Legal Subcommittee. In particular, they raise the need for an open discussion on whether the Moon Agreement still holds valid solutions for those issues, or whether it would need to be revised in order to adapt it to the new frontiers of space law.

This is in line with the proposal made by the delegation of Austria at the forty-seventh session of the Legal Subcommittee that an interdisciplinary seminar on issues related to the Moon Agreement should be organised before the Subcommittee holds its session next year.

Note: United Nations documents quoted in this article are available from the website of the United Nations Office for Outer Space Affairs at www.unoosa.org and from the Official Document System of the United Nations at documents.un.org.

Disclaimer: The views expressed in this article are purely those of the author and do not necessarily reflect the position of the United Nations and its Office for Outer Space Affairs.

[23] A/63/20, paras 177-180.

IAC-08-E8.4.A

OUTER SPACE TRAFFIC MANAGEMENT: SPACE SITUATIONAL AWARENESS REQUIRES TRANSPARENCY

By Paul B. Larsen*)

Introduction

The purposes of outer space traffic management are to avoid intentional and unintentional collisions, to make outer space traffic safe and to assure the freedom of access, egress and uses of outer space granted by the Outer Space Treaty (OST), Art I.[1] China's 2007 anti-satellite exercise followed by the 2008 US shoot-down of its reconnaissance satellite awakened states to the importance of space situational awareness. This paper describes current space traffic laws and regulations It concludes that the main stakeholders are beginning to realize that in order to assure safety of their space objects from collisions by others, they must disclose their locations. Dialogues at the highest levels have begun between private space operators and the military. Furthermore US Senator Barack Obama's recent policy statement expressed that he would negotiate international agreements on rules of the road for outer space. [2] Outer space traffic management is now in a state of change which may eventually lead to standard-setting by an international organization. The paper also concludes it would be impractical at this time to encumber traffic management with military disarmament talks.

A. Issues:

 The author teaches space law at Georgetown Univ. Law Center, Washington DC.

[1] Treaty on Principles Governing the Activities of States in the Exploration and Use of Outer Space, Including the Moon and Other Celestial Bodies, 610 UNTS 2005 (1967).

[2] Policy Statement released Aug. 16. 2008 http://www.spaceref.com/news/viewsr.htma?pid=28880
Senator Obama recognizes the importance of an international approach towards "enhancing capabilities for space situational awareness." He favors an international agreement on rules of the road and plans to enter into negotiations with other countries "to ensure all nations have a common understanding of acceptable behavior."

Existing traffic in outer space is mostly concentrated in identifiable orbits: The many communication satellites in Geostationary Orbit (GSO); the International Space Station and many observation satellites in Low Earth Orbit (LEO); the GNSS satellites in Mid Earth Orbit (MEO). The traffic is a mix of civilian and military space objects of various national registrations. The traffic is mainly tracked by the United States and by Russia. The Europeans have plans to establish their own tracking systems. Only large space objects are being tracked. Several hundred thousand objects, like flecks of paint and droplet of fuel, are not tracked. Traveling at orbital speeds in excess of 25,000 miles per hour, even small space objects can cause great damage. The increasing amount of traffic since the beginning of the space age, in particular the great increase of space debris, and the great economic investments in outer space objects, make collision avoidance an urgent issue. How can outer space traffic be effectively and safely managed, and ultimately how can access, egress and use of outer space be assured? What traffic rules and guidelines are the spacepowers willing to accept?

Outer space traffic management is an ongoing concern among thinkers and writers.. Lubos Parek wrote groundbreaking assessment of the issues in his 1982 paper.[3] In 2007 Corinne Jorgenson, Kai-Uwe Schrogl and Petr Lala studied the issues in their IAA Space Traffic Management Study, taking into account the most recent developments.[4] Several studies have been published in the International Space Law Colloquium. In view of the plans to send humans to the Moon by 2020, planned suborbital flights of humans, the increasing magnitude of plans for economic exploitation of the Earth's orbits, and the increasing military uses of outer space, the study of outer space traffic management must go on. Particular attention-grabbing events during the last year and a half were the intentional shoot-down of the Chinese satellite and the US reconnaissance satellite shoot-down, as well as unilateral demands for freedom of action in outer space for reasons of national security.

B. The Stakeholders

The stakeholders in safe unimpeded outer space flight include the following:

1. The commercial satellite operators, including INTELSAT, the Direct Broadcast Satellite operators, the remote sensing satellite operators, the manufacturers of satellites and satellite equipment, the launch operators, and the operators of spaceports.
2. The military users including the military establishments that operate military reconnaissance satellites, dual purpose satellite services such as Global Navigations Satellite Services

[3] L. Parek, Traffic Rules of Outer Space, 1982 Colloquium on the Law of Outer Space, 37.

[4] Corinne Jorgenson, Kai-Uwe Schrogl, and Petr Lala, Space Traffic Management, IAA Cosmic Study, 2007 Colloquium on the Law of Outer Space, 580. See also discussion of AIA Report in S. Aoki, Space Traffic Management for the Prevention of Weaponization of Outer Space, 2008 Colloquium on the Law of Outer Space. The AIA Report concludes that transparency is essential for outer space traffic management.

(GNSS), as well as secret satellite operations.

3. The countries that do not have current space capability but who want to be assured that they can use outer space when they do obtain space capability individually or in cooperation with other countries.
4. Governmental and nongovernmental international organizations interested in preserving unimpeded access to and use of outer space.

I. Magnitude of Space Traffic Congestions.

A. Keeping Track of Outer Space Traffic

Simply tracking outer space objects does not mean that the traffic is managed and that order is established, but tracking is important in order to know how to circumvent space objects. The course of non-functional space debris cannot be changed, but data collection about location identifies traffic hazards. Thus the few space objects such as the International Space Station (ISS) that can change course, have to be moved out of the way of debris. The ISS experiences at least one major debris encounter each year. The following are important sources of information about the location of objects in outer space; they are all imperfect but all contribute towards a comprehensive traffic data collection

1. *The Registration Convention*:[5] Under the Registration Convention countries are required to register the location of their space objects including "component parts of a space object as well as its launch vehicle and parts thereof." This registry is now electronically accessible. Countries may file information about changes in location after launch. The registry is maintained by the United Nations Office of Outer Space Affairs. Some countries, like the United States, register both their functioning satellites as well as major non-functioning debris known to stem from their satellites.
2. *The International Telecommunication Union*[6] (ITU) registers and recognizes orbital locations (as well as radiofrequencies) of satellites, in particular satellites in the scarce GSO. ITU also registers changes in location of satellites. The information is used to avoid interference with registered satellites. Military satellites do not have to be registered with ITU, but they most often are registered in order to avoid radio interference.
3. *U.S. Commercial and Foreign Entities Program:* Large objects over 10 centimeters are being tracked by the United States Air Force. Each known object is assigned an identity marking that indicates what is known about its history and characteristics and location. Thus the course of

[5] Convention on Registration of Objects Launched Into Outer Space, 1023 UNTS 15 (1976).

[6] Instrument Amending the Constitution of the International Telecommunication Union, (ITU), Project 2001, Legal Framework for Commercial Satellite Telecommunications, 325.

satellites can be changed to avoid known non-functional debris. Orbital data are distributed by the US Air Force in the Commercial and Foreign Entities (CFE) program. Russia maintains a similar tracking system. This valuable information is available to operators and prospective operators of space objects. Debris objects smaller than 10 centimeters are not tracked. They are therefore exceptionally dangerous.

4. *EU/ESA:* In 2008 the EU and ESA decided to establish a space surveillance network in order to identify outer space objects. The new tracking space object network will create an independent tracking network for the EU/ESA countries and will vastly improve their space situational awareness capability. The network is designed to be dual purpose and will track both military and civilian space objects. [7]
5. *Cape Town Convention Registry*: A new comprehensive registry of space assets is being established under the Space Protocol to the Cape Town Convention.[8] This is also an electronic registry which will be publicly accessible. It will be a secondary source of information about location.
6. *National military authorities as well as private operators* keep track of and control their own satellites in order to protect their existence and operation. For that purpose they also track, to the best of their capability, extraneous space objects that may interfere with the operation of their satellites. Military information is often secret and not available to the public.
7. *NASA*: High value individual flights are tracked very carefully by their operations centers. Flights of the Space Shuttle to the ISS are anxiously watched by NASA. Each operational center guides the mission and indeed retains some operational control over the spacecraft.
8. *Private observers* track satellites and report unusual events. Their observations often find their way into the news.

B. Projections into the Future

Astronauts describe outer space as being a very dangerous environment which does not allow for mistakes in operation of space objects. Collisions are dangerous for astronauts, and can frustrate commercial and military objectives; the malfunctions of several satellites are believed to be due to debris strikes. It is important to keep in mind that the concern is not only with objects launched into outer space but also with the cascading effect of successive collisions causing fragments that collide with other satellites and fragments, which in turn cause further escalating fragmentations. The ultimate result could be virtual denial of access to outer space, somewhat like the rings of icy debris that encircle the planet Saturn. Accidental or deliberate military confrontations potentially portends a

[7] DeSelding, European Ministers Endorse Military Role for Galileo, Space News, 29 Sept. 2008, at 8.

[8] Convention on International Interests in Mobile Equipment, www.UNIDROIT.org (2001).

virtual denial of movements into, from, and within outer space. [9]

II. Existing Laws and Regulations on Outer Space Traffic

Existing laws and regulations provide a good legal foundation for space traffic management. The existing regime is mixed international and national. However, the existing legal authority has not really been developed to establish as effective space traffic management system as exists for air, sea and road and rail traffic.

a. International Laws and Regulations.

1. The Space Law Treaties

In their outer space activities States are bound by international law and the UN Charter. Pursuant to the Outer Space Treaty, Art. VI, states have assumed "international responsibility" for the movements of their national space objects in outer space. The Outer Space Treaty provides the most significant legal foundation for management of space traffic. All countries have a right of free access to outer space; which is characterized as the "province of all mankind" (Art 1). States may not claim sovereignty over outer space, therefore individual states may not, without their permission, direct the space objects of other states to travel in specified ways (Art II). Their nationally registered space objects are subject to continuing national oversight and are under national jurisdiction and control (Art VIII). In their outer space activities States must pay due regard to the interests of other states (Art IX), and shall inform other states "to the greatest extent feasible and practicable, of the nature, conduct, locations and results of such activities." [10]

Outer space traffic collisions may cause liability. The Liability Convention, Article III, provides that, when space objects collide in outer space, the states can be liable for the damages caused by their fault or the fault of persons for whom they are responsible. The Convention also provides for joint and several liability of several states acting in unison; they would be jointly liable for damages caused by them to innocent third parties. [11]

The Registration Convention requires launching states to register the location and nature of their space objects. The Convention's definition of space objects is very broad. The definition includes component parts as well as the launch vehicle and parts of launch vehicles. This broad definition makes it difficult for states to register space debris constituting components of space objects because debris is hard to identify, particularly debris that has collided with other debris causing cascading collisions. Consequently, debris objects are often not registered in the UN registry. The sheer mass of debris

---

[9] See Stimson Center, Model Code of Conduct for the Prevention of Incidents and Dangerous Military Practices in Outer Space, www.stimson.org. See P. Larsen, Guidelines for Military Activities in Outer Space, 2007 Colloquium on the Law of Outer Space. Also see S. Aoki, supra n. 4.

[10] Outer Space Treaty, supra n. 1

[11] Convention on International Liability for Damage caused by Space Objects, 961 UNTS 187 (1972)

complicates registration in the UN registry.[12]

2. ITU Registry

The ITU registers satellite orbits and radiofrequencies for effective, interference-free communication. The ITU Radiocommunications Sector ensures rational, equitable, efficient and economical use of the radio-frequency spectrum. ITU therefore registers radiofrequencies as well as orbits used by satellites.[13] ITU does not require military space objects to be registered. But orbits of military space objects are usually registered and their orbits publicly known anyway. The reasons are that military satellites need cleared radio frequencies for their operation just like civilian communication satellites do. The military authorities disclose the orbits of their satellites so that other users can avoid them. The ITU registry is therefore an excellent source of information about the location of current space objects. It is even a good primary source of information about orbits of future space objects because states often register orbits of space objects planned for future use. Assured location of satellites in GSO orbits is particularly sought after because of the density and scarcity of slots in the GSO. Special ITU diligence in placing satellites in GSO orbit is therefore required.[14] However, it should be emphasized that efficient and economic distribution of orbital resources is the main ITU objective; the main objective is not collision avoidance. Users of the ITU Registry must take this into consideration[15]

3. Environmental Regulation of Space Flight

a. Space Flight Restrictions

OST, Art IX, provides that State activity in outer space shall be conducted so as not to cause harmful contamination of the Earth. The UN Principles on use of Nuclear Power Sources (NPS) in Outer Space[16] provide detailed traffic rules for NPS-powered satellites in outer space. Launching states may use NPS to propel outer space flights only when non-nuclear power sources are not available. When used in outer space, NPS may not expose people and the biosphere to radiation in excess of minimum standards set by the NPS Resolution. While NPS may be used in restricted cases, they must be stored in "in sufficiently high orbits after the operational part of their mission." Furthermore, the launching state must inform other states of malfunctions of the NPS space object The Nuclear Test Ban Treaty bans testing of nuclear devices in outer space.[17]

b. Outer Space Debris: The Inter-Agency Debris Committee (IADC)

Traffic in outer space would be much safer if space debris were eliminated. Debris in outer space constitutes the

[12] Id. The registry is maintained by the UN Office of Outer Space Affairs (UNOOSA).

[13] Supra n. 6, ITU Constitution, Art 12, also see Arts 44 and 45.

[14] Id. ITU Constitution, Arts. 44 and 45

[15] Stefan Kaiser, Rules of the Road for Space Traffic, 2003 Colloquium on the Law of Outer Space, 351

[16] Note that the NPS Principles, UNGA Res. 47/68 (Dec. 1992), are not a treaty obligation.

[17] Treaty Banning Nuclear Weapons Tests in the Atmosphere, In Outer Space and Under Water, 480 UNTS 43 (1963)

greatest danger of collision. Removal of all existing debris is not feasible, but generation of new debris can be diminished. That is the work objective of the IADC, [18] which is a committee composed of the major national space agencies (Italy, UK, France, China, Germany, India, Japan, USA, Ukraine, Russia and ESA). The IADC's guidelines for debris mitigation are important because they constitute the internationally agreed policy of the participating governments to prevent in orbit collisions. The IADC voluntary guidelines were endorsed by COPUOS and by the United Nations General Assembly in 2007. [19] The guidelines are a step towards greater outer space traffic safety. They limit the release of further debris, minimize the potential for break-up during operations, as well as post mission break-up and intentional destruction of satellites. Non-operational space objects must be removed from congested areas by post-mission disposal into graveyard orbits. In view of the recent debris-causing shoot-down of their satellites in outer space by China and the United States, it is significant that both countries support the IADC rules. All the spacefaring states have promised to abide by the IADC rules. The rules will continue to be improved.

Individual nations, most importantly the space powers represented on the IADC, have adopted the IADC guidelines on debris mitigation domestically and require that the guidelines be observed both by governmental and by non-governmental launch and satellite operators. National enforcement of the IADC guidelines is also primarily through the national process of authorization and continuing oversight.

B. National Space Traffic Management

1. National Space Traffic laws and Regulations

Outer space is not sovereign.[20] Air space is sovereign.[21] National air traffic separation rules apply in sovereign airspace. They conform with the ICAO separation rules which apply directly in non-sovereign airspace. However, States maintain national jurisdiction and control over their own space objects in outer space,[22] but not over the space objects of other countries. A defined delimitation of air space from outer space would be a significant benefit in outer space traffic management because the launching states would know when their space objects are in their national sovereign air space and when they enter foreign air space during launch but particularly during the long flat landing trajectory of their space objects. Passing though the sovereign air space of other states has to be carefully coordinated. [23] Some countries, like Australia, have domestic laws that define space above 100 kilometers as constituting non-sovereign outer space in which they do not assume

[18] www.iadc-online.org See S. Mirmina, Reducing the Proliferation of Orbital Debris: Alternatives to a Legally Binding Instrument, 99 Amer. J. Int'l L. 649 (2005)

[19] Id. UN DOC.RES/62/217 (Jan 2008).

[20] Outer Space Treaty, supra n. 1, Art. II

[21] Convention on International Civil Aviation, Art. 1, 15 UNTS 295 (1944) (Chicago Convention).

[22] Outer Space Treaty, supra n. 1, Arts. VI and VIII

[23] Note case of aerial intrusion into USSR air space by Korean Air Lines, see Chan v. Korean Air Lines, 490 U.S. 122 (1989)

responsibility for traffic separation.[24] Major spacepowers, like the United States, have declined to delimit outer space by international or by national legislation.

National tracking is the main source of information about the location of space objects. The United States constantly, but unilaterally, tracks all space objects regardless of whether they are of US registry of other registry. The US also tracks unregistered space debris and shares that information. Information about the location of both registered and unregistered space objects is used to calculate probability of possible collisions. For example, recently when a debris object was calculated to be on a collision course with the International Space Station (ISS), the European Automated Transfer Vehicle was used to place the ISS in a lower orbit in order to avoid the debris. [25]

The national states, pursuant to the Outer Space Treaty, Art VI, are obligated to authorize and exercise continuing oversight over activities of private operators in outer space. This is done though national licensing of launches and of spaceports. Operators must show that they are safe operators and that their launch vehicles and payloads are safe enough to navigate in outer space. Not only do the operators have to prove the safety of their space vehicles, but also that they are able to navigate safely. States have one great incentive to insure the safety and navigability of authorized space activities: That is to preclude activities that may cause state liability under the Liability Convention.

2.National Jurisdiction and Control.

Space vehicles, including astronauts on board, while in outer space are subject to the jurisdiction and control of the state on whose registry the space vehicle is carried. Therefore, the state of registry can and does direct the space objects on its registry to operate in accordance with its domestic laws and international obligations. US Space vehicles are built to NASA safety specifications. NASA subjects astronauts to very extensive physical examinations and trains them for specific tasks. The astronauts are also subject to the astronauts code of conduct. The ISS Agreement, Art 11, provides that each ISS partner state shall provide only qualified personnel to serve as crew members. ISS flight assignments are made by special agreements. The partner states agree to establish a code of conduct for the ISS crew. A partner state must approve the code of conduct before it can provide crew members. Each member state must ensure that its crew members observe the crew code of conduct. [26]

The US FAA requires private space vehicles and payloads to comply with specified safety standards. The personnel of suborbital space flights must be qualified pilots who passed medical examinations. They must receive FAA specified training before they are

[24] Australian Space Activities Act, Section 8. See excellent review of Australian legislation in R. Lee, The Australian Legal & Regulatory Framework for Space Launches, (2004/05 ed.)

[25] www.esa.int/SPECIALS/operations/SEM640SAKF_0

[26] International Space Station Agreement, Art 11, www.nasa.gov The states using the International Space Station(ISS) have by separate agreement stipulated how to exercise jurisdiction and control over the jointly operated ISS.

permitted to pilot suborbital space vehicle. The training is intended to protect the health and safety of the public whether on Earth or traveling in space. [27]

III. Current Trends in Outer Space Traffic Management

A. Analogies to International Sea, Road, Rail and Air Traffic

Road, rail, sea and air traffic is by necessity extensively regulated. Such regulations resulted in huge improvement in road, rail, maritime and aviation safety. There are international treaties on road traffic.[28] One need only think of congested city traffic in Glasgow or New York. Accidents such as the sinking of the Titanic, the maritime collision of the Andrea Doria with the Stockholm led to new international regulation of maritime traffic.[29] The collision of two planes, one a passenger plane and the other a military plane over the Grand Canyon in the United States led to the creation of the US Federal Aviation Agency and to improved international air traffic rules in ICAO. These analogies to other modes of transportation do not fit very well to outer space traffic because the other modes of traffic assume that all the moving objects can be directed and can maneuver. Different traffic rules are required for outer space objects.

The most rudimentary air traffic separation scheme is for each operator to see and be seen. To put it bluntly: that is still the traffic separation scheme for outer space. This separation system works as long as outer space is so vast. But when outer space is crowded with space objects, then this rudimentary system is inadequate. This system does not work properly when most of the objects in outer space are incapable being maneuvered. Non-functional outer space debris causes the see and be seen traffic management scheme to fail. Basically nothing much can be done about uncontrolled outer space debris except to minimize future debris. What remains then is the possibility of better traffic management of the space objects that can maneuver and change course.

B. Military Outer Space Objects

Unilateral claims of authority to deploy military space objects in outer space cause the problems of space traffic management to escalate. Uncoordinated intentional or unintentional approaches to space objects of other states may create a danger of military confrontation in outer space. Military engagements in outer space would cause intense debris accumulation and would disrupt civilian uses of outer space. Nuclear war in outer space would be disastrous for future uses of outer space. Unilateral exertion of authority in outer space by China in 2007 to shoot down one of its own 'dead' satellites greatly added to the volume of outer space debris. US claims in its 2006 Presidential policy statement

[27] Commercial Space Launch Act, 49 USC 70101. See excellent article by Timothy Hughes and Esta Rosenberg, 'Space Travel Law (and Politics): The Evolution of the Commercial Space Launch Amendments of 2004, 31 J. Space L. 1 (2005).

[28] Convention on Road Traffic, 125 UNTS 22 (1949)

[29] Convention on International Regulations for Preventing Collisions at Sea, 28 UST 3459 (1972), International Convention for the Safety of Life at Sea, 536 UNTS 27 (1965).

to act unilaterally,[30] that is, to do whatever it considers necessary for national security objectives, followed by the 2007 shoot-down of a non-functional reconnaissance space object, also affected outer space traffic. These and similar claims have increased vulnerability of outer space traffic and called into question their safety and security. The danger of accidental and even intentional collisions leading to claims that sovereign rights have been violated, led to apprehensions of aggressive activities in outer space. Such apprehensions were confirmed in 2008 by then-Secretary of the US Air Force, Michael Wynne, saying that outer space is expected to be a future field of military engagements. Outer space "will shape the American way of warfare well into the 21st Century." [31] Weaponization of outer space and military activities in outer space could restrict access to, egress from, and traffic in outer space.

C. Current Efforts at Transparency of Military and Civilian Outer Space Traffic

Competition of the military space objects of friendly and competing space powers in the 'outer space commons' is dangerous. The tendency to claim and assert unilateral authority to do whatever is in the interest of their national security creates a possibility of military confrontations among the space powers. Some of those confrontations may be accidental for lack of information about the locations and intentions of competing space powers, but some confrontations may be deliberate. Second, increasingly assertive private commercial operators need assurance that they can do business in outer space. [32] Third, private operators may be performing military tasks and can be considered military objectives. Fourth, some military and commercial space operations are indistinguishable; they are dual service space objects, such as GNSS satellites.

Possible conflicts in space traffic have resulted in proposals for maximum transparency of all outer space traffic in order to avoid both accidental and deliberate collisions in outer space. The Stimson Center has put forward draft guidelines for outer space navigation proposing greater transparency of military uses of outer space and urging reduced probability of accidental collisions of military satellites in outer space. The proposed traffic rules are well within the framework of the Outer Space Treaty. [33] The proposal has

[30] US National Space Policy Statement, Oct 6, 2006, www.spaceref.com/news/views.html.

[31] Michael Wynne, Secretary of the US Air Force, Crafting the Next Generation of Military Space, Space News, June 2, 2008, at 19.

[32] P. Larsen, Guidelines for Military Activities in Outer Space, supra n. 8.

[33] Supra n. 8. Stimson Center model code provides: 1. States shall not simulate attacks and shall avoid maneuvers that increase the risk of collision. 2. States shall not use directed energy devices, such as lasers to impair satellites in outer space. They shall not use anti-satellite weapons or space weapons to impair a satellite. 3. States shall follow the IADC debris mitigation rules. 4. States shall keep each other informed about launches and legitimate approaches to each other's satellites. 5. States shall adopt ITU's international traffic management regulations and recommendations. 6. States shall permit verification. 7. States may establish caution zones in order to avoid collision; and States shall maintain mandatory communication systems and engage in consultations to resolve problems. The main purpose is to establish transparency of all space traffic activities so as to create predictability and safety.

started a discussion of and a search for the common ground rules that all states can live with. The proposal can be modified to suit the needs of the countries involved. The result could be space traffic guidelines for simple rules of the road, providing advance notice to identify space activities so that collisions can be avoided. The objective would be to create more certainty, predictability and safety of space flights. Ultimately the traffic rules could reduce the danger of war in outer space

The concept of transparency of location and of movements can be applied not only to accidental and warlike military encounters in outer space; transparency of space traffic maneuvers can also serve the needs of civilian space operators who need to coexist with the military satellites. In pursuit of transparency, a process of coordination and information has begun. A joint government-space industry commercial Satcom mission assurance working group was formed in 2003 to improve communication between the US Department of Defense (DOD) and the commercial satellite operators. The working group is chaired by high level DOD officials (Secretary of the US Air Force) and civilian leaders of the space industry (the President of INTELSAT). The working group is meeting " to begin hammering out basic guidelines, or rules of the road, for space operations." The goal is to discuss fundamentals: how to track debris and how to exchange information. The military and the civilian authorities must initially develop common terminology for sharing traffic data, such as satellite positions. The parties participating in the working group acknowledge that outer space, in particular GSO, is becoming crowded. They are finding the discussions to be fruitful. Communications between the two classes of operators are becoming clearer. The consequence of the military-industry dialogue is that the US Defense Department is making a major effort to teach "senior military leaders about the need for guidelines to bring transparency and predictability to the U.S. and allied space operations.[34] DOD's position is that" China's January 2007 antisatellite test sparked interest in the idea and the Pentagon's recent destruction by missile of a failed spy satellite helped demonstrate the benefits for a space power of communicating its actions and intentions to the world to lessen the likelihood of misunderstanding." [35]

In the same vein, Michael Vickers, the DOD Assistant Secretary of Defense for Special Operations, Low-intensity Conflict and Interdependent Capabilities informed the US Congress on Feb. 27, 2008, that DOD is interested in development of space operating guidelines. DOD "seeks to promote compliance with existing legal regimes, acceptance of international debris mitigation guidelines and development of additional voluntary guidelines for safe and responsible space operation."[36] DOD acknowledges that the military-space industry coordination of traffic in space will improve conditions for space commerce. It will also lead to greater safety, protection and military security in outer space. Knowledgeable commentators believe that the prospective improvement in outer space traffic management "could serve as a foundation for better orbital data sharing

[34] C. Clark, Industry, Pentagon Eye Operating Guidelines, Space News, March 10, 2008, at 5.
[35] Id.
[36] Id.

between the United States and other spacefaring governments - something that will become critical as space gets increasingly crowded." Theresa Hitchens, director of the Center for Defense Information stated that "space situational awareness is critical to any strategy for keeping space assets safe."[37]

Focus on a kind of space traffic management that the stakeholders, from practical experience, conclude is required, is realistic. Such traffic management may result only in minimal order in space, but it is what the stakeholders find they need and can accept. This is constructive movement away from previous policy claim of unilateral freedom of action in space, claim of right to deny such freedom to adversaries, combined with US policy view opposing new international space regimes [38] The new policy initiative is moving in the direction of recognition of the benefits and indeed the need for international co-operation among governments and the space industry. From a US point of view, such co-operation "will be required to ensure the level of space situational awareness (SSA) needed to protect U.S. and the rest of the world's orbital assets." Consequently governments are now discussing potential data sharing plans in order to create a socalled "Neigborhood Watch" network. "While nearly everyone agrees on the need for better data sharing to increase transparency in space, to avoid collisions and monitor debris, the question of how that will be done is as yet unclear." The

[37] Id.

[38] US National Space Policy Statement, Oct 6, 2006, supra n. 30.

US Air Force CFE tracking program [39] is being revised to allow for the greater transparency. The commercial satellite industry is working independently on its own data collection program using not only the US Air Force data but also their own operator data. The objective is to create a new independent voluntary but trustworthy industry collection of traffic data.[40]

Because of the inherently international nature of outer space traffic, there will have to be international cooperation among those parties generating data for greater transparency. Therefore space traffic data collection and traffic management must naturally involve all the spacepowers. [41] Towards that purpose, ESA proposes to organize European tracking resources in order to create an independent European organization for surveillance and data collection that will be made available to both military and civilian users. Furthermore, ESA wants to create a voluntary outer space code of conduct for outer space activities.[42]

CONCLUSION

There is growing realization among all the satellite operators that space situational awareness is essential and that it requires transparency of outer space traffic. The operators are beginning to acknowledge that they

[39] T. Hitchens, 'BuildingTransparency in Space', Space News, 28 July, 2008, at 19.

[40] Id.

[41] Rouge, Interview with Director, DOD National Space Office, Space News, June 30, 2008, at 12

[42] DeSelding, France to Keep Pushing for Ambitious New EU Space Policy, Space News, July 7, 2008, at 7.

must be willing to exchange information about the location of their space objects so that they can 'see' other space objects and themselves be 'seen' and thus be avoided. To avoid accidental military confrontations, the locations of military space objects also need to be disclosed. The DOD Director of the National Security Space Office is right that monitoring of locations can best be done through international cooperation in order to get a full picture of all moving space objects. Unilateral monitoring is not sufficient. Some international organizational structure is necessary for independent monitoring and communication. The unilateral space tracking organizations that now monitor space objects, can coordinate directly with each other the way the GNSS providers began coordination to establish GNSS interoperability. Prof. Aoki correctly concludes that "it seems more important to reach the pertinent gentlemen's agreement among like-minded countries."[43] The international coordination would eventually widen and could involve international organizations able to set standards.[44] Using again the GNSS experience, such standard-setting could involve the COPUOS. It would not be difficult to extend COPUOS functions the way the COPUOS mandate was extended to include GNSS.[45]

Finally, stakeholders agree that outer space traffic safety is the key objective. Single focus on safety can lead to minimal traffic order in outer space. Combination of the outer space traffic management objective with the objective of disarmament would greatly enlarge the issue of space traffic management, place talks in the different forum of the Conference on Disarmament, and burden space traffic management with national security concerns. It is preferable to resolve these issues one at a time.

[43] S. Aoki, supra n. 4

[44] The AIA Report, supra n. 4, contains several recommendations for standards, including rules of the road, orbits, safety and navigation standards.

[45] See International Committee on Global Navigation Satellite Systems (ISG) http://www.unoosa.org/oosa/en/sap/gnss/icg.html

THE IMPORTANCE OF THE RULE OF LAW FOR SPACE ACTIVITIES

Prof. Dr. Stephan Hobe, LL.M.
Board of Directors, International Institute of Space Law
Director, Institute of Air and Space Law, University of Cologne
Albertus-Magnus-Platz
D-50923 Cologne, Germany
stephan.hobe@uni-koeln.de

ABSTRACT

Beginning with the rise of the modern sovereign State and particularly in the 20th century the observance of the rule of law for States – later for all subjects of international law - has become increasingly important. It was basically the idea that all subjects of international law should be subject to (international) law because the law would guarantee that not the economically or militarily superior power would govern, but that all States would be treated - more or less - equally. The norms of international law are so to say rules for the international power game. Only if States undertake to solve all their conflicts by abstaining from using force and instead observe the rule of law there is a chance for international law to be accepted and for the weaker States to maintain and survive. The observance of the rule of law is thus also crucial for the future of the international community as a whole.

Space law as a branch of international law takes part in the need to closely follow the rule of law. Regrettably we can observe with an increasing speed a decreasing respect for the legal rules. Already the five international space law conventions enjoyed an ever decreasing number of ratifications. Moreover, in more recent times, the space law community even starts to totally abstain from any legally binding rules, but to resort to the so called "soft law" of United Nations General Assembly resolutions. The current peek of this development is that by way of an unbinding UN General Assembly resolution, hard international law of the five space treaties has been (re-)interpreted.

This paper will have a look into the consequences of the decreasing observance of the rule of law for outer space activities, but also for the international community as a whole. In this context it will be asked which negative consequences the denial of this observance will have for the space law community as well as for the international legal community as a whole.

INTRODUCTION

The rule of law is an achievement of modern times.[1] Regardless of the various approaches to international law – be they informed by rather positivist thinking like *Herbert Hart*[2] or *Hans Kelsen*[3], be they of a more natural law's point of view,[4] or be they policy-oriented like the famous New Haven School of *Prof. Myres McDougal* and Associates[5], there is no question among international legal scholars as well as in the community of States that to honour the rule of law in international relations is a considerable achievement. Law delimits the competence of States. It is more or less the agreement of States, and of course of other subjects of international law, that their international relations shall be governed by legal rules and not by the use of force. This is, of course, the result of a considerable development of more than 300 years of modern international law.[6] Whereas with the inception of the modern international system after the Westphalian Peace in the 17th century, still the use of force was the dominant means for international relations, this has considerably changed after 300 years. Namely the United Nations Charter in 1945 almost abandoned any use of force reducing it to matters of self-defence (Art. 51 of the UN Charter) or allowing the use of force only after a decision of the Security Council in accordance with Chapter VII of the Charter has been taken.[7] As will be described later, making thus the use of force the exception rather than the rule has important consequences for the rule of law.

After 1945, new legal rules were introduced concerning the use of territory or with regard to the common spaces. This basically meant a refraining from the exercise of sovereignty in so-called common spaces like the Deep Sea Bed or the High Seas or Antarctica, or, most importantly in this respect, outer space and on celestial bodies. In the UN era, States and other subjects of international law have pledged to settle all their international relations by legal, thus peaceful means and by doing so to honour international law. It is therefore of utmost importance to closely observe the rule of law in that the solutions for all main conflicts are incorporated in the legal rules determining the behaviour of States and international organizations.

It is that aspect that we are interested in the following paper. What will be described is a development with regard to the legal relations of outer space activities that clearly shows many crucial rules for outer space activities being not governed any longer by hard international law. Rather, rules are implemented that aim to avoid any legally binding character. All this is seen as a considerable deviation from the observance of the rule of law. And this deviation asks for consequences of such non-observance. It is thus submitted that the current development actually means a deviation from the rule of law and answers shall be given with regard to future prospects for the rules for outer space activities. This will finally allow for an assessment of where we stand with regard to international relations with regard to outer space activities.

MAIN PART: DEVELOPMENTS

1. The Function of Law in the International Community

After the establishment of the United Nations Organisation in 1945 it was clear that a much stricter observance of the rule of law with regard to international relations should take place.[8] The prohibition of the use of force with the few exceptions mentioned above had a signal function for international relations. Moreover, the UN era gave rise to many more areas to become subject of international legal regulation. To mention only a few, there was the vast field of human rights[9], of international environmental law[10], of international economic law[11], of the law of the common spaces like the High Seas and the Deep Sea Bed[12], of Antarctica[13] and outer space[14], and a considerable part of international humanitarian law through the Geneva Conventions of 1977[15] that came into being only after 1945. In general we can say that in an enormous way international relations were made subject to legal regulation. This was the case notwithstanding the fact that even the United Nations Organisation could not overcome one major systemic weakness of international law. Of course, any question of enforcement of international law is still problematic due to the missing of a central enforcement agent leaving enforcement of international legal rules in a decentralized way to the States.[16] Still, a growing variety of international tribunals symbolize in an intensity that cannot be overlooked any longer a general tendency to create agents for the enforcement of international legal rules. Besides the International Court of Justice and some regional courts like the European, African and American Courts of Human Rights, and the European Court of Justice, we find at the universal level the Law of the Sea Tribunal, the WTO Panels and the Appellate Body, the

International Criminal Court as well as, with a more limited scope the Yugoslav and Ruanda Tribunals. Also a discussion on the creation of an international court on environmental matters is taking place.[17] This clearly shows that the tendency towards a "legalisation" of international relations is precisely giving governments the opportunity to settle their conflicts by legal means. And this all is a development within the UN era after 1945. Of course, there are problems, but this shows very clearly the general tendency of framing international relations against the possibility to honour international law. Of course, one may also ask the question how much judgements of these tribunals are fully enforceable. But as a whole, the tendency of a grater "legalisation" just described holds true for international relations.

2. Law-Making with Regard to International Space Activities: From Hard Law to "Soft Law"

As is well known international legal relations with regard to space activities started relatively early after the inception of the space age with the launch of the first artificial satellite Sputnik 1.[18] Already in 1957, major decisions were taken by the international community. The first decision was that, of course, the major space powers of that time, the then Soviet Union and the United States, wanted to have legal rules governing outer space activities. Moreover, they wanted to have those rules, although leaving the most crucial decisions to their bilateral decision-making, being decided in a multilateral scenario, namely the United Nations. This led to the creation of the United Nations Committee on the Peaceful Uses of Outer Space as a special committee to the United Nations General Assembly in 1959.[19]

2.1. The work of this Committee had at the beginning been remarkably successful. Five international conventions were negotiated by this Committee and later adopted by the United Nations General Assembly. We can thus speak of a first phase of law-making for outer space activities that lasted from the late 1950ies to 1979. Here, firstly the very important Magna Charta for outer space activities, the Outer Space Treaty of 1967, was adopted. This important international agreement as is well known lays down all the main principles for outer space activities.[20] It does so in a rather general way giving leeway for States and other actors to develop on these legal rules. It has so far been proven to be flexible enough to basically serve the purpose of governing international space activities. Major principles as the freedom for outer space activities in terms of research, exploration and use (Art. 1 OST), the prohibition of the making of claims of sovereignty to areas of outer space and on celestial bodies (Art. 2 OST), the principles of registration, liability and assistance (Art. 5, 7, 8 OST), some rather rudimentary rules on the protection of the environment (Art. 9 OST) make this international agreement really a very important legal contribution. It is, moreover, widely accepted that having now 98 ratifications though this is only a little more than half of the members of the international community, is a considerable number because all the main space powers as most of the States whose interests are mostly affected are member of this club of treaty parties.

Although the three following international conventions, namely the Rescue Convention of 1968, the Liability Convention of 1972, and the Registration Convention of 1975 were not so successful in terms of membership, they can nevertheless be regarded as concretizing general principles of international space law. In this regard they are also major contributions to the governance of the rule of law. In the Rescue Agreement[21], major principles of rendering assistance to astronauts in distress are laid down. Moreover, the Liability Convention of 1972[22] has laid down rather innovative principles in that it contains the principle of strict liability for States for damages caused by space objects on the Earth or in the air (Art. 2 LiabC). Finally, the Registration

Convention of 1995[23] makes a first attempt to come up with the major rules for the registration of space objects in an international as well as a national register (Art. 2 and 3 RegC). As of 1 January 2008, the Rescue Agreement has got 90 ratifications whereas the Liability Convention had 86 and the Registration Convention 51 ratifications. This is, of course, particularly in case of the Registration Convention, not a very high number, but it is with regard to the very special focus of outer space activities still considerable.

The crisis of international law-making with regard to outer space activities began with the negotiation of the Moon Agreement that was eventually agreed upon in 1979.[24] Here, space legislation became heavily affected by the general discussion about a new international economic order.[25] This international discussion that was the result of a decolonization process bringing about the numerical majority for developing countries in the UN General Assembly and their - sometimes violent – request for more participation in international law-making. This led to the result that also international common spaces like the High Seas and the Deep Sea Bed or even outer space were at the beginning subjected to principles that aimed at contributing to the interests of so-called developing countries. One could very clearly see the consequences in the Law of the Sea Convention that was negotiated between 1973 and 1982.[26] Its Part XI on the exploitation of the Deep Sea Bed as well as the respective Annex VII to the Law of the Sea Convention contained in its original feature the famous provision that the Deep Sea Bed and its resources were the "common heritage of mankind". This basically meant that developed States were put under a legal obligation to transfer technology to developing countries in order to enable them to actively participate in Deep Sea Bed mining activities and moreover that an International Deep Sea Bed Authority was established for the distribution of mining rights for manganese noodles in the Deep Sea Bed. This rather rigid approach, putting enormous restrictions on the freedom of States and private enterprises with regard to mining activities, was however later considerably reduced by an Amending Protocol to the Montego Bay Convention in 1994.[27] The language of the common heritage of mankind was just a symbol for the motives of developing countries and their effort for participatory rights.[28] It consequently also affected the negotiations on the Moon Agreement and became an integral part of this Agreement in its Article 11. The Moon and its resources are considered to be the common heritage of mankind whereby, interestingly enough, only very rudimentary details about the consequences of this declaration to the common heritage of mankind are contained in the Moon Agreement.[29] Nevertheless, this was reason enough for major industrialized countries to totally disagree and refrain from their signature as well as the ratification of this Agreement. And it also meant the end of the 20 years phase of international treaty-making for outer space activities. Until now, the Moon Agreement has got only 13 ratifications, none of them being the ratification of a major space power.

2.2. What followed then as a <u>second phase</u> approximately from 1980 to 1995 was the refraining from the adoption of international agreements and the orientation towards the adoption of United Nations General Assembly Resolutions. That was the case in 1982 with regard to satellite broadcasting, i.e. on the field of international direct telecommunications where a set of legal principles was established.[30] It was, furthermore the case in 1986 with regard to the United Nations Principles on Remote Sensing.[31] And it was, thirdly, the case in 1992 with regard to the Principles on the Use of Nuclear Power Sources.[32] The idea behind all these sets of principles was certainly that they wanted to cover specific areas of outer space activities and describe the legal rules for such activities. These rules are less general than the international agreements and they are all focussed on specific fields of outer space activities. Nevertheless, it is well known that United Nations General Assembly Resolutions do not possess legally binding character.[33] This is so because the United Nations General Assembly has no law-making power. A resolution is a recommendation that can certainly reflect some State practice. But it has certainly not the same character as international treaty law. For example, one can now very seriously doubt that the direct broadcasting by satellites' legal principles of 1982 still reflect international State practice with regard to these satellites in the era of global communications that we have in the early 21st century. Moreover, major parts of the Principles on Remote Sensing cannot live up any longer to the era of commercialization that is presently characteristic for our times. We can thus summarize that this second phase brought a certain deviation from the strict observance of the rule of law.

2.3. A third phase that started approximately around 1995 and lasts until today is now characterized by a development that can give rise to even more serious concern. Starting in the early 1990ies in the aftermath of the struggle for the new international economic order, an effort for a reinterpretation of Article I para. 1 of the Outer Space Treaty had been made.[34] The principle - known as the common benefit principle of Art. I para. 1 of the Outer Space Treaty - is rather difficult to interpret because it lacks preciseness.[35] This was deliberately done in that the Outer Space Treaty in its Article I para. 1 says that the exploration and use of outer space shall be made for the benefit of all mankind. It was thus not clear whether this rule would contain an obligation to share military and particularly economic advantages derived from outer space activities. Under the request of developing countries, the United Nations Committee on the Peaceful Uses of Outer Space deliberated for quite some time and concluded in 1996 in a resolution some major principles for the interpretation of Article I para. 1 of the Outer Space Treaty.[36] These interpretations contained as a general direction the reiteration of the freedom of States to determine which States they would like to cooperate with and how they would distribute the benefits and the results gained by their own space activities. But the interesting point is that for the first time, the United Nations General Assembly adopted a resolution to reinterpret international treaty law, namely Article I para. 1 of the Outer Space Treaty.

Moreover, in 2004, after lengthy and complicated deliberations, a resolution was adopted that came up with a reinterpretation of one of the key notions of outer space legislation, namely the notion of "launching state".[37] It had become clear that the notion of "launching state" as is contained in the Liability Convention as well as in the Registration Convention did not fully pay tribute to all arising problems, namely with regard to private space activities. And again it is important to see that the United Nations General Assembly resorted to a legally non-binding resolution for the reinterpretation of this key notion of international space legislation.

Finally, in 2007, the United Nations General Assembly adopted a resolution on restating some major principles for the registration of space objects.[38] That was considered necessary because such State parties did not pay tribute to the major obligation of international space law to register each of their objects launched into outer space in a national as well as in an international register.[39] Moreover, problems with regard to private space activities concerning registration gave also rise to make this new move.

What is now interesting with regard to this investigation is that this third still continuing phase is a totally new attempt to (re-)interpret international treaty law for outer space activities by non-binding United Nations General Assembly resolutions. And this exactly gives rise to further concerns with regard to the rule of law as will be shown in the following.

3. Observations

With regard to these developments, some observations are necessary. We see that there is a slow but clear deviation from hard international law via the making of legally non-binding rules for specific space activities towards the reinterpretation of legally binding rules by non-binding legal rules. We must thus first ask the question whether this in fact means such a weakening of the rule of law.

In this paper the submission is made that this in fact means such a weakening. It is different whether strict treaty law obligations are incumbent upon States or whether there is the adoption of United Nations General Assembly resolutions that are precisely not legally binding. In the realm of legally non-binding rules, no State is under a strict obligation to honour international law. Even if one would strengthen the argument like is very often done that the consecutive and repetitious mentioning of certain rules in legally non-binding UNGA resolutions could amount to the creation of (new) international custom,[40] there is no doubt about the fact that the deviation from a certain custom is much more difficult to prove than a deviation from strict treaty rules. Thus, the point is that not only the clear preference of adopting legal principles in the form of the United Nations General Assembly resolutions covering various fields like direct broadcasting satellites, remote sensing or nuclear power sources is a weakening rule of law. More severely, the (re-)interpretation of binding international treaty law by non-binding UNGO resolutions is a very serious deviation from the rule of law. The statement of people involved in the work of the UNCOPOUS that nothing else than non-binding resolutions could be successful is significant in itself for this regrettable development.

It can also be compared with other developments that shall be briefly mentioned here. First of all, the observance of the international legal duty to register international objects and to furnish information about the object and purpose of such space object with regard to the national and international registers is very weak.[41] This gave precisely rise to the fact that the United Nations General Assembly tried to come up with some precision as contained in UNGA resolution 62/191 of 17 December 2007.

The second development of interest is that, until now, regrettably, there is no governmental statement concerning the unilateral claims of individuals to plots on the Moon. It was only a – obviously non-binding – declaration of the Board of Directors of the International Institute of Space Law[42] that made it clear that these individual claims are totally unjustified under and are even against international law. But it is interesting that the respective State whose individual has made such claim did not object to this behaviour although it had infringed deeply upon international legal rules.

And finally, it is interesting to observe that according to Article 18 of the Moon Agreement, 10 years after the entry into force of the Moon Agreement which took place in 1984 – the question of the review of the Agreement should have been included on the Agenda of the United Nations General Assembly in order to consider whether the Agreement would require revision. Although much opposition was heard in international legal circles against the existence of the Moon Agreement, the natural means for expressing this concern, namely the putting of the Agreement on the Agenda of the United Nations General Assembly, was never undertaken.

One can, therefore, clearly say that today the observance of the international rule of law is in a severe crisis with regard to outer space activities.

CONCLUSION

As had been shown in the beginning, the observance of the rule of law is crucial for international relations. This is also true for outer space activities. All countries, be they small or be they big, should observe international agreements of binding character.

In the future, at least two developments give rise to strongly appeal to a more close observance of the international rule of law: These developments are the growing trend to more private space activities that should encourage more international and particularly more national space legislation. It should be up to the United Nations as has been done very generally by its Legal Subcommittee to take up this development and to encourage the drafting of more national space legislation. Also in the field of national space legislation, States can in their own interest demonstrate that they prefer legal regulation over a non-regulated area. A drafting of a model law would e.g. give the possibility to States to draft their national laws against the background of such a model as a good example.[43]

Moreover and perhaps even more importantly, the United Nations Committee on the Peaceful Uses of Outer Space has taken up the extremely important question of space debris on its Agenda with a view to consider its legal regulation as well in the course of these deliberations some time later. It is clear from any scientific report that the solution of the problem of space debris also by legal means will be decisive for the future of all outer space activities because such regulation can preserve the outer space environment for such activities. It would thus be of utmost importance to end the negotiations on space debris with a legally binding instrument and not just with a non-binding United Nations General Assembly resolution.

With such a move, the international community could prove that it would be willing to live up again to the observance of the rule of law which is probably the most important development of modern international law also in the 21st century.

---

[1] For an overview of the modern influence of the rule of law on international relations and international law see e.g. I. Brownlie, The Rule of Law in International Affairs, The Hague et al. 1998.

[2] H. L. A. Hart, The Concept of Law, 2nd. ed. Oxford, 1994.

[3] H. Kelsen, The General Theory of Law and State, New York 1945, 325 et seq.

[4] Classic naturalist approaches are the ones of H. Grotius, A. Gentili or F. Suarez, see on their theoretical considerations T. Meron, Common Rights of Mankind in Gentili, Grotius and Suarez, AJIL 85 (1991), 110 et seq.

[5] M. S. McDougal/H.D. Lasswell/W. M. Reisman, The World Constitutive Process of Authoritative Decision, Journal of Legal Education, 19 (1967), 253 et seq.

[6] For a description of this development see inter alia W.G. Grewe, Epochen der Völkerrechtsgeschichte (Epochs of the History of International Law) Baden-Baden 1984, 323 et seq; K.-H. Ziegler, Völkerrechtsgeschichte History of Public International Law), 2. ed. 2007, 142 et seq.; Stephan Hobe/Otto Kimminich, Einführung in das Völkerrecht (Introduction to Public International Law), 8th ed. Tübingen/Basel 2004, 36 – 63.

[7] See for a description Hobe/Kimminich, note 6, 305 et seq.

[8] This is the main conclusion of Brownlie, op. cit., note 1, passim.

[9] As is well known this development started with the adoption of the Universal Declaration of Human Rights on 10 December 1948, Res. A (III), GAOR Doc. A/810, 71 and was continued by the two universal international agreements , the International Covenant on Civil and Political Rights and the International Covenant on Social, Economic and Cultural Rights, both of 19 December 1966, 999 UNTS 171 resp. 993 UNTS 3.

[10] Starting with the Stockholm Declaration on the Human Environment of 1972; for a good description of the development of international environmental law see P. Birnie/A. Boyle, International Law and the Environment, 2nd ed. 2002.

[11] The start of this development was made by the installation of the Bretton Woods Institutions World Bank and International Monetary Fund whereas the foundation of an International Trade Organisation failed directly after 1945, see for a concise description C. Herrmann/W.Weiß/C. Ohler, Welthandelsrecht (International Economic Law), 2nd ed. München 2007, 28 – 82.

[12] The four Geneva Conventions of 1958 were followed by the Law of the Sea Convention of 10 December 1982, reprinted in 21 ILM 1982.
[13] Antarctic Treaty of 1 December 1959, 402 UNTS 71,
[14] Starting with The Outer Space Treaty of 1967, 610 UNTS 205, and the description infra 2.
[15] Geneva Conventions of 1949 with Additional Protocols of 1977, 75 UNTS 135; 75 UNTS 287; 1125 UNTS 3; 1125 UNTS 605.
[16] This question had been addressed in the symposium J. Delbrück (ed.), "Allocation of Law Enforcement Authority in the International System", Berlin 1995.
[17] On the problem of fragmentation of international law though the growing diversity of tribunals see inter alia B. Rudolf, Unity and Diversity of International Law in the Settlement of International Disputes, in: A. Zimmermann/ R. Hofmann (eds.), Unity and Diversity in International Law, Berlin 2006, 389 et seq.
[18] See for a very good description of the beginning of the space age W.A. McDougall, ... the Heavens and the Earth – A political History of the Space Age, New York 1985, 20 - 73.
[19] By UNGA res.1472 (XIV) of 12 December 1959.
[20] Source: 610 UNTS 205.
[21] Agreement on the Rescue of Astronauts, the Return of Astronauts and the Return of Objects Launched into Outer Space of 22 April 1968, source: 672 UNTS 119.
[22] Convention on International Liability for Damage Caused by Space Objects of 29 March 1972, source: 961 UNTS 187.
[23] Convention on Registration of Objects Launched into Outer Space of 14 January 1975, source: 1023 UNTS 14.
[24] Agreement Governing the Activities of States on the Moon and Other Celestial Bodies of 18 December 1979, source: 1363 UNTS 3.
[25] For a description of this development see S. Hobe, Die rechtlichen Rahmenbedingungen der wirtschaftlichen Nutzung des Weltraums (Legal Framework for Economic Space Activities), Berlin 1992, 93 et seq, 239 et seq.
[26] On this negotiating process see R. Wolfrum, Die Internationalisierung staatsfreier Räume (The Internationalisation of Common Spaces), Heidelberg et al. 1984, 371 et seq.
[27] Agreement on the Implementation of Part XI of the Law of the Sea Convention of 10 December 1982, adopted on 28 July 1994, text reprinted in: 33 ILM 1309.
[28] On the principle see inter alia R. Wolfrum, The Principle of the Common Heritage of Mankind, ZaöRV 43 (1983), 312 and for a current account of its influence S. Hobe, Was bleibt vom gemeinsamen Erbe der Menschheit?, in: Liber Amicorum Jost Delbrück, Berlin 2005, 328 et seq.
[29] See on more current problems F.G. von der Dunk, The Moon Agreement and the Prospect of Commercial Exploitation of Lunar Resources, AASL XXXII (2007), 91 et seq.
[30] Principles Governing the Use of States of Artificial Earth Satellites for International Direct Television Broadcasting, UNGA res. 37/92 of 10 December 1982.
[31] Principles Relating to Remote Sensing of the Earth from Outer Space, UNGA res. 41/65 of 3 December 1986.
[32] Principles Relevant to the Use of Nuclear Power Sources in Outer Space, UNGA res. 47/68 of 14 December 1992.
[33] The is the absolute leading opinion and practice, see only Hobe/Kimminich, op. cit., note 6, 196 et seq.
[34] Declaration on International Cooperation in the Exploration and Use of Outer Space for the Benefit and in the Interest of All States, Taking into Particular Account the Needs of Developing Countries, UNGA res. 51/122 of 13 December 1996.
[35] See inter alia Hobe, op. cit., note 25, 93 et seq.
[36] See Hobe, op.cit., note 28, 339.
[37] Resolution 59/115 of 10 December 2004: Application of the concept of "launching State".
[38] Resolution 62/101 of 17 December 2007: Recommendation on enhancing the practice of States and international intergovernmental organizations in registering space objects.
[39] See for a description of the previous State practice S. Mick, Registrierungskonvention und Registrierungspraxis, Cologne et al. 2007.
[40] See for an assessment of the arguments e.g. Hobe, op. cit. note 6, 196 et seq.
[41] See the striking description of the State practice in leaving aside this obligation Mick, op. cit., note 39, 70 et seq.
[42] However a statement of the Board of Directors of the International Institute of Space Law on this matter of 2005 can be found under http://www.iislweb.org/docs/IISL-_Outer_Space_Treaty_Statement.pdf
[43] In the Space Law Committee of the International Law Association such work on a Model Law will be undertaken during the period between the Rio (2008) and the Hague (2010) Conference; see also for important work in preparing the basis the Berlin Workshop of the Cologne Institute of Air and

Space Law and of DLR in S. Hobe/ B. Schmidt-Tedd/ K.-U. Schrogl (eds.), Towards a Harmonised Approach for National Space Legislation, Cologne 2004.

IAC-08-E8.4.15

SPACE LAW STUDENTS PARTICIPATION IN REAL ENGINEERING RELATED PROJECTS AND RESEARCH

Amalio Monzón
Laboratory for Space and Microgravity Research, Spain
amalio.monzon@leem.es

Andrés Russu
Laboratory for Space and Microgravity Research, Spain
andres.russu@leem.es

Héctor Salvador
Laboratory for Space and Microgravity Research, Spain
hector.salvador@leem.es

ABSTRACT

Space Law is a recognized way of collaboration, promotion of sustainable development and international cooperation since it started, with the launch of world first artificial satellite in 1957. Likewise education and research in a scientific, technical and juridical way constitute a key factor in the development of Space Law. Laboratory for Space and Microgravity Research (LEEM), a Spanish non-profit space students and young professionals association dedicated to motivate and support students to develop space related projects has created a Juridical Committee whose principal objectives are to divulgate this field between science and law students, and to collaborate in Space Law development. Through direct participation in real space projects, LEEM provides to students the opportunity of having a real hands-on the topics of these activities. Students research about all related aspects of Space Law concerning a specific project at the same time they collaborate in its development, making possible a wide knowledge in technical and legal aspects. These research activities support the legal aspects of LEEM and the development of more ambitious projects, assuring all applicable laws. Our initial research line is concerned about launch of sounding rockets and picosatellites. Spain has not a clear regulation about them and in some cases; activities related with these fields are not specifically covered by the present legislation. Our Juridical Committee finds ways where our activities are adapted to Spanish legislation and ways to motivate a fluent contact with authorities in order to promote a specific regulation. On the other hand, students and young professionals who develop these Space Law researches have the possibility of networking with professional and experts, presenting their results, and discussing about new ideas during the association annual congress, the Spanish Space Students (S3). This innovative way of research makes possible to study the current status of local applicable legislation in space from a practical point of view, to promote Space Law field among students of different disciplines and collaborating to its development and progress. In addition, it constitutes an opportunity to include practical ideas and knowledge in a future national law.

I. INTRODUCTION

Space Law is one of the disciplines with more future work in the Space related activities due to the requirement of higher detail in the current Law and new jurisprudence for the incoming problems. The new objectives of this discipline of the Law are related with the current goal of the Space business. The commercial spaceflights or the international collaboration to avoid Near-Earth object impacts, are examples of the necessity of having an adequate juridical environment to solve these problems. The development of the Space Law requires lawyers, engineers and scientifics that working together serve like an engine of such development. For that reason, these objectives require educational programs, which provide the required space specialization of all of them. In such framework, the Laboratory for Space and Microgravity Research (LEEM), association with the objective is to motivate and to support students to develop their own Space related projects, has created a Juridical Committee. This committee has the aim of the Space Law promotion among the students of different disciplines at the same time that provides legal support for all the association activities. From the creation of this Juridical Committee in the beginning of 2008, there are two main lines of work. On one hand, there is the aim to promote the Space Law, as referred before, among the different university disciplines. On the other hand, the association projects and conferences, lectures and congresses are the perfect framework for the introduction to space student of the Space Law basis. The two work packages have been already started. Firstly, the different projects of the association as Sounding rockets, International Space Facilities for experimentation and Satellites projects have been studied from a juridical point of view. This paper will present the main conclusions acquired. Secondly, the first Space Law Workshop has been done among the LEEM members. The purpose of the training is to equalize the knowledge in space topics among the different disciplines required for the understanding of the several points of view of the Space. This paper will detail the actions items, conclusions and lessons learned. The main goal of the LEEM

Space Law Committee is to create the basis for the development of more ambitious projects in the future with more students and Space Law related Institutions.

II. SPACE LAW THROUGH ENGINEERING RELATED PROJECTS

The Juridical Committee recent activities have been focused in the support of the engineering projects developed by the LEEM members. Depending on the experimentation platforms different juridical considerations are required. Until today, the Juridical Committee has support the sounding rocket activities carried out by our association during the present year. The committee has identified the main problems and risks derived from these activities and it has supported the search of ways to develop them. The experimentation at international space facilities and the development of a picosatellite constitute the new activities, which allow students to introduce them to the Space Law through the participation in real engineering projects.

Sounding rockets

The First International LEEM CanSat Competition was organized during 2008 with the sponsorship of the Technical University of Madrid. It took place from 10th to 12th of April 2008, and had a participation of ten teams from different Spanish universities and two teams from Malaysia, with a total of 85 students participating on the event.

A CanSat (1) is a scale satellite integrated within a soft drink can, that it is launched up to 500 meters onboard an amateur rocket. Inside this reduced volume, all subsystems must be fit (power, telemetry & command, recovery, payload). Once it is built and launched the CanSat should accomplish a full mission as to come back to a previous defined position or to obtain telemetry data. This "scale" project undergoes all the same phases as a real satellite mission, including the launch campaign and the data analysis, having an important educational value.

In addition to the organization of the competition, LEEM decided to develop the rockets needed to launch the device. At the same time that this project offered to student the opportunity to participate on real hands-on engineering activities, it has been used as a way to introduce students into the Space Law discipline through the collaboration with the Juridical Committee in the support of this activity.

During the preparation phase of the competition the Juridical Committee identified a lack of a clear law about the experimentation with sounding rockets in Spain. These problems are caused mainly by the absence of a specific law, which regulates the operation of sounding rockets as experimental devices. In consequence, a lot of difficulties arise focused on these aspects:

- Purchase of materials.
- Transport of the rockets.
- Authorization to use the launch camp.
- Closing of their surrounding airspace.
- Management of a risk insurance, which would cover the damage and responsibilities of a potential accident. The difficulty is higher if we try to get an insurance for an on board or deployable (CanSat) payload within the sounding rocket.
- Lack of support of the local authorities (police, fireman, and so on).
- Socially confused vision, which relates Sounding rockets to military activities.

This gap in the Law confuses the local governments and institutions at the moment of designating the required organism, which would be on charge of the activities supervision to guarantee the safety. In addition it is necessary to clarify if this activity should be developed under the military supervision (use of military facilities to launch the rockets) or not (use of civil aerodromes). On the other hand, this lack of definition difficulties the support to the experimentation from government and institutions.

The solution applied by LEEM has been to cover their activities by different legal frameworks. For instance, the rocket engines were bought abroad Spain as educational material, the igniters were acquired as pyrotechnic material, the event was managed as aero model aircraft flying exhibition and the civil responsibility was assured until an specific quantity. It is not the proper one and has the risk to involve sounding rocket activities in a non defined law framework if an accident occurs. In addition this lack of legal frame means the assumption of real risk from the organizers side, because if the material imported (mainly the engines) is understood as explosive it could means the arrest of the purchase responsible.

Based on the experience in the operation of sounding rockets, we think necessary that the following measures will be considered in order to develop the current state of the law:

- The clarification of a clear framework for the sounding rocket experimentation and educational activities, which facilitates them.
- The establishment or the designation of an organism, which provides the certification of engines, rockets and operators. Nowadays the sounding rockets activities in Europe are developed under the American Tripoli license. While, this certification is recognized by vendors and buyers, it is not by the competent national authorities. Into this situation the best solution is to cover the product and its operation through the American certification.
- The specification of the upper airspace law related aspects for sounding rockets. In case of a high altitude rocket, the Law should be clear, in order to avoid critical problems as the invasion of different airspace from the one where the rocket is lifted off.

- The clarification of the Civil Responsibility on the rocket landing. The lack of a governmental organism able to certify and supervise the activities makes that all the assurances do not cover the damage caused by aircrafts or devices destined to the aerial navigation.

We understand that these measures contribute to the promotion of the sounding rockets as experimentation platform and will facilitate the support to these activities from institutions and the government. Furthermore we consider that it is especially important to develop this framework to make possible the operation of high altitude sounding rockets, which will allow the microgravity experimentation on board of them, constituting a low cost platform for the research in conditions closer to the weightlessness.

Further projects to introduce to Space Law

Members of LEEM have participated in projects that use international space facilities for their experimentation. Examples of these platforms are the International Space Station or the Foton Satellite. They are used as a research platform where high-level educational experiments are operated by an astronaut or automatically at unmanned spacecrafts. From this experience we have identified, several points to pay attention if any collaboration between our association and a third party is established to develop similar experimentation projects. Firstly, it is necessary to clarify the rights of the data obtained from the experimentation by students on behalf of a university, association or other, who normally have the supervision of a teacher or a person from the institution on behalf of they work. Secondly, there is the possibility of an error of the students experiment which could be propagated to the facility system and, consequently, to other experiments. Because of this, it is necessary to establish previously all the responsibilities of each party involved in the experimentation activities.

On the other hand, the development of a picosatellite is one of the main projects that LEEM will carry out for the coming years. Currently, the project is in the conceptual phase and the main objectives have been established. They are the creation of know-how needed to develop further more ambitious projects, the organization of an operational network based on different universities, infrastructure, partnership with industry and institutions, and the entirely development by students of the satellite, covering the design, the manufacturing, the test and the operation. The Juridical Committee will foresee the possible legal handicaps that this project can lead.

Through these lines that will be developed by the Juridical Committee in a short term, the students will have the opportunity to collaborate with real engineer projects at the same time that it introduces themselves to the Space Law.

III. SPACE LAW THROUGH EDUCATION ACTIVITIES

The second line of work during the present year of the Juridical Committee has been focused on the development of educational activities to students related to Space Law. Due to Space Law is a very specialized discipline, LEEM does not expect to develop educational activities, which covers in depth all related aspects of the Space Law. LEEM has the intention of performing a training whose objective is to level the knowledge of student from different disciplines (law, technical, scientific, etc).

Through this training is expected to carry out an introduction to the basic concepts and the recent history of the Space Law, providing to the student a compilation of the resources available on the net in order to make possible their development in an autonomous way. The last objective of the committee related to education is to bring the Space Law to students.

Space Law Workshop

As previous step to develop more ambitious educational activities with the collaboration of other third parties specialized in this discipline, it was decided to organize a pilot Space Law Workshop to be performed in a small training scale in Space Law. The purpose of this pilot is to obtain the lessons learnt necessaries to develop trainings with a major scope, open to university students. LEEM members, coming from technical and law studies, carried out the pilot workshop. In this way we simulate the heterogeneity that has a standard group of training in Space Law. Likewise, this pilot workshop was used to identify the expectative of the participants in a Space Law training to adapt the contents and the methodology of our training.

The process of development had a preparation phase to search the information and the pedagogic material to use, a phase to check with the participant if the initial objectives of the workshop fulfil their expectatives, a phase to adjust the contents, a phase to give the pilot workshop and finally a lesson learnt phase.

It was remarkable the inputs provided by the students concerning their expectations on the training. It reflected that beyond a traditional training, there are very valued factors as the repercussion of the training in their future career, the network of people interested in the discipline and professionals that the training provides or the flexibility of the training.

According to these expectations the objectives of the workshop were defined as:

- To know the history, the current status and the future challenges of the Space Law.
- To know the institutions and organizations which collaborate in the development of the Space Law and their opportunities.

- To provide the basic concepts and the information sources which make able the student to develop their own knowledge.
- To establish a base to develop those Space Law aspects which are related to experimentation activities.
- To organizes a network between students and young professionals interested in this discipline, which allows to collaborate in the development of the Space Law.

In a transversal manner to these objectives, it was agreed to provide the enough flexibility to the workshop to make it compatible with other activities at the same time that the aim of the project will be reached. With this proposal, it was decided to develop a 60% of the training on line supported by on site sessions which facilities the contact with the other participants and their interaction.

Once the objectives and the format were established, the pedagogic tools were developed. These tools were designed to balance contents, teamwork, and reflection activities which helps to obtain the capabilities to the further development of the discipline by the student.

Based on this scheme five sessions were designed. The first three sessions were given on line through an Internet platform with a restricted access area destined to the student communication and interaction. The first session was dedicated to learn about the platform, to solve the problems and doubts and to meet the other participants. The second session was used to read the theoretical contents focused on the history, the principal agreements, the challenges and the institutions, solving by mail all related doubts. The third session was composed by a forum in which the student answered to an open question. The fourth and fifth sessions were given on site. In the fourth one the theoretical concepts were explained in detail. In addition a round table was carried out to perform a brainstorming about student projects to develop related to Space Law with the objective to facilitate the formation of a student collaboration network in this discipline. Finally, the fifth session was dedicated to a highlight session by a member LEEM in which it was exposed the experience related to the legal difficulties that the association has found during the development of student space related projects.

From this pilot experience we have obtained the following conclusions:

- Due to the knowledge heterogeneity among technical, scientific and law students, it is necessary to focus the contents on those basic aspects of each discipline, which are essential for the Space Law. It is also important to level the knowledge of the participants to start from an equal base.
- The on line sessions provide a great flexibility at the same time that facilitates the interaction with all the participants of the workshop.
- One remarked expectative of the participants is that the workshop will have a positive repercussion in their careers. A practical focus of the workshop is essential to reach this objective. At the same time it is important to promote the creation of ideas, projects and collaborations, which contribute to the development of the participants.

These conclusions joint to the experience developed during this pilot Space Law Workshop is the base to prepare more ambitious projects which promotes the Space Law, collaborating with the Spanish institutions and making possible to bring Space Law to students in an effective way.

IV. SPACE LAW THROUGH RESEARCH

The Spanish Space Students Congress is the annual congress of the association in which students from all over the country meet to share their projects status, their experiences, their difficulties and their results. Furthermore, the Spanish Space Students Congress is a meeting point for students, industry and institutions facilitating the network among them, and the scene of the general meeting of all the members of the association.

As support activity to the two lines of action of the judicial committee previously referred to, and with the objective of collaborating in the promotion of the Space Law among the students, a section into the congress dedicated to it will be organized. In this section, the students will have the opportunity to present their works in Space Law in front of the institutions and organizations of reference of this discipline in our country. On the other hand a series of highlight sessions finish each day of this section of the congress. Besides, a reduced workshop of introduction to Space Law will be given to the attendants, which includes a round table to promote the collaboration in further projects.

The final objective of this section of the Spanish Space Students Congress dedicated to the Space Law is to collaborate in the aim of bringing the Space Law to students.

V. CONCLUSIONS

The Juridical Committee has supported the association to avoid any conflict between our educational projects and the absence of a clear framework and to prevent any potential risk of our activities, and on the other hand has anticipated the potential problems derived of further projects.

At the same time we have identified a series of problems, which are not covered by the current law, that we consider a practical point of view to the legal aspects of this activities. It constitutes a practical contribution to a future national law. Also, it has provided to students the opportunity to introduce to Space Law through the participation in real hand on projects.

Besides, we have promoted the Space Law among the students through educational activities.

After its first activities, the LEEM Space Law Committee has reached the initial experience necessary to develop more ambitious projects. This step makes possible a stronger support to further LEEM activities as well as a wider promotion of the Space Law among the Spanish students. On the other hand, it facilitates the collaborations with the reference institutions to work for the promotion of this discipline.

All these aspects make the Juridical Committee an essential part of the mission of our association as well as an important actor in the development of the Space Law trough the students' community.

REFERENCES

A. Russu, B. Hidalgo, A. Monzón, A. Gutiérrez, C. Peregrín, E. González, V. Albert, J. Marín-Yaseli, J.A. Plaza, D. Zurita, J. Fernández-Cabrera, F.J. Robles (2008); Laboratory for Space and Microgravity Research: One Year of Student Projects; Proceeding IAC-08-E1.3.103 International Astronautical Congress 2008 (International Astronautical Federation)

(1) The CanSat Concept was proposed by Prof. Bob Twiggs of the Stanford University Space Development Laboratory, at the Japan-U.S. Science, Technology and Space Applications Program (JUSTSAP) conference in November 1998.

SPACE AND THE COMPLEXITY OF EUROPEAN RULES AND POLICIES: THE COMMON PROJECTS GALILEO AND GMES – PRECEDENCE FOR A NEW EUROPEAN LEGAL APPROACH?

Dr. Annette Froehlich, LL.M., DESS
Senior Research Associate,
Institute of Air and Space Law, University of Cologne
annette.froehlich@uni-koeln.de

Albertus-Magnus-Platz
D-50923 Cologne, Germany

ABSTRACT

The two European space projects, Galileo and GMES, clearly show that the current existing legal rules of the two organisations involved (European Union and ESA) are not compatible. Moreover, it is quite impossible to implement a common project if every single organisation insists on the application of its own rules *strictu sensu*. Nevertheless, due to the political desire to advance these projects rapidly and to bring them to success, legal obstacles were overcome. Consequently, recently concluded agreements between ESA and the EU-Commission concerning the financial and governmental matters of the GALILEO and GMES implementation feature a new approach of cooperation between these two organisations. However, the question remains if they can be taken as precedence for a future institutionalized cooperation? It follows that the agreements have to be analysed in order to understand how a mutually acceptable agreement was reached despite the differences in the rules of both organisations. In this regard, especially the financial decision agreement concerning GALILEO in December 2007 shows a very interesting, unique way in applying EU-competition law. In the same way, the GMES-Delegation Agreement of spring 2008 is a good example how two different legal systems can be applied to bring a project to success. Additionally, the reasons and arguments of both organisations have to be considered, especially once the Treaty of Lisbon will be in force. As these two main projects of the European Space Policy are characterized by the desire for a successful European cooperation, they can be regarded as an important step forward to a new legal approach. A new system emerges which could be taken into consideration for further European common projects.

INTRODUCTION

Initially, the EU defined itself as an economic community and left the matters concerned with space to the ESA, which was expressly created for this purpose.[1] However, by now the EU has (despite lacking express competence in matters of space) already taken over Galileo and GMES (in co-operation with ESA) and thereby begun its first operations in the matter of space. Even though these projects

were declared pilot projects only, they will bind the EU for the long term.[2] Keeping in mind that the EU has included a separate budget for Galileo in the financial prospective for 2007-2013 and that the Treaty of Lisbon includes a shared competence in the area of space, one can conclude that there is an increased interest in this area by the EU, combined with a raised presence. Consequently, one can expect the EU Commission to put before the Council and the Parliament suggestions on how to realise the potential within the new EU competences.
Furthermore, the Treaty of Lisbon will break with the traditional "three pillar structure" of the EU, so that the EU will receive further competences in the area of foreign policy (such as a common defence strategy).[3] From that point onwards the EU may act in the military sphere and even award projects of a military nature to ESA. This would naturally lead to an extension of tasks of ESA which up till now may only carry out projects *for peaceful purpose* (art. II ESA-Convention)[4].
Therefore, one has to determine exactly the kind of competences belonging to the EU according to the Treaty of Lisbon and how they might reflect on the ESA-EU agreements already in force. This is especially relevant when bringing to mind the experiences already made in relation to the implementation of Galileo and GMES.

TREATY OF LISBON

For the first time ever, the Treaty of Lisbon, which was supposed to come into force in 2009, contains a codification of the competences of the EU in the area of space.[5] Following the preamble, a new title setting down the "Categories and Areas of Union Competence" was added to the treaty (Title I). Within this new title, article 2 b enumerates the areas in which the EU has exclusive competence (the customs union, the establishment of the competition rules, the common monetary policy, the conservation of marine biological resources under the common fisheries policy, the common commercial policy and the conclusion of an international agreement when its conclusion is provided for in a legislative act of the Union or is necessary to enable the Union to exercise its internal competence).
According to article 2 c) the EU will have only shared competence in areas of internal market, economic, social and territorial cohesion, agriculture and fisheries (excluding the conservation of marine biological resources), environment, consumer protection, transport, trans-European networks, energy, area of freedom, security and justice, further in the areas of common safety concerns in public health matters.
Additionally article 2 c) para. 3 expressly states a shared competence between the EU and the member states in the area of space: "*In the areas of research, technological development and space, the Union shall have competence to carry out activities, in particular to define and implement programmes; however, the exercise of that competence shall not result in Member States being prevented from exercising theirs.*"
Moreover, the title XVIII „Research and Technological Development" was amended to include the words "and space" as well as a new article 172 a): "*To promote scientific and technical progress, industrial competitiveness and the implementation of its policies, the Union shall draw up a European space policy. To this end, it may promote joint initiatives, support research and technological development and coordinate the efforts needed for the exploration and exploitation of space.*"[6]
Lastly, article 3 states: "*The Union shall*

establish any appropriate relations with the European Space Agency".[7]
Therefore, art. 172 a) of the Treaty of Lisbon contains the legal basis for the EU to act in the area of space.[8] This new competence was desired by all member states, including those also belonging to ESA,[9] (even though research and development in the area of space traditionally belong to the domain of ESA).[10] Accordingly, the Resolution on the European Space Policy[11] from March 25, 2007 reaffirms "*the roles and responsibilities of the European Union, ESA and Member States, as identified in the Orientations of the second meeting of the 'Space Council'.*"[12] The Orientations from the Space Council[13] (2nd meeting on June 7, 2005, point 3-2) set the framework for the relationship between the EU and ESA; particularly each party's role and responsibilities. As a result, the EU is obliged to ensure the availability and continuity of operational services of the European space infrastructure (at the moment Galileo and GMES), whereas ESA will develop space technologies and systems, supporting global competitiveness. "*Their activities will focus on exploration of space and on the basic tools on which exploitation and exploration of space depend: access to space, scientific knowledge and space technologies".*[14]
This division of competence between ESA and the EU as set down in the Orientations from the Space Council of 2005 and affirmed in the Resolution on the European Space Policy should be continued in the future. Even though the Treaty of Lisbon, as a superior source of law compared to the common ESA-EU agreements, states that the EU, *"may promote joint initiatives, support research and technological development and coordinate the efforts needed for the exploration and exploitation of space".*[15] In light of the prior agreement this text should be interpreted in a way as to say that the EU will support space research which will, as previously agreed upon, be initialised by ESA.

RESOLUTION on the EUROPEAN SPACE POLICY

In Chapter II "Further Steps – Programmes and Implementation", point F.12 "Governance", the Resolution on the European Space Policy describes the role of ESA when realising common projects. It envisages that ESA will play the role of a technical expert in relation to the EU-Commission, whereas the realisation of common projects will follow the EC-procedure. "*Such ESA role should also include: supporting the European Commission as technical expert in the elaboration of European Community initiatives involving space-related activities and relevant work programmes, and in the selection and monitoring of relevant work contractors, and the management by ESA of European Community space-related activities in accordance with the rules of the European Community".*[16]
In this context, the "rules of the European Community" mean that the EU competition law would be applicable to the area of space. The competition law itself can be found in art. 81-86 of the Treaty of the European Community, its purpose is to prevent a distortion of competition by member states or private entities.[17] Even though this represent one of the most basic aims of the Treaty,[18] it presupposes the existence of an effective competition within the relevant market,[19] implying "*the existence of the market of workable competition, that is to say the degree of competition necessary to ensure the observance of the basic requirement and the attainment of the objectives of the treaty, in particular the creation of a single market achieving*

conditions similar to those of a domestic market".[20]
Still, the competition is not an end in itself; instead it is only a means to achieve the goals of the EC as laid down in art. 2. It follows that in certain constellations the protection of the pure competition may step back in favour of other important aspects.[21]
Considering that the market on space is not a market as traditional understood, but instead one that is strongly institutionalised, representing higher state policies which include interventions for the protection of the civilian population und the interest to keep the European space industry in competition with foreign countries,[22] the EU competition law has to give way to these other considerations at hand. Additionally, art. 2 of the Treaty of the European Community states that the implementation of the common market is not the only aim of the EC, the introduction of common policies which also include the advancement of research and development (art. 3, para. 1 lit. n of the Treaty) is also listed. Furthermore, of these declared aims neither one enjoys prominence over the other (art. 3 of the Treaty of the European Community).[23]

These considerations have been officially adopted by the Council in the Resolution on the European Space Policy from May 25, 2007. The Council declares in Chapter II "Further Steps – Programmes and Implementation" in point G.14 "Industrial Policy" explicitly an exception to the general competitive law, it: "*EMPHASISES in this context in particular the political and economic dimension of ESA's 'fair return' principle; and the importance to assess and improve, when necessary, the implementation of the 'fair return' principle in view of the future challenge for industry to remain competitive in a changing environment worldwide while maintaining, and possibly increasing, Member States' motivation to invest in space*".[24]
Even the Orientations from the European Space Council (2nd meeting on June 7, 2005) in its key principles (point 3-4) contain the statement that "*the implementation of the European Space Policy requires an industry policy tailored to the specificities of a sector subject globally to government influence. This policy should provide all stakeholders in Europe with the motivation (...) to make the necessary investment to maintain (...) a globally competitive space industry. This is central to the achievement of Europe's economic and political objectives, thus contributing to Growth and Employment*".[25]

Furthermore, the EU already has made extensive concessions in the application of its competitive law when negotiating the financing of its two projects, Galileo and GMES.

EU-DECISION regarding GALILEO 2008

The common project of Galileo and its Public-Private-Partnership has opened new ways of cooperation between ESA and the EU. The regulations regarding the implementation of the European Satellite Navigation Programs (EGNOS and Galileo)[26] from July 9, 2008 determine in article 17 that the "*following principles for the procurement of the activities of the deployment phase of the Galileo programme shall apply*" in order to guarantee transparency and equality among the competitors. These principles are meant to "*promote balanced participation of industry at all levels*"[27] in order to "*avoid possible abuse of dominance*"[28]. In order to achieve this aim, the tender proceedings declare that the "*procurement of the infrastructure shall be split into a set of six main work packages"*[29]. Even more, a rule was established according to which *"one independent legal entity, or a group represented for this purpose by a legal entity belonging to that group, may bid for the role of prime contractor for a maximum of two of the six main work packages"*[30]. Furthermore, *"at least 40 % of the aggregate value of the activities shall be subcontracted (...) to companies other than those belonging to groups of entities that are prime contractors for any of the main work packages"*[31].

In establishing these rules, the Commission has taken into consideration how a public tendering procedure may lead to certain member states being privileged over others. Accordingly, the Commission has accepted a division of its market and certain flexibility in the application of its competition law. This resolution leads the way into a division of the Galileo project and its market, supported by an extensive reading of the EU competition law, even though art. 81, para. 1, lit 1 of the EU Treaty itself declares that to „*share markets or sources of supply*" was incompatible with the goal of a common market.

ESA-EU AGREEMENT concerning GMES FINANCING

In the course of implementing the common GMES project it turned out that the set of rules of each organisation couldn't simply be transferred to a co-project without alterations. Therefore, on February 28, 2008, the Commission interpreted the agreement with ESA as consistent with the EU financial regulation in order to allow for a continuation of the project.

CONCLUSION

When looking at the agreement at hand, it becomes obvious that where the EU Commission desires to have the EU Competition Law adapted to a specific situation, it may very well do so. Special agreements for further common projects in the area of space industry are especially prone for such an adaptation, due to the particularities of the subject matter. This approach gains even more relevance in light of other possible common projects like the project of Space Situation Awareness, where there remains plenty of room for actions.

---

[1] Currently the area of satellite communication is the one most commercialised.

[2] Neither the Galileo nor the GMES project will be finished at the end of the FP7 (Seventh Research Framework Programme) in 2013, this may lead to issues of liability and other long time problems.

[3] Legal basis: art. 11 (a) Treaty of Lisbon: "The Union's competence in matters of common foreign and security policy shall cover all areas of foreign policy and all questions relating to the Union's security, including the progressive framing of a common defence policy that might lead to a common defence". This means that the competence of the EU in the area of common foreign and security policy now also includes the task of implementing a common defence, (cf. *Fischer*, Der Vertrag von Lissabon, p. 147 in relation to the article' cryptic phrasing, the author doubts whether the new art. 11 of the treaty may lead to a more efficient and effective common foreign and security policy). It follows that the competences of the EU in the area of common foreign and security policy now include all matters related to it, there is no inherent limitation. (cf. *Fischer*, Der Vertrag von Lissabon, p. 146).

[4] Convention on the European Space Agency, 1975, came into force on October 30, 1975.

[5] These terms were adopted unaltered from the Constitutional Treaty, cf. article I-14 of the Constitutional Treaty, "Categories and Areas of Union Competence".

[6] Para. 2 "To contribute to attaining the objectives referred to in paragraph 1, the European Parliament and the Council, acting in accordance with the ordinary legislative procedures, shall establish the necessary measures, which may take the form of a European space programme, excluding any

harmonisation on the laws and regulations of the Member States", Text according to C 306/43, Official Journal of the European Union, from December 17, 2007.
[7] Para. 4 „This Article shall be without prejudice to the other provisions of this Title".
[8] Cf. *Fischer*, Der Vertrag von Lissabon, p. 318.
[9] This anomaly leads to questions concerning the coherency of member states in matters of a "European Space Policy", especially regarding those member states with a double membership in the EU and ESA. However, one has to keep in mind that the member states did not discuss the matter of space during the negotiations to the Treaty of Lisbon, leading effectively to a dissent among those present.
[10] Cf. art. II and art. V ESA-Convention and the decision concerning the European Space Policy.
[11] The Resolution on the European Space Policy was preceded by the Communication from the Commission to the Council and the European Parliament, European Space Policy on April 26, 2007. An identical document was presented to the Council of the European Space Policy by its director-general. The Resolution on the European Space Policy can therefore be seen as the member states' reply to the Commission's Communication, whereby they recognise the outstanding work done by ESA and endorse its current institutional framework.
[12] Council of the European Union, Resolution on the European Space Policy, May 25, 2007, Chapter II.F "Governance", pt. 12.
[13] The Space Council consists of the Council of the European Union and the Council of the European Space Agency (ESA).
[14] Orientations of the second meeting of the Space Council, June 7, 2005, pt. 3-2.
[15] Art. 172 a, para. 1 Treaty of Lisbon.
[16] Chapter II Further Steps "Programmes and Implementation", pt. F.12 "Governance".
[17] Art. 3 para. 1 lit. g Treaty of the European Community demands a "system ensuring that competition in the internal market is not distorted".
[18] Cf. ECJ, Case 6/72, Continental Can, report 1973, 215, para. 25, *Lenz/Borchardt* (ed.), EU- und EG-Vertrag, Kommentar, 4. Edition 2006, p. 944.
[19] Cf. *Lenz/Borchardt* (ed..), EU- und EG-Vertrag, Kommentar, 4. Edition, 2006, p. 944.
[20] ECJ, Case. 26/76, Metro I, report 1977, 1875, para. 20.
[21] *Lenz/Borchardt* (ed.), EU- und EG-Vertrag, Kommentar, 4. Edition, 2006, p. 944.
[22] Cf. the statement by the EU-Commission, Communication from the Commission to the Council and the European Parliament, European Space Policy, April 26, 2007, pt. 5, p. 10 "A competitive European space industry is of strategic importance".
[23] *Oppermann*, Europarecht, 3. Edition, 2005, p. 311, at. 5.
[24] Decision of the European Space Policy from May 25, 2005, Chapter II, pt. G.14.
[25] Orientations from the second Space Council, 7 June 2005, pt. 3-4.
[26] Regulation (EC) No 683/2008 of the European Parliament and of the Council of 9 July 2008 on the further implementation of the European satellite navigation programmes (EGNOS et Galileo), L 196/1.
[27] Art. 17, pt. 2 a).
[28] Art. 17, pt. 2 b).
[29] Art. 17, pt. 3 a).
[30] Art. 17, pt. 3 b).
[31] Art. 17, pt. 3 c).

IAC – 08 – E8.5.2

NETWORK OF EUROPEN REGIONS USING SPACE TECHNOLOGIES AN UPDATE ON THE NEREUS CONSTITUTION

Marianna Morelli
CORILA Consortium for Coordination of Research Activities Concerning the Venice Lagoon System, Venezia, Italy, marianna_morelli@yahoo.it ; morelli@corila.it

Pierpaolo Campostrini
CORILA Consortium for Coordination of Research Activities Concerning the Venice Lagoon System, Venezia, Italy, campostrini@corila.it

ABSTRACT

The Outer Space research and development programmes give a consolidated model for international cooperation. They represent a good framework to develop and enhance different initiatives at global, regional and local level. The impact of space activities covers different areas of interest such as: the development of Earth observation applications for environment monitoring and risk prevention and management, the development of satellite communications and information systems, the exploitation of global satellite positioning and navigation systems, the knowledge of the Universe, the utilisation of ground segment engineering. These activities involve also the regional and local governments. This is particularly evident if we look at the competences of the European regional governments. Several European Regions deal with a number of issues in areas linked to space activities. In line with these reflections a number of European regional governments supported by the Committee of the Regions launched the idea to create a Network of European Regions Using Space technologies – NEREUS. This Network is an International non-profit Association (AISBL) under the Belgian Law. The aim of this paper is to review the course of the initiative since its launch in April 2006, giving an update of the work done for its constitution.

INTRODUCTION

The main purpose of this paper is to analyse and assess the legal basis for the cooperation among European regional governments, which has been established in recent years. In particular, this article offers an update[1] of the work done for the constitution of the Network of European Regions Using Space technologies (NEREUS).
The idea of implementing a network of European regions to promote the use of space services has been presented, for the first time, by the Midi-Pyrénées Region, during the Conference on the Global Monitoring for Environment and Security (GMES)[2], held in Graz in April 2006.
A significant advancement has been performed since the end of 2006, with the conclusion of the NEREUS' Political Charter, signed in Toulouse on 18 December 2007 by the interested European regions, and the enter into force of the NEREUS Statute in June 2008.
Today NEREUS is an International non-profit Association under the Belgian law, located in Brussels.
The Political Charter and the Statute of the Network offer the basis for future activities of the new International Association.

The present article reviews the course of the initiative since its launch in April 2006. First, we will recall the evolution of the role of States in the exercise of Space activities taking into particular account the Outer Space

Treaty of 1967. Then we analyse the role of the European regional governments within the EU Space Policy. Therefore we will focus the paper on the main aspects of the Network, giving an update of the work done for its constitution: the definition of the organizational structure, the objectives and the methods of work, the definition of the cooperative activities and the common projects within the Network.

1. THE EVOLUTION OF THE ROLE OF STATES IN SPACE ACTIVITIES

States are the primary and major subjects of international law.

The role of States was predominant during the "Cold War", which was characterized by the rivalry between the USA and USSR, on the level of prestige as well as on the level of military capabilities[3]. In that period, space activities were largely either military or scientific[4]. This was reflected in the beginning of space law: the Outer Space Treaty[5] of 1967 was largely established to prevent extension of the arms race into outer space, and to preserve the freedom of exploration of outer space[6].

The progress of science and technology in space activities as well as the changes in global politics and international relations led to a new dimension of space activities.

This is particularly evident if we look at the new applications of space technologies in the field of telecommunications, satellite navigation systems and Earth observation.

With the rising profitability of space activities in these fields, the increasing interest of the private sector leads to a gradually evolving tendency towards the relocation of parts of the activities to the non-state players.

However, States remain the major and predominant actors of space law. States form custom, treaties and other legal instruments, they create international organisations, regulate, supervise and license private operators within their national legislation, which must conform with international obligations[7].

The role of States in space activities is expressly foreseen by Article VI of the Outer Space Treaty of 1967.

According to Article VI of the Outer Space Treaty, States bear international, political and legal responsibilities for all national space activities – whether they are carried out by private or public entities - and for assuring that national activities are carried out in conformity with the provision set in the Treaty[8].

Article VI also affirms that States have to authorize and supervise the space activities of non governmental entities placed under their jurisdiction.

Finally, Article VI expressly recognises the role of international organisations in performing space activities. In order to assure that the international organisations comply with the provision of the Treaty, Article VI provides that when activities are carried out by an international organisation, the responsibility for these activities shall be borne both by the international organization and by the States Parties to the Treaty participating in such organization[9].

Therefore States are the actors of the exploration and use of outer space, but their role is now decreasing. We can observe an increasing importance of the non-governmental sector and the consequent emergence of various new actors on the space scene, e.g. private entities or Non Governmental Organisations (ONG)[10].

In line with these considerations, the major States define a national space policy and thus develop space programmes to implement this policy.

2. THE EU SPACE POLICY

The deepening of integration in the framework of the European Union implies new questions also for space-related undertakings. The European Space Agency (ESA) and the European Union jointly work on an intensification of their relationship, in order to create a coherent and highly competitive environment for the European strategy for space[11].

"ESA and the EU are two distinctive organisations with their own legal bases, different mandates and membership and different funding mechanisms"[12].

ESA has provided for more than 30 years an efficient structure for European cooperation on joint space projects. The independence and reliability of ESA, thanks to the support of its Member States, contribute to the increasing role of Europe in space, through the successful development of the European space sector, and to the strong position of the European space industry on world markets.

For many years, cooperation between ESA and EU regarding specific projects has been carried out on a case-by-case basis. Today, the EU considers the space sector as a strategic domain which can directly contribute to the implementation of a large group of European policy objectives, such as the Europe's Sustainable Development Strategy, the Common Foreign and Security Policy, the Lisbon Strategy for growth and employment.

In April 2007 the European Union adopted the document on the "European Space Policy"[13], presented as a proposal of the ESA Director General and as a Communication of the European Commission.

The implementation of the "European Space Policy" represents the major step in establishing a coordinated and effective European space effort of ESA and EU.

The basis for the institutional relationship between ESA and EU has been established in the EC-ESA Framework Agreement[14], signed in 2003 and officially entered into force on 28 May 2004.

The Preamble of the Framework Agreement states that ESA and EC recognize that they have "specific complementary and mutually reinforcing strengths" and "commit themselves to cooperate in an efficient and mutually beneficial manner", to avoid "any unnecessary duplication of effort"[15].

The final aim of this Framework Agreement is to address the coherent and progressive development of an overall European Space Policy able "to link demand for services and applications using space system in support of the Community policies with the supply of space systems and infrastructures necessary to meet that demand"[16].

The document on the "European Space Policy" has been elaborated taking into account the "White Paper on Space Policy"[17], titled "Space a New European Frontier for an Expanding Union", which was adopted by the European Commission on 11 November 2003. It provides the overall political framework for the development of a viable and strong European space sector which will seek:

- to develop and exploit European space applications serving Europe's public policy objectives and the needs of European enterprises and citizens
- to establish appropriate funding arrangements for the operational phase of EU programmes such as the Global Monitoring for Environment and Security (GMES) and the European Satellite Navigation System (GALILEO) to increase synergy between defence and civil space programmes and technologies
- to contribute to the knowledge based society by investing strongly in space-based science
- to ensure a strong and competitive space industry which foster innovation, growth and the development and delivery of sustainable, high quality, cost-effective services
- to develop a more coordinated and coherent approach to international relations in space

The European Space Policy will be implemented through the establishment of a common European Space Programme, serving as a basis for coordination of national and European level space activities.

The European Space Programme will be the main instrument to implement the most important space related initiatives, in particular the Global Monitoring for Environment and Security (GMES), or the GALILEO programme.

The success and sustainability of these space programmes can only be achieved through a user-driven approach[18]. The main target of the European Union nowadays is to set-up a more open and permanent dialogue with the largest possible users' community for the new services which will be developed within the GMES and GALILEO.

The document on the European Space Policy restates the role and the responsibilities of the European Union, ESA and the Member States. The European Union has the task to define the priorities, aggregate the political will and users' demand and ensure the availability and continuity of services. ESA and its Member States have the task to develop space technologies and systems according to the objectives of the European Space Policy. Finally, Member States, under the coordination of ESA and in close cooperation with the EU, should provide the best expertise for European Space Programmes and increase synergy between national, ESA and EC contributions to space programmes.

The final aim is to progressively achieve an integrated approach respecting national sovereignty. Therefore the European Space Programme should be elaborated through a process of dialogue including all stakeholders: EU, ESA, Members States with their national space agencies, national research centres, European industries, European organisations such as EUMETSAT, EEA, etc.

Within this framework, the involvement of the regional and local dimension is essential in order to collect users' requirements and to ensure new space services. The active participation of European regions is also requested to enhance the awareness of potential advantages coming from future space-based services.

Regional governments can represent the new players in outer space in an environment previously ruled only by national governments, the European Union and the European Space Agency.

3. THE ROLE OF REGIONS IN THE EU SPACE POLICY: THE PRINCIPLE OF SUBSIDIARITY

At the beginning of the 21st century the development of the EU is faced with a number of major tasks of historic importance: the enlargement of Europe to include the new Member States, the establishment of political, economic, social and monetary Union, the consolidation of EU's position as a global player.

Technological changes, the information and communications revolution, globalization and the integration of markets are all elements of an increasingly interdependent world.

Consequently, there is a growing number of issues that cannot be dealt within the boundaries of national States. There is a large area of issues having an European dimension, whereas for other issues the EU must act more on the international level.

There is a growing number of public and private decisions taken at European and international level, which affect directly the daily life of citizens.

In this context the principle of subsidiarity - in the sense of proximity, with decisions taken at the institutional and operational level closest to the citizens, ensuring the recognition, enhancement and involvement of private individuals and their social groupings – becomes more and more important.

The principle of subsidiarity, as defined in Article 5 of the EC Treaty, applies exclusively to the relations between the Community and the Member States, and not to the relations between sub-national bodies and the Member States. The relations between sub-national bodies and Member States are regulated by the constitutional systems of each Member States.

However, the Committee of the Regions[19], created in 1994 as a consultative body of the EU, is based on the subsidiarity principle to ensure that the decisions are taken as close as possible to the citizens.

The Committee of Regions acknowledges that EU actions, in accordance to the subsidiarity principle, not only concern the Member States but also their regional and local governments.

Concerning the concept of region there is a problem of definition. The concept of region considerably varies in Europe.
Local and regional authorities in the EU Member States have different structures and denominations: Länder, Regions, Counties, Provinces, Autonomous Communities[20].
Generally, in the context of the Committee of the Regions, the word "region" refers to the territorial administration which is directly subordinated to the central administration, and endowed with an independent political representation[21].

The increasing role of the regional governments within the Member States as well as in the EU policies, for bringing citizens closer to decision making centres, has also affected the space sector.
Regions are legitimate stakeholders in the EU Space Policy. Despite the different size and territorial characteristics of European regions and the different functions granted to the regional entities, most European regional governments deal with a number of issues linked with space activities.
European regions are involved in space activities from infrastructures to the use of applications. They are involved in the installation, development and maintenance of space infrastructures, the so-called "Ground Segment". They play a major role in the process of innovation and promote the process of transferring technologies from research sector to commercial sector. They give a significant contribution to the creation and support of clusters and competitiveness zones that bring together manufacturers (also Small and Medium Enterprises - SME), higher education and scientific research. They are users of satellite data having competences in the field of environment, territorial management, agriculture, emergencies and security. They are in charge of the development of territorial and marine spatial planning and they provide different services to the citizens.
Therefore the regional dimension is essential to the definition and use of space services.
The involvement of regional entities makes it possible to open up the space sector to players, others than those of the traditional space sector: service providers, content providers, public and private users. This phenomenon considerably enhances the potential of the space sector in terms of benefits for European citizens.

In this spirit, and following the example of existing regional associations, such as the Peripheral Maritime Regions Conference (PMRC), a number of European Regions, supported by the Committee of the Regions, launched the idea of creating a Network of European Regions Using Space technologies (NEREUS), in order to strengthen the dialogue among the regional authorities.

4. A NEW FORM OF COOPERATION: THE NETWORK OF EUROPEAN REGIONS USING SPACE TECHNOLOGIES (NEREUS)

The idea of establishing a Network of European Regions Using Space technologies has been presented by the Midi-Pyrénées Region during the GMES Conference, "A Market for GMES in Europe and its Regions – The Graz Dialogue", organised in Graz on 19-20 April 2006, by the Austrian Government during its EU Presidency.
From the extensive discussion held during the Conference, the Austrian EU Presidency prepared a document including recommendations for the further development of GMES in Europe, titled "The Graz Roadmap for GMES services development"[22].
The Graz Roadmap highlights and recommends the creation of a Network of European Regions Using Space technologies to reinforce the dialogue among all concerned actors.
From December 2006, a group of European regions started the discussion on the actual implementation of this Network of European Regions, defining the objectives, the structure and the future activities of the Network.
The focus of the Network shall be on the use of space technologies, in order to federate new users and their information needs and to promote the development of solutions responding to the various needs.

The process for the constitution of the Network of European Regions Using Space technologies (NEREUS)[23] has been carried out for more than two years.
The Network was inaugurated in Toulouse at the Midi Pyrenees Regional Council on 18 December 2007. On that occasion 22 Regions coming from 9 European Member States signed the NEREUS Political Charter[24].
The Network is governed by a Statute[25] which entered into force on June 2008. According to the Statute, NEREUS is an International non-profit Association (AISBL).
The Association is open to the access of regions of the European Member States.
At present, 21 regions have finalised their membership procedure, becoming full members of the Networks: the City of Vienna (Austria), Wallone, Brussels Capital Region (Belgium), Midi Pyrenees (France), Abruzzo, Basilicata, Lombardia, Molise, Puglia, Veneto (Italy), Brandenburg, Bremen, Hessen, Bayern, Baden-Wurttemberg (Germany), Mazovia (Poland), Açores (Portugal), Andalucia, Castilla y Leon, City of Madrid (Spain) and East Midlands (United Kingdom). Others regions have already expressed their intention to join the network.

The first General Assembly of NEREUS took place on 7 July 2008 at the Committee of the Regions in Brussels.

4.1 OBJECTIVES AND ACTIVITIES OF THE NETWORK

The Network supports the strategies set up by the European Union in its European Space Policy, articulating the common interest of Regions within the EU space programs. The initiative aims at strengthening the dialogue between the regional authorities and the main actors in the space sector such as the European Commission, the European Space Agency, the Member States and the European space industry.

NEREUS is independent from the EU institutions, it should be a representative voice of the regions in order to identify their joint concerns and suggestion with regards to the EU Space Policy.

The NEREUS Political Charter establishes the main objectives of the Network[26]:

- To introduce the regional level in the elaboration and development of European space programmes and activities linked to infrastructures and applications.
- To promote and implement partnerships and cooperation between European regions, in order to develop common or complementary approaches and policies.
- To sustain the fulfilment of the end-users' needs in the space services provided by the European Union programs.
- To ensure that space services are used across all the regions for ensuring a balanced development of the European Union.
- To support a better promotion of the European space dimension in a context of increasing worldwide competition.
- To increase the citizens' participation in the European policy construction and in the developing of the Space services market.

In order to accomplish these objectives, the Network could pursue the following activities[27]:

- To carry out research projects and programs in all major fields of space technologies
- To spread the information concerning the available sources of financing in Europe, both public and private, in order to allow NEREUS members to build up projects on space technologies
- To organize workshops, study seminars, educational activities and supervise scientific studies, plans, in order to enhance the dissemination of knowledge on space technologies
- To carry out activities such as identifying final users, mapping common issues and needs, sharing problems related to space applications and federating users' demand

- To express the opinion of the regions involved about EU publications having an impact on matters concerning NEREUS' aims.

All the activities of the Network should be in harmony with the activities of the concerned institutions in Europe and open to international cooperation.

4.2 STRUCTURE OF THE NETWORK

The Network NEREUS has the legal status of an international non-profit association (AISBL) under the Belgian Law. The Secretariat of the Association is located in Brussels.

Article 5 of the Statute states that the Association has two categories of members: full members and associate members.
Full membership is open to regions or territorial entities from all EU Member States formally acknowledged as "regions". As mentioned in paragraph 3, the word "region" refers to the territorial administration with a measure of autonomy which is directly subordinated to the central administration and endowed with an independent political representation.
When regions, as above defined, do not exist, public institutions representing territorial areas comparable to regions can be considered as full members. In the countries where regional authority is not in force, the admission as full members of lower level governmental authorities will be taken into consideration, on condition that they are not municipalities.

Associate membership is open to local authorities (other than regions), industries, Small and Medium Enterprises (SME), associations, foundations, banks, universities, as long as they have competences, knowledge and interest within the NEREUS aims.

The membership of the Association is linked to the payment of the annual subscription fees.

The organisation of the Network includes the following organs:

- The General Assembly, composed of representatives of all NEREUS members. The Assembly, where each region has one vote[28], is the political body of NEREUS. The General Assembly shall meet at least once a year. The Assembly elects the Management Board, the President, the Vice-President and the treasurer of the Association. It shall approve the political guidelines and the budget of NEREUS
- The Management Board, elected by the General Assembly, for a period of two years. The members should be representatives of regional authorities. The Management Board proposes the political guidelines for NEREUS and implements the decisions adopted by the General Assembly.
- The General Secretariat, headed by a Director who is the permanent working organ of the Association.
- The Working Groups, set up by the Management Board. Their aim is to carry out the work programme of the Network. The Working Groups are organised around specific themes of cooperation.

4.3 COMMON PROJECTS WITHIN THE NETWORK

The Working Groups will be the real "core" of NEREUS' activities. In fact, the definition and the implementation of a strategic development plan for NEREUS depend on the actual activities of the Working Groups.

At present the NEREUS members have identified four initial Working Groups covering a broad range of space activities and programmes:

1. **GMES and Earth Observation**: GMES services are closely linked with the daily life of citizens, providing immediately usable terrestrial, oceanographic and atmospheric data for environmental protection and security of citizens. End users operating at the regional level can take advantage of GMES core

services, and will contribute to develop downstream services.

2. The European Satellite Navigation System – GALILEO: Many sectors managed by regional administrations are affected by the development of satellite navigation, e.g. transport and communication, land survey, agriculture, fisheries and waterways, tourism, waste disposal and logistic domain. Navigation systems also provide instruments for civil protection, emergency management and humanitarian aid.

3. Telecommunication: Distance learning, remote medicine, remote health and remote administration are key fields of application for the future in every region. Information and telecommunication technologies play a key role in improving the quality of services delivered at the regional level

4. Acquisition of Knowledge, Education, Training and Communication with the Public: the need to develop new applications and technologies for present and future space programmes requires improvement of the current knowledge. Research centres and universities in Europe are able to meet this challenge, supporting industry activities in a competitive world. Regional administrations shall involve the education process at all levels.

The organisation of the Working Groups is in progress. Nowadays, the main challenge is to nominate the Working Groups' leaders, to identify and motivate, by the end of 2008, specific sub-themes for each Working Group on the basis of clear criteria and stakeholders consultations.
The first "Kick-off meeting" for each Working Group is scheduled for autumn 2008.

CONCLUSIONS

Since 2006, when the idea to establish a Network of European Regions Using Space technologies emerged, discussions have taken place between the interested European regions, regarding the actual implementation of the Network. The initial discussion concerned the aspects of the role, necessity and functioning of the Network.
Now the Network of European Regions Using Space technologies (NEREUS) has become a reality.
The General Assembly, held on 7 July 2008 at the Committee of the Regions in Brussels, elected the organs of the Association: the Management Board, the President and the Vice-President. The next step to complete the establishment of a fully operational Network includes the implementation of the Working Groups, mentioned in the paragraph 4.3.
For the successful development of the Network in space activities, it is now necessary to identify programmes and projects in which the members of the Association can cooperate and effectively undertake joint activities.

The European objective to involve regions in the construction of the European Space Policy aims to consider the final users needs in the realization of space-based services, following the subsidiarity principle.
This new emerging form of cooperation in Europe opens a new dimension of space activities that look at the space from a different point of view taking into consideration the new emerging concept of subsidiarity in the European governance as well as the European environmental policy and law, the economic development and the scientific community.
Today EU considers space applications as an important tool at the service of numerous European objectives and policies: such as transport and mobility, information society, environmental protection, land use planning, agriculture and sustainable development.
In this context the involvement of Regional Authorities can give a significant contribution to support the development of European space applications, exploiting new emerging markets.

[1] M. Morelli, P. Campostrini, "The Role of European Regions in the EU Space Policy", Proceedings of the 58th IAC Congress, Hyderabad, 24-29 September 2007.
[2] GMES Conference, "A Market for GMES in Europe and its Regions – The Graz Dialogue", Graz, 19 - 20 April 2006.
[3] M. Bourély, "The Institutional Framework of Space Activities in Outer Space", Proceedings of the 9th European Summer Course on Space Law and Policy of ECSL, Nice, 27 August – 8 September 2001, pp.9-14.
[4] F. von der Dunk, "National Space Legislation", Proceedings of the 10th European Summer Course on Space Law and Policy of ECSL , Nice, 27 August – 8 September 2001, pp.33-36.
[5] Treaty on Principles Governing the Activities of States in the Exploration and Use of Outer Space, Including the Moon and Other Celestial Bodies, 610 UNTS 205, 27 January 1967.
[6] Ibidem Art.I: "The exploration and use of outer space, including the Moon and other celestial bodies, shall be carried out for the benefit and in the interests of all countries, irrespective of their degree of economic or scientific development, and shall be the province of all mankind.
Outer space, including the Moon and other celestial bodies shall be free for exploration and use by all States without discrimination of any kind, on a basis of equality and in accordance with international law, and there shall be free access to all areas of celestial bodies.
There shall be freedom of scientific investigation in outer space, including the Moon and other celestial bodies, and States shall facilitate and encourage international cooperation in such investigation."
[7] E. Back Impallomeni, "Sources of Space Law", Proceedings of the 11th European Summer Course on Space Law and Policy of ECSL, La Rochelle, 2-13 September 2002, pp.4-14.
[8] Treaty on Principles Governing the Activities of States in the Exploration and Use of Outer Space, Including the Moon and Other Celestial Bodies, 610 UNTS 205, 27 January 1967, Art. VI: "States Parties to the Treaty shall bear international responsibility for national activities in outer space, including the Moon and other celestial bodies, whether such activities are carried on by governmental agencies or by non-governmental entities, and for assuring that national activities are carried out in conformity with the provisions
set forth in the present Treaty. The activities of non-governmental entities in outer space, including the Moon and other celestial bodies, shall require authorization and continuing supervision by the appropriate State Party to the Treaty(...)"
[9] Ibidem Art. VI: "(...)When activities are carried on in outer space, including the Moon and other celestial bodies, by an international organization, responsibility for compliance with this Treaty shall be borne both by the international organization and by the States Parties to the Treaty participating in such organization."
[10] S. Hobe, J. Hettling, "Challenges to Space Law in the 21st Century – Project 2001 Plus", Proceedings of the 45th Colloquium on the Law of Outer Space of IISL, Houston, October 2002, pp. 51-55.
[11] Ibidem.
[12] C. Baudin, "Cooperation and International Agreements: Article XIV of the ESA Convention", Proceedings of the International Colloquium "Legal Aspects of Cooperation between the European Space Agency and Central and Eastern European Countries", Prague, 11-12 September 1997, Edited by ECSL 1998, pp.11-25.
[13] COM(2007) 212 final, "European Space Policy", Brussels, 26 April 2007.
[14] Framework Agreement between the European Community and the European Space Agency, entered into force on 28 May 2004.
[15] Ibidem, Preamble: " the Parties recognise that they have specific complementary and mutually reinforcing strengths and are committed to cooperating in an efficient and mutually beneficial manner and to avoiding any unnecessary duplication of effort".
[16] Ibidem Art.1.
[17] COM(2003) 673, White Paper "Space: a New European Frontier for an Expanding Union. An Action Plan for Implementing the European Space Policy", Brussels, 11 November 2003.
[18] Report from the Presidency to the Council, "A Market for GMES in Europe and its Regions- the Graz Dialogue 19 – 20 April 2006", acknowledged in Brussels on 15 May 2006.
[19] The Maastricht Treaty, 7 February 1992, Art.4 : "The Council and the Commission shall be assisted by an Economic and Social Committee of the Regions acting in an advisory capacity."
[20] F. Massart , "Panorama des missions et competences régionales au travers de l'Europe", Atelier GMES " Le role des Régions européennes dans GMES", Toulouse, 9 mars 2006. http://recherche.midipyrenees.fr/IMG/pdf/gmes-session2-massart.pdf
[21] COR Studies I-1/2004, "The selection Process of COR Members, Procedures in the Member States", Brussels, September 2004.
[22] Report from the Presidency to the Council, "A Market for GMES in Europe and its Regions- the Graz Dialogue 19 – 20 April 2006", acknowledged in Brussels on 15 May 2006.
[23] The Network is named NEREUS - Network of European Regions Using Space technologies. www.nereus-regions.eu
[24] Charter on the Creation and the Implementation of the Network of European Regions Using Space Technologies-NEREUS, signed in Toulouse on 18 December 2007.
[25] Statute of the Network of the European Regions Using Space Technologies, entered into force on 13 June 2008.

[26] Charter on the Creation and the Implementation of the Network of European Regions Using Space Technologies-NEREUS, signed in Toulouse on 18 December 2007, Paragraph 1, p.2

[27] Statute of the Network of the European Regions Using Space Technologies, entered into force on 16 June 2008, Title I Art. 3.

[28] Ibidem Title II Art.5. Only full members have the right to vote at General Assembly meetings, whereas associate members have a status of observers.

International Institute of Space Law
51st International Colloquium on the Law of Outer Space
Glasgow, Scotland
Friday, October 3, 2008, 10H00, Room Alsh 1, E8, Session 5

Bilateral legal framework agreements governing international cooperation in outer space*

Michael C. Wholley**
General Counsel, NASA

Steven A. Mirmina***
Senior Attorney, NASA

Abstract

Over the past decade, NASA has increasingly utilized bilateral legal framework agreements to establish the legal parameters for the conduct of cooperative space missions with the space agencies of other countries. The paper explains from a legal perspective why these framework agreements are both useful, and in some cases necessary. NASA's framework agreements generally contain a core set of legal clauses including liability, intellectual property rights, and the transfer of goods and technical data. Through practice, these clauses are now well recognized throughout the community of space faring nations. Successful conclusion of legal framework agreements facilitates the resolution of legal issues and makes scientific and technical cooperation between the Parties proceed more smoothly. By resolving in advance all of the legal issues that routinely arise when negotiating an international agreement to conduct an activity in outer space, significant time and resources are saved, thereby allowing the space agencies to focus on performing their underlying missions more efficiently and effectively.

** Retired Marine Brigadier General Michael Wholley formerly served as the Staff Judge Advocate to the Commandant of the Marine Corps/Director, Judge Advocate Division. He managed the active and reserve Marine legal community of more than 1,500 attorneys, paralegal and civil service support staff. He was commissioned through the Naval ROTC program at Harvard University, Cambridge, Mass., after receiving his B.A. degree, magna cum laude, in History and Literature. He holds a J.D. degree from the University of Virginia Law School; an LL.M degree from George Washington University in Environmental Law and Land Use; and a master's degree in National Security and Strategic Studies from the College of Naval Warfare.

*** Steve Mirmina is the Lead Counsel for NASA's Space Operations Mission Directorate. Previously, he served nine years in the International Law Practice Group of NASA's Office of the General Counsel. He has been NASA's primary attorney negotiating these framework agreements. Steve has LL.M. degrees from both Georgetown University in Washington, DC and Leiden University in the Netherlands. He is a graduate of Brandeis University and received his law degree from the University of Connecticut School of Law.

Overview

Over the past decade, attorneys from NASA's Office of the General Counsel have worked on an increasing number of bilateral legal framework agreements to facilitate cooperation in the civil uses of outer space. Working with colleagues from NASA's Office of External Relations, attorneys from the Office of the Legal Adviser of the U.S. State Department, and attorneys from other space agencies and foreign ministries around the globe, NASA has concluded these framework agreements to establish the legal parameters for NASA to conduct cooperative space missions with the space agencies of other countries. The article explains why these agreements are both useful and necessary from a legal perspective, and it outlines the general content of these framework agreements.

Purpose of Legal Framework Agreements

NASA's agreements regarding cooperation in outer space generally contain a number of core legal clauses concerning, *inter alia*, liability, intellectual property rights, and the transfer of goods and technical data. By resolving in advance the myriad legal issues that routinely arise when negotiating an international agreement to conduct an activity in outer space, precious time and resources are saved allowing the space agencies to focus on performing their underlying scientific and technical missions.

Authority to Conclude the Agreement

Authority of NASA to conclude international agreements

NASA may cooperate with other nations and groups of nations in work done pursuant to its organic statute, the National Aeronautics and Space Act.[1] Section 205 of the Space Act authorizes NASA to engage in a program of international cooperation to fulfill its mission. In carrying out these activities, the Space Act specifically provides:

> 42 USC 2475: International Cooperation Sec. 205: **The Administration**, under the foreign policy guidance of the President, **may engage in a program of international cooperation** in work done pursuant to this Act, and in the peaceful application of the results thereof, **pursuant to agreements** made by the President with the advice and consent of the Senate. (emphasis added)

An accompanying signing statement by President Eisenhower on July 29, 1958, underscored NASA's authority to conclude international executive agreements on behalf of the United States: "I regard this section merely as recognizing that international treaties may be made in this field, and as not precluding, in appropriate cases, less formal arrangements for cooperation. To construe the section otherwise would raise substantial constitutional questions." Since NASA's founding 50 years ago, NASA and the State Department have consistently interpreted and applied this provision to require

[1] The National Aeronautics and Space Act of 1958, Pub. L. No 85-568, 72 Stat. 426-438 (Jul. 29, 1958), as amended, 42 U.S.C. § 2451 *et seq.*, hereinafter "the Space Act."

"legally binding agreements."[2] President Eisenhower's reference to "less formal arrangements for cooperation," refers to international arrangements less formal than Treaties requiring the advice and consent of the Senate, yet which are, nevertheless, legally binding agreements.[3] In U.S. parlance, these "less formal arrangements for cooperation" are known as "executive agreements,"[4] and under the Case Act, executive agreements must be notified to the Congress within 60 days of their entry into force.[5]

<u>Authority of other space agencies to conclude international agreements</u>

NASA's legal framework agreements establish the legal provisions under which future cooperative activity could take place. Generally, such agreements are structured in one of two ways, contingent on the legal authority *of the other Party*. NASA, with the assistance of the State Department, must ascertain what entity is the appropriate contracting Party to conclude the agreement. NASA conducts international cooperative activities pursuant to binding agreements under international law. A fundamental question NASA regularly encounters is whether the other prospective Party to the agreement has the juridical capacity to sign agreements that are binding under international law. Certain countries' space agencies apparently have such authority. For example, the German Space Agency (Das Deutsche Zentrum für Luft- und Raumfahrt (DLR)) and the Russian Federal Space Agency (Roscosmos) have such legal capacity. International agreements between NASA and DLR or between NASA and Roscosmos can be concluded readily by NASA, subject to consultation with the Secretary of State.[6]

In other countries, space agencies sometimes do not have the authority to conclude agreements that are legally binding under international law. In such cases, in order for NASA to ensure that its framework agreement is legally binding under international law, such agreements generally must be concluded by the foreign ministry of the other country. The U.S. Government can accomplish this in two basic ways. One way is for NASA to request that the State Department conclude the legal framework agreement on a Government-to-Government level. The other option involves NASA requesting the State Department to exchange Diplomatic Notes with the corresponding Ministry of Foreign Affairs (or other appropriate Ministry) of the other country authorizing the cooperation described in the underlying agency-agency legal framework agreement. In this

---

[2] Parallel NASA authority to conduct activities with U.S. entities through binding agreements under U.S. law is found in Section 203(c)(5) of the Space Act, which states that, in the performance of its functions, NASA is authorized: "to enter into and perform such contracts, leases, cooperative agreements, or other transactions as may be necessary in the conduct of its work and on such terms as it may deem appropriate, with any agency or instrumentality of the United States, or with any State Territory, or possession, or with any political subdivision thereof, or with any person, firm, association, corporation, or educational institution."

[3] *Youngstown Sheet & Tube Co. et al v. Sawyer*, 343 U.S. 579; 72 S. Ct. 863 (1952), concurring opinion of Frankfurter, J.

[4] For a discussion of U.S. practice with executive agreements, see ANTHONY AUST, MODERN TREATY LAW AND PRACTICE, 157-158 (2002).

[5] See generally, 22 U.S.C. § 2656d, the Case Act, 1 U.S.C.112a, 112b, and regulations implementing the Case Act, 22 CFR Part 181.

[6] Ibid.

circumstance, the international agreement is not the underlying legal framework agreement itself, but rather is found in the exchange of diplomatic notes.

Contents of Legal Framework Agreements

Preamble

Most legal framework arrangements for civil cooperation in outer space are structured similarly. They often start with a Preamble, in which the countries or the space agencies recall successful cooperation that they have shared in the past. For example, a few illustrative preambular paragraphs of the U.S.-Canada legal framework agreement are as follows:[7]

> Recognizing a mutual interest in the exploration and use of outer space for peaceful purposes; Considering the desirability of enhanced cooperation between the Parties in human space flight, space science and exploration, Earth science, civil aeronautics research and other activities; ... [and] Recalling their long and fruitful cooperation since 1959 in the exploration and peaceful use of outer space, through the successful implementation of cooperative activities in a broad range of space science and applications areas; ...

The Preamble will customarily conclude with a closing similar to: "the Parties have agreed as follows..." (then setting forth the substance of the Agreement).

[7] The Framework Agreement between the Government of the United States of America and the Government of Canada for Cooperation in the Exploration and Use of Outer Space for Peaceful Purposes is currently under review by the Government of Canada and has not yet entered into force, although the Parties reached *ad referendum* agreement on the text in April, 2007.

Purpose and Scope

The Preamble may be followed by a "Purpose" article explaining that the Parties wish to establish a legal framework to govern the terms and conditions for their future cooperation. After the Purpose, when needed, agreements contain an article with any necessary definitions of specific terms for the agreement. Following that, one finds the Scope article explaining the broad range of activities covered by the agreement.

Implementing Agencies

When the agreement is a Government-to-Government agreement, it usually contains an article explaining that NASA would be the "Implementing Agency" on behalf of the United States, and the foreign country's space agency would be the Implementing Agency for that country. What does the Implementing Agency implement? Specifically, the Government-to-Government framework agreement establishes the general legal structure for the cooperation, while Implementing Arrangements concluded under the Government-to-Government agreement set forth the specific roles and commitments of the space agencies to conduct missions in outer space (or in air space, or on Earth, as appropriate). Generally, the Implementing Arrangements refer to and are subject to the legal framework agreement, and should there be any inconsistency between the Implementing Arrangements and the framework agreement, then the framework agreement prevails.

Financial Arrangements

NASA's legal framework agreements customarily contain a clause on Financial Arrangements generally stating that the obligations of the parties (or Implementing Agencies in the case of a Government-to-Government agreement) are subject to the availability of appropriated funds, and that if one party should encounter budgetary difficulties that may affect its performance under the agreement, then it would consult with the other as soon as possible.

Customs Duties and Taxes

NASA has special legal authority to exempt from customs duties and taxes articles that are used in international space cooperative activities.[8] Thus, legal framework agreements routinely contain a clause obliging the other party to waive customs duties and taxes if they have authority to do so, or to pay those duties or taxes should they be imposed.

Intellectual Property

Intellectual property (IP) rights are also typically addressed in a legal framework agreement. In short, patents, trademarks, and copyrights are all covered. Generally speaking, any IP created by one party before or outside the scope of the agreement belongs to that party; allocation of any IP rights solely created by one party or its contractors during the course of the implementation of the agreement is determined by that party's national laws. For any IP jointly created

[8] In order to encourage and facilitate the use of NASA's launch services for the exploration and use of space, section 116 of Public Law 97–446 provided for the duty-free entry into the United States of certain articles imported by NASA for its space-related activities or articles imported by another person or entity for the purpose of meeting its obligations under a launch services agreement with NASA. This exemption from duty was provided for in Subheading 9808.00.80, Harmonized Tariff Schedule of the United States (HTSUS) (19 U.S.C. 1202). Also, HTSUS, Chapter VIII, U.S. note 1, pursuant to the same law, provided that return of articles by NASA from space to the United States would not be considered an importation, and similarly not be subject to a duty. As a result of the Uruguay Round agreements of the 1994 General Agreement on Tariffs and Trade, this authority was revised and expanded in scope. It now provides that imports of articles for NASA's use and articles imported to implement NASA's international programs, including articles to be launched into space, parts thereof, ground support equipment, and uniquely associated equipment for use in connection with NASA's international programs and launch service agreements would be eligible for duty-free customs entry upon certification by NASA to the Commissioner of Customs. The revised authorities also provided, in U.S. note 1 to subchapter VIII of chapter 98 of the HTSUS, that articles brought into the customs territory of the United States by NASA from space or from a foreign country as part of a NASA's international programs would not be considered imports or subject to customs entry requirements. The provision of the HTSUS: 9808.00.80 00 lists the following in a table annotating their rate of duty as "free:"

> Articles for the National Aeronautics and Space Administration and articles imported to implement international programs between the National Aeronautics and Space Administration and foreign entities, including launch services agreements;
> Goods certified by it to the Commissioner of Customs to be imported for the use of the National Aeronautics and Space Administration or for the implementation of an international program of the National Aeronautics and Space Administration, including articles to be launched into space and parts thereof, ground support equipment and uniquely associated equipment for use in connection with an international program of the National Aeronautics and Space Administration, including launch services agreements.

during performance of the agreement (which is relatively unusual), the parties would be obliged to consult to determine: the allocation of rights to, or interest in, such joint invention; the responsibilities, costs, and actions to be taken to establish and maintain patents or other forms of protection for each joint invention; and the terms and conditions of any license or other rights to be exchanged between the parties or granted by one party to the other.

Transfer of Goods and Technical Data

NASA is committed to compliance with export control laws, including the Export Administration Regulations (EAR) and the International Traffic in Arms Regulations (ITAR).[9] For this reason, NASA's legal framework agreements routinely contain a clause concerning the transfer of goods and technical data. While the specific clause is very detailed, it essentially states a few fundamental principles: that only those technical data and export-controlled goods necessary to fulfill the agreement are transferred; and that, in performance of the agreement, the Parties will follow all applicable laws and regulations, particularly those concerning export control and control of classified information. Additionally, the clauses provide that if export-controlled goods or data are transferred, they must be marked appropriately, and that such goods and data are to be used solely for purposes of fulfilling that Party's responsibilities under the agreement, and then either returned or destroyed.

Cross-Waiver of Liability

NASA's liability provisions are extremely complex and they contain precise definitions.[10] Nevertheless, they are well recognized over decades of practice as the standard of risk allocation for cooperative space activities throughout the globe. Succinctly stated, the fundamental principle of NASA's liability article states that each Party assumes its own risks inherent in the cooperative activity. The article contains a mutual promise by both Parties not to sue each other for losses caused by any of the activities that take place under the agreement, subject to a few exceptions. The article also requires that each Party flow this cross-waiver of claims down to any "related entities" to ensure that they also waive claims against the other Party and the other Party's related entities. With each entity assuming its own risks under the agreement, the clause functions to encourage the collaborative exploration of outer space by lowering the costs of cooperation by reducing the risk of exposure to liability. Sec. 309 of the Space Act confirms and clarifies NASA's authority to waive claims of the U.S. Government in cooperative

[9] The U.S. Department of Commerce controls exports of dual-use commodities, software, and technologies, and administers the Export Administration Act of 1979, as amended (EAA, 50 U.S.C. app. 2401 et seq.), through its Export Administration Regulations (EAR, 15 CFR 730-774). The U.S. Department of State has responsibility for the control of the permanent and temporary export and temporary import of defense articles and services, governed by section 38 of the Arms Export Control Act (AECA, 22 U.S.C. 2778) and Executive Order 11958, as amended. The AECA, among these other requirements and authorities, provides for the promulgation of implementing regulations, the International Traffic in Arms Regulations (ITAR, 22 CFR 120-130).

[10] A detailed discussion of the changes NASA recently made to 14 CFR Part 1266 (NASA's cross-waiver clauses for international cooperative agreements) is found at 73 FR 10143 (February 26, 2008).

agreements in exchange for a reciprocal waiver of claims from the foreign cooperating party. Since NASA's waiver of claims includes waiving claims of other U.S. Government agencies, NASA looks to its partner space agencies (or their governments) to ensure that the scope of their waiver is reciprocal.

Amicable Settlement of Disputes

NASA's legal framework agreements also contain provisions on dispute settlement. These provisions state unequivocally that the parties will resolve disputes through negotiations and consultations. NASA aims to resolve disputes at the lowest possible technical level – for example at the level of the program managers. Failing that, disputes can be resolved at a higher managerial level, such as at the level of the Associate Administrator. Only in rare cases would a dispute be elevated as high as the NASA Administrator or his designee for joint resolution.

Final Clauses

NASA's legal framework agreements conclude with relatively standard final clauses, generally comprising the procedures for amendments, entry into force, duration, and termination of the Agreement.[11]

[11] Final clauses "owe their designation as *final clauses* not merely to the fact that they are normally placed at the end of a treaty but to the fact that they cannot be negotiated until work on the substantive provisions of the treaty itself is essentially completed." Shabtai Rosenne, *When is a Final Clause Not a Final Clause,* 98 Am.J.Intl.L.546.

Conclusion

As of the date of this paper, NASA has concluded ten of these bilateral legal framework agreements for cooperation in outer space either at the space agency level or at the Government-to-Government level.[12] At the agency level, NASA has concluded legal framework agreements with the space agencies of Argentina, India, and Israel. At the Governmental level, framework agreements have been concluded with: Brazil, France, Hungary, Norway, Russia, Sweden, and Ukraine. In fact, NASA's agreement with the Ukraine was signed on March 31, 2008, during the visit of President Bush and Secretary of State Rice to Kiev.

Each of NASA's legal framework agreements has varying levels of detail and the scope can sometimes vary.[13] Nevertheless, successful conclusion of a framework agreement establishing the legal regime for potential cooperation between space agencies has proven to facilitate scientific and technical cooperation in a variety of areas, often on an expedited basis. By resolving legal

[12] Not all of these agreements remain in force. Some have expired by virtue of their own terms.

[13] For example, the U.S. concluded an agreement with Japan in 1995 solely in the area of cross-waiver of liability. See the Agreement between the Government of the United States of America and the Government of Japan Concerning Cross-Waiver of Liability for Cooperation in the Exploration and Use of Space for Peaceful Purposes, signed at Washington, on April 24, 1995, as amended, and the Exchange of Notes of the same date between the two Governments concerning subrogated claims. Less formally, however, there have been preliminary discussions between NASA and the Japan Aerospace Exploration Agency concerning the possibility of a more comprehensive framework agreement on civil space cooperation between the U.S. and Japan.

questions in advance, NASA and its partner space agencies around the globe can focus their attention on what they do best: exploring the universe, understanding our own planet, and inspiring the next generation of explorers.

IAC-08-E8.5.4

THE GLOBAL EXPLORATION STRATEGY: LEGAL PERSPECTIVES

Karl W. Abendschein, Esq.
Berliner, Corcoran & Rowe, L.L.P., Washington DC, USA

Corinne C. Jorgenson
Advancing Space Consulting Group, Fairfax, VA, USA

I. Abstract

The Global Exploration Strategy ("GES") is an initiative by 14 space agencies that establishes a collaborative strategy for exploring our solar system. In order to foster an environment that is conducive to private sector investment in space exploration, the GES acknowledges the need for the development of a common understanding with respect to difficult legal and policy issues such as: (1) national export control regimes and (2) intellectual property rights ("IPR") related to industrial inventions (e.g., patents, trademarks, and trade secrets). The participants in the GES established a coordination committee to implement the GES, the International Space Exploration Coordination Group ("ISECG"). The ISECG operates on a consensus basis and will focus on developing non-binding findings, recommendations, and other outputs to address critical issues for the space agencies involved in exploration activities, including the assessment of relevant legal issues. In order to start a discussion regarding the efficacy of ISECG, this paper will (1) identify and flesh out the legal challenges posed by national export control regimes and IPR protection; and (2) suggest that the ISECG explore modest and practical solutions that take full advantage of the structure established for the ISECG.

II. The Global Exploration Strategy

A. Overview

In May 2007, fourteen space agencies released *The Global Exploration Strategy: The Framework for Coordination* (the "GES"), which establishes a collaborative strategy for the robotic and manned exploration of our solar system.[1] The GES focuses on those destinations where humans will one day live and work (*e.g.,* the Moon, Mars, and near-Earth Asteroids) and recognizes that "[s]ustainable space exploration is a challenge that no one nation can do on its own."[2] Therefore, the GES seeks to build a global framework that will (1) articulate a compelling case for globally coordinated space exploration, and (2) set the stage for future international discussions on coordination mechanisms and architectures.[3]

---

[1] Participants in the development of the GES included representatives from Australia, Canada, China, the European Space Agency ("ESA"), France, Germany, Great Britain, India, Japan, Russia, South Korea, Ukraine, and the United States.

[2] GES, at p. 2.

[3] *See The NASA-ESA Comparative Architecture Assessment*, (July 16, 2008)

The GES does not establish a single program but rather calls for the development of international exploration coordination tools to enhance mutual understanding among partners and identify areas for potential cooperation. As the GES explains, "[a] coordinated strategy will help individual nations with shared objectives to engage in joint projects that will maximize their return on investment."[4]

The GES indentifies the following five general themes in which space exploration provides a benefit to society:

- Theme 1: New Knowledge in Science and Technology
- Theme 2: A Sustained Presence – Extending Human Frontiers
- Theme 3: Economic Expansion
- Theme 4: A Global Partnership
- Theme 5: Inspiration and Education

It is participation of private industry in these themes that is the focus of this paper, especially with respect to Themes 1-3. Undoubtedly, industry will play an important role in future exploration from the development and manufacture of the spacecraft and equipment that will enable the sustained presence of man beyond the Earth to the exploitation of mineral resources on the Moon, Mars, and near-Earth asteroids. Therefore, implementation of the GES will require developing new and innovative technologies, many of which will come from the private sector, such as space-based resource extraction, safe habitats, human-robot cooperation, and efficient power generation and energy storage. It is also envisioned that commercial providers of crew and cargo transportation services, and telecommunications and navigation systems will play an important role in future space exploration activities.

B. Potential impediments to the full realization of the goals of the GES must be addressed

In order to foster the an environment that is conducive to private sector investment in space exploration, the GES acknowledges the need for the development of a common understanding with respect to difficult legal and policy issues such property rights and technology transfers. Within property rights and technology transfers, two principal areas of concern are: (1) national export control regimes and (2) intellectual property rights ("IPR") related to industrial inventions (e.g., patents, trademarks, and trade secrets). It is incumbent on participants in the GES to seek the consistent treatment of technology with respect national export control regimes and to help ensure that the intellectual capital of commercial participants is protected.

Recent history has shown that the inconsistent treatment of technology with respect to the application national export control regimes can lead to unnecessary delays and complications with respect to international cooperative space programs. Most notably the continued insistence by the United States Government to treat civil spacecraft as "munitions" for export control purposes has damaged the ability of U.S. companies to participate in international missions and spurred international partners to avoid the use of U.S.-origin technology. Building upon efforts that began with the International Space Station ("ISS") and other international cooperative space ventures, the GES can provide a forum with which to discuss the harmonization of export control regimes as they relate to space exploration.

---

www.nasa.gov/pdf/259237main_NASA_ESA_CAA-Report.pdf

[4] GES, at p.6.

The legal issues related to IPR in space are being brought into the limelight due to the increased participation of the private sector in space activities but there remains a great deal of ambiguity. For example, none of the major international legal regimes concerning IPR protection or the international legal treaties governing the use of outer space deal directly with the issue of IPR in space.[5] Building upon the practical solutions developed for other international cooperative programs, most notably ISS, the GES is well placed to offer innovative solutions until the appropriate international legal framework can be developed.

C. <u>The GES Coordination Mechanism can work towards practical solutions to export controls and IPR concerns</u>

The GES called for the creation of a "coordination mechanism," now known as the International Space Exploration Coordination Group ("ISECG"), that will be responsible for the implementation of the GES. On July 10, 2008, the ISEGC held its first meeting in Montreal. During this meeting, the ISECG established a Secretariat, initially to be hosted by the European Space Agency ("ESA"), and took the initial steps towards identifying critical space infrastructure interfaces, such as between spacecraft, lunar rovers, and lunar habitats.[6]

The Terms of Reference ("ToR") for ISECG state that ISECG is intended to meet "the need [identified in the GES] to establish a voluntary, non-binding international coordination mechanism through which individual agencies may exchange information regarding interests, objectives, and plans in space exploration with the goal of strengthening both individual exploration programs as well as the collective effort."[7] The ISECG is designed to ensure that international space activities are not duplicative of each others work/investment; that results and lessons learned are shared; participation in cooperative missions (*i.e.*, bilaterally, trilaterally, and so on) is encouraged; and interoperability standards are agreed upon.[8]

The framers of the GES and the ISECG rejected the adoption of a binding ISS type of Intergovernmental Agreement ("IGA"), which could have dealt with many of the legal challenges up front in a binding manner.[9] The ISECG, instead, operates on a consensus basis and will focus on developing non-binding findings, recommendations, and other outputs to address critical issues for the space agencies involved in exploration activities, including the assessment of relevant legal issues. The ISECG is open to all space agencies which have or are developing space exploration capabilities for peaceful purposes.

While the creation of comprehensive binding IGA was rejected as unfeasible by the participants in the drafting of the GES, it is anticipated that the ISECG will work towards the development of consensus based solutions that can be readily applied to the IGAs that will be negotiated in the future between countries that anticipate cooperative ventures. This can include, as appropriate, the

---

5 Sasikkumar, *Space Commercialization: Addressing Intellectual Property Issues*, Proceedings of the 48th Colloquium on the Law of Outer Space, AIAA (2006), p. 263.

6 *NASA: Global Exploration Strategy talks continue*, Dow Jones Factiva, July 16, 2008.

7 ISECG ToR at 1.

8 *See* Gibbs, Kirkham, *The Global Exploration Strategy: Developing A Framework for International Coordination and Cooperation*, Proceedings of the 50th Colloquium on the Law of Outer Space (2007), IAC-07-B3.1.08 (pp. 13).

9 *Ibid.* at p. 12.

establishment of specialized committees, working groups or workshops that address key issues and challenges to the coordinated exploration of our solar system. With respect to national export controls and IPR issues, the ISECG could establish committees or working groups supported by appropriate experts to develop a forum in which to address these issues and develop practical solutions, such as template provisions for future IGAs. Given the broad participation of stakeholders in the ISECG, any practical solutions or recommendations developed by these committees or working groups are likely to gain wide acceptance and provide a positive influence on future exploration ventures.

III. Export Controls

A. There is a need for the rationalization of national export controls with respect to the space exploration activities contemplated by the GES

There will always be a delicate balance between the need for the free flow of ideas and technology with respect to the commercialization of space and the legitimate needs of nations to protect their national security and foreign policy needs. If a nations export controls are overly strict, such rules risk stifling those industries responsible for developing the next generation of technology necessary to enable space exploration without an appreciable gain in national security.

All of the space faring nations participating in the GES have export controls regimes that require licenses for sensitive technologies involved in civil space activities. The United States, however, is the only one of these countries to treat civil spacecraft as "munitions" subject to control as military items through its International Traffic in Arms Regulations ("ITAR"). The other nations control sensitive civil space hardware and technologies under regimes that control items designed for civil/commercial uses but also have military applications (*i.e.,* dual use items). This disparity in the level of controls for U.S. civil space hardware and technology versus those controls for civil space hardware and technology from the other GES participants has contributed to a hesitancy among international partners to use U.S. technology and/or engage with the U.S. Government in cooperative space ventures. The ITAR is further blamed, at least in part, for the recent decline in the market share of U.S. companies in the international space market.

B. What is the ITAR?

The ITAR, which implement the Arms Export Control Act, 22 U.S.C. §§ 2778 *et seq.*, control the transfer to foreign persons of U.S. commodities and technologies listed on the U.S. Munitions List, (the "USML") 22 C.F.R. § 121. In addition to traditional military items, the USML also includes many commodities and technologies related to civil space programs, such as: (a) spacecraft (including all remote sensing satellites), (b) space launch vehicles and rocket engines; (c) missile tracking systems; (d) systems or subsystems, components, parts, accessories, and associated equipment specifically designed or modified for the foregoing; and (e) U.S.-origin technical information directly related to the foregoing. The ITAR also regulates retransfers, by foreign persons, of the foregoing items outside of the United States. Under the ITAR, almost all civil spacecraft items (and related technology) on the USML require licenses for export, unless a limited number of narrow exemptions apply.

C. Why does the ITAR apply to civil space programs?

In 1998, U.S. congressional investigations discovered that during 1996 Hughes Electronics and Loral Space and Communications illegally transferred technology to China that may have improved the capabilities of China's intercontinental ballistic missiles. The assistance was provided in the wake of the February 1996 Intelsat 708 launch failure. The most publicized investigation was that of the U.S. House of Representatives Select Committee on U.S. National Security and Military Commercial Concerns with the People's Republic of China, chaired by Representative Christopher Cox. A similar investigation led by Jesse Helms was conducted in the Senate. The Select Committee concluded that Hughes and Loral deliberately transferred technical information and know-how to China during the course of the accident investigations.

In response to the investigations and their conclusions, in Section 1513 of the 1999 National Defense Authorization Act ("NDAA") the House and Senate transferred jurisdiction over commercial satellites and satellite components exports from the Commerce Department to the State Department, and tightened restrictions allowing U.S. made satellites, satellite components, equipment, and technical information (*e.g.,* launch failure analysis).

The 1999 transfer reversed the 1996 transfer of commercial communications satellites and related items from the ITAR to the Commerce Control List ("CCL") of dual-use commodities and technology controlled by the U.S. Department of Commerce under the Export Administration Regulations ("EAR"), 15 C.F.R. §§ 730-774. (The 1996 change had put U.S. export controls in accord with those of the EU, Japan, and other Wassenaar Arrangement members, which treat commercial satellites and related items as predominantly commercial items subject to less stringent export controls than those imposed on Munitions List items.)

D. <u>The Negative Impact of the ITAR on International Space Activities</u>

D.1. The ITAR has negatively impacted international commercial cooperation and innovation

The reversion had a significant impact on international space cooperation and satellite programs. Because nearly every U.S. manufacturer's spacecraft involves foreign subcontractors, a foreign launch service provider (primary or back-up), foreign insurers, and/or foreign customers, multiple export authorizations are required from the outset of a spacecraft program, through launch, and continuing into investigation and resolution of on-orbit anomalies that a spacecraft may experience. Lengthy processing times and inconsistent application of the ITAR by the U.S. Department of State's Directorate of Defense Trade Controls ("DDTC") tend to dissuade non-U.S. customers from purchasing U.S. spacecraft, especially given that the time from contract execution to launch is extremely important for spacecraft customers. Further, the U.S. Government policy of not permitting the export of certain technology, such as detailed design data and concepts regarding physical engineering implementation of design methodology, engineering analysis, and manufacturing know-how, makes procurement of many U.S. spacecraft components frustrating for foreign spacecraft manufacturers.

There is evidence that the ITAR has had a significant negative impact on the financial health of second and third tier spacecraft component suppliers in the United States. A recent Center for Strategic and International Studies report estimates that

U.S. spacecraft component manufacturers lose approximately USD 600 million per year due to ITAR restrictions.[10] This has led to questions concerning the financial security of these companies, especially manufactures of traveling-wave tubes, optical coatings, imagers, and solar cells, which are traditional sources of innovation for spacecraft development in the United States.[11] These suppliers are likely to lose more money as the European Union, citing concerns over U.S. export controls, announced that U.S. components are to be avoided wherever possible in the €3.4 billion procurement for the Galileo global satellite navigation system.[12]

The Galileo procurement is just the most recent example of the trend over the past several years for European manufacturers to actively design out U.S. subsystems and components due, at least in part, on the negative impact of the ITAR. Thus, it appears that the ITAR is speeding up the development of advanced technology outside of the United States. For example, Thales Alenia offers an ITAR-free satellite bus. The ITAR has also spurred the development of an ITAR-free European apogee motor, thruster valves, and star tracker. In addition, countries of concern to the United States (*e.g.,* China), are getting the technology they require to develop their space programs despite U.S. controls. As Mike Gold of Bigelow Aerospace stated to The Economist, "if the purpose of the ITAR is to lose billions of dollars of business, ship jobs overseas, and the Iranians and the Chinese get the same technology anyway, then mission accomplished."[13]

Despite a challenging regulatory environment, there are interesting and successful examples of cooperation in the space transportation industry. Of course, the first one that comes in mind is Atlas V, as it incorporates the Russian-made RD-180 engine, the Swiss-made fairing (the same as Ariane V), as well as a the Spanish-made vehicle equipment bay and payload adaptors. Another example is the Delta IV with its the RL-10B2 upper stage engine incorporating a French-made carbon-carbon nozzle.

D.2. Government exploration programs have likewise been negatively affected by the ITAR

The reversion has had an impact on civil government space programs as well. For instance, the $110-million 2005 Demonstration of Autonomous Rendezvous Technology ("DART") mission failure was attributed to, among other things, the ITAR. NASA failure-analysis report concluded, in part, that "insufficient technical communication between the project and an international vendor due to perceived restrictions in export control regulations did not allow for adequate insight," referring to Orbital Sciences Corporation's, the prime contractor communications with Surrey Satellite Technology, the U.K. supplier of a miniature GPS receiver. In 2005, ESA scaled back its plans to cooperate with the United States in the development of NASA's Mars rover due, at least in part, over fears that U.S. export controls could make the project "too complicated to be feasible."[14]

[10] *Briefing of the Working Group on the Health of the U.S. Space Industrial Base and the Impact of Export Controls*, Center for Strategic and International Studies (Feb. 2008), p. 33.
[11] *Ibid.* at 10.
[12] *Gravity is not the main obstacle for America's space business*, The Economist (Aug. 21, 2008).
[13] *Ibid.*
[14] *Ibid.*

Due in large part to the ITAR, NASA has adopted a "clean interface" approach, which encourages the development of standard interfaces that minimize technical interactions with respect to the procurement of components and instruments from international partners. Unfortunately, this approach to integration in international cooperative programs has not proved as successful as earlier efforts. Thus, a Cassini-Huygens-type mission, starting today, might not be as successful today.

The ITAR has also led to a reshuffling of the respective roles of governments participating in international civil space programs. One example is the James Webb Space Telescope. NASA scrapped initial plans for European suppliers to develop the bus in Europe and, instead, opted to use a U.S. developed bus in order to lessen the various levels of ITAR burden on the program. Thus, the integration of sensitive U.S. instrumentation would take place in the United States, limiting the transfer of U.S. technology.

The ITAR will also have severe consequences for any future missions that involve China. China is subject to an arms embargo by the United States and, therefore, it is the policy of the United States to deny any licenses for the export of any items to China or Chinese nationals under the ITAR.[15]

The U.S. proved willing to provide some import concessions with respect to export controls in connection with ISS largely due to the presence of significant political will. Article 19 of the ISS IGA contains detailed provisions with respect to the ISS partners' responsibilities with respect to the transfer of technology related to the development, manufacture, and operation of the ISS. Consistent with its obligations under the ISS IGA, the U.S. Government determined that the ISS itself should be subject to the "dual use" controls under the EAR. While this has lessened tensions with respect the building and operation of the ISS, the U.S. Government reserved certain sensitive technology under the ITAR, leading some to question the safe operation of the ISS given the continued need to obtain export authorizations under the ITAR.[16]

E. Recent reform efforts have provided some procedural relief but more work is needed

The past year has seen intense industry lobbying for export control reform led principally by the Coalition for Security & Competiveness (whose members include the Aerospace Industries Association and the Satellite Industries Association). In response, the U.S. President issued National Security Presidential Directive ("NSPD") 56 in January. The Directive builds upon efforts by the DDTC to improve the defense licensing process.[17] DDTC efforts in support of the Directive have significantly improved license processing times. A year ago technical assistance agreements were taking about 6 months to process, which caused an enormous to industry. Recently, DDTC's Missiles and Space Division reported that technical assistance agreements are now being processed in approximately 35-47 days.

Unfortunately, the political climate in the United States at the moment does not

[15] *See* 22 C.F.R. § 126.1(a).

[16] *Gravity is not the main obstacle, supra.*, note 12.

[17] NSPD 56 mandates: (1) Additional financial and intelligence resources to support the licensing process; (2) that license application must be acted upon within 60 days, unless there is a strong reason for additional time; (3) that DDTC update its electronic licensing system, D-TRADE; (4) removes the requirement that third country nationals sign NDAs; (5) the creation of an interagency dispute resolution mechanism for the commodity jurisdiction process.

appear ripe for dramatic substantive reform. There does, however, appear to be a movement spear headed by the U.S. National Security Space Office to update the USML to remove spacecraft technologies that are not considered critical to national security.[18] Even this limited change, could go a long way in making the U.S. export controls on spacecraft more consistent with its international space partners.

F. The ISECG should consider the creation of permanent committee or working group to address potential impediments caused by national export controls

National export control regimes have the potential to impede international space exploration activities, especially when such regimes are not applied consistently among partner nations. As described above, the ITAR is one such regime that has been applied to civil spacecraft in a manner that is inconsistent from that of the other participants in the GES.

While export controls are national security laws that remain the province of national policy, the ISECG can provide a forum to discuss the unique issues related to international space exploration. For example, discussions could be held to develop a consensus with respect to the level of control appropriate to emerging space exploration technologies. In such a forum the burden would be on the United States or other nations proposing the imposition of military technologies to articulate the reasons for doing so. This would be complementary to the reform efforts underway in the United States focused on defining the military critical technologies warranting strict controls. Such a forum could also help stimulate political will that has proved critical in other projects, such as ISS and the Soyuz-launcher at the Guiana Space Center.[19] Such a forum could also work toward the creation of template IGA provisions related to export controls (e.g., based on past projects such as Article 19 of the ISS IGA) that could be adopted, rejected, or modified for future project specific IGAs. Although not binding, such a forum could prove useful to discuss these issues and develop potential solutions.

[18] *Gravity, supra.*, note 12.

IV. Intellectual Property Rights Issues

A. IPR protection requires the application of national jurisdiction

A.1. International space law and IPR protection

Commercial participants may be discouraged from participating in international space exploration activities, unless an adequate legal framework exists that offers protection of their intellectual property. Under current IPR regimes, this protection comes in the form of a state grant of title, under which the grantee enjoys the exclusive right, though limited in scope, duration, and geographical reach, to exploit and benefit from its invention.[20] This protection can come in the form of patents, trademarks, trade secrets, etc., which permit the holders of such rights to prevent others from using them without authorization.

Not only do existing international space agreements not directly address the issue of IPR in space,[21] they leave open

[19] See Ejova, *Euro-Russia Cooperation in Space and Export Control: Policies and Practices*, Proceedings of the 48th Colloquium on the Law of Outer Space (2005), IAC-05-E6.4.13 (p. 399).
[20] L. Malager, M. Malagar, *International Law of Outer Space and the Protection of Intellectual Property Rights*, Boston Univ. Int'l. L. J.L. (Fall 1999), at p. 350.
[21] *See* Sasikumar, *supra.*, note 5.

important questions concerning national sovereignty. Without some degree of sovereignty there can be no jurisdiction over IPR. The Treaty on Principals Governing the Activities of States in the Exploration and Use of Outer Space, including the Moon and other Celestial Bodies (the "1967 Outer Space Treaty"), the principal treaty governing space exploration, provides that "[o]uter space, including the moon and other celestial bodies, is not subject to national appropriation by claim of sovereignty, by means of use or occupation, or by any other means."[22]

The Agreement Governing the Activities of States on the Moon and Other Celestial Bodies (the "Moon Treaty") also provides a similar prohibition against claims of sovereignty and goes even further by prohibiting all forms of property rights and stating that "the moon and its natural resources are the common heritage of mankind."[23] The concept of the common heritage of mankind is not well defined but appears to imply the creation of a legal regime to make sure that the proceeds from space exploration are shared equally between space faring and non-space faring nations.[24] Uncertainty concerning the application of this principal and its potential chilling effect on private investment, contributed to the failure of the Moon Treaty to become generally accepted.

[22] Article II of the 1967 Outer Space Treaty.

[23] *See* Article XI of the Moon Treaty. Only nine nations ratified the Moon Treaty, none of whom are space faring nations, which calls into question its validity. *See* P. Tobias, *Opening the Pandora's Box of Space Law,* 28 Hastings Int'l & Comp. L. Rev. 299 (Winter 2005), at p.305.

[24] *See* S. Nandakumar, *"Common Heritage of Mankind" – Property Rights, in the Wake of Commercial Use of the Moon and Other Celestial Bodies,* Proceedings of the 48th Colloquium on the Law of Outer Space (2005), IAC-05-E6.4.02.

Even though national appropriation of outer space is prohibited by the 1967 Outer Space Treaty, the 1967 Outer Space Treaty does provide States with support in connection with the application of jurisdiction in connection with the application of national IPR laws. First, Article VI provides that States remain liable and responsible for national actions undertaken in space, whether or not such activities were conducted by governmental or non-governmental actors. This means that States have a duty to control and supervise the activities of private actors in space undertaken of spacecraft or facilities that are carried on the registry of such nation States under the 1975 Convention on Registration of Objects Launched in Outer Space. Second, Article VIII provides that States retain control and jurisdiction over spacecraft launched on their registries.

A.2. The U.S. Patents in Space Act provides for quasi territorial effect in space on space craft carried on the registry of the United States

Efforts have been made on a national level to address the unique nature of IPR protection in outer space. Consistent with the 1967 Outer Space Treaty, the United States became the first country to enact a law that specifically links national IPR laws to activities in outer space. In 1990, the United States Congress enacted the Patents in Space Act in an effort to encourage private investment in space activities by ensuring that the investments of U.S. inventors will be protected.[25] The Patents in Space Act purports to do this by extending U.S. jurisdiction with respect to patents to space

[25] 35 U.S.C. § 105 (Inventions in Outer Space). *See* J. Shoemaker, *The Patents in Space Act: Jedi Mind Trick or Real Protection for American Investors on the International Space Station,* 6 J. Intell. Prop. L. 395 (Spring 1999), at p. 398.

objects that are carried on the registry of the United.

The Patents in Space Act provides that inventions made, used, or sold in outer space on objects under the jurisdiction or control of the United States, shall be considered made, used, or sold in the United States, unless otherwise agreed to by an international agreement. Thus, the United States has attempted to extend the jurisdiction of its IPR laws, including protecting investors from infringement, to activities in outer space by in essence making U.S. registered space objects U.S. territory for this limited purpose.

At best, however, the Patents in Space Act represents a partial solution for U.S. national missions in that it does not directly address the issues that arise in international space exploration missions. A comprehensive approach that address the unique aspects of IPR in outer space and that stitches together the various national IPR and international IPR regimes is needed.

A.3. The IPR provisions of the ISS IGA represent a practical approach to IPR protection in the absence of a comprehensive legal framework

Article 21 (Intellectual Property) of the ISS IGA contains several provisions intended to protect existing intellectual property from infringement and to provide some clarity over jurisdiction with respect to inventions made aboard the ISS. In addition, Article 16 specifically excludes IPR claims from the cross-waiver of liability requirements of the IGA making infringement claims possible.

The ISS IGA provides that each partner retains jurisdiction over the elements that it provides to ISS.[26] Therefore, each ISS partner that contributes research modules to the ISS retains jurisdiction over IPR claims that result from activities undertaken in such modules. Any claims for infringement would be made according to that Party's national legal regime for intellectual property.

Article 21 of the IGA addresses the issue of the country of invention for discoveries made onboard ISS. The second paragraph of Article 21 provides that inventions are generally deemed to have been made in the country that retains jurisdiction over the module in which the invention was made.[27] For example, an invention made on the Kibo Laboratory will be deemed to have been made in Japan. The purpose of this provision is to determine the location of invention and does not prevent the filing of patents in multiple countries. The county of invention provisions of Article 21 are qualified by the third paragraph of Article 21 that prohibits partner States from applying any invention secrecy laws to prevent nationals from another nation to seek patent protection in other partner State, so long as such other partner State has adequate laws to protect information that is either classified or protected for other national security reasons (e.g., export controls).

The ISS IGA also contains marking procedures for identifying proprietary data that are intended to aid in the identification of and protection of IPR.[28]

---

[26] Article 5 (Registration, Jurisdiction, and Control), ISS IGA.

[27] Article 21 provides that for ESA registered modules "any European Partner State may deem the activity to have occurred within its territory." Paragraph 4 of Article 21 also limits the recovery for the IPR infringement claim in multiple European partner countries.

[28] Article 19 (Exchange of Data and Goods), Paragraph 3(b), ISS IGA.

B. Until a comprehensive legal frame work is developed that takes into account the unique aspects of IPR is outer spacc, the ISECG can provide an important forum to advance the protection of IPR

Until a comprehensive legal regime can be developed to address the protection of IPR in space exploration activities, space faring nations will need to continue to use existing legal regimes and IGAs to assure the adequate protections for commercial participants. The ISECG can assist in this process by establishing a permanent committee or working group to address IPR issues. Appropriate experts from participating space agencies should be invited to participate in such a committee or working group.

While the creation of a binding IGA was rejected in the establishment of the ISECG, the IPR committee or working group could work towards the drafting of template IPR provisions for future project specific IGAs. These template IPR provisions could be based on the lessons learned from previous IGAs (e.g., the ISS). The participation of all ISECG members in the development of these template provisions would contribute to consensus based solutions that would address the interests of all members. Space agencies would then be free to accept, reject, or modify these template provisions to suit the needs of individual space exploration missions. Each project would not be forced to reinvent the wheel.

The IPR committee or working group could also provide a forum to identify more comprehensive reforms. While the non-binding nature of the ISECG does not appear to hold much promise towards resolving these issues on its own, such a committee or working group could prove useful in making recommendations concerning potential solutions and identifying the appropriate forums in which to address IPR issues in a binding manner.

V. Conclusion

The establishment of the GES presents a wonderful opportunity to address the potential legal impediments to the exploration of our solar system. The ISECG creates an important, although non-binding, forum in which to address the negative impact of national export control regimes and uncertainties with respect to IPR protection in outer space. Given the non-binding nature of the ISECG, we believe that the ISECG should concentrate on those potential solutions that are best suited to its structure. This includes the development of practical tools (e.g., template IGA provisions) and providing forums to identify potential solutions.

50 YEARS OF EARTH OBSERVATION FROM SPACE AND SPACE LAW

José Monserrat Filho *
Brazilian Association of Air and Space Law (SBDA)
Brazilian Society for the Advancement of Science (SBPC)
Rio de Janeiro, SP, Brazil
monserrat@alternex.com.br

"There is still time to address this dramatic disconnect between institutions and reality. Doing so requires leadership, vision and faith that our future is not merely work of destiny but ours to shape."
Inge Kaul, Isabelle Grunberg and Marc A. Stern[1]

Abstract

Over the past five decades, space-faring States and international organizations have developed impressive capabilities for observing Earth from satellite platforms. The way humankind view and live in our home planet has fundamentally transformed for the good and for the bad. These fantastic advances are shown in rich details in the excellent book "Earth Observation from Space – The First 50 Years of Scientific Achievements"[2], prepared by a special committee of the National Research Council of the United States National Academies, and published in 2008.

Their conclusions provide ample and rich information, as well as a strong motivation to assess the relationship existing between the technological development of Earth observation systems and the development of legal instruments regarding these very intense and essential activities. They permit us to accomplish the quite useful task of establishing to what extent the remarkable development of Earth Observation technologies has been internationally regulated, and determine the legal deficiencies and insecurity such activities face today and will be facing tomorrow.

One of the conclusions states that "the full benefits of satellite observation of Earth are realized only when the essential infrastructure, such as models, computing facilities, ground networks, and trained personnel, is in place."

The clear awareness of this situation raises some legal issues: at which level are these necessities and demands looked after today by the existing instruments of International Space Law? And what kind of effective commitments should be set up in order to accelerate the dissemination of the full benefits of Earth observation satellites all over the world? Earth observations from space demonstrate successful synergy between science and technology. Is there such a synergy between science and technology, on one hand, and their legal implications, on the other hand?

The present paper aims at giving an answer to such questions.

Introduction

The US National Academies carried out a "decadal survey" entitled "Earth Science and Applications from Space: National Imperatives for the Next Decade and Beyond"[3] (NRC 2007a), considered an "essential study", the first of its kind in the Earth Science. This study recommends the US Administration "a renewal of the national commitment to support on going observations from space in order to face scientific and societal challenges over the next decades and to understand and manage natural resources".

To complement this decadal survey, the US National Aeronautics and Space Administration (NASA) asked the National Academies to draft a report on the value of the first 50 years of Earth observations from space. This led to preparation of thc document entitled "Earth Observation From Space – The First 50 Years of Scientific Achievements", issued by the US National Academy of Sciences in 2008.

The report highlighted important contributions of Earth observation from space to the current understanding of our planet. It points out: "The global view obtained routinely by observations from space is unmatched in its ability to resolve the dynamics and variability of Earth processes. Ship-based observations, for example, cannot provide the spatial and temporal information to detect the dynamic nature of the ocean. Similarly, aircraft and weather balloon measurements alone cannot resolve the details required to understand the complex dynamics of ozone depletion. Space observations provide detailed quantitative information on many atmospheric, oceanic, hydrologic, cryospheric, and biospheric processes. Because satellite information is gathered at regular intervals, it provides, like a movie, a view of changes over time. For the first time, satellites make it possible to track a tropical cyclone from its gestation over the ocean to landfall and to observe the ever-fluctuating intensity of the storm."

One of the central objectives of the National Academies report is to convince the American authorities and public opinion that the United States must continue to support the national programs of satellite observations, which has "revolutionized the Earth Science".

It is amazing that the National Academies consider it necessary to convince the authorities of the country that made such remarkable advances in these essential activities during last half century of the usefulness of these observations.

Earth observation from space is really an international issue. Since its inception, it has been an activity of great interest to the community of States because of its national security impact, and also as an indispensable tool for the improvement of everyday life, the development of all nations and people, the protection of the environment, and the management of natural disasters.

Earth observation from space takes place by remote sensing satellites, and is – or should be – governed by international space law.

Conclusions of the National Academies Report

First of all, the report's seven conclusions are presented, as they are the main topics of reference for the reflexions developed in this paper:

"1. The daily synoptic global view of Earth, uniquely available from satellite observations, has revolutionized Earth studies and ushered in a new era of multidisciplinary Earth sciences, with an emphasis on dynamics at all accessible spatial and temporal scales, even in remote areas. This new capability plays a critically important role in helping society manage planetary-scale resources and environmental challenges.

2. To assess global change quantitatively, synoptic data sets with long time series are required. The value of the data increases significantly with seamless and inter-calibrated time series, which highlight the benefits of follow-on missions. Further, as these time series lengthen, historical data sets often increase in scientific and societal value.

3. The scientific advances resulting from Earth observations from space illustrate the successful synergy between science and technology. The scientific and commercial value of satellite observations from space and their potential to benefit society often increase dramatically as instruments become more accurate.

4. Satellite observations often reveal known phenomena and processes to be more complex than previously understood. This brings to the fore the indisputable benefits of multiple synergistic observations, including orbital, suborbital, and *in situ* measurements, linked with the best models available.

5. The full benefits of satellite observations of Earth are realized only when the essential infrastructure, such as models, computing facilities, ground networks, and

trained personnel, is in place.

6. Providing full and open access to global data to an international audience more fully capitalizes on the investment in satellite technology and creates a more interdisciplinary and integrated Earth science community. International data sharing and collaborations on satellite missions lessen the burden on individual nations to maintain Earth observational capacities.

7. Over the past 50 years, space observations of the Earth have accelerated the cross-disciplinary integration of analysis, interpretation, and, ultimately, our understanding of the dynamic processes that govern the planet. Given this momentum, the next decades will bring more remarkable discoveries and the capability to predict Earth processes, critical to protect human lives and property. However, the nation's commitment to Earth satellite missions must be renewed to realize the potential of this fertile area of science."

To begin with, it is important to note that, indeed, thanks to Earth observation from outer space, Earth sciences have developed as never before, but this accelerated scientific and technological progress, in fact, has helped only part of human society in managing natural resources and environmental challenges. To be truly effective, this new capability needs to reach all human society.

Legal framework

Some of the main provisions of the international legal regime regulating Earth observation activities from outer space are cited, and compared with the technological advances these activities have reached today.

The international legal regime consists basically of the 1967 Treaty on Principles Governing the Activities of States in the Exploration and Use or Outer Space, including the Moon and Other Celestial Bodies (Outer Space Treaty)[4], and the 1986 United Nations Resolution on Principles Relating to Remote Sensing of the Earth from Outer Space (UN Remote Sensing Principles), adopted by consensus at the United Nations Committee on the Peaceful Uses of Outer Space (COPUOS). This Resolution currently is the only specific international reference in existence on the regulation of remote sensing. It is considered to reflect a customary law.[5]

Carl Q. Christol noted as early as 1988: "The Principles were the product of consensus, but there were numerous compromises along the way. This suggests that the consensus was, if not unsubstantial, at least so thin that some members of COPUOS have viewed the consensus as conditional and subject to review. It cannot be said that the agreement was a temporary one, but the commentary reviewed (...) suggests a grudging acceptance of some provisions, with the view on the part of developing countries in particular, that there is a need to reinforce the Principles beneficial to them in order to buttress existing assurances and commitments."[6]

Part of the problem is that the 1986 Principles have been overtaken by the swift technological advances in the sector that have occurred in the last 22 years. Consequently, remote sensing activities are far from being sufficiently regulated.[7]

If, as the first conclusion of the National Academies report states, "the daily synoptic global view of Earth, uniquely available from satellite observation, has revolutionized Earth studies and ushered in a new era of multidisciplinary Earth sciences, with an emphasis on dynamics at all accessible spatial and temporal scales, even in remote areas", and if "this new capability plays a critically important role in helping society manage planetary-scale resources and environmental challenges", then it cannot be said that the current legal framework of Earth observations activities is equally developed.

The new era of multidisciplinary Earth sciences began just in the middle of the 1980s. And despite the extraordinary remote sensing technological advances registered since that time, the 1986 UN Remote Sensing Principles remain the same, without any updating amendments. Not by chance, it is seen today as an ineffective instrument.

The UN Remote Sensing Principles is an instrument focused on the intent to reconcile the rights and duties of sensing States and those of sensed States. But it would seem that the necessary and fair balance in this relationship was not reached. The technological superiority of sensing States gave them a solid advantageous position in relation to the sensed States. And these have not been able to negotiate the adoption of more guarantees for their legitimate interests.

Principle IV states: "Remote sensing activities shall be conducted in accordance with the principles contained in Article I of the Treaty on Principles Governing the Activities of States in the Exploration and Use of Outer Space, including the Moon and Other Celestial Bodies, which, in particular provides that the exploration and use of outer space shall be carried out for the benefit and in the interests of all countries, irrespective of their degree of economic or scientific development, and stipulates the principle of freedom of exploration and use of outer space on the basis of equality."

These are the two fundamental pillars of the Outer Space Treaty, usually referred to as the principles of "common benefits" and of "freedom of exploration". The issue is that it is much easier to interpret and implement the second principle than the first one, which is not yet legally defined, despite the great demand for some definition.

Principle IV also states: "These activities shall be conducted on the basis of respect for the principle of full and permanent sovereignty of all States and peoples over their own wealth and natural resources, with due regard to the rights and interests, in accordance with international law, of other States and entities under their jurisdiction. Such activities shall not be conducted in a manner detrimental to the legitimate rights and interests of the sensed State."

Thus, the Remote Sensing Principles formally incorporate both the principle of "freedom of sensing" the territory of any country, at any time, and the principle of full and permanent sovereignty of all States and peoples over their own wealth and natural resources. But they do not specify in which cases and how each of these – in large extent – contradictory principles should be applied.

As Atsuyo Ito noted, the conditions for sensing from space in terms of respecting the rights of sensed States are not explicit: "Hence the data collection policy provides: 1) no restrictions based on geography; objects observed including natural resources and the sea surface; and territories beyond national jurisdiction; 2) no prior consent, consultation, or notification is required before sensing; therefore, no veto right is granted to sensed States with regards to their territories being sensed; 3) no conditions are imposed for sensing capabilities with varying degrees of spatial resolution, or the type of sensor such as radar or optical."[8]

The new Earth Observation from outer space capability effectively plays a critically important role in helping society manage planetary-scale resources and environmental challenges, but it could much wider and better done if a more just and beneficial international legislation on the matter were in place.

It is important to stress that, under Principle VI, remote sensing activities "shall not be conducted in a manner detrimental to the legitimate rights and interests of the sensed State".

This Principle certainly is the strongest provision concerning the protection of the rights and interests of sensed States; however, these are still not defined. Bin Cheng made a remarkable comment on that issue: "Those who are apprehensive that data gathered from outer space by others might work to their detriment or that the data gathered from outer space might be misused by third parties to their detriment can probably find only scant comfort from United Nations Principles."[9]

As to the issues of access and distribution of remote sensing data, there is only the provision formulated in Principle XII: "As soon as the primary data and the processed data concerning the territory under its jurisdiction are produced, the sensed State shall have access to them on a non-discriminatory basis and on reasonable cost terms. The sensed State shall also have access to the available analyzed information concerning the territory under its jurisdiction in the possession of any State participating in remote sensing activities on the same basis and terms, particular regard being given to the needs and interests of the developing countries."

On the one hand, this Principle affirms the positive right of sensed States to access the data concerning their territory, as well as the obligation of sensing State to provide these data to the sensed States. On the other hand, it does not establish the appropriate conditions under which this right shall be respected. It is stated that access to the data and the obligation to provide them shall be carried out on "a non-discriminatory basis and on reasonable cost terms". Both seeming requirements may be questionable, as they are not clearly defined, and thus, do not offer any certainty of being fulfilled.

The concept of "non-discriminatory basis" means that the sensing States should provide the data to the sensed States under the same conditions as other States wanting to have access to the data. Sensed States have no preference or priority in relation to other States when they wish to access the data on their own territory. Even in this situation there is no clear regulation or mechanism guaranteeing them this simple access.

The concept of "reasonable cost terms", in turn, is vague, ambiguous, not defined and can be interpreted in different ways. "It could mean marginal costs or market price insofar as it is reasonable for the particular data in question. It does not serve as a general guideline for setting different prices, as could be expected for different types of products or uses by the divergent data generators", as Atsuyo Ito rightly observes.[10]

The new capability of Earth Observation from outer space effectively plays a critically important role in helping society manage planetary-scale resources and environmental challenges, but it could be widely and better done if there were a more just and beneficial international legislation on the matter.

In view of such an uncertain legal framework, it is worth citing the sixth conclusion of the National Academies report. It states: "Providing full and open access to global data to an international audience more fully capitalizes on the investment in satellite technology and creates a more interdisciplinary and integrated Earth science community. International data sharing and collaborations on satellite missions lessen the burden on individual nations to maintain Earth observational capacities."

As can be seen, there currently is an increasing awareness of the crucial issue of access to data produced by Earth observation from space. This awareness provides strong reasons to reflect this positive vision in a new, updated instrument governing remote sensing activities. This instrument must be based on the most important concepts contained in the UN Remote Sensing Principles, such as free and responsible remote sensing from outer space all over the world.

Space co-operation

The fifth conclusion of the National Academies' report cautions: "The full benefits of satellite observation of Earth are realized only when the essential infrastructure, such as models, computing facilities, ground networks, and trained personnel, is in place".

If the existence of such an essential infrastructure in all countries is absolutely necessary, it is indispensable to have great international cooperative action, so that all countries may have access to models, computing facilities, ground networks and to training personnel. Of course, this is a complex undertaking, requiring the active participation of many countries, organizations and institutions. That is why in 2006, the Brazilian delegation to COPUOS presented to the Plenary a three-year plan to study the need for an ample international cooperation program to enable the establishment of a geospatial data infrastructure in all countries, or at least in the majority of them. The plan was approved by COPUOS.

Principle V recommends: "States carrying out remote sensing activities shall promote international co-operation in these activities. To this end, they shall make available to other States opportunities for participation therein. Such participation shall be based in each case on equitable and mutually acceptable terms."

This is certainly a laudable premise, but in reality, it is not sufficient. More than bilateral efforts, multilateral ones are required. What is needed is an international program, supported by a great number of nations, intergovernmental and non-intergovernmental organizations, as well as involvement by all interested enterprises. An emblematic example of this kind of international undertaking is the purpose and the work of the Group of Earth Observation (GEO) and its ten-year program, the Global Earth Observation System of Systems (GEOSS).[11]

According to Principle VI, "in order to maximize the availability of benefits from remote sensing activities, States are encouraged, through agreements or other arrangements, to provide for the establishment and operation of data collecting and storage stations and processing and interpretation facilities, in particular within the framework of regional agreements or arrangements wherever feasible". And according to Principle VII, "States participating in remote sensing activities shall make available technical assistance to other interested States on mutually agreed terms".

Encouraging States to provide for the establishment and operation of data collecting, storage stations, processing and interpretation facilities, without doubt, is a good intention. Yet, today the community of States has to go much further.

Collecting and storage stations, as well as processing and interpretation facilities become a crucial necessity for all States in any program directed towards national and regional development, as well as in any efforts to cope with natural disasters. Technical assistance is just a first and modest step; the majority of States need much more. True progress in this area requires technology transfer as well as capacity building.

International cooperation should aim, *inter alia*, at "fostering the development of relevant and appropriate space capabilities in interested States", as is recommended in point 5 (b) of the 1996 Declaration on International Cooperation in the Exploration and Use of Outer Space for the Benefit and in the Interest of all States, Taking into Particular Account the Needs of Developing Countries.[12]

Earth observation as a global common good

It is clear that all seven conclusions of the National Academies' report strongly support the idea of Earth observation as a global common good.

The concept of public good has its roots in 18th century scholarship[13]. The Scottish philosopher David Hume[14] (1711-1776) and the British economist Adam Smith[15] (1723-1790) wrote about this subject. Since then, a vast and varied literature has been published on the topic in many countries. In recent years, the concept of global public goods has become an increasingly important part of international policy making. The concept appears in the agendas of United Nations agencies, the IMF/ World Bank, and many non-governmental organizations. Everyone depends on public goods; they are indispensable for all peoples and their activities. A clean environment, health, knowledge, property rights, peace and security, are among the examples of public goods that could be made global[16].

According to Inge Kaul, Isabelle Grunberg and Marc A. Stern, "the ideal public good has two main qualities: its benefits are non-rivalrous in consumption and non-excludable". Non-rivalrous means that there is no additional cost incurred in providing the good to an additional person (its marginal cost is zero). Non-excludable means that one cannot exclude others from deriving benefit from the good once it has been produced.

Kaul, Grunberg and Stern stress that the global public goods are public goods whose benefits reach across borders, generations and population groups. They form part of the broader group of international public goods, which include as other sub-group, regional public goods. "Their benefits are quasi universal in terms of countries (covering more than one group of countries); people (accruing to several, preferably all, population groups); and generations (extending to both current and future generations, or at least meeting the needs of current generations without foreclosing development options for future generations). This property makes humanity as a whole the *publicum*, or beneficiary, of global public goods."[17]

Based on the concept of satellite imagery as a public good, in 2004 Brazil adopted a policy of free distribution of data obtained from the China-Brazil Earth Resources Satellites – CBERS.

In 2004-2008, through the Brazilian National Institute of space Research (INPE) more than half a million satellite images had been distributed free of charge through the Internet (www.dgi.inpe.br/CDSR/).

In this way, Brazil became a worldwide example in the Earth Observation field, making remote sensing activities an easy and accessible tool. From these half million images distributed in the last 4 years, 435 thousand came from CBERS.

This free distribution has increased the availability of CBERS-2B satellite images launched in September 2007, and from LANDSAT-5. Since January 2008 Brazil has been distributing about 800 images per day, according to Luiz Geraldo Ferreira, from INPE's Image Generation Division (DGI), where the Remote Sensing Data Center (CDSR) is allocated.

Free distribution through the Internet started with CBERS-2 satellite images and LANDSAT-1, 2, and 3. Since its inception, this initiative has represented important progress in the use of Remote Sensing resources in all the country.

Later, the benefit was extended to South America and now it is extending to Africa. In additon to data from CBERS satellites, the user can access images from the LANDAT 1, 2, 3, 5 and 7 satellites.

The CBERS turned Brazil into the biggest distributor of satellite images from all over the world.

In Brazil alone around 430 thousand CBERS images were distributed to about 15 thousand users from several public and private institutions, which proves the economic and social benefit of offering data free of charge. In China, after adopting a policy similar to Brazil's more than 200 thousand images were distributed, with the Earth and Natural Resources Ministry being its main user.

At the end of 2007, Brazil and China decided to offer the CBERS images free of charge to all the African Continent. The free distribution policy will allow institutions and authorities in Africa to monitor natural disasters, deforestation, and threats to agricultural production and public health risks.[18]

Hence the commentary of the Brazilian researchers Hilceia Santos and Gilberto Camara: "The trend towards openness and free access to remote sensing data worldwide is gaining momentum. The next decade will likely see the emergence of a global land-imaging consortium, which would provide data access from a constellation of satellites. The land imaging satellite constellations will provide free 10-30 meter global land cover multispectral images available worldwide at least once a week, and if possible, every two days. This timely data will meet the needs for fast-response applications, which are critical in all areas. Thirty years of experience using land-imaging satellites shows that timely, free and high-quality geospatial data provide significant societal benefits. There is a high likelihood that this policy will become widespread."[19]

At the COPUOS Plenary Meeting, held in June 19, 2008, in Vienna, Austria, Brazil explained its policy of free access to geospatial data. On this occasion, the COPUOS was considering the item "International cooperation in promoting the use of space-derived geospatial data for sustainable development", a Brazilian proposal adopted in 2006. During this 2008 session, Brazil presented its experience in setting up proper national infrastructure for space-derived geospatial, data collection, processing and application.

It stated that space-derived data can make a large contribution to the national spatial infrastructures of the developing nations, given some conditions. First, they should have access to spatial data as public goods. Secondly, there should a strong investment in capacity building. Thirdly, developing countries should have access to open source software, scientific literature and suitable hardware. However, concerted action is necessary to achieve these goals.

The most important parts of the

Brazilian statement at COPUOS meeting,[20] formulated several relevant proposals:

"Developing nations present both a challenge and an opportunity for spatial data infrastructures (SDI). These countries have significant challenges in handling their natural resources and SDI can play an important role in their management. Most decision-making in developing nations requires spatial data. For example, planning a new hydroelectric power plant needs an assessment of its potential impact on communities and the environment. This leads to a need for building different scenarios based on acceptable data and analysis techniques.

"The first problem for setting up an SDI in a developing nation is to create the data sets, since in most countries they are outdated or nonexistent. Much important information for a spatial data infrastructure can be obtained from satellite imagery. Land imaging data provided by remote sensing satellites provides significant societal benefits associated with applications in areas such as agriculture, deforestation assessment, disaster monitoring, drought relief, and land management. Land use and land cover maps, hydrologic features, the road network, and the urban structures can be extracted from remote sensing data.

"There are grounds for moderate optimism on this matter. Brazil and China have agreed that their CBERS satellite imagery is available as open access in China, Latin America and Africa. The United States has announced that all its LANDSAT archive will be openly accessible. Japan has committed to provide a digital terrain model of the Earth at 30-meter resolution, based on its ASTER sensor. Together, the initiatives of Brazil, China, the United States and Japan show that much can be done to provide developing nations with much-needed data.

"Given that the Earth is a finite place, an ideal configuration of land imaging satellites is possible, covering different ends of the spectrum (optical, infrared, thermal and microwave) and with complementary spatial and temporal resolutions. Many countries, some with existing space program and some with emerging space activities, can contribute to this global, concerted effort. By combining compatible satellite specifications, open data access, open source software, and extensive capacity building, there will be optimal global benefits of land imaging data.

"Satellite data is useful to build spatial data infrastructures in developing countries, but other data sources and technologies are needed. Developing nations need a land survey and registry, where each parcel of land is registered. Fortunately, the costs of building a land registry have decreased enormously in recent years, due to the use of global navigation satellite systems (GNSS). The GNSS technology allows countries to produce road network maps and cadastral maps at a fraction of the previous cost.

"Finally, the technologies for organizing and sharing spatial data obtained from satellites and GNSS-based surveys are also available as open source. There are open source spatial databases that handle large sets of geographical data. These open source solutions are extremely efficient and cost-effective.

"All of these promising technologies need trained personnel to make use of them. How can we make good use of such diverse data for sustainable development? Open access to land imaging data has to be combined with capacity-building on the use of space-based geospatial data. For developing nations, multi-tiered solutions are needed. Short term approaches include training programs through international institutions such as the United Nations Regional Centres for Space Science and Technology. Longer term knowledge building requires international support for undergraduate and graduate education, complemented by open source and low-cost software for processing remote sensing data, which are becoming widely available.

"The global potential benefits of spatial data infrastructures in developing nations should motivate developed nations to contribute significantly to efforts in capacity building. The world needs trained technicians, engineers, and researchers who know how to build and make use of spatial data infrastructures. This requires continual commitment of resources for at least 10 years.

We need sustained investment in education in this area.

"To achieve the goal of free and open distribution and use of land imaging data, a global governance structure is needed. Fortunately, such governance structure exists today. The Group on Earth Observations (GEO) and its associated space-branch, the Committee on Earth Observation Satellites (CEOS) have established plans for a Land Surface Imaging (LSI) Constellation. GEO has a membership of 73 countries, plus the European Commission. Most land imaging satellite operators are members of CEOS. As soon as the member countries of GEO decide to adopt an open data policy and actively commit resources to the proposed LSI constellation of CEOS, the governance structures of GEO and CEOS can organize an effective action.

"Combining open data access and open source software is the best way to combine the efforts of developed and developing nations to promote the use of space-derived geospatial data for sustainable development. Brazil is working hard with its partners worldwide to support and achieve these goals."

Future of Earth observation from space, and some conclusions

"Over the past 50 years, space observation of the Earth has accelerated the cross-disciplinary integration of analysis, interpretation, and, ultimately, our understanding of the dynamic processes that govern the planet", as the seventh conclusion of the National Academies report states. Its expectation is that "the next decades will bring more remarkable discoveries and the capability to predict Earth processes, critical to protect human lives and property".

However, for a bright future to Earth observations from outer space we need to reach agreement on some measures as *conditia sine qua non*:

1) To establish a solid universal commitment to peaceful and constructive exploration and use of outer space, in order to guarantee sustainability for all space activities.

2) To adopt an ample, updated and contemporary legal international regulation of Earth observation from outer space.

3) To promote an international cooperation program to construct in all countries a national infrastructure for space-derived geospatial, data collection, processing and application as essential tools for national sustainable development, disaster management and environment protection.

4) To discuss the creation of a global governance structure to deal with the goal of free and open distribution and use of land imaging data (or at least at a minimum, symbolic cost).

5) To commit all countries to realize the potential of Earth satellite missions science.

The National Academies' report provides an extensive sampling of important accomplishments enabled by Earth satellite data. However, many scientific questions and societal challenges remain unresolved, including improving 10-day weather forecasts, more accurately forecasting hurricane intensity, increasing resolution of earthquake fault systems and volcanoes to detect precursors of events, mitigating climate change impacts, and protecting natural resources.

It is extremely important to sustain the rate of scientific discovery and progress, committing to the maintenance of long-term observing capabilities and to innovation in observation technology. The Earth is a highly dynamic system and not as predictable as initially thought. Thus, long-term observations are required if we wish to understand and predict future changes. Future advances must be associated with fundamental societal benefits, given the current challenges presented, for example, by climate change and loss of biodiversity. We can envision the availability of regional annual climate predictions to assist in water resource management, in control of infectious diseases, early warning systems, operational use of air pollution maps, and improved ability to foresee volcanic eruptions or earthquakes, according to the National Research Council (NRC) Reports 2001a and 2007a.

If we can fulfill these conditions, a report in 50 years will be able to highlight many more valuable scientific achievements and discoveries, assuring prosperity and health for all countries and peoples.

References

1) *Global Public Goods – International Cooperation in the 21st Century*, Edited by Inge Kaul, Isabelle Grunberg and Marc A. Stern, published by the United Nations Development Programme (UNDP), USA, New York: Oxford University Press, 1999, p. XXXVI.

2) The National Academies Press, Washington, 2008.

3) The National Academies Press, Washington, 2007.

4) The Treaty on Principles Governing the Activities of States in the Exploration and Use of Outer Space , including the Moon and Other Celestial Bodies, adopted by the General Assembly in its Resolution 2222 (XXI), opened for signature on 27 January 1967, entered into force on 10 October 1967, 98 ratifications and 27 signatures (as of 1 January 2008).

5) The Principles Relating to Remote Sensing of the Earth from Outer Space (United Nations General Assembly Resolution 41/65 of 3 December 1986). See Joanne Irene Gabrynowicz, *Expanding Global Remote Sensing Services*, in Proceedings of the Workshop on Space Law in the Twenty-First Century, New York, 2000.

6) Christol, Carl Q., *Remote Sensing and International Space Law*, Journal of Space Law, Number 16, 1988; *in Space Law, Past, Present and Future*, Kluwer, 1991, pp. 73-95.

7) Monserrat Filho, José, *A Remote Sensing Convention for the Advancemente of Space Law*, Proceedings of The Forty-Sixth Colloquium on The Law of Outer Space, International Institute of Space Law (IISL), 29 September-3 October 2003, Bremen, Germany. Published by the American Institute of Aeronautics and Astronautics (AIAA), pp. 63-73.

8) Ito, Atsuyo, *Improvement to the Legal Regime for the Effective Use of Satellite Remote Sensing Data for Disaster management and Protection of the Environment*, *in* Journal of Space Law, University of Mississippi School of Law, Vol. 34, Number 1, Spring 2008. p. 49.

9) Cheng, Bin, *Studies in International Space Law*, Clarendon Press Oxford, 1997, p. 596.

10) Idem Ibid, pp. 49-50.

11) See <www.earthobservations.org/>

12) See <www.unoosa.org/oosa/en/SpaceLaw/gares/html/gares_51_0122.html>

13) Kaul, Inge; Grunberg, Isabelle; and Stern, Marc A., *Defining Global Public Goods, in Global Public Goods International Cooperation in the 21st Century*, Edited by Inge Kaul, Isabelle Grunberg and Marc A. Stern, published bythe United Nations Development Programme (UNDP), USA, New York: Oxford University Press, 1999, pp. 2-19.

14) David Hume discussed the concept of "public good" in his *Treatise of Human Nature*, first published in 1739.

15) Adam Smith analyzed the idea of "public good" in his *Inquiry into the Nature and Causes of the Wealth of Nations*, first published in 1776

16) See <www.globalpolicy.org/socecon/gpg/index.htm>.

17) See article quoted in reference number 10, pp. 2-3.

18) See <www.inpe.br>.

19) Ferreira, Hilcéia Santos; Câmara, Gilberto, *Current Status and Recent Development in Brazilian Remote Sensing Law*, *in* Journal of Space Law, University of Mississippi School of Law, Vol. 34, Number 1, Spring 2008. p. 17. Gilberto Câmara is the General Director of the Brazilian National Institute of Space Research (INPE).

20) See <www.cbers.inpe.br/?hl=en&content=imprensa_inpe> The statement of Gilberto Câmara, General Director of INPE, was read by Fernando Ramos, International Cooperation Adviser of INPE, as member of the Brazilian Delegation at COPUOS Plenary Meeting.

IAC-08-E8.5.8

A SUPPORT OF IMAGINATION TO BECOME REALITY: RELATION OF PATENT LAW AND SPACE LAW

Ms. Fruzsina Tari
Member of MANT – Hungarian Astronautical Society
and
Associate of Faludi Wolf Theiss Attorneys at Law

Székesfehérvár, Hungary
fru.tari@gmail.com

ABSTRACT

Human mankind has had a huge dream of leaving the Earth, and challenges himself in space. To achieve this dream, several legal issues came up. Some of them had already been solved before the technology got to the point to refer to it (mostly in the first 3 decades of space age). Though the advance of law, in last decades more and more legal issues are coming up. To support the dreams, legal background is to be ensured. One corner stone of these questions is related to intellectual property and more precisely to patents.

The aim of the paper is to present the basic conflict of patent law and space law. By studying the territoriality and legal monopoly of patent law and disclosure of sovereignty and non-appropriation in space law, the issues developments will be considered, though no definitive solutions are yet achieved.

As a conclusion to keep the space industry going on a less risky way, a kind of harmonization is needed in patent law. This paper wishes to propose some solutions (e.g. a global regulation through WIPO, or directly by a new and separate international organization), while keeping in mind, that the field of application is special, so the rules should adapt to the special conditions.

INTRODUCTION

In the beginning of Space Age the law was ahead of technological possibilities. The basis of space law, as part of international public law was settled rather soon after the first satellite was launched.

From that time many dreams in space technology have come true. Not only the state institutes and organizations are now actors of space industry, but private companies are also shown up with new aims and new solutions.

International cooperation in any level is stronger than ever before.

In this quickly changing and developing environment intellectual property becomes more and more a key issue, though its regulation and specification to outer space is still pending.

As in space the dominant intellectual property right is patent right, the present document will focus on the relation of patent law and space law. However the reader is encouraged to apply analogy between patent law and other intellectual property rights.

PATENT LAW

Patent represents a legal monopoly to get interested inventors to develop innovations. Patents are awarded by states and are enforceable under the state's jurisdiction. The owner of patent right may own, use, licence or exploit by any other means the patented invention, and may forbid any third party from such an exploitation. Patent rights are different from traditional property rights (among others) in their limitation of territory and time.

During the last twenty years, many development were achieved in patent law and especially in harmonization of patent rights.

The main agreements to harmonize patent rights are the following:

- 1883: Paris Convention for the Protection of Industrial Property;
- 1967: Convention establishing the World Intellectual Property Organization (as a specialized organization of the United Nations) ("WIPO"), which has already 180 member States;
- 1970: Patent Cooperation Agreement (carried out by WIPO and was amended in 1979 and 2001);
- 1995: Trade Related Aspects of Intellectual Property Rights (signed in the framework of World Trade Organization as part of the General Agreements of Tariff and Trade) ("TRIPS");
- 2000: Patent Law Treaty (open for members of WIPO or the Paris Convention) it entered into in 2005 and has already 18 contracting members, while another 42 parties have already signed the agreement.

There are two main structures of patent rights according to the protection awarded. The more spread system is the first-to-file system, while the other solution is the first-to-invent system.

<u>First-to-file</u>

Most of the world follow the first-to-file system, which means that the basis to grant a patent is the first application filed to the competent authority. No further requirements are needed regarding the inventor or the place of the invention in this system.

<u>First-to-invent</u>

Though only two countries in the world follow the first-to-invent system, considering it is needed, one of them is a key actor of space industry. So considering these jurisdictions –the United States and the Philippines[1]– is rather important.

The person who can improve to be the first to invent has a right to the patent. In the US, one more provision interfered with international trends, namely the place of an invention, where the invention made abroad is not acknowledged for the time of an invention.

With signing the TRIPS agreement, latter clause is not any more applicable for other members of the agreement, but still in force for invention out of these States.

SPECIALITY OF SPACE LAW

A very young branch of international law was launched almost in the same time as the Sputnik reached successfully the outer space. Soon, in 1967, the first and still the most important international agreement entered into, which is the Treaty on Principles Governing the Activities of States in the Exploration and Use of Outer Space, including the Moon and Other Celestial Bodies ("OST"). It declares main principles while using outer space, and among other states that no appropriation is possible in

---

[1] Meeting of consultants on inventions made or used in outer space, International Bureau of WIPO, Genova, 1997

outer space. The research, exploration and exploitation shell be conducted for the benefit of all mankind.

The other important international treaty from the point of view of patent law is the Convention on Registration of Objects Launched into Outer Space in 1975 ("Registry Convention").

The enforceability of the patent is usually depends on the jurisdiction of the spacecraft, which refers to the registration State. However, defining State of registry may be uncertain as any of the launching States can be the State of registry in the meantime.

OST is ratified by 98 States and another 27 State signed it[2], the global application may be pending. The first treaty to apply for outer space is become however stronger than a simple international agreement. It entered into custom law as well, which means that all of its principles are valid for all and any of the States (being member or not).

Some of the experts are considering all 5 main international agreements as part of custom law, although the other four (beside OST) are accepted by less and less States. While the Rescue Agreement[3] and the Liability Agreement[4] were ratified by many States, the Registry Convention is in force only in 51 States and another 4 States have signed it and further 2 accepted the obligations and duties rising from.

Accordingly the Registry Convention remains some doubts if it shall be considered as a custom law. Some of the States have not yet accepted it even if they may be considered as launching State (e.g.: Israel)

[2]http://www.unoosa.org/oosa/en/SpaceLaw/treaties.html (date of download: 30 August 2008)

3 Agreement on the Rescue of Astronauts, the Return of Astronauts and the Return of Objects Launched into Outer Space

[4] Convention on International Liability for Damage Caused by Space Objects

TYPOLOGY OF INVENTIONS

The inventions in question may be divided into three groups:

(1) Invention invented on Earth and used in outer space (too);
(2) Invention invented in outer space and used either in outer space or on Earth (or both);
(3) Invention, which is applicable only in outer space.

As of the first two types of invention, the first-to-file system allows all inventors, to file a patent application under the selected jurisdictions. Though awarded patents are providing protection on Earth, only the flagship principle enables patent protection in outer space on the board of a spacecraft. In other parts of outer space, no applicable patent protection is available.

In a first-to-invent system the patent application is obtainable with different basis. The place of invention was also part of this principle and it is still in force in the US against countries out of the TRIPS agreement. The importance is emphasized in the second case, when invention is carried out in outer space, and so the patentability according to the US legislation is only available on the board of US registered parts of spacecrafts (and now other TRIPS agreement members' parts of spacecraft are also considered as valid place of invention)

The third category of invention rises the most doubt since the invention is not applicable, owned and other was used on Earth and so not subject of enforcement under any jurisdiction.

The only State which however grants patent is the US for such an invention. US Space Bill and patent regulation are extended to outer space too.

The invention considered in this category are orbits for different purpose or frequencies able to transmit signals, but may be any solution, which are not applicable on Earth.

CONFLICTS OF PATENT LAW AND SPACE LAW

The two systems are difficult to use in the same time. The nature of patent law requires the ownership, the limitation in the territory and supports mostly the international private law. In the meantime space law forbids any appropriation in outer space, which means also that an orbit or frequency should not be appropriated. So the granted patent should not be valid according to space law.

Space law has no limitation in territory, while in traditional patent law the patent is applicable in a restricted territory (in the territory of the jurisdiction).

The enforceability of patent is attached to the jurisdiction, which granted the patent. All infringement of patent rights may be challenged at court. However the uncertainty of outer space jurisdiction or even more, the neutrality of outer space questions the possibility to enforce a patent (even if it is granted by one or another jurisdiction).

The patent by its nature ensures monopoly to the entitled of patent right. Monopoly is contrary with the main principles of space law, namely the Art. 2 of OST.

> *"Outer space, including the Moon and other celestial bodies, is not subject to national appropriation by claim of sovereignty, by means of use or occupation, or by any other means."*

Outer space should used according to OST on an equal, non discriminatory basis, and to the benefit of all countries (irrespective whether the country managed to get out to outer space, or not)

JURISDICTION

As already mentioned above, as a main rule, there is no jurisdiction settled in outer space. However on board of spacecrafts a certain regulation is needed, for which the flagship principle is the most accepted. It means that the applicable law is that of the State of registry.

State of registry can be any of the launching States according to the Art. 1 of the Registry Convention:

> *"The term "launching State" means:*
> *(i) A State which launches or procures the launching of a space object;*
> *(ii) A State from whose territory or facility a space object is launched;"*

In the practice, more launching States are available for the same spacecraft, as such a mission is usually carried out by more nations. Accordingly the most favourable legal environment is chosen for a State of registry.

International Space Station

The International Space Station ("ISS") is nowadays biggest challenge regarding jurisdiction. The Intergovernmental Agreement ("IGA") was signed in 1998 and foresees the meaning of "intellectual property" as it is laid down in the Convention establishing the WIPO.

Patents and other intellectual property rights are regulated under Article 21 of IGA. As the US contributed an important value in the ISS, the applicable regulation is closer to US solution. The place of the activity is dominant when determining the jurisdiction to apply. Each element of ISS is registered separately and so the plurality of State of registry is at stake. In this aspect the word "international" in the name of the ISS

reflects less to the reality, as several jurisdictions are present in parallel.
The territoriality is applied to the crew as well. Even if an astronaut comes from a certain State, the applicable law remains that of the element of the ISS.

FUTURE TRENDS

Today the patent law's future is not decided in relation to space law. The conflicts are more likely to stay though some recent developments.
A close future challenge is waiting for patent law in space. The State of registry is not changeable at the moment, which means that even if a company purchases a satellite, the applicable law will not change at all.
In the Author's point of view, with the more and more intense presence of private sector, it will demand the change of registry according to their State.
On the other hand the State of registry will also want to change the registration, as the liability remains there, while the owner of the satellite is already placed in another State.
We can state that the Registry Convention arrives soon to the point to be changed.
Such a change will effect on the patent as well. If the State of registry is changed, the patent granted by the original State of registry will not be in force in the new one (unless the application was filed in the new State at the same time, or within a short time after filing the original patent.). So the patented invention becomes free for the satellite's new owner, as the documentation has to be handed over during the sale and purchase procedure.

SOLUTIONS

To maintain the enforceability of space law and to switch the patent law in space to a less contested form, a new solution or at least a certain level of harmonization is needed.

World without patent

As a first direction, according to OST, the total deny of patent law would be a simple solution. In that way no more questions would come up related to the non-appropriation clause or related to the extraterritoriality. As no jurisdiction would be necessary, the OST and other treaties related to outer space could be amended without effect on patent law in space.
Although this solution seems to be simple, probably it will not be supported by private sector and not even by State institutes, as the intellectual property and patent rights are more and more valuable and better and better recognized.

Harmonization

A certain level of harmonization would already help in application of patent law in space.
Recently many changes were carried out by WIPO and by some regional organizations (such as European Patent Office), which shows that even on the territories of Earth the harmonization is needed.
A patent law, which is unified in all countries would solve the most of the questions occurring in patent law in outer space. However the conflict with the principle of non-appropriation would certainly remain.

Braking the rules

The most difficult to carry out, but the most effective solution would be to work out a

new way of patentability in space. It would exclude appropriation in a first step.
As a second step, the inventions in space would be used on an equal and non-discriminatory basis, which means that even if a space patent is awarded a free of charge licence or a reasonable price licence would be transferred to any parties willing to use the invention in outer space.
This kind of protection would be effective in all countries and would be enforceable on the Earth according to the legislation of the State where it is applied or infringed.
Such a solution would break up with the elements of the traditional patent definition. It should be paired with an intergovernmental, independent organisation, which would register the patent and would follow up it. In case of infringement of such a patent, this organization would be competent to make a judgement.

CONCLUSION

Recently there were developments on patent law, but still some of the nations are extending their jurisdictions to space (e.g.: US and Germany), which is contradictory of space law.
In case we wish to stay inline with technical and economical development, new solutions are needed.
To obtain the best solution (which is durable as well), we need to define in a new manner the patent law in space. We should not be afraid of do so, but should refer to space law in global, as it was also a pioneer way of approach by rejecting territorial sense of regulations.
For new challenges new answers would provide the best solutions.

BIBLIOGRAPHY

Gyula Gál: Space Law, Budapest (1969)

Bradford Lee Smith: Evolution of Intellectual Property Law in Outer Space In: Marietta Benkő and Walter Krell (editors): Air and Space Law in the 21st Century, Karl Heymanns Verlag KG (1999)

S. G. Sreejith: Intellectual Property Clause of the International Space Station Agreenebt: Damp Squib?

Bradford Lee Smith: Intellectual Property Rights in Outer Space Activities: Aid or Impediment? In:: Bringing Space Benefits ot the Asian Region, Bangalore (2005)

Intellectual Property and Space Activitites, Issue paper prepared by the International Bureau of WIPO (April, 2004)

Meeting of Consultants on Inventions Made or Used in Outer Space, WIPO, Genova (1997)

Évolution du Droit de l'Espace en France, Ministere Délégué á la Recherche et aux Nouvelles Technologies, Ovation, Paris (2003)

Texts of international treaties of outer space:

http://www.unoosa.org/oosa/en/SpaceLaw/treaties.html (downloaded on 10 August 2008)

Texts of agreements on intellectual property:

http://www.wipo.int/treaties/en/ (downloaded on 24 August 2008)

FORMALISM, INFORMALISM & INNOVATION IN SPACE LAW: LENSES TO VIEW, ASSESS, AND GUIDE THE DEGREE OF FORMALISM IN THE REGULATION OF SPACE ACTIVITIES

Professor Matthew Schaefer*
Director, Space and Telecom Law Program
University of Nebraska College of Law

Abstract: Too often outer space regulation has been viewed only through the lenses of public international law and the related international relations school of regime theory (or institutionalism) in which states are the primary, if not exclusive actors, and in which formal international treaties and international organizations are the means by which regulation occurs in a non-zero sum game environment (i.e. gains from cooperation can be achieved). However, it is equally important now to begin viewing the regulation of space through private international law and liberal international relations theory lenses if we are to fully understand how space activities can be and are regulated for the current "system" of space regulation that involves international, national and local government regulation but also a degree of private regulation too. From a less positivistic view and more normative perspective, it is important to continue to consider, where appropriate, possibilities for private regulation, including through the insurance market and industry self-regulation, as gap fillers and supplements for governmental regulation in the space arena if we are to maximize global space-created welfare.

 The author would like to thank Julie Jiru, UNL Space and Telecom LLM Class of 2008-09, for her outstanding research assistant work on this project.

I. INTRODUCTION

Outer space actors, both governmental and private, operate in an environment in which their actions are guided and constrained by finances, technology, policy, and regulation. Outer space regulators, given the global commons nature of space, must also be acquainted with and guided by finances, technology, and policy, but also diplomacy and politics. And, of course, it is important to realize that regulators of space activity are not only governments but indeed private actors in the space arena as well, including both the space industry and insurance companies. Governments used to be the only actors in space, but now private commercial actors and operators are well established within outer space activities. Similarly, private actors are well enshrined as regulators of the outer space regime. Too often outer space regulation has been viewed only through the lenses of public international law and the related international relations school of regime theory (or institutionalism)[1] in which states are the primary, if not exclusive actors, and in which formal international treaties and international organizations are the means by which regulation occurs in a non-zero sum game environment (i.e. gains from cooperation can be achieved). However, it is equally important now to begin viewing the regulation of space through private international law and liberal international relations theory[2] lenses if we are to fully understand how space activities can be and are regulated for the current "system" of space regulation that involves international, national and local government regulation but also a degree of private regulation too. From a less positivistic view and more normative perspective, it is important to continue to consider, where appropriate, possibilities for private regulation as gap fillers and supplements for governmental regulation in the space arena if we are to maximize global space-created welfare.

II. THE TRADITIONAL PUBLIC INTERNATIONAL LAW & REGIME THEORY INTERNATIONAL RELATIONS VIEW OF SPACE REGULATION: STATES AS THE EXCLUSIVE OR PRINCIPAL ACTORS

At the time the seminal Outer Space Treaty (OST) was concluded in 1967, governments, principally the United States and the Soviet Union, were the exclusive actors in the space domain. Regulation of space activities was primarily viewed as a means to constrain governmental behavior, governments being the only actors in space. Traditional public international law being viewed as regulation the relationship between states, the OST fell comfortably within this framework. The OST also fell comfortably within regime

[1] For a fuller description of institutionalism or regime theory, see Kenneth W. Abbott, "Modern International Relations Theory: A Prospectus for International Lawyers," 14 Yale J. Int'l L. 335 (1989); Kenneth W. Abbott, "Elements of a Joint Discipline," 86 Am. Soc'y Int'l L. Proc. 167 (1992); Stephen Krasner, "Structural Causes and Regime Causes: Regimes as International Variables," 36 Int'l Org. 185 (1982).

[2] For a fuller description of liberal international relations theory, see Anne Marie Slaughter, "A Liberal Theory of International Law," 94 Am. Soc'y Int'l L. Proc. 240 (2000); Anne Marie Slaugher, "Liberal International Relations Theory and International Economic Law," 10 Am. U. J. Int'l L. & Pol'y 717 (1995).

theory[3], which posits, much like realist theory, that states were the principal actors in international relations, but distinguishing itself from realist theory in seeing international relations as a non-zero sum game where gains from formal regimes could be had through treaty making. Principles of non-interference and free access found in the OST could be seen as increasing the size of gains that could be derived from space.

Many of the other "Cold War"-era treaties on space negotiated over the ensuing 12 years, including the1968 Rescue and Return agreement, the 1972 Liability Convention, the 1976 Registration Convention and the 1979 Moon Agreement, also paid primary attention to the behavior of governmental actors. To be sure, private activity in space was not unforeseen even in the OST, as Art. VI of the OST requires governments to provide continuing authorization and supervision of non-governmental actors within their jurisdiction. Nevertheless, with little private activity in space in the 1960-70's, much of the regulation of space activities was principally through these treaties. Later still, United Nations General Assembly (UNGA) Resolutions, such as the 1985 Resolution on Direct Broadcast Satellites and the 1992 Resolution on Remote Sensing Principles, became a key regulatory tool. However, this form of regulation was still of an intergovernmental, negotiated nature, albeit legally non-binding unless later acquiring the status of customary international law. If regulation of space activities did not transform itself, the debate between so-called "hard law" (e.g. treaty) v. "soft law"[4] (e.g. UNGA resolution) would continue to be the principal debate in the regulation of space. It is still an important debate in space regulation, but certainly not the principal one, especially since formal amendment of the OST, although occasionally proposed, or even formal new treaties on space activities do not appear feasible anytime soon given the international political landscape. This is why for instance the issue of mitigation of space debris is handled principally through voluntary soft-law instruments.[5]

III. THE PRIVATE INTERANTIONAL LAW & LIBERAL INTERNATIONAL RELATIONS THEORY VIEW OF SPACE REGULATION: DOMESTIC CONSTITUTIONAL STRUCTURES AND LEGISLATION, NON-GOVERNMENTAL ORGANIZATIONS (NGOs), MULTINATIONAL CORPORATIONS (MNCs) AND NETWORKS

Private international law has rapidly grown throughout the past several decades, principally through efforts undertaken in the Hague Conference on Private International Law. While historically, in the pre-WWII era, references to private international law were basically synonymous with a reference to conflict of law rules (those rules that told you which nation's substantive law would guide resolution

[3] See generally Abbott, supra note 1.

[4] For an excellent discussion of hard law v. soft law in the context of space, see Comments by Geraldine Goh at University of Nebraska College of Law conference on "Formalism, Informalism, and Innovation in Space and Telecommunications Law," May 1-2, 2008.

[5] See, e.g., Steve Mirmina, "Reducing the Proliferation of Orbital Debris: Alternatives to a Legally Binding Instrument," 99 Am. J. Int'l L. 649 (2005).

of the dispute at hand), private international law has rapidly expanded in the past several decades through the traditional public international law making method of treaties. As private actors became more engaged in space are even efforts in the UNIDROIT (which continue today) regarding security interests in mobile assets in outer space (or as described by one of the lead UNIDROIT officials, "mortgages" in space) so as to create more possibilities for financing space activities. But even the traditional view of public international law as only regulating interaction between government actors has come under increasing pressure, for much of public international law has very tangible impacts on private actors (e.g. WTO/GATT tariff limits, human rights norms, etc.). Thus, the distinction between public international law and private international law is breaking down, and what is more critical is to realize that there are a range of actors that participate in international space relations and in regulating space activities.

Liberal international relations theory posits that states are not monolithic entities, that their preferences and behavior are influenced by domestic constitutional structures and legislation, and that international relations are not the exclusive domain of governments, with NGOs, MNCs, and networks (educational or otherwise) also playing key roles.[6]
This view of international relations, although only gaining popularity in the last decade plus in terms of joint international relations-international law scholarship, certainly has some descriptive weight even in the early origins of space law. For example, the OST borrowed language from (and was guided and constrained by) the original NASA Act of the US Congress.[7] In current environment, the liberal international relations lens has even more explanatory force for what we are witnessing. In today's world, we see competition and experimentation among arious nations in terms of their approaches to space legislation, in particular the different models of national space legislation adopted by space active nations. We see the explanatory force of the theory in the US delegation to the International Telecommunications Union (ITU) World Radio Communication Conferences (WRC). At the most recent WRC-07, the US delegation was comprised of 157 persons, over half of whom where private sector advisors.[8] Indeed, the body sponsoring the technical session at which this paper was delivered, the International Institute of Space Law (IISL), is an example of the influence of NGO's in international relations too, through its recently acquired permanent observer status at the UNCOPUOS. But the impact of non-governmental private entities in the regulation of space activities is not limited to input at intergovernmental negotiations and conferences. Private regulation through the insurance market and industry self-regulation are already

[6] See generally Slaugher, supra note 2.

[7] See, e.g., Jonathan Galloway, "Revolution and Evolution in Outer Space Law," forthcoming in Nebraska Law Review.

[8] See comments by Ambassador Richard Russell, US Ambassador to WRC-07, made at University of Nebraska College of Law conference on "Formalism, Informalism, and Innovation in Space and Telecommunications Law," May 1-2 (2008), Lincoln, Nebraska.

existing components of the regulatory domain of outer space too.

IV. PRIVATE REGULATION: THE INSURANCE MARKET AND INDUSTRY SELF-REGULATION

A. Insurance Market as a Regulatory Vehicle

The insurance market is able to act as an institution of governance because it can stimulate or control behavior and uses many of the same techniques such as surveillance, information systems, investigators, and inspectors as government does when it regulates.[9] Indeed, "competitive pressures among insurers [can] result in continual improvements in the art of risk analysis."[10] For example, the fire insurance industry is widely viewed as playing a key regulatory role and greatly reducing the number of fire losses that would otherwise occur.[11] The satellite launch insurance market can be analyzed as an example of this type of regulation.[12]

The historical failure rate of launches is 8-12%, although in recent years it had fallen to 6%, leading to profitable years for launch insurers. However, in the past year or so, numerous launch failures have occurred, including the Russian Proton/AMC-14 Comsat; Sea Launch/Dutch NSS-8; Proton Rocket/JCSat 11 plus satellite damage to the QAF-1 Comsat.[13] One can ask whether certain launch companies or vehicles might be squeezed out of the launch market due to "regulation" of such activities by the insurance market. "Regulation" of launch companies through the insurance market can occur through a variety of means, including the cost of premiums, level of deductibles, the denial of insurance altogether, inspections, and even definitions of coverage terms such as "material change" and "due diligence" provisions.[14] Indeed, the predictions of lower premiums for manned flights than unmanned flights[15] might be suggestive of regulatory power of the insurance market, namely manned flight operators will exercise more care, utilize more technology, and include more redundant systems in order to keep insurance costs low and thereby keep the viability of the private space flight industry.

The launch industry is not the only "transport" industry in which regulation is thought to occur through the insurance market. In the maritime industry, private insurance companies play an important regulatory role. Marine underwriters may for example ask ship owners with "bad loss records" to implement specific

---

[9] See generally, Ericson, Doyle, & Barry, Insurance As Governance (2003)

[10] Martin Katzman, "Environmental Risk Management Through Insurance," 6 Cato Journal, No. 3 (1987).

[11] See id.

[12] For a description of space insurance law, see generally, Pamela Meredith "Space Insurance Law, With A Special Focus on Satellite Launch and In-Orbit Policies," 21 Air & Space Law 13 (2008).

[13] See Ron Panko, "Falling Stars: A String of Launch Failures Has Brought the High-Flying Satellite Insurance Market Back to Earth," 109 Best's Review 82 (July 1, 2008).

[14] For similar type of regulation in the maritime industry, see Franco Furger, "Accountability and Systems of Self-Governance: The Case of the Maritime Industry," 19 Law & Policy 445, 459 (October 1987),

[15] See Meg Green, "Reach for the Stars: Insurers are Playing a Critical Role in Transforming the Dream of Affordable Space Tourism," 105 Best's Review 20 (March 1, 2005)(quoting Jean Michel Eid, managing director of Aon).

safety recommendations and those ship owners with enviable safety records may ask for premium discounts.[16] Marine insurers also rely in part on classification societies to certify the seaworthiness of ships prior to deciding on whether to offer insurance coverage. One problem with the combined regulatory power of the insurers and the classification societies is that no single society has a monopoly, instead they compete for members as self-financed organizations, and thus have some incentive to be lenient in their certifications in order attract or at least not lose members.[17] Moreover, it is important to realize the limits on insurance market acting as regulator so as to not accord too great of a role to it in the "regulatory mix" of space regulation.

A significant limit is the so-called "insurance cycle" relating to periodic fluctuations in insurance capacity. The insurance cycle related to the maritime industry is described as follows:

> "Industry insiders divide the cycle into two phases, "hard" and "soft." In a hard market situation, insurance capacity is scarce, premiums are rising and underwriters will be able to charge adequate premiums and to impose tight conditions. Rising premiums attract additional insurance capacity and new underwriters into the market. As a result, competition for fleets among underwriters sets in, and the ability of underwriters to dictate premiums, deductibles and conditions sharply diminishes. Eventually capacity will be too high and the market becomes soft. In a soft market, premiums are falling and insurers compete, often desperately, for market shares."[18]

Hard markets are thought to enhance operational safety in the maritime industry; the reverse is thought to be true in soft markets. The launch and satellite insurance markets appear to be subject (at least to some degree) to a similar cycle as well. From 2002-2006 many new insurance players entered the marketplace as insurers were making lots of money and operators gradually obtained cheaper insurance.[19] However, the recent launch failures of the past year and half may have tipped the insurance market from soft to hard. Indeed, an official of the insurance company with the highest market share in space insurance has said the recent launch failures " 'have reduced the pressures on terms and conditions…and for the first time in a number of years, we were able to increase prices by some 25-30%... So the competitive pressures are a little more in our favor.' "[20] Similar insurance cycles for satellites and launches can be seen as far back as the beginnings of the satellite and launch insurance markets in 1965 and 1968 respectively. Many insurers continued to enter these markets until the 1977 "destruction of the OTS-1 communications satellite caused by the explosion of [its] launch vehicle…[the damage] wiped out the prior 12 years combined international premium

[16] See Furger, supra note 14, at 458.
[17] See id. at 462-63.

[18] See id. at 464.
[19] See Panko, supra note 13.
[20] See id. (quoting Ernst Steilen, head of space underwriting at Munich Re)

income."[21] To the extent the insurance cycle leads to sub-optimal standards periodically, steady, stable, non-fluctuating government regulation may be preferable, although government regulation tends to go in cycles as well depending on events and shifts in the control of government by different political parties.

B. Industry Self-Regulation

Industry self-regulation is the second branch of private regulation that is important to examine from a liberal international relations theory view of space regulation. The US has adopted a relatively "hands off" approach in its space tourism and transport regulations.[22] For instance, with respect to reusable launch vehicles, the FAA currently only has authority to restrict design features that have a high risk or have caused injury or death.[23] For space flight participants (or space tourists) the FAA does not regulate medical clearance. However, an industry association has been created, the Personal Spaceflight Federation, and that body has taken steps towards establishing medical clearance standards for space tourists. The possible motivations and benefits of self-regulation are many[24], including but not limited to warding off restrictive government regulation, deferring to those with potentially the greatest knowledge, establishing quicker and more easily adjustable regulation,[25] and, of no small significance in the space sector, more easily becoming effective at the global/international level (as the market and the industry is not limited to a particular geographic boundary). Further, it does appear that the space tourism industry is a "community of fate,"[26] in which companies collectively have an incentive to fill in regulatory gaps left by the government, in the sense that a major accident by one operator is likely to damage the fortunes of all as demand for flights would likely decrease significantly or onerous government regulation would likely appear. While each space tourist operator wants to out-compete the other in terms of the quality of the experience, none of them want to see a disastrous accident by one of their competitors.

Self-regulation, of course, causes many potential concerns as well, including self-serving standards that do not adequately protect the public,[27] standards that are potentially not backed with sufficient enforcement mechanisms, problems of potential free-riders[28] (referring to those companies that simply do not join or do not follow the industry standard yet benefit from the public's sense that standards are in place), and

[21] See Munich Re, Re-View A Magazine: 07 A Space Odyssey, available at www.munichre.com/en/corporate/history/re-view_a_magazine/magazine_07.aspx

[22] See comments by Rachel Yates at University of Nebraska College of Law conference on "Formalism, Informalism, and Innovation in Space and Telecommunications Law," May 1-2, 2008.

[23] 49 U.S.C.A. Sec. 70105(c)(2).

[24] For a discussion of benefits and drawbacks of industry self-regulation, see generally, Margot Priest, "The Privatization of Regulation: Five Models of Self-Regulation," 29 Ottawa L.Rev. 233, 268-275 (1997-98)

[25] See Neil Gunningham & Joseph Rees, "Industry Self-Regulation: An Institutional Perspective," 19 Law & Policy 363, 366 (October 1997).

[26] See id. at 395.

[27] See id. at 373-75.

[28] See id. at 391-93.

possible anti-trust concerns[29] as well. Given these potential drawbacks, it is important to realize that private regulation, be it through the insurance market or industry self-regulation, need not become a full replacement to government regulation, but can rather fit within a system of larger government regulation. For example, the government can mandate certain insurance coverage levels (rather than leaving it exclusively to a business decision) or can mandate self-regulation (or provide penalties for lack of compliance with self-regulatory schemes). These types of mixed governmental and private regulation are referred to as "co-regulation" or "regulated self regulation."[30] Co-regulation already exists in the form of the FAA requirements that private space operators maintain 3rd party liability insurance for the maximum probable loss, up to a maximum of $500 million dollars.

V. MULTI-LEVEL, MULTI-TYPE REGULATORY MIX: THE MOST EFFICIENT PATH TO SPACE-CREATED GLOBAL AND NATIONAL WELFARE ENHANCEMENTS?

The business of space is an industry subject to the influences of complex technology, diplomacy, politics, and finance. It is no surprise then that a complex regulatory mix already exists for space activities, one that involves governmental regulation at international, national, and even state, regional and local levels and one that involves elements of both governmental and private regulation. For example, OST Article VI requires national governments to authorize and supervise private space actors under their jurisdiction. The US Congress passed the Commercial Space Launch Act Amendments Act (CSLAA) of 2004 and the FAA issued regulations providing a baseline of regulation for space tourism operations, including requirements that space flight participants sign written informed consents, but no waiver of liability was required between the space flight operator and the participants (tourists). The industry is proceeding with developing contractual waivers and even lobbying state legislatures to pass laws providing tort liability protection should a space flight participant be injured or die during a voyage (because of worries that the written informed consent and the contractual waivers might not provide watertight protection in the absence of such legislation).[31] Indeed, it is possible that in a search for even further protection from legal uncertainty, the space flight operators (or even the participants) might seek insurance to cover any remaining risk.[32] Another existing example of a multi-level, multi-type regulatory mix is OST Article VI's direction leading to the CSLAA and accompanying regulations requiring

[29] See Robert Pitofsky, "Self-Regulation and Anti-Trust," Prepared Remarks to DC Bar Association Symposium, February 18, 1998 (noting such concerns are raised "in the limited group of cases in which rivals are foreclosed from the market without justification").

[30] See generally Hans Bredow Institute, "Regulated Self-Regulation As a Form of Modern Government," (October 2001); Gunningham & Rees, supra note 25, at 365-366 (for "co-regulation").

[31] I adapted this example from comments made by Rachel Yates, Holland & Hart, at the UNL Conference on "Formalism, Informalism, and Innovation in Space and Telecommunications Law," May 1-2, 2008.

[32] See, e.g., Meredith, supra note 12, at 13.

insurance be obtained by reusable launch vehicle permit applicants for maximum probable loss (up to a maximum of $500 million) and then the insurance industry regulating further through insurance policies issued to companies.

This paper should not be read as a call for more private regulation or less government regulation. Rather, as the title indicates, this paper explores lenses through which to view and assess the degree of formalism in the regulation of space activities. It acknowledges as a positive matter the complex regulatory mix already existing for space activities and this is why we should most appropriately views space regulation as a system of rules, at various governmental levels and by various private bodies.[33] As a normative matter, it simply calls for all regulatory options to be considered and recognizes that a mix is likely to continue to be necessary as we seek the most efficient path to space –created global and national welfare enhancements. It also hopes to serve as a reminder that sometimes private regulation –often having international scope and buy-in --may partially fill regulatory voids left by the inability to achieve international consensus for regulation among governments. For example, the potential problem of "flags of convenience" in space regulation (i.e. industry moving to states with weak regulatory schemes)[34], and the related problem of the large difficulty of negotiating "hard law" harmonized standards among governments, might be partially cured through private regulation.

In sum, we might seek to visualize each space activity issue demanding regulation as a search for the appropriate regulatory points in a space bounded by a Y axis going from local regulation upward to national regulation and up further to international regulation and a X axis moving from pure government regulation further outward to co-regulation and further still toward true private regulation. The concept of subsidiarity (that decisions should be made as closely to those affected having regard for the scope of the issue at hand) and the benefits of federalism (including the benefits and drawbacks of experimentation among jurisdictions v. harmonization across jurisdictions) will likely guide us to appropriate regulatory points as we move along the Y axis. The benefits and drawbacks of private regulation through the insurance market and industry self-regulation can guide us along the X axis.

[33] For a related argument of viewing international trade constraints as a system of rules, see Matthew Schaefer, "U.S. States, Sub-Federal Rules, & the World Trading System," in New Directions in International Economic Law 525 (Bronckers & Quick eds., 2000).

[34] The potential problem of flags of convenience was raised during the question and answer session at the plenary session hosted by IISL in Glasgow on Tuesday September 30, 2008. In response, the panel did not discount the problem ever arising but thought the Isle of Mann was not such a case as they had strong regulation but rather recruited space industries through favorable tax treatment.

IAC-08-E8.5.11

ALCÂNTARA CYCLONE SPACE: A NEW JOINT VENTURE IN THE SPACE MARKET

Álvaro Fabricio dos Santos
Brazilian Association of Air and Space Law (SBDA)
São José dos Campos, SP, Brazil
alvaro.santos@agu.gov.br

ABSTRACT

On 05/Dec/2007, the Brazilian-Ukrainian Company "Alcântara Cyclone Space" (ACS) was officially inaugurated. ACS is in charge of developing the launching vehicle Cyclone-4 that will launch satellites from the Alcântara Launching Center in the Northwest of Brazil. The Company was created according to the articles 2 and 3 of the Treaty signed on 21/Oct/2003, between the Governments of Brazil and Ukraine entitled: "Long-Term Cooperation in Utilization of the Cyclone-4 Launch Vehicle at the Alcântara Launching Center". This Treaty established the first international legal framework for the commercialization of space launches from the Alcântara Launching Center. Under the agreement, ACS was defined as "*an international entity for economic and technical purposes*". It was not the first bi-national endeavor of Brazil, because in 1975 a Treaty between Paraguay and Brazil for the construction of a hydroelectric dam was signed. At the 47th Colloquium on the Law of Outer Space, held in Vancouver / Canada, from 4 to 8 October 2004, the Brazilian Professor José Monserrat Filho presented his paper "Brazilian-Ukrainian Agreement on Launching Cyclone-4 From Alcântara: Impact on Brazilian Legislation", where the legal issues derived from the Treaty were duly analyzed. The purpose of this paper is to analyze the main articles of the ACS Statute, which was approved by the Directive # 559, of 31/Aug/2006, from the Brazilian Ministry of Science and Technology. There are many legal issues in the ACS Statute, such as applicable law, liability to third parties, and intellectual property rights, which will be presented in this paper. Current activities and future projects of ACS will also be presented.

INTRODUCTION

On 21/Oct/2003, the Ukraine and Brazil signed a "Long-Term Cooperation in Utilization of the Cyclone-4 Launch Vehicle at the Alcântara Launching Center"[1] which established the first international legal framework for the commercialization of space launches from the Alcântara Launching Center. Under the agreement, a joint venture company, "Alcantara Cyclone Space" was created. On the same occasion, the Brazilian Space Agency and the National Space Agency of Ukraine signed a "*Memorandum of Understanding on Future Bilateral Projects*", expressing "*their commitment to expand their cooperation further, through the exploration of additional fields of collaboration and the commitment to the joint development of new technological endeavors and projects, such as those related to the areas of liquid propulsion both for satellites and launchers, systems of guidance and control and upgrading of launch vehicles*".

The Treaty states in its Article 5 that Brazil is in charge of developing the general infrastructure of the Alcântara Launching Center (ALC), in order to provide technical conditions for the launching of Cyclone-4. On the other hand, Ukraine shall develop Cyclone-4, its assembly, integration, and tests procedures, especially its electrical injection fuel system, as well as the qualification of the first flight-model.

According to Article 3 of its Statute, the objective of ACS is to develop and operate the place of launching of Cyclone-4, established in the Alcântara Launching Center, in the State of Maranhão, Brazil, including the infrastructure for the launching of Cyclone-4, as well as the preparation and integration of spaceship, auxiliary systems and equipments, in order to carry out launching services for Brazil, Ukraine, and other public or commercial customers.

The Cyclone-4's technology is based on the Cyclone-3 rocket, which has been widely and successfully used by the Russian Federation for the launching of several satellites.

On the other hand, the Alcântara Launching Center has a privileged geographical location, placed at 3 degrees in the south of the Equator Line. Experts say that launchings from the Alcântara Launching Center, due its strategic position, can provide 30% of fuel saving. Beyond that, favorable weather conditions, which permit launchings all over the year, and the low demographic density in the Region, are also other factors that contribute to choose the Center as one of the best places for carrying out space launchings.

Currently, Ukraine has launched its rockets from plants located in the Russian Federation or in maritime platforms of the Sea Launch Company, a joint venture among companies from Ukraine, Russian Federation, United States and Norway.

The General-Director of ACS nominated by the Ukrainian Government, Mr. Olexander Serdiuk, states that the partnership between Ukraine and Brazil will be very fruitful for both sides: for Ukraine, which has the opportunity to put its new and promising Cyclone-4 vehicle into the world market; and for Brazil, which is advancing in its strategic plan to introduce the Alcântara Launching Center as an attractive and competitive site for launchings into outer space. He is convinced that ACS is a new highly competitive choice in the space market, because it may provide launchings in a safe and economic way.[2]

ACS's STATUTE

The ACS's Statute was approved by the Directive # 559, on 31/Aug/2006, from the Brazilian Ministry of Science and Technology.

At the United Nations/Ukraine Workshop on Space Law, held in Kiev, Ukraine, in 2006, Prof. José Monserrat Filho made a presentation entitled

"Development of Brazilian-Ukrainian Space Cooperation", where he mentioned that there was an impasse for thc approval of ACS's Statute. It is important to remember that the Treaty between Brazil and Ukraine was signed on 21/Oct/2003 and ACS's Statute was only approved on 31/Aug/2006. Prof. Monserrat reports that "*there was a long legal discussion between two points of view: 1) By-Laws has to be approved by the National Congress*[3]*, as it has financial implications; 2) By-Laws has not to be approved by the National Congress, since the Congress had approved the Treaty, in accordance to which the Company was created*". As the end, it was decided that the approval of the Brazilian National Congress was not needed.

The main dispositions of ACS's Statute will be presented in this topic.

ACS is a Bi-national company, established in Brasília, Brazil, which is represented, by the Brazilian Ministry of Science and Technology, the Brazilian Space Agency, the Ukrainian Fund for State Property, and the Ukrainian National Space Agency.

The Company has the following structure: a) General Assembly, which is the company's summit; b) Administration Board, which is composed by the Administrative Council; General Directors, Executive Directors, and Deputy Directors; and c) Survey Board, which is composed by Fiscal Council and Internal Audit. The Statute also foresees the need of an independent Auditor, who will prepare a final report to be submitted to the Administrative Council within 30 days after the end of the fiscal year[4].

The General Assembly should have one annual ordinary meeting but it can be installed extraordinarily whenever it is necessary or by a request from the Administrative Council.

Eight members compose the Administrative Council (4 Brazilians and 4 Ukrainians) and it is in charge of supervising the company's activities, as well as to fulfill the guidelines of the Treaty established between Brazil and Ukraine.

There are two General Directors, one Brazilian and one Ukrainian, and they are in charge of coordinating, organizing, and managing the Company's usual activities, as well as to represent it on the administrative and judicial levels.

The Company also has four Executive Directors and four Deputy Directors as follow: for the Brazilian side – Executive Administrative Director, Executive Financial Director, Deputy Technical Director, and Deputy Supply and Quality Director. For the Ukrainian side – Executive Technical Director, Executive Supply and Quality Director, Deputy Administrative Director, and Deputy Financial Director. The Executive and Deputy Directors shall implement the General Directors' decisions.

Two auditors nominated by the General Directors compose the Internal Audit. It is subordinated to the Administrative Council.

Four effective members and four substitutes compose the Fiscal Council. Each part – Brazil and Ukraine – shall nominate 2 effective members and 2 substitutes who will act in the absence of the effective member. The Fiscal Council shall survey the acts of the Directors and evaluate financial reports of the Company.

According to Article 6 of ACS's Statute, the Company has an initial social capital of US$ 4,500,000.00. It is also established in the Statute (Article 21) that the Company will use the currency of the United States of America – US dollars - in its commercial and financial operations.

The decision making process within ACS is regulated by Article 13. The following steps should be observed: a) Any member of the Administrative Council or Directorate may submit a proposal to be considered by the Administration Board; b) The General Directors present their conclusive opinion on the proposals to the Administrative Council in ten days; c) The proposal is approved only when both General Directors agree with it and there are at least five favorable votes of the Administration Board; d) In case of a dissenting opinion of a General Director, six favorable votes of the Administration Board are required to approve the proposal; and e) Proposals on modification of the number and responsibilities of the Executive Directors and Deputy Directors shall be approved by eight votes of the Administration Board.

Article 22 deals with profits, losses and Reserve Fund of ACS. Brazil and Ukraine have an equal share in the net profits and losses of the Company. Unless otherwise decided by the General Assembly, share in profits shall be paid out in 30 days upon approval of the Annual Report. The Company shall establish a Reserve Fund composed by up to 10% of the Company annual profit, until the investment made by the Parties into development of the Launch Site is fully repaid.

The Company will be solved, as stated in Article 23, in two cases: (a) extinction of the Treaty between the Brazilian and Ukrainian Governments; and (b) by the unanimous decision of the General Assembly.

Regarding intellectual property, Article 26 states that the Company will ensure the ways to protect previous rights of the Parts, as well as new inventions jointly developed.

Article 28.1 foresees that the Brazilian Law will be applicable in cases between the Company and: (a) Brazilian citizens and residents in Brazil; (b) Brazilian entities; and (c) foreigners, residents abroad, and international entities. On the other hand, the Ukrainian Law will be applicable in cases involving the Company and: (a) Ukrainian citizens and residents in Ukraine; and (b) Ukrainian entities.

In the case of conflict between the Brazilian side and the Ukrainian side, the Statute establishes in its Article 29 that the divergences should be solved in a good will manner, attending the principles of good faith and transparency. Beyond that, the practices and experiences of other international joint ventures or bi-national companies should also be considered in such situations.

Article 30 states that in case of disputes either regarding the Statute or the Company's activities, both Parts should do the best to solve them in a friendly way. If an agreement could not be reached, the dispute will be submitted to the Brazilian and Ukrainian Governments.

According to Article 34, the employees may receive their salaries in the currency of the Parts, depending on where the working-contract was signed. There is also the possibility of receiving salaries in US dollars, through a decision of the Administrative Council. The principle of wage equality, related to works with the same nature, time, and efficiency, will be applicable despite of nationality, sex, race, religion, or civil status.

The last disposition, i.e. Article 36.3, establishes that the Administrative Council is in charge of solving doubts regarding the application and interpretation of ACS's Statute.

ACS'S LEGAL NATURE

As a Bi-national company, ACS has many particularities that require a different legal approach. On this regard, it is interesting to know how Brazilian jurists have deal with another bi-national company called "Itaipu", a hydroelectric dam constructed in the seventies as a partnership between Brazil and Paraguay.

Luiz Rafael Mayer, former Brazil's General Legal Consultant, states that Itaipu is a new legal person in the field of International Public Law and its specific goal is to carry out commercial activities. He considers that Itaipu is submitted to the International Law regime and, as a consequence, constitutional or administrative domestic laws are not applicable to it.[5]

Prof. Miguel Reale, talking about Itaipu's employees, states that Brazilians and Paraguayans carry their personal rights, as well as their civil and penal responsibilities, according to the laws of their respective countries. The Brazilian or Paraguayan laws, depending on how it is established in the working-contract, will regulate the rights of foreigner employees.[6]

Prof. Paulo Salvador Frontini draws the attention to the fact that there are international institutions created by the initiative of particulars. The Red Cross is an example of such institutions. It is international because it has a worldwide range, but it is not an international organism, because it was not created by sovereign States. It is submitted only to the laws of one State, which is the one where it was established. Itaipu was created by Brazil and Paraguay and has legal and financial capacity. Despite its bilateral origin, its legal capacity is not limited to Brazil and Paraguay. Itaipu is Bi-national Company due to its origin, but it is international in its legal capacity.[7]

Turning to ACS, there are no doubts that it is an international company with a particular legal regime, where neither Brazilian nor Ukrainian laws are applicable, except in the cases foreseen in Article 28. Otherwise, ACS could not have fiscal benefits, for instance, because the Brazilian Constitution does not allow such privilege. Beyond that, if Brazilian domestic laws were applicable to ACS, the Statute could not have established that the currency would be the US dollars, or that the employees could receive their salaries in a foreigner currency. It should be emphasized that domestic laws will only be applicable to persons, employees or members of the Company's Board, where the subject is their civil or penal responsibilities, or even their particular rights.

ACS's PERSPECTIVES

The development of ACS's activities is currently linked to the solution of two main difficulties: 1) the improvement of the Alcântara Launching Center's (ALC) infrastructure; 2) the establishment of a technological safeguard agreement between the Governments of Brazil and the United States of America, for the utilization of ALC. Both difficulties have to be solved by the Brazilian Government.

According to the Treaty between Brazil and Ukraine, Brazil is in charge of providing a suitable infrastructure in ALC, in order to enable launchings of space objects from that Center. The Brazilian Space Agency has been putting its best efforts to become ALC operational as soon as possible. ALC is not property of ACS, so it is subjected to the Brazilian laws, especially the Law # 8.666, of 21/Jun/1993, which regulates biddings and contracts of the Brazilian Federal entities. The Brazilian Space Agency is committed to promote public biddings related to ALC. The winners of these biddings will provide equipments and services for the improvement of ALC's infrastructure. Unfortunately, the legal procedures established in Law # 8.666 are not as expedite as expected. Furthermore, competitors may present complains to judicial or even administrative courts, which may decide to interrupt biddings. Another impasse is the existence of a "quilombola" community in ALC. "Quilombolas" are descendents of African slaves who live in communities protected by the Brazilian law. The Brazilian Government has already identified another area for replacing the one that is in ALC. However, some "quilombolas" are reluctant to leave ALC. The Brazilian General Advocacy has created a working-group to define, in cooperation with "quilombola" communities, areas – not only in ALC but also in other regions of Brazil - where they might be placed. Despite of these obstacles, it should be affirmed that the efforts of the Brazilian Space Agency have been well succeeded.

As a company, ACS is obviously interested in catching the North-American market, because it is the leader in the field of space activities. However, this goal can only be reached when Brazil and the United States signed a technological safeguard agreement for the utilization of ALC as the launching base of North-American space objects. In 2001, Brazil and the United States had their first attempt of signing an agreement of this nature. At the time, many voices against the agreement were heard. Favorable and unfavorable opinions were well reported by Wálteno Marques da Silva, current Head of Legal Department of the Brazilian Space Agency, and Prof. Altair Stemler da Veiga, in their article "Technological Safeguard Agreement between Brazil and the United States: Why Do Not Sign It?" [8].

According to the draft presented by the United States, the agreement's most polemical points were the follows: 1) Brazilians, civil or military, are not allowed to enter in the North-American ground station placed in ALC; 2) Revenues from the launchings in ALC could not be used by the Brazilian Government for the development of a space vehicle; and 3) ALC could not be used by States that have not adhered to the Missile Technology Control Regime – MTCR.

According to those who were against the agreement, the Brazilian authorities could and should enter into the North-American ground station, since it was placed in the Brazilian territory. They considered that, in theory, the United States could use ALC as its base for military or spy operations and, in this case, Brazil would connive with such actions. They also considered that Americans did not have the power to control the Brazilian revenue and, furthermore, one of the aims of the Brazilian Space Program was the development of a space-launching vehicle. Finally, those criticizers remembered that China, the most important international partner of Brazil in its space endeavors, had not adhered to MTCR. So, if the technological safeguard agreement was signed, the China-Brazil Earth Resources Satellite – CBERS – would never be launched from ALC.

However, favorable opinions draw the attention to the nature of the agreement: it was not an instrument of cooperation, but a technological safeguard instrument. In this context, it was obvious that North-American space technology should be protected. Hence, the restricted access to the North-American ground station in ALC was fully justified. They also consider normal that revenues from launchings of North-American space objects in ALC could not be used for the development of a Brazilian space-launching vehicle. Just a few groups of States in the world have the technology of developing a space-launching vehicle. Of course, the United States do not want to support other potential competitors in this narrow market. But this revenue could be used for other purposes, such as the improvement of ALC's infrastructure. Besides that, Brazil had a specific budget for the development of its space-launching vehicle, and the project would not be cancelled. Regarding the fact of China had not adhered to MTCR, the enthusiasts for the agreement stated that according to the CBERS Program, all satellites would be launched from the Chinese Center, in Taiyuan.

In their article, Da Silva and Da Veiga advocates the signature of technological safeguard agreement between Brazil and the United States due the following reasons: a) the agreement does not hurt neither the Brazilian sovereignty nor the Brazilian Space Program; b) the agreement's dispositions are in accordance with the international legal framework, especially with the Outer Space Treaty, and the Agreement on the Rescue of Astronauts; c) the strategic position of ALC as well as its infrastructure should be better exploited by Brazil; and d) the agreement will bring financial resources and social benefits to the region.

At the end, the Brazilian Government decided not to send the agreement to the National Congress, because it was well known that it would not be approved there.

Dr. Joao Azevedo, Director of ACS, nominated by the Brazilian Government, has emphasized that the Brazilian Government, through its Ministry of Foreign Affairs and the Brazilian Space Agency, has already started discussions with the North-American authorities, in order to establish a new technological safeguard agreement for the utilization of ALC[9]. Of course, at this time, the Brazilian representatives, during the discussion process, will try to avoid the difficulties that were faced at the first occasion.

CONCLUSION

ACS has the necessary skills to be well succeeded in its activities. On one side, it has the technological experience in developing space-launching vehicles, provided by the Ukrainian expertise; on the other side, it has the privileged location of ALC, where launchings may be carried out practically all year over.

The Company's objectives, decision-making process, administrative and financial procedures, and the Brazilian and Ukrainian responsibilities are perfectly defined in the Treaty and in ACS's Statute.

Regarding ACS's legal nature, there is no doubt that the Brazilian and Ukrainian domestic laws, even its respective Federal Constitutions, are not applicable to the Company, except in those cases foreseen in Article 28.

The partnership with Ukraine strengths the Brazil's position as a space-faring country, as well as endorses the feasibility of cooperation between two developing countries. On this regard is always valid to remember that the first Brazil's partnership with a developing country, i.e., the one established with China for the development of CBERS satellites, has been considered worldwide as an example of a well succeeded cooperation.

Brazilians and Ukrainians expect that current impasses involving ACS, such as "quilombolas", biddings promoted by the Brazilian Spacc Agency, and the signature of a technological safeguard agreement with the United States, can be solved soon. Then, hopefully, ACS will become one of the most important suppliers of space-launchings around the world.

---

[1] The Brazilian Decree # 5.436, on 28/Apr/2005, approved the Treaty.

[2] Data obtained from "Jornal da Ciência", published by the Brazilian Society for the Advancement of Science – SBPC – on 13/Nov/2006.

[3] According to the Brazilian Constitution, any agreement celebrated between Brazil and a foreigner country, which involves financial resources from the Brazilian Treasury, should be approved by the National Congress.

[4] According to Article 21.1 of the Statute, the fiscal year starts on January 1st and ends on December 31st of each year.

[5] In his Advice L-208, issued on 22/Sep/1978, approved by the Brazilian President on 17/Oct/1978.

[6] In his article "The Legal Structure of Itaipu", published in the Brazilian Magazine "Problemas Brasileiros", Aug/1974, pages 3 to 8.

[7] In his article "Itaipu Bi-national: A New Kind of Company", published in the Brazilian Magazine "Direito Mercantil" #15/16, 1974.

[8] Published by the Brazilian Association of Air and Space Law – SBDA – in its website www.sbda.org.br/revista/Anterior/1725.htm

[9] The author would like to express his gratitude to Dr. Joao Azevedo, Director of ACS, for the Brazilian side, which kindly provided information regarding ACS's projects. Notwithstanding, the views expressed in this paper reflect exclusively the author's opinion.

China's Space Activities Licensing Regime

Xiaodan Wu
Universita degli Studi di Milano / Harbin University, China

Abstract

According to 1967 Outer Space Treaty and Liability Convention to which China has acceded respectively in 1983 and 1988, China is required to authorize and continuously supervise and take international responsibility for national space activities. In order to implement these international obligations and arrange domestically for the liability between the State and non-governmental entities, China National Space Administration has adopted the "Interim Provisions on Licenses for Civil Space Launching Projects" on November 21, 2002. Its main provisions include the scope of application; the licensing agency, criteria for issuing licenses; procedure for license application and its review; modification, cancellation and revocation of the license; administrative penalties and criminal liabilities, etc. The licensing regime established by the Interim Provisions demonstrates Chinese government's firm efforts in carrying out international obligations on space issues and commitment to achieving legal transparency in outer space. Ever since 2000,aiming at promoting the healthy development of space activities, one of the major policies and measures of China's space industry is to encourage industrial enterprises, scientific research institutions, commercial corporations and social organizations to participate in space activities. On the other hand, these entities have shown great interest in launching civil satellites and have become involved in space activities. So, the relevant practices can provide further evidence that China's intend to create a regulatory climate for commercial space activity.

1. INTRODUCTION

The provisions of responsibility and liability enunciated in Outer Space Treaty and Liability Convention requires that the state parties shall authorize, supervise and bear international responsibility for national activities in outer space, whether such activities are carried on by governmental agencies or by non-governmental entities. The most transparent and effective instrument of implementing these international obligations is to regulate the activities of private entities through the license procedure. Therefore, many states parties have elaborated domestic space regulation on the licensing of outer space activities in order to arrange the liability between the State and non-governmental entities. This paper intends to evaluate China's licensing regime for space activities and highlight its significance in promoting space commercialization.

Due to various reasons, especially China's circumstances in the past that all the programs of exploring and using outer space is carried out by the government or its authorized organs, justified that there is no space activities licensing regime for three decades since its first satellite Dongfanghong-1 was launched in 1970. However, in recent years, the importance of establishing a licensing system has been realized by China's government because of following changes:

1.1 The Rapid Development of Market Economy.

Since the adoption of reforming and opening-up policy in 1978, China's economy enjoyed average annual growth rates in excess of 9 percents for three decades[1] and a market economy system has already been established. China's enterprises, no matter state-owned or non-state owned sector, have made breakthrough in development and has shown great interests in becoming involved in space activities, especially in the application of space technologies.

1.2 The Extensive and Rapid Development of Space Activities.

As one of the major space powers, China's space industry has made eye-catching achievements and ranks among the world's most advanced countries in some important fields of space technology. Ever since 2000, in order to maintain long-term, stable development in this field, one of the major policies and measures for China's space industry is to muster strength in implementing key space scientific and technological projects and encourage industrial enterprises, scientific research institutes, commercial corporations and social organizations to take an active part in space activities under the guidance of national space policies.[2]

1.3 Improvement of Legal System.

The importance of national space legislation has been acknowledged by individual nations and international society. As a space-faring country, China's space activities need to be regulated by law, which is the basic requirement of a nation ruled by law. In China, there are now frequent appeals for strengthening space legislation to solve a series of problems arising from the rapid development of space activities in a market economy.

1.4 The International Climate of Space Commercialization and Privation.

Space commercialization and privation is the logical consequence of the progress of space activities because this trend contributes to the social welfare of all mankind. This trend has accelerated and many space-faring countries have adopted some measures to encourage non-governmental entities to invest in space industry. For example, the United States Congress has declared that "the general welfare of the United States requires that the NASA shall seek and encourage, to the maximum extent possible, the fullest commercial use of space."[3] Although there is great distinction in the developing strategy of space industry between the United States and China, American policy of space commercialization and privation and its prominent effects has established a good example for China.

In December 1999, Ministry of Education, on behalf of Tsinghua University, submitted an application to the National Space Agency(NSA) for launching the micro satellite, Tsinghua-1, built as a collaborative project between Surry Satellite Technology Limited(SSTL) and Tsinghua University. NSA issued the license after reviewing the report of Tsinghua University. Based the above mentioned reasons and this practice, NSA adopted the "Interim Provisions on Licenses for Civil Space Launching Projects"(hereinafter referred as the "Interim Provisions") on November 21, 2002.

2. UNDERSTANDING ON INTERIM PROVISIONS

The basic purpose of the Interim Provisions is to regulate civil space launch project; promote the healthy development of civil space industry; maintain the state security and the public benefits and implementing the international obligations of outer space conventions. The Interim Provision, 28 articles in total, contains the basic framework of licensing regime for civil space launching projects.

2.1 Scope of Application

The Interim Provisions regulate all the entry of space objects inside the territory of China into the outer space not for military purpose or the entry of such space objects into outer space from outside of the territory of China for which the natural persons, legal persons or other organizations of China have had property rights including by means of in-orbit delivery. NSA shall be responsible for review, approving and supervising civil space launch projects.[4]

2.2 Criteria for Issuing Licenses The contractor of the space projects or the final owner of the satellites or other space objects shall be the applicant of the license, who should meet the following conditions: the project shall be operated according to the laws and regulations; shall maintain the state security and public interests; obtain the relevant permission documents issued by the other sectors of the government; have the technological, financial ability for carrying out the projects.[5] The holder of the license shall comply with the relevant requirement of third party liability insurance and other insurances for launching a space object.[6] Furthermore, contract of foreign-related projects shall be entered and organized by an international trade company designated by Chinese government and shall not entered into force until it is approved by NSA and other governmental sectors.[7]

2.3 Review; Modification, Cancellation and Revocation of the Licenses.

Within 30 days upon receiving the application documents, NSA shall review and issue a license to the qualified applicant. For the unqualified ones, the applicants and relevant sectors of the government shall be notified in writing.[8] If any content of the license needs to be modified, the holder shall file an application to the NSA. Until it has been approved upon examination, the license shall not be modified. For the project unable to be accomplished due to inappropriate management of the license holder, NSA shall nullify the license.[9] NSA can suspend the license under the following severe circumstances: violation of the relating laws or regulations or the agreements on confidentiality during operation of the projects; jeopardizing national security or national interests; action against the national diplomatic policy or international conventions concluded or acceded by China; launching activities carried out beyond the scope approved by the license, etc.[10]

2.4 Administrative Penalties and Criminal Liabilities.

Any natural person, legal person or other organizations shall not be carried out space activities without the license. The involved parties shall be imposed administrative penalties or criminal punishment in accordance with the law.[11] During the application or operation of the launching projects, if the parties conceal the truth, practice frauds or infringe the national interests, it shall be imposed administrative penalties or criminal punishment in accordance with the law.[12] When NSA or its staff neglects duties or abuses powers during the review or approval of the license and caused losses to the state, it/he shall be punished according to the law.[13]

3. RELEVANT PRACTICES

Soon after the launch of Tsinghua-1, Aerospace Tsinghua Satellite Technology Co. Ltd (ATST) was established with an official sanction from the State Administration of Industry. The financial backing came from the China Aerospace Machinery and Electronics Corporation, Tsinghua University Enterprise Group and Tsinghua Tongfang Co.Ltd. In 2002, China YinTai Holdings Co., Ltd, a private enterprise, invested in ATST by the means of risk capital. ATST focuses on developing micro satellites and detector technologies, and marketing their applications. In China, this is the first time the high-tech aerospace industry, a well-established university collaborate with a private enterprise to fund and share resources in a joint business venture. This company has changed the situation that the State and the government is the exclusive investor for the satellite developing and operation. In 27th October 2005, after been issued the license in accordance with the Interim provisions, Beijing-1 was launched successfully, which is collaborated by Beijing Landview Mapping Information Technology Co., Ltd (BLMIT) and SSTL. Beijing-1 is the first satellite that Beijing posses the whole right of controlling and can provide remote sensing images periodically, which can service as the data base for the city planning, the ecological environment monitoring, important projects and land use projects. This year it has made contribution for the decision making of Wenchuan earthquake relief and the 2008 Olympics. Now, under the guidance of the national space policy and the Interim Provisions, inspired by the successful launches of the two micro satellites, several other commercial entities are setting their sight to compete in the micro satellite market and made clear indication to establish their presence in this booming market both at national and international level.

4. EVALUATION AND RECOMMENDATION

Legal society is governed by law, with the aim to improve legal certainty and transparency. The licensing regime established by the Interim Provisions demonstrates Chinese government's efforts in carrying out international obligations on space issues and commitment to achieving legal transparency in outer space. The promulgation of the Interim Provisions has also indicated that China's firm determination to encourage the commercialization and privation of space activities and the level of space legislation has reached a new stage. And most importantly, Interim Provisions has led to some remarkable achievements after it entered in force in 2002. However, there are some limitations in the articles that run against national space policy. In order to encourage commercial entities to invest in space, a comprehensive regulatory scheme must be enacted to provide private enterprise with a minimum of regulatory interference and long-term predictability. [14] But Interim Provisions require special approvals for the foreign-related projects contract. In practice, the above-mentioned micro satellites both needs communications with the Ministry of Foreign Affairs and obtained special approvals from State Council because they are joint developed with a British company and were launched in Russian. Consequently, it is very necessary for China to study other existing commercial space-faring nation's legislation and make the licensing regime perfect when it is incorporated into national space legislation. For example, China can learn from U.S space policy about how to expand private-sector investment and involvement in civil space and space related activities. Therefore, although the space commercialization and privation of space is developing in China, but it is developing slowly because there are limitations and interferences for private capital to invest in space ventures. The role of the government is over-emphasized in the management of space activities. As a conclusion, Interim Provision has provided a testing bed for a higher level of legislation to make further improvement and there is much that China's government can do to encourage private space activities through national legislation.

References

1 See Report given by the Director of National Development and Reform Commission, available at http://news.xinhuanet.com/politics/2007/05/03/content_6055744.htm(viewed on August 12, 2008).

2 See White Book on China's Space Activity in 2000 and 2006, available at http://www.cnsa.gov.cn(viewed on August 12 2008).

3 National Aeronautics and Space Act, 42 U. S. C. § 245 (c) (1984).

4 Article 2 of the Interim Provisions; see http://www.cnsa.gov.cn (view on 13th August 2008).

5 Article 5 of the Interim Provisions.

6 Article 19 of the Interim Provisions.

7 Article 9 of the Interim Provisions.

8 Article 7 of the Interim Provisions.

9 Article 13 of the Interim Provisions.

10 Article 15 of the Interim Provisions.

11 Article 25 of the Interim Provisions.

12 Article 24 of the Interim Provisions.

13 Article 26 of the Interim Provisions.

14 Fred Kosmo, The Commercialization of Space: A Regulatory Scheme that Promotes Commercial Ventures and International Responsibility, 61 Southern California Law Review, p. 1058 (May 1988).

IAC-08-E8.5.16

LEGAL FRAMEWORK IN SUPPORT OF COMMERCIALIZATION OF OUTER SPACE: THE CASE OF THE ISLE OF MAN

T. Aganaba (aganaba@isu.isunet.edu)
International Space University, Strasbourg, France

Abstract

This paper looks at the continued growth of the Isle of Man as the financial hub for the international space community. Particularly in light of the proposed UNIDROIT Convention on International Interests in Mobile Equipment as Applied to Space Assets, the effect of the Isle of Man's intentions to adopt elements of UNIDROIT to establish the world's first commercial space register on the Island will be discussed. With the increasing surge towards increased commercialization of outer space, "New Space" entities are emerging. The profile of space actors is moving from strictly governmental, intergovernmental and multi national companies to include smaller entrepreneurial companies with ambitious ideas. For many new actors, access to capital and lack of certainty of the law are proving major challenges to the progression of the commercial space industry. As such, the choice of regulatory environment is crucial to the success and profitability of business. A little Island at the heart of the British Isles, the Isle of Man is adopting measures to tackle these challenges and attracting established satellite communications as well as New Space business through the establishment of a tight regulatory environment specifically tailored to the needs of the industry. The latest strategy, the creation of an Isle of Man financial register of space assets, will mark a symbolic maturing of the space sector in the Isle of Man. The proposed registry will seek to provide a platform for commerce as no other space registry has. Already a renowned jurisdiction for shipping and aircraft registration, this initiative may be the next logical step for the Island to facilitate strong regulatory environment for the financing of space assets while providing a platform for the Island in its continuing work to establish itself as a leading international finance centre dedicated to space.

1. INTRODUCTION

Under the auspices of the United Nations Committee for the Peaceful Uses of Outer Space (COPUOS), the international legal framework governing space activity is set out in the following conventions: The Outer Space Treaty 1967[1] (OST); The Rescue Agreement 1968[2]; The Liability Convention 1972[3]; The Registration Convention 1975[4]; The Moon Agreement 1979.[5] Whilst the Moon Agreement has not received wide international support, the Outer Space Treaty (OST) can be described as the Magna Carta of space law, which sets out the general framework the other treaties elaborate upon. Many of the principles therein have customary status and are binding on all Nations. Under Art. VI and Art. VII OST, States must authorize and supervise national space activities, including the activities of nongovernmental entities such as commercial and non-profit organizations and are liable and responsible for any damage that occurs as a result of space activity. The advent of commercial space activities has led more countries to consider how to regulate private space activities in a manner that does not hinder or preclude investment, while still ensuring that commercial activities comply with international law. Following the space initiative in the Isle of Man, it is necessary to determine the legal framework applicable to space activity on the Isle, bearing in mind its unique status, not as a sovereign State but as an internally self governing British Crown dependency. This paper introduces the Isle of Man and its space strategy. The special arrangement between the UK and Isle of Man with respect to space activity will be discussed as well as the Isles latest strategy to create a space register; bearing in mind the proposed UNIDROIT Convention on International Interests in Mobile Equipment as Applicable to Space Assets. The paper concludes with recommendations for the Isle of Man space register.

2. THE ISLE OF MAN

The Isle of Man is located at the centre of the British Isles in the middle of the Irish Sea. Made up of 4 designated towns with Douglas as the capital and numerous village and parish districts, The Isle is roughly 52km by 22km with a population of approximately 81,000 people. The Isle of Man is a British crown dependency, but is not part of the UK or the EU. According to Schedule 1 to the Interpretation Act 1978 (c.30), the Isle is a "British Possession" but not a colony.

As an internally self governing nation, the Isle of Man has its own legal system, which is based on principles of English common law, with political and legislative independence through Tynwald - the oldest legislature in the world. There is no enshrined constitutional document defining the relationship with the UK and whilst proud of its internally self governing status, it is accepted by the Isle of Man Government that in certain specific limited circumstances the UK has the power to legislate on its behalf in matters of mutual concern, as the crown claims ultimate responsibility for the good government of the Island. This is mostly with respect to foreign affairs and matters of defence, for which the Island pays an annual contribution.

UK law is not directly applicable to the Isle of Man, but Acts of the British parliament may be applicable to the Isle of Man, after consultation and prior agreement, through extension by Order-in-Council. An Order-in-Council is a statutory instrument made formally in the name of the Queen through the Privy Council. Over the years, the Isle of Man has obtained significantly greater self determination as evidenced by the latest 2006 declaration signed by the UK and Isle of Man Government, which set out a framework of principles for the development of the International identity of the Isle of Man.[6]

Despite such agreements to clarify the relationship between the two countries, the Isles strategy is to align itself with the UK, but differentiate itself where it can be more effective as a service provider. The Isle of Man recognizes that the relationship with the UK is one of its economic strengths and it is clear that the Isle will not deviate from what is acceptable or is not objectionable to the UK Government. In that light, from a pragmatic perspective it is unlikely that the Isle would look to full independence from the UK. Instead the Isle works hard to ensure that all business and dealings with respect to or from the Isle of Man add value to the UK, maintaining a hub and spoke relationship. The Isle is considered part of the UK for VAT collection purposes imposing, save in certain circumstances, an identical VAT rate to that of the UK. Economically, there is free trade and increasing value has flowed from the Isle of Man to the UK, particularly with respect to capital flow and access to the international community.

Why the Isle of Man

The principal benefit for managing or locating business on the Isle of Man is based on the tax structure system of the Isle, which provides a competitive regime for companies and individuals. Keen to distance itself from the negative connotation of being tagged as an unscrupulous tax haven, the Isle of Man established a tax strategy that fully complies with the EU Code of Conduct on Business Taxation[7] and is in line with the OECD standards on transparency, exchange of information and the removal of perceived harmful tax practices.[8] This effort, as well as its strict enforcement of Anti Money Laundering controls and wholehearted abidance to best international standards has made the Isle a 'participating partner' of the OECD and contributed to its impressive credit rating as a jurisdiction of class and choice.

To take advantage of the tax and many other benefits of incorporation on the Isle of Man, the Isle completely overhauled its companies legislation and introduced the "New Manx Vehicle" ("NMV") company structure by way of The Companies Act 2006 (c.13). Before 2006, the principal Act was the Companies Act 1931. Amending Acts were enacted in 1960, 1961, 1968, 1974, 1982, 1986, 1992 and 1993 and are construed as one with the principal Act. The 2006 Act (a stand alone Act) was drafted by Isle of Man lawyers with a brief to create a new corporate vehicle which was: flexible, simple and inexpensive to administer. Since its inception, the NMV has proved very attractive to international businesses.

The principal benefits of company incorporation under the new Manx Vehicle are with respects to:[9]

- Capital arrangements
- Proceedings of the company
- Accounting and Annual Returns

The types of company structures that the Isle of Man offers are both simple and creative. It is the strength and adequate regulation of the structures that make incorporation on the Isle beneficial for many businesses. Confidentiality is guaranteed through greater regulation and responsibility on corporate service providers through the Corporate Service Providers Act 2000 (c.13), rather than on bureaucratic and stringent company requirements. Creative structuring includes establishment of structures such as the Protected Cell Company (PCC) known in some jurisdictions as a segregated accounts company. A PCC is a corporate body that can be segregated into legally distinct portions, known as the "core" and underlying "cells". Each cell has its own separate segment of the PCC's overall share capital, allowing shareholders to maintain sole ownership of an entire cell whilst owning only a small proportion of the PCC as a whole. Though a cell is not a separate legal entity in its own right; its revenue streams, assets and liabilities are kept separate from all other cells and from the core.

Other benefits to setting up business on the Island include:

- Neutral jurisdiction
- International reputation - no red flag and no black list
- Proactive & collaborative regulatory regime with minimal bureaucracy
- Access to decision makers
- Business development initiatives
- Regime based on English law
- Stability and Independence – longest continuous parliament in the world
- Capacity and space for sustaining growth
- Low operating cost environment
- No restrictions on new residents and open property market
- Good skill set with top rated professional services sector

3. OFFSHORE CENTERS AND SPACE

In several articles by American tax attorney William Andrews, tax relief on space commerce is promoted as well as the establishment of a new type of company structure, referred to as the International

Space Company (ISC). He proposes that an offshore entity could take the lead in forming, servicing and managing ISC's, particularly jurisdictions with "a well educated, stable, moral populace with nimble, effective leadership and an adequate communications system."[10] He adds that the Offshore financial centers could aggressively target space related support functions offering corporate and fiscal incentives, human resources packages and specialized training and education initiatives and tailor made space company legislation. Under this structure, there would be no income tax liability for space ventures and if an ISC generates 51% of its business in space, its non space income would also be tax free. The Isle of Man is way ahead of other offshore jurisdictions through direct promotion of itself as the "spaceisle" – the international finance center dedicated to space. The spaceisle represents a flexible and nimble, yet highly regulated and open base for space and space related business.

Isle of Man Space Strategy

Seeking to diversify the economy further and on a mission to develop its international identity, the Isle of Man embarked on developing a Space strategy to attract the space industry to the Isle. Whilst focusing on the established satellite industry, the Isle also differentiates itself as a jurisdiction that will set the way for the efficient and successful growth of the emerging New Space industry.

The uniqueness of the Isle of Man space strategy is that it is essentially commercial in nature and enjoys whole hearted support at the highest levels within Government. The Isle of Man Space strategy is spearheaded in the Treasury Department by the E-business/Space Commerce Division. This is a unique and strong statement that ensures that involvement in the space arena in the Isle is seen as a commercial and financial endeavour. Also playing a role is the Communications Commission, which is a statutory board of the Government, with responsibility for the licensing and regulation of telecommunications and broadcasting in the Isle of Man.

With a focus on encouraging commercialization, the Government works hand-in-hand with the private sector through a public/private Space Industry Group and took the novel step to fully outsource the commercialization of its space activity by contracting the services of ManSat Limited. The effect of this fusion between Government and the private sector, through both ManSat and the space industry group is that the law and policy framework for space activity on the Isle can be guided by the space industry and the relevant stakeholders, increasing the Isles potential as a global centre for private activities in Space. Set to attract established satellite companies as well as modern New Space companies, including the general benefits discussed in section 2 above, the Isle of Man space and satellite "package" comprises the following benefits[11]:

- Access to Orbital Slots through ManSat Ltd
- 0% Corporate Tax
- No Insurance Premium Tax for satellite insurance
- Space aware/Supportive Government
- Friendly Commercial Legislation
- Ability to Source Funding through the Isle of Man
- Professional Services Expertise in Space Related Matters
- State-of-the-Art Telecommunications Infrastructure

4. LEGAL FRAMEWORK

Due to the status of Outer Space as the "province of mankind" (Art.1 OST), outer space activity is primarily regulated by international law as well as under national law. As a British Crown dependency, the UK is responsible for the international relations of the Isle of Man. To that end it is necessary to asses the UK legal and regulatory framework for space activity to determine the framework that the Isle of Man operates under.

The UK Regime

The UK has two separate legal/regulatory frameworks for space activity, particularly:

- The granting of licenses for activities in outer space by virtue of the Outer Space Act 1986 (c.38)
- The management of orbital filings by the Office of Communications (Ofcom) with the International Telecommunications Union (ITU) and by virtue of The Communications Act 2003 (c.21)

Specifically, the UK Outer Space Act 1986, set out to:

1. Provide regulation / licensing of UK space activities through the UK Secretary of State and the British National Space Centre (BNSC)

2. Facilitate compliance with International Obligations, specifically that:
   - UK Secretary of State to maintain a register of space objects as set out under the Registration Convention
   - UK to bear international responsibility and liability for activities in outer space which it has licensed as set out in the Outer Space Treaty and Liability Convention.

3. Ensure that activities do not jeopardise public health or the safety of persons or property

4. Provide for indemnification of UK Government against any claims in respect of damage or loss arising out of the licensed activity.

The main provisions of the Outer Space Act apply to the launch, the procuring of the launch, the operation of space objects and to "any activity in outer space". The breath of activity covered is extremely wide but space activity licenses are not required for leasing of space segment capacity (transponders) or space activities governed by separate international agreements securing compliance with the international obligations of the UK.

With respect to satellite communications, the first step is gaining the right to orbital slots and frequencies. An orbital slot is a function of intended use, orbital location and frequency band. The ITU allocates all Geostationary orbital positions on a first come, first served basis, only to States. This is because the geostationary orbit is seen by States as a natural resource that is the province of all mankind. It is a highly regulated area with rules of notification, and coordination established under the ITU Radio Regulations. The purpose of such notice, coordination and registration is to eliminate harmful interference between earth and space-based stations of different countries, to improve the use of the radio frequency spectrum and the geosynchronous satellite orbit and to accommodate, to the extent possible, a country's needs. This is managed in the UK by Ofcom, the regulator and competition authority for the UK communications industries.

The Isle of Man Regime

Space activities can be carried out on and from the Isle of Man and are governed by the Outer Space Act 1986 (Isle of Man) Order 1990 (SI 1990/596), which extends to the Isle of Man all the provisions (as amended) of the UK's Outer Space Act 1986. The UK Communications Act 2003, extended to the Isle by Communications (Isle of Man) Order 2003 (SI 2003 No. 3198), provides the legal basis for activities in the satellite communications sector in the Isle. Due to the Isle of Man's status as a Crown dependency, Ofcom has the power to represent the Isle of Man at international bodies such as ITU. When Ofcom exercises these functions, it acts on behalf of the Isle of Man Government. In January 2005, and at the request of the Isle of Man, the Secretary of State formally requested Ofcom to extend its ITU representation role to the Channel Islands, the Isle of Man and the British Overseas Territories.

Uniquely, the Isle of Man Government has granted, via commercial contract, sole and exclusive rights to ManSat Ltd for all its orbital applications and ManSat Ltd offers a bespoke filing and consultancy service for Isle of Man registered space companies undertaking or wishing to undertake any space activity. As such, ManSat became the first commercial company in the world to have filing rights at ITU.

As a Crown dependency, Isle of Man registered companies wishing to undertake in any type of space activity must apply to UK Secretary of State via the British National Space Centre for a launch and operations licence. Here, Isle of Man companies are treated exactly the same as UK companies. To minimize any additional administrative burden on satellite operators or launch service providers wishing to operate from the Isle of Man, the Isle of Man Government has appointed a commercial due diligence and risk management service provider to manage space activity licensing processes as they directly affect the Isle of Man.

Contingent Liability

Section 10 of the Outer Space Act requires all persons carrying on space activities to indemnify the UK Government against any claim for loss or damage, if such person is one to whom the Act applies. Currently, the BNSC requires the licensee to provide evidence of third party liability insurance of £100 million and for the UK Government to be an additional insured to the insurance policy. The insurance policy should cover both the launch and the in-orbit phase for the operational lifetime of the satellite. With respect to this issue of liability, the Isle of Man and the UK have reached a sharing arrangement with respect to Isle of Man space activity and the UK's contingent liabilities. The process of this unique agreement is indicated in Figure 1 below.

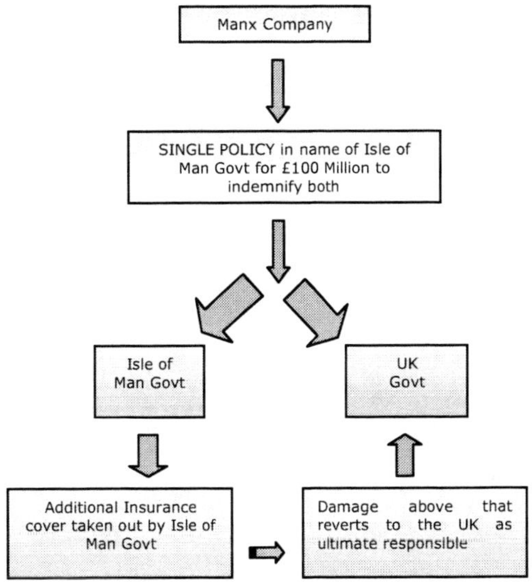

Figure 1: Isle of Man and UK Contingent Liability Arrangement

The effect of this arrangement is that ultimate responsibility and liability for damage caused by space activities of Isle of Man registered companies rest on the UK Government, but the Isle of Man has made adequate provision to ensure that the UK is adequately indemnified in the event of any damage.

New Enabling Law

Despite adequate agreement with the BNSC, the Isle of Man seeks increased parity with other jurisdictions with respect to the ability to grant licenses, which currently is decided by the UK Secretary of State. To that end, the Isle of Man has requested a new Order-in-Council from the UK Government to effect the Isle of Man Space and Telecommunications Bill 2007. The Bill aims to vest regulatory powers for space activities for the Isle of Man on the Communications Commission.

Under the Bill, the Communications Commission may, subject to the provisions of the Act, and after consultation with the Secretary of State, regulate the following activities which are carried on in the Isle of Man or elsewhere:
(a) The launching or procuring of the launching of a space object;
(b) The operation of a space object;
(c) Any activity in outer space

Under Section 1, the Commission shall undertake the functions vested in it from time to time by the Outer Space Act 1986 as it has effect in the Island. The Commission shall also in exercising its functions under the Act, take such action as it considers appropriate to comply with the international obligations of the United Kingdom in the field of the exploration and use of outer space.

The effect of the Space and Telecommunications Bill is that the Isle of Man will be able to enact licensing procedures for its space activities and maintain its own space register, in any format as long as it complies with UK international obligations. Bearing in mind that the new order and enabling law simply seeks parity with other jurisdictions, the Isle of Man is essentially satisfied with its connection to the UK Outer space licensing system. Where it seeks to differentiate itself is through the registration system, whereby under the Manx process, a one stop approach is taken to all business aspects of space projects in a similar fashion to its corporate jet and super yacht registration.

5. REGISTRATION

There are two principal obligations under Art. 2 Registration Convention 1975:

- As a "Launching State" with respect to a specific space object, a party should register that space object in its own national registry; and
- As "State of Registry" it should provide the Secretary-General with specific information on the space object, for inclusion in United Nations Register

Loosely defined in Art. 1 of the Liability Convention, a space object "includes component parts of a space object as well as its launch vehicle and parts thereof". The main purposes of Registration of space objects are:

- To know what is in space and record national activity
- To give notice to other States to aid in avoidance of collision
- To aid in identifying the launching state for liability purposes under the Liability Convention 1972

The UK currently has 2 registers managed by the BNSC. As of August 2008;

1. The UK Registry of Space Objects - currently contains 37 objects and 7 decayed objects not in the register
2. A supplementary Register - where the UK was not a launching state or was a joint launching state. Currently contains 27 objects

The registers contain the following information about space objects as set out in Table 1 below.

Table 1: Contents of UK Space Objects Register

Name of Space Object:	Perigee:
Owner/Operator:	Orbital position:
Date of launch:	General function:
Location of launch:	Date notified to United Nations:
Designation:	Disposed or decayed:
Catalogue Number:	Disposal Notification to UN:
Nodal period:	Date of acceptance onto UK Register of Space Objects:
Inclination:	Date of Licence issued under the Outer Space Act 1986:
Apogee:	Additional information:

There is no commercial application of this register as that is not its purpose. Under the new enabling law, the Isle of Man intends to establish the World's first commercial space registry. It is proposed therefore that the register would include not only the requirements as specified under the Registration Convention but would include a financial register of security in space assets registered in the Isle of Man. With the enactment of this financial register, it is believed that more business will flow to the Isle of Man and though it is an idea that will benefit the satellite communications industry - looking to the future - it will greatly benefit the New Space industry.

UNIDROIT Cape Town Convention

The idea of a space assets register is not a new one. Under the auspices of the International Institute for the Unification of Private Law (UNIDROIT), a space assets register has been proposed. The Cape Town Convention[12], read alongside the proposed Space Assets Protocol to the Convention,[13] aims to facilitate asset-based financing and leasing of space assets through the creation of an "international interest" recognized in all contracting States, and an electronic international register for the registration of these international interests.

The Space Assets Protocol is currently at drafting stage by the UNIDROIT Committee of governmental experts, in co-operation with the Space Working Group and a Steering Committee. At the last meeting of the group held in May 2008, the following outstanding issues were and continue to be deliberated[14]:

- Sphere of Application
- Default Remedies
- Identification of space assets for the purposes of registration
- Public service Issues
- How to increase the value of the future Space Protocol
- Issue of Insolvency
- Salvage interests

Following the May 2008 meeting, Space assets are currently defined as:

"any man-made uniquely identifiable satellite, satellite transponder, payload, space station, space vehicle, launch vehicle, reusable space capsule or other object capable of independent control, in or intended to be launched in or into space or used or intended to be used as a launch vehicle, including any such asset in course of manufacture or assembly, together with all modules and other installed, incorporated or attached accessories, parts and equipment and all data, manuals and records relating thereto''.[15]

Acknowledging that assets in space go beyond simply hardware and materials; included are also *related rights* and *debtors rights*. Related rights means any permit, licence, authorisation, concession or equivalent instrument that is granted or issued by, or pursuant to the authority of, a national or intergovernmental or other international body or authority, when acting in a regulatory capacity, to manufacture, launch, control, use or operate a space asset, *or* relating to the use of orbits positions ***or*** the transmission, emission or reception of electromagnetic signals to and from a space asset. Debtor's rights refers to all rights to payment or other performance due to a debtor by any person with respect to a space asset.[16]

The Convention also incorporates "proceeds" from the asset as part of the assets. These proceeds are limited to money or non money proceeds arising from the total or partial loss or physical destruction of the object or its total or partial confiscation, condemnation or requisition.[17]

The notion that lies at the heart of the Convention is the concept of an international interest. An international interest is a proprietary interest in mobile equipment and associated / related right, which is either:

1. granted under a security agreement;
2. vested in a conditional seller under a title reservation agreement;
3. or vested in a lessor under a leasing agreement.[18]

These transaction types are in line with the legal systems that distinguish between security and title type interest's e.g. as in the US for the former and Europe for the latter. Though these categories of international interests are dealt with slightly differently in the text, particularly with respect to default remedies, the importance of establishing a written agreement is that the focus is on the effect of the transaction and not on the form – that is to say that any type of non possessory interest in specific equipment is protected through the transaction document established specifically for the purpose of securing the performance of an obligation.[19]

Once created, the International interest must be preserved. To preserve the interest, creditors must have the ability to centrally register their security interest and search for competing interests. This notion of registration goes to the core of the Convention because publication regimes not only ensure protection of third parties and give creditors the opportunity to determine whether assets are charged by security, but also provides an objective means of resolving dispute. In essence, increased access to credit market can be achieved through easier evaluation of risk.

To that effect the Convention/Protocol provides for the establishment of an international registry for the registration, amendment, extension or discharge of: [20]

- International interests
- Prospective international interests – interest intended to be made in the future or dependent upon the occurrence of a certain event
- Sales and Prospective sales
- Registrable non consensual rights and interests – priority interests as determined by the State
- Assignment and prospective assignments of international interests
- Acquisition of international interests
- Notices of national interest
- Subordinations of above mentioned interests

To protect an interest in an asset from other creditors, the registry is a notice based electronic registration system that gives priority to the interest/creditor first to register. Clear rules establishing priority must be established to ensure that subsequent creditors and potential purchasers are aware of to their status in relation to others. As such, under the Convention,[21] a registered interest has priority over any interest subsequently registered and over an unregistered interest. This rule is applicable despite knowledge of competing interests. This clearly brings certainty to the regime, but the provisions may also cause a problem for commerce as they do not adequately provide for priority rules governing competition between unregistered interests.

The Convention provides for administrative and regulatory functions of a supervisory authority and registrar[22] as well as for registration requirements.[23]The registration is complete on entry of the required information into the database and following receipt by the Registry of the debtor's and creditor's consent to the registration.

Any person may request or make a search of the register upon which a registry search certificate will be issued stating: all registered information related to the asset with a statement indicating the date and time of registration, or that there is no information on the registry.[24] The certificate is prima facie proof of the fact included in it and the time and date of the registration.[25]

To build confidence in the registration system, the registrar shall be liable for compensatory damages for loss suffered directly resulting from any error, or omission or from a malfunction of the system.

Such a register would prove significant to space asset insurers, financiers and lenders, and the space industry, particularly "New Space". Though the Isle of Man could have been a suitable jurisdiction to house such an international register, it is focused on creating its *own* register for Isle of Man registered entities using positive aspects of the UNIDROIT Convention Framework.

6. CONCLUSIONS AND RECOMMENDATIONS

The choice of regulatory environment is crucial for the profitability and success of business. The Isle of Man has realized just in time that the commercial space industry is an industry in need of a home. A home where enterprise and innovation are encouraged; finance can flow and best practices are maintained! The Isle is well on its way to meeting its goal – to "create a legal environment that will encourage, nurture, and secure the growth of commercial space industries that choose to operate in the jurisdiction." [26] The Isle is doing so by continuously providing incentives to attract space companies, such as 0% tax rate for companies engaging in space and space related activity, easy access to orbital slots and a space aware and supportive Government .

Currently the UK Outer Space Act 1986 governs licensing and registration of space objects in the UK and for all dependencies that the law extends to. In that light all space activity for the Isle of Man is currently licensed by the UK. The UK maintains an active register of space objects, which contains Manx satellites but as part of the drive to diversify its economy, the latest strategy on the Isle is the establishment of a licensing and registry scheme to give the Isle of Man increased autonomy with respect to space activities. Whilst keeping in line with the international obligations of the UK, as an offshore jurisdiction, it can easily create an environment that takes into account the specific needs of the industry.

Under the UN Registration Convention, Registers of Space Objects, whether national or International, are primarily for record keeping purposes. With this objective in mind, existing registers in their current form play a limited role in the drive towards increased commercialization of Outer Space. The UNIDROIT Convention on International Interests in Mobile Equipment and the Space Assets Protocol on the other hand, aim to facilitate asset based finance for the space industry through the creation of international security interests in space assets and the establishment of an International registry for recording those security interests. Creditors can record their interests and establish priority vis-à-vis competing creditors within and outside the insolvency of the debtor. It is proposed that when in force, access to finance in the space industry will improve as creditors can greater ascertain their risk; will be better protected and hence will lead to increased flow of investment.

The Isle of Man Space Register would seek to emulate the positive aspects of the UNIDROIT Space Assets Register. Worthy of mention is that the establishment of an assets registry itself would not and can not establish any rights. It also can not guarantee that investors will suddenly start financing projects. Such is still conditional on ordinary investment principles such as proof of concept, evidence of viable market and excellent return on investment potential. The New Space industry particularly, will have to continue working on strengthening the many business plans that exist. But, what the register can guarantee is that it will help ease the minds of potential investors in that if a contest arose to have priority in an interest, they will be in good standing to claim on the asset if he is first to be file. The domestic law must therefore be prepared to accept the register as evidence of standing.

Ospina states that a lacuna in all current space registration processes is that "no one single national or international entity seems to have a centralized registry of space objects or activity. Nor is any entity involved in planning the use of or allocation of non geostationary orbits, where the vast majority of the privately owned and operated G Mobile Personal Communication services satellite systems are/will be located."[27] Alongside the establishment of a financial register for space assets, there is great scope for the Isle of Man registration system to be the most comprehensive registration system.

Suggestions for Isle of Man Register:

- The registry should contain not only space assets but also interests in assets
- A suitable definition of assets should be given to include component parts if they can be independently controlled
- Related rights must be accounted for
- internet notice based registry - recorded data should give notice to third parties that an interest has been vested in a specific space asset
- Register should accept notification of interests in objects still in manufacture or in transit
- 24 hour accessibility
- Data Input should be possible from private and public bodies and registered individuals
- Decisions should be made on technical capacity to identify, file and search, and whether the search results meet credit industry needs
- The checking of the registration applications, the registrations themselves and the responses to searches should be processed automatically
- Clear publication rules must be developed, especially priority rules governing competition between unregistered interests
- As much as possible must be registered prior to launch
- Parameters that are not yet known must be followed up to ensure timely inclusion as soon as available
- The registry should be all encompassing including information about all data and information applicable to the space asset (subject to privacy and intellectual property rights)

It has been proposed[28] that a comprehensive register can be separated into 4 general categories:

1. Technical data
2. Intended use/services to be provided
3. Financial/economic data
4. Contact information

References

1 Treaty on Principles Governing the Activities of States in the Exploration and Use of Outer Space, Including the Moon and Other Celestial Bodies, Jan. 27, 1967, 610 U.N.T.S. 205 [hereinafter Outer Space Treaty (OST)].

2 Agreement on the Rescue of Astronauts, the Return of Astronauts and the Return of Objects Launched Into Outer Space, Apr. 22, 1968, U.N.T.S. 119.

3 Convention on International Liability for Damage Caused by Space Objects, Mar. 29, 1972, 961 U.N.T.S. 187 [hereinafter Liability Convention].

4 Convention on Registration of Objects Launched into Outer Space, Nov. 12, 1974, 1023 U.N.T.S. 15 [hereinafter Registration Convention].

5 Agreement Governing the Activities of States on the Moon and Other Celestial Bodies, Dec. 5, 1979, U.N. Doc. A 34 46, 1979. [hereinafter Moon Agreement]

6 Isle of Man Government, " International Identity of the Isle of Man Recognized," 2007. Available: http://gov.im/lib/news/cso/internationalide.xml, Last accessed 20 August 2008.

7 ECOFIN Council, "Conclusions of the ECOFIN Council Meeting on 1 December 1997 Concerning Taxation Policy", Official Journal of the European Union, 1997, No. 98/C 2/01, C002, 6/1/98, p.0001.

8 OECD, Towards Global Tax Co-operation: Progress in Identifying and Eliminating Harmful Tax Practices, OECD Publications, Paris, 2000.

9 For an overview of the differences between a 1931 company and a 2006 company see West Corporation Limited, "Flexible Offshore Companies Isle of Man Companies Act 2006", 2007. Available: http://www.westiom.com/plugins/p13_download_manager/getfile.php?categoryid=37&p13_sectionid=1&p13_fileid=1&p13_versionid=1, Last Accessed 01 September 2008.

10 Andrews, W., "A Mighty Stone for David's Sling: The International Space Company". *Regent Journal of International Law*. 2003, Volume 1. (Electronic Version)

11 See Stott, C. et al, "Isle of Man: Opportunities for Space and Satellite Business", *Satellite Finance*, May 18 2005, Issue 80, pp 44 -46.

12 The Convention on International Interests in Mobile Equipment , Nov. 16, 2001, UN Doc. No. A/AC.105/C.2/2002/CRP.3

13 UNIDROIT, Preliminary Draft Protocol On Matters Specific To Space Assets, 2004, UNIDROIT 2004 Study LXXIIJ - Doc. 13 rev.

14 UNIDROIT, Summary Report, 2003, UNIDROIT 2003 Study LXXIIJ – Doc. 14.

15 UNIDROIT, "Proposal For An Alternative Text Of The Preliminary Draft Protocol To The Convention On International Interests In Mobile Equipment On Matters Specific To Space Assets", 2008 , UNIDROIT 2008 Study LXXIIJ – Doc 14 Appendix III.

16 Ibid.

17 Convention on International Interests in Mobile Equipment 2001, Article 29(6) .

18 Convention on International Interests in Mobile Equipment 2001, Article 2.

19 Hazan, N., "The UNIDROIT Preliminary Draft Protocol on Matters Specific to Space Assets," *Annals of Air and Space Law*, Volume 28, pp. 219-255.

20 Convention on International Interests in Mobile Equipment 2001, Article 16.

21 Convention on International Interests in Mobile Equipment 2001, Article 29.

22 Convention on International Interests in Mobile Equipment 2001, Article 17.

23 Convention on International Interests in Mobile Equipment 2001, Chapter V.

24 Convention on International Interests in Mobile Equipment 2001, Article 22.

25 Convention on International Interests in Mobile Equipment 2001, Article 24.

26 Stott, C., "Making Space Profitable: the Role of Law and Policy", *Proceedings of the Forty-third Colloquium on the Law of Outer Space*, Rio de Janeiro, 2 October - 6 October 2000.

27 Ospina, S., "Revisiting the Registration Convention: A Proposal to Meet the Need to Know "what is up there", *Proceedings of the Forty-third Colloquium on the Law of Outer Space*, Rio de Janeiro, 2 October - 6 October 2000, vol. 43, pg. 199-209.

28 ibid

SESSION REPORT
"PRIVATE INTERNATIONAL LAW REGARDING SPACE ACTIVITIES"

Chairmen: Lesley Jane Smith
Francis Lyall

Rapporteur: C. Doldirina

Session 1 was held on September 30, 2008 and opened the IISL activities within the 51st Colloquium on the Law of Outer Space. The chairs of the session were Professors Francis Lyall and Lesley Jane Smith, with Ms. Catherine Doldirina as rapporteur.

Ten out of thirteen announced papers were presented during the session. One of the submitted papers could not be presented as the author Ms. Mariam Yuzbashyan ("Regarding Formation of the International Space Private Law) was unable to attend the Congress because of visa issues. Two other papers, one by Ms. Axelle Cartier ("The Need for Commercial Risk Reduction in Private Manned Activities in Outer Space: A Proposal for a Regulatory Evolution") and one by Mr. Muhamed Mustaque ("Interoperability of GNSS, Legal Issues and Implications") were withdrawn.

The first paper was presented by Ms. Diane Howard with the title "Achieving a Level Playing Field in Public Private Partnerships: Can Sovereign Immunity Upset the Balance?". It addressed the international and domestic legal responsibilities of both public and private parties in jointly administered space projects with the special emphasis on the doctrine of sovereign immunity in both the United States and the European Union. Professor Smith was the next speaker with a joint paper with Mr. Christian Bank, "Capability and Fair Return in European and International Space Cooperation: the Example of Columbus". It described the challenges of using barter agreements in the context of space cooperation projects (like the ISS) for a crew transport vehicle from an industrial point of view, such as the high frequency of deliveries, liability and risk issues, and the fair share in the utilization of a system from a commercial perspective.

The paper "Space tourism: private law implications" by Dr. Peter Haanappel was presented by Dr. Peter van Fenema. It highlighted the existing areas of law that influence the development of space tourism, alongside discussing whether new private law mechanisms need to be developed. This was followed by Dr. Hanneke van Traa-Engelman's paper "The Need for a Uniform Law System Protecting Intellectual Property Rights in Outer Space" on the ongoing significance of intellectual property in space contracts and the feasibility of the establishment of an international intellectual property framework enforced by a single body. Professor Smith's second paper entitled "Jurisdiction and Applicable Law in cases of Damage from Space – the Advent of the Most Suitable Choice in Europe - Rome II", co-authored by Ms. Catherine Doldirina, covered the European solution to the issue of the conflict of laws for non-contractual liability for space activities.

Professor Souichirou Kozuka presented his paper "Liability from the Launch of a Space Object as a Matter of Private Law" that dealt with the liability issues related to the launch of a space object in the light of domestic private law rules. The next presentation was made by Dr. Fabio Tronchetti who addressed “The future of the UNIDROIT Draft Space Protocol: legal proposals to encourage the participation of private financiers in space activities”. His paper analysed the limits of the Draft Protocol illustrated its ability to support participation of financiers in current and future space activities. This topic was further addressed in the presentation by Dr. Bernhard Schmidt-Tedd entitled “The UNIDROIT Draft Of A Space Assets Protocol – A Civil Law Instrument Under A Public Framework”. This paper, co-authored by Ms. Isabelle Arnold, discussed the difficulties stemming from the interdependency of the civil law financing instruments and the international regulation of space activities, both merged within the Protocol.

Mr. Michael Mineiro presented his paper “CISG and the Final Frontier: Contracting for the Sale of Goods and the Risk of Loss in Outer Space” elaborating on issues of international sales within space contracts. The session was concluded by a presentation made on behalf of Professor Edythe Weeks “Which Laws Will Govern Private Space Travel, Commercial Space Mining and Commercial Space Settlements?”. The paper addressed the issue of applicability of public space law to private space activities like spaceflight, space hotels and spaceports.
The session was held for the first time last year, and attracted considerable attention both from potential participants (the chairpersons received 18 abstracts to choose from), as well as from the International Astronautical Congress attendees. The session was visited in all by around 40 people.

SESSION REPORT

THE 40TH ANNIVERSARY OF THE RESCUE AGREEMENT: LOOKING AHEAD/ WEAPONISATION OF OUTER SPACE IN THE LIGHT OF ARTICLE 4 OF THE OUTER SPACE TREATY, PART A

Chairmen: Prof. Sergio Marchisio
Prof. Vladimir Kopal

Rapporteur: Dr. Fabio Tronchetti

In their paper on **Controlling ballistic missile defense and anti-satellite missiles: should we look elsewhere or rather amend Art. IV OST?** Mr. A. Agarwal, K. Vuppala and R. Mukherjee (India) analyzed the issue of weaponisation of outer space by pointing out that international cooperation and confidence building represent the means for securing outer space from military use and from becoming an area of conflicts for States. This process of international cooperation must lead to the development of an international legal regime regulating military applications in the space environment, in particular the use and impact of ballistic missile defense and anti-satellite missiles. First of all, the need for such a legal regime stems from the fact that the interpretations on Article IV of the Outer Space Treaty, dealing with military uses of outer space, vary and it is rather unclear whether its provisions applies to the utilization of anti ballistic missiles and ballistic missile defense. Secondly, the use of these categories of missiles has been regulated through specific arm treaties, mainly on a bilateral rather than multilateral basis. Therefore, a multilateral effort aimed at clarifying what military activities are allowed in outer space is required so as to keep the space environment peaceful.

Prof. S. Aoki (Japan) presented a paper entitled: "**Space traffic management for the prevention of weaponization of outer space**". Looking at the increasing number of military operations in space and at the difficulties of arriving at an agreed international instrument dealing with military uses of outer space, Prof. Aoki suggests an alternative strategy to deal with this issue, namely the development of space safety measures to be crystallized into a soft law instrument.
The paper starts with a detailed analysis of the work of the Conference on Disarmament on the issue of militarization of space, by showing the failure of the Conference to reach any widely accepted agreement or definition. The lesson to be learn from it is that when dealing with weaponization of outer space a definition of "space weapons" shall be avoided if consensus is seriously sought out. The paper goes on with the analysis of the ASAT weapons and of the international efforts made to regulate their use.
Afterwards, Prof. Aoki deals with the concept of Space Traffic Management (STM) by focusing on the activities of the IAA in such respect. The paper focuses on the STM model in 2020 prepared by the IAA and suggests to use it as a model to regulate the issue of weaponization of outer space. Hence, space safety measures should be developed to regulate that issue. The paper concludes by stressing that these safety measures would be successful only upon the fulfillment of 5 conditions: 1) emphasis is placed on safety rather than security; 2) definition of "space weapons" is avoided; 3) the rules are contained in a soft law instrument; 4) such soft law instrument is, at first, implemented by the

like-minded States; 5) such soft law instrument is supported by the major space-faring nations.

The paper of Prof. C. Christol (USA) entitled: "**The joinder of telecommunications and globalization**" was presented by Prof. J. Galloway (USA). In his paper Prof. Christol discusses about the characteristics and relations between telecommunication and globalization. He argues that telecommunication was the force leading to the existence of globalization. Indeed, without the advanced communication process, the gathering of data through remote sensing and the transmission via satellite the evolvement of multifaceted areas of discourse that contributed to the emergence of globalization would have not happened. The paper shows the importance of instant communication for national security and for contributing to harmonious relations among States which, as a result, contribute to the success of globalization. Later, the paper deals with the negative impact of global terrorism on globalization and on the fact that terrorists make use of space technologies to pursue their criminal acts. The paper also analyzes the efforts of International Organizations in combating and preventing terrorism. Prof. Christol concludes by stressing that terrorists cannot be prevented form seeking access to advanced communication equipments. However, one response could be that terrorists be brought under such detailed scrutiny that they would be denied opportunities to make use of all of the hardware that is essential for their operations.

The paper: "**A new indemnification policy spacecraft that rescue astronauts in need**" was presented by K. Corner (USA). The paper focuses on the issue of rescuing spacecraft and astronauts in need and argues the need for the setting up of detailed rules dealing with such an issue. After pointing out the differences between today's space environment and the one existing at the time the Rescue Agreement was drafted, Mr. Corner touches upon the issue of the definition of astronauts by arguing that all humans on board a spacecraft must be considered astronauts with regard to their need of being rescued while in danger. The paper goes on with the analysis of the Rescue Agreement's provisions on the rescuing of astronauts. It is argued that the Agreement provides only a skeletal framework and that a more sophisticated system is needed. In order to set up such system a useful example is provided by maritime law where the duty to help sailors in distress is well established.

Mr. Corner argues that these new rules on rescuing astronauts are very likely to be developed at national level first, although they should be acceptable to space-faring nations. These rules should establish clear procedures and standards to be followed in case of emergency. The author also stresses the importance of an insurance system to compensate companies for any loss resulting from a rescue, due to the risky nature of rescuing in space.

Mr. J. de Faraminan Gilbert and Ms. C. Munoz Rodriguez (Spain) presented the paper entitled: "**The return of space objects launched into outer space some legal questions**".

The paper analyses in detailed the provisions of the Rescue Agreements. The paper begins with the study of the drafting history of the agreement containing the discussions held within the COPUOS. Then, the analysis switches to the provisions of the Rescue Agreement by paying particular attention to those on notification of accidents, search and rescue in a contracting State's territory, landing in areas not under the jurisdiction of any State, returning of astronauts, returning of space objects, elimination of possible danger or harm and payment of expenses. The paper concludes with listing some of the problems and danger being faced by astronauts and space object nowadays, such as space debris.

The paper: "**Anti satellite missile testing: a challenge to Article IV of the Outer Space Treaty**" was presented by Mr. K. Mohan, Mr. A. Agarwal and Mr. Bhat (India).
It analyzes the relation between anti satellite weapons and Article IV of the Outer Space Treaty and pointes out some of the loophole of the latter. After describing the different typologies of anti satellite weapons and their most recent uses by States, the paper turns to the study of the Outer Space Treaty's provisions, particularly those of Articles III, IV and IX due to their relevance for the issue of anti satellite weapons. Specific attention is paid to the provisions of Article IV which leave room for doubts regarding whether or not anti satellite missile tests are prohibited, what is the meaning of the terms "peaceful purposes" and "military activities".
The authors conclude their paper arguing that a new treaty dealing with the use of anti satellite weapons is far from being a feasible option. They suggest that States parties to the Outer Space Treaty should take as position on this matter as well as the United Nations General Assembly.

Ms. Z. O'Brien (Ireland) presented a paper entitled: "**The Rescue Agreement and private space carriers**".
Looking at the current outer space environment, where private enterpries have become a major actor, the paper addresses whether the duties set forth in the Rescue Agreement should be extended to bind commercialo entitites engaged in space carriage.
The paper considers the Rescue Agreement a great achieviement in space law because of its duties upon contracting States to protect astronauts. However, the lack of clarity of some of its provisions has meant that the duties of private parties have not been clarified and certain situations involving private party rescuerses may not come within the scope of the Agreement. The author argues that for primarily humanitarian reasons, it is preferable to impose duties to rescue on the commander of spacecraft akin to those on the mastes of vessels wiht some modifications suitable to space law. Information-sharing obligations, requiring commanders to inform their launching authority of any distress call recieved should be imposed in addition to more substantive duties to rescue the lives of those aboard. No distinction should be made between personnel and non-personnel and the burden for reimbursing for the rescue should fall either on the rescuee or the launching authorithy of the rescue. Such duty to rescue should be subject to requirements of practicability. The author concludes recommending a new multilateral convention requiring States to ensure that commander of spacecraft comply with the above duties.

The paper: "**Verification mechanism for peaceful uses of outer space**" was presented by Ms. Y. Takaya-Umehara (France).
The paper analyzes the meaning of the term "peaceful purposes" of outer space and suggests to ensure security in space by means of a verification mechanism.
The paper begins with the study of the term "peaceful purposes" as defined in the legal regime regulating activities in Antarctica, the high seas, the seabed and the ocean floor. Each of these regimes has different definition and mechanism for ensuring peace and security in the respective areas. Outer space is not an exception in such respect. The paper argues that due to the impossibility of arriving at an agreed definition of "peaceful uses", the legal regime in outer space fails to establish an effective verification mechanism. Considering the tendency of States to cooperate in verification mechanism through international organization, the author suggests that it is suitable for space

security to establish an organization responsible for verification with the aim of enhancing transparency in outer space.

E8. 51st INTERNATIONAL INSTITUTE OF SPACE LAW (IISL)

E8.4. Legal Aspects of Natural Near Earth Objects (NEOs) / Other Legal Matters, Part A

Chairmen: Richard J. Tremayne-Simith, Maureen Williams
Rapporteur: Yuri Takaya-Umehara

IAC-08-E8.4.1

International Responsibility vis-à-vis natural near-Earth objects (NEOs) and their possible implications

Professor Dr. Maureen Williams / University of BuenosAires / Conicet / Chair of the ILA Space Law Committee (HQ London)

Prof. Maureen Williams examined the legal aspects of NEOs, clarifying some similarities between the guidelines for NEOs and space debris under space law and customary international law. Her analysis indicates that effective dispute settlement mechanisms are essential and the ILA International Instrument on Space Debris appears applicable to the NEOs. She pointed out the questions on the decision-making process, technical capability to avoid space debris, and political will, which is indispensable for further consideration.

IAC-08-E8.4.2.

Natural Near Earth Objects and the International Law of Outer Space

Dr. Luis F. Castillo Argañarás / National Council of Scientific and Technical Research of Argentina (CONICET) / Universidad Argentina de la Empresa – Instituto de Ciencias Jurídicas y Sociales (UADE)

Dr. Luis F. Castillo Argañarás examined the existing law on NEOs hazard that is a global issue to be effectively addressed only through international cooperation and coordination. He emphasized the importance of international cooperation particularly between States who have space capability, concluding that a suitable legal framework on NEOs exists already but the public awareness for NEO should be enhanced.

IAC–08–E8.4.3.

Defining Subject Matter Under Space Law: Near Earth Objects Versus Space Objects

Prof. Frans G. von der Dunk / University of Nebraska, College of Law, Space and Telecommunications Law Program

Prof. Frans G. von der Dunk analyzed the legal terms "outer space objects" and "Near Earth Objects" through the draft treaty elaborated by Russia and China on de-wepaonization of space. Clarifying the definition of the terms "objects," "objects in outer space" and "space objects" being linked with man-made artifacts, but not with NEO of any sort, he concluded that, apart from space weapon which is designed specifically to attack, "objects" are not necessarily linked with weapons.

IAC-08-E8.4.4

Pella Vilya: Near Earth Objects – Planetary Defence Through the Regulation of Resource Utilisation
Dr. Gérardine Meishan Goh / Deutsches Zentrum für Luft- und Raumfahrt e.V. (DLR), Bonn, Germany, Institute of Air and Space Law, University of Cologne, Germany

Drawing from recent analogies and examples from international environmental law, Dr. Gérardine Meishan Goh proposed a legal framework for the regulation of NEOs' resource utilisation, including a mechanism to ensure the political will and economic investment necessary for technological advances in planetary defense.

IAC-08-E8.4.7
Japanese Space Fundamental Law
Prof. Yasuaki Hashimoto / National Institute for Defense Studies, Tokyo, Japan

Prof. Yasuaki Hashimoto introduced Japanese Space Fundamental Law passing in the Diet on 21 May 2008, adding the new perspective of national security to space activities in Japan. He explained the shift in interpreting the term "peaceful" from "non-military" to "non-aggressive" in Japan, allowing military uses of outer space to the extent lawful under the Japanese Constitution.

IAC-08-E8.4.11
The Moon Treaty in the 21st Century
Ms. Antonella Bini / United Nations Office for Outer Space Affairs (UNOOSA), Vienna, Austria

Ms. Antonella Bini analyzed the Moon Agreement from the perspectives of commercialization and privatization of space activities. She addressed questions raised in the LSC of the UN COPUOS in 2008 and emphasized the importance of feasibility and political will in the evaluation of the Moon Agreement.

IAC-08-E8.4.12
Outer Space Traffic Management
Prof. Paul B Larsen / Washington DC, United States, OUTER SPACE TRAFFIC MANAGEMENT

Prof. Paul B Larsen explained the purposes of outer space traffic management as: to avoid intentional and unintentional collisions; to make outer space traffic safe; and to assure the freedom of access and uses of outer space under space law. He emphasized the importance of having better situational awareness and of enhancing transparency in outer space by the exchange of information among operators of space vehicles under international cooperation, including the co-operability of GNSS.

IAC-08-E8.4.14
The Importance of the Rule of Law for Space Activities

Prof. Stephan Hobe

Prof. Stephan Hobe examined the rule of law under space law by focusing on "soft-law." Considering UN General Assembly resolutions providing non-binding but binding character for the privatization of space activities and mitigation of space debris, he concluded that the UN COPUOS plays an important role in observing the rule of law which might contribute to the development of modern international law.

IAC-08-E8.4.15
Space Law Students Participation in Real Engineering Related Projects and Research
Mr. Amalio Monzón, Ms. Andrés Russu, Mr. Héctor Salvador / Laboratory for Space and Microgravity Research, Spain

Considering the importance of education and research in the field of science, technology and law constituting a key factor in the development of space law, Mr. Amalio Monzón, Ms. Andrés Russu, Mr. Héctor introduced Salvador Laboratory for Space and Microgravity Research (LEEM), a Spanish non-profit space students and young professionals association.

IAC-08-E8.4.16
Decision Program on Asteroid Threat Mitigation
Dr. Russell L. Schweickart / Chairman, Committee on Near-Earth Objects, Association of Space Explorers, Sonoma, California

Dr. Russell L. Schweickart introduced the Association of Space Explorers Committee on Near-Earth Objects (NEOs) and the Panel on Asteroid Threat Mitigation, preparing a decision program to aid the international community in organizing a coordinated response to asteroid impact threats, which would be under consideration of the UN COPUOS in 2009.

SESSION REPORT
"OTHER LEGAL MATTERS, PART B"

Chairmen: Joanne Gabrynowycz
Joanne Wheeler

Rapporteur: Sylvia Ospina

A total of 11 papers were presented, ranging over a variety of subjects and regions. Also included in the presentations were some papers that were moved from other sessions, to accommodate the presenter.

The over-all quality of the papers was very high, and several were outstanding: "Space And The Complexity Of European Rules And Policies: The Common Projects Galileo And GMES – Precedence For A New European Legal Approach?", by Dr. Annette Froehlich, and "Network Of European Regions Using Space Technologies: An Update On The NEREUS Constitution", co-authored by Dr. Marianna Morelli and Dr. Campostrini. Both of them addressed issues specific to the European Union / European space sector, and presented in an understandable form the complexity of issues faced by the European Union Member States.

Dr. Froehlich describes the increasing role of the European Union (EU) in space activities, and in space policy, and points out that the legal rules of the EU and ESA are not quite compatible at this time. Another issue that is likely to be contentious in future is the question of financing and ESA's policy of fair return. It is likely to cause some difficulties, if not at present, in future.

The authors of the paper on NEREUS provide much valuable information on this ambitious network, which involves or will involve European regional institutions and enterprises of all kinds, big and small, academic and business-oriented. (The definition of "region" provided by the authors seems to be all-encompassing.) The author's intent was to make a short presentation on "a new form of cooperation" that seems to be open to any kind of institution in Europe. Their objective was well met.

(It is worth noting that Froehlich's and the Morelli-Campostrini papers were selected for publication in "Acta Astronautica.")

"Fifty years of Earth observation from Space and Space Law", by José Monserrat Filho, and "Alcântara Cyclone Space: A New Joint Venture in the Space Market", by Mr. Alvaro Fabricio Dos Santos provided valuable information on the Brazilian space program, in particular Brazil's involvement in earth observation / remote sensing activities. (Dr. Dos Santos' paper was summarized by Dr. Monserrat.) The pairing of these two presentations provided a good synopsis of some of the aspects of Brazilian space activities.

"A Support of Imagination to Become Reality: Relation of Patent Law and

Space Law", by Ms. Fruzsina Tari provided interesting and valuable information on patent law and its relation to space law, topics that are not usually spoken about together.

Space activities on the other side of the globe, in particular in China, were also addressed by Ms. Xiaodan Wu, who gave an account on the Chinese licensing regime of space activities.

A paper on the "Legal framework in support of commercialization of outer space: the case of the Isle of Man", was presented by Ms. Timiebi Aganaba, a graduate student at the International Space University in Strasbourg, France. Its content aroused a fair amount of discussion, as it seemed to be more of a "sales pitch" for the Isle of Man than an exposé of legal issues.

In brief, the quality of this session's papers was very good, as was their presentation. Unfortunately, there was little time left at the end of the session for any in-depth discussion of the many issues that were written about. Interestingly, when the session began, there were about 30 people present; at the end of the session, more than 40 were there!

Report of the IISL Plenary 2008: "Real Space, Real Law, Real Progress"

By Scott Hatton

Moderator: Robert Richards (CEO, Odyssey Moon)

Panelists:

- Walter Peeters (Dean of ISU)
- Clayton Mowry (President, Arianespace USA)
- Christian Sallaberger (MDA, Canada)
- Axelle Cartier (General Counsel, Excalibur Almaz Europe)
- Tom Shelly (Space Adventures USA)
- Martin Stanford (Secretary General UNIDROIT)

New economic activities have entered the space arena and their legal and regulatory needs could be seen to be in conflict with those of the existing space agencies and satellite markets.

These new entrants pose both challenges and opportunities to the space law community and in turn the space industry in general. Their issues need to be understood to encourage this new growth, but also to negate any harmful impacts they may or may not have.

This was a discussion panel, focusing on International Space Law and its role in the modern world of commercial space exploitation. As one panel member put it "Where explorers go, lawyers follow." Bob Richards started the session off by asking each of the panelists to comment on what their occupation is and how space law impacts on it, how stable it is and how they see it will develop.

Walter Peeters works in education, and he finds it difficult to teach the next generation of space leaders about the Outer Space Treaty, knowing that it may not survive in its current form into the future. He wonders what will a new treaty put in place? For example, the treaty attaches liability to the concept of a 'launching state', but with the rapid commercialisation of space, will this make any sense in the future?

Clayton Mowry's role is head of a company that carries out a large number of launches of satellites and spacecraft of all varieties, and often has to deal with a large amount of regulation. However, it was often regulation closer to home that was more difficult to deal with (for example ITAR, the International Traffic in Arms Regulations). This was a recurring comment made during the session, as many of the panelists found that local or national laws, rather than the international treaty, were most limiting in the commercialisation of space.

These sentiments were echoed by Christian Sallaberger, who spoke of the problems of having an international company work on projects restricted by such legislation. It made transfer of technology or personnel within the company, but crossing national boundaries, legally problematic.

Axelle Cartier worked for a private space company registered in the Isle of Man, which has special legal status, and this has alleviated many of these problems.

Governments can do this to help such commercial interests to flourish.

Tom Shelly runs Space Adventures, a company responsible for sending six private individuals to the ISS. Originally the ISS regulations had no provision for private citizens to visit. The help of the Russian government helped get past this, and now the USA are also behind them.

Martin Stanford works for the International Institute for the Unification of Private Law (UNIDROIT). His organisation was involved in not trying to improve the legal situation of space law, but helping people through the rules.

It seems then, from what was being said, that the major barrier to commercial exploitation of space was not the Outer Space Treaty. Many companies had managed to make money despite the treaty and found the laws closer to home to be more of a problem.

However, as the expansion of Earth's economic sphere expands to include the Moon, this may change. There is already an agreement on the peaceful use of outer space, but will governments respect the rights of private companies, and vice versa? Should the sites of the Moon landings be considered hallowed ground, and action be taken to stop commercial companies disturbing them? (Selling pieces of Apollo 11 on eBay was one example.)

The panel stated that the commercialisation of space is demonstrating that the Outer Space Treaty established several good fundamental principles that remain valid today, but could not recognise or anticipate the extent of the involvement of private commercial players in space. They spoke of the probablity that common law and case law will come into play as private players increase their activity in space and on the Moon. The panel concluded that the Outer Space Treaty has important and valid principles to build upon. (for example ITAR, the International Traffic in Arms Regulations). This was a recurring comment made during the session, as many of the panelists found that local or national laws, rather than the international treaty, were most limiting in the commercialisation of space.

These sentiments were echoed by Christian Sallaberger, who spoke of the problems of having an international company work on projects restricted by such legislation. It made transfer of technology or personnel within the company, but crossing national boundaries, legally problematic.

Axelle Cartier worked for a private space company registered in the Isle of Man, which has special legal status, and this has alleviated many of these problems. Governments can do this to help such commercial interests to flourish.

Tom Shelly runs Space Adventures, a company responsible for sending six private individuals to the ISS. Originally the ISS regulations had no provision for private citizens to visit. The help of the Russian government helped get past this, and now the USA are also behind them.

Martin Stanford works for the International Institute for the Unification of Private Law (UNIDROIT). His organisation was involved in not trying to improve the legal situation of space law, but helping people through the rules.

It seems then, from what was being said, that the major barrier to commercial exploitation of space was not the Outer Space Treaty. Many companies had managed to make money despite the treaty and found the laws closer to home to be more of a problem.

However, as the expansion of Earth's economic sphere expands to include the Moon, this may change. There is already an agreement on the peaceful use of outer space, but will governments respect the rights of private companies, and vice versa? Should the sites of the Moon landings be considered hallowed ground, and action be taken to stop commercial companies disturbing them? (Selling pieces of Apollo 11 on eBay was one example.)

The panel stated that the commercialisation of space is demonstrating that the Outer Space Treaty established several good fundamental principles that remain valid today, but could not recognise or anticipate the extent of the involvement of private commercial players in space. They spoke of the probablity that common law and case law will come into play as private players increase their activity in space and on the Moon. The panel concluded that the Outer Space Treaty has important and valid principles to build upon.

24th IAA–IISL Scientific-Legal Roundtable

23rd IAA/IISL Scientific-Legal Roundtable
Glasgow, UK, October 2nd, 2008

"Paper Satellites" - Problems of Policy, Regulation and Economics

Chairpersons: Luboš Perek (Czech Republic), Kai-Uwe Schrogl (Austria)
Rapporteur: Nicola Rohner-Willsch (Germany)

At the 2008 joint IAA/IISL Roundtable, the issue "Paper Satellites" - Problems of Policy, Regulation and Economics was discussed among lawyers, scientists and engineers. The Roundtable was opened by the IISL President Mrs. Tanja Masson-Zwaan, the IAA Secretary General Dr. Jean-Michel Contant and the Chairman of the IAA/IISL Scientific-Legal Liaison Committee Dr. Kai-Uwe Schrogl.

The term "paper satellites" refers to proposals for space communication networks submitted to the International Telecommunication Union but not yet notified in the Master International Frequency Register. The lively debate about this topic exceeded the topic of paper satellites and touched on the general problem of management of outer space, in particular in the geostationary satellite orbit, GEO.

According to the Convention on Registration of Objects Launched into Outer Space, the launching of all space objects, including those launched into the GEO have to be announced to the United Nations. However, about 15% of objects have not been registered. Improvement of the situation can be expected after the recent UN General Assembly Resolution 62/101 on enhancing the practice of States and international intergovernmental organizations in registering space objects becomes widely used. In order to provide quick and unlimited access to information provided in launching announcements of Member States, The Office of Outer Space Affairs has developed an electronic on-line searchable Index to the Register containing substantial information also on objects not announced to the UN. The Index assists in locating information contained in governmental announcements and it provides outside complementary information.

The European Space Operation Centre (ESOC) in Darmstadt provides a yearly status report on satellites in and near the GEO. It is based on observations provided by a worldwide network of observatories and listed in the US Space Surveillance Network Catalogue. There are 934 objects in the list, out of which 318 objects are controlled in longitude, 148 objects are in libration orbits, and 462 are drifting above the GEO ring. The internationally accepted recommendation to re-orbit retired objects into disposal orbits was not followed universally. Only 41% were re-orbited properly, 33% were left in orbits too close to the GEO and 26% were abandoned in the GEO. The failure to re-orbit properly will lead to a steady increase in the collision risk of operational spacecraft with inactive objects.

International management of the use of the radio-frequency spectrum and orbits is entrusted to the International Telecommunication Union (ITU). A legal regime, codified in the ITU Constitution and Convention, complemented by the Radio Regulations, contain main principles and regulations governing the registration of space network frequency assignments. Administrations are submitting every year an extremely high number of space networks, including speculative filings, so called "paper satellites". For a single real network, there could be ten or more filings for different orbital positions. Independent information on the real use of the orbit/spectrum resource shows some divergence from information submitted by administrations to the ITU.

A Comparative Table of notified space networks (i. e. transmission of information) on one hand and of satellites (i.e. solid space vehicles) on the other hand, at or near a specific orbital position was set up in order to find out the real state of affairs in the GEO. There are about twice as many notified space networks (about 880) than there are active satellites (about 370). There are 165 networks at 83 different orbital positions with no satellites close enough. It means that at least 19% of notified networks do not transmit and that the GEO spectrum/orbit resource is not as much overcrowded as the lists of notified space networks suggest. The opposite case, satellites with no corresponding notified networks, appears in the Comparative Table as well. There are 23 such satellites. If radio stations on these satellites are transmitting, and it is likely that they do, it is contrary to the ITU rules.

The current situation in the orbit/frequency resource is affected by the trend of countries to use their own satellites for their domestic satellite telecommunication system. The GEO is going to be overloaded. New opportunities arise in using slots not used by "veterans of the space club". Another novelty in using the GEO resource is to place a network in service before the completion of the coordinating process, provided that no harmful interference is caused. It is also possible to use parts of the frequency bands which are not used by the original operator.

The problem of developing countries in establishing new national satellites was raised in the discussion. It is difficult and time consuming to find frequencies and orbital slots which do not interfere with existing space networks, in particular in heavily used geographical longitudes. If requirements of a country allow it, the possibility exists to rent a transponder on an existing commercial satellite.

Authors of Papers	**Titles of Papers**
Petr Lála	Registration of Geostationary Satellites by the United Nations
Heiner Klinkrad	Status of the Geosynchronous Orbit
Yvon Henri	Long-Term Efficiency of the Space Regulatory Framework: "Paper Satellites" issue
Luboš Perek	Paper Satellites and Space Networks
Victor S. Veshchunov	Statement on behalf of Intersputnik

23rd IAA/IISL Scientific-Legal Roundtable
Glasgow, UK, October 2nd, 2008

Registration of geostationary satellites by the United Nations

Petr Lála
Czech Board for Space Activities, Czech Republic

ABSTRACT

The geostationary satellites should be registered by the United Nations according to the Convention on Registration of Objects Launched into Outer Space. Based on the registration documentation from the United Nations Office for Outer Space Affairs and other sources, effectiveness of this process is evaluated for all geostationary satellites launched by the end of 2007. It is concluded that about 15% of geostationary satellites is not registered – mainly those belonging to States non-party of the Convention and to commercial operators. Improvement of the situation might be expected from the new UN General Assembly resolution 62/101 on enhancing the practice of States and international intergovernmental organizations in registering space objects.

The registration of artificial objects launched into outer space is governed by the United Nations Convention on Registration of Objects Launched into Outer Space. It was adopted by the UN General Assembly on 12 November 1974, opened for signature on 14 January 1975 in New York and entered into force on 15 September 1976. The depository of the Convention is the Secretary-General of the United Nations and practical issues are dealt with by the UN Office for Outer Space Affairs (OOSA), currently located at Vienna, Austria. Of course, the geostationary satellite is a special case of artificial satellite – it is moving on the unique geostationary orbit (GSO), so that for terrestrial observer it appears motionless. In this paper, only aspects of the Convention important for GSO satellites will be discussed.

As of 1 January 2008, there were 51 ratifications and 4 signatures of the Convention[1].
In addition, two international organizations have declared their acceptance of rights and obligations provided for in the Convention[2]. Hence, the Convention is ratified by the majority of main space players, but in spite of the continuous efforts of the United Nations Committee on the Peaceful Uses of

[1] Algeria, Antigua and Barbuda, Argentina, Australia, Austria, Belarus, Belgium, Brazil, Bulgaria, Burundi (Signature only), Canada, Chile, China, Cuba, Cyprus, Czech Republic, Denmark, France, Germany, Greece, Hungary, India, Indonesia, Iran (S), Italy, Japan, Kazakhstan, Lebanon, Liechtenstein, Mexico, Mongolia, Montenegro, Netherlands, Nicaragua (S), Niger, Norway, Pakistan, Peru, Poland, Republic of Korea, Russian Federation, Saint Vincent and the Grenadines, Serbia, Seychelles, Singapore (S), Slovakia, Spain, Sweden, Switzerland, Turkey, Ukraine, United Arab Emirates, United Kingdom, United States of America, Uruguay

[2] European Space Agency (ESA) and European Organization for the Exploitation of Meteorological Satellites (EUMETSAT)

Outer Space (COPUOS), not all of its members participate in it. The full text of the Registration Convention can be found at[3], or at OOSA webpage[4].

The Registration Convention requires that the launching State should furnish to the United Nations, as soon as practicable, the following information concerning each space object:

- Name of launching State or States;
- An appropriate designator of the space object or its registration number;
- Date and territory or location of launch;
- Basic orbital parameters, including:
  Nodal period (for GSO it should be around 1436 minutes);
  Inclination (inclination of the orbit – for equatorial orbit it is 0 degrees);
  Apogee (highest altitude above the Earth's surface – for GSO 35,850 km);
  Perigee (lowest altitude above the Earth's surface – for GSO 35,850 km);
- General function of the space object.

The fundamental weakness of the Convention is the formulation "as soon as practicable" which is open to different interpretations, because there is no given deadline. Also the selected set of "basic orbital parameters" is not complete – it does not provide enough information for computation of the satellite orbit in space. This is particularly true for satellite on GSO, since no clue is provided about the geographic longitude of the sub-satellite point at the equator, above which the satellite is stationed. The parameters requested, such as inclination, nodal period, perigee and apogee, are almost the same for every GSO satellite.

As requested by the Registration Convention, the official United Nations Register of objects launched into outer space has been established at OOSA on behalf of the Secretary General. The first document ST/SG/SER.E/1 was issued on 14 April 1977 and it contains information on space objects launched by the United States of America as of 31 December 1976. As of 31 July 2008, OOSA has issued 540 documents containing registration data on 7245 space objects (including non-functional). In addition, the Office continues to maintain registration information furnished by member States under the previous resolution 1721 B (XVI) of 20 December 1961 and on a voluntary basis. Such information appears in document series A/AC.105/INF.1-416 and covers 5185 objects (most of them were launched before the Registration Convention went into force). A voluntary registration information was provided by Egypt, Israel, Luxembourg, Malaysia and Philippines and also by some states which later acceded to the Convention (Algeria, Brazil and Italy).

In order to provide quick and unlimited access to the information contained in the Register, OOSA has developed an electronic on-line searchable Index to the Register[5]. It contains information received from member States and also complementary information collected from external sources on all functional objects launched into outer space since 1957 (space debris and non-functional objects are not included). The search could be performed using different parameters (name, international identification, State of registry, date of launch, orbital status, etc.). In addition, it provides links between space objects and their relevant documents of registration. This way, every user can

---

[3] United Nations, *Treaty Series*, vol.. 1023, No. 15020

[4] http://www.unoosa.org/oosa/SpaceLaw/gares/html/gares_16_1721.html

[5] http://www.unoosa.org/oosa/osoindex.html

download and print any registration document. At thc same time, additional unofficial information is clearly marked as such.

Table 1: UN registration of geostationary satellites launched by 31.12.2007 (R – ratification, S – signature, A – Acceptance, N – non-party)

State of registry	Registration treaty status	Registered GSO objects	Non-registered GSO objects
Arabsat	-		6
Arabsat (to India)	-		1
Argentina	R	1	
Australia	R	7	1
Brazil	R	5	3
Canada	R	13	
Canada (to Argentina)	R	1	
Canada (to Brazil)	R	1	
Canada and USA	R	2	
China	R	19	8
China (from UK)	R	4	
ESA	A	5	
ESA (for Eumetsat)	A	2	
ESA (for Eutelsat)	A	2	
ESA (for Inmarsat)	A	2	
Egypt	N		2
Eumetsat	A	3	1
Eutelsat	-		4
France	R	11	1
France (for ESA)	R	3	
France (for Eutelsat)	R	20	
Germany	R	5	
Germany and France	R	2	
Greece and Cyprus	R	1	
India	R	19	
India and USA	R	2	
Indonesia	R		9
Inmarsat	-		9
Intelsat	-		34
Intersputnik	-		1
Israel	N		2
Italy	R	3	1
Japan	R	45	3
Kazakhstan	R	1	
Korea, Republic of	R	4	
Luxembourg (for Astra)	N	15	
Malaysia	N	3	

Mexico	R	2	4
Netherlands (for SES New Skies Sat.)	R		6
Nigeria	N		1
Norway	R		2
Philippines	N	1	
Russia	R	139	
Russia (to Rimsat)	R	3	
Singapore and Taiwan	S		1
Spain	R	3	3
Sweden	R	4	
Sweden (from UK)	R	1	
Thailand	N		5
Turkey	R	3	
UK	R	13	1
USA	R	260	12
USA (for Canada)	R	2	
USA (for Indonesia)	R	3	
USA (for Inmarsat)	R	3	
USA (for Intelsat etc.)	R	27	
USA (for NATO)	R	5	2
USA (to Australia)	R	1	
United Arab Emirates	R	2	
TOTAL		673	123

This on-line Index had been used to obtain statistical information on registration of functional objects in GSO, launched by the end of 2007. The results are given in Table 1. From the total number of 796 relevant satellites, 123 (15%) is not registered by the United Nations. In general, States which participate in the Convention are doing well in fulfilling their obligations. More important are problems with international organizations – some of them (e.g. Eutelsat and Inmarsat) register their objects through other States or entities, but not always. Sometimes, the satellite was registered by two States procuring the launch, even though only one should be the State of registry. In addition, some satellites have changed the ownership/operators during their operational period, but their UN registration have not changed.

At its 75th plenary meeting on 17 December 2007, the GA adopted resolution 62/101[6] prepared by COPUOS on "Recommendations on enhancing the practice of States and international intergovernmental organizations in registering space objects" . Some of the recommendations would be very beneficial particularly for registration of satellites in GSO. Further studies should verify whether substantial improvements occur in the future.

---

[6] Document A/RES/62/101, 10 January 2008

In paragraph 2, the Resolution deals with recommendations to provide important information on geostationary satellite location and any changes of this location, particularly in case of moving a space object to a disposal orbit:

Also recommends, with regard to the harmonization of practices, that:

2 (b) Consideration should be given to the furnishing of additional appropriate information to the Secretary-General on the following areas:

(i) The geostationary orbit location, where appropriate;

(ii) Any change of status in operations (inter alia, when a space object is no longer functional);

(iii) The approximate date of decay or re-entry, if States are capable of verifying that information;

(iv) The date and physical conditions of moving a space object to a disposal orbit;

(v) Web links to official information on space objects.

Finally, in Paragraph 4, the Resolution proposes submission of information regarding the change of ownership/supervision of a space object which in the past had not been requested at all:

4. Recommends that, following the change in supervision of a space object in orbit:

(a) The State of registry, in cooperation with the appropriate State according to article VI of the Outer Space Treaty, could furnish to the Secretary-General additional information, such as:

(i) The date of change in supervision;

(ii) The identification of the new owner or operator;

(iii) Any change of orbital position;

(iv) Any change of function of the space object.

CONCLUSIONS

- Registration of geostationary satellites is a special case of the UN registration. Till the end of 2007, 673 GSO objects was registered (84.5% of existing);
- There is usually a long delay in registrations;
- Non-registered satellites belong mainly to States or organizations which are not parties of the Convention;
- Only 3 States (China, Sweden and UK) have officially registered a change of ownership; Changes of orbital position or transfer to disposal orbit are generally not registered (good examples: France, Luxembourg, Sweden).
- New recommendations in Resolution 62/101 are expected to improve the efficiency of the registration process.

Long-Term Efficiency of the Space Regulatory Framework
"Paper satellite" issue

Yvon Henri, Chief, Space Services, Radiocommunication Bureau, ITU

Satellite applications

Satellite technology offers effective technical and economic solutions for the establishment of state-of-the-art telecommunication networks, providing trunk telephony and data, direct digital radio and TV and broadband services to fixed and mobile user terminals. Satellites are also key elements in emergency telecommunications, meteorology, global positioning systems, environmental monitoring and communication services that ensure safety of life on land, at sea and in the skies. Accordingly, the demand for satellite capacity has grown, and with it, the demand for use of the orbit/spectrum resource. As the options for such use are limited by a combination of technical and economic factors, the problem of resource scarcity has arisen, along with the concern that the development of new satellite applications and networks may be held back.

The successful introduction of any radio system thus depends on the availability of radio frequency spectrum as a valuable common resource. The development of satellite communication has led to the appearance of a new, invisible international resource – the geostationary-satellite orbit (GSO). At present, most communication satellites describe a circular orbit in the plane of the Earth's equator at an altitude of about 36000 km, resulting in a 24-hour period of revolution around the centre of the Earth. They are synchronous with the Earth's rotation and would appear to be motionless in relation to a reference point on the Earth's surface. This characteristic enables the satellite to provide permanent coverage of a given area, which simplifies the design of earth stations as they are not required to track such satellites. The latter are thus located on the GSO and are designated as geostationary satellites. The very characteristics of the GSO mean that it is finite, and subject to possible congestion but not depletion as a common resource.

Various other satellite systems use non-geostationary (non-GSO) satellites, mainly for space research, earth exploration and radionavigation applications. They mostly involve low-Earth orbit (LEO) satellites, designed to operate at altitudes between 400 and 1500 km, and to a more limited extent medium-Earth orbit (MEO) satellites, orbiting at altitudes between 7 000 and 12 000 km. For non-GSO satellites the problem of orbit scarcity is less crucial, as greater flexibility exists in designing the orbital characteristics of the constellations. This flexibility would nevertheless decrease in the case of systems with numerous non-GSO satellites providing applications in for example, the mobile- or fixed-satellite services.

International space regulatory framework

International management of the use of the radio-frequency spectrum and orbits is entrusted to the International Telecommunication Union (ITU). The ITU Member States have established a legal regime, which is codified by the ITU Constitution and Convention, complemented by the Radio Regulations (RR) which enshrine the main principles and lay down the specific regulations governing the registration of satellite network frequency assignments. The Radio Regulations are revised partially or, in exceptional circumstances, completely by world radiocommunication conferences (WRCs) and constitute a binding international treaty (Radio Regulations, 2008).

The Radio Regulations constitute the unique international basis for achieving an interference-free - or rather interference-controlled – environment for satellite operation and guaranteeing equitable access to use of the natural resources of the frequency spectrum and GSO. Two major mechanisms for the sharing of orbit and spectrum resources have been developed and implemented: the *a priori* planning procedures which include the allotment plan for the fixed-satellite service and the plan for the broadcasting-satellite service along with the associated plan for feeder links (planned services); and the coordination procedures (unplanned services). The right to use a satellite position and associated frequency bands in unplanned services is acquired through negotiation ("coordination before use") with the administrations concerned by actual use of the same portion of the orbital segment and associated spectrum. Completion of the registration procedure, including the coordination and recording of frequency assignments in the Master International Frequency Register (MIFR), establishes the rights and obligations of administrations in the domain of orbit/spectrum management, the aim being to prevent possible loss of investment, customers and revenue by minimizing the capacity that remains unusable due to potential interference.

Concerns regarding application of the registration procedures (paper satellites)

The current regulatory procedures were developed back in the 1970s, and they have successfully served their purpose since then. However, as occupancy of the resources increased and demand for satellite services grew, a somewhat abnormal behaviour on the part of administrations emerged in international satellite network registration in the second half of the 1990s. Indeed, administrations submitted an extremely high number of satellite networks to ITU each year, most probably including some speculative filings, or so-called "paper satellites". This resulted in unacceptable delays in processing space notices at ITU, pushing administrations to file more even systems in order to be sure of securing a position at the end of the registration process. Indeed, for a single real network, there could be multiple filings for different orbital positions (up to ten and more in some cases).

Additional resources, coupled with improvements in the Radiocommunication Bureau's workflow and processes, adjustments to the regulatory procedures and the provision of software packages to administrations, have resulted in the complete elimination of the backlog in processing satellite network notices (see attachment 1).

However, even this will not suffice in the long term, given the magnitude of the challenge.

As a matter of fact, independent information available today on the real use of the orbit/spectrum resource shows some divergence from the corresponding information submitted by administrations to ITU. This means that "paper satellite" issues – or, more precisely, fictitious frequency assignments recorded in the MIFR - still exist, with the majority of such assignments recorded with the indication that they have been brought into regular operation in accordance with the notified satellite network characteristics (see attachment 2).

One reason for this is that administrations have no real incentive to give up underused orbit/spectrum resources or update their satellite network parameters at the stage of notification and recording of assignments in the MIFR in order to accurately reflect their planned operations. Rather, there is a fairly strong incentive to reserve (and thus freeze) spectrum regardless of real future needs, thus *de facto* denying access to new entrants. The enforcing mechanisms that exist at present to ensure that a satellite system is operating in accordance with recorded parameters are based mainly on the goodwill of administrations. When goodwill is linked to financial consequences, enforcing mechanisms of this kind tend to be disregarded.

WRC-11 and beyond

The future development of satellite communications is closely bound up with the international regulatory procedures, which as things stand may be seen as placing some limitations on the development of new satellite projects or as being jeopardized by certain behaviour. There is thus a pressing need to take steps to guarantee and increase efficiency in the use of the orbit/spectrum resource.

ITU is 'committed to connecting the world' and satellite communication systems have an enormous potential to offer promising, high-capacity transmission capabilities in this regard. To achieve that goal, however, governments, international organizations and the private and public sectors must continue to enable ITU to carry out its vital work of recording frequency assignments in the MIFR, thus ensuring that frequencies and orbital positions are compatible and do not result in radio interference.

In this regards, WRC-11 will be a unique opportunity for the improvement of the international regulatory framework for registering satellite networks.

Attachment 1

CR/C filings and processing trends for the years 1997-2008.09

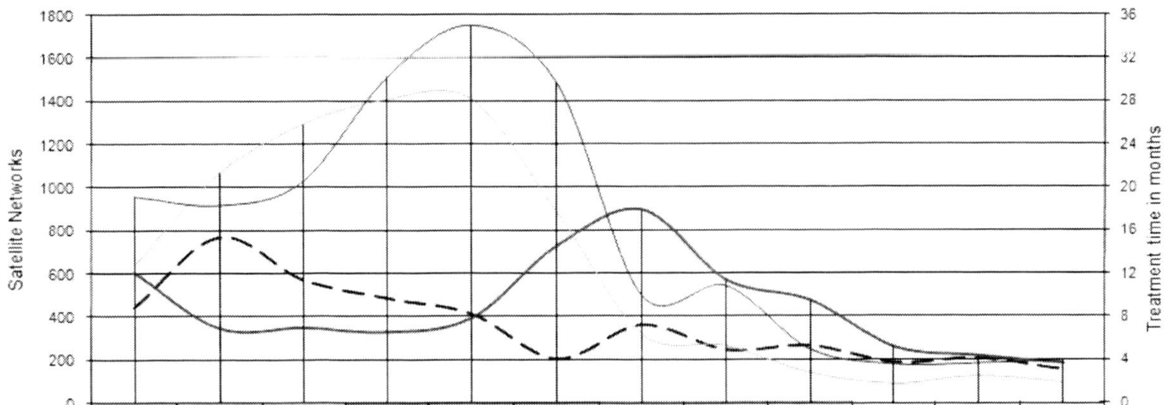

	1997	1998	1999	2000	2001	2002	2003	2004	2005	2006	2007	2008.09
Filings Received	447	773	576	489	416	208	365	249	268	190	209	154
Filings Published	605	346	350	328	394	728	898	570	475	258	217	182
Under Treatment	642	1069	1295	1410	1409	890	324	269	143	91	127	97
Treatment Time (Months)	19.2	18.4	20.6	30.3	35.1	29.8	10	10.9	5	3.6	3.7	4

Attachment 2

Number of GSO positions per administration recorded in the ITU International Master Frequency Register (MIFR), for which frequency assignments have been declared brought into use

Administration	Organization	Number of GSO orbital positions
ARG		2
ARS	ARB	4
AUS		9
B		7
BEL		4
BLR	IK	5
CAN		8
CHN		30
CTI	RAS	4
CYP		1
D		4
E		2
F		12
F	ESA	4
F	EUT	26
G		28
GRC		1
HNG		1
HOL		9
I		1
IND		6
INS		8
IRN		3
ISR		1
J		19
KOR		5
LBY		1
LUX		6
MEX		4
MLA		4
NOR		2
PAK		2
PNG		1
RUS		53
RUS	IK	1
S		2
S	NOT	1
SEY		3
SNG		1
THA		5
TON		5
TUR		3
UAE		4
URG		1
URS		12
USA		138
VTN		3
Total		*456*

23rd IAA/IISL Scientific-Legal Roundtable
Glasgow, UK, October 2nd, 2008

PAPER SATELLITES AND SPACE NETWORKS

Luboš Perek[1]
Astronomical Institute
Academy of Sciences of theCzech Republic, Prague

Abstract

A Comparative Table of notified space networks and satellites present at or near nominal orbital positions illustrates the occupation of the geostationary orbit. Clear distinction has to be made between a network as transmission of information on one hand and a satellite as a solid body orbiting the Earth on the other hand. Paper satellites do not exist but there is an overabundance of over 2500 separate requests for transmission frequencies. The situation has been made transparent by showing which networks are fit to transmit by having both, a notification by the ITU as well as an active satellite in a correct orbital position. There are 802 notified networks and at least 84 networks in the Broadcasting Satellite Service operated on 372 satellites. More than 165 notified networks do not have a satellite at relevant orbital position.

1. INTRODUCTION

The term "paper satellites" gives the impression that some satellites are not real but exist on paper only. In fact, the number of active satellites in the most populated region, the geostationary orbit, **GEO**, at the end of June 2008 was 372.

On the other hand, the number of requests for frequency bands for radio transmissions from satellites to ground stations exceeds 5000. The requests exist only on paper until they have passed successfully through a coordinating process initiated by the International Telecommunication Union (ITU) and became "notified" space networks enjoying international recognition and protection. The ITU sieve is strict. In June 2008 only 802 networks in SNL, Section A1, made it and were supposed to transmit. The coordinating process is an attempt to prevent possible harmful interferences and eliminate interferences detected at a later stage.

In order to understand the problem at hand better, it is necessary to compare notified networks with reality, i.e. with active satellites in orbit. For that reason a Comparative Table has been set up listing side by side the "paper situation" of space communication networks with the "physical situation" of satellites in the geostationary orbit.

Also, the distinction between a network and a satellite has to be well understood. A space network is an electromagnetic transmission of information between radio stations in space and on the ground. A satellite is a solid body in outer space, in an orbit around the Earth. Official terminology, as well as its popular counterpart, evolved historically in such a way that the word "satellite" stands sometimes for the material body, sometimes for its function as the space component of a communication network. The present paper provides understanding. It is not a proposal for changing a widely used terminology.

2. THE COMPARATIVE TABLE

The operation of a space network depends on the presence of a satellite near the nominal orbital position of the network. There is tolerance in position because a transmission lobe covers with sufficient intensity a certain area. The ITU passed a rule that a satellite operating a space network has to be within an angle of 0.1 degree in latitude as well as longitude from the nominal orbital position.

The Comparative Table[1] shows the relation between **"notified space networks"** which enjoy international recognition and protection, and **active satellites** which are close to the nominal orbital position. Networks appear on the left, satellites on the right. The reference date for Networks is 3 June 2008, i.e. the date of issue of the ITU Space Network

List. For satellites it is the end of June 2008.

The main parameter in the table is the nominal position **(Nom. long.)** shown in the first column in bold characters. All notified networks operating at the same nominal position are contained within horizontal lines. On the right, between the same horizontal lines, are satellites which could operate networks on the left.

Data on space networks have been taken from the ITU Quarterly Publication of the Space Network List[2] **(SNL)**, Section A1, issue of 3 June 2008. The columns show a symbol for the notifying administration **(Adm)**, for the organization operating the network **(Ntw org)**, and the **Space Network Name**. The heading of that column in the SNL is **"Sat_name"**. As explained in the preface to the SNL, it identifies the space network, not the satellite. Names of networks are sometimes changed by operators, and some networks are known under more than one name. The last column shows the symbol for **frequency bands** according to table 1 of the preface to the SNL. E.g., the C band corresponds approximately to numbers 43 - 49, the K_u band to 53 – 65, the K_a band to 75-80.

For satellites in the Broadcasting Satellite Services, BSS, Sections A10 and A11 of the SNL were used to supplement data of Section A1, in particular in cases when a satellite operates mostly networks in the BSS.

The right-hand half of the Comparative Table shows spacecraft launched before the end of June 2008. The first column gives the COSPAR international designator **(COSPAR Int. Desig.)** indicating the year of launch, the serial number of the launch in the year, and the serial letter of the object in the launch. The designator identifies the satellite in a unique way and for its entire lifetime.

The remaining columns show the **Satellite Name,** which sometimes is subject to a change[3], and **Status of Orbit** (see section 5). Notes concerning functions performed by the satellite, printed in red, appear on the same line or the line below, or in the left part of the table, wherever convenient. The notes are very brief in order not to extend the length of the table. More details on the satellite appear in the Encyclopedia of Satellites and Probes of the Czech Academy of Sciences[4].

3. SPACE NETWORKS

The total number of entries in the **SNL,** Section A1, has grown rapidly in the past. It attained around 3000 entries back in 1998, it slightly exceeded 4000 entries in 2001 and attained 5770 in March 2007. Since that time, possibly reflecting new measures by the ITU, the number declined to 5164 entries in December 2007. It increased to 5507 in June 2008. Some networks are listed separately at different stages of coordination, thus the number of separate networks is smaller than the number of entries. In the December 2007 issue of the SNL, there are 2940 separate networks only, a reduction of the "paper population" by 43%.

An administration wishing to bring a network into use has to submit an advance announcement to the ITU. The announcement appears in the SNL as class **A**. When a coordinating process to avoid possible harmful interference with existing networks is initiated, the network is denoted as class **C**. Its entry as class **A** is, in most cases, retained. If and when the process is successfully completed, the network is notified in the Master International Frequency Register and denoted as category **N**. As before, entries of the same network in classes **A** and **C** are frequently retained. A possible reason may be to preserve a record of changes which the network was subjected to in the course of the coordinating process. Thus, in the June 2008 issue of the SNL, Section A1, 802 networks were notified in the

Master International Frequency Register and received international recognition and protection. From Sections A10 and A11, 84 networks in Broadcasting Satellite Service have been added, making thus a total of 886 Networks.

4. DUE DILIGENCE

The ITU has adopted measures, termed Due Diligence, to restrict the growth of space networks. Without these steps, the present number would have been even larger. A significant reduction is, however, in the hands of launching and operating agencies.

Administrations have to provide evidence of seriousness of intentions to establish a space network[5] according to the ITU Resolution 49, Annex 2, of the World Radio Conference 2003. Among the many data required are (i) Identity of the satellite network, (ii) Name of the satellite, and (iii) Orbital characteristics.

Item (i) refers to the term "space network" used in the title of the SNL.

Item (ii) refers to a satellite as a material body. Its unique identification is desirable. Launching names have the disadvantage of rather frequent changes. The COSPAR international designation is the best option for objects in orbit. It is unique, does not change but it is rarely used. It is assigned after a successful launch, therefore it is not available as pre-launch announcement.

Item (iii) refers, in the case of a GEO, to the geographic longitude of the nominal orbital position.

A summary of the data appeared in a table provided by the courtesy of the ITU[6]. The table lists the "ITU Name" as well as the "Commercial Name". These two names are in most cases identical with the name of the space network and only rarely with the name of the satellite. The table lists networks in all three classes, A, C, N. That means, that only some of the networks have completed the coordinating process at the date of publication of the list.

The Administrative Due Diligence, together with the fees introduced in the Financial Due Diligence and Cost Recovery, and with the time limit of bringing into use, are very efficient measures adopted by the ITU for limiting the excessive number of applications for transmission frequencies.

5. SATELLITES IN THE GEO

For a meaningful comparison of space networks and satellites a reliable list of spacecraft and their positions in the GEO is required. The most authoritative information on objects launched into outer space is provided by launching states. It appears in the UN Register in the form of governmental announcements[7]. Unfortunately it is not complete. An online searchable index listing all objects announced to the UN, as well as those not announced, appears at the same website. Additional information is, however, needed for computing actual positions of space objects at a given time. The best available and most reliable source of actual positions of objects in the geostationary orbit is the "Classification of Geosynchronous Objects" referring to situation at the end of each year[8]. It is based on the DISCOS database[9] which uses as source of information the NASA Two-Line Elements[10]. These, in their turn, are based on observations of physical presence of satellites in orbit. Only objects larger than 1m can be tracked on a regular basis at the distance of the geostationary orbit. All active satellites can be tracked but small inactive objects cannot be routinely detected. The most recent issue of the "Classification" refers to the end of 2007. Results can be summarized as follows, showing the complexity of the population of the geostationary orbit:

• The total number of objects in and reasonably close to the geostationary orbit was 1150, 29 more than the year before, 68 more than two years before.

Table1, Objects with recently updated orbital elements:

• **Status C1**: 243 objects are controlled in longitude and latitude. These are suitable for transmissions to fixed antennas. Five satellites launched towards the end of 2007 were supplemented with data obtained in 2008 and have been added to that group. Launchings in the 12 years, 1996-2007, averaged 19 per year. In years 1993-1995 the rate was lower.

• **Status C2**: 75 objects are controlled in longitude only, requiring movable ground antennas. These objects have been launched in the 23 years 1985-2007. One satellite, ATS 3, survives from 1967.

• **Status D**: 458 objects are in a drift orbit passing through all longitudes. These could obviously not been associated with a nominal orbital position, just as the following group,

• **Status L1, L2, L3**: 145 objects are at a libration orbit, oscillating around the Eastern (92), Western (36), or both (17) stable points in the orbit,

Table 2, Objects without orbital elements determined during the last 6 months of 2007:

• **Status C**: 47 objects under control. Their recent orbital elements are not available in the NASA Two-Line Elements. Orbits of 36 objects out of those 47 can be determined from amateur observations. Most of these objects are used for governmental uses, such as reconnaissance or verification of international treaties. There are thus at least 11 active satellites in controlled orbits at unknown positions,

• **Status D, L1, L2**: a total of 7 objects, mostly retired satellites.

• **Status U**: Uncontrolled: 99 objects, mostly rocket stages and debris, a few payloads, without available orbital elements. These objects are old. The most recent three objects have been launched in 1998, 1994, and 1992 respectively.

• **Status UU**: uncontrolled un-catalogued: 67 objects, all of them debris, such as covers and restraint cables, or rocket bodies,

• **Status UI**: 146 unidentified objects which have been repeatedly observed but not correlated to a specific launch. 22 of these objects have provisional identification coinciding with objects of status C. and finally,

• **Status Ind**: 9 objects launched or maneuvered close to the end of 2007 of indeterminate status. Positions of five of the objects, all launched in November and December 2007, have been computed by P. Lála[11]. ,from one or more orbital elements determined in 2008. In the Comparative Table they appear under Status of Orbit **PL**.

Of interest to the problem of paper satellites are active objects at their nominal positions within ITU tolerances (Status C1), those controlled in longitude only (Status C2), objects of Status C, and objects of Status Ind.

Also considered were objects launched in the first half of 2008. Positions have been computed and listed by P. Lála, Status **PL,** as above, and/or by A. Vítek and published in the Encyclopedia (see ref. 3), Status **Enc**. The physical presence of all 372 satellites listed in the Comparative Table has been confirmed by data derived from observations.

The number of notified space networks in the ITU tables is larger than the number of active spacecraft in orbit. Allowance, however, has to be made to the fact that more than one space network can be operated on one satellite.

6. RELATING NOTIFIED SPACE NETWORKS WITH SATELLITES

Operators know which satellites operate which of the networks but do not always publish the fact. If such information is available, e.g., on the web, or if the name of the satellite is sufficiently close to the name of the network, a bullet • was put near the central line of the Comparative Table. An example is at nominal position 0.00 E. An ESA Meteosat network is operated by satellite 2006-049B, Meteosat 9. Very helpful in this respect was, e.g., the List of Satellites Approved to Provide Fixed-satellite Services in Canada[12].

In other cases, a possible match is a matter of conjecture. An example is at nominal position 4.00 E. Satellite 1997-049A, Hot Bird 3 = Eurobird 4, providing regional services in Europe in band K_u, may be operating networks EUTELSAT 2-4E or EUTELSAT 3-4E or both. Without more detailed information we cannot say if it operates also network EUTELSAT-KA-4E transmitting in band K_a or network TELECOM-4E transmitting in band C.

In case there is no satellite at a nominal position of a network, the network cannot transmit. Possibly a satellite is being prepared to take up the position and will be launched at a future time. An example is at nominal position 5.70 E, where network MEASAT-SA1 has no satellite to transmit from. In the Comparative Table are 83 nominal positions with a total of 165 networks with no satellite at the relevant position. In other words, almost 19% of all notified networks could not operate at the end of June 2008.

There are also cases of satellites without a notified network at its nominal position. Such a satellite may have terminated its activities, or it may be a spare to be brought into use at a later date. In fact, most of satellites having no corresponding notified network, have a pending application in the coordinating process. An example is satellite 1997-071A, Sirius 2, at 31.5 E. It was moved from its previous position at 4.80 E, where it had its network SIRIUS 2. Its services at 4.80 E have been taken over by 2007-057A, Sirius 4, in May 2008. At the same time Sirius 2 was moved to its new position at 31.5 E, where corresponding networks have classes A and C, but have not been "notified" at the time of the shift of position. The annexed Comparative Table shows that a total of 23 satellites had no notified network at the critical date. Evidently, commercial reasons are in some cases stronger than discipline in complying with agreed rules.

In general, the table restricts considerably the number of possibilities of matching networks with satellites. In some cases, however, more detailed information is desirable.

7. CONCLUSIONS

There are no "Paper Satellites". There is, however an overabundance of over 2500 separate requests for transmission frequencies. Many requests appear in the lists repeatedly, according to their status in processing and coordinating with existing space networks. The ITU instituted administrative and financial measures, called Due Diligence, to curb the excessive number of applications. Only networks which do not interfere with earlier comers receive international recognition and protection. The work of the ITU is essential for preventing harmful interference among space telecommunication networks and for providing a forum for removing interference whenever it appears. A Comparative Table listing both, space networks and satellites, illustrates the situation at occupied nominal orbital positions. It is proposed to maintain and regularly update the Comparative Table.

At the end of June 2008, there were 372 satellites in orbit operating not more than 886 networks duly notified or planned. The actual communication traffic is considerably less than the numbers would suggest because 165 networks are at nominal positions with no satellite present. Moreover, even if a satellite is present, there is no confirmation that it operates **all** networks at the relevant position.

There are 23 satellites without a notified network, some of them very likely to transmit without international recognition and protection

The situation is in constant change and evolution for several reasons. New ventures have been undertaken, some networks or satellites have been sold to other operators of networks or owners of satellites and acquired new names in the process.

The term "satellite" should be used in reference to a material body only. The heading "Sat_name" in tables listing space networks may lead to confusion.

References

[1] The Comparative Table is too long to annex it here. It is available on the DVD distributed to participants at the IAC 2008 under the title of the Session. It is available also on request to perek@ig.cas.cz.
[2] ITU Space Network List, Section A1, Editions of 3 December 2007, 3 March and 3 June 2008, Sections A1, A10 and A11 for Broadcasting SatelliteServices. See at www.itu.int/ITU-R/space/snl.
[3] E.g., the satellite at 121W, EchoStar 9, is also known as Telstar 13, or Intelsat Americas 13, or Galaxy 23, or G 23.
[4] Encyclopedia of Satellites and Probes of theCzech Academy of Sciences, author A. Vítek, at www.lib.cas.cz/space40.
[5] Ram Jakhu: Legal Issues of Satellite Telecommunications, the Geostationary Orbit, and Space Debris, Astropolitics, 5, p.173-208.
[6] ITU Space Network Systems Online, Special Query System: Administrative Due Diligence Information (Res, 49), List of Geostationary and Non-geostationary Satellite Networks.
[7] See at www.unoosa.org/oosa/SORegister/index.html.
[8] Classification of Geosynchronous Objects, Issue 10 by R. Choc and R. Jehn, February 2008, European Space Operations Centre, Darmstadt, Germany. Preceding issues cover years 1999-2007. For years 1990-1999 see Log of Objects Near the Geostationary Ring, Issues 1 to 20, also at ESOC, Darmstadt.
[9] Established at the European Space Operations Centre in Darmstad, Germany.
[10] See at ghrc.msfc.nasa.gov/orbit/tleformat.html.
[11] P. Lála, Private communication. Computed from CelesTrak elements and published in the Encyclopedia, ref. 3.
[12] See www.ic.gc.ca/epic/site/smt-gst.nsf/en/sf02104e.html.

23rd IAA/IISL Scientific-Legal Roundtable
Glasgow, UK, October 2nd, 2008

Statement by Victor S. Veshchunov
Director, International and Legal Department
Intersputnik Organization of Space Communications

When assessing the current condition of the geostationary orbit/frequency resource and prospects of its utilization one can not ignore the relatively recent trend where countries that were never members of the 'space club' are now willing to have domestic satellite telecommunications systems using their own satellites. Over the last few years alone, such projects have been successfully launched by Kazakhstan, Nigeria and Vietnam that have already deployed their domestic spacecraft and are planning to add more. Azerbaijan's domestic satellite telecommunications project is about to come off with the Azersat 1 satellite. Similar plans have been announced by Ukraine, Angola and other countries.

Certainly, in the nearest future this trend will result in the growth of the number of geostationary spacecraft. Accordingly, the orbit is going to be overloaded. In other words, the GSO frequency resource is going to become even scarcer for new satellites.

Formally, any new states that are willing to launch their own spacecraft but lack coordinated orbital slots for such satellites have to follow the routine of filing, coordinating and notifying their new satellite networks in the GSO. However, they clearly understand that on the one hand this would take too much time and require much effort. On the other hand, considering the current 'population density' in the GSO, the final conditions of frequency coordination will be quite stringent and would noticeably limit the opportunity of deploying a satellite that would be up to snuff.

In these conditions, there comes up a new opportunity that could hardly be imaginable just several years ago, where states that are joining the 'space club' turn for assistance to 'space club' veterans that already have several coordinated orbital positions. For example, for its Kazsat-1 satellite Kazakhstan uses a Russian slot. Azerbaijan's Azersat-1 will be located in the Turkish position at 50°E. Can you imagine anybody criticizing Russia, Turkey or any other country that have available slots for offering such slots to other countries as was the case twenty years ago with the Kingdom of Tonga? (although one must admit that Tonga filed its slots for re-sale only while Russia, Turkey or similar countries filed their slots for their own telecommunications satellites).

It is noteworthy that it is not only countries that are looking for suitable orbital positions for their domestic satellite systems but also satellite operators that have only recently entered the market of satellite telecommunications services. Our international organization is just one example. In the mid-90s Intersputnik decided to procure its own space segment using its own orbit/frequency resource and filed with the ITU a number of orbital slots. Regretfully, for a number of objective reasons Intersputnik had to somehow reduce its plans. Instead of deploying a fleet of three or four satellites we had to confine ourselves to just one satellite in one orbital position. Here I mean the well-known Lockheed Martin Intersputnik project. Therefore, today Intersputnik has satellite networks filed in several orbital slots that can be offered for joint use to interested government or private satellite operators. For example, recently we have launched a project to operate a Protostar satellite at 98.5°E. Frankly speaking, when this project was announced certain operators that were using adjacent positions became somewhat concerned. This is quite natural and understandable because competition on the Asian satellite market is getting fiercer. Here I do not want to go into greater detail technically, but I can assure you that the conditions of Intersputnik's cooperation with ProtoStar just like with all other potential partners explicitly exclude any harmful interference to any satellite networks that have ITU time priority.

This example illustrates another 'novelty' in using the GSO resource that used to be an exception just several years ago, but quite obviously may in the nearest future become much more widespread. I mean that spacecraft are placed in service in a given position before the international frequency coordination of the corresponding satellite network is completed. You will recall that under section II, Article 9 of the ITU Radio Regulations a satellite network can be brought into use after its international frequency coordination is completed. However, as I already mentioned, both today and in the nearest future new satellite networks can only be finally coordinated with very serious constraints. It is beyond any doubt – and this was mentioned during this round table discussion – that the problem can be resolved by technologically upgrading spacecraft and terrestrial equipment. At the same time, more and more often spacecraft tend to be placed in service before the satellite network is formally brought into use provided that no harmful interference is caused to satellite networks that have ITU time priority. This procedure conforms to sections 11.44.1, 11.41 and 4.3 of the Radio Regulations. Experience shows that this can be done because quite a few satellite networks already filed use satellites with frequency plans that do not cover all frequencies filed in a given network. Formally, the entire bandwidth of such frequencies belongs to the operator having time priority, but actually this bandwidth includes 'cantles' that can be used by a 'foreign' satellite that, if properly configured, will never cause interference to the 'first-priority' satellite. It may be expected that against the background of the growing scarcity of the orbit/frequency resource and the growing number of geostationary satellites such situations will occur more often.

United Nations

Important notice

Non-Delegates

In order to attend the Symposium please submit your name to the Office for Outer Space Affairs and present this invitation, together with an appropriate identification document at the entrance of the Vienna International Centre.

UNITED NATIONS
OFFICE FOR OUTER SPACE AFFAIRS

United Nations Office at Vienna
5 Wagramer Strasse
A-1220 Vienna, Austria

Telephone: + 43-1-260 60 4950
Fax: + 43-1-260 60 5830
E-mail: oosa@unvienna.org
Web site : http://www.unoosa.org

International Institute of Space Law (IISL) and
European Centre for Space Law (ECSL)

SPACE LAW SYMPOSIUM 2008

Legal Implications of Space Applications for Global Climate Change

Monday, 31 March
&
Tuesday, 1 April

Conference Room III
Building C
Vienna International Centre

On the occasion of the 47th Session of the Legal Subcommittee
of the United Nations Committee on the Peaceful Uses of Outer Space

PROGRAMME

Day 1
31 March 2008
(16h-18h)

Legal Implications of Space Applications for Climate Change: Principles and Rules

Chairman
Ambassador Peter Jankowitsch, Chair, Advisory Board, Austrian Aeronautics and Space Agency. Former Chairman of COPUOS

Legal Features of the Climate Change Convention: from Kyoto to Bali
Gerhard Loibl, The University of Vienna, Austria

Space Applications and Climate Change
Francesco Sarti, Centre for Earth Observation of the European Space Agency (ESA/ESRIN), Coordination and Planning Office (EOP-SA)

Legal Aspects of Cooperation for Space Monitoring of Climate Change and Sustainable Development
Monserrat Filho, Brazilian Society of Space Law, Brazil

Promoting Access to, and Exchange of, Data and Information related to Climate Change: the Legal Perspective
Joanne Irene Gabrynowicz, The University of Mississippi School of Law, United States

Coordinators of the Symposium:
Corinne Jorgenson (Secretary, IISL)
Kai-Uwe Schrogl (Board of Directors, IISL)
Sergio Marchisio (President, ECSL)

Day 2
1 April 2008
(16h-18h)

Legal Implications of Space Applications for Climate Change: Institutions and Instruments

Chairman
Sergio Marchisio, President, European Centre for Space Law (ECSL)

Coordination Instruments and Satellite Observation of the Climate System: the Contribution of CEOS
Evangelina Oriol Pibernat, Centre for Earth Observation of the European Space Agency (ESA/ESRIN)

Monitoring the Environment for Climate Change: the Case of GMES
Gisela Süss, European Space Agency (ESA)

Monitoring the Kyoto Protocol: Greenhouse Gases Observation & the Global Forest Carbon Monitoring System
Masami Onoda, Japan Aerospace Exploration Agency (JAXA)/the University of Kyoto, Japan

Legal Aspects of Climate Monitoring by Means of Treaty Law
Frans von der Dunk, the University of Nebraska, United States

Concluding Remarks
Vladimír Kopal, the University of Pilsen, Czech Republic

Legal Implications of Space Applications for Global Climate Change

A Symposium jointly organised by the
International Institute of Space Law and the European Centre for Space Law

On the occasion of the
47th Session of the Legal Subcommittee
of the United Nations Committee on the Peaceful Uses of Outer Space

31 March – 01 April 2008

Report of the Symposium

Dr. Gérardine Meishan Goh, LL.M.
Assistant to the Secretary, Board of Directors, International Institute of Space Law
German Aerospace Center (DLR) / University of Cologne, Germany
Email: Gerardine.Goh@dlr.de

Climate change and sustainable development have given rise to a growing number of symposia and conferences, most notably in the field of international environmental law. The role of space applications in the management of climate change has been increasingly acknowledged. What is striking however is the almost exclusive discussion of this role within the international environmental arena; space applications and space law are considered generally a subset of measures available for environmental specialists. Deliberations on State practice in regard of international environmental regulatory mechanisms and the global nature of environmental cooperation have generally been rooted in environmental protection studies rather than international space law.

In this context, the Symposium jointly organised by the International Institute of Space Law (IISL) and the European Centre for Space Law (ECSL) points to the usefulness and originality of considering climate change issues from a different perspective. The Symposium, entitled "Legal Implications of Space Applications for Global Climate Change", was held on the occasion of the 47th Session of the Legal Subcommittee of the United Nations Committee on the Peaceful Uses of Outer Space (COPUOS). As has been the tradition from previous years, the Symposium was held on the afternoons of the first two days of the Legal Subcommittee sitting from 31 March – 01 April 2008. The topic was exceptionally proposed to the IISL and the ECSL in a nod to the role of international space law in addressing environmental, climate change and sustainable development issues from the viewpoint of space technology and space applications.

The Symposium was divided into two sessions. The session on Day 1 (31 March 2008) focused on "Principles and Rules", presenting the legal framework and interaction between international space law and international environmental law. Four presentations were given by legal experts in the field, followed by an open discussion with delegates and observers of the Legal Subcommittee. Day 2 saw a shift in the emphasis of the Symposium towards "Institutions and Instruments" – the implementation, compliance and enforcement issues related to the legal implications of space applications for global climate change. The delegates and observers of the Legal Subcommittee heard another four expert presentations on this topic. The floor was then open again for queries and comments. Concluding remarks were made by ***Professor Dr. Vladimír Kopal***, Chairperson, Legal Subcommittee, COPUOS / University of Pilsen, Czech Republic. Thereafter the Symposium was closed by the chair and all participants were invited to a cocktail reception hosted by the IISL and the ECSL.

Day 1: Legal Implications of Space Applications for Global Climate Change – Principles and Rules

Chairperson: ***Ambassador Dr. Peter Jankowitsch***, Chair, Advisory Board, Austrian Aeronautics and Space Agency, Austria

Speakers: ***Prof. Dr. Gerhard Loibl***, University of Vienna, Austria ("Legal Features of the Climate Change Convention: from Kyoto to Bali"); ***Mr. Jorge Lafourcade*** on behalf of ***Ambassador Raimundo Gonzalez Aninat***, Chile ("Space Applications and Climate Change" / "Implicaciones jurídicas de las Aplicaciones de la Tecnología Espacial para el Cambio Climático Mundial"); ***Prof. José Monserrat Filho***, Brazilian Society of Space Law, Brazil ("Legal Aspects of Cooperation for Space Monitoring of Climate Change and Sustainable Development") and ***Prof. Joanne Irene Gabrynowicz,*** University of Mississippi Law School, United States of America ("Promoting Access to, and Exchange of, Data and Information related to Climate Change: the Legal Perspective")

Ambassador Dr. Peter Jankowitsch opened the Symposium with words of thanks to the delegates of the COPUOS Legal Subcommittee for inviting the IISL and the ECSL to jointly organise the Symposium. He noted that past Symposia have contributed very constructively to the deliberations of the Subcommittee delegates, especially since the topics addressed by the Symposia have regularly been adopted as a single-item agenda issue for the next sitting of the Legal Subcommittee. *Ambassador Jankowitsch* noted that the first session would give a clear overview of the legal principles relating to the use of space technology and space applications for global climate change, providing a strong basis for discussions in the Subcommittee.

The first expert speaker was ***Professor Dr. Gerhard Loibl,*** University of Vienna, Austria. His presentation, entitled "Legal Features of the Climate Change Convention: from Kyoto to Bali" laid the groundwork for the context of the Symposium. *Prof. Loibl* discussed the historical development of the United Nations Framework Convention for Climate Change (UNFCCC), including the negotiations of the Kyoto Protocol and the recent conference in Bali, Indonesia. First looking at the schedule of negotiations, *Prof. Loibl* noted the lengthy negotiations needed to finalise the Kyoto Protocol, as specific details relating to the compliance and financing systems, in particular with regard to the developing States, had to be drawn up. This process carried on until the conference in Marrakech, Morocco, and only then did the Kyoto Protocol become ratifiable for all Annex I State parties. *Prof. Loibl* then discussed the current situation of the UNFCCC negotiations, drawing the attention of the delegates to the fact that the Kyoto Protocol was further stalled by the U.S. refusal to ratify the Protocol. He noted that it was only the 2004 Russian Federation ratification that triggered the Protocol, when more than 55% of the Annex I Parties' emissions were covered by ratifications. *Prof. Loibl* turned thereafter to the discussion relating to the situation after 2012. The Bali Conference aimed to set out the future process of negotiations, including a plan of meetings to decide upon the actions after 2020. With the Bali Action Plan, *Prof. Loibl* believes that there is now a roadmap for international and national mitigation mechanisms, such as technology development, technology transfer, and the support of development countries. *Prof. Loibl* further discussed the proposals to review Article 9 of the Kyoto Protocol, and the issues of whether and how new States parties may join the commitments as listed in Annex B to the Protocol. The second issue of Article 9 reform focuses on voluntary commitments of developing States to reduce their emissions, and how these can be incorporated into the UNFCCC framework. *Prof. Loibl* also delved into the compliance system of the UNFCCC and Kyoto Protocol, particularly noting that the Protocol mechanism broke new ground with its two-part compliance system of facilitative and enforcement branches. The latter branch in particular becomes more significant this year, as with 1 January 2008, data submitted by Annex I parties will be decisive in the trade of emission certificates, participation in the clean development mechanism, and the progress of the joint development mechanisms.

The broad overview of the UNFCCC framework and international environmental law was put in context by the next speaker,

Mr. Jorge Lafourcade, who made the presentation on behalf of ***Ambassador Raimundo Gonzalez Aninat***, Chile. Entitled "Space Applications and Climate Change" / "Implicaciones jurídicas de las Aplicaciones de la Tecnología Espacial para el Cambio Climático Mundial", the presentation aimed at focusing the principles and rule of the UNFCCC framework with regard to the specific purpose of the Symposium at hand. It was noted that the 1967 Outer Space Treaty (OST) is the most important international space law convention, and defines briefly the elementary aspects of outer space as the common heritage of Mankind. The OST also marked a qualitative change to traditional international law since it sought out new challenges in the global context, such as the peaceful uses of outer space in consideration of the global public good and human environment as provided for in Article 9 OST. In the opinion of the author, the mandate in Article 9, in the context of the principle of the common heritage of Mankind, is mandatory, and possibly a non-derogable norm. This is especially so in light of the various ambient developments since 1967, including the political changes in strategic alliances, commercialisation, new actors and novel space technology. However, two issues in particular complicate the implementation of space applications for the global public good: terrorism and intellectual property rights. In the first instance, the author is of the opinion that terrorism has exacerbated the trend towards secrecy over data due to national security legislations. The 1986 Principles Relating to Remote Sensing (RS Principles), in particular Principles 1, 4, 10 and 12, relate to the provision and use of earth observation data for climate change issues. However, the author contends that these Principles are vague and therefore weak, and in particular lacking, especially since there is no mention of global warming. The author warns of a watering-down of multilateralism to a matrix in which commercial parties and certain States seek to protect unilateral interests. Against the background of the 1996 Space Benefits Declaration, this would be contrary to the fundamental principles of international space law. The author advocates international legislation in step with the realities and needs of present and future scenarios, and proposes a resolution by the United Nations General Assembly to provide technological parameters for States to apply space technology in the face of global climate change.

Prof. José Monserrat Filho, Brazilian Society of Space Law, Brazil, made the next presentation, entitled "Legal Aspects of Cooperation for Space Monitoring of Climate Change and Sustainable Development". *Prof. Monserrat* noted that climate change is the single largest scientific issue in the world, and that cooperation for space monitoring of climate change for sustainable development is a new issue facing international space law. Space monitoring of climate change for sustainable is a complex of space infrastructure that integrate earth observation and telecommunications assets. However, *Prof. Monserrat* noted that space monitoring is not a legal concept, and also did not find expression in the 1986 RS Principles or in the UNFCCC. He argued that since space monitoring is a remote sensing activity however, it should be conducted according to the 1986 RS Principles, in particular Principle 12. Principle 12 provides that all sensed States should have open access to data on their territories as soon as the data is produced, and *Prof. Monserrat* is of the opinion that this is applicable also to data relating to climate change and sustainable development. In April 2007, the United Nations Security Council recognised the threat of climate change and its environmental, social and economic impacts, in particular its threat to the maintenance of international peace and security. In this context, space monitoring is increasingly crucial, and should be subsumed under the "systematic observation and development of data archives" for sustainable development as envisaged under the UNFCCC and the Kyoto Protocol. Pursuant to Article 4 UNFCCC and Article 10 Kyoto Protocol, and against the background of the 1996 Space Benefits Declaration, *Prof. Monserrat* moots that there is a strong commitment to cooperation. He then recommends that the concepts of "space monitoring", "climate change" and "sustainable development" be defined for greater clarity and consistency, especially with respect to earth observation activities. Due to the complexity of the legal aspects underlining cooperation in this field, *Prof. Monserrat* proposes that the COPUOS Legal Subcommittee should consider this issue under a 3-year Workplan.

The last presentation on Day 1 of the Symposium was given by ***Prof. Joanne Irene Gabrynowicz***, University of Mississippi Law School, United States of America on the issues of “Promoting Access to, and Exchange of, Data and Information related to Climate Change: the Legal Perspective”. Taking the perspective of a practising lawyer in advising a client, *Prof. Gabrynowicz* noted that the technologies and practices relating to climate change develop generally quicker than the law, especially since law is reactive while technology development is proactive. She also noted that issues relating to the climate transcends all organisational elements of legal institutions, jurisdictions and competences, and as such is very complex. The collective data sources for climate change also span a multitude of mechanisms, ranging from space-based assets to aerial and other in-situ data sources. *Prof. Gabrynowicz* described the three levels in the legal landscape for data procurement: International, national and sub-national. With so many principles and regulations, *Prof. Gabrynowicz* then proceeds to question what should be taken from the law. In the short term, she advocates the strengthening of coordination and communication among data providers such as world data centres, as well as to build upon scientific conventions of openly publishing and certifying data sets. In the medium-term, she proposes distinguishing between scientific and commercial data. The long-term goal as stated by *Prof. Gabrynowicz* is the establishment and maintenance of open archives for the repository of data. *Prof. Gabrynowicz* argues that to promote access to and exchange of data, there should be a presumption of openness, which should be institutionalised with some recognition of the need for clearly worded legitimate exceptions. There is also a need to harmonise and clarify inconsistent definitions and to publish relevant legislation in open source format.

Delegates and observers of the Legal Subcommittee were then called upon by the Chairperson to make their queries or comments. The delegate of ***Colombia*** started with a comment that the methodology of the Kyoto Protocol leads to an extremely complex mechanism for the exchange of data on climate change. He noted that the re-definition of terms is particularly necessary, and then asked the panel of speakers as to the description of the ideal international legal framework for data provision. ***Prof. Gabrynowicz*** answered that her proposal of a presumption of openness at the international level would be the starting point in the generation of an ideal international legal framework.

Prof. Dr. Frans G. von der Dunk then commented that various speakers referred to the 1986 Remote Sensing Principles, and asked for an opinion on how the RS Principles could be made applicable to climate change, especially in light of the fact that the Principles were drafted specifically for the earth observation of land for economic use and not for climate change. ***Prof. Monserrat Filho*** answered that the concept of remote sensing is general and therefore there should be an identification of those remote sensing activities that relate to climate change. ***Prof. Gabrynowicz*** noted that the word “environment” appears in the RS Principles, and proposed that the scientific relationship between “environment” and “climate” should be resorted to, so as to clarify the law.

The delegate of ***France*** then pointed to the 2007 Directed issued by the European Union (EU), which obliges member States that collect earth observation data to free up databases according to international standards. It was noted that satellite data providers, EU member States and some data users have funded this operation, and there is access to information that is freely available.

Mrs. Tanja Masson, *President, International Institute of Space Law*, then asked the panel as to their views relating to the Chilean proposal to review the 1986 Remote Sensing Principles, whether the Principles should be adapted to current demands, whether a new treaty would be the solution, or whether working agreements between agencies (e.g. INSPIRE) would be the best path. ***Prof. Gabrynowicz*** pointed to the Disasters Charter as a good point of reference, noting that the activities may be different, but the value of the Charter is that there is practice evolving based on actual events. ***Prof. Monserrat Filho*** noted that the present level of remote sensing activities deserves a better document than the RS

Principles and it is necessary to update the Principles.

The delegate of ***Algeria*** then noted that there is a difference in the nuances between States. He noted that it was a question of negotiation after a sound understanding of what other States are doing for climate change. In particular, developing States should define precisely their negotiating positions so as to understand the global scenario better. Spatial remote sensing data is particularly crucial, and in this context the COPUOS should provide a forum for States to work towards the benefit of all.

Prof. Sergio Marchisio mentioned that one of the most controversial issues in the UNFCCC is with regard to its Article 4 precautionary approach. This approach fixes scientific, social and economic fields as one of the main objectives of the convention, including a specific reference to data archives. He noted that planet monitoring is one of the most important tools to constantly monitor climate changes and global temperature. He then tabled the question as to whether this reference in the UNFCCC is sufficient, or whether there is room still for critique. ***Prof. Loibl*** answered that the question of data collection has not played a prominent role in the negotiation so far. The issues relating to making the Kyoto Protocol operative, as well as post-2012 actions were focused on, with data collection taken for granted. ***Prof. Gabrynowicz*** further noted that the interesting point about data access and climate change is that different types of data are necessary. As such, political will to cooperate is necessary.

The delegate from ***Belgium*** commented thereafter of his uncertainty after listening to the Symposium that the debate is really of a legal nature. He mentioned that if it is, there are questions as to whether the issues at hand relate to the 1986 RS Principles, and whether at all the focus should be on space applications, when it should be on climate change. Space monitoring worried the delegate because there may well be developing regimes of law which are contradictory when seen from the perspective of space applications. The delegate instead proposed that this should be left up to the experts on climate change, since in itself climate change is not a legal problem. The only legal issue at and is that access to available data may be prevented due to a lack of knowledge and the lack of a coherent approach to data provision and use. ***Prof. Gabrynowicz*** noted that the lack of knowledge as to the availability and use of data poses the greatest challenge to the use of space applications for global climate change.

Day 2: Legal Implications of Space Applications for Global Climate Change – Institutions and Instruments

Chairperson: ***Prof. Sergio Marchisio***, President, European Centre for Space Law

Speakers: ***Dr. Evangelina Oriol Pibernat***, Centre for Earth Observation, European Space Agency (ESA/ESRIN), Italy ("Coordination Instruments and Satellite Observation of the Climate System: the Contribution of CEOS"); ***Dr. Gisela Süss***, European Space Agency (ESA/HQ), France ("Monitoring the Environment for Climate Change: the Case of GMES"); ***Ms. Masami Onoda***, Japan Aerospace Exploration Agency (JAXA) / University of Kyoto, Japan ("Monitoring the Kyoto Protocol and Beyond: Greenhouse Gases Observation & the Global Forest Carbon Monitoring System"); ***Prof. Dr. Frans G. von der Dunk***, University of Nebraska, United States of America ("Legal Aspects of Climate Monitoring by Means of Treaty Law"); ***Prof. Dr. Vladimír Kopal***, Chairperson, Legal Subcommittee, United Nations COPUOS and University of Pilsen, Czech Republic ("Concluding Remarks")

Day 2 of the Symposium was based on the theme of "Legal Implications of Space Applications for Global Climate Change – Institutions and Instruments". ***Prof. Sergio Marchisio***, President, European Centre for Space Law, chaired the session. Prof. Marchisio opened the session with a look at the 4th Assessment Report of the Intergovernmental Panel on Climate Change (IPCC), noting that the Earth could warm by as much as 6 degrees Celsium by 2030. He concluded therefore that space agencies and other mechanisms for international cooperation must understand the seriousness and urgency of climate change, and to begin work on this field.

Dr. Evangelina Oriol Pibernat, Centre for Earth Observation, European Space Agency (ESA/ESRIN), Italy launched the second session with her presentation on "Coordination Instruments and Satellite Observation of the Climate System: the Contribution of CEOS". *Dr. Oriol Pibernat* gave the historical development of the Committee of Earth Observation Satellites (CEOS). CEO was formed in 1984 with a view to harmonising and coordinating earth observation activities worldwide. It now has twenty-six members, mostly consisting of space agencies; and twenty associate members, which are national and international organisations that do not own any space assets. The European Space Agency (ESA) is a founding member of CEOS. Next *Dr. Oriol Pibernat* turned to the discussion of the Group on Earth Observation (GEO), and the Global Earth Observing System of Systems (GEOSS), which was launched in 2003 to integrate not only space-based systems, but also air-, land- and ocean-based systems that monitor the climate through a data management system that assessing various societal benefit areas affected by earth observation. *Dr. Oriol Pibernat* also discussed the Global Climate Observing System (GCOS), in particular focusing on its Implementation Plan. This Plan has several crosscutting actions of integrated products, reprocessing, re-analysis and archiving, so as to enable a complete atmosphere and climate monitoring database. She also considered the role of the various space agencies throughout the world, the CEOS virtual constellations, ESA, and the European Global Monitoring for Environment and Security (GMES) Programme in the GCOS.

"Monitoring the Environment for Climate Change: the Case of GMES" was the title of the presentation given by ***Dr. Gisela Süss***, European Space Agency (ESA/HQ), France, which flowed seamlessly from the earlier presentation. *Dr. Süss* described in detail the purposes, operations and architecture of the GMES system. She noted that the GMES is an EU-led initiative, with three elements: services, in-situ and space. The services component produces information services in response to the European policy priorities in environment and security, and relies on data provided by the in-situ and space components. The in-situ component is mainly a national responsibility with coordination at the European level, and can be ground-, air-, sea- or space-based. The space component sees ESA with a two-fold role. ESA is the development agency for dedicated space infrastructure, including earth observation missions developed specifically for GMES ("Sentinels"). ESA also has a coordinating function with its Member States, EUMETSAT and other intergovernmental organisations, as well as private and commercial actors. The latter function sees ESA providing its long experiences in third-party mission schemes to the GMES programme. *Dr. Süss* then spoke on the cooperation between the European Commission (EC) and ESA, both generally and specifically on the GMES initiative. She traced the timeline on the EC-ESA cooperation, noting that the December 2005 ESA Conference of Ministers in Berlin provided 255 million Euro for Phase 1 of the GMES Space Component Prgamme, and that January 2007 saw the approval of the EC Framework Programme 7, with 1.2 billion Euro allocation to GMES. Funding was further augmented in September 2007 with an investment of 501 million Euro into Phase 2 of GMES. The overall indicative EC budget for the GMES Space Component is 710 million Euro. *Dr. Süss* then turned to the ESA-EC Agreement on GMES, indicating the legal issues arising from conflicting procurement rules. She also discussed governance and data policy issues relating to GMES, and noted that the GMES is an on-going project with great promise.

Ms. Masami Onoda, Japan Aerospace Exploration Agency (JAXA) / University of Kyoto, Japan, was the next speaker. Her presentation addressed the topic of "Monitoring the Kyoto Protocol and Beyond: Greenhouse Gases Observation & the Global Forest Carbon Monitoring System". *Ms. Onoda* introduced the early forms of use of earth observation data in international institutions, and explained the relationship between earth observation data and the arms control regime. She then turned to the present institutional model of information gathering, decision making and risk assessment, considering the structure in light of the legal context and its contribution to climate change issues. *Ms. Onoda* noted that earth observation is generally used for both environmental and compliance monitoring, and made references to various international treaty regimes in this regard. She then focused on the UNFCCC, emphasising the change in terminology from space "monitoring" to "systematic observation", a change that originated in the Vienna Convention for the Protection of the Ozone Layer. The Japanese GOSAT and American OCO missions were then discussed in the context of the UNFCCC obligations and the Kyoto Protocol mechanisms. *Ms. Onoda* discussed the potential of geo-information data in realising the goals and obligations of the UNFCCC framework, in particular with regard to the IPCC Guidelines. She illustrated this potential with an example from a project helmed by the Japanese Ministry of Environment on forest carbon monitoring, where satellite data is used to derive information on the status of forest cover mapping and the identification and quantification of changes. *Ms. Onoda* then discussed the integration of political requirements and technological solutions, mooting that it is essential to achieve optimal institutional procedures and space systems design. She noted that international space law is silent on these issues and proposed that greater transparency and cooperation between States may be the best way forward.

The last presentation of the Symposium was given by ***Prof. Dr. Frans G. von der Dunk***, University of Nebraska, United States of America, on the topic of "Legal Aspects of Climate Monitoring by Means of Treaty Law". In his presentation, *Prof. von der Dunk* took a bird's eye view of the topic, presenting four essential aspects on the theme: Dispute settlement, reference to satellite data in treaties, authenticity guarantee of satellite data, and the inclusion of quantifiable parameters. On the issue of dispute settlement, *Prof. von der Dunk* argued that any dispute settlement mechanism in this regard should be tailor-made to the topic in view of its technical complexities. Referring to the International Tribunal on the Law of the Sea and other specialised tribunals, he argued that a specialised tribunal is necessary for this theme. Secondly, *Prof. von der Dunk* mooted that satellite data should be specifically referred to in treaties, as was done in the 1998 Aarhus Convention. Thirdly, he discussed the necessity to certify data authenticity so as to establish a clear audit trail. Lastly, he mooted the determination of objective, measurable parameters by a recognised body such as CEOS or GEOSS, so as to provide standards for data provision and use. in looking to the future, he noted that treaties would be useful for reinforcing use of earth observation data in combating climate change if they are properly developed, with a special focus on enforcement issues.

The Chairperson then opened the floor for comments and questions. ***Dr. Peter van Fenema***, the observer from the *International Astronautical Federation*, asked ***Dr. Süss*** whether, with the European GMES system specifically addressing both environmental and security needs, one use may interfere with the other through the guise of national security. *Dr. Süss* acknowledged the conflict. However, she replied that at present it is difficult for her to make a proper statement as the GMES data policy is being elaborated. Further, ESA should only comment on this issue together with the EC.

The delegate from ***France*** then noted that there should be an establishment of a database with data from all available earth observation satellites. He then asked as to the actions undertaken to take into account the interests of the developing States, which level of data will be made available, and how such data would be made available. ***Dr. Süss*** explained that the international exchange of data is being considered within the elaboration of the GMES data policy. She said that the time is not yet right to deal with this issue, since the coordination of the existing satellite data is still being discussed. ***Dr. Oriol Pibernat*** further added that the GEO is currently examining the aspects raised with regard to developing States, and that this will also be raised at all levels with the GMES initiative when the time is correct. ***Prof. Monserrat Filho*** then asked the panel and the delegates whether they would agree to this proposition being discussed by the Legal Subcommittee on its regular agenda. ***Prof. Marchisio*** indicated that ***Prof. Kopal*** will address this last issue in his "Concluding Remarks".

Prof. Dr. Vladimír Kopal, Chairperson, Legal Subcommitteee, United Nations COPUOS and University of Pilsen, Czech Republic, then proceeded with his "Concluding Remarks" to the Symposium.

Prof. Kopal opined that the Symposium was very substantive and rich in advice and legal observations. Some contributions were very instructive and academic, such as that from ***Prof. Loibl***, which was very instructive on the principles and regulations of environmental law. The reports presented by ***Ms. Onoda, Dr. Oriol Pibernat*** and ***Dr. Süss*** showed how complex the situation with regard to environmental monitoring and climate change is, and how complicated the developments in this field have become. The presentations given by ***Prof. Monserrat Filho, Prof. Gabrynowicz*** and ***Prof. von der Dunk*** were more oriented towards legal conclusions and towards pinpointing the real legal issue at hand that should be discussed at the Legal Subcommittee of the COPUOS. ***Prof. Kopal*** noted that the subject matter is very innovative, and it is a new field with new horizon highlighting problems in the development of the law in this respect. He then advised speakers for the next Symposium to formulate in specific suggestions to the Legal Subcommittee for consideration.

Prof. Marchisio welcomed the insight and advice of ***Prof. Kopal***, and after thanking the speakers, delegates and observers, closed the Symposium.

The 2009 IISL-ECSL Symposium will be held on the first two afternoons of the 48th Session of the Legal Subcommittee of the United Nations Committee on the Peaceful Uses of Outer Space.

The papers and presentations contributed by the speakers may be obtained from the website of the United Nations Office for Outer Space Affairs (UN OOSA) at http://www.unoosa.org/oosa/COPUOS/Legal/2008/symposium.html (last accessed: 23 April 2008).

Legal Features of the Climate Change Convention: from Kyoto to Bali

Gerhard Loibl, University of Vienna, Austria

Introduction

In December 2007 the 13th session of the Conference of the Parties (further: COP13) to the United Nations Framework Convention (further: UNFCCC) and the third session of the Conference of the Parties serving as the meeting of the Parties (further: COP/MOP) to the Kyoto Protocol (further: KP) was held in Bali. These meetings got intensive attention by public as they were regarded as decisive for the further evolution of the international regime addressing climate change.

Let me remind you of the international legal framework addressing climate change.

The UNFCCC, which was adopted and opened for signature at the United Nations Conference on Environment and Development (held in Rio de Janeiro in 1992), contains, inter alia, general principles and institutional arrangements. Its Article 2 sets out the objective of the international climate regime. It reads as follows:

"The ultimate objective of this Convention and any related legal instruments that this Conference of the Parties may adopt is to achieve, in accordance with the relevant provisions of the Convention, stabilizing greenhouse gas concentrations in the atmosphere at a level that would prevent dangerous anthropogenic interference with the climate system. Such a level should be achieved within a time-frame sufficient to allow ecosystems to adapt naturally to climate change, to ensure that food production is not threatened and to enable economic development to proceed in a sustainable manner."

It should also be underlined that UNFCCC contained commitments – from a legal point it may be discussed whether they are politically or legally binding – for the "industrialized countries", in the climate change regime they are known as Annex I countries, to reduce their emissions of CO2 equivalent to their 1990 levels. By doing so they would demonstrate that they take the lead in combating climate change.

The Kyoto Protocol, which was adopted in 1997, was based on the so-called "Berlin mandate" adopted by COP1, held in Berlin in 1995. The Berlin mandate established a (negotiating) process as the CP.1 after having reviewed Article 4 (a) and (b), of UNFCCC concluded that these subparagraphs were not adequate.[1] It stated, inter alia, will "not introduce any new commitments for Parties not included in Annex I, but reaffirm existing commitments in Article 4.1 and continue to advance the implementation of these commitments in order to achieve sustainable development, taking into account Article 4.3, 4.5 and 4.7".[2]

Why is there a need to move now?

The Kyoto Protocol established in Article 3 paragraph 1 a commitment period from 1 January 2008 to 31 December 2012 to reduce or limit CO2 emissions. The commitments for Annex I Parties are set out for each of them in Annex B, ranging from reduction commitments of 8 % to a limited increase of emissions by 10 %. Thus, if there is the need for any further commitments to be agreed before the expiry of the first commitment period set out in the KP.

---

1 See preamble of Decision 1/CP.1.
2 Decision 1/CP.1, paragaph 2 (b).

Article 3 paragraph 9 takes this already into account. It reads: "Commitments for subsequent periods for Parties included in Annex I shall be established in amendments to Annex B to this Protocol, which shall be adopted in accordance with Article 21, paragraph 7. The Conference of the Parties serving as the meeting of the Parties to this Protocol shall initiate the consideration of such commitments at least seven years before the end of the first commitment period referred to in paragraph 1 above."

The Kyoto Protocol took much longer to enter into force than had been expected. That was due to the fact that the Kyoto Protocol only contained very general rules e.g. concerning the mechanisms (joint implementation, clean development mechanisms and emissions trading), a compliance mechanism and financial mechanisms. Only at CP.7 in 2001 agreement could be reached on controversial questions, inter alia, on the specific rules for the operation of the mechanisms, the functioning of the compliance mechanism, financial issues. These compromises reached in Marrakesh, the Kyoto Protocol became "rectifiable" for Annex I Parties, as they knew then the conditions under which they would have to implement their commitments, i.e. the reduction and/or limitation of their emissions. The Kyoto Protocol only entered into force in 2005 after the Russian Federation ratified it in 2004.

At the first COP/MOP, held in 2005 in Montreal, the future of the Kyoto Protocol became the essential issue. A number of different tracks were discussed and set up:

- The Convention dialogue was established to discuss the future of the Convention. Ideas and suggestions were to be collected on how the objective of the Convention is to be achieved by the international community. The decision establishing the dialogue acknowledged that "the global nature of climate change calls for the widest possible cooperation and participation in an effective and appropriate international response"[3] and stated that "the dialogue will take the form of an open and non-binding exchange of views, information and ideas in support of enhanced implementation of the Convention, and will not open any negotiations leading to new commitments."[4]
- The Ad Hoc Working Group on Further Commitments for Annex I Parties under Kyoto Protocol was established in accordance with Article 3.9 (as mentioned above).

Two other issues were raised:

Article 9 review: Article 9 provides that a first review of the Kyoto Protocol in the light of the best available scientific information and assessments on climate change and its impacts, as well as relevant technical, social and economic information shall take place at the second session of the COP/MOP.

The so-called "Russian proposal": the Russian Federation presented orally a proposal at COP/MOP1 that in the future a simplified procedure should be established for joining Annex B. The first review under Article 9 was held at COP/MOP2 in Nairobi 2006. No substantive discussion took place. Neither was there a substantive discussion of the Russian proposal, but it was agreed that during the meeting of the subsidiary bodies in spring 2007 in Bonn the Russian Federation would make a presentation about its proposal.

The presentation in Bonn made it clear that the Russian proposal consisted of two parts:
The first part addresses simplified procedures to join Annex B of the Protocol. According to the provisions of the KP any amendment of Annex B needs to be adopted by consensus, or, as a last resort, by a three-fourths majority of the parties present and voting, as well as the written consent of the party concerned. Moreover, the amendment enters into force ninety days after three-fourths of the parties to the Protocol have deposited their instruments of ratification. This may in

[3] Decision 1/CP.11, Dialogue on long-term cooperative action to address climate change by enhancing implementation of the Convention, UN Doc. FCCC/CP/2005/5/Add.1.

[4] Recital 7 and paragraph 2.

practice lead to a number of legal questions, such as a "fragmentation of commitments" between parties.

The second part deals with options to recognise contributions towards achieving the ultimate objective of the Convention by way of voluntary commitments. The Russian proposal suggests that if countries take commitments based on the implementation of national policies and measures. A party should be rewarded if it reaches its voluntary commitments, but will not be "punished" – as Annex I parties – if they do not. Such "possible types of benefits", could be the following:

- Emissions trading;
- Technology transfer options;
- Finance mechanisms (such as insurances); and
- Other types.[5]

IPCC reports predict an increase in global temperatures. The fourth assessment report predicts that measures have to be taken within the next years in order to slow down the increase of temperatures. The question is how much temperature increase is "acceptable" to the global eco-system without threatening the objective that is set out in Article 2 UNFCCC.

The Bali Sessions and the future negotiations

A post-2012 arrangement has to address four main building blocks:

- Mitigation (measures against climate change)
- Adaptation (measures to adapt to climate change)
- Technology transfer
- Financial resources.

At the Bali meetings an institutional set-out for the future negotiations has been agreed as well as a time-table.

- The Bali Action Plan addresses the further process under the Convention. The process is to be conducted in an Ad Hoc Working Group on Long-term Cooperative action under the Convention (AWG). The decision sets out those issues that should be discussed by this new AWG. Paragraph 1 lists as items:
    - Shared vision for a long-term action: i.e. how is the objective of the Convention to be reached taking into account the provisions and principles of the Convention, in particular the principle of common, but shared responsibilities;
    - Enhanced national/international action on mitigation;
    - Enhanced action on adaptation
    - Enhanced action on technology development and transfer to support action on mitigation and adaptation
    - Enhanced action on the provision of financial resources and investment to support action on mitigation and adaptation and technology cooperation.
- The Ad Hoc Working Group on Further Commitments for Annex I Parties under the KP is to discuss as the title already states the future commitments of Annex I Parties. These discussions focus on the question of mitigation measures of Annex I Parties, i.e. what should they do to reduce emissions.
- Second Review of the Kyoto Protocol under Article 9 of the Kyoto Protocol.
    - Article 9 reads as follows:

"1. The Conference of the Parties serving as the meeting of the Parties to this Protocol shall periodically review this Protocol in the light of the best available scientific information and assessments on climate change and its impacts, as well as relevant technical, social and economic information. Such reviews shall be coordinated with pertinent reviews under the Convention, in particular those required by Article 4,

[5] Presentation of the Russian Federation at the Workshop on the Russian Proposal, 11 May 2007, Slide 10, available at http://unfccc.int/files/meetings/workshops/other_meetings/application/pdf/russian_presentation.pdf

paragraph 2(d), and Article 7, paragraph 2(a), of the Convention. Based on these reviews, the Conference of the Parties serving as the meeting of the Parties to this Protocol shall take appropriate action.

2. The first review shall take place at the second session of the Conference of the Parties serving as the meeting of the Parties to this Protocol. Further reviews shall take place at regular intervals and in a timely manner."

It was agreed in Bali that the second review will address specific issues, such as:

- Extending the share of proceeds to assist in meeting the costs of adoption to joint implementation and emissions trading;
- Relevant procedural elements for inscribing commitments for Annex I Parties in Annex B;
- Privileges and immunities for individuals serving on constituted bodies established under KP
- Scope, effectiveness and functioning of the flexibility mechanisms, including ways and means to enhance an equitable regional distribution of CDM projects;
- Minimization of adverse effects, including the effects of climate change, effects of international trade, and social, environmental and economic impacts on other Parties.

- On the Russian proposal it was agreed that elements of this proposal should be introduced during the second review of the KP pursuant to Article 9 and in the Ad Hoc Working Group on Long-term Cooperative Action under the Convention.

In Bali it was agreed that the post-2012 arrangement concerning climate change is to be agreed at COP15 and COP/MOP5, which is to be held in Copenhagen. It will be for COP14 and COP/MOP4 in Poznan/Poland to evaluate the progress made in the various fora and to propose the work-programme until December 2009.

Promoting Access to, and Exchange of, Data and Information Related to Climate Change: a Legal Perspective

Executive Summary

by

Prof. Joanne Irene Gabrynowicz

The context for data sharing includes the fact that technologies and practices related to climate change are developing much more quickly than the law. Law is, by design, reactive, whereas technology is, by intent, proactive. Practices are, of necessity, responsive and evolutionary. Climate transcends all organizational elements of legal institutions, jurisdictions, and competencies. In short, it is extremely complex.

Complexity is compounded by the fact that data sources are numerous. Space – based platforms include environmental, weather, remote sensing, and navigation satellites. Aerial platforms include aircraft, balloons, sondes, etc. There are also a wide variety of other in-situ platforms: ground-based sensors, buoys, CTDs, and human-made measurements.

On the legal side, the landscape for data is equally complex and involves many legal systems. At the international level there are various relevant treaties; principles; and, binding and nonbinding agreements. At the national level there are constitutions, laws, regulations, and codes. Many nations have smaller political units like states, cantons, and provinces, etc. that have additional local municipal laws that can also govern data acquisition and access.

At the international level, some of the most important relevant instruments include the Kyoto Protocol; the U.N. Principles Relating to Remote Sensing of the Earth from Outer Space (Principles); WMO Resolution 40; the Montreal Protocol; the Agreement Between US NOAA and EUMETSAT on Joint Transition Activities Regarding Polar-orbiting Operational Environmental Satellite

Systems; the Disasters Charter; and, the Group on Earth Observations Data Principles (in Progress).

Access and exchange strategies ought to be considered in the short-term, medium-term, and long-term. They are not mutually exclusive, and ought to be considered as complements to one another. Short-term strategies ought to seek to strengthen coordination and communication among data providers. They can build on scientific conventions such as openly publishing and certifying data sets. Medium – term strategies ought to distinguish between the scientific and commercial (however defined) value of data. Typically, commercial value decreases over time as scientific value increases. The most important aspect of long-term strategies is to establish and maintain open archives. Examples of institutions that employ these strategies include the World Data Centers, WMO, and the U.S. National Satellite Land Remote Sensing Data Archive.

The single most important component of any policy that is intended to promote data access and exchange is a presumption of openness. That is it should be presumed that all data is subject to open access. Generally, this is starting point of existing laws and policies. To promote access, the presumption ought to be institutionalized. Agreements should include specific open access language and, if necessary, recognize the need for some legitimate exceptions, such as, personal privacy; indigenous rights; conservation and protection of sensitive ecological, archaeological, or cultural resources; national security; and, propriety interests. However, these exceptions ought to be clear and limited and should not outweigh the presumption of openness. Precise "sunset clauses" should be set for each and every period of restricted access. There should be a specific end date applicable for each and every restriction.

To encourage participation in data sharing, there should be

attribution of all significant data and information sources. Inconsistent definitions for critical terms like "remote sensing", "raw data", "value-added", etc. should be clarified and harmonized to the degree possible. All laws and policies regarding data sharing should be published in open, accessible, public sources.

Regarding satellite data there are some very specific current trends. Overall, access to data is the presumed norm with exceptions for national security. For high-resolution systems the number and kind of exceptions are growing worldwide. However, for medium to low-resolution data, access is growing and data from them are more openly available. The Principles and the nondiscriminatory access policy are routinely acknowledged in various laws, policies, and agreements. However, regarding high-resolution systems, the Principles and the nondiscriminatory access policy are being more narrowly construed in national laws.

Treaty Law in Support of Climate Monitoring

Prof. Dr. Frans G. von der Dunk
University of Nebraska-Lincoln, College of Law

1. Introduction

The other papers presented at the present IISL-ECSL Symposium have all offered extended and detailed elaborations on various aspects of climate change monitoring and efforts to combat it, with a view *inter alia* to their legal aspects and parameters. The present paper, following upon the heels of those, presents an effort to tie some of the strands developed in the course of the two sessions of the symposium together.

Thus, it takes a bird's eye view of the contribution treaty law as the most specific source of international law available[1] can make to efforts to solve or mitigate this crucially important problem, in particular – with a view to the context offered by the symposium as part of the opening sessions of the Legal Subcommittee of UNCOPUOS – focusing on the benefits that remote sensing[2] and other attendant uses of space technology[3] could bring to such an effort.

The task facing the international law-making community in this respect was aptly summarised by Dr. K. Kasturirangan, former Head of the Indian Space Research Organisation (ISRO): "Thus, the great challenge is to develop this space-web for (...) bringing more clarity to inputs on measurements and international treaties and conventions, thus bringing in transparency to global climate

[1]. Famously, the International Court of Justice (ICJ) is charged to solve disputes between states by basing itself on the sources of international law as enumerated by its Statute (Statute of the International Court of Justice, San Francisco, done 26 June 1945, entered into force 24 October 1945; 156 UNTS 77; USTS 993; 59 Stat. 1031; UKTS 1946 No. 67; ATS 1945 No. 1); Art. 38(1) pointing to "international conventions, whether general or particular, establishing rules expressly recognized by the contesting states" as the first of such sources. This clause in the ICJ's Statute, which is an integral part of the United Nations Charter (Charter of the United Nations, San Francisco, done 26 June 1945, entered into force 24 October 1945; USTS 993; 24 UST 2225; 59 Stat. 1031; 145 UKTS 805; UKTS 1946 No. 67; Cmd. 6666 & 6711; CTS 1945 No. 7; ATS 1945 No. 1), has since been commonly understood as stating, indeed, the sources of international law also outside of the specific setting of an ICJ case, even if later developments in international law may have led to additional sources being recognised.

[2]. It should be noted that in writings the terms "remote sensing" and "earth observation" are often used interchangeably, alternatively the latter term is used by others to refer to a somewhat narrower version of the former, focusing on 'the earth' and leaving out observation of the skies and perhaps even the oceans (separately headed "meteorology" in many cases). For those reasons, the term "remote sensing" is used in the current paper, as it may indeed be the comprehensive use of space remote sensing technology for monitoring any of those phenomena which may contribute to the monitoring and, thereby, mitigation of climate change. Cf. already M. Ferrazzani, The Status of Satellite Remote Sensing in International Treaties, in *'Project 2001' – Legal Framework for the Commercial Use of Outer Space* (2002), 179; C.Q. Christol, Remote Sensing in an Era of Global Warming, *Proceedings of the Fiftieth Colloquium on the Law of Outer Space* (2008), 405-10.

[3]. Though the focus is, and should be, on remote sensing technology as referred to *supra*, n. 2, it should be noted that space technologies increasingly become mixed, e.g. for effective use of remote sensing data telecommunications, including satellite communications, is indispensable, and the value of such data in enormously enhanced by attendant usage of satellite positioning systems such as the US GPS.

convention management."[4] Since treaties and conventions have the specific advantage over other sources of international law of being on the face of it clear, or at least less ambiguous and debatable than for example customary international law, their specific benefit here might well be to ensure that such "inputs" and "measurements" are more easily and formally accepted.

From such a bird's eye perspective then, four closely interlinked aspects seem to be key: (1) treaties should somehow include, at least from the theoretical vantage point, a dedicated and specialised dispute settlement system in one form from another; (2) treaties should include references to satellite data as tools for monitoring; (3) as following from the foregoing an effort should be made through relevant treaties to somehow guarantee the authenticity of satellite data for the intended purposes; and (4) in order *inter alia* to support the previous points treaties should include, as clearly as possible, quantifiable and measurable parameters upon which to base decisions as regarding the violation of treaty obligations and the appropriate measures to take. I will address each of those four aspects in somewhat more detail.

2. Include appropriate dispute settlement clauses in climate change conventions

The main problem in the application or establishment of any system of settlement of disputes over interpretation, application and/or perceived violation of obligations under treaties with respect to climate change concerns the technical complexities involved in climate change – both legally and otherwise. Effectively therefore, the dispute settlement system needs to be tailor-made for the cause. This makes reference to existing courts and tribunals, which are either of a general nature or focused on a different area of international law and activities, less obvious, and possibly even counter-productive.

Climate change issues are not unique in this sense, however, and there have been other areas where the subject matter was considered to be specialised enough to warrant the establishment of separate, specialised judicial structures. Consequently, it would be worthwhile for further studies on climate change treaty development to review and analyse some of those precedents.

A first example thereof that comes to mind is the International Tribunal for the Law of the Sea[5], as this "Tribunal shall be composed of a body of 21 independent members, elected from among persons enjoying the highest reputation for fairness and integrity and *of recognized competence in the field of the law of the sea.*"[6]

Similarly, the Statute of the International Court of Justice allows for the possibility for parties to a dispute to have that dispute settled by a special chamber, formally subject to the authority of the Court. Thus, chambers of a more permanent character may be formed "for dealing with particular categories of cases", specific examples referred to being "labour cases and cases relating to transit and communications".[7] The

[4]. K. Kasturirangan, Space technology for humanity: A profile for the coming 50 years, 23 *Space Policy* (2007), 162-3.

[5]. See Artt. 186-191, 285, also Annex VI, United Nations Convention on the Law of the Sea, Montego Bay, done 10 December 1982, entered into force 16 November 1994; 1833 UNTS 3 & 1835 UNTS 261; UKTS 1999 No. 81; Cmnd. 8941; ATS 1994 No. 31; 21 ILM 1261 (1982).

[6]. Art. 2(1), Annex VI, United Nations Convention on the Law of the Sea (emphasis added).

[7]. Art. 26(1), Statute of the International Court of Justice.

idea behind this was to have those judges on the Court with specific knowledge of the non-legal aspects of a specialised subject matter adjudicate relevant cases. In addition, the possibility of using the instrument of *ad hoc* chambers is also offered: "The Court may at any time form a chamber for dealing with a particular case", an option which again can be used amongst others to allow substantive specialists on the Court to adjudicate a relevant case.[8]

The area closest to climate change in terms of subject matter, if not actually and effectively (at least under some definitions) encompassing climate change itself, obviously is that of environmental protection broadly speaking. Here, as a matter of fact, many interesting precedents may be found, worthy of further scrutiny.[9]

One interesting example on a national level for instance is offered by the case of India, where the Supreme Court of India has established a Central Empowered Committee in order to deal with the technical intricacies of the 1980 Forest Conservation Act in terms of the admissibility as evidence of certain satellite data in court.[10] Other examples refer to the United States and Australia (Queensland in particular).[11]

Of course, such national developments from the perspective of global climate change raise the question of desirability of (potentially widely varying) national judgements. Although it will perhaps be inevitable that certain countries take the lead in this respect (and such leadership is even to be lauded), at the end of the day we need a global system also in this respect, or more to the point a global framework within which national regulations and national dispute settlement systems may still have an important role to play – just not a completely independent one. And this in turn, of course, calls for an international treaty providing precisely for such a dispute settlement framework.

Also, in Europe in this regard once more the potential of harmonisation of national laws of the – now twenty-seven – member states of the European Union by means of EC law may be noted.[12] Indeed, the Union has since a decade and a half decisively moved into general environmental protection

[8]. Art. 26(2), Statute of the International Court of Justice.

[9]. Cf. N. Peter, The Use of Remote Sensing to Support the Application of Environmental Treaties, *Proceedings of the Forty-Sixth Colloquium on the Law of Outer Space* (2004), 74 ff.; M. Onoda, Monitoring Greenhouse Gases from Space and the Kyoto Protocol, *Proceedings of the Forty-Ninth Colloquium on the Law of Outer Space* (2007), 204 ff.

[10]. See R. Purdy, Satellites: A New Era for Environmental Compliance?, 3 *Journal for European Environmental Planning Law* (2006), 409, 412.

[11]. Cf. Purdy, 409, 411; and on the United States specifically H. Ginzky, Satellite Images as Evidence in Legal Proceedings Relating to the Environment – A US Perspective, 51 *Droit et Ville* (2001), 41-68.

[12]. The European Union was essentially created by means of the Treaty on European Union, Maastricht, done 7 February 1992, entered into force 1 November 1993; 31 ILM 247 (1992); OJ C 191/1 (1992); which *inter alia* incorporated the former EEC Treaty, now re-christened EC Treaty, properly speaking; see Treaty Establishing the European Community (Consolidated Version); OJ C 325/33 (2002). In particular the Treaty of Amsterdam (Treaty of Amsterdam Amending the Treaty on European Union, the Treaties Establishing the European Communities and Certain Related Acts), Amsterdam, done 2 October 1997, entered into force 1 May 1999; OJ C 340/73 (1997)) and the Treaty of Nice (Treaty of Nice amending the Treaty on European Union, the Treaties establishing the European Communities and certain related acts), Nice, done 26 February 2001, entered into force 1 February 2003; OJ C 80/1 (2001)) further amended both treaties, including some rearrangement and renumbering of Articles.

legislation, to ensure that for example economic motives of individual member states would not be allowed to undercut a high, and progressively higher, level of environmental protection throughout the Union.[13] Satellites, moreover, are indeed gradually becoming involved in that process.[14]

Further to this, at least as far as EC law is concerned an extended judicial dispute settlement system exists, spearheaded by the European Court of Justice (ECJ) and further comprising the Court of First Instance as well as, essentially, the comprehensive judicial systems of the member states, to address any violations of relevant law.[15] As a consequence, at least some down-to-earth environmental disputes have been adjudicated by the ECJ.

At the same time, whilst EC law contains a wide range of environment-related regulations, as yet none of them specifically refer to global climate change, meaning that the aforementioned judicial system can not yet be used for legal action and/or dispute settlement in relevant cases.[16] Also, the European case is no different from others in that the need for sufficient specialised know-how on climate change issues might not be sufficiently taken care of by this general system of adjudication. Even more importantly, also a Europe-wide effort falls short of the global one required. Still, the European case may serve as an interesting precedent for sovereign states accepting some measure of supranational adjudication on environmental issues.

Finally, the Global Monitoring for Environment and Security (GMES) project, recently renamed Kopernikus, should be mentioned here.[17] It represents a concerted European effort, led by the European Commission on behalf of the European Union and the European Space Agency (ESA)[18], to realise "an operational and autonomous European capability for global monitoring for environment and security" by 2008, involving satellites.[19] This capability in substance was to take the form of databases to be filled with relevant data, partly self-generated by the key GMES players, notably ESA and (likely) EUMETSAT[20], which was

---

[13]. See further e.g. R.H. Folsom, Principles of European Union Law (2005), 190-6, esp. 191.

[14]. Cf. Purdy, 408-9.

[15]. See Artt. 225-240, EC Treaty; cf. further Folsom, 70-93.

[16]. Jurisdiction of the ECJ by definition is limited to issues somehow – even if sometimes somewhat implicitly or indirectly – falling within the scope of EC law.

[17]. See Council Resolution on the launch of the initial period of global monitoring for environment and security (GMES), of 13 November 2001; OJ C 350/4 (2001); Communication from the Commission to the European Parliament and the Council – Global Monitoring for Environment and Security (GMES): Establishing a GMES capacity by 2008, COM(2004) 65 final, of 3 February 2004; Communication from the Commission to the Council and the European Parliament – Global Monitoring for Environment and Security (GMES): From Concept to Reality, COM(2005) 565 final, of 10 November 2005.

[18]. For ESA, GMES/Kopernikus comprises a set of activities undertaken as "optional activities" in accordance with Art. V(1.b), Convention for the Establishment of a European Space Agency (ESA Convention), Paris, done 30 May 1975, entered into force 30 October 1980; 14 ILM 864 (1975); Space Law – Basic Legal Documents, C.I.1. As a consequence of such characterisation, individual member states are entitled to opt out of those activities, and/or determine their individual levels of financial commitment to them; cf. also Art. XIII(2), ESA Convention.

[19]. Para. (3), Council Resolution of 13 November 2001.

[20]. EUMETSAT is the European Organization for the Exploitation of Meteorological Satellites established by the Convention for the Establishment of a European Organization for the Exploitation of Meteorological Satellites (EUMETSAT), Geneva, done 24 May 1983, entered into force 19 June 1986; as amended 14 July 1994, entered into force 27 July 1994;

considered another major stakeholder even though not (so far) an 'official' partner in the project, partly generated from 'outside providers'.

There can be little doubt about the comprehensiveness of the approach taken under GMES/Kopernikus as to what constitutes data relevant for environment and security, and its consequent contribution to climate monitoring: "GMES will provide the EU with a tool for participating in the international efforts (…) to strengthen the global climate observing system"[21].

Even GMES/Kopernikus, in spite of its broad scope and comprehensive character, however, does not in any way provide for a dispute settlement system itself. Once more, this testifies to the necessity to generate a first level of global dispute settlement mechanisms at the global level.

3. Include explicit references to satellite data in climate change conventions

In the above, reference was already made to GMES/Kopernikus, as amongst others crucially bent upon using satellite data (in addition to any other data considered useful for the purpose) for the purpose of combating climate change. More generally, of course also a 'system' such as the GEOSS, as 'combining' the various existing national and multi-national remote sensing systems, could play a role here.[22] As indicated, this raises the issue of the extent to which satellite data can become formally engaged and acknowledged as helpful, in certain cases even indispensable, tools in that context.

Also here other areas of international law show interesting examples where such references to satellite data are already incorporated in major legal regimes. One early example concerns the so-called MARPOL Convention[23], where an obligation is included for states parties to the Convention to "co-operate in the detection of violations and the enforcement of the provisions of the present Convention, *using all appropriate and practicable measures of detection and environmental monitoring*, adequate procedures for reporting and accumulation of evidence."[24] Subsequent practice has shown that indeed satellites and satellite data were seen to be included in that rather comprehensive formulation, and have been used occasionally to build a case for violation of relevant rules on marine environmental pollution, such as most notably in the Song San case.[25]

Furthermore, with reference to the previously discussed issue of dispute settlement, it may be noted that the MARPOL Convention calls for any dispute as regarding its interpretation, including issues concerning the interpretation and/or validity of

Cmnd. 9483; Space Law – Basic Legal Documents, C.III.1; 44 ZLW 68 (1995).

21. Communication of 10 November 2005, 5.

22. The Global Earth Observation System of Systems (GEOSS) is a loosely formed group of major remote sensing players with the aim of enhancing the coherent contributions to global benefits to be derived from their respective satellite systems to global issues such as related to the environment – including global warming and climate change issues.

23. International Convention for the Prevention of Pollution From Ships, 1973 as modified by the Protocol of 1978 (MARPOL Convention as Modified), London, done 17 February 1978, entered into force 2 October 1983; ATS 1988, No. 29.

24. Art. 6(1), MARPOL Convention as Modified (emphasis added). Under Art. 17(b) furthermore, states parties are even encouraged to support "the supply of necessary equipment and facilities for reception and monitoring". See further Purdy, 409; *Applications of Earth Observation to the legal sector*, Final Report, BNSC Sector Studies Programme, August 2001, para. 7.2.1.

25. Cf. e.g. N.J. Brehon, L'Utilisation des Satellites d'Observation pour la Détection des Déballastages en Mer, 51 *Droit et Ville* (2001), 102-3; Ferrazzani, 192.

satellite data in any given case, "if settlement by negotiation between the Parties involved has not been possible, and if these Parties do not otherwise agree, be submitted upon request of any of them to arbitration as set out in Protocol II to the present Convention."[26]

Another area where the potential to use satellites for verification purposes has already been addressed is that of disarmament and arms control. Most notably the (now defunct) ABM Treaty[27] of 1972 included a reference to non-interference with so-called 'national technical means' for verification of compliance of the parties to the treaty obligations, a clause widely agreed to include (national) satellites.[28]

From another angle, the International Court of Justice itself has already made use of satellite data in trying to solve boundary disputes – albeit with mixed success to the extent that the parties' varying interpretations and conclusions could not be verified or falsified by the use of the data as such.[29]

Thus, whilst on the one hand the use of satellite data is slowly becoming more acceptable, on the other hand there does not seem to be a clear-cut and/or widespread understanding on the side of the (rather dispersed) entities charged with monitoring – mostly, in any case, still the states parties to the treaty at issue themselves – of how, what and where satellites might or even should come into the picture with a view to their special advantages as compared to other monitoring means. There would seem to be sufficient reason to try and develop a more explicit and detailed system at the international level, for the time being perhaps developed as per individual treaty, for involving satellites in the task of monitoring treaty obligations of the parties.

By way of possible example, in the context of the European Union certain instruments of EC law authorise the use of satellite data for monitoring potential fraud in the case of farm subsidies, as well as fish catches with a view to quota limitations established under EC law.[30] This evolves from such legal instruments as Regulation 2371/2002[31], Regulation 2244/2003[32], and Regulation 796/2004[33].

---

[26]. Art. 10, MARPOL Convention as Modified.

[27]. Agreement Between the United States of America and the Union of Socialist Soviet Republics on the Limitation of Anti-Ballistic Missile Systems (ABM Treaty), Moscow, done 26 May 1972, entered into force 3 October 1972, no longer in effect 13 June 2002; TIAS 7503; 23 UST 3435.

[28]. See Art. XII(2), ABM Treaty. Further e.g. B. Cheng, Legal and commercial aspects of data gathering by remote sensing, *The Highways of Air and Outer Space Over Asia* (1992), 60-1; Ferrazzani, 188-9; S. Ushioda, Recent Developments in Multilateral Satellite Monitoring Systems, in *Issues in International Air and Space Law, and in Commercial Law* (1994), 387.

[29]. See e.g. Ferrazzani, 193-4; Purdy, 409-10, incl. n. 28; *Applications of Earth Observation to the legal sector*, paras. 8.3, 8.5.

[30]. Cf. e.g. Purdy, 408-9.

[31]. Council Regulation on the conservation and sustainable exploitation of fisheries resources under the Common Fisheries Policy, No. 2371/2002/EC, of 20 December 2002; OJ L 358/59 (2002); see Art. 23(3).

[32]. Commission Regulation laying down detailed provisions regarding satellite-based Vessel Monitoring Systems, No. 2244/2003/EC, of 18 December 2003; OJ L 333/17 (2003); see in particular Art. 4, providing for the requirement of "an operational satellite-tracking device installed on board" of Community ships subject to the regime.

[33]. Commission Regulation laying down detailed rules for the implementation of cross-compliance, modulation and the integrated administration and control system provided for in of Council Regulation (EC) No 1782/2003 establishing common rules for direct support schemes under the common agricultural policy and establishing certain support schemes for farmers, No. 796/2004/EC, of 21 April 2004;

4. Certify the authenticity of satellite data in climate change conventions

The next step in any evolution of legal instruments and frameworks for maximising the benefits satellite remote sensing could bring to climate change monitoring activities, which has already been alluded to, inexorably moves into the realm of the future. It would lie in a very fundamental discussion on how such data would qualify as evidence in judiciary proceedings.[34] Such a discussion might arise in any adversarial context, but the scope and extent of it could considerably be limited in case, somehow, the credibility and reliability of satellite data could be established in an *a priori*, objective and legally underwritten fashion.

As the (limited extent of) research undertaken on this matter has shown, so far in fact such evaluation and valorisation of satellite data has indeed largely been a matter of adversary testing.[35] Apart from the fact that the few instances where satellite data have so far been actually used in court have mainly seen those data being used in conjunction with other – *in situ* – data to verify and validate preliminary conclusions drawn from satellite data, this means that both sides in a dispute are likely to contest such value by means of their own technical experts and 'expert witnesses'.

OJ L 141/18 (2004); see in particular Art. 32(5.a).

[34]. See e.g. Ferrazzani, 191-4; and for the United States specifically, R.J. Rychlak, J.I. Gabrynowicz & R. Crowsey, Legal Certification of Digital Data: The Earth Resources Observation and Science Data Center Project, 23 *Journal of Space Law* (2007), 195-219.

[35]. Cf. e.g. R. Macrory & R. Purdy, The Use of Satellite Images as Evidence in Environmental Actions in Great Britain, 51 *Droit et Ville* (2001), 73-88; also Purdy, 410-1; Rychlak, Gabrynowicz & Crowsey, 198-202.

This may not altogether be a negative development, as over time it will certainly build a certain familiarity with, and understanding of, the value of satellite data, of what they can prove and what not, as well as of where they might be tempered with, on the part of the courts and arbitral tribunals, but this will be a long process – given the relative scarcity of cases involving satellite data, when compared example for instance to such recent 'new' types of evidence as faxes and e-mails.

Thus, there would seen to be a special interest in establishing an 'audit trail' system to certify authenticity,[36] in view of the highly-technical character of satellite data and their generation, the multiple processing going on from binary data to useful information[37] and the international character of most satellite operations – in many cases, satellite data necessary to help decide a

[36]. See already Macrory & Purdy, 81-4; Purdy, 411-2; *Applications of Earth Observation to the legal sector*, para. 4.5.1.

[37]. An interesting reference here would be to Principle I of the Principles Relating to Remote Sensing of the Earth from Outer Space, UNGA Res. 41/65, of 3 December 1986; UN Doc. A/AC.105/572/Rev.1, at 43; 25 ILM 1334 (1986); generally considered to contain customary law; see e.g. Ferrazzani, 182. Principle I namely makes a threefold distinction between data generated by remote sensing satellites: "(*b*) The term "primary data" means those raw data that are acquired by remote sensors borne by a space object and that are transmitted or delivered to the ground from space by telemetry in the form of electromagnetic signals, by photographic film, magnetic tape or any other means; (*c*) The term "processed data" means the products resulting from the processing of the primary data, needed to make such data usable; (*d*) The term "analysed information" means the information resulting from the interpretation of processed data, inputs of data and knowledge from other sources". This clearly indicates that already back in 1986 fundamental activities with regard to raw data such as 'processing' and 'interpretation' were necessary, were remote sensing satellites to deliver on their promises.

(legal) dispute may have to come from a jurisdiction quite different from the one where the dispute is being played out.

Taking into account the many differences that the special character of satellite operations and data bring with them, there are nevertheless interesting precedents that may serve as useful indicators of how to approach such an audit trail approach. Reference has been made already by a number of authors to such other new developments in the electronic area as automatic speed cameras – which may be especially valuable perhaps as illustrations, since remote sensing is basically also about cameras taking pictures, often in an automated mode.[38] Ultimately, scientific and technical experts may be called upon to help the lawyers in drafting the correct and feasible audit trails, but the call for development thereof should clearly emanate from the latter.

5. Insert quantifiable parameters into climate change conventions

That last remark also pertains to the final issue to be briefly addressed in this introductory paper on remote sensing satellites contributing to the monitoring and mitigation of climate change under appropriate treaty law. Quantifiable parameters should be included, wherever feasible, into any international treaties dealing with climate change issues, so as to allow a reasonably objective determination in specific cases as to whether a particular party has complied with the relevant obligations or not. Such quantification, obviously, cannot be achieved without substantial input from scientific and technical experts.

Next, such parameters – from a procedural perspective – would best be included in an annex or protocol to the primary treaty, yet forming an inseparable part thereof – much like in the field of frequency allocation the Radio Regulations, being constantly reshaped, form an inseparable and equally binding part of the international regime developed within the context of the International Telecommunication Union (ITU).[39] That way, regular updates of such parameters, as technology or other developments may require, may be provided for – by a body generally recognised to have the competence to do so within the treaty's framework. In the context of the ITU for example, that body is the Radio Regulations Board, consisting of individual experts [40].

As mentioned, there is a major role to play in developing such a system for scientific experts, which makes one expect that the Scientific and Technical Sub-Committee of COPUOS would have to play a key role to play in this respect, perhaps assisted by the GEOSS platform and/or the GMES/Kopernikus programme.

6. Concluding remarks

It will be clear that a considerable amount of research would have to be

[38]. See e.g. R. Purdy & R. Macrory, Satellite photographs: 21st Century evidence?, *New Law Journal*, 7 March 2003, 338.

[39]. The ITU, since 1992, most fundamentally bases its competencies and activities upon two intergovernmental agreements, the Constitution of the International Telecommunication Union (ITU Constitution), Geneva, done 22 December 1992, entered into force 1 July 1994; 1825 UNTS 1; UKTS 1996 No. 24; Cm. 2539; ATS 1994 No. 28; Final Acts of the Additional Plenipotentiary Conference, Geneva, 1992 (1993), at 1) and the Convention of the International Telecommunication Union (ITU Convention), Geneva, done 22 December 1992, entered into force 1 July 1994; 1825 UNTS 1; UKTS 1996 No. 24; Cm. 2539; ATS 1994 No. 28; Final Acts of the Additional Plenipotentiary Conference, Geneva, 1992 (1993), at 71). See esp. Art. 4(3), ITU Constitution.

[40]. See Art. 14, ITU Constitution; Art. 10, ITU Convention.

undertaken in supporting, and where necessary further refining, the above set of four conclusions on what needs to be done to make satellite activities contribute (even) more to combating climate change through legal instruments, notably international treaties. Part of that research should be directed at a few precedents that, though not dealing with climate change strictly speaking, nevertheless may provide interesting clues to such further refinements.

In this context, mention should also be made of the research project of Columbia University that culminated in a workshop dealing with remote sensing in support of ecosystem management treaties and transboundary conservation.[41] Similarly, there is the major research project undertaken under the auspices of the University College of London that deals with satellite monitoring as a legal compliance tool in the environmental sector.[42] Somewhat more removed from the core topic, some interesting examples available for comparative analysis have already been mentioned, such as in the arms control and human rights areas, and the jurisprudence of such courts as the ICJ and the Supreme Court of India – but likely there would be more that are worthy of attention.

What should hopefully be clear is that treaties could be even more helpful as tools for combating climate change if they somehow (re)enforce the use of satellite earth observation data in such contexts, from monitoring to, in the end, enforcing compliance of relevant obligations and agreements.

---

41. See *Applications of Earth Observation to the legal sector*, para. 7.1.

42. See http://www.ucl.ac.uk/laws/environment/satellites/index.shtml?aims.

MONITORING THE KYOTO PROTOCOL & BEYOND: Greenhouse Gases Observation & the Global Forest Carbon Monitoring System

IISL and ECSL Space Law Symposium 2008
Vienna, 1 April 2008

Masami Onoda

JAXA Industrial Collaboration Department/

Kyoto University Graduate School of Global Environmental Studies

onoda.masami@jaxa.jp

1. Introduction

As we are to talk about "Legal Implications of Space Applications for Global Climate Change: Institutions and Instruments" in this symposium, I would like to try to identify in my presentation the exact meaning of "monitoring" by taking up the cases of space application s for the Kyoto Protocol and beyond. Early forms of the use of satellite Earth observation data in international institutions can be seen in the arms control regimes, where satellite imagery was used as National Technical Means of verification, and then as "open source information" for verification by the International Atomic Energy Agency. After the cold war, there has been an emerging trend of international cooperation to achieve international common interests, leading to the increasing need of transparency. "Environmental monitoring" is thus a general and core requirement to achieve transparency and reassurance, in decision-making and implementation or compliance of the regime.

This could be shown in an institutional model of environmental treaties, where policy is based on scientific information in a manegerial framework based on treaty procedures: Information is gathered on a certain risk; States may acknowledge the risk and make political decisions; and measures (e.g. emissions reduction) will be carried out. Monitoring can broadly mean two functions in this model: (a) Environmental monitoring, or "systematic observations" in the Vienna Convention for the Protection of the Ozone Layer[1] and UNFCCC[2], or monitoring compliance or verification of treaty obligations by States. We will now look into these functions of "monitoring", and the role satellite data could play therein.

2. Satellite Earth Observation as "Systematic Observation"

Most multilateral environmental treaties call for improvement of information in both functions of "monitoring", i.e. provided in the treaty text in terms such as "research and systematic observation" (including "monitoring" in general) and compliance monitoring as "self-reporting" or in other terms in the treaty text. Those treaties include the 1985 Vienna Convention (Arts 2.2 (a), 3, Annex II for the former; and Arts 5 and 6 for the

---

[1] Vienna Convention for the Protection of the Ozone Layer, Mar. 22, 1985, T.I.A.S. No. 11,097, 1513 U.N.T.S. 293 [hereinafter Vienna Convention].

[2] Framework Convention on Climate Change, 31 ILM (1992), 851[hereinafter UNFCCC].

latter), the 1987 Montreal Protocol (Art 6; Art 7), [3] the 1992 UNFCCC (Arts 4.1(g), 5; Art 12), and its Kyoto Protocol[4] of 1997 (Art 10 (d); Arts 7 and 8), the 1979 LRTAP (Arts 4, 6, 7, 9; Art 8),[5] the 1982 UNCLOS (Arts 200 and 204; Arts 205 and 206),[6] and the 1992 Biodiversity Convention (Arts 7, 12, 17, 18; Art 14).[7] While this is not a complete nor comprehensive list of relevant treaty provisions, it indicates the extensive use of provisions on environmental monitoring in treaty texts.

The Vienna Convention, adopted in 1985, was the first multilateral legal instrument to adopt the term "systematic observations". In the former draft Convention, the definitions for "monitoring" was: "a system of observations, collation of the results of these observations, and assessment and forecasting of change in the amount and vertical distribution of ozone and substances having a significant impact on the state of the ozone layer on the basis of factual data." [8] During the negotiations,[9] it was agreed that throughout the draft convention and associated texts, "monitoring" would be replaced by "systematic observations."

The UNFCCC Preamble recalls the Vienna Convention and its Montreal Protocol, and provides that all parties shall "[p]romote and cooperate in scientific, technological, technical, socio-economic and other research, systematic observation and development of data archives ...."[10] It also provides that parties shall support and further develop international and intergovernmental programs and networks or organizations, taking into account the need to minimize duplication of effort.[11]

The Kyoto Protocol provides that: "[p]arties ... shall ... cooperate in scientific and technical research and promote the maintenance and the development of systematic observation systems and development of data archives ... and promote the development and strengthening of endogenous capacities and capabilities to participate in international and intergovernmental efforts, programs and networks on research an systematic observation, taking into account Article 5 of the Convention."[12]

"Research and Systematic Observation" has been a regular agenda item of the UNFCCC' Subsidiary Body for Scientific and Technological Advice (SBSTA)[13] since its 17th session. The 9th Conference of Parties (COP 9) (Milan, December 2003) adopted a decision on global observing systems for climate.[14] This decision calls for the preparation of an implementation plan for global climate observations to be coordinated by GCOS in collaboration with the Group on Earth Observations (GEO). [15] COP 10

---

3 Montreal Protocol on Substances That Deplete the Ozon Layer, 26 ILM 1541, 1989.
4 Protocol to the Framework Convention on Climate Change (Kyoto), 37 ILM, 1998 [hereinafter Kyoto Protocol].
5 Convention on Long-Range Transboundary Air Pollution, 18 ILM 1442, 1979.
6 United Nations Convention on the Law of the Sea, 21 ILM 1261, 1982.
7 Convention on Biological Diversity, 31 ILM, 818, 1992.
8 *Revised Draft Convention for the Protection for the Ozone Layer, with Additional Commentary,* UN Doc. UNEP/WG/78/10 (1983), p.3.
9 *Second Revised Draft Convention for the Protection of the Ozone Layer, with Additional Commentary,* UN Doc. UNEP/WG.94/3 (1983), p.2.
10 UNFCCC, Art. 4(1)(g).
11 *Ibid.*, Art. 5.
12 Kyoto Protocol, Art. 10.
13 See UNFCCC, Research and Systematic Observation, at http://unfccc.int/methods_and_science/research_and_systematic_observation/items/2312txt.php (last visited 25 Mar. 2008).
14 Conference of the Parties, Milan, Italy, Dec. 1-12, 2003, *Report of the Conference of the Parties on its Ninth Session, Addendum,* FCCC/CP/2003/6/Add1, 11/CP.9, (Apr. 22, 2004), pp.20-22.
15 See generally, Group on Earth Observations, *Homepage, at* http://earthobservations.org/ (last visited Mar. 25

also adopted a decision on research and systematic observation.[16] SBSTA 22 (Bonn, May 2005) welcomed the endorsement of the 10-year Implementation Plan at the third Earth Observation Summit in February 2005, which establishes the Global Earth Observation System of Systems (GEOSS) as an important development in systematic observation to contribute to the enhancement of climate change research.[17]

The potentials and implications of systematic observation is that firstly, satellite Earth observation is an integral part of systematic observation, as an international obligation in the Vienna Convention and UNFCCC. Secondly, the fundamental issue is the coordination of national law and policies, including programme planning and data policy. Although there has been almost 30 years of history of international harmonisation of space programmes and data policy, what has been achieved is a common largest denominator at best. It is yet to be clear if GEOSS could provide effective coordination tools for national policies or successfully link satellite Earth observation to environmental treaties. Thirdly, synergy between space and environmental law and policy is essential for the effective application of space technologies.

3. Potential Compliance Monitoring

Next, I will argue the potentials and issues of compliance monitoring in two cases, atmospheric greenhouse gases monitoring, and monitoring emission sinks by land remote sensing.

3.1 Atmospheric Greenhouse Gases Monitoring

Today, two missions exist with the aim of atmospheric greenhouse gases observation from space: Japan is developing the Greenhouse gases Observing SATellite (GOSAT)[18] and the U.S. is developing the Orbiting Carbon Observatory (OCO)[19]. GOSAT mission has a certain political focus in relation to the Kyoto Protocol,[20] while OCO objectives are mainly scientific.

2008).

[16] Conference of the Parties, Buenos Aires, Arg., Dec. 6-18, 2004, *Report of the Conference of the Parties on its Tenth Session, Addendum*, FCCC/CP/2004/10/Add1 (Apr. 19, 2005), p.2.

[17] Subsidiary Body for Scientific and Technological Advice , Bonn, Germany, May 19-27, 2005, *Report of the Subsidiary Body for Scientific and Technological Advice on its twenty-second session,* at 14, FCCC/SBSTA/2005/4 (Jul. 29, 2005). Also see *The Global Earth Observation System of Systems (GEOSS) 10-Year Implementation Plan* (Feb. 16, 2005) *available at* http://earthobservations.org/docs/EOS%20III/10YR%20IMPLEMENTATION%20PLAN.doc (last visited 25 Mar. 2008).

[18] Greenhouse Gases Observing Satellite (GOSAT) , *Homepage, at* http://www.jaxa.jp/missions/projects/sat/eos/gosat/index_e.html (last visited Mar. 25 2008).

[19] Orbiting Carbon Observatory, *Homepage, at* http://oco.jpl.nasa.gov/ (last visited Mar. 25 2008).

[20] Japan Aerospace Exploration Agency (JAXA), Plan to Achieve the Mid-term Goal of the Japan Aerospace Exploration Agency, at 11, available (only in Japanese) at http://www.jaxa.jp/missions/plan/pdf/mid-term_plan.pdf (last visited Sept. 20, 2006) ; National Institute for Environmental Studies (NIES) Center for Global Environmental Research, GOSAT brochure, *available at* http://www-cger.nies.go.jp/cger-e/e_pub/e_pamph/pamph_index-e.html (last visited Mar. 25 2008).

(1) GOSAT Mission

GOSAT is a joint project of the Japan Aerospace Exploration Agency (JAXA) and the Ministry of Environment (MOE), and the National Institute for Environmental Studies (NIES). It will be the first satellite to observe the concentrations of CO_2 and CH_4 from space. The results will not only contribute to a deeper scientific understanding of global warming, but also provide fundamental information for refining climate change prediction and formulating global warming countermeasures.[21]

The mission targets are: (i) to observe CO_2 and CH_4 column density at 100-1000km spatial scale, with 1% relative accuracy for CO_2 and 2% for CH_4, during the Kyoto Protocol's first commitment period; and (ii) to reduce sub-continental scale CO_2 annual flux estimation errors by half[22]. This challenging mission certainly has many technical issues to overcome, but scientifically, there would be an important value of global repetitive data collected over a continuing period. Currently, the profile of greenhouse gas concentration is mapped by some 300 ground and airborne observations, while GOSAT will be capable of providing 56,000 observation points. Such measurement by satellite is expected to significantly enhance the current observation network capability.

(2) Monitoring Procedures of the UNFCCC & Kyoto Protocol

If we understand "compliance" as the degree of implementation by Parties of the treaty obligations, we must first examine the treaty obligations in order to identify the potential role of satellite Earth observation in treaty compliance. The ultimate objective of the UNFCCC being to achieve stabilisation of greenhouse gases concentrations, in the atmosphere at a level that would prevent dangerous anthropogenic interference with the climate system, it sets forth differentiated commitments to Parties, including systematic observation, and reporting and review. Specifically, it requires as commitments of all Parties, to make available national inventories; formulate programmes; develop technologies; sustainable management of sinks and reservoirs; cooperate in preparing for adaptation; research and systematic observation; information exchange; education and training; and communication.[23] It also provides that Annex I Parties[24] shall adopt national policies and measures on mitigation of climate change, and communicate detailed information[25]. Annex II Parties[26] shall provide new and additional financial resources; and transfer of technology.[27] The Conference of Parties shall review the adequacy of national policies and measures communicated.[28]

The Kyoto Protocol to UNFCCC provides regulations for specific numeric goals for emissions reduction by Annex I Parties of the UNFCCC,[29] as well as the procedures for

---

[21] Ministry of Environment, National Institute for Environmental Studies and Japan Aerospace Exploration Agency, *GOSAT Project*, published by the Center for Global Environmental Research, NIES, Nov. 2007.

[22] Akihiko Kuze, Kayoko Kondo, Yutaka Kaneko & Takashi Hamazaki, Greenhouse Gases Observation from the GOSAT Satellite, A14C-03, presented at the American Geophysical Union Fall Meeting, San Francisco, 2005, p.3.

[23] UNFCCC, Art 4.1.

[24] *Ibid.*, Annex I; Annex I countries include the industrialized countries that were members of the OECD (Organisation for Economic Co-operation and Development) in 1992, plus countries with economies in transition (the EIT Parties), including the Russian Federation, the Baltic States, and several Central and Eastern European States.

[25] *Ibid.*, Art. 4.2.

[26] *Ibid.*, Annex II;OECD Members.

[27] *Ibid.*, Art.4.3.

[28] *Ibid.*, Art 4.2.

[29] Kyoto Protocol, Art. 3.1.

Parties to conduct research and systematic observation,[30] conduct emissions estimation, reporting and review,[31] and cooperation between developed and developing countries, and commercial entities through the market mechanism, to achieve emission reduction goals.[32] These provisions of the UNFCCC and Kyoto Protocol constitute an elaborate set of rules and procedures of political goals, regulations and economic incentives, based on scientific information, to which satellite Earth observation may partly contribute, if not for all the treaty requirements.

(3) Potentials and implications

The GOSAT mission, in addition to the scientific goals, has a certain political focus. In the usages of terms in the political rationale of the GOSAT satellite mission, there seems to be a mixture of focus in "validating (verifying)" "understanding (clarifying, estimating)" and "predicting". The role satellite Earth observation can play in the Kyoto Protocol procedures is, however, still limited at the present stage, especially with regard to compliance monitoring. The Kyoto Protocol calls to ensure that "their aggregate anthropogenic carbon dioxide equivalent emissions of the greenhouse gases… do not exceed their assigned amount". The estimation guidelines for establishing national inventories are based on calculations of CO_2 equivalence of anthropogenic emissions using global warming potentials as accepted by Intergovernmental Panel on Climate Change and agreed upon by the Conference of Parties.[33] On the other hand, it is not possible to identify and measure anthropogenic emissions by satellite observation. Because of the technical limitations of inverse modelling using satellite data, as opposed to forward calculation methods that is adopted by the treaty procedures, it is not realistic to expect satellite observation to be able to contribute directly in the reporting and review process of national inventories, or to replace such data in this process in the Kyoto Protocol.

The most significant role of greenhouse gases observations by satellite is, in the context of implementation of Kyoto Protocol obligations, that they can provide an integral part of systematic observation on greenhouse gases (CO_2 and CH_4) by enhancing observation capacities and providing accurate estimates of the flux of greenhouse gases on a subcontinental basis. This would improve the quality of information and understanding on the long-term and general status of climate change, thus reducing the uncertainty of the scientific basis of the treaty.

3.2 Forest Carbon Monitoring System

(1) The Kyoto Mechanism and Earth observation

As the Kyoto Protocol requires quantification of afforestation, reforestation, and deforestation (ARD), [34] revegitaton and land used management, [35] and to establish a baseline of carbon stocks for 1990,[36] existing and historical remote sensing data could

[30] *Ibid.,* Art 10 (d).

[31] *Ibid.,* Art 5.1, 7.1&2, Art 8.

[32] *Ibid.,* Art 6, 12, 17 on the Kyoto Mechanism.

[33] Intergovernmental Panel on Climate Change, *Revised 1996 Guidelines for National Greenhouse Gas Inventories, available at* http://www.ipcc-nggip.iges.or.jp/public/gl/invs1.htm (last visited 25 Mar. 2008).

[34] Kyoto Protocol, Art 3.3.

[35] *Ibid.,* Art 3.4.

[36] *Ibid.,* Art 3.1.

play and important role in supporting the establishment of these information. As the negotiations of the COP has defined specific arrangements for the so-called "Kyoto Mechanisms", it turned out that in the Clean Development Mechanism (CDM) projects on emission sinks, remote sensing or GIS data is not being fully utilised. On the other hand, the IPCC Good Practice Guidance[37] encourages the use of spatial data and the use of GIS for monitoring and evaluation of sinks.

In 2005 at COP11, "Reducing Emissions from Deforestation & Degradation" (REDD) was proposed by Papua New Guinea, leading the Coalition of Rainforest Nations. A formal decision was adopted at the recent COP13 in Bali. [38] REDD is a concept in which reduction of emissions from deforestation and forest degradation can be compensated through payment of carbon credits generated from deliberate actions and policies to reduce or avoid deforestation and degradation. While only afforestation and deforestation are applicable under the Kyoto Protocol CDM, avoidance or deduction of deforestation is not. It is hoped that starting after 2012, REDD can start as an international mechanism where "reduced" emissions can be certified and transacted in carbon markets. It is expected to provide a sustainable source of financing for forest conservation, by giving incentives to preserving forests. This poses another demand for satellite data. In order to quantify the amount of carbon "conserved", and credits duly issued, it is necessary to establish means of accurately quantifying the baseline data and to monitor deforestation and degradation.

(3) The Forest Carbon Monitoring System (FCMS) Research Project

The Ministry of Environment of Japan, JAXA and Tokyo University, together with international partners, has initiated a 3-year research project for the development of a Forest Carbon Monitoring System (FCMS) based on the synergetic use of in-situ measurements and Earth Observation data, for quantitative estimates of CO_2 emissions induced by deforestation and degradation. The prime data source for baseline mapping and annual change monitoring is the L-band Synthetic Aperture Radar (PALSAR) onboard ALOS. The aim of the FCMS project is to develop a system for mapping and monitoring of forest carbon stocks and of changes therein, through the synergetic use of in-situ networks and Earth Observation data. In particular, the project aims to assess the utility of using ALOS PALSAR data to derive information about the status of the forest cover mapping and for identification and spatial quantification of changes in the forest cover as a result of deforestation, forest degradation and regeneration. The project also aims to develop operational methodologies for forest monitoring and quantitative carbon accounting – an FCMS system – which, depending on the outcome of UNFCCC post 2012 regime agreements, can support the potential implementation of international climate change regulations for the REDD as a part of a post 2012 regime. In this sense, the FCMS research project is a precursor initiative that is foreseen to lead to the development of an operational international monitoring system for quantitative assessment of deforestation and degradation.

In this project, a 3-step approach will be adopted for developing carbon accounting methods through the following 6 subtopics:

---

[37] IPCC, Good Practice Guidance for Land Use, Land-Use Change and Forestry, 2003, at *http://www.ipcc-nggip.iges.or.jp/public/gpglulucf/gpglulucf_contents.htm* (last visited 26 Mar. 2008)

[38] UNFCCC, Decision 2/CP.13, *Reducing emissions from deforestation in developing countries: approaches to stimulate action*, Un Doc. FCCC/CP/2007/6/Add.1, pp.8-11.

(1) Developing carbon accounting methods
(2) Analysis of PALSAR data for identifying deforestation and degradation
(3) Analysis of forest inventories
(4) Estimating forest structure parameters using PALSAR
(5) PALSAR characterisation of vegetation disturbance
(6) Evaluating carbon balance with ecosystem models

Quantitative estimates of carbon emissions will not be possible to be derived from satellite data alone. Ancillary data will be required from other data sources such as *in-situ* observations. SAR data can provide information on *where the changes occurred*, *when they occurred* and the *spatial extent of the changes*. The SAR images can provide cloud-penetrating, periodical data on forest covered areas, which is especially useful for observation of the tropical region with persistent cloud cover.

(3) Potentials and Implications

The FCMS initiative is a challenging project integrating political requirements and technological solutions in to a operational system. Easily said than done, there are a number of issues to be overcome before it proves operational. While the ALOS SAR is a powerful tool in detecting land use and land cover change, it must be identified what can be delivered, in response to the political framework of international monitoring requirements. Data availability, as well as cost, at present and near-long term is crucial. In this sense, a single system is not sufficiently reliable as the source of information in the implementation phase. It is thus essential to achieve optimal institutional procedures and space system design; and collaborative approach between developed and developing countries through treaty mechanisms for sustainable forest conservation and emissions reduction. Although REDD is still in a very early stage of negotiations, it should never be too early to start such institutional coordination.

4. Legal Implications

As the first day of this symposium session extensively discussed "Principles and Rules", here, I will just point out a couple relevant points with relevance to international space law. With regard to the 1967 Outer Space Treaty[39], 3 principles are of relevance : free exploration and use of outer space (Art. 1); international cooperation (Art. 1,3,9); and due regard to interests of all other States (Art. 9).

The UN Remote Sensing Principles of 1986[40] attempts to coordinate the benefit and interests of all countries and national sovereignty issues. However, the following 3 points may be specified as remaining open for clarification: (1)Access rights: especially the issue of property position of holder of data (IV); (2)Equality and equity (promotion of international cooperation "shall be based in each case on equitable and mutually acceptable terms.") (V), while the OST states the norm of equality, and specific conditions for "non-discriminatory basis" (XII) ; or "cost terms" are left open; (3) As to

[39] Treaty on Principles Governing the Activities of States in the Exploration and Use of Outer Space, including the Treaty on the Non-Proliferation of Nuclear Weapons, 729 UNTS 161, 1967.
[40] UN, *Principles Relating to Remote Sensing of the Earth from Space*, UN Doc. A/RES/41/65, annex.

the disclosure of remote sensing environmental "information", there is no mention of cost, leading that information obligations are established at no cost, and environmental info is given special status of "public good."

As to the relevance to international environmental law, the improvement of environmental information is an established objective of international environmental law.[41] Various supervisory techniques are adopted, including reporting, inspection, non-compliance procedures, and "preventive global monitoring". In the Vienna Convention and UNFCCC, satellite data is referred to as part of systematic observation, which countries must promote and cooperate in, as treaty obligations.

Most of these legal implications pointed out above are of very general nature, regarding normative principles rather than specific State obligations or procedures. The ambiguity concerning the Remote Sensing Principles could be subject to future discussions when the procedures and legal issues become clearer, as instrumental frameworks on satellite Earth observation applications for global climate change further develop.

4. Conclusion

Satellite Earth observation is an integral part of *systematic observation* required in the UNFCCC and Kyoto Protocol. For *compliance monitoring* satellite data is not fully utilized because of the gaps between technical capabilities and the treaty methodology; thus, it is essential to generate political procedures and scientific capability in a compatible and consistent manner. For enhanced transparency with *cost-effective, independent & reliable data*, there is a need to design environmental-space systems and political institutions involving *non-state actors,* where *developed and developing countries* work together in treaty mechanism with joint economic incentives; and with *coordinated data rights and accessibility.*

However, space law is silent on these issues. It is thus expected that future research and discussions in this field will clarify the incompatibilities between technical capability and legal or policy procedures, so that satellite Earth observation can provide an effective observation basis for environmental data and scientific knowledge, upon which future regimes could be developed and implemented.

---

[41] See Sands, Philippe, *Principles of International Environmental Law,* Second Edition, Cambridge: Cambridge University Press, 2003.

The 3rd Eilene M. Galloway Symposium On Critical Issues In Space Law

The National Center for Remote Sensing, Air and Space Law
at the University of Mississippi School of Law,
the JOURNAL OF SPACE LAW,
the International Institute of Space Law
and
Arianespace, Inc.
Present

The 3rd Eilene M. Galloway Symposium on Critical Issues in Space Law
Art. VI of the Outer Space Treaty: Issues and Implementation
December 11, 2008
8:30 AM - 4:00 PM
Cosmos Club, Washington D.C.

Agenda

08:30 - 09:00 Registration Sign-In

09:00 - 09:15 Welcome

Prof. Joanne Irene Gabrynowicz, Director, National Center for Remote Sensing, Air, and Space Law, Univ. of Mississippi; Director, IISL

Mrs. Tanja Masson-Zwaan, President, IISL, International Institute of Air & Space Law, Leiden University

Mr. Clayton Mowry, President, Arianespace, Inc.

Introduction

Prologue to Article VI

Dr. Eilene M. Galloway, Honorary Director, IISL

Presented by Dr. Jonathan F. Galloway, Professor of Politics, Emeritus on the Irwin L. & Fern D. Young Presidential Chair, Lake Forest College

09:15 – 11:00 Article VI: The Legal Landscape

Overview of State Responsibility in a Global Commons

Prof. Joanne Irene Gabrynowicz, National Center for Remote Sensing, Air, and Space Law, Univ. of Mississippi

Article VI of the Outer Space Treaty in the European Context

Prof. Frans von der Dunk, Member, Board of the European Centre for Space Law; Professor of Space Law, University of Nebraska

Implementation of Art VI of Outer Space Treaty in North America

Prof. Ram Jakhu, Institute of Air and Space Law, McGill University

Art. VI of the Outer Space Treaty and Private Human Access to Space

Tanja Masson-Zwaan, International Institute of Air & Space Law, Leiden University

11:00 – 11:15 Break

11:15 – 12:30 National Implementation of Article VI by Governments

Philipe Clerc, Head of the Legal Service, CNES, Paris, France

Margaret Roberts, Senior Attorney, Office of General Counsel, NASA, Washington, DC

Glenn E. Tallia, Senior Counselor for Atmospheric and Space Services, and Research, U.S. Dept. of Commerce/NOAA

Richard Crowther, Science & Technology Facilities Council United Kingdom

12:30 - 1:30 Lunch and Presentation

Selected Findings from the Andrew G. Haley Archive at National Center for Remote Sensing, Air and Space Law: the Work Product of the World's First Space Law Practitioner

Prof. Joanne Irene Gabrynowicz

Michael S. Dodge

1:30 - 2:45 Operating a Space Business Under National Laws that Implement Article VI

Richard DalBello, VP Govt. Affairs, INTELSAT General

Michael Gold, Bigelow Aerospace, Wash., DC

Franceska Schroeder, Fish & Richardson P.C., Wash., DC

2:45 – 3:00 Break

3:00 - 4:00 Discussion

Moderator: Prof. Joanne Irene Gabrynowicz

REPORT OF THE 3RD EILENE M. GALLOWAY SYMPOSIUM ON CRITICAL ISSUES IN SPACE LAW

'Article VI of the Outer Space Treaty: Issues and Implementation'

It was already for the third time, that the last month of the year in Washington, D.C., witnessed a Symposium on Critical Issues in Space Law being held in honour of Dr. Eilene M. Galloway. This time discussing "Article VI of the Outer Space Treaty: Issues and Implementation", the Symposium (organised by the National Center for Remote Sensing, Air and Space Law at the University of Mississippi School of Law, the Journal of Space Law, the International Institute of Space Law (IISL) and Arianespace, Inc.) took place 11 December 2008 at the Cosmos Club in Washington, attracting some 60 participants in the process.
Dr. Eilene M. Galloway is generally considered to be one of the founding giants of space law, as one of the leading participants in the legislative process on the national US level leading to the establishment of NASA in 1958 and subsequently one of the great contributors to the evolution of space law at the international level through the UN space treaties of the 60's and 70's and beyond.

The Symposium kicked off by opening remarks on behalf of the various organizing entities. Thus, **Prof. Gabrynowicz** welcomed the participants on behalf of the National Center for Remote Sensing, Air and Space Law and the Journal of Space Law, explaining the history and background of the Symposium. She pointed to the fact that the first symposium had been held two years ago to celebrate the hundredth birthday of Dr. Galloway, and had been such a success that the Center and the Journal had been stimulated to organise a second, and now a third symposium.
Mrs. Masson-Zwaan, President of the International Institute of Space Law (IISL), highlighted in her welcoming remarks the reflection of international cooperation in the Symposium and its themes over the past few years, pointing in particular to the presence of participants also from other parts of North and South America, as well as Europe – notably from the European Centre for Space Law (ECSL). Also this could be judged reflective of Dr. Galloway's career, as she had always laid the greatest importance in her work and writings on international cooperation for the peaceful uses of outer space.
On behalf of the third main sponsor of the Symposium, Arianespace Inc., **Mr. Mowry** provided an overview of some current developments in space activities and space law. He noted that, while 'change' was a very popular word these days in Washington, the question for space would still be whether such change would be going in the right direction. In terms of today's issues with space activities, and in particular (with an eye to the theme of the symposium) those resulting from the increasing commercialisation and privatisation of many space activities, he noted problems with the allotment of frequencies and slots with the ITU (which in one case has some 400 applications filed for the same slot), and with the total lack of international coordination of the launch services sector now that the launch services agreements of the early 90's with Russia, China and the Ukraine had all expired as the global environment has changed profoundly.

As Dr. Galloway herself had not been able to join the Symposium, her son, **Prof. Galloway**, presented a message of welcome and thanks on her behalf. He noted in particular the appropriateness of the Symposium's theme with a view to the career of his mother, with one of the key features of Article VI being the requirement of authorisation and continuing supervision of non-governmental entities' 'national activities' in outer space.
In this context, the discussion of Article VI of the Outer Space Treaty – providing essentially for international responsibility of states for their national activities in outer space – was a very appropriate theme in the context of the obligations of mankind, in particular the leading spacefaring nations, to use outer space responsibly, which have always been defended staunchly by Dr. Galloway. During the negotiations leading to the final text of this Treaty, perhaps the most salient question

raised was whether both governments and nongovernmental actors could be recognized as legitimate space actors. Article VI, stating that "activities of non-governmental entities shall require authorization and continuing supervision by the appropriate State Party to the Treaty" is the compromise that recognizes both as legitimate space actors, as part of the broader focus on responsible usage of outer space.

Since the text of Article VI was repeatedly put up on the powerpoint screen during various presentations in order to remind the audience and the discussants of the exact wording, it is appropriate also to reproduce that text here:

> States Parties to the Treaty shall bear international responsibility for national activities in outer space, including the Moon and other celestial bodies, whether such activities are carried on by governmental agencies or by non-governmental entities, and for assuring that national activities are carried out in conformity with the provisions set forth in the present Treaty. The activities of non-governmental entities in outer space, including the Moon and other celestial bodies, shall require authorization and continuing supervision by the appropriate State Party to the Treaty. When activities are carried on in outer space, including the Moon and other celestial bodies, by an international organization, responsibility for compliance with this Treaty shall be borne both by the international organization and by the States Parties to the Treaty participating in such organization.

The first speaker of the first substantive session, on ***"Article VI: The Legal Landscape"***, was **Prof. Gabrynowicz**, providing an *"Overview of State Responsibility in a Global Commons"*. She noted that, whilst the concept of a 'global commons' was nowhere authoritatively defined in international law treaties or other documents, few would dispute that this was what the 1967 Outer Space Treaty referred to in an indirect manner. In that context, usually concepts such as *res communis* or *terra communis* were used, but they essentially amounted to the same. She then noted that currently there would be three areas falling under that heading: Antarctica, the high seas and space; and proceeded to analyse the application of the general concept of 'responsibility' in its various incarnations of 'state responsibility' and 'international responsibility' as per Article VI of the Outer Space Treaty in those three different areas. In response to a question from the audience bringing up the concept of the 'common heritage of mankind' finally, the speaker clarified that that concept should not be seen as interchangeable with that of the 'province of all mankind' – a concept indeed officially referred to in the Outer Space Treaty.

The second speaker of the session was **Prof. Von der Dunk**, who addressed the subject of *"Article VI of the Outer Space Treaty in the European Context"*. Whilst noting at the outset that there was no such thing as a European context for – let alone a European approach to – Article VI, he proceeded to discuss the various European levels at which outer space activities were addressed (the European Space Agency, the European Community, then Union, and the 'geographical Europe' as this included Russia and the Ukraine). In addition, noting the fundamental absence of any comprehensive authority at any of those levels to replace the individual sovereignty of (member) states in dealing with Article VI at their national discretion, he also discussed those states where Article VI had given rise to establishment of a more or less comprehensive national space law – in chronological order: Norway, Sweden, the United Kingdom, Russia, the Ukraine, Belgium, the Netherlands and France. Comparing those pieces of national law using five aspects – their scope in terms of (space) activities *ratione materiae*; their scope in terms of license obligation *ratione personae*; how they dealt with issues of liability and insurance; the licensing and registration authority(/ies); and their respective implementation experience – the speaker came to the somewhat ironic conclusion that the two European national space laws that were most identical to each other were the Russian and Ukrainian ones, for obvious historic reasons.

The third speaker of the session was **Prof. Jakhu**, who spoke about *"Implementation of Article VI of the Outer Space Treaty in North America"*. Addressing, from this perspective, essentially developments in the United States and Canada, he defined implementation as incorporation in, alternatively giving effect under, national law to a provision of international law. He furthermore noted in this regard that the traditional requirement of state responsibility under general international law, of imputability / attributability of a certain act as a 'state act' to a certain state, has been done away with in space law as per Article VI. From the international perspective therefore, there is not so much a difference between public and private activities, but between "national activities" and 'non-national', in other words 'international' activities. He highlighted in this respect that 'national activities' at the time of drafting the clause was meant to include everything linked or connected to a state or its national or its territory or its facilities. Coming to the issue of implementation proper finally, he analysed that Canada had followed very much the US approach in terms of enunciating national space legislation.

The last speaker of that session was **Mrs. Masson-Zwaan**. She discussed *"Article VI of the Outer Space Treaty and Private Human Access to Space"*, by starting to refer to Article I of the Outer Space Treaty as that declared space to be the "province of all mankind", and wondering what the legal consequences thereof would be for, for example, space tourism. The speaker considered Article VI from that vantage point to provide substance to the general obligation of due diligence of a state, and proceeded to analyse, using 'space tourism' as the test case, whether current space law would still be up to the task of guaranteeing such due diligence. The fundamental answer to this should be 'yes', although adaptation and further implementation would certainly be necessary – at the national, regional (European) as well as international levels. She finally remarked that Article VI only was to apply to international spaceflight (as long as the label of 'international' of course included effects as opposed to merely who the partners were), and proposed to apply space law to space tourism primarily on a functional basis.

After a coffee break, the second session dealt with the ***"National Implementation of Article VI by Governments"***. The first speaker in the session was **Mr. Clerc**, tackling the subject from his position as Head of the Legal Service of the French space agency CNES. He essentially dealt with the brand new French national space law, explaining first of all that there had existed a *sui generis* legal framework all along for the operations of Arianespace. This was not only a matter of *de facto* control; there were also international agreements operative in the 'triangle' of key stakeholders France-ESA-Arianespace. As for Eutelsat, since until relatively recently it was an intergovernmental organisation, it had only been affected by national electronic communications regulations. With the fundamental reorganisation of the French 'spacescape', however, CNES now was no longer allowed to undertake commercial activities, and a law was necessary to deal with future private space activities falling within French jurisdiction. Since Arianespace is already authorised through the aforementioned international agreements, it does not require authorisation under the new law, whereas Eutelsat was also provided with a general license extending the existing situation, but any other space activities should be covered by the new law, the provisions of which he finally analysed in considerable detail.

The second speaker of the second session was **Mrs. Roberts**, sharing her perspective as Senior Attorney at the Office of General Counsel of the US National Aeronautics and Space Agency (NASA) with the audience. She went through an extensive analysis of how, in the various areas concerned, the US government had implemented its general obligations to authorise and continuously supervise under Article VI of the Outer Space Treaty, from the Communications Act and follow-on acts on telecommunications as they applied to satellite communications, to the various Acts and policies on remote sensing, the Commercial Space Launch Act in various incarnations and the Commercial Space Act of 1998. Finally, pointing to a key issue in any implementation exercise as far as the international responsibility of Article VI is concerned, she noted that within the United States the Crew Code of Conduct for the International Space Station was implemented through the Code of Federal Regulations, in order to ensure enforceability of the relevant provisions.

The third speaker was **Mr. Tallia**, who spoke from his perspective as Senior Counselor for Atmospheric and Space Services, and Research with the US Department of Commerce, more precisely with NOAA as the national authority responsible *inter alia* for licensing private remote sensing operators in the United

States. Such licensing, speaker pointed out, also took Articles III and VII of the Outer Space Treaty into account. He discussed amongst others the Kyl-Bingaman Amendment National Defence Authorization Act of 1997, placing limits in dissemination of imagery of Israel, and the licensing regulations enunciated in the Code of Federal Regulations. Then he addressed the question whether participants to the Google Lunar Prize potentially making pictures of the earth would require a NOAA license. In his view, if they would orbit the earth for a week and do so, the answer would likely be 'yes', whereas if they would just be blasting off in the direction of the moon, the answer would probably be 'no'. Speaker's final remark pertained to the requirement under the UN Principles on Remote Sensing to allow foreign governments access to unenhanced data regarding its territory on reasonable (cost) terms, which was included in any license granted: so far such a request has never occurred.

The fourth and final speaker of that session was **Mr. Crowther**, from the Science and Technology Facilities Council of the British National Space Centre (BNSC) in the United Kingdom. Going through the UK Outer Space Act as it entered into force in 1986, he highlighted several aspects regarding how it implemented key aspects of the space treaties, including Article VI of the Outer Space Treaty. This concerned especially the scope and further relevant parameters of the licensing obligation under the Act, down to such important practical details as applicable time limits, fees, sanctions and environmental impact assessments. The speaker also, however, pointed out the same overarching issues that all national implementation activities with respect to Article VI have to deal with, such as liability arrangements and insurance requirements, where he noted in particular a need for international standards applicable to in-orbit activities, to avoid the potential negative impact of widely varying national regulations.

During lunch, the participants were informed by **Mr. Dodge** of the establishment of an Andrew G. Haley Archive at the National Center for Remote Sensing, Air and Space Law, announced as "The Work Product of the World's First Space Law Practitioner", and he proceeded to present the audience with a number of highlights from the documents thus archived.

After the lunch break, the third and final session focused on the private sector, ***"Operating a Space Business Under National Laws that Implement Article VI"***. The first speaker here was **Mr. DalBello**, in his capacity as Vice-President of Government Affairs with Intelsat General. He started out by saying that Article VI implies an ongoing responsibility, through the concept of "continuing supervision". The speaker further analysed what this meant in the context of Article VI: could failure to comply with such an ongoing responsibility, especially if over time more sophisticated technical means would have become available and hence would have raised the applicable standards of reasonableness, invoke the application of strict liability under the Liability Convention, if relevant damage were to result? Finally, speaker went in depth into the various technical aspects involved in the possibility for governments to actually control any licensees, to ensure continuing conformity of their activities with the pertinent international responsibility of those states.

The second speaker was **Mr. Gold**, the Legal Counsel for Bigelow Aerospace, who amongst others went into considerable detail when sharing his experience with the audience with national implementation of Article VI in the US context *vis-à-vis* his private company, bent on developing new technology for space habitats – and contracting with a Russian launch service provider for the launch of the first two space objects involved in that development process. His talk thus also brought the US International Trade in Arms Regulations, the much-feared and -loathed ITAR's, into the discussion, as these certainly were one particular form of national authorisation and continuing supervision considered a pain in the neck by most (US) private enterprise. He finally noted that, in the United States, currently no regulatory authority exists for crewed on-orbit activities; in his view the establishment of any relevant authorisation and continuing supervision here would need a specific Congressional statutory act.

The final speaker of that session was **Mrs. Schroeder**, of Fish & Richardson in Washington, who succinctly remarked that the United States had taken the lead globally speaking in implementing domestically not only the Outer Space Treaty but also the Liability Convention and the Registration Convention, the two other international treaties relevant here. Speaker considered all of them sufficiently broadly drafted to allow for ongoing relevance and to allow individual states to tailor their national regulations to their national activities as they develop – and consequently suggested that changes to be made to that regime should only be undertaken at the national level.

Thus, another interesting Symposium lived up to the expectations as following from the link to the Galloway-name, and as it seemed from various comments either during the sessions or during the informal gatherings around drinks and food immediately following, this time in particular the relatively substantial international involvement could be seen as another plus. Which made yours truly ever more pleased with having been able to participate!

Frans von der Dunk
Member of the Board, ECSL

ARTICLE VI OF THE OUTER SPACE TREATY AND PRIVATE HUMAN ACCESS TO SPACE

Tanja Masson-Zwaan

President, International Institute of Space Law (IISL) and Deputy Director, International Institute of Air and Space Law, Leiden University, The Netherlands

1. Introduction

The theme of the third Eilene Galloway Symposium on Critical Issues in Space Law was "Article VI of the Outer Space Treaty: Issues and Implementation".[1] One of the main fields where Article VI needs to be analysed is that of private human spaceflight, commonly called 'space tourism'.[2] This will be *the* private commercial space activity of the next era of spaceflight, and although the industry is not taking off as quickly as was thought until last year, expectations are still that in the next few years the first commercial flights for wealthy individuals seeking the thrill of going on a sub-orbital flight will take off. Therefore this paper analyses Article VI in the context of space tourism, to see if it will be able to adequately accommodate this activity while at the same time preserving the framework of state responsibility and liability designed in the Sputnik-era.

2. Article I of the Outer Space Treaty

Before turning to Article VI, it is important to set the scene by recalling the words of the first article of the first space treaty: Article I of the Outer Space Treaty. It states in relevant part that:

> *The exploration and use of outer space [...] shall be carried out for the benefit and in the interests of all countries [...] and shall be the province of all mankind.*[3]

The question this article raises is whether it contains a firm and enforceable legal principle or 'merely' a moral obligation. Of course it is hard to realise it in practice, because it is hard to concretely conceptualise 'the benefit and interests of all countries' or the 'province of all mankind'. Would it mean for instance that space tourism should become 'Spaceflight for Mankind' in order to be a legally acceptable space activity? Looking at current price tags this hardly seems realistic. Does this mean that space tourism is illegal under the Outer Space Treaty? That would be pushing it too far, and after all, Article VI does open the door for private activities, which by nature are 'for profit'. Perhaps it is then best to see whether space tourism will eventually make access to space cheaper, which will create a spinoff that is 'beneficial' for 'Mankind'?

3. Article VI of the Outer Space Treaty

Discussing Article VI in the context of space tourism necessitates a full quote of its provisions:

> *States Parties to the Treaty shall bear international responsibility for national*

[1] This paper is based on the presentation made by the author at the 3rd Eilene Galloway Symposium on Critical Issues in Space Law, held on 11 December 2008 in Washington DC. All websites have been accessed in May 2009.

[2] Even though I believe that the term 'space tourism' is not the ideal one, because the word tourism implies large groups of people going on a trip, and that will not be the case for space travel for individuals for the foreseeable future, I will use the term for the sake of convenience.

[3] Treaty on Principles Governing the Activities of States in the Exploration and Use of Outer Space, including the Moon and other Celestial Bodies, 18 UST 2410 (1967) (Outer Space Treaty).

activities in outer space, including the Moon and other celestial bodies, whether such activities are carried on by governmental agencies or by non-governmental entities, and for assuring that national activities are carried out in conformity with the provisions set forth in the present Treaty.

The activities of non-governmental entities in outer space, including the Moon and other celestial bodies, shall require authorization and continuing supervision by the appropriate State Party to the Treaty.

When activities are carried on in outer space, including the Moon and other celestial bodies, by an international organization, responsibility for compliance with this Treaty shall be borne both by the international organization and by the States Parties to the Treaty participating in such organization.

Many eminent scholars have written about the history of the drafting of Article VI, so I will just recall here that the agreed text was a compromise between States that wished not to allow private activities at all, and those that wished to permit such activities. The compromise was that private activities are allowed, on the condition that the appropriate state exercises authorisation and continuing supervision over the activities of its non-governmental entities.

It implies in fact an obligation of due diligence; the state must verify for instance that activities by its nationals do not present a danger to the public health or to the safety of persons and goods, that they are not inconsistent with the State's treaty obligations, do not pose any risk for national security or have a negative environmental impact, and that they do not pose significant financial risks.

Since private space activity is still in its infancy, this system has worked well. What can be observed in recent years is that a constantly growing number of States has come to the conclusion that it is wise for them to adopt national legislation to bring the activities of non-governmental entities under their control. After all, they bear responsibility for those activities and can be held liable in case of damage. Even though the adoption of a licensing system under national law is not an obligation flowing from Article VI, it is certainly the recommended method of protecting the State's interests and exposure to liability risks.

4. Case study: space tourism

As mentioned in the introduction, space tourism was selected to provide a case study to test the suitability of Article VI to present and future needs. Space tourism is still considered one of the most promising private space activities in the short- and mid-term.

Evaluating the applicability and suitability of Article VI to space tourism is an opportunity to verify if the space treaties in general are still valuable and can be maintained as the basis for space activity in the future. It will also show if and to what extent there is a need to clarify and/or supplement them, for instance through national law, codes of conduct or via other means. Some authors suggest that the treaties are hopelessly outdated and should be binned altogether; others have suggested major amendments that would almost certainly mean the end of the space treaties as we know them. My belief is that the legal framework created after the launch of Sputnik is still extremely valuable and that it should be preserved.[4]

---

[4] This was also the firm belief of Eilene Galloway, who sadly passed away a few months after the Symposium. She wrote to me in August 2008: "We need to explain what needs to be done to bring private space activities under the control of the space system we have. We need a code of conduct for the [100] nations that have ratified the '67 Treaty. I think the framers of Article VI were foresighted in anticipating the possibility of private ventures. If we can suggest the legal language for implementing Article VI, it would be a real contribution for the future."

Categories of space travel

There are basically three categories of space travel which are briefly outlined below:

Suborbital spaceflight

In sub-orbital spaceflight, orbital velocity is not achieved. After engine shutdown, 3 to 6 minutes of microgravity is achieved, after which the vehicle falls back to Earth and re-enters the atmosphere. Most current projects will offer this kind of space travel. Vehicles usually attain an altitude of around 100 km.[5]

Orbital spaceflight

In orbital spaceflight, orbital velocity must be achieved for the vehicle to keep flying along the curvature of Earth and not fall back to Earth. Orbital space flight is technically highly complex and therefore expensive. Providing orbital spaceflight for private paying clients is much more demanding than suborbital flight, both in terms of technology and cost, but is nevertheless envisaged by several ventures. This is also what the six tourists who went up to the ISS so far have experienced.

Intercontinental rocket transport

Intercontinental rocket transport implies a transit through space in order to substantially shorten the travel time from one point on earth to another. It is not a new idea[6], but the technical challenges are sky high in terms of the velocity and the amount of propellant required, and the need for robust thermal protection for re-entry. Cost is therefore prohibitive, at least for the mid-term.[7]

In the present paper I will address only sub-orbital space tourism, since the other two categories seem still so far away that it would not be wise, and premature, to try to fit them into a regulatory framework now.

A brief overview of providers of sub-orbital space tourism

In order to have a good understanding of the potential impact of 'space tourism', it is useful to give a broad overview of projects underway. On October 4, 2004, the birthday of the launch of Sputnik 1, the first private manned spacecraft exceeded 328,000 ft two times in 14 days, thus winning the 10M$ Ansari X-Prize. The list of enterprises venturing into this new and promising market is surprisingly long, but many of them may never see the light of day and I will concentrate on the most well-known ones.

Space Adventures' trips to the ISS

In April 2001, the first commercial space tourist Dennis Tito spent six days in the Russian section of the ISS, after extensive training at the Star City complex.[8] After him, five others were launched to the ISS on a Russian Soyuz; Mark Shuttleworth in 2002, Gregory Olsen in 2005, Anousheh Ansari in 2006, Charles Simonyi in April 2007 (and again in 2009) and Richard Garriott (son of a NASA astronaut) in October 2008. The price for a flight brokered by Space Adventures to the ISS on board Soyuz is now around $35 million. Recent reports claimed that space tourism seats will be unavailable on Soyuz

[5] Numerous different technologies are under consideration. Some concepts involve a horizontal take-off or 'launch' (sometimes from an aircraft), while others take off vertically. For landing, they can vary from aircraft to parachute, the main technology challenge being thermal protection during re-entry.

[6] For an overview of legal issues of aerospace planes, see Tanja L. Masson-Zwaan, *The aerospace plane: an object at the cross-roads between air and space law*, in: Air and Space Law: *de lege ferenda* (Essays in Honour of Henri A. Wassenbergh), (T. Masson-Zwaan / P. Mendes de Leon, eds.), 247-261 (Nijhoff 1992).

[7] See David Hoerr, *Point-to-point Suborbital Transportation: Sounds Good on Paper, But...*, at http://www.thespacereview.com/article/1118/1.

[8] See http://www.spaceadventures.com/.

spacecraft for the next few years, since the International Space Station doubled its crew size up to six people in May 2009. However it is now reported that Cirque du Soleil founder Guy Laliberté may fly to the ISS in September 2009, as Kazakhstan cancelled its plans to send a trained cosmonaut.[9] It is expected that after that, this form of space tourism will also continue. Spaceports are being planned in the UAE and Singapore

Virgin Galactic's SpaceShipTwo

The most famous space tourism enterprise is without doubt Virgin Galactic, the project of the flamboyant Sir Richard Branson. Despite earlier reports, Virgin Galactic is now expected to take off with commercial flights in 2012 at the earliest. The concept involves a launch of SpaceShipTwo in midair at 50,000ft from the mothership, an aircraft called WhiteKnightTwo. Its homebase will be Spaceport America in New Mexico, but flights are also planned from Kiruna in Sweden. The 2½ hour journey into space sells for US$200,000 a seat, and hundreds of people have reportedly already signed up.[10]

EADS' Spaceplane

EADS unveiled its plans for a spaceplane at the 2007 Paris Airshow. EADS, unlike Virgin Galactic, is not planning to act also as the operator of the plane, it is just planning to build it. However due to funding problems it is now on hold.[11]

Xcor's Lynx

XCOR Aerospace is developing the two-seat Lynx suborbital spaceship. The spaceship intends to take off in 2010 and will be capable of flying several times each day.[12] Some say it may beat Virgin Galactic in being the first to start commercial operations.

Spaceports

The first commercial spaceport is Spaceport America, which is under construction in New Mexico.[13] The US Government provides licences to build a number of spaceports across the country, and there are proposals for spaceports in various states. Plans for spaceports in the UAE and Singapore have also been announced. In Europe, a planned spaceport in Kiruna, Sweden has made an agreement with Virgin Galactic. Other spaceports are being planned in Scotland and France, although the latter (in Montpellier) is on hold. A spaceport is also being planned in the Netherlands Antilles.

5. Realities

It is certain that space tourism is going to happen. And, at the risk of sounding pessimistic, accidents are also going to happen.

The basic legal framework for private commercial space activity is in place, although the extent to which humankind would one day engage in commercial space tourism activities was not anticipated.

---

9 See http://www.space.com/missionlaunches/090403-space-adventures-future.html and http://uk.reuters.com/article/scienceNewsMolt/idUKTRE54C48520090513

10 See http://www.virgingalactic.com. Up to 300 Virgin Galactic 'astronauts' could venture to space in the foreseeable future.

11 See http://www.astrium.eads.net/en/families/space-plane-tourism-flight-shuttle and http://www.flightglobal.com/blogs/hyperbola/2009/03/eads-astrium-puts-its-space-je.html.

12 http://www.xcor.com/press-releases/2008/08-03-26_Lynx_suborbital_vehicle.html.

13 For information on various planned spaceports: http://www.spaceportamerica.com/, http://www.spaceportsingapore.com/, http://emiratesupdate.wordpress.com/2008/01/15/uae-spaceport/, http://www.ssc.se/?id=9500, http://www.spaceportscotland.org/, http://spatioportfrance.free.fr/, http://caribbeanspaceport.com/.

More and more licensing systems are being put in place under national law, complementing the international legal framework, which will help to provide legal certainty and harmonised rules. Legal certainty is good for new industries as well as for passengers and third parties.

Mass tourism is probably still several decades away. When ticket prices come down to $20-40.000, the numbers of passengers will certainly increase, as the prospect of experiencing weightlessness and observing the 'Blue Planet' from outer space is very attractive to many people.

For the immediate future, it will be only for the rich few, at considerable risk, liability for which they will be requested to waive, while insurance will not yet be available.

6. Needs

The multilateral space treaties elaborated within UNCOPUOS were formulated in the 'Cold War' era, when only a small number of countries had space-faring capability. They could not fully anticipate the extent to which humankind would one day engage in commercial space tourism activities. The Outer Space Treaty foresaw that private entities would one day engage in space activities, yet one of the most essential topics for private operators, namely their exposure to second- or third-party liability,[14] is not addressed. Instead, the Treaty, as well as the Liability Convention[15], only addresses liability at the level of the States involved. There is no cap on liability of operators, and no opportunity for passengers or third parties to present direct claims for compensation.

[14] Second-party or contractual liability refers to liability of the operator vis-à-vis passengers and cargo, while third-party or non-contractual (tort) liability refers to liability for damage to persons or property on the ground, who have no contractual relations with the activities of the operators.

[15] Convention on International Liability for Damage Caused by Space Objects, 672 UNTS 119 (1968) 24 UST 2389 (1972) (Liability Convention).

Thus, even though the Treaties maintain their relevance even after several decades, the existing international legal regime needs to be supplemented with additional and more specific rules.

A balance must be found between commercial and technological opportunities on the one hand and principles of international space law on the other, and between the interests of the State and those of private enterprise and passengers and third parties.

In essence, it is necessary to protect the legitimate interests of states and to ensure the safety of crew, passengers and third parties in a satisfactory way, without creating a regulatory overkill.

7. Does international space law or air law apply to space tourism?

A lot has been written about the legal aspects of space tourism.[16] The UN space law treaties apply to relations between states in carrying out space activities. International air law conventions (Chicago, Warsaw, Montreal) deal with international carriage by air.

Many of the currently planned space tourism projects plan to operate from one

[16] See for instance S. Hobe, G. Goh & J. Neumann, *Space Tourism Activities – Emerging Challenges to Air and Space Law,* in 33 Journal of Space Law 359 (2007), F. von der Dunk, *Passing the Buck to Rogers : International Liability Issues in Private Spaceflight,* in 86 Nebraska Law Review 400 (2007), S. Hobe, *Legal Aspects of Space Tourism, ibid.* at 439 and S. Freeland, *Up, up and ... Back: The Emergence of Space Tourism and its Impact on the International Law of Outer Space*, in 6 Chicago Journal of International Law 1 (2005). These articles contain extensive references to the majority of earlier publications in this field. See also T. Masson-Zwaan & S. Freeland, *Private Human Access to Space: legal challenges and possibilities*, presented at the 1st IAA Symposium on Private Manned Access to Space, May 2008, Arcachon, France (an elaborated version is to be published in Acta Astronautica under the title *Between Heaven and Earth: the legal challenges of human space travel*).

territory only. And as long as the vehicles 'take off' and 'land' in that territory, the likelihood of cross-border damage is limited, so in principle that State's *national* law will apply. Most ventures are planned to take place in the USA, for instance the Mojave desert. Here, no international element will be involved a state may determine whether it will consider the activity as an aviation or space activity under its national law.

The USA has developed a substantive body of rules governing private human spaceflight that is applicable until the end of 2012, possibly longer.[17] A 'light touch" legal approach has been taken, and licenses from the FAA's Office of Commercial Space Transportation are mainly concerned with public safety, not so much with the safety of passengers (who are voluntarily engaging in a risky activity).

However there are other cases where the probability of an *international* element, and thus the applicability of international air or space law, is much less remote, for instance if and when Virgin Galactic would launch from Kiruna in Sweden. Countries in Europe are much smaller and so the risk of cross-border damage is much bigger. This could then lead to damage being caused by (the private entity of) one state to persons or property of another.

Before turning to a discussion of whether air or space law would apply in such cases, it must be noted that even though more and more states in Europe have enacted national space legislation creating a licensing system, most of these do not contain any specific rules on space tourism.[18] Interestingly, the Dutch Act contains a provision stating that it can be declared 'wholly or partly applicable to the organization of outer space activities by a natural or juridical person from within the Netherlands' (Sec.2.2.b). The explanatory note states: 'This might include the commercial organization of space tourism activities'.

What is worrying is that efforts at harmonisation are only just starting, and the risk is that we end up with a patchwork of rules that may lead to flags of convenience and forum shopping. Recently, the European Aviation Safety Agency (EASA), the European counterpart of the FAA, has declared its competence in regulating space tourism, limited to horizontal take-off concepts such as SpaceShipTwo, and has taken a less flexible approach than the FAA, basically considering these craft as aircraft requiring full certification at the start of commercial operations.[19]

Returning to the situation where an international element warrants the application of international law, the challenging question is whether we should apply air or space law to damage resulting from space tourism (what's in a name...!).

As is well known, space law is based on 'freedom of use', in the absence of sovereignty, whereas air law is based on the sovereignty of states over the airspace above their territory. This results in major differences in both regimes. In air law, passenger liability and liability of the operator towards third parties on the ground is laid down in an elaborate system

[17] Code of Federal Regulations (CFR), Title 14, Chapter III, based on the CSLAA of 2004. Relevant parts are 14 CFR 401, 415, 431, 435, 440 & 460, at http://ecfr.gpoaccess.gov/cgi/t/text/text-idx?sid=6a5153b45a2675c8b05adfd8d7195483&c=ecfr&tpl=/ecfrbrowse/Title14/14cfrv4_02.tpl#300.

[18] E.g. Norway, Sweden, the UK, Belgium, the Netherlands and France. An overview and texts are available at http://www.unoosa.org/oosa/en/SpaceLaw/national/index.html. . For the French law ('Loi no. 2008-518 du 3 juin 2008 relative aux opérations spatiales', JO (Official Journal) 129, 4 June 2008), see http://www.senat.fr/dossierleg/pjl06-297.html. Its full implementation is expected in mid-2010.

[19] See *Accommodating sub-orbital flights into the EASA regulatory system*, by J.B. Marciacq *et.al.*, at www.congrex.nl/08a11/presentations/day1_S09/S09_05_Marciacq.pdf. See also Masson-Zwaan/ Freeland, *supra* note 16.

of rules tested and clarified extensively by jurisprudence, while space law is based on a rudimentary state-based system of liability that has moreover never been put to the test in a court case.

So, should air law apply for part of the journey and space law at some (as yet undefined) point during the activity? Is the case different for suborbital flights and for orbital flights? And what about horizontal (aircraft) take-off as opposed to vertical (rocket propelled) take-off?[20] The application of two totally different regimes to one suborbital flight may be the result; this would be both unsatisfactory and impractical. Of course, ideally, a comprehensive and uniform 'hybrid' legal regime ('aerospace law'?) should cover the complete launch and return journey of private individuals. Such a regime may well require a new treaty, but we all know that the probability of being able to agree on a new treaty is not high nowadays and might well take decades.

Since there is a need to provide clarity to today's space tourism entrepreneurs and to safeguard the interests of all players involved, the second best, or interim solution would seem to apply space law to the entire suborbital flight, on the basis of the function of the vehicle or mission. Since the purpose of space tourism is to go to space, space law should be applied to the entire mission.

But, there is a but, and that is that appropriate clarifications and additions (perhaps based on the model of the US CFR/FAA) are made to supplement the provisions of the space treaties. And national laws should be harmonised as much as possible. This will require action at the international level, probably steered by UNCOPUOS, relying on the expertise and experience of ICAO, and at the regional level, e.g. by the EU.

[20] One can also wonder if the case is different for suborbital flights and for orbital flights, but as stated earlier, this article only addresses sub-orbital flights.

8. Article VI and space tourism

Having thus established a preference for the applicability of space law, there is a need to discuss some of the elements of Article VI in the context of space tourism, and to see what needs to be done to supplement or clarify the regime.

'National activities in outer space'

The state bears international responsibility for 'national activities in outer space'. It is not very clear what is meant by the term 'national activities'. Is reference made to activities of nationals, also if they are abroad? Does it refer to private companies registered or headquartered in its territory? Do the activities concerned need to take place in outer space, or have an effect in outer space? What about a ticket sales office for space tourism? Does that make the state where the sales office is registered or headquartered responsible? This would seem farfetched, as it would just concern the activity of selling tickets to go into outer space; the activity has no direct effect in outer space, it is not an activity 'in space'. Would the operation of a spaceport make the country where it is located or headquartered internationally responsible? Operating a spaceport also is not an activity 'in' space, so its operation would most probably not fall under Article VI, and thus a license under national space legislation would also not seem necessary under the various national space legislations. Operating a spaceline on the other hand could be construed as an activity 'in outer space' and Article VI would probably make the state of incorporation responsible. For operating a spaceline therefore, national space legislation would probably request a license. Nevertheless, we can observe that the term would benefit from clarification.

'Appropriate State Party'

The 'appropriate' state party has to authorise and supervise, however the question is why the term 'appropriate state' is used instead of another. This is again a complicated question.[21] Is it the state of nationality of a spaceline that is meant, or is it the launching state? But in that case, why use a different name? Can there be several 'appropriate states', like there can be several launching states? Can the appropriate state change, for instance if a spaceline is sold?

What must be kept in mind is that the purpose of Article VI is to provide for control by the state that is responsible, which is the main subject of this article. Again, clarification would be beneficial.

'Authorisation'

In order to authorise an activity, the company carrying out the activity must have a link with the authorising state. The establishment of a licensing system is one way of establishing an authorisation mechanism. But it is not the only way. The state decides, this is an internal affair. Lately many states have come to the conclusion that the establishment of a national licensing system is the most logical way of authorising space activities, but for instance France, a major space power, has functioned quite well without an explicit licensing system for many decades. And besides, as already mentioned above, even states that have enacted national space legislation did not always address space tourism or create a licensing system.

'Continuing supervision'

The issue of continuing supervision is of course closely linked to authorisation, and the best way seems for a state to include a mechanism of for instance bi-annual reviews once a license has been granted in order to fulfil this obligation under Article VI. But again, this is not the only way, and states may also choose not to provide for a supervision audit under their national schemes. A license definitely facilitates supervision as it creates a direct link between the state and the private entity. Also, nothing is foreseen in Article VI about what 'continuous' implies, so for instance the frequency and the extent of the supervision are left for states to decide. Should the potential risk exposure be taken into account? Or new technological developments? Should the nature of the activity have an influence on the extent of supervision? For instance for telecommunications satellites in the geostationary orbit once every three years, and for the operation of a spaceline once every year? Again, we can see that Article VI imposes a rather general obligation on states and it is left to their discretion to give meaning to it – which may lead to a patchwork of differing national regulations. Here as well, harmonisation and guidance on how to fill in the obligation of Article VI seems desirable.

9. Problems with implementing Article VI

The question for states where private space activities, including space tourism, are (or will be) carried out, is how they should realise an adequate authorisation and supervision process, so that they can fulfil their obligations under Article VI of the Outer Space Treaty in a satisfactory way, avoiding uncontrollable exposure to international responsibility and liability. As stated above, not all states have implemented their obligations under Article VI in the same manner. The absence of objective evaluation criteria and standards does not facilitate the work, and in some countries the authorities in charge of the licensing process lack the necessary know-how to properly evaluate the activity, especially in smaller countries without a long history of space activities. For the US

[21] See K.H. Böckstiegel, *The Term "Appropriate State" in International Space Law,* Proceedings of the 37th Colloquium on the Law of Outer Space 77 (1994).

government to assess whether an activity should obtain a license or not will thus be much easier than for the Dutch or Belgium government for example. These will often be obliged to hire external experts in order to carry out the necessary safety, financial and insurance audits. This is amplified by the fact that space activities in such countries will remain the exception rather than the rule for the foreseeable future, and therefore it is not economically feasible for them to build up and maintain such hi-tech expertise in-house.

The lack of harmonisation in legislations interpreting Article VI may lead to a wide variety of implementation methods.

Another problem that may severely impact the capacity of states to carry out a proper evaluation of a proposed space activity is the long arm of ITAR.[22] The purpose of ITAR is to prevent international proliferation of US technology that may be critical to national security, and non-adherence to its provisions can result in huge fines. This may lead to the undesirable effect that a third state is unable to properly verify the implications of a proposed space activity under its responsibility because ITAR prevents the applicant from providing the information that will assure the government of the compliance of the activity.

And then of course we have not even started to address the problems that may arise for passengers of space tourism flights or third parties on the ground; the space treaties do not allow the victims to present a claim directly to the operator, claims must be presented by one state to another. Moreover, nationals of the launching state may be excluded altogether from presenting a claim under Article VII of the Liability Convention. This provision was designed for astronauts on board of the launch vehicle of their own national state, but is inadequate for a paying passenger on board a commercial flight – even though for the time being passengers are required to sign an 'informed consent' letter waiving their right to claim for damages.[23]

10. Solutions?

Efforts to harmonise national legislations are undertaken by the UNCOPUOS and this is a very good initiative, as ideally harmonisation must be a global effort.[24] An agenda item called "General exchange of information on national legislation relevant to the peaceful exploration and use of outer space" is considered under a multi-year work plan for the period 2008-2011, and a new Working Group on National Legislation Relevant to the Peaceful Exploration and Use of Outer Space has been established. The gathered information will allow all States, in particular developing States, to gain an understanding of existing national regulatory frameworks and could assist States in their efforts to establish their own national regulatory frameworks in accordance with their specific needs and level of development.

At the regional level as well such efforts should be continued, e.g. in Europe.[25]

---

[22] International Traffic in Arms Regulations, ITAR (22CFR120-130). see http://ecfr.gpoaccess.gov/cgi/t/text/text-idx?sid=a5d12a23a5dfe0e4495181703bdae79a&c=ecfr&tpl=/ecfrbrowse/Title22/22cfrv1_02.tpl. See for instance P.J. Blount, *The ITAR Treaty and its implications for US space exploration policy and the commercial space industry*, 73 JALC 705 (2008)

[23] See T. Knutson, *What is "Informed Consent" for Space-Flight Participants in the Soon-To-Launch Space Tourism Industry?* 33 JSL 105 (2007). It is questionable whether legal representatives of deceased space tourists will be bound by such letters.

[24] See for instance the Report of the Legal Subcommittee on its forty-eighth session, held in Vienna from 23 March to 3 April 2009, http://www.oosa.unvienna.org/pdf/reports/ac105/AC105_935E.pdf at 22.

[25] Initial steps are being taken by EASA, see *supra* note 19, and ESA has issued a position paper and a press release on space tourism, but the EU has not yet expressed itself. See the ESA press release at http://www.esa.int/esaCP/SEM49X0YUFF_index_0.html, and the position paper at

As regards ITAR, a solution may be to conclude TAA's (Technical Assistance Agreements) that may allow the government to have access to the information, but this generally takes several months and in some states, a shorter term applies for applicants to receive a decision on their license application.[26]

Apart from the necessary clarifications of the component concepts of Article VI ('national activities', 'appropriate state', authorisation', 'continuing supervision'), it is also necessary to complement the state-based responsibility and liability system of the space treaties.

Detailed rules governing second- and third-party liability are needed, and lessons may be learned here from the extensive liability provisions governing the international carriage by air under the Warsaw/ Montreal/ Rome system.[27]

The Warsaw Convention of 1929, as amended, provides for upper limits for liability in relation to the carriage of passengers and of baggage and cargo as well as dealing with areas of responsibility and insurance. Article 20(1) exonerates the carrier from liability where it or its servants and agents 'have taken all necessary measures to avoid the damage or that it was impossible for him or them to take such measures'.

The Montreal Convention of 1999 was designed to supersede the Warsaw Convention and removed the system of arbitrary limits on air carrier liability, by providing that the carrier was liable for the full amount of the damages, unless it could demonstrate that it was not negligent or that a third party was solely responsible for the damage: Montreal Convention, art VI.

The Rome Convention on Damage Caused by Aircraft to Third Parties on the Surface and its Montreal Protocol of 1978, deals with third party liability in respect of commercial air activities. It is intended 'to ensure adequate compensation for persons who suffer damage caused on the surface by foreign aircraft, while limiting in a reasonable manner the extent of the liabilities incurred for such damage in order not to hinder the development of international civil air transport'. However the Rome Convention is not widely ratified because of its low limits and its relevance is negligible. In most cases, national law will determine the compensation to third parties on the ground.

Some other areas where space law might use provisions from air law are vehicle safety, the status of the crew, commander, and passengers, navigation (traffic, transit), security (criminal law), and of course liability (collisions, third parties, passengers). With these at its disposal, states will have the necessary tools to fulfil their obligations under Article VI.

The unintended extra-territorial effect of ITAR must also be reduced, but this will probably take time. Perhaps the conclusion of TAA's will help in the meantime, but it is also a responsibility of the US government to reduce the global negative effects of ITAR.

11. Conclusions

Safe, efficient private human access to space at reasonable cost will boost space activity, the global economy, and thus will benefit Mankind as a whole. Article I of the Outer Space Treaty therefore does not

---

http://esamultimedia.esa.int/docs/gsp/Suborbital_Spaceflight_ESA_Position_Paper_14April08.pdf.

[26] Under the Dutch Space Activities Act, the authorities are expected to give their decision within six months.

[27] Convention for the Unification of Certain Rules Relating to International Carriage by Air, Warsaw, 12 October 1929; Convention for the Unification of Certain Rules for International Carriage by Air, Montreal, 28 May 1999; Convention on Damage Caused by Foreign Aircraft to Third Parties on the Surface, Rome 7 October 1952. See for texts http://www.mcgill.ca/iasl/research/treaties/air_law/private/.

stand in the way of seeing space tourism as a legitimate use of space.

Article VI Outer Space Treaty sets the scene for state control of space tourism by posing a due diligence obligation on the state and imposing a state-based system of responsibility and liability.

Clarifications to this system are needed, as is harmonisation at the global and regional level. Also, additional rules are to be derived from air law in order to better equip the state to fulfil its obligations of authorisation and continuous supervision under Article VI, and to develop an adequate system of second- and third party liability.

A clear, functional, harmonized legal framework for space tourism is essential to safeguard the interests of the state and those of the private entities engaging in this great new leap for Mankind

Article VI of the Outer Space Treaty 'in the European Context'

Prof. Dr. Frans G. von der Dunk
Member of the Board, European Centre for Space Law (ECSL)
University of Nebraska-Lincoln, College of Law

1. Introduction

The theme of the 3rd Eilene M. Galloway Symposium on Critical Issues in Space Law clearly was a very appropriate one in view of the extraordinary career of Dr. Galloway. One of the key features of Article VI of the Outer Space Treaty[1] being the requirement of authorisation and continuing supervision of non-governmental entities' "national activities in outer space", Article VI represents a major aspect of the obligations of mankind, in particular the leading space-faring nations, to use outer space responsibly that have always been defended so staunchly by Dr. Galloway. It states that "activities of non-governmental entities shall require authorization and continuing supervision by the appropriate State Party to the Treaty", and thus recognizes their proper place as part of the overarching paradigm of the responsible usage of outer space, to be guaranteed by one state or another. The present paper tries to analyse how this issue has been tackled in the specific European context, and whether perhaps something akin to a 'European approach' can be discerned here.

2. A 'European context' for Article VI?

When reference is made to 'the', or even 'a', 'European context', the preliminary question always is: "*which* 'Europe'"? For even if merely confining ourselves to outer space and activities of private entities in that sense, there would be a number of 'Europes' that could likely be at issue.

First amongst these would be the European Space Agency (ESA)[2], as the general standard-bearer for European-wide space activities. Activities of ESA, itself an intergovernmental organisation, have amongst many other things given rise to the establishment of two further intergovernmental space organisations: EUTELSAT[3] for satellite communications – which has meanwhile been privatised[4] – and EUMETSAT[5]

[1]. Treaty on Principles Governing the Activities of States in the Exploration and Use of Outer Space, including the Moon and Other Celestial Bodies (hereafter Outer Space Treaty), London/Moscow/Washington, done 27 January 1967, entered into force 10 October 1967; 610 UNTS 205; TIAS 6347; 18 UST 2410; UKTS 1968 No. 10; Cmnd. 3198; ATS 1967 No. 24; 6 ILM 386 (1967).

[2]. ESA was established by the Convention for the Establishment of a European Space Agency (hereafter ESA Convention), Paris, done 30 May 1975, entered into force 30 October 1980; 14 ILM 864 (1975); Space Law – Basic Legal Documents, C.I.1; currently, it has 18 member states.

[3]. EUTELSAT was established by the Convention Establishing the European Telecommunications Satellite Organization (EUTELSAT), Paris, done 15 July 1982, entered into force 1 September 1985; Cmnd. 9069; Space Law – Basic Legal Documents, C.II.1; and the Operating Agreement Relating to the European Telecommunications Satellite Organization (EUTELSAT), Paris, done 15 July 1982, entered into force 1 September 1985; Cmnd. 9154; Space Law – Basic Legal Documents, C.II.2. During its heydays, shortly before privatisation, it had more than 45 member states.

[4]. See Convention Establishing the European Telecommunications Satellite Organization (EUTELSAT), done 15 July 1982, entered into force 1 September 1985, as amended 20 May 1999, amended version not yet entered into force but applied provisionally 2 July 2001; Space Law – Basic Legal Documents, C.II.1.

[5]. EUMETSAT was established by the Convention for the Establishment of a European Organization for the Exploitation of Meteorological Satellites (EUMETSAT), Geneva, done 24 May 1983, entered into force 19 June 1986; as amended 14 July 1994, entered into force 27 July 1994; Cmnd. 9483; Space Law – Basic Legal Documents, C.III.1; 44 ZLW 68 (1995); currently it has 21 member states.

for satellite meteorology. For the sake of brevity, however, these two entities will be left out of the analysis at this point.

A second 'Europe' that, though coming from a quite different angle, also meets the eye as a potential flag-bearer for Europe in the present context is the European Community (EC), since 1992 part of the overarching European Union.[6] The Community essentially developed as an intergovernmental organisation *sui generis*, with certain competences that could be qualified as 'supranational' and the overriding aim to establish a liberalised economic level playing field by means of a common market.

Thirdly, from a yet different perspective 'Europe' simply means the geographical concept, a continent running from the Arctic waters in the north to the Mediterranean Sea in the south and from the Atlantic Ocean in the west to the Ural mountains in the East. From this point of view, however, Europe really still consists of a number of sovereign individual states, where then – with a view to the theme of the present Symposium – the focus would logically be on those states that have, somehow, developed a system for "authorization and continuing supervision" as is required by Article VI of the Outer Space Treaty, as opposed to those who are as of yet without such a system.

3. The 'Europe' of ESA

The role and modes of operation of ESA, as an intergovernmental organisation essentially pooling the scientific, technical and financial resources of its member states, have been dealt with *in extenso* by many other books, contributions to books and articles, and will not be reiterated here.

Article VI of the Outer Space Treaty, in conjunction with Article XIII, provides for a somehow secondary legal status of intergovernmental organisations active in outer space such as ESA. Whilst such organisations under the former share international responsibility for such activities with their member states[7], the latter makes clear that ultimately international organisations merely present a "framework" for these individual member states to conduct their activities within[8].

For the present purpose, it is noteworthy firstly that ESA has declared its willingness, as a major player in space, to be bound by the rights and obligations of those follow-up treaties to the Outer Space Treaty where that option was offered: the Rescue Agreement[9], the Liability Convention[10] and the Registration Convention[11]. The Moon

[6]. The European Community was originally based on, especially, the Treaty of Rome, or Treaty establishing the European Economic Community (hereafter EC Treaty, where reference should be had to its currently ruling version), Rome, done 25 March 1957, entered into force 1 January 1958; 298 UNTS 11; after a number of important revisions most fundamentally changed by the Treaty on European Union, Maastricht, done 7 February 1992, entered into force 1 November 1993; 31 ILM 247 (1992); OJ C 191/1 (1992); its present legal status however is determined as per the Treaty of Nice amending the Treaty on European Union, the Treaties establishing the European Communities and certain related acts, Nice, done 26 February 2001, entered into force 1 February 2003; OJ C 80/1 (2001). The EC currently comprises 27 member states.

[7]. Art. VI, Outer Space Treaty, provides in relevant part: "When activities are carried on in outer space, including the Moon and other celestial bodies, by an international organization, responsibility for compliance with this Treaty shall be borne both by the international organization and by the States Parties to the Treaty participating in such organization."

[8]. Art. XIII, Outer Space Treaty, makes reference to "the activities of States (...) including cases where they are carried on within the framework of international intergovernmental organizations".

[9]. See Art. 6, Agreement on the Rescue of Astronauts, the Return of Astronauts and the Return of Objects Launched into Outer Space, London/Moscow/Washington, done 22 April 1968, entered into force 3

Agreement, though also offering such an opportunity, did not enjoy such treatment as only a few ESA member states are parties to that treaty themselves.[12]

Secondly, however, whilst the formulation of Articles VI and XIII of the Outer Space Treaty and the acceptance by ESA of rights and obligations under three other space treaties might leave open possibilities for ESA to be substantially involved in authorisation of private space activities, it should be realised that ESA is mainly an instrument for organising joint space programmes of the member states and for trying to provide some overall direction to the various individual and joint space programmes abounding in Europe.

The key concept of the ESA structure with a view to such space programmes, including any involvement of private actors, is the two-pronged approach of mandatory *versus* optional activities. Mandatory activities are those in which all member states participate, such as basic activities, science programmes potentially giving rise to actual space programmes, and information collection and dissemination tasks, to the financing of which all member states have to contribute as per a predetermined scale.[13] Optional activities, comprising *inter alia* all space operations themselves, by contrast allow for opt-out by individual member states as well as individual determination of their respective contribution for those who decide not to opt out.[14]

In either case, private industry only contributes as subcontracted developers and builders of software and hardware, and possibly integrators thereof. As soon as such soft- and/or hardware has to go into outer space, however, this is either taken care of by ESA itself or contracted out to other (launch service) providers. Whilst it is true that such launches are often contracted out to Arianespace, which is a European private company (often furthermore using technology developed by ESA and cooperating with ESA in many other ways); the relationship in a formal sense is still a purely contractual one, not that between a licensor and a licensee.[15]

December 1968; 672 UNTS 119; TIAS 6599; 19 UST 7570; UKTS 1969 No. 56; Cmnd. 3786; ATS 1986 No. 8; 7 ILM 151 (1968); Declaration of 31 December 1975.

10. See Art. XXII, Convention on International Liability for Damage Caused by Space Objects (hereafter Liability Convention), London/Moscow/Washington, done 29 March 1972, entered into force 1 September 1972; 961 UNTS 187; TIAS 7762; 24 UST 2389; UKTS 1974 No. 16; Cmnd. 5068; ATS 1975 No. 5; 10 ILM 965 (1971); Declaration of 23 September 1976; Space Law – Basic Legal Documents, A.III.2, p.1.

11. See Art. VII, Convention on Registration of Objects Launched into Outer Space (hereafter Registration Convention), New York, done 14 January 1975, entered into force 15 September 1976; 1023 UNTS 15; TIAS 8480; 28 UST 695; UKTS 1978 No. 70; Cmnd. 6256; ATS 1986 No. 5; 14 ILM 43 (1975); Declaration of 2 January 1979; Space Law – Basic Legal Documents, A.IV.4.2, p.2.

12. Agreement Governing the Activities of States on the Moon and Other Celestial Bodies, New York, done 18 December 1979, entered into force 11 July 1984; 1363 UNTS 3; ATS 1986 No. 14; 18 ILM 1434 (1979); see Art. 16. Amongst the current 18 member states, only Austria, Belgium and the Netherlands are parties to the Moon Agreement; France in addition has signed but not ratified the Agreement.

13. See Artt. V(1) sub (a), XIII(1), ESA Convention.

14. See Artt. V(1) sub (b), XIII(2), ESA Convention.

15. Interestingly, in some measure the relationship between ESA and Arianespace in terms of substance does deal with issues typically included in licenses and other authorisations, such as liability and reimbursement thereof, but it is illustrative of the fundamental lack of competence of ESA in these respects that that relationship also involves the individual member states of ESA and in particular France, the state of nationality of Arianespace as well as the holder of sovereignty over the territory from which Arianespace operates; cf. e.g. the author's *Private Enterprise and Public Interest in the European 'Spacescape' – Towards Harmonized National Space Legislation for Private Space Activities in Europe* (1998), 155-61, 167-70, for an analysis of this construction.

Thus, no issue arises of a national activity in outer space conducted by a non-governmental entity, with reference to the phrasing of Article VI, requiring authorisation and/or continuing supervision by ESA – which in addition, as an organisation focused on scientific, technical, operational and financial cooperation, has no competences under the ESA Convention to license or otherwise exercise state-like authorising or supervising powers *vis-à-vis* private operators in any event.

4. The 'Europe' of the European Community and the European Union

Whilst ESA was not at all about pooling regulatory resources and competences at an international level, the European Community, also as later subsumed within the European Union, certainly was. From the beginning the founding treaties were not only meant to provide substantive rules and guidelines for individual member states and other actors in terms of market regulation and behaviour, but also to provide for the mechanisms to continuously enlarge and update such substantive regimes through the set-up of an elaborate institutional structure.

Thus, a Commission, a Council of Ministers and a European Parliament were created by the treaties, which in a complicated interplay were able to draft Regulations, Directives and Decisions overriding as relevant any national member state legislation and regulation on the point.[16] Furthermore, the European Court of Justice was created to adjudicate cases where EC-law rules were, presumably, violated by any relevant actor – including, as appropriate, individual member states.[17]

Whilst such regulatory power over economic activities – which, in principle, encompasses economic activities involving outer space – could have included establishment of an EC-level licensing process, and the possibility to thus deal with the authorisation and supervision of private activities in the terminology of Article VI of the Outer Space Treay including such activities in outer space, that has remained largely theoretical even outside the domain of outer space.

The only area where space activities have obtained a clear-cut commercial character, being satellite communications, has indeed seen a half-hearted effort at the EC level to ensure that the licensing process of such activities – still exclusively a domestic matter – would not result in major distortions in the level playing field which was to result from the establishment of an Internal Market.

The process took off with a first Green Paper in 1987, calling for liberalisation of the environment for telecommunications,[18] as of yet excluding satellite communications in view of the special character of that sector; an omission which was quickly repaired when a second Green Paper three years later addressed satellite communications specifically along the same lines[19]. It took another four years, however, before the

---

[16]. The role and competences of each of these main institutions are outlined in the EC Treaty; Artt. 211-219 for the Commission, Artt. 202-210 for the Council and Artt. 189-201 for the Parliament. The two main law-making mechanisms involving all three institutions are provided for in Art. 251 and Art. 252; the concepts of Regulations, Directives and Decisions in Art. 249.

[17]. See for the role and competences of the ECJ Artt. 220-245, EC Treaty, where Artt. 226, 227, 230, 232 & 234 essentially determine the scope of its jurisdiction.

[18]. Towards a Dynamic European Economy – Green Paper on the Development of the Common Market for Telecommunications Services and Equipment, Communication from the Commission, COM(87) 290 final, of 30 June 1987; OJ C 257/1(1987); as per Council Resolution on the development of the common market for telecommunications services and equipment up to 1992, of 30 June 1988, OJ C 257/1 (1988).

[19]. Towards Europe-wide systems and services – Green Paper on a common approach in the field of satellite communications in the European Community, Communication from the Commission, COM(90) 490 final, of 20 November 1990.

general economic principles asserted by that Green Paper led to a first piece of proper EC legislation, the 1994 Satellite Directive[20], applying as its full title indicated some older Directives on the introduction of certain competition rules to the satellite communication sector.

Soon, the Commission in particular, driving this process of liberalising the European satellite communication markets, started to call for a harmonisation of the national processes of licensing the operators. This was considered an important prerequisite for realising an Internal Market for satellite communications without regulatory distortions following from the rather divergent conditions and fees imposed by the various relevant national licensing authorities. For a number of reasons, however, it did not manage to move beyond a half-way effort to harmonise national licensing processes; its original aim of an EC-wide licensing process principally conducted at the EC level rapidly disappeared behind the horizon.[21]

As a consequence, so far the increasing interest of the European Community in outer space and the benefits it could bring to economic activities on earth beyond the aforementioned applicability and application of competition law to satellite communications has mainly manifested itself through the lead which the Commission took in two major European space projects.

The first of these is Galileo, a joint programme to develop Europe's own, second-generation satellite navigation system together with ESA.[22] The cooperation with ESA in this context was institutionalised through the creation first of the Galileo Joint Undertaking[23], then of the European GNSS Supervisory Authority[24], and also gave rise to several international agreements with non-European states on cooperation in its context[25].

However, interestingly enough the original aim to establish a concession allowing a private consortium to operate the Galileo satellite system and market its services – which would have been somewhat akin to a license, and would certainly have come to encompass authorisation and supervision mechanisms, including arrangements on

---

[20]. Commission Directive amending Directive 88/301/EEC and Directive 90/388/EEC in particular with regard to satellite communications, 94/46/EC, of 13 October 1994; OJ L 268/15 (1994).

[21]. Cf. e.g. Commission Directive amending Directive 90/387/EEC with regard to personal and mobile communications, 96/2/EC, of 16 January 1996; OJ L 20/59 (1996); Decision of the European Parliament and of the Council on a coordinated authorization approach in the field of satellite personal communications systems in the Community, No. 710/97/EC, of 24 March 1997; OJ L 105/4 (1997); Directive of the European Parliament and of the Council on a common framework for general authorizations and individual licenses in the field of telecommunications services, 97/13/EC, of 10 April 1997; OJ L 117/15 (1997); Decision of the European Parliament and of the Council on the selection and authorisation of systems providing mobile satellite services (MSS); No. 626/2008/EC, of 30 June 2008; OJ L 172/15 (2008).

[22]. See e.g. Council Resolution on the European Contribution to the Development of a Global Navigation Satellite System (GNSS), of 19 December 1994; OJ C 379/2 (1994); and Green Paper on Satellite Navigation Applications, COM(2006) 769 final, of 8 December 2006.

[23]. See Council Regulation setting up the Galileo Joint Undertaking, No. 876/2002/EC, of 21 May 2002; OJ L 138/1 (2002).

[24]. See Council Regulation on the establishment of structures for the management of the European satellite radio-navigation programmes, No. 1321/2004/EC, of 12 July 2004; OJ L 246/1 (2004).

[25]. Such as the People's Republic of China (Cooperation Agreement on a Civil Global Navigation Satellite System (GNSS) – Galileo between the European Community and its Member States and the People's Republic of China, of 30 October 2003; Doc. Council of the European Union, 13324/03) and Israel (Cooperation Agreement on a Civil Global Navigation Satellite System (GNSS) between the European Community and its Member States and the State of Israel, of 2 June 2005; Doc. Council of the European Union, 9482/04).

liability, with the European GNSS Supervisory Authority in the supervisory role – had to be shelved as the commercial possibilities for operating a satellite infrastructure were insufficient for the time being to attract a concessionaire.[26]

The second major European space project driven by the Commission with the technical and operational support of ESA was the development of GMES (Global Monitoring for Environment and Security), which was recently re-christened Kopernikus and is also to result ultimately in a series of satellites to be launched, this time primarily for remote sensing purposes in the service of European policies on the environment and civil protection of various kinds.[27]

Finally, it should be noted that neither the European Community nor the European Union has ever declared its acceptance of rights and obligations under any space treaty, nor has it – as far as can be judged – ever seriously contemplated doing so.

5. The geographical Europe: a number of individual states

That last conclusion, therefore, brings us to the geographical Europe and those states part of it that on an individual level have decided to implement in some comprehensive manner the obligation to authorise and continuously supervise space activities by non-governmental entities.

This is not the place to undertake an extensive and comprehensive analysis of the way in which the various states concerned have sought to implement the obligations arising under Article VI of the Outer Space Treaty *vis-à-vis* private enterprise by means of a national law, act or set or regulations; the present effort will confine itself to comparison on a select number of key issues which in a sense may serve as a yardstick with reference to such implementation.

These issues are basically fivefold: (1) the scope of the national law and the licensing obligation established by it in terms of (space) activities, which refers back to Article VI's applicability to "activities *in outer space*"; (2) the scope thereof in terms of persons and entities requiring a license under the national law, which refers back to the concept of "*national* activities" as propounded by Article VI; (3) how the specific issue of dealing with international third party liability and the related one of insurance thereof is dealt with, as the most tangible and directly *valuable* subjects to be addressed by authorisation and supervision; (4) the national entity responsible for the licensing process, hence for proper authorisation and supervision, which includes – from an international law perspective – registration of space objects involved; and (5) to the extent possible, the actual practice of application. Each of seven states geographically part of Europe (all being parties to the Outer Space Treaty moreover) will thus – summarily – pass scrutiny.

---

[26]. Cf. Communication from the Commission to the European Parliament and the Council – Galileo at a Cross-Road: The Implementation of the European GNSS Programmes, COM(2007) 261 final, of 16 May 2007; Council Resolution on GALILEO, 2805th Transport, Telecommunications and Energy Council Meeting, Luxembourg, 6-8 June 2007; and Amended proposal for a Regulation of the European Parliament and of the Council on the further implementation of the European satellite radionavigation programmes (EGNOS and Galileo), Brussels, 4 April 2008; Interinstitutional File: 2004/0156 (COD), 8046/08.

[27]. Cf. Council Resolution on the launch of the initial period of global monitoring for environment and security (GMES), of 13 November 2001; OJ C 350/4 (2001); Communication from the Commission to the European Parliament and the Council – Global Monitoring for Environment and Security (GMES): Establishing a GMES capacity by 2008, COM(2004) 65 final, of 3 February 2004; and Communication from the Commission to the Council and the European Parliament – Global Monitoring for Environment and Security (GMES): From Concept to Reality, COM(2005) 565 final, of 10 November 2005.

6. Norway

Norway, a state member of ESA but not of the European Union (having twice in the past voted not to accede), was the first European state to enact a national act on space, the Act on launching objects from Norwegian territory into outer space in 1969.[28] As the title already discloses, the scope of the Act *ratione materiae* is limited to launching activities; *ratione personae* the license obligation resulting from it applies to such activities conducted from Norwegian territory, vessels and aircraft, as well as conducted by nationals if operating from global commons.[29]

On arrangements for reimbursing the Norwegian state for liability claims resulting from authorised activities under the international liability regime, the Norwegian Act is silent, but that should not be surprising in view of the fact that the Act was enunciated before the 1972 Liability Convention offered any further detail on how the general principles on liability provided by Article VII of the 1967 Outer Space Treaty were to be applied. Presumably, experience with application of the Act has remained limited to the case of the *Norsk Romsentr*, being authorised to launch from the Andøya rocket launch facilities. Both are, however, more or less government-owned and -controlled entities, so that one might actually question whether such an authorisation was really necessary for the Norwegian government to fulfil its obligations under Article VI of the Outer Space Treaty.

7. Sweden

Sweden, as the second European country historically speaking developing its national regime for authorising and supervising private space activities, was the first state both member of ESA and (now) the European Union[30] to do so: in 1982, both an Act on Space Activities and a Decree on Space Activities saw the light of day.[31]

In terms of the Act's scope *ratione materiae*, most noteworthy was the exclusion of the launch of sounding rockets of the otherwise comprehensive set of space activities requiring a license,[32] which were apparently not considered either to be likely undertaken with private involvement, or to present much of an international risk in liability terms. The license obligation was applied rather comprehensively also *ratione personae*, in being imposed both upon those undertaking any of the relevant activities from Swedish soil and upon those with the Swedish nationality.[33]

As to the issue of liability, the Swedish Act and Decree being enunciated a full ten years after the Liability Convention, in principle the Swedish government is entitled to full reimbursement of any international liability claim, although no obligatory insurance is provided for[34] – and hence in an actual case the Swedish government may end up

---

[28]. Act on launching objects from Norwegian territory into outer space, No. 38, 13 June 1969; National Space Legislation of the World, Vol. I (2001), at 286.

[29]. See Sec. 1, Act on launching objects from Norwegian territory into outer space.

[30]. Sweden became a member of the EU in 1995 only, that is 13 years after it drafted its national legislation on space activities.

[31]. Act on Space Activities, 1982: 963, 18 November 1982; National Space Legislation of the World, Vol. I (2001), at 398; Space Law – Basic Legal Documents, E.II.1; 36 Zeitschrift für Luft- und Weltraumrecht (1987), at 11; and Decree on Space Activities, 1982: 1069; National Space Legislation of the World, Vol. I (2001), at 399; Space Law – Basic Legal Documents, E.II.2; 36 Zeitschrift für Luft- und Weltraumrecht (1987), at 11.

[32]. See Sec. 1, Act on Space Activities.

[33]. See Sec. 2, Act on Space Activities.

[34]. Cf. Sec. 6, Act on Space Activities.

with less than full reimbursement, simply because the licensee has run out of funds to compensate from.
The licensing regime is detailed to some degree by the Decree, which *inter alia* provides the National Board for Space Activities (NBSA) with the authority to monitor licenses granted by the Swedish government, as well as the obligation to run the national register of space objects.[35] The most import thing to note about implementation of the Swedish Act and Decree, however, is a matter for the future: Virgin Galactic's business-deal to start launching space tourists from the Kiruna launch facilities in 2012 will present the major test for the actual application of the Act, as it still leaves considerable leeway in its throughout concise formulation to determine the terms of an individual license.

8. The United Kingdom

In 1986, the United Kingdom, another state member of both ESA and the European Union, enacted its Outer Space Act.[36] In terms of its application to a range of relevant activities, most notable was the explicit reference to "procuring the launch of a space object" as being included.[37] Also, the broad definition of "any activity in outer space" where a person is considered to conduct such an activity as soon as he "causes it to occur or is responsible for its continuing" leads to a rather sweeping scope of application – at least *ratione materiae*.[38] *Ratione personae*, namely, the scope is considerably less comprehensive, as the license obligation only extends to nationals undertaking the relevant activities, not for example to foreigners conducting them from British territory.[39]
As for relevant liability arrangements with a view to the international treaties, in principle full indemnification of the UK government is called for, and this time, contrary to the Swedish case, the licensing authority is expressly authorised to require insurance to cover such liability.[40] That licensing authority, in fact, is the British National Space Centre (BNSC), which also keeps the national register further to the relevant provisions of the Act under a mandate from the Secretary of State.[41]
The opportunity to obtain a license under the Outer Space Act has been made good use of, as since 1989 more than twenty licenses have been issued by the BNSC, in addition to at least ten more by the Governors of UK Overseas Territories.[42] Was the fee requested for licensing originally £ 1,000, in the following years it rose to £ 6,500, showing – if anything – a realisation that the process of licensing turned out to be more expensive than originally envisaged, or was getting more complicated as time moved on and more specific requirements were added.

---

35. See Sec. 3, Act on Space Activities & Sec. 2, Decree on Space Activities; resp. Sec. 4, Decree on Space Activities.
36. Outer Space Act, 18 July 1986, 1986 Chapter 38; National Space Legislation of the World, Vol. I (2001), at 293; Space Law – Basic Legal Documents, E.I; 36 Zeitschrift für Luft- und Weltraumrecht (1987), at 12.
37. Sec. 1(a), Outer Space Act. It may be noted that Art. I(c), sub (i), Liability Convention, refers to "A State which (…) procures the launching of a space object" as part of the definition of the "launching State".
38. Secc. 1(c), resp. 13(2), Outer Space Act.
39. See Secc.1, 2 & 3(1), Outer Space Act; see for the definition of 'UK national' Sec. 2.
40. See Sec, 10(1), resp. 5(2) sub (f), Outer Space Act.
41. Cf. Secc. 3, resp. 7(1), Outer Space Act.
42. Cf. Secc. 2(2) sub (a), 13(1) sub (a), (b), Outer Space Act.

As for the insurance issue, the general policy developed by the UK authorities was to oblige the licensee to take out an insurance policy covering reimbursement of the UK government up to £ 100,000,000 – noting again that under the Liability Convention liability is essentially unlimited. Noting furthermore that for example in the case of US launch licenses the maximum insurance requirement could be capped to the extent third-party liability coverage would be available at reasonable rates,[43] the UK government at the time noted that such coverage was available up to some US$ 250,000,000 for the launch phase and some US$ 400,000,000 for in-orbit operations.

9. The Russian Federation

Russia is neither a member state of ESA nor of the European Union, so it is one of the two European states considered here that are only part of geographic Europe. In 1993 it enunciated the Law of the Russian Federation on Space Activities,[44] providing the basic framework for dealing with the newly introduced principles of privatisation and market economy as they were also impacting the space industrial complex.

The Russian Law is characterised by its general and broad scope, as well as by certain ambiguities. For example, the definition of "space activities" as constituting the substantive scope of the Law, encompasses amongst others all "kinds of activities performed with the aid of space technologies" and the "creation (including development, manufacture and test) of, as well as using and transferring space technics, space technologies, other products and services necessary for carrying out space activities".[45] Similarly, the license obligation extends to all such activities "under the jurisdiction of the Russian Federation", further detailed as "the space activities of organizations and citizens of the Russian Federation or the space activities of foreign organizations and citizens under the jurisdiction of the Russian Federation, if such activities include tests, manufacture, storage, preparation for launching and launching of space objects, as well as control over space flights".[46] In addition, it should be noted, jurisdiction is expressly extended to space objects duly carried on the Russian registry.[47]

The precise arrangements on handling liability and reimbursement issues are yet to be clarified, but as far as the Law itself is concerned, the indemnification of the Russian government is in principle unlimited, coupled to compulsory insurance coverage up to a level to be decided – so far, presumably, on a case-by-case basis.[48] It is the Russian Space Agency *Roskosmos* which is charged with running the licensing process, as well as the national register of space objects.[49]

A few words finally on Russian practice. While not many details are readily available, already a number of years ago a major insurance company, Megaruss, could report on having been involved in over 60 contracts in the decade or so following 1992, having worked with a rough schedule for the limits to *de facto* reimbursement of liability

---

43. See Sec. 70112(a)(3)(B), Commercial Space Transportation – Commercial Space Launch Activities, 49 U.S.C. 70101 (1994).

44. Law of the Russian Federation on Space Activities, No. 5663-1, 20 August 1993, effective 6 October 1993; National Space Legislation of the World, Vol. I (2001), at 101.

45. Art. 2(1) & (2), Law of the Russian Federation on Space Activities.

46. Artt. 1(1), 9(2), Law of the Russian Federation on Space Activities.

47. See Art. 17(2), Law of the Russian Federation on Space Activities.

48. Cf. Art. 30(2) & (4), Law of the Russian Federation on Space Activities.

49. See Artt. 6(2), 17(1), Law of the Russian Federation on Space Activities.

claims which in terms of US$ ran between 80,000,000 for a Start launch vehicle and 300,000,000 for a Proton (the heaviest launch vehicle in the Russian market).

10. The Ukraine

For essentially the same reasons as Russia – because the Ukraine after Russia was the largest heir of assets and know-how of the previous Soviet space complex – the Ukraine drafted its national Law of the Ukraine on Space Activity[50] in 1996. Also the result was very much in line with the Russian Law which had been announced three years earlier: the definition of "space activity" was similarly broad[51], the license obligation equally pertained to activities undertaken "in the Ukraine or, under jurisdiction of the Ukraine, abroad"[52], and liability reimbursement of the government in case of an international claim is unlimited in principle, coupled to a compulsory insurance subject to a limit[53] – all more or less like in the case of Russia, though generally speaking formulated in a more straightforward fashion.

In this case, it is the National Space Agency of the Ukraine (NSAU) which grants, monitors and if necessary suspends and cancels licenses,[54] thus implementing the obligations resting upon the Ukraine under Article VI of the Outer Space Treaty. It may be said, quite comprehensively so; again like Russia, as following from the scope discussed above, the phrase "national activities" for example is apparently taken to refer to all activities falling within Ukrainian jurisdiction, hence those subject to legal and factual possibilities of control and supervision.

A final point where the Ukraine does largely follow the Russian approach concerns the registration: the obligation to register is expressly subject to exceptions where another state already takes care of registration as relevant.[55] The reason behind this is the involvement of Ukrainian space industry partners in the international launch consortium Sea Launch, led by the US company Boeing, which was being developed at the time – just as Sea Launch also comprised Russian partners.

11. Belgium

The next state in geographical Europe, member of ESA as well as of the European Union, creating its national system of authorisation and continuing supervision of private space activities was Belgium, with its Law on the activities of launching, flight operations or guidance of space objects in 2005.[56]

The license obligation established by the Law as the major instrument for control extends principally to activities conducted from Belgian territory; in addition it applies to activities conducted by Belgian nationals outside of Belgium if such is provided for by special agreement – presumably with the state from whose territory such activities

---

50. Law of the Ukraine on Space Activities, No. 502/96-VR, 15 November 1996: National Space Legislation of the World, Vol. I (2001), at 36.

51. Cf. Art. 1, 1st bullet, Law of the Ukraine on Space Activities.

52. Art. 10, Law of the Ukraine on Space Activities.

53. See Artt. 24, 25, Law of the Ukraine on Space Activities.

54. See Art. 10, also Art. 20, Law of the Ukraine on Space Activities.

55. Cf. Artt. 13, 14, Law of the Ukraine on Space Activities. It may be noted here that the Registration Convention does not allow for double registration; cf. Art. II(2). In the case of Russia, a more ambiguous provision referred to the possibility to deal with such an issue under an international agreement; see Art. 17(4), Law of the Russian Federation on Space Activities.

56. Law on the activities of launching, flight operations or guidance of space objects, 17 September 2005, adopted 28 June 2005.

would take place.[57] *Ratione materiae*, the obligation covers "the activities of launching, flight operations and guidance of space objects", where "'flight operation' and 'guidance' mean any operation relating to the flying conditions, navigation or evolution in outer space of the space object, such as the control and correction of its orbit or its trajectory".[58]
Regarding reimbursement by the licensee of the government in case of international liability claims addressed towards the latter, as the Law is the first European one making explicit reference here both to Article VII of the Outer Space Treaty and to the Liability Convention, it is in principle unlimited.[59] As far as the Law itself is concerned, there is no obligation to take out insurance, but the appropriate Minister may, in granting a license, "create an obligation for insurance to be taken out in favour of third parties to cover the damage that may result from the activities authorised by him".[60] That Minister is the Minister "with responsibility for space research and its applications in the framework of international cooperation", holding the general authority to grant authorisations, suspend and/or cancel them, as well as to take care of the "National Register of Space Objects".[61]

12. The Netherlands

Where Belgium has preceded, its Northern neighbour the Netherlands, ESA and EU member as well, can not stay far behind, and in 2007 its national Law Incorporating Rules Concerning Space Activities and the Establishment of a Registry of Space Objects entered into force.[62] In terms of its scope *ratione materiae*, it is noteworthy that it may come to include the "organization" of space activities as such[63] – which especially refers to the plans to conduct space tourism flights from the 'Dutch' Caribbean, organised from the European part of the Netherlands.[64] *Ratione personae*, the licensing obligation pertains to those conducting such activities from Dutch territory, which includes for this purpose ships and aircraft registered in the

---

57. See Art. 2(1), resp. (2), Law on the activities of launching, flight operations or guidance of space objects.

58. Art. 2(1), resp. Art. 3(5), Law on the activities of launching, flight operations or guidance of space objects.

59. See Art. 15(1), Law on the activities of launching, flight operations or guidance of space objects. Art. 15(3) allows the state to cap such reimbursement liability, although § 4 requires the licensee "to comply with the conditions attached to his authorization" in order to be able to enjoy the benefits of such a cap.

60. Art. 5(2), Law on the activities of launching, flight operations or guidance of space objects.

61. Art. 3(6), resp. Art. 14(1), Law on the activities of launching, flight operations or guidance of space objects. Art. 14(2) & (3) spell out further details of the registration, *inter alia* faithfully copying the exact terms of Art. IV(1), Registration Convention, but then going well beyond those to take care of many potential scenarios regarding private involvement in space objects thus registered.

62. Law Incorporating Rules Concerning Space Activities and the Establishment of a Registry of Space Objects, 24 January 2007; 80 *Staatsblad* (2007), at 1.

63. See Sec. 2(2.b), Law Incorporating Rules Concerning Space Activities and the Establishment of a Registry of Space Objects. Space activities themselves are defined as "the launch, the flight operation or the guidance of space objects in outer space", Sec. 1(b).

64. It should be noted here that the Kingdom of the Netherlands, which ratified the five space treaties, also includes a few small remnants of its colonial empire in the Caribbean region that internally hold a special status, including some measure of autonomy on a number of issues. Whilst externally the Netherlands is the state for example responsible and liable also for activities conducted from those Caribbean islands, internally it is not automatically entitled to impose domestic legislation regarding space activities to that extent.

Netherlands.[65] Furthermore, that obligation *can* be made applicable to Dutch nationals if active in the territory of states not parties to the Outer Space Treaty[66] – in other words, where no other state may be an obvious "appropriate State" to undertake the authorisation and continuing supervision required by Article VI of the Outer Space Treaty.

As for the applicable liability arrangement, the licensee is required to offer redress to the Dutch government up to the value of the sum insured,[67] whereas in this respect "the prospective holder shall have and maintain what Our Minister considers to be the maximum possible cover for the liability arising from the space activities for which a licence is requested. Account is taken here of what can reasonably be covered by insurance."[68]

Finally, it is the Radiocommunications Agency, part of the Dutch Ministry of Economic Affairs, which serves as the entity taking care of the licensing applications and procedures, as well of registration of space objects involved as appropriate with a view to the Registration Convention.[69]

13. Concluding remarks

Whilst the above analysis has addressed seven European states having taken fundamental steps to implement relevant obligations under the space treaties, as a matter of fact there is an eighth: France, first amongst a few equals in terms of space activities at least amongst the ESA and EU member states, which has even more recently than the Netherlands enunciated a comprehensive national law dealing with these issues *vis-à-vis* private entities. Since the new French law however is covered in much more detail by the paper of Mr. Philippe Clerc presented to the Symposium, it will not be addressed here.

Without discussing the French law as such therefore, there is nevertheless one point for discussion which the French case gives rise to that is of interest for the present analysis. The seven states referred to above, and since recently also France, have all implemented the obligations stemming from Article VI of the Outer Space Treaty by means of a comprehensive and – at least at the highest level transparent – law. At the same time it may be noted there is a principled alternative to such approach to exercising authorisation and continuing supervision: that of *de facto* control – and this was the road France has taken until recently. Arianespace and SPOTImage so far are the two main private entities with French nationality and operating from French territory conducting space activities – and both had CNES, the French national space agency, as their largest single shareholder. The almost day-to-day control which CNES could consequently exercise over these companies' activities took care of implementation of the international obligations at issue satisfactorily (at least until recently).

Whilst France has now, finally, taken the step towards a comprehensive national space law, such step is still missing in the three other major space-faring nations in Western

---

65. See Sec. 2(1), Law Incorporating Rules Concerning Space Activities and the Establishment of a Registry of Space Objects.

66. See Sec. 2(2.a), Law Incorporating Rules Concerning Space Activities and the Establishment of a Registry of Space Objects.

67. See Sec. 12(2) & (3), Law Incorporating Rules Concerning Space Activities and the Establishment of a Registry of Space Objects; following § 1 which provides for the generic reimbursement obligation.

68. Sec. 3(4), Law Incorporating Rules Concerning Space Activities and the Establishment of a Registry of Space Objects.

69. Cf. Secc. 1(a), 3(1), 4-8, resp. Sec. 11, Law Incorporating Rules Concerning Space Activities and the Establishment of a Registry of Space Objects.

Europe: Germany, Italy and Spain. Where Germany at least has created a national act focusing on important security aspects of one space activity, that of remote sensing[70], Italy and Spain currently have not moved beyond the realm of academic discussion in this area. One can really wonder, whether any 'European' approach to application and implementation of Article VI of the Outer Space Treaty without those major space-faring states would be a sensible concept.

Such a European approach would not likely result either from ESA policies, or even from developments in the context of the European Union which does not seem ready yet to 'implement' that Article in practice.[71] Consequently, the European context for application and implementation of Article VI will remain a mixed landscape for the time being, driven predominantly by national concerns and competences, where, as briefly analysed, a number of considerable divergences exist even between those states that have, at least, established a national space law.

Such divergences are clearest and most pronounced in such areas as the scope of the licensing obligation, both *ratione materiae* and *ratione personae*, and the handling of insurance – whether an obligation to take out insurance is imposed on the licensee or not; whereas in areas such as liability reimbursement (which is usually unlimited, at least in principle) and the creation of a dedicated entity for licensing and registration purposes there, apparently, seems to be a larger measure of harmonisation. Perhaps the ultimate irony is that precisely those two states *only* part of geographical Europe, and not of the integrative 'platforms' of ESA and the European Union, show the largest measure of commonality: the Russian Federation and the Ukraine – for obvious historic reasons.

In any event, it will take considerable more time and effort before anything like a substantive and detailed European *approach* to dealing with Article VI's obligations will arise, as beyond the general high-level understanding that issues like liability, insurance, licensing procedures and registration requirements should be dealt with in a sensible fashion. The European *context*, consequently, is precisely that: a haphazard, fractured and almost organic process whereby many forces of a more traditional sovereign nature are sometimes in agreement, sometimes not, on the need for more European coherence, sometimes out of necessity or by default, sometimes also stemming from some higher ideals. That conclusion, however, for better or worse is not confined to matters of outer space, space activities and the Outer Space Treaty alone.

---

[70]. Act Protecting Against the Endangerment of German Security Through the Proliferation of High Resolution Aerial Imagery of the Earth, 23 November 2007, effective 1 December 2007; Federal Gazette (BGBl.) Year 2007 Part I No. 58, of 28 November 2007.

[71]. Whilst the newest treaty trying to further develop the legal basis for the European Union, the Lisbon Treaty, does contain some clauses on EU competence in matters of space, those are far too vague to expect them to give rise rapidly to a competence at the EU level to 'authorise' and 'continuously supervise' private space activities. Furthermore, the fate of the Lisbon Treaty itself as of this writing is still uncertain as a consequence of the recent Irish 'no'.

17th Manfred Lachs Space Law Moot Court Competition

THE 2008 MANFRED LACHS SPACE LAW MOOT COURT COMPETITION

CASE CONCERNING THE CONTINUED PROVISION OF
LIFELINE SATELLITE SERVICES TO COUNTRIES
IN THE FACE OF SATELLITE OPERATOR INSOLVENCY

Concordia and Landia v Usurpia

PART A: INTRODUCTION

The 17th Manfred Lachs Space Law Moot Court Competition was held during the Glasgow IISL Colloquium. The *Case concerning the Continued Provision of Lifeline Satellite Services to Countries in the Face of Satellite Operator Insolvency (*Concordia and Landia v Usurpia*)* was written by Ricky Lee. Preliminaries were held at regional level in Europe, North America and in the Asia Pacific region. The Finals were judged by two Judges of the International Court of Justice, Judge A. Koroma and Judge P. Tomka, and by IISL Board Member Professor F. Lyall.

The final was hosted by the Glasgow City Council Shettleston, City Chambers.

The organizations that graciously supported the World Finals are:
EADS Astrium
European Centre for Space Law (ECSL)
Scott Group Limited
Glasgow Caledonian University
Glasgow University
LOC of the 2008 International Astronautical Congress
Japan Aerospace Exploration Agency (JAXA)
Martinus Nijhoff Publishers
National Aeronautics and Space Administration (NASA)/ Association of US Members of the IISL, AUSMIISL
SpaceIsle
University of Aberdeen
University of Strathclyde

Results of the world finals:

Winner: University of New South Wales, Australia (Madeleine Ellicott, Tamara Phillips, Katrina Taylor; Coach: Mr Pouyan Afshar)

Runner-up: University of Augsburg, Germany (Christian Odehnal, Melanie Ortlieb, Maximilian Widmann; Coach: Mrs. Sarah Schumann)

2nd runner-up: Georgetown University, USA (Jonathan Reisman, Robert Gajarsa; Coach: Prof. Paul Larson)

Eilene M. Galloway Award for Best Written Brief: University of Augsburg, Germany

Sterns and Tennen Award for Best Oralist: Ms Madeleine Ellicott (University of New South Wales)

Lee Love Award for members of the Wining team: University of New South Wales, Australia

Participants in the regional rounds

In North America:

1. George Mason School of Law
2. University of Mississippi School of Law
3. Catholic University of America School of Law
4. McGill University
5. University of Nebraska College of Law
6. Georgetown University Law Center
7. University of Virginia School of Law
8. Drexel University Earle Mack School of Law

In Europe:

1. University of Leiden, The Netherlands
2. University of Inner Temple, London, UK
3. Riga Graduate School of Law, Latvia
4. John Paul II Catholic University of Lublin, Poland
5. University of Strathclyde, Glasgow, Scotland
6. University of Augsburg, Germany
7. Catholic University of Leuven, Belgium
8. University of Paris XI, Sceaux, France

In the Asia Pacific:

1. Amity Law School, New Delhi (India)
2. Army Institute of Law, Mohali (India)
3. Bangalore University Law College, Bangalore (India)
4. Beijing Institute of Technology, Beijing (China)
5. China University of Political Science and Law, Beijing (China)
6. Dr Ram Manohar Lohiya National Law University, Lucknow (India)
7. Government Law College, Mumbai (India)
8. Government Law College, Ernakulam (India)
9. Gujarat National Law University, Gandhinagar (India)
10. Indian Law Society Law College, Pune (India)
11. Murdoch University, Perth (Australia)
12. National Academy of Legal Studies and Research, Hyderabad (India)
13. National Law Institute University, Bhopal (India)
14. National Law University, Jodhpur (India)
15. National University of Juridical Sciences, Kolkata (India)
16. National University of Singapore (Singapore)
17. Padjadjaran University, Bandung (Indonesia)
18. Parahyangan Catholic University, Bandung (Indonesia)
19. Tamil Nadu Dr Ambedkar Law University, Chennai (India)
20. Trisakti University, Jakarta (Indonesia)
21. Pelita Harapan University, Banten (Indonesia)
22. University College of Law, Dharwad (India)
23. University of Kyoto, Kyoto (Japan)
24. University of New South Wales, Sydney (Australia)
25. University of Sydney, Sydney (Australia)
26. University of Tokyo, Tokyo (Japan)

Judges for written briefs:

- Mr. Ian Awford , Australia
- Dr. Peter van Fenema, The Netherlands
- Dr. Martha Mejia-Kaiser, Mexico/Germany
- Mr. Larry Martinez, USA
- Mr. Fabio Tronchetti, The Netherlands
- Dr. Yun Zhao, Hong Kong

Judges for semi finals:

- Prof. Dr. Stephan Hobe, Germany
- Prof. Dr. Maureen Williams, Argentina
- Mr. K.R. Sridhara Murthi, India

Judges for finals:

- H.E. Judge Abdul Koroma, ICJ
- H.E. Judge Peter Tomka, ICJ
- Prof. Francis Lyall, UK

Contact details regional rounds:

- *North America:* Milton (Skip) Smith SSMITH@sah.com
- *Europe:* Melanie Vincent Melanie.Vincent@esa.int
- *Asia Pacific:* Ricky Lee asiapacific@spacemoot.org

PART B: THE PROBLEM

STATEMENT OF FACTS

1. The year is 2010. Landia, a landlocked and geographically isolated country, is surrounded by uninhabitable terrain on all of its borders, with few natural resources and limited economic means. Its Gross Domestic Product places it in the lowest 5% of national GDPs in the world.

2. Given its isolated condition, Landia is totally dependent on satellites to meet its basic telecommunications requirements, both for international telecommunications links connecting it to the rest of the world and for providing a critical basic domestic telecommunications infrastructure within Landia. In order to fulfill these basic requirements, Landia recently entered into a long-term, non-preemptible lease with Satelsat, Inc. ("Satelsat"), a private global satellite operator incorporated in the country of Concordia. Pursuant to this lease, Landia, through its state-owned Landia Telecommunications Authority ("LTA"), leases three transponders from Satelsat on the Satelsat-

18 satellite. These transponders are used for the following purposes:

(a) to provide links from Landia to all other countries in the world;

(b) to provide backbone internet connectivity within the country, including to more than 250 remote and isolated villages located throughout the Landia countryside and access to which, according to the Constitution of Landia, is recognized as a fundamental right of all of its citizens; and

(c) to provide critical infrastructure used to support various of its important governmental activities and functions, including e-government, distance learning and telemedicine.

3. Satelsat operated a fleet of 25 geosynchronous satellites providing satellite services and connectivity on a global basis, operating in the conventional C and Ku-band frequencies available for use by the Fixed Satellite Service. Satelsat is incorporated and has its principal place of business in Concordia, which also serves as the notifying administration with the International Telecommunication Union ("ITU") on behalf of Satelsat, although Satelsat does have a major business presence in other countries, including the location of a number of satellite control facilities in the Kingdom of Usurpia. All of Satelsat's satellites are licensed by the Concordia Communications Commission ("CCC") and are deployed at orbital locations that Concordia has notified to the ITU on Satelsat's behalf. All of these satellites were launched from the Concordia Space Center by commercial launch services providers based in Concordia and licensed by the government of Concordia.

4. Over the past 15 years, Satelsat has undergone a number of corporate reorganizations and transformations, having on multiple occasions been successively sold to differing groups of private investors, with the effect of significantly increasing the overall debt level of the company. In 2010, it has debt obligations in excess of $25 billion with annual debt service of approximately $3 billion and annual revenues of approximately $4.5 billion. The bulk of Satelsat's debt is held by banks located in Usurpia and is secured by the assets of Satelsat, including the entire Satelsat satellite fleet and its satellite control facilities located in Usurpia.

5. Usurpia, Concordia and Landia are also all parties to an international intergovernmental agreement pursuant to which each party commits to provide affordable satellite services to those countries of the world, each having a GDP in the bottom quartile (a "Lifeline Dependent Country"). The agreement, known as the Global Legacy International Telecommunications Satellite Organization Agreement (the "GLITSO Agreement"), was established in 2009 to supersede a number of other international agreements that had previously been in place with respect to the privatization of former international satellite organizations. Pursuant to the GLITSO Agreement, each State party thereto has committed to the principles of maintaining global connectivity and global coverage to all countries of the world on a non-discriminatory basis and supporting the provision of affordable services to all Lifeline Dependent Countries requiring such services, in order to meet their international or domestic telecommunications services.

6. While GLITSO has overall responsibility for overseeing the adherence to these principles by its member states, it does not possess any binding enforcement authority to compel adherence or to impose remedies in the event that a member state breaches these principles. Moreover, the GLITSO Agreement does not specify any particular means by which a State party thereto must honor its obligations, this being left to the discretion of each State party. In ratifying the GLITSO Agreement, each State party undertakes to issue a Declaration indicating how it intends to adhere to these objectives. In the case of the various satellite licenses that Concordia has issued to Satelsat regarding the Satelsat fleet, Concordia has imposed the affirmative obligation on Satelsat that it must adhere to the principles set forth in the GLITSO Agreement and abide by the conditions set forth in Concordia's ratification Declaration, whenever providing services to any Lifeline Dependent Country.

7. Due to a major downturn in the global economy, a number of Satelsat's major customers have either become insolvent or fallen significantly in arrears in their payments to Satelsat for space

segment capacity leased from Satelsat. Consequently, Satelsat has been unable to meet the interest payments on its debt for the past six months, resulting in the breach of a number of covenants in its various debt instruments. Given concerns by the banks holding Satelsat's debt that the prospects for rectifying the situation at any time in the foreseeable future were dim, the banks felt they had no recourse but to place Satelsat under the protection of a bankruptcy proceeding, choosing to do so in their home country of Usurpia. This petition was filed with the Usurpia Bankruptcy Court on June 1, 2010.

8. The petition sought to restructure Satelsat so as to maximize the likelihood that it could continue in business on a profitable basis and meet its debt obligations as restructured through the bankruptcy process, while avoiding a potentially much more disruptive total liquidation of the company. The reorganization plan put forward would keep Satelsat largely intact, but contemplated redeployment of certain Satelsat satellites to different orbital locations, all of which had previously been notified by Concordia to the ITU. The objective was to be able to achieve utilization levels (and revenue generation) at these new locations that would be significantly higher than achievable at current locations.

9. In particular, one potential customer was prepared to commit to a long-term lease of an entire Satelsat satellite at premium rates, if Satelsat could quickly redeploy one of its satellites to a particular portion of the orbital arc that presently was unserved by any Satelsat satellite. The revenues that would be generated by this transaction would significantly improve Satelsat's future financial prospects. Fortuitously, Concordia happened to have a currently unoccupied, registered orbital slot within the required portion of the orbital arc and which would be acceptable to the potential customer. If, however, a Satelsat satellite could not be redeployed to such a location within a three-month period (by the end of August 2010), the potential customer has indicated that it would make alternate arrangements to provide the service, instead utilizing a new fiber optic cable that had been recently activated.

10. Of all of the satellites in the Satelsat fleet, the one that would be easiest to relocate and have the necessary configuration of transponders to meet this customer's requirements was the Satelsat-18 satellite. However, if the Satelsat-18 satellite were moved to this new orbital location, Landia's current leases could not be maintained. This was both because the Satellite- 18 satellite would be fully dedicated to this new customer and would be unable to provide adequate coverage of Landia from the new orbital location. To address the situation, the banks proposed that Landia's current services be reapportioned among three other Satelsat satellites serving the same region. These satellites, however, were older and less powerful than the Satelsat-18 satellite. As such, the effect of dispersing Landia's services among these three satellites would be to force Landia, at great expense, to modify its current ground segment infrastructure. Even with these changes, Landia was of the view that the substitute services would be markedly inferior to the current levels of service that it was receiving on the Satelsat-18 satellites. In particular, Landia's ability to operate its internal domestic networks and its external international links on an integrated basis would be substantially impeded.

11. Based on an expedited order issued by the Usurpia Bankruptcy Court approving the proposed reorganization, Satelsat applied to the CCC in Concordia for the necessary authority to relocate the Satelsat 18 satellite to this new orbital location.

12. When notified of these developments, Landia sent a strong diplomatic note to Concordia, protesting the relocation of the Satelsat-18 satellite. In that note, Landia contended that it was entitled to special consideration as a Lifeline Dependent Country, since this measure would significantly harm the interests of all Landian citizens. Landia's plea struck a responsive chord with certain portions of the Concordian public, resulting in public demonstrations in support of Landia throughout Concordia. Following these demonstrations, the CCC issued an interim order on July 1, 2010 withholding authority for Satelsat to relocate the Satelsat-18 satellite until the CCC could further consider the situation.

13. Fearful that any delay in the relocation of the Satelsat-18 satellite would imperil the entire reorganization plan, the banks devised a revised plan that was submitted to the Usurpia Bankruptcy Court on July 8, 2010. This revised plan sought authority to create a new subsidiary of Satelsat, to be known as New Satelsat, which would take title to certain Satelsat assets, including the Satelsat-18 satellite. This subsidiary would be established under the laws of Usurpia. Without intending to affect the licensing status of the other Satelsat satellites, the banks proposed that the Satelsat-18 satellite be re-licensed by the Usurpian Telecommunications Authority ("UTA") as an Usurpian satellite and requested that redeployed to a new, but currently unoccupied orbital location that was currently notified to the ITU by Usurpia, and which was also fully acceptable to the new customer. This revised plan was approved by the Usurpia Bankruptcy Court on an expedited basis on July 15, 2010. Satelsat immediately notified the CCC of its intent to relinquish its license to operate the Satelsat-18 satellite and any rights it had to locate the satellite at its current orbital location, and simultaneously applied on an emergency basis to the UTA for licensing authority for the satellite. The UTA granted the license request on August 15, 2010, based upon which Satelsat immediately commenced the relocation process for the Satelsat-18 satellite.

14. Landia and Concordia strongly protested these actions, claiming that this was a sham transaction intended to circumvent commitments that previously had been made by Concordia and that national responsibility for the satellite could not be transferred from Concordia to Usurpia without the express consent of Concordia. Usurpia responded by arguing that its actions were entirely appropriate, in that it was acting on the proper application of an Usurpian commercial enterprise to license a satellite in accordance with standard Usurpian procedures. For that reason, it asserted that the prior status of the satellite as having been licensed by Concordia was completely irrelevant to the actions now requested by Newtelsat as a Usurpian company. And while Usurpia is also a member of GLITSO, its licensing procedures only contain a "best efforts" provision with respect to the furnishing of services to any Lifeline Dependent Country.

15. Landia, having now lost the use of the Satelsat-18 satellite and dissatisfied with what it viewed as a wholly inadequate alternate arrangement offered by Satelsat, contacted a second satellite operator, Orbitsat, to determine if Orbitsat could accommodate its requirements. Orbitsat, also licensed by Concordia, did have capacity available on its Orbitsat SpaceStar satellite to meet Landia's requirements, although the cost of such capacity would be five times the cost of the capacity that Landia has previously obtained from Satelsat. Without knowing how it would be able to handle these additional costs, Landia entered into a provisional leaseagreement with Orbitsat, to take effect on September 1, 2010, subject to Landia's ability to obtain emergency funding from the World Bank or a similar international organization.

16. In light of Landia's and Concordia's protests and concerned about what impact they might have on Usurpia, New Satelsat decided to speed up the relocation of the Satelsat-18 to the new orbital location licensed by Usurpia. Unfortunately, as a direct result of this effort, the Satelsat-18 satellite collided in geosynchronous orbit on August 25, 2010, with the Orbitsat Space Star satellite, completely destroying both satellites.

17. Following the collision, Landia found itself not only lacking the ability to continue to receive services from the Satelsat-18 satellite, but also deprived of the ability to secure appropriate replacement capacity on the Orbitsat SpaceStar satellite. In Landia's view, it was now totally deprived of any suitable means for meeting its internal and external telecommunications requirements, especially given the inferiority of the alternate arrangements that had previously been proposed by the banks.

18. Estimating that it would take at least three years to get adequate replacement capacity from another satellite operator and that, during the interim, Landia would suffer more than $2 billion in losses to its economic welfare as a result of the disruption of its telecommunications infrastructure, Landia

submitted demands for compensation to both Concordia and Usurpia for this amount, contending that both countries were ultimately liable for the loss. Usurpia rejected this demand, disavowing any breach of international law or obligations owed to Landia. Moreover, Usurpia denied that there was any basis under international law for recovery of the type of damages allegedly incurred by Landia. Concordia, which has its own claim for compensation from Usurpia for loss of both the Satelsat-18 and Orbitsat SpaceStar satellites, did not directly deny Landia's claim for compensation, but rather took the position that, to the extent it would be held liable for compensation, it was entitled to indemnification from Usurpia.

19. In an effort to resolve this impasse, Landia, Concordia and Usurpia have agreed to submit this dispute for resolution to the International Court of Justice, which has accepted jurisdiction over the matter. Concordia's damages claim against Usurpia relating to the loss of the Orbitsat SpaceStar satellite has been resolved by negotiation and is not presented for further consideration. However, Concordia's damages claim against Usurpia relating to the loss of the Satelsat-18 satellite has not been resolved. Because of the overall commonality of many of their respective positions, Landia and Concordia have joined forces in opposition to Usurpia in the submission of the dispute to the International Court of Justice.

20. Landia seeks declarations from the International Court of Justice to the effect that:

(i) Usurpia's decision to license and then authorize the relocation of the Satelsat-18 satellite over the objections of Landia is contrary to applicable principles of international law, including, *inter alia*, the 1967 Outer Space Treaty, the 1975 Registration Convention and the GLITSO Agreement; and

(ii) Landia is entitled to compensation for economic consequences of its loss of basic satellite telecommunications services from Usurpia for the relocation of the Satelsat-18 satellite and from both Concordia and Usurpia as a result of the collision destroying the Satelsat-18 and Orbitsat Space Star satellites, pursuant to, *inter alia*, the 1972 Liability Convention and the GLITSO Agreement.

21. Concordia seeks declarations from the International Court of Justice to the effect that:

(i) Usurpia's decision to authorize relocation of the Satelsat-18 satellite over its objections is inconsistent with applicable principles of international law, including, *inter alia*, the 1975 Registration Convention and the GLITSO Agreement;

(ii) Usurpia is liable to Concordia for the loss of the Satelsat-18 satellite under, *inter alia*, the 1972 Liability Convention and the GLITSO Agreement; and

(iii) Usurpia is obligated to indemnify Concordia for any liability Concordia might owe to Landia for the economic consequences of Landia's loss of basic satellite telecommunications services arising from the collision of the Satelsat-18 and Orbitsat SpaceStar satellites, pursuant to, *inter alia*, the 1972 Liability Convention and the GLITSO Agreement.

22. Usurpia seeks declarations from the International Court of Justice to the effect that:

(i) Usurpia's decision to license the Satelsat-18 satellite and to permit it to be deployed at an Usurpian orbital location over the objections of both Landia and Concordia is consistent with applicable principles of international law, including, *inter alia*, the 1967 Outer Space Treaty, the 1975 Registration Convention and the GLITSO Agreement;

(ii) Landia is not entitled to compensation from Usurpia as a result of the collision that destroyed the Satelsat-18 and Orbitsat SpaceStar satellites, pursuant to, *inter alia*, the 1972 Liability Convention and the GLITSO Agreement;

(iii) Concordia is not entitled to compensation for the loss of the Satelsat-18 satellite, pursuant to, *inter alia*, the 1972 Liability Convention and the GLITSO Agreement; and

(iv) Concordia is not entitled to indemnification from Usurpia for any financial obligation owed to Landia, as a result of the collision destroyed the Satelsat-18 and Orbitsat SpaceStar satellites, pursuant to, *inter alia*, the 1972 Liability Convention and the GLITSO Agreement.

23. All three countries are members of the United Nations and the ITU and are parties to the 1967 Outer Space Treaty, the 1972 Liability Convention and the 1975 Registration Convention. Concordia and Usurpia are

members of the World Trade Organization but Landia is not.

24. Both the Satelsat-18 and Orbitsat Space Star satellites were registered with the Secretary General of the United Nations in accordance with the 1975 Registration Convention, with Concordia listed as the "launching State" and the "State of registry." Usurpia has placed the Satelsat-18 satellite on the registry it maintains for such purposes and had commenced the process of notifying the Secretary-General of the United Nations in accordance with the 1975 Registration Convention of its status as the State of registry for the Satelsat-18 satellite but had not completed the process at the time of the collision.

25. Concordia and Usurpia are both parties to the Convention on International Interests in Mobile Equipment. However, to date, negotiations regarding a specific Protocol to the Convention on Matters Specific to Space Assets are ongoing, and therefore no such Protocol has yet been opened for signature.

26. For purposes of this problem, participants are to assume that there are no technical coordination matters associated with any of the orbital locations referenced therein.

Relevant Provisions of the GLITSO Agreement and Party Declarations Made Pursuant Thereto GLITSO Agreement

Preamble:

The State Parties to this Agreement, Considering the principle set forth in Resolution 1721(XVI) of the General Assembly of the United Nations that communication by means of satellites should be available to the nations of the world as soon as practicable on a global and non-discriminatory basis,

Considering the relevant provisions of the Treaty on Principles Governing the Activities of States in the Exploration and Use of Outer Space, including the Moon and Other Celestial Bodies, and in particular Article I, which states that outer space shall be used for the benefit and in the interests of all countries, and

Considering the importance of continuing to assure that, in today's modern era of satellite telecommunications, all countries of the world, including those that may be uniquely pendent on satellite telecommunications to meet their domestic and international telecommunications requirements, which for purposes of this Agreement are specified as all countries comprising the bottom quartile of countries in the world as determined by level of Gross Domestic Product ("GDP") and hereinafter referred to as a "Lifeline Dependent Country", have reasonable access to the satellite telecommunications services they require on fair and equitable terms and conditions,

Agree as follows:

. . . . .

Article II: Purposes and Means for Achievement

Each Party to this Agreement hereby commits to adhere to the following objectives:

(a) To maintain global connectivity and global coverage, available to all countries on a non-discriminatory basis; and

(b) To support the provision of affordable satellite service to all Lifeline Dependent Countries so requiring such services, in order to meet their international or domestic telecommunications requirements Each Party to this Agreement shall take such action as it determines to be appropriate, consistent with its national regulatory regime, to achieve the objectives set forth above. In ratifying or acceding this Agreement, each Party shall issue a Declaration indicating the specific measures by which it intends to abide by its commitment to the achievement of these objectives.

Party Declarations

In connection with its ratification of the GLITSO Agreement, Concordia issued the following Declaration, in which it stated:

Concordia views these obligations to be of paramount importance and will include in all licenses issued for satellites licensed by our national regulatory authority, the Concordia Communications Commission, the specific requirement that licensees are obligated to adhere to these principles and must not take any actions inconsistent therewith; moreover, to the extent that any licensee sells or otherwise disposes of any particular satellite asset, as a condition of that sale or transfer, any successor in interest holding that satellite license shall similarly be obligated to adhere to such obligations.

In connection with its ratification of the GLITSO Agreement, Usurpia issued the following Declaration, in which it stated:

Usurpia is fully committed to supporting the objectives of the GLITSO Agreement, while recognizing that such measures must be harmonized with the realities of the commercial nature of the satellite telecommunications business. Consistent therewith, Usurpia will require all satellite operators to accommodate the objectives in Article II of the GLITSO Agreement on a "best efforts" basis consistent with prudent business practices.

In connection with its ratification of the GLITSO Agreement, Landia issued the following Declaration, in which it stated:

Landia, as a Lifeline Dependent Country, lacks the resources to launch its own satellite and does not expect to have such resources for many years to come. In light of our geographic and economic circumstances, Landia is uniquely dependent on satellite telecommunications services to meet its international and domestic telecommunications requirements and is therefore totally dependent on the commitments made by other Parties to the GLITSO Agreement, and their continuing good will in adhering to their commitments, in order to be able to provide basic telecommunications services to the citizens of our country.

Statement of Additional Facts

1. After New Satelsat was incorporated on 16 July 2010, the Board of Directors of this new company, could not decide on the name for the company and so for some time the company was known as Newtelsat. The two names belong to the same company.
2. Orbitsat is licensed by Concordia and is 100% owned by Concordian private interests.
3. None of the States referred to are parties to the Vienna Convention on the Law of Treaties.
4. Satelsat-18 has 11 transponders on board, of which only 10 were used at all relevant times.
5. The front cover to the present *compromis* has been corrected.

PART C: FINALISTS BRIEFS

MEMORIAL FOR THE APPLICANTS LANDIA AND CONCORDIA

University of New South Wales, Australia (Madeleine Ellicott, Tamara Phillips, Katrina Taylor; Coach: Mr Pouyan Afshar)

ARGUMENT

1. Usurpia's decision to license and then authorise the relocation of Satelsat-18 over the objections of Landia and Concordia breached international law

1.1 Usurpia breached its obligations under *GLITSO* to provide Landia with affordable satellite services in a non-discriminatory manner

Landia, Concordia and Usurpia are all States parties to *GLITSO*,[1] under which each party commits to providing affordable satellite services to lifeline dependent countries ('LDCs').[2] Article II of *GLITSO* requires States parties to 'maintain global connectivity and global coverage available to all countries on a non-discriminatory basis and; to support the provision of affordable satellite services to all LDCs so requiring such services.'[3]

Landia's Gross Domestic Product ('GDP') is in the lowest 5% of national GDPs in the world, making it an LDC under *GLITSO*. In addition, Landia's landlocked and geographically isolated position makes it totally dependent on satellite services to meet its basic telecommunications requirements. Usurpia's authorisation of Satelsat-18's relocation, and subsequent failure to offer Landia a viable alternative service constituted a breach of its *GLITSO* obligations.

(a) Usurpia circumvented the Concordian Declaration to *GLITSO*

1 *Compromis* ¶5.
2 *Compromis* ¶1.
3 *Compromis*, Appendix A.

by which New Satelsat was bound under international law

GLITSO requires each State party to issue a Declaration upon ratification of the treaty specifying how it intends to adhere to these objectives.[4] In connection with its ratification of *GLITSO*, Concordia issued a Declaration stating that for all satellites licensed by the CCC, 'licensees are obliged to adhere to these [*GLITSO]* principles and must not take any actions inconsistent therewith.' The Declaration provides that to the extent that any Concordian licensee disposes of its satellite asset, 'as a condition of that sale or transfer, any successor in interest holding that satellite license shall similarly be obligated to adhere to such obligations.'[5]

Therefore, New Satelsat's acquisition of Satelsat-18 was conditional upon its strict compliance with the principles in *GLITSO.* The restructuring of Satelsat and Usurpia's re-licensing of Satelsat-18 were part of a sham transaction designed to circumvent the stringent obligations imposed on New Satelsat by Concordia's Declaration. This act was contrary to the international law doctrine of abuse of rights as Usurpia exercised its rights 'in such a way as to cause damage to another States, or in a way which may impair the rights of other States'.[6] This Court stated in the *Case of the Free Zones of Upper Savoy and the District of Gex* that if a State attempted to avoid its contractual obligations by resorting to measures that had the same effect as a specifically prohibited act, an abuse of rights would result.[7] In this case, Usurpia deliberately licensed Satelsat-18 as a means of avoiding Satelsat's international obligations under Concordia's *GLITSO* Declaration. This act constituted an abuse of rights, as it enabled New Satelsat to act in breach of Concordia's *GLITSO* commitment by depriving Landia of satellite services.

(b) In the alternative, Usurpia failed to adhere to its obligations under *GLITSO* on a 'best efforts' basis in accordance with its own Declaration

Usurpia's Declaration in connection to its ratification of *GLITSO* commits it to requiring its satellite operators 'to accommodate the objectives in Article II of *GLITSO* on a 'best efforts' basis consistent with prudent business practices.'[8] Although this Court has not had the occasion to interpret the meaning of 'best efforts', it may refer to the practice of States[9], as well as to judicial decisions and the writings of the most highly qualified publicists.[10] These authorities establish that the 'best efforts' standard requires a high standard of performance.

The 'best efforts' standard is commonly employed in space contracts.[11] Bernhard Schmidt-Tedd has stated that it is 'not a green light for the launch service provider's non-performance of its contractual obligations.'[12] Rather, the phrase requires that parties behave 'with the greatest commitment and the highest quality standards' in performing their obligations.[13] This is also reflected in domestic case law. For example, American Courts have stated that 'best efforts' will not prevent a party from giving reasonable consideration to its own

---

4 *Compromis* ¶6.

5 *Compromis,* Appendix A.

6 Ian Brownlie, *Principles of Public International Law* (5th ed, 1999) 444. See also L. Oppenheim, *International Law: A Treatise* (8th ed, 1955) 345.

7 *Case of the Free Zones of Upper Savoy and the District of Gex* (*France* v *Switzerland*) (1932), PCIJ (ser. A/B) No 46, 167.

8 *Compromis* ¶14, Appendix A.

9 *Vienna Convention*, opened for signature 23 May 1969, 1155 UNTS 331 art 31(3)(b), (entered into force 27 January 1980).

10 *Statute of the International Court of Justice*, art 38(1)(d).

11 *Hughes Communications Galaxy, Inc., v United States*, (2000) 47 Fed Cl 236, *American Satellite Co. v United States*, (1993) 998 F 2d 950; See also K.H. Bockstiegel, 'The Law Applicable to Contracts' (1982) 25 *Proceedings of the 25th Colloquium on the Law of Outer Space* 207; Julian Hermida, *Legal Basis For A National Space Legislation* (2004), xxii.

12 Bernhard Schmidt-Tedd, 'Best Efforts Principle and Terms of Contract in Space Business'(1988) 31 *Proceedings of the 31st Colloquium on the Law of Outer Space* 330, 330.

13 Ibid, 336.

interests but imposes an obligation to act in 'good faith and to the extent of its own total capabilities'. [14] The Australian High Court interpreted the term as being measured by what is reasonable in the circumstances, having regard to the nature, capacity, qualifications and responsibilities of the contracting party in the light of the particular contract.[15] Similarly, Canadian[16] and English Courts[17] have held that 'best efforts' means doing all that can reasonably be done in the circumstances.

In determining whether Usurpia acted reasonably in the circumstances, it is necessary to have regard to the prudent business practices employed by other major satellite operators.

Intelsat, one of the largest satellite service providers is a party to the ITSO Agreement[18] which provides for affordable satellite services to developing countries under Article III. Intelsat is required to provide services to developing countries on an affordable, non-discriminatory basis. These obligations are accorded such importance that even in the event of the company falling into receivership 'its insolvency would not excuse it from fulfilling its public service obligations to developing countries'.[19] EUTELSAT, another large satellite service provider, is bound by Article III(d) of its Convention to provide non-discriminatory telecommunications services.[20] EUTELSAT's management report of 30 June 2006 highlights its respect for the principle of non-discrimination, stating that it must be factored into the operational activities of the company.[21] The commercial practices of these satellite operators reveal that the provision of services to lifeline countries and a practice of non-discrimination is an essential element of prudent business practice in the satellite industry.

Usurpia failed to meet the high standard imposed by its best efforts declaration. Whilst Satelsat had a large fleet of 25 satellites,[22] Usurpia specifically chose to relocate Satelsat-18 to meet the high-paying customer's demands. It did so with the knowledge that Landia relied entirely on Satelsat-18 to meet its telecommunications requirements. Rather than exercising its best efforts, Usurpia simply adopted the '*easiest*' solution to Satelsat's financial difficulties, unreasonably failing to consider any alternatives which would not have infringed its obligations to Landia. Given the large number of satellites in Satelsat's fleet, Usurpia should have at least explored alternative restructuring options involving various satellites, before approving a solution which it knew would adversely compromise Landia's vital communications services.

While Usurpia did offer Landia an alternative service, this was inadequate and practically unusable by Landia. This proposed substitute would have been markedly inferior to

---

14 *Bloor v Falstaff Brewing Corp*, 601 F.2nd 609 (2nd cir. 1979); *Pips (Leisure Productions) Ltd v Waltons* (1980) 43 P & CR 415.

15 *Transfield Pty Ltd v Arlo International Ltd* [1998] HCA 15; for a similar interpretation see *Hospital Products Ltd v U.S. Surgical Corporation* (1984) 156 CLR 41.

16 *Atmospheric Diving Systems Inc v International Hard Suits Inc* [1994] 89 BCLR (2d) 356.

17 *Terrell v Mabie Todd & Co Ltd [1952]* 69 RPC 234.

18 *Agreement Relating to the International Telecommunications Satellite Organization* <http://www.itso.int/htmldocs/agreement.htm> at 10 June 2008 ('*ITSO Agreement*').

19 Kenneth Katkin, 'Communication Breakdown? The Future of Global Connectivity after the Privitisation of INTELSAT' (2005) 38 *Vanderbilt Journal of Transnational Law* 1323, 1373.

20 *Convention Establishing the European Telecommunications Satellite Organization EUTELSAT* <http://www.eutelsatigo.int/en/docs/Amended_Convention.pdf> at 19 July 2008.

21 *Management Report of Consolidated Accounts at 30 June 2006* (2006) EUTELSAT Communications < http://www.eutelsat.org/investors/docs-0506.html> at 19 July 2008.

22 Making it one of the largest satellite telecommunications providers in the world, following only Intelsat and SES who in 2007 possessed 51 and 36 satellites respectively; Peter de Selding, 'The List: Top Fixed Satellite Service Operators', *Space News* (New York), 25 June 2007, 12.

the services Landia had received from Satelsat-18, and would have impeded Landia's ability to operate its internal domestic networks and its external international links on an integrated basis.[23] Further, in order to utilise this substitute service, Landia would have had to modify its ground segment infrastructure at great expense, which it could not afford given its limited economic means.[24]

Thus, in authorising the relocation of Satelsat-18, and failing to provide Landia with an appropriate replacement service, Usurpia adopted the most convenient and profitable course of action, and clearly failed to exercise its 'best efforts' to support the principles in *GLITSO.*

1.2 Usurpia breached the customary international law principle not to discriminate against developing countries

In addition to its *GLITSO* obligations, Usurpia is concurrently bound by a principle of customary international law requiring it not to discriminate against developing countries in the provision of telecommunications services.[25] This customary principle is evidenced by the requisite State practice and *opinio juris.*[26]

Evidence of widespread and consistent state practice[27] is found in the large number of States parties to treaties and international organisations espousing the principle of non-discrimination. For example, the *Treaty on Principles Governing the Activities of States in the Exploration and Use of Outer Space, Including the Moon and Other Celestial Bodies* has been ratified by 98 States and provides in Article I that the exploration and use of outer space should be carried out 'for benefit and in the interests of all countries, irrespective of their degree of economic or scientific development.'[28] Publicists have described the benefit of all mankind principle as a 'central tenet of international space law.'[29]

The principle of non-discrimination also forms an integral part of the *Principles Relating to Remote Sensing of the Earth from Space.*[30] Principles II, IV and XII require States to 'take into particular consideration the needs of developing countries' and to supply data on a 'non-discriminatory basis and on reasonable cost terms.'

Furthermore, the International Telecommunications Union ('ITU') (with 191 member States) enshrines the principle of equitable access to the orbital spectrum in Article 33(2) of the *ITU Convention*, requiring member States such as Usurpia to conduct space activities 'taking into account the special needs of developing countries and the geographical situation of particular countries.'[31] Moreover, the *ITU Constitution* requires telecommunications service providers to give priority to all telecommunications concerning safety of life and government functions.[32] Also, the World Trade Organisation ('WTO') (of which Usurpia is a member) promotes the assistance of developing countries in regard to the provision of telecommunications.[33]

Additionally, the importance of the non-discrimination principle has been confirmed by many resolutions of the United Nations General

23 *Compromis* ¶10.

24 Ibid.

25 *Case Concerning Military and Paramilitary Activities in and against Nicaragua (Nicaragua v United States) (Merits)* [1986] ICJ Rep 14, ¶177.

26 *North Sea Continental Shelf Cases (Germany v Denmark; Germany v Netherlands)* [1969] ICP Rep 3.

27 *Asylum Case (Columbia v Peru)* [1950] ICJ Rep 266.

28 Opened for signature on 27 January 1967, 610 UNTS 205(entered into force 10 October 1967) ('*OST*').

29 Nandasiri Jasentuliyana, 'Ensuring Equal Access to the Benefits of Space Technologies for All Countries' (1994) 10(1) *Space Policy* 7, 7.

30 UN Doc GA Res 41/65, UN GA, 95th plen mtg, UN Doc A/Res/41/65.

31 *International Telecommunication Convention*, opened for signature Nairobi, 6 November 1982 (entry into force 1 January 1984).

32 *ITU Constitution* arts 40-41.

33 See, eg, *Fourth Protocol to the General Agreement on Trade in Services,* opened for signature 30 April 1996, art IV (entered into force 5 February 1998).

Assembly,[34] which constitute evidence of a customary rule.[35] In particular, Resolution 1721(XVI) recognised that 'satellites should be available to the nations of the world as soon as practicable on a global and non-discriminatory basis.'[36] International support for this objective culminated in the conclusion of the *GLITSO Agreement* in 2009.

Although it is difficult to ascertain whether a customary rule has crystallised mandating the active provision of satellite services to developing countries, there is significant state practice and *opinio juris* in support of the principle of non-discrimination against developing countries in the provision of telecommunications services. Landia is a developing country completely reliant on satellite services to satisfy its telecommunications requirements, including e-government and telemedicine.[37] Thus, Usurpia's approval of the relocation of Satelsat-18 to a customer with greater economic capabilities breached this customary principle.

1.3 Usurpia breached international law by registering Satelsat-18 and seizing jurisdiction and control of the satellite

Under the *Convention on Registration of Objects Launched into Outer Space*, a space object may only be registered on the registry of one State at any given time.[38] Whilst the *Registration Convention* does not explicitly provide for the transfer of a space object between State registries, State practice suggests that this is possible only where States have agreed to transfer registration.[39] Moreover, States are required to 'jointly determine' the registration of satellites pursuant to Article II of the *Registration Convention.*

In the present case, Concordia as the original State of Registry for Satelsat-18 in accordance with the *Registration Convention*, registered the satellite with the UN Secretary General.[40] Subsequently, Usurpia unilaterally decided to place Satelsat-18 on its national registry, and commenced the process of notification of its status as a State of Registry to the United Nations, in the absence of any agreement with Concordia. Indeed, Concordia unequivocally protested against Usurpia's actions, arguing that national responsibility for the satellite could not be transferred from Concordia to Usurpia without the express consent of Concordia.[41] Therefore, Usurpia's actions in registering Satelsat-18 breached Article II of the *Registration Convention.*

Additionally, Article VIII of the *OST* states that '[a] State party to the treaty on whose registry an object launched into outer space is carried shall retain jurisdiction and control over such object.'[42] As Concordia remained the lawful State of registry for Satelsat-18, and did not agree to Usurpia's registration of the

---

[34] *The Declaration on International Cooperation in the Exploration and Use of Outer Space for the Benefit and in the Interest of All States, Taking into Particular Account the Needs of Developing Countries,* GA Res 51/122, UN GAOR, 51st sess, 83rd plen mtg, UN Doc A/Res/51/122 (1996).

[35] *Case Concerning Military and Paramilitary Activities in and against Nicaragua (Nicaragua v United States) (Merits)* [1986] ICJ Rep 14, ¶188.

[36] *International Co-operation in the Peaceful Uses of Outer Space*, GA Res 1721 (XVI) UNGAOR, 16th sess, 1085th plen mtg, UN Doc A/Res/1721 (1961).

[37] *Compromis* ¶2.

[38] *Convention on Registration of Objects Launched into Outer Space*, opened for signature 14 January 1975, 1023 UNTS 15, art I(c) (entered into force 15 September 1976) ('*Registration Convention*').

[39] See Julian Hermida, 'Transfer of Satellites in Obit: An International Law Approach' (2003) *Proceedings of the 46th Colloquium on the Law of Outer Space* 189, 190; *Practice of States and International Organizations in Registering Space Objects*, UNCOPOUS, UN Doc A/AC.105/C.2/L.266 (2007) 7 which recommends that with regard to the transfer of ownership of space objects in orbit, both States involved in the transfer should furnish information to the UN Secretary General).

[40] *Compromis* ¶24.

[41] *Compromis* ¶14.

[42] *OST*, opened for signature on 27 January 1967, 610 UNTS 205, art VIII (entered into force 10 October 1967).

satellite, Concordia remained entitled to exercise jurisdiction and control over the satellite. Usurpia's interference with Concordia's legitimate jurisdiction and control over Satelsat-18 by way of licensing the satellite and authorising its redeployment to an Usurpian orbital location therefore breached Article VIII of the *OST*.

1.4 Usurpia breached the international obligation to cooperate and consult with respect to activities in Outer Space

In expeditiously authorising the relocation and re-licensing of Satelsat-18 in the face of the Applicants' protests, Usurpia violated the principle of international cooperation and consultation. Cooperation is a fundamental principle of international law, enumerated in Article 74 of the *UN Charter* which emphasises that States are to take into account the 'interests and well-being' of other States.[43] The *OST* also stresses the importance of international cooperation, mentioning it no less than six times.[44]

The requirement to consult forms part of this cooperation obligation. In particular, Article IX of the *OST* stipulates that where a State has reason to believe that an 'activity planned by it or its nationals in Outer Space...would cause potentially harmful interference with activities of other States', then it is required to 'undertake appropriate international consultations before proceeding with any such activity'. Failing to consult where that is required constitutes a material breach of Article IX.[45] The duty to consult extends to activities which may cause 'physical interference' to another object in outer space, as well to activities which might interfere with the 'earthly ingredients' of a State's space activities. It is submitted that the on-earth use of a satellite signal is also covered by this Article.[46]

In light of the Applicants' diplomatic protests to the revised plan, Usurpia clearly had reason to believe that the relocation would harm first Landia's interests, by impeding its ability to operate its telecommunications services, and second Concordia's interests, by undermining the CCC's interim order. Usurpia's failure to cooperate or consult with Landia or Concordia in relation to the possible impacts of the satellite relocation clearly violated international law.

1.5 Creditors of space assets are required to act in accordance with international space law

The Banks' status as creditors of Satelsat does not preclude Usurpia's conformity with its obligations under international law.

The International Institute for the Unification of Private Law ('UNIDROIT') recently completed a *Preliminary Draft Protocol on Matters Specific to Space Assets*[47] to the *Convention on International Interests in Mobile Equipment.*[48] When the *Space Protocol* enters into force, it will provide a legal regime for the international financing of space assets. Until the *Space Protocol* is ratified however, the *Mobile Equipment Convention* has no binding force with respect to the creation of rights for creditors of space assets.[49] Moreover, even when the *Space Protocol* does become effective, creditors of space assets will remain bound by international space law. Article XXI *bis* of the

43 *Charter of the United Nations*, art 74.

44 *OST,* opened for signature on 27 January 1967, 610 UNTS 205, arts I, II, IX, X, XI.(entered into force 10 October 1967).

45 Jerzy Sztucki, 'International Consultations and Space Treaties' (1975) *Proceedings of the 17th Colloquium on the Law of Outer Space* 147, 159. Istvan Herczeg, 'Introductory Report: Provisions of the Space Treaties on Consultation' (1975) *Proceedings of the 17th Colloquium on the Law of Outer Space* 141; The intended consultation is to take place on a diplomatic level, ideally through the vehicle of an international organisation.

46 Sztucki, above n 45, 159.

47 *Preliminary Draft Protocol on Matters Specific to Space Assets* UNIDROIT Doc Study LXXIIJ (2004) *('Space Protocol')*

48 *Convention on International Interests in Mobile Equipment*, opened for signature 16 November 2001, UN Doc A/AC.105/C.2/2002/CRP.3 (entered into force 1 March 2006 excl art 49) ('*Mobile Equipment Convention*').

49 Ibid, art 49.

current *Space Protocol* provides that the 'Convention as applied to space assets does not affect State party rights and obligations under the existing United Nations Space Treaties or instruments of the International Telecommunications Union.' Thus, neither the Banks nor Usurpia may escape their international obligations under the *OST* or *Registration Convention* by virtue of any rules of international insolvency law.[50] Landia is entitled to compensation for the economic consequences of its loss of basic satellite communications services from Usurpia for the relocation of Satelsat-18, and from both Usurpia and Concordia as a result of the collision destroying Satelsat-18 and SpaceStar

1.6 Usurpia is internationally responsible and liable to compensate Landia for the unlawful relocation of Satelsat-18

It is a well established principle of international law that every internationally wrongful act of a State entails its international responsibility.[51] Article 2 of the *State Responsibility Articles* provides that 'there is an internationally wrongful act of a State when conduct consisting of an action or omission (a) is attributable to the State under international law; and (b) constitutes a breach of an international obligation of the state.'

As discussed above in section 1, Usurpia breached the international obligations it owed to Landia in approving the relocation of Satelsat-18. This unlawful removal of lifeline satellite services from Landia is also attributable to Usurpia. The UBC is a State organ of Usurpia, and its authorisation of the redeployment of Satelsat-18 is imputable to Usurpia pursuant to Article 4 of the *State Responsibility Articles.*[52] Moreover, as an entity empowered to exercise governmental authority to grant satellite licenses, the UTA's conduct in licensing the relocation of Satelsat-18 is also attributable to Usurpia according to Article 5 of that treaty. Finally, in response to Landia and Concordia's diplomatic protest to the actions taken to relocate Satelsat-18, Usurpia responded that 'its actions were entirely appropriate, in that it was acting on the proper application of an Usurpian commercial enterprise.'[53] Thus, Usurpia 'acknowledge[d] and adopt[ed] the conduct in question as its own,' so that it should be considered an act of that State pursuant to Article 11 of the *State Responsibility Articles*. Therefore, the licensing of Satelsat-18 and the authorisation of the relocation was unlawful and attributable to Usurpia, incurring its international responsibility for these actions.

Accordingly, Usurpia is liable to make full reparations for the damage caused by its breaches of international law.[54] Article 36 of the *State Responsibility Articles* provides that a 'State responsible for a wrongful act is under an obligation to compensate for the damage caused thereby, insofar as such damage is not made good by restitution' and that 'the compensation shall cover any financially assessable damage.' The relocation of Satelsat-18 clearly caused damage to Landia, by depriving it of services it was entitled to receive under its long term non-preemptible lease.[55] As Satelsat-18 has been destroyed, restitution is impossible. It is predicted that it will take three years for Landia

---

50 Paul B Larsen, 'UNIDROIT Space Protocol: Comments on the Relationship Between the Protocol and Existing International Space Law' (2001) *Proceedings of the 44th Colloquium on the Law of Outer Space*, 187, 191.

51 *Articles on Responsibility of States for Internationally Wrongful Acts*, GA Res 56/83, UN GAOR, 56th sess, Supp No 10, art 1, UN Doc A/Res/56/83 (2001) ('*State Responsibility Articles*').

52 In *Difference Relating to Immunity from Legal Process of a Special Rapporteur of the Commission on Human Rights* [1999] ICJ Rep 62 the ICJ held that according to a rule 'of a customary character,' courts are State organs and their decisions are attributable to States.

53 Compromis ¶14.

54 *State Responsibility Articles,* GA Res 56/83, UN GAOR, 56th sess, Supp No 10, art 31, UN Doc A/Res/56/83 (2001).

55 Compromis ¶2; A non-preemptible lease affords the lessee the highest priority service offered by a satellite company; Timothy Bonds et al, *Employing Commercial Satellite Communications* (2000) 87.

to obtain adequate replacement capacity from another satellite operator, and that in the meantime, Landia will suffer more than an estimated $2 billion in losses to its economic welfare resulting from the disruption of its telecommunications infrastructure.[56] Usurpia is therefore required to compensate Landia for this economic loss, placing it in a position as though it had retained its long term lease of premium satellite services.[57]

1.7 Concordia and Usurpia are liable to Landia for the satellite collision as launching States under the *Liability Convention*

(a) Usurpia is a launching State of Satelsat-18

A launching State is defined in Article I of both the *Liability Convention* and the *Registration Convention* as a State that 'launches or procures the launch of a space object' or 'a State from whose territory or facility a space object is launched.'

As the *Liability Convention* was intended to be victim-oriented, it is appropriate that the definition of the 'launching State' is interpreted broadly, as to enable recovery of compensation from a large number of States.[58] Publicists have argued that 'a plurality of launching States is an advantage for the victims'[59] and that 'no real harm is caused by there being a host of launching States because...States can make provision amongst themselves as to how to apportion damage.'[60] Conversely, a narrow conception of the term would limit the number of States from which a victim of damage caused by space activities could claim.

First, Usurpia should be considered a launching State due to its provision of satellite control facilities to Satelsat. Wirin has suggested that 'launches' or 'facility' could be given a broad interpretation to include any State which assists in any way, for example by providing telemetry support.[61] Ground control facilities are integral in ensuring that a launched satellite achieves its desired orbital trajectory. As Satelsat-18 could not have been launched without the use of the Usurpian control facilities, Usurpia should be considered a launching State for the satellite.[62]

Additionally, Usurpia should be considered a launching State as it procured the launch of Satelsat-18. According to the most highly qualified publicists, procurement by a State occurs when it or its nationals are involved in 'acquiring, securing, or bringing about the launch'.[63] Professor Cheng has argued that 'a State would be one of the launching States of the space object if the launch was effected or procured by a private person, whether natural or corporate, that bears its nationality.'[64] Similarly, Professor Hurwitz has proposed that a State which procures a launch can be a 'State...whose nationals have financed or ordered the launching.'[65] As the ongoing activities of Satelsat have been financed by Usurpian

---

56 *Compromis* ¶18.

57 *Factory at Chorzów (Merits)*[1928] PCIJ Series A No 17, 47.

58 Article 31(1) of the *Vienna Convention* requires treaties to be interpreted in light of their object and purpose; Even though the States in this case are not party to the *Vienna Convention,* it has been established that *Vienna Convention* arts 31-32 reflect customary international law; *Case Concerning Sovereignty over the Islands of Litigan and Sipadan (Indodensia v Malaysia)*(2002) ICJ 23-24.

59 Armel Kerrest, 'Remarks on the Notion of the Launching State' (1999) *Proceedings of the 42nd Colloquium on the Law of Outer Space* 308, 311.

60 William Wirin 'Practical Implications of Launching State and Appropriate State Definitions' (1994) *Proceedings of the 37th Colloquium on the Law of Outer Space* 109, 113.

61 Ibid, 111-2.

62 Schmidt-Tedd and Gerhard, 'How to Adapt the Present Regime or Registration of Space Objects to New Developments in Space Applications' (2005) *Proceedings of the 48th Colloquium on the Law of Outer Space* 353, 359.

63 Wirin, above n 60, 111.

64 Bin Cheng, *Studies in International Space Law* (1997) 627.

65 Bruce Hurwitz, *State Liability for Outer Space Activities* (1992) 22.

Banks;[66] Usurpia should be considered to have procured the launch of Satelsat-18.

Moreover, there is support for the proposition that a State which registers a satellite is considered a launching State under the *Liability Convention.* Therefore, even if Usurpia is not considered to be a launching State due to its involvement with Satelsat-18 at the time the space object was launched, it should be deemed to have acquired launching State status due to its registration of the satellite.[67]

Relevant rules of international law applicable between the parties to a dispute may be used in ascertaining the meaning of a treaty provision.[68] The launching State concept also appears in the *Registration Convention*, which provides in Article 1(c) that a 'State of registry means a launching State on whose registry a space object is carried.'[69] The *Registration Convention* was concluded after the *Liability Convention*. The concept of launching State in the later convention thus assists in interpreting the definition of the 'launching State' under the *Liability Convention.*[70] Professor Christol has noted that the launching state referred to in the *Registration Convention* was intended to be the same State as that defined under the *Liability Convention,* in order to achieve consistency and uniformity in the international law of outer space.[71] Moreover, Professor Kerrest has advocated an interpretation of the *Registration Convention* which allows a 'non-original launching state to register a spacecraft… [and] to consider this state as one of the liable launching states'.[72] The Applicants respectfully request this Court to adopt this interpretation.

Interpreting the term 'launching State' broadly also accords to the realities of the space industry. In today's commercial space industry, the launching entity will generally not continuously operate and control the satellite, as predominantly occurred in the past where States were the key space actors. It has been noted that 'if a launching State is required to be an original launching State, this could mean that none of the States subject to international liability under the *Liability Convention* have the ability to prevent damage being caused by the space object in question.'[73] These considerations will become even more important when the *Mobile Equipment Convention* enters into force in relation to space assets, and the transfer of possession of space objects becomes even more frequent. A narrow interpretation of the term 'launching State' has been described as 'a serious flaw in the logic of the liability regime'[74] and as 'an injustice' to the launching State which continues to incur liability when it no longer has any control or influence over the operation or control of the satellite.[75]

In light of these concerns, in meetings of UNCOPUOS, a number of States have recommended that a State which comes to own or operate a space object should be liable for the damage caused by it, even if that State is not directly involved in the launch phase.[76] Indeed, agreements concluded as part of State practice support the notion that a State which operates a

66 *Compromis* ¶4.
67 *Compromis* ¶24.
68 *Vienna Convention*, opened for signature 23 May 1969, 1155 UNTS 331 art 31(3)(c), (entered into force 27 January 1980); see above n 58.
69 *Registration Convention*, opened for signature 14 January 1975, 1023 UNTS 15, art 1(c) (entered into force 15 September 1976).
70 *Vienna Convention*, opened for signature 23 May 1969, 1155 UNTS 331 art, 30(2) (entered into force 27 January 1980).
71 Carl Q Christol, *The Modern International Law of Outer Space* (1982) 240.
72 Kerrest, above n 59, 309.

73 Kai-Uwe Schrogl and Charles Davies, 'A New Look at the Launching State – The results of the UNCOPUOS Legal Subcommittee Working Group "Review of the concept of the launching state (2000-2002)"' (2002) *Proceedings of the 45th Colloquium on the Law of Outer Space* 286, 295.
74 Frans von der Dunk, 'The Illogical Link: Launching, Liability and Leasing' (1993) *Proceedings of the 36th Colloquium on the Law of Outer Space*, 349, 354.
75 Ricky J Lee, 'Liability Convention and National Licensing Regimes' (Paper presented at the United Nations/Republic of Korea Workshop on Space Law, Daejeon, Republic of Korea, 4 November 2003) 443.
76 *Review of the concept of the 'launching State,'* UNCOPOUS Legal Subcommittee, UN DOC A/AC.105/768 (2002) 16.

satellite should be liable for any damage caused. For example, in order to define the respective liabilities of parties involved in the 1990 launch of the ASIA-I satellite from China, a liability agreement was concluded between China and the UK.[77] Under this arrangement, China was made liable for damage caused to third countries during the launch phase of the satellite, whilst the UK assumed liability for any damage occurring after the satellite's successful launch, as the State which owned and operated the space object whilst it was in orbit. Similar agreements have been since employed by China, as well as other space actors including the European Space Agency, reflecting international acceptance that a State which controls the activities of a space object should be held liable for any losses which arise from its actions.[78] Further, State practice in the sale of satellites also supports the idea that a State of registry should bear liability. The satellite BSB-1A was originally registered with the UN by the United Kingdom. However, following Sweden's purchase of the satellite in orbit in 1996, it was listed as Sirius-1 on the Swedish national register and this was conveyed to the UN.[79] According to Swedish law, its national registry is to be used for space objects for which Sweden is considered the launching State in accordance with Article I of the *Registration Convention.*[80] The UN's acceptance of Sweden's registration reflects recognition of the fact that a State which registers a satellite is a launching State for that space object.

Thus, there is international support for the proposition that a state which registers a satellite is one of the liable launching States under the *Liability Convention.* This Court should respectfully adopt the view that a 'launching State is not required to be an original launching State'[81] and that a State which purchases a payload already in orbit 'should be considered a launching State even though this status occurs after the launch.'[82] Usurpia unlawfully seized jurisdiction and control of Satelsat-18. It then placed Satelsat-18 on its national register and commenced the process of notifying the UN of its status as a State of registry. This was a unilateral act which evinced Usurpia's intention to be recongised as a State of registry and a launching State under both the *Liability Convention* and *Registration Convention.* Usurpia should accordingly be bound by its actions.[83]

Thus, Usurpia should incur liability as a launching State either due to its involvement with Satelsat-18 prior to its launch, or its subsequent registration of the satellite.

(b) Concordia is a launching State of Satelsat-18 and SpaceStar

Concordia is a launching state of Satelsat-18, as the satellite was launched from a Concordian facility, in Concordian territory.[84] Additionally, Concordia is the launching State of SpaceStar, as it licensed or 'procured' the launch of this satellite.[85] The Applicants do not dispute Concordia's status as a launching State, but rather submit below in section 4 that any liability Concordia incurs in this capacity should be indemnified by Usurpia.

(c) Landia's loss is recoverable damage under the *Liability Convention*

Landia is claiming $2 billion for the economic loss it predicts it will suffer because of the loss of its lease of satellite services.[86] Damage is defined under Article I(a) of the *Liability Convention* as 'loss of life, personal injury or other impairment of health; or loss of or damage to property.'[87] However, compensation under the *Liability Convention* is to be determined 'in accordance with

77 Ibid, 13.

78 Ibid, 14.

79 Ibid, 16.

80 Section 4, *Decree on Space Activities* (1982:1069) at <http://www.unoosa.org/oosaddb/showDocument.do?documentUid=319> at 1 June 2008.

81 Schrogl and Davies, above n 73, 295.

82 Wirin, above n 60, 313.

83 For the binding nature of a unilateral undertaking in international law, see eg, *Nuclear Tests (Australia v France)* [1974] ICJ Rep 253, 270.

84 *Compromis* ¶3.

85 *Compromis* ¶15.

86 *Compromis* ¶18.

87 *Liability Convention,* opened for signature 29 March 1972, 961 UNTS 197, art I(a) (entered into force 1 September 1972).

international law and the principles of justice and equity, in order to provide such reparation in respect of the damage as will restore the...State... to the condition which would have existed if the damage had not occurred.'[88]

Under international law, 'property' is not confined to tangible assets, but rather extends to any right which can be subject to a commercial transaction.[89] The satellite collision resulted in Landia's loss of its provisional lease of segment capacity on SpaceStar, which clearly constitutes a 'loss of property' within the threshold definition of damage.

Indirect economic damage arising from this loss of property is therefore recoverable under the *Liability Convention* insofar as it is required to effect reparation.[90] Article 36(2) of the *State Responsibility Articles* confirms the availability of compensation for prospective financial loss by expressly providing for compensation for future loss of profits. It is acknowledged that the *travaux preparatoires* of the *Liability Convention* reveal that negotiating States were divided as to whether indirect damage should be explicitly included in the Treaty. However, the majority of States held the opinion that the issue should be left open to be dealt with as individual cases arose.[91] To date, there has been only one international claim made pursuant to the *Liability Convention*, following the Cosmos 954 incident. This claim was presented by Canada in 1979, for damage caused when a Soviet satellite deposited nuclear debris on Canadian territory.[92] Although this dispute was eventually settled through negotiation, Canada's inclusion of 'clean-up costs' in its claimed amount suggests that compensation for indirect losses arising from the destruction or contamination of property does fall within the scope of recoverable damage under the *Liability Convention.*[93] Significantly, the most highly qualified publicists have since supported the view that the *Liability Convention* covers the 'additional consequences produced as a result of the initial hit [of a space object].'[94]

Such indirect damage will be compensable under the *Liability Convention*, as in general international law, so long as causation is established. In order for damage to be recoverable in Space law, it must be 'caused by' a space object.[95] The requisite causal link under both the *Liability Convention* and general international law is 'proximate causation,' a general principle of international law[96] which is supported in international case law[97] and by the most highly qualified publicists.[98] The rule of 'proximate causation' limits recoverable damage to that which arises as a normal consequence of an act and is reasonably foreseeable.[99] A normal

88 *Liability Convention,* opened for signature 29 March 1972, 961 UNTS 197, art XII (entered into force 1 September 1972).

89 See eg, discussions of 'property' in *Amoco International Finance Corp v Iran (US v Iran)*(197) USCTR 189, ¶108; *Shufeldt Claim (US v Guatemala)* (1930) 2 RIAA 1083, 1097.

90 *Chorzow Factory Case (Merits)* [1928] PCIJ Series A., No 17, 47; *State Responsibility Articles* GA Res 56/83, UN GAOR, 56th sess, Supp No 10, art 31, UN Doc A/Res/56/83 (2001).

91 WF Foster, 'The Convention on International Liability for Damage caused by Space Objects' (1972) 10 *Canadian Yearbook of International Law* 137, 159.

92 Christol, C, 'International Liability for Damage Caused by Space Objects' (1980) 72 *American Journal of International Law* 346.

93 Ibid 362.

94 Foster, above n 91, 159. See also, Hurwitz, above n 65, 16.

95 *Liability Convention,* opened for signature 29 March 1972, 961 UNTS 197, arts II-III (entered into force 1 September 1972); Foster, above n 91, 157-158.

96 To which recourse may be had pursuant to Article XII of the *Liability Convention.*

97 *Administrative Decision No II (US v Germany)* (1923) 7 RIAA 23, 29-30; *Lake Lanoux Arbitration (France v Spain)*(1957)12 RIAA 281; *Trail Smelter Arbitration (United States v Canada)* (1949)33 AJIL 182, 183.

98 Bin Cheng, *General Principles of Law as Applied by International Courts and Tribunals* (1953) 241-56. ; Hurwitz, above n 65, 15–17; Christol, above n 71,109.

99 For example, in *Administrative Decision No II (US v Germany)* (1923) 7 RIAA 23 it was held at 29-30: 'the proximate cause of the loss must have been in legal contemplation the act of Germany... all indirect losses are covered...provided

and foreseeable consequence of a satellite collision is that entities reliant on the services of the satellites involved will incur grave losses. The collision of Satelsat-18 and SpaceStar caused Landia losses of such a nature. Landia was deprived of the opportunity to lease services from SpaceStar – an opportunity it most likely would have benefitted from had the satellite not been destroyed. It is highly likely that had the collision not occurred, Landia would have received funding to secure its provisional lease over SpaceStar, given the high value Landia places on internet connectivity as a constitutional right,[100] and its declaration under *GLITSO* of its total dependence and reliance on the commitments of other countries in enabling it to meet its basic telecommunications needs.[101] Following the collision, Landia lacked the capacity to operate its basic telecommunications infrastructure, and as a result will suffer enormous economic damage.[102] Accordingly, regardless of whether Landia's loss is categorised as direct or indirect, this damage is recoverable under the *Liability Convention* as it was legally caused by a space object.

(d) Therefore, Landia may claim full compensation from either Usurpia or Concordia under the *Liability Convention*

Article IV(1) of the *Liability Convention* provides that 'in the event of damage being caused elsewhere than on the surface of the Earth to a space object of one launching State...by a space object of another launching State,' these states shall be jointly and severally liable to a third state that suffers damage. The collision of Satelsat-18 (for which Usurpia and Concordia are joint launching States) with SpaceStar (for which Concordia is a launching State) caused Landia to suffer damage on the surface of the Earth, in the form of economic losses due to disruption to its telecommunications infrastructure. Pursuant to Article IV(2) of the *Liability Convention*, Landia is entitled to seek the entire compensation due to it 'from any or all of the launching States which are jointly and severally liable' – that is, from either Concordia or Usurpia.

1.8 In the alternative, Usurpia is liable to compensate Landia because it bears international responsibility for the collision under Article VI of the *OST*

Even if Usurpia is not held liable as a launching State under the *Liability Convention*, the Applicants may still claim compensation from Usurpia as the State internationally responsible for the collision.[103]

Article VI of the *OST* provides that State parties 'shall bear international responsibility for national activities in outer space...whether such activities are carried on by governmental agencies or non-governmental entities.'[104] According to the *lex specialis* rule, Article VI displaces general international law principles of State responsibility and attribution, automatically imposing international responsibility on States for their national activities in outer space.

The 'national activities' referred to in Article VI comprise of those space activities within a State's effective jurisdiction or control.[105] This encompasses both activities 'conducted by a company enjoying the nationality of [a] State'[106] and 'activities

that Germany's act was the efficient and proximate cause from which they flowed.'

100 *Compromis* ¶2(b).

101 *Compromis,* Appendix A.

102 *Compromis* ¶18.

103 Frans von der Dunk, 'Liability versus Responsibility in Space Law: Misconception or Misconstruction' (1991) *Proceedings on the 34th Colloquium on the Law of Outer Space* 363, 368; Ricky J Lee, 'Liability Arising From Article VI of the Outer Space Treaty: States, Domestic Law and Private Operators'(2005) *Proceedings of the 48th Colloquium on the Law of Outer Space* 216.

104 *OST,* opened for signature 27 January 1967, 610 UNTS 205, art VI (entered into force 10 October 1967).

105 Bin Cheng, 'Article VI of the Space Treaty Revisited: "International Responsibility," "National Activities," and the "Appropriate State"' (1998) 26 *Journal of Space Law* 7, 19.

106 Kerrest, A, 'Remarks on the Responsibility and Liability for Damages Other Than Those Caused by the Fall of a Space Object' (1997) *Proceedings of the 40th Colloquium on the Law of Outer Space* 134, 138.

undertaken from within the territory of the State in question.'[107] The relocation of Satelsat-18 was performed by New Satelsat, a company incorporated in Usurpia bearing Usurpian nationality.[108] In addition, the relocation was likely to have been performed using Satelsat's satellite control facilities located in Usurpian territory.[109] Thus, the relocation was within the national and territorial jurisdiction of Usurpia, and constituted a Usurpian 'national activity.'

The effect of Article VI is to impute to a State international responsibility for all private space activities under its control, including those that it has a duty to authorize and supervise as the appropriate State.[110] Professor Kerrest has stated that '[in] the case of violation by a private entity of any international regulation or principle, the State is responsible without having the possibility to avoid liability by proving ignorance of such a violation, nor even by showing it had [done] its best...to control the activity.' This represents a stringent application of the principle of international law established in the *Trail Smelter Arbitration* that a 'State owes at all times a duty to protect other States against injurious acts by individuals from within its jurisdiction.[111] In space law, States are not able 'to dodge a potential duty for damage by claiming they have taken due care.'[112] Usurpia breached its duty as a responsible State to supervise the activities of New Satelsat, resulting in the company rushing the satellite relocation and the subsequent collision.[113] Regardless of whether Usurpia had knowledge of the company's proposed speedy relocation, Usurpia's failure to adequately monitor New Satelsat's actions causes it to incur international responsibility.

As the State responsible for its national activities in outer space under Article VI, Usurpia is internationally liable to pay compensation for any damage caused by such activities.[114] Therefore, Usurpia is obliged to fully compensate Landia for its economic losses arising from the collision.[115]

2. Concordia is entitled to compensation from Usurpia for the loss of Satelsat-18

2.1 Concordia may claim for the destruction of Satelsat-18 on behalf of Satelsat

Concordia is entitled to claim compensation on behalf of its national, Satelsat[116] under both Article VIII(1) of the *Liability Convention* which provides that 'a State which suffers damage or whose natural or juridical persons suffer damage, may present...a claim for compensation for such damage'[117] and customary international law principles of diplomatic protection.[118] This is the case even though New Satelsat was the company which owned Satelsat-18 at the time it was destroyed, and Satelsat's loss occurred by virtue of its shareholding in New Satelsat.

In the *Barcelona Traction Case,* this Court stated that in some situations, 'considerations of equity might call for the possibility of protection of shareholders by their national state.'[119] The present case requires such considerations of equity. In particular, the *Barcelona Traction Case* recognised the

107 von der Dunk, above n 103, 367.

108 *Barcelona Traction, Light and Power Company Limited (Belgium v Spain) (Judgment)* [1970] ICJ Rep 3, ¶42 ('*Barcelona Traction Case*').

109 *Compromis* ¶4.

110 Lee, above n 103, 218.

111 3 RIAA 1905, 1963; See also, *Corfu Channel Case (Merits)*(1949) ICJ Rep 4, 22.

112 von der Dunk, above n 103, 367.

113 *Compromis* ¶16.

114 *Chorzow Factory Case (Merits)* [1928] PCIJ Series A., No 17, 47; This rule applies under Article VI of the *OST*; von der Dunk, above n 103, 367; Kerrest above n 106, 139.

115 *State Responsibility Articles,* GA Res 56/83, UN GAOR, 56th sess, Supp No 10, art 36, UN Doc A/Res/56/83 (2001).

116 Due to its incorporation in Concordia: *Barcelona Traction Case* [1970] ICJ Rep 3, ¶42.

117 *Liability Convention,* opened for signature 29 March 1972, 961 UNTS 197, art VIII(1) (entered into force 1 September 1972).

118 *Draft Articles on Diplomatic Protection,* Report on the 55th Sess ILC (2003) UN GAOR, 58th Sess, Supp 10, arts 1-4.

119 *Barcelona Traction Case* [1970] ICJ Rep 3, 93.

applicability of the municipal company law doctrine of 'lifting the corporate veil' in international law.[120] This doctrine, recognised in both common law[121] and civil law[122] jurisdictions, allows the corporate veil to be pierced where the legal personality of a company is misused. For example, where the dominant intention of forming a company is to avoid an existing obligation, the corporate structure may be disregarded. It is submitted that the creation of New Satelsat was part of a sham transaction in order to evade the stringency of Concordia's *GLITSO* declaration and circumvent the interim order of the CCC. As such, the Court should look beyond New Satelsat's corporate veil, allowing Concordia to claim compensation on behalf of Satelsat, notwithstanding that New Satelsat incurred the loss.

2.2 Usurpia is liable compensate Concordia as the State at fault under the *Liability Convention*

Article III of the *Liability Convention* imposes fault-based liability on States in the event of damage being caused in outer space 'to a space object of one launching State...by a space object of another launching State.'

The damage suffered by Concordia does not precisely fall within this provision, as Concordia's claim is for the destruction of Satelsat-18, which was caused by Usurpia's relocation of that same satellite. However, an imposition of liability as provided for exactly by the article has become impractical and anachronistic with respect to today's commercial realities.[123] With the development of the commercial space industry resulting in an increasing number of launching States for any given satellite, a variety of losses in outer space are possible other than through a collision between different space objects belonging to different launching states. The purpose of the *Liability Convention* is 'to establish effective international rules and procedures concerning liability for damage caused by space objects...and a full and equitable measure of compensation.'[124] In particular, Professor Christol states that the imposition of fault-based liability in Article III constitutes 'an inducement to launching entities to exercise care.'[125] Article III should therefore be interpreted as providing a general regime of fault-based liability for damage caused in outer space. To allow Usurpia to escape liability simply because Concordia was involved in the launch of both satellites would contradict the victim-oriented nature of the *Liability Convention* and would also curtail any inducement of creditors which take possession of space objects from exercising due care. Thus, even though the present situation does not precisely meet the parameters of Article III, fault-based liability should be imposed upon Usurpia.

In this case, the destruction of Satelsat-18 was entirely due to the fault of Usurpia. Fault is not defined in the *Liability Convention.* However, in international law the principle of 'fault' refers not to culpability or malice, but rather a failure to comply with a legal duty or obligation.[126]

As discussed above, Usurpia breached international law in discriminatorily depriving Landia of the services of Satelsat-18 and failing to supervise New Satelsat's relocation of the satellite. These breaches of international law constitute fault on Usurpia's part. Baker has also proposed that in general, a State would be at fault if it failed to maintain the required spacing between satellites in the geostationary orbit.[127] Thus, fault for the collision and destruction of Satelsat-18 lies solely with Usurpia, as the collision was caused by its unlawful act of failing to prevent New Satelsat from speeding up the relocation.[128]

120 *Barcelona Traction Case* [1970] ICJ Rep 3, 57.

121 *Gilford Motor Co v Horne* [1993 Ch 935].

122 *German Civil Code.*

123 Bruce Hurwitz, 'Liability for Private Commercial Activities in Outer Space' (1990) *Proceedings of the 33rd Colloquium on the Law of Outer Space, 37.*

124 See Preamble to *Liability Convention.*

125 Christol, above n 71, 107.

126 Cheng, above n 98, 218.

127 H Baker, 'Liability for Damage Caused in Outer Space by Space Refuse'(1988) 12 *Annals of Air and Space Law* 183, 192.

128 *Compromis* ¶16.

In addition, Concordia bore no fault in the satellite collision. Professor Cheng states that according to the international law principle of *nemo tenetur ad impossibile*, launching States cannot be held liable if a space object happens to be under the effective jurisdiction of another State, and it proves impossible to bring them within its effective jurisdiction.[129] In light of Usurpia's seizure of control and jurisdiction over Satelsat-18 in spite of Concordia's protests, Concordia had no control over the satellite at the time of the collision. Therefore, Usurpia's total fault requires it to pay complete compensation for the destruction of the satellite in accordance with a broad interpretation of Article III of the *Liability Convention.*

2.3 In the alternative, Usurpia is liable to compensate Concordia as the State responsible for the collision

As discussed above in sections 2.1 and 2.3, Usurpia bears international responsibility for the satellite relocation and subsequent collision under general principles of State responsibility and Article VI of the *OST*. It is thus liable to compensate Concordia for its losses arising from these acts.

3. Usurpia is obligated to indemnify Concordia for any liability Concordia might owe Landia

3.1 Usurpia is a joint launching State of Satelsat-18

Article V of the *Liability Convention* allows a launching State which has paid compensation for damage to present a claim for indemnification to other participants in the joint launching, that is, the other launching States.[130]

Where the launching States have not concluded an agreement regarding the apportionment of their liability, this is to be determined in accordance with their relative fault.[131] As discussed above in section 3.2, the satellite collision was entirely due to Usurpia's fault. On the other hand, Concordia bore no fault. Therefore, as the State solely at fault, Usurpia must completely indemnify Concordia to the extent that the latter is liable to Landia.

3.2 In the alternative, Concordia is entitled to indemnification under general principles of international law

Usurpia's unilateral seizure of Satelsat-18 violated international law and deprived Concordia of any opportunity to conclude agreements with either New Satelsat or Usurpia, transferring to them the burden of its liability under the *Liability Convention.*

It is almost uniform State practice for States to require private satellite operators and their successors in title to indemnify them for any liability the private company's activities may cause the State to incur under the *Liability Convention.*[132] Such indemnification agreements protect States from the burden of liability arising from space activities which they are not in a position to control. For example; agreements between China and the United Kingdom concerning the AsiaSat-1, AsiaSat-2, Apstar-1 and Apstar-2 satellites contained an indemnification clause in favour of the United Kingdom assigning all liability to China in the launch phase when a Chinese corporation had complete control over the satellite.[133]

Had Usurpia not breached its international obligations, especially the duty to cooperate, Concordia would have concluded appropriate indemnification agreements. Therefore, Usurpia should be liable to indemnify Concordia to the extent that Concordia is required to pay compensation to Landia.

---

129 Cheng, above n 98, 227.

130 Although the *Liability Convention* expressly specifies that 'a State from whose territory or facility a space object is launched shall be regarded as participant in a joint launching,' joint launching States can bear any of the definitional features of the launching State as defined in Article I; Cheng, above n 64, 329.

131 Kerrest, above n 106, 312.

132 *Report of the Secretariat, 'Review of the Concept of the 'launching State'* UN Doc A/AC.105/768 (2002) at 11.

133 See Exchange of Notes between the United Kingdom and China concerning Liability for Damage arising during the Launch Phase of the Asiasat Satellite, Peking, 26 March 1990 and 2 April 1990, UKTS No 7(1993) Cm. 2138.

SUBMISSIONS TO THE COURT

For the foregoing reasons, the Governments of Landia and Concordia, Applicants, respectfully request to the Court to adjudge and declare that:

1. Usurpia's decision to license Satelsat-18 satellite and permit it to be deployed at an Usurpian orbital location was inconsistent with applicable principles of international law;

2. Landia is entitled to compensation from Usurpia for the economic consequences of its loss of basic satellite telecommunications as a result of the collision that destroyed Satelsat-18 and SpaceStar;

3. Concordia is entitled to compensation from Usurpia for the loss of Satelsat-18; and

4. Usurpia is obligated to indemnify Concordia for any liability Concordia might owe to Landia as a result of the satellite collision.

MEMORIAL FOR THE RESPONDENT THE KINGDOM OF USURPIA

University of New South Wales, Australia (Madeleine Ellicott, Tamara Phillips, Katrina Taylor; Coach: Mr Pouyan Afshar)

ARGUMENT

1. Usurpia's decision to re-license and authorise the relocation of Satelsat-18 was consistent with its international obligations

1.1 Usurpia acted in accordance with its *GLITSO* obligations

The *Global Legacy International Telecommunications Agreement* ('*GLITSO*') sets out an international regime for the provision of satellite services.[134] Usurpia, Concordia and Landia are all States parties to this Agreement. Under Article II of *GLITSO*, parties commit to maintaining global connectivity and coverage to all countries on a non-discriminatory basis and to supporting the provision of affordable satellite services to Lifeline Dependent Countries ('LDCs'). *GLITSO* does not specify any particular means by which States must honour their obligations. Rather, each State party has the discretion 'to take such action as it determines to be appropriate' to achieve the objectives set out in *GLITSO*.[135] Each party to *GLITSO* is required to issue a Declaration indicating how it intends to adhere to these objectives.[136]

Accordingly, the Declaration issued by Usurpia in connection with its ratification of *GLITSO* requires satellite operators to accommodate the objectives of *GLITSO* on a 'best efforts' basis, consistent with prudent business practice and in light of the commercial realities of the satellite industry.[137]

134 *Compromis* ¶5.
135 *Compromis* Appendix A.
136 Ibid.
137 Ibid.

(a) Usurpia's Declaration is consistent with the object and purpose of *GLITSO*

Usurpia's Declaration is a statement of how it intends to adhere to the principles of *GLITSO*. The rules regarding reservations in the *Vienna Convention on the Law of Treaties*,[138] which codify customary international law,[139] apply to Usurpia's *GLITSO* Declaration, which, like a reservation, purports to modify the effect of a treaty provision.[140] Under Article 19 of the *Vienna Convention*, States are entitled to formulate a reservation provided this does not defeat the object and purpose of the treaty. Article 20 provides that a State is considered to have accepted another State's reservation if it does not raise any objection to the reservation within a stated time.

Usurpia's 'best efforts' declaration is entirely consistent with the object and purpose of *GLITSO*, when viewed in the context of space industry practice. Due to the high-risk, capital-intensive and volatile nature of the satellite services industry, it is common for operators to include a 'best efforts' clause in their service agreements due to the substantial risk of satellite service failure in the event of insolvency.[141] Satellites are subject to significant operational risks whilst in orbit, including control system failures, in-orbit malfunctions, transponder failures and power system failures.[142] Given the demonstrated volatility of the satellite industry, and the directive in *GLITSO* that a State may 'take such action as it deems appropriate,' Usurpia's declaration is clearly valid. Further, Landia and Concordia did not raise any objection to this Declaration, thus they may be deemed to have accepted its validity.

(b) Usurpia complied with its 'best efforts' Declaration

While there is an absence of international law jurisprudence interpreting the term 'best efforts', this Court is entitled to apply general principles of law, judicial decisions and writings of publicists in deciding disputes submitted to it.[143]

An examination of the case law of national courts reveals that a 'best efforts' standard only obliges the declarant to do what is reasonable in the circumstances. This principle has been upheld by the Australian High Court.[144] Similarly, the Canadian Supreme Court has held 'best efforts' means taking, in good faith, all reasonable steps to achieve an objective.[145] United States case law also requires a party to do what is reasonable, and notes that a party with a best efforts obligation is not required to disregard its own interests or spend itself into financial ruin.[146]

Further, publicists have noted that States frequently employ a 'best efforts' clause in relation to their high-risk activities in outer space in order to 'refrain from promising the accomplishment of their respective obligations, committing themselves only to using their best efforts to achieve success… [t]his is associated with both a reduction and a waiver of liability'.[147] Usurpia's commitment to

---

138 *Vienna Convention on the Law of Treaties*, opened for signature 23 May 1969, 1155 UNTS 331 art 31(3)(b) (entered into force 27 January 1980) ('*Vienna Convention*').

139 *Reservations to the Genocide Convention* [1951] ICJ Rep 15, 24.

140 *Vienna Convention*, opened for signature 23 May 1969, 1155 UNTS 331, art 2 (entered into force 27 January 1980).

141 Julian Hermida, *Legal Basis for a National Space Legislation* (2004) xxiv.

142 *Telesat Form F-1* (2006) US Securities and Exchange Commission <http://www.sec.gov/Archives/edgar/data/1375450/000090956706001910/o33788fv1za.htm> at 11 July 2008.

143 *Statute of the International Court of Justice*, arts 38(1)(c)-(d).

144 *Transfield Pty Ltd v Arlo International Ltd* (1980) 54 ALJR 323, 329.

145 *Atmospheric Diving Systems Inc v International Hard Suits Inc* [1994] 89 BCLR (2d) 356.

146 *Bloor v Falstaff Brewing Corp*, 601 F.2d 609 (2d Cir 1979); *Hughes Communications Galaxy, Inc v United States* (2000) 47 Fed Cl 236, which interpreted the term in the context of a satellite contract.

147 Bernhard Schmidt-Tedd, 'Best Efforts Principle and Terms of Contract in Outer Space

performing its obligation on a best efforts basis is therefore not an absolute guarantee of performance, but only required it to do what was reasonable in the circumstances, in accordance with 'prudent business practice'.

In assessing the nature of 'prudent business practice' as referred to in Usurpia's Declaration, it is necessary to examine the practice of commercial satellite operators. This reveals that satellite companies often resort to selling or redistributing their assets as a means to avoid insolvency. For example, in 2000, Iridium LLC was forced to sell its satellite assets (previously valued at US$5 billion) for US$25 million after falling into significant debt.[148] Also, in 2003, Loral Space and Communications was forced to sell five in-orbit satellites to Intelsat to repay its US$959 million debt.[149] In light of Satelsat's $25 billion debt, the UBC made a prudent and necessary decision to approve the restructuring plan in order to save the company.

Industry practice reveals that satellite companies are highly dependent on the lease rates of their transponders to remain profitable,[150] and often rely on the patronage of a few major clients to maintain acceptable revenue levels.[151] Satelsat's potential customer was able to lease *all* of the Satelsat-18 transponders at premium rates (as opposed to only 10 of the 11 transponders which were being utilised at its existing location), and this would significantly aided the prevention of Satelsat's insolvency.[152] Prudent business practice necessitated action by Usurpia to secure the customer quickly in order to save Satelsat and ensure that the company could continue to provide satellite services, including lifeline services, into the future.

Furthermore, the *Convention on International Interests in Mobile Equipment* which was concluded eight years before *GLITSO* provides useful guidance as to what is 'prudent business practice' for creditors in the space industry.[153] Whilst the *Mobile Equipment Convention* is not yet in force,[154] the remedies available to creditors under it have been accepted by over 40 contracting States which have signed or ratified the treaty. This includes countries heavily involved in the financing of space assets such as the United States and United Kingdom. The *Space Protocol* aims to promote the financing of space activities though the creation of a clear legal framework for the securitisation of space assets, in order to instil greater confidence in financiers who wish to loan money for space ventures.[155] Both Usurpia and Concordia are parties to the *Mobile Equipment* Convention and have acceded to the procedures within it, and they have an obligation

Business' (1988) *Proceedings of the 31st Colloquium on the Law of Outer Space* 330, 330.

148 *Iridium Satellite LLC* (2001) <http://www.spaceandtech.com/spacedata/constellations/iridium_sum.shtml> at 21 July 2008.

149 *Space News Business Report* (2003) <http://www.space.com/spacenews/archive03/loral_102003.html> at 21 July 2008.

150 *RSCC Feels Effects of Transponder Rates, Insurance Costs* (2005) Russian Satellite Communications Company <http://rscc.ru/en/company/publ/2005.04.25.html> at 19 July 2008.

151 *Telenor ASA Form 20-F* (2004) US Securities and Exchange Commission <http://www.sec.gov/Archives/edgar/data/1126113/000115697305000475/u48549e20vf.htm#107> at 19 July 2008.

152 *Compromis,* Statement of Additional Facts, 4.

153 *Convention on International Interests in Mobile Equipment*, opened for signature 16 November 2001, UN Doc A/AC.105/C.2/2002/CRP.3 (entered into force 1 March 2006 excl art 49)('*Mobile Equipment Convention*').

154 It will only enter into force with respect to space assets when the *Draft Protocol on Matters Specific to Space Assets ('Space Protocol')* is acceded to: *Preliminary Draft Protocol on Matters Specific to Space Assets*, UNIDROIT Doc Study LXXIIJ (2004).

155 Rolf Olofsson and Mark Bisset, 'The UNIDROIT Convention on International Interests in Mobile Equipment, the Aircraft Protocol and the Draft Space Protocol' (2002)3 *Business Law International* 307, 307; Paul Larsen, 'Critical Issues in the UNIDROIT Draft Space Protocol' (2002) *Proceedings of the 45th Colloquium on the Law of Outer Space* 2, 2.

not to act contrary to the object and purpose of the Agreement before it enters into force.[156]

Under the *Mobile Equipment Convention* and the *Space Protocol*, a creditor may apply for a court order authorising various remedies upon a satellite operator's default, including orders allowing the creditor to repossess, sell, lease or manage the use of the space object, as long as this is done in a 'commercially reasonable' manner, in line with industry practice.[157] Here, Satelsat defaulted on its interest payments on its debt to the Banks over a six month period.[158] Accordingly, in applying for the necessary authority to restructure Satelsat and relocate Satelsat-18, the Banks acted consistently with the rights of creditors which have been internationally recognised in the *Mobile Equipment Convention*.

Usurpia's acts in re-licensing and authorising the relocation of Satelsat-18 were entirely consistent with its best efforts declaration with regard to the standard of prudent business practice set out above. Given Satelsat's huge debt obligations, the Banks were forced to urgently adopt a commercially viable solution to prevent Satelsat's insolvency. Had the Banks not seized the opportunity to take on a customer who was able to lease all of the transponders on Satelsat-18 at premium rates, Satelsat would have been in grave and imminent danger of insolvency, which would have led to many countries dependent on Satelsat's 25 satellites (including Landia) becoming entirely deprived of Satelsat's satellite services. The only option available in these circumstances was to offer Landia an alternative service on three other Satelsat satellites. While this may not have been an ideal alternative for Landia, it was the only commercially reasonable option available in the circumstances. In light of Satelsat's impending insolvency, acceding to Landia's demands to maintain their satellite services on Satelsat-18 would have been financially unsound and may have resulted in Landia being deprived of Satelsat satellite services altogether. Usurpia's decision represented the exercise of its best efforts to support the provision of affordable satellite services to Landia given the exigencies of the situation.

(c) New Satelsat was not bound by Concordia's *GLITSO* obligation

Concordia's Declaration in connection with its ratification of *GLITSO* requires that licensees of the CCC adhere to the principles of *GLITSO*, and that as a condition of sale of any satellite assets, a 'successor in title holding that satellite license shall be similarly obligated to adhere to such obligations.'

Immediately upon receiving approval for the revised restructuring plan, Usurpia notified the CCC of its intent to relinquish its license to operate Satelsat-18 and any rights it had to locate the satellite at its Concordian orbital location.[159] As the Concordian license was relinquished before the transfer of Satelsat-18, New Satelsat was not a 'successor in title holding' a Concordian license to operate Satelsat 18. Thus, New Satelsat was at no time bound by the Concordian *GLITSO* Declaration.

1.2 Usurpia's actions were consistent with the *Outer Space Treaty*

(a) Usurpia did not breach Article I

Article I of the *Treaty on Principles Governing the Activities of States in the Exploration and Use of Outer Space, including the Moon and Other Celestial Bodies*[160] stipulates that the exploration of space should be carried out for the benefit and interests of all countries 'irrespective of their degree of scientific and economic development'.

---

156 *Vienna Convention,* opened for signature 23 May 1969, 1155 UNTS 331, art 18 entered into force 27 January 1980).

157 *Mobile Equipment Convention*, opened for signature 16 November 2001, UN Doc A/AC.105/C.2/2002/CRP.3, art 8 (entered into force 1 March 2006 excl art 49).

158 *Compromis* ¶4.

159 *Compromis* ¶13.

160 *Treaty on Principles Governing the Activities of States in the Exploration and Use of Outer Space, including the Moon and Other Celestial Bodies*, opened for signature 27 January 1967, 610 UNTS 205 (entered into force 10 October 1967) ('*OST*').

However, this provision remains non-binding for two reasons.

First, the principle is simply too vague to create a binding legal obligation as the language used is too general and lacks the minimum precision required in legal documents.[161] The most highly qualified publicists have stated that Article I(1) is merely an expression of intention conferring no real rights and imposing no real obligations on states.[162] Gorove has stated that the 'common interests' clause in Article I(1) 'is not self-executing, but rather a kind of imperfect legislation in that it expresses an aspiration couched in very general terms which could not be specifically implemented without further elaboration.'[163] Furthermore, Jenks has seen it as merely a 'point of departure' requiring further definition and conceptualisation before making the transition from executory to executed international law.[164] Therefore, Usurpia had no legal obligation towards Landia to ensure commercial activities in outer space were conducted for its benefit.

Second, subsequent state practice in the application of the *OST* (which the Court is entitled to take into account in its interpretation of the Treaty under Article 31(3)(b) of the *Vienna Convention*)[165] reveals that States do not view Article I as a binding legal obligation. Since the *OST* entered into force no over 40 years ago, no State has sought to rely on Article I to assert a claim to benefits obtained by another country through its space programs.[166] Further, most States have not implemented domestic laws which accommodate the principles embodied in Article I, reflecting that States merely perceive Article I as a statement of aspirational principle rather than a legal obligation.[167]

In any event, even if Article I is binding, Usurpia acted consistently with the principle by seeking to ensure Satelsat remained solvent. An obligation to carry out space activities for the benefit and interests of all countries does not require States to accord preferential treatment to any one country. With 25 satellites in its fleet, Satelsat serviced many members of the international community. Usurpia acted in the common interest of this international community by restructuring Satelsat and relocating Satelsat-18.

(b) Usurpia's acts were consistent with Article IX

Article IX of the *OST* imposes an obligation on States to conduct their activities in outer space with due regard to the interests of other States.[168] Usurpia did consider the interests of Landia when it decided to authorise the relocation of Satelsat-18. Usurpia complied with Article IX as Landia's position was duly considered, and it was offered the best possible alternative service available.

Moreover, Usurpia did not breach any duty to consult with Landia or Concordia under Article IX. The duty of consultation in this provision only applies to a State's 'potentially harmful interference' in other States' use of outer space, which refers to environmental harm such as contamination and pollution.[169] This

---

161 Carl Q Christol, *The Modern International Law of Outer Space* (1982) 42.

162 Bin Cheng, *Studies in International Space Law* (1997) 404-405.

163 Stephen Gorove, 'Implications of International Space Law in Private Enterprise' (1982) VII *Annals of Air and Space Law* 319, 322.

164 C. Wilfred Jenks, *Space Law* (1965) 310. See also Cheng, above n 162, 235.

165 *Vienna Convention,* opened for signature 23 May 1969, 1155 UNTS 331, art 31(3)(b) entered into force 27 January 1980).Even though the States in this case are not party to the *Vienna Convention,* it has been established that *Vienna Convention* arts 31-32 reflect customary international law; *Case Concerning Sovereignty over the Islands of Litigan and Sipadan (Indonesia v Malaysia)* (2002) ICJ Rep 23-24.

166 Christol, above n 161.

167 Nandasiri Jasentuliyana, 'Ensuring Equal Access to the Benefits of Space Technologies for All Countries' (1994) 10(1) *Space Policy* 7,10.

168 *OST* opened for signature 27 January 1967, 610 UNTS 205, art IX (entered into force 10 October 1967).

169 Jerry Sztucki, 'International Consultations and Space Activities' (1975) *Proceedings of the 17th Colloquium on the Law of Outer*

clearly does not extend to the present situation involving the proposed relocation of a satellite.

Further, even if the duty to consult in this Article did apply to the proposed relocation of a satellite, the exchanges between Usurpia and Landia constituted the requisite consultation. The consultation requirement is fulfilled if a 'reasonable time was left for an exchange of communications and if this exchange took place in whatever form, even if it only showed that the positions of the parties involved were irreconcilable.'[170] Usurpia took Landia's interests into account by offering it an alternative service which it rejected. To fulfil the consultation requirement, Usurpia was not obliged to act in accordance with Landia's wishes, but was merely required to take Landia's interests into account, which it did by offering an alternative service and responding to Landia and Concordia's protests.[171] Hence, Usurpia did not breach any duty under Article IX of the *OST*.

1.3 Usurpia's actions were consistent with the *Registration Convention*

The *Convention on Registration of Objects Launched into Outer Space*[172] does not require States to notify the UN of changed orbital parameters for spacecraft.[173] Although the UN General Assembly has suggested that following the change in supervision of a space object in orbit, information regarding any change in its orbital position *could* be furnished to the Secretary General,[174] this is merely a *de lege ferenda* recommendation which has not crystallised into a binding obligation.[175] Indeed, State practice reveals that notification of changed orbital parameters to the UN is not common.[176] Thus, Usurpia was not bound to notify of the changed orbital parameters of Satelsat-18.

In any event, Usurpia's acts in placing Satelsat-18 on its national registry and commencing notification to the UN occurred 'as soon as practicable', in conformity with its obligations under the *Registration Convention*.[177] State practice under the *Registration Convention* reveals that delays of nine months or longer are common in States' implementation of their notification obligations. For example, in September 2007 the United States completed the process of notifying the UN of its registration information for a satellite launched in December 2006.[178] Further, in June 2007 the United Kingdom notified the UN of registration information for satellites launched between March and November 2005.[179] Thus, Usurpia's commencement of applicable notification procedures at some point between the approval of the restructure on 15 July 2010 and the collision on 25 August 2010 fell far short of the delays noted above, and was thus in conformity with the requirement to notify as soon as practicable. **Landia is not entitled to** compensation from Usurpia for loss suffered due to the collision of Satelsat-18 and SpaceStar

---

Space 159, 159. See also Cheng, above n 162, 256.

170 Sztucki, above n 169, 165.

171 Ibid.

172 *Convention on Registration of Objects Launched into Outer Space*, opened for signature 14 January 1975. 1023 UNTS 187 (entered into force 15 September 1976)('*Registration Convention*').

173 David Enrico Reibel, 'Registration of Space Objects: Beyond Conventional Compliance'(1989) *Proceedings of the 32nd Colloquium on the Law of Outer Space* 391.

174 *Recommendations on enhancing the practice of state and international intergovernmental organizations in registering space objects*, GA Res 62/101, UN GAOR 62nd sess, adopted 17 December 2007, UN Doc A/RES/62/101 (2008).

175 This is emphasised by the preamble to the resolution noting that 'nothing in the conclusions of the working group or in the present resolution constitutes an authoritative interpretation of or a proposed amendment to the *Registration Convention.*'

176 *United Nations Registry of Space Objects* <http://www.unoosa.org/oosa/en/SORegister/index.html> at 23 July 2008.

177 *Registration Convention*, opened for signature 14 January 1975, 1023 UNTS 187 art IV(1) (entered into force 15 September 1976).

178 UN DOC ST/SG/SER.E/514, accessed at <http://www.unoosa.org/oosa/en/Reports/Regdocs/ser514.html> at 23 July 2008.

179 UN DOC ST/SG/SER.E/518, accessed at <http://www.unoosa.org/oosa/Reports/Regdocs/ser518.html> at 23 July 2008.

1.4 Usurpia is not liable to pay compensation to Landia under the *Liability Convention.*

Landia, Concordia and Usurpia are all States parties to the *Convention on International Liability for Damage Caused by Space Objects.*[180] This treaty acts as the *lex specialis* for determining issues of liability for damage caused by space objects.[181] The *Liability Convention* provides that only 'launching States' can incur liability in space law.[182] Article I(c) defines a launching State as a 'State which launches or procures the launching of a space object' or a 'State from whose territory or facility a space object is launched'.

Usurpia is not liable to Landia because: the launching State regime remains applicable, Usurpia is not a launching State and Concordia is a launching State for Satelsat-18 and is therefore liable for any damage caused to Landia.

1.5 The *Liability Convention* only provides for the liability of launching States

Even though the regime of holding launching States liable was formulated before the proliferation of commercial space activities, it remains applicable today. State practice in the application of the *Liability Convention* (which may be referred to as an interpretive aid pursuant to Article 31(3)(b) of the *Vienna Convention*)[183] reflects continued international acceptance of the current space liability regime exclusively based upon the launching State. This is evident in the domestic legislation which has been enacted by the majority of spacefaring nations, requiring private operators to indemnify the launching State for any liability it incurs under the *Liability Convention*. For example, Japan, Russia, the United States and the United Kingdom all require private operators to indemnify the State for damage caused by their space activities.[184] Moreover, these States legally require that commercial satellite operators obtain compulsory third party liability insurance, to ensure that the private entities can actually cover the payment of any potential liability they incur.[185]

Furthermore, the increasing sale and purchase of satellites in orbit has not affected launching State liability, which subsists after a transfer of satellite ownership.[186] The *Liability Convention* was designed to simply put up a 'safety net' to ensure that victims may always identify a state from which to claim compensation.[187] Upon the sale of satellites in orbit, launching States can disburse the burden of their liability through their domestic laws or private agreements.[188] For example, in the 1990 launch of the ASIA-I satellite for Hong Kong from China, a liability agreement was concluded between the governments of China and the United Kingdom.[189] This agreement privately distributed the liability of the parties for damage caused in the various stages of the operation of the satellite for which the countries were responsible. China has subsequently invoked this agreement many times in its space activities. Similar agreements have also been concluded between the French government and the European Space Agency in relation to use of the Guiana Space Centre.[190]

---

[180] *Convention on International Liability for Damage Caused by Space Objects*, opened for signature 29 March 1972, 961 UNTS 197 (entered into force 1 September 1972) ('*Liability Convention*').

[181] *Rights of Passage over Indian Territory (Portugal v India)(Merits)*[1960] ICJ Rep 6, 44.

[182] *Liability Convention,* opened for signature 29 March 1972, 961 UNTS 187, arts II- V (entered into force 1 September 1972).

[183] *Vienna Convention,* opened for signature 23 May 1969, 1155 UNTS 331, art 31(3)(b) entered into force 27 January 1980); see above n 165.

[184] For summary of national laws see, *Review of the concept of the 'launching State,'* UN DOC A/AC.105/768 (2002), 11-12.

[185] Ibid, 10-11.

[186] Ricky Lee, 'Effects of Satellite Ownership Transfers on the Liability of the Launching States' (2000) *Proceedings of the 43rd Colloquium on the Law of Outer Space* 148.

[187] Armel Kerrest, 'Remarks on the Notion of the Launching State' (1999) *Proceedings of the 42nd Colloquium on the Law of Outer Space* 308, 308.

[188] Ibid, 309.

[189] Review of the concept of the 'launching State,' UN DOC A/AC.105/768 (2002) 13.

[190] Ibid, 14.

The conclusion of the *Mobile Equipment Convention* and the negotiation of the *Space Protocol* reinforce the continued relevance of launching State liability in situations where a satellite changes ownership. The *Space Protocol* specifically considers the scenario of a creditor taking control of a secured space asset following a debtor's default, which is exactly the situation which occurred in the instant case. Article XXI *bis* of the *Space Protocol* provides that the rights and obligations of States under the existing international space treaties should not be affected by the *Protocol* or *Mobile Equipment Convention.*[191] Thus, even where a satellite is repossessed whilst in orbit, the rules in the *Liability Convention* attributing liability to the launching State still apply.

In 2002, the Legal Subcommittee of the United Nations Committee on the Peaceful Use of Outer Space ('UNCOPUOS') reviewed in detail the concept of the launching State, and explicitly declined to change either the use or interpretation of the term.[192] The Review noted that States have expressed the opinion that it would be inappropriate to extend the definition or interpretation of the term launching State.[193] This would hinder the development of the space industry, and result in space activities becoming excessively expensive for both governments and the private sector, as States would be 'hesitant to approve participation by their private entities if they would be subject to liability even for tangential connections to the launch.'[194] A broad interpretation of the 'launching State' would also increase the 'red tape costs' of the space sector.[195] Instead of recommending changes to the existing liability regime, the Working Group recommended that States continue to implement national laws to apportion the burden of their obligations under the *Liability Convention.* This view was also adopted by the UN General Assembly in Resolution 59/155, which encouraged States to enact national laws and conclude private agreements to fulfil their international obligations under the *Liability Convention.*[196]

Thus, the political consensus of a majority of States has been to maintain the current regime of launching State liability, and to adapt to lacunae in the treaty with the enactment of national legislation, rather than wholesale amendment or novel interpretation of the *Liability Convention.* Therefore, liability in this case must only be borne by the launching State of the satellites involved in the collision.

1.6 Usurpia is not a launching State of Satelsat-18

Usurpia did not participate in the launch of Satelsat-18. The satellite was not launched from its territory or its facility. While there are Satelsat satellite control facilities located in Usurpia, there is no evidence to suggest that these facilities were employed in the launch of Satelsat-18. Moreover, even if these were utilised, it is submitted that the provision of satellite control facilities without more does not qualify a State as a launching State.[197]

In addition, Usurpia did not 'procure the launch' of Satelsat-18. Whilst the term 'procurement' is not defined in the *Liability Convention,* the *travaux préparatoires* of the treaty explain that a State procures a launch if it 'actively and substantially participate[s]' in the launch.[198] Professor Hurwitz has confirmed that a State which procures a launch is 'the State which specifically requests the launch' or the 'State whose nationals have financed or ordered

---

191 *Space Protocol*, UNIDROIT Doc Study LXXIIJ (2004).

192 *Review of the concept of the 'launching State,'* UN DOC A/AC.105/768 (2002).

193 Ibid, 17.

194 William Wirin, 'Practical Implications of Launching State – Appropriate State Definitions' (1992) *Proceedings of the 36th Colloquium on the Law of Outer Space* 109, 113.

195 Kai-Uwe Schrogl and Charles Davies, 'A new look at the launching State:' the results of the UNCOUOS Legal Subcommittee Working Group' (2002) *Proceedings of the 45th Colloquium of the Law of Outer Space* 286, 293.

196 *Application of the Concept of the 'Launching State,'* GA Res 59/155, UN GAOR, 59th sess, 71st plenary mtg UN Doc A/Res/59/115 (2005).

197 *Review of the concept of the 'launching State,'* UN DOC A/AC.105/768 (2002), 18.

198 *Report of the Legal Subcommittee on its Sixth Session* UNCOPUOS, Legal Subcommittee, UN Doc A/AC.105/C.2/SR.77 (1967).

the launch.'[199] Procurement has also been limited to States which have 'direct control over the launch'[200] or those that 'organise' the launch.[201]

Usurpia did not request, finance or actively participate in the launch of Satelsat-18. Indeed, there is no indication that Usurpia was at all involved in the launch of Satelsat-18. Thus, Usurpia is not a launching State, and cannot be liable under the *Liability Convention* for the losses arising from the satellite collision.

Furthermore, Usurpia did not become a launching State as a result of its placement of Satelsat-18 on its national registry. The drafters of the *Registration Convention* 'generally believed registration of a space object by itself, to be an insufficient basis for linking a State to damage caused by that Space object.'[202] The launching State concepts within the two Conventions have a different purpose, scope and effect. Whilst the *Liability Convention* aims at the recovery of compensation by injured parties, the *Registration Convention* is directed to the identification and management of space objects. A State's liability arises from its status as a launching State, not as a State of registry, such that 'ascribing liability to a state solely because it has registered a space object is not only legally questionable, but also unnecessary and counter-productive.'[203] Therefore, even though Usurpia placed Satelsat-18 on its register, this does not make it the launching State of Satelsat-18.

It has been stated that if the 'satellite of the national A is assigned to the national B of a different nationality, the State of the national B will *not* become the launching State retroactively.'[204]The *Liability Convention* does not impose liability on States in the position of Usurpia, which are not involved in the launch of a satellite, but whose national later purchases the satellite whilst in orbit.[205] As Usurpia is not a launching State, it is therefore not liable for any damage caused by Satelsat-18 under the *Liability Convention.*

1.7 Concordia is liable for damage caused by Satelsat-18 as it is the launching State

Satelsat-18 was launched from the Concordia Space Center by commercial launch services providers based in Concordia and licensed by the government of Concordia.[206] As Satelsat-18 was clearly launched from both Concordian 'territory' and a Concordian 'facility', Concordia is the launching State and is liable for any damage caused by the satellite.[207] Usurpia is not internationally responsible for the satellite collision

Article VI of the *OST* provides that 'State parties to the Treaty shall bear international responsibility for national activities in outer space…and for assuring that national activities are carried out in conformity with the provisions set out in the present Treaty.'[208] The term 'responsibility' in this provision is used in the same sense as responsibility in general international law. That is, responsibility under the *OST* also arises as a result of an internationally wrongful act by a State.[209] Professor von der Dunk has stated that 'international State responsibility, in space law as much as elsewhere, therefore arises in [the] case of activities being in violation of relevant

---

199 Bruce Hurwitz, *State Liability or Outer Space Activity* (1992) 22.

200 Wirin, above n 194, 113.

201 Schrogl and Davies, above n 195, 293.

202 Edward A Frankel 'Once a Launching State, always *The* Launching State?' (2001) *Proceedings of the 44th Colloquium on the Law of Outer Space* 32, 36.

203 Ibid, 37.

204 Bernhard Schmidt-Tedd and Michael Gerhard, 'Registration of Space Objects: What are the Advantages for States Resulting from Registration,' in Marietta Benko and Kai-Uwe Schrogl (eds) *Space Law: Current Problems and Perspectives for Future Regulation* (2005) 121, 131 (emphasis added).

205 Kerrest, above n 187, 308.

206 *Compromis* ¶3.

207 *Liability Convention*, opened for signature 29 March 1972, 961 UNTS 197, art I(c)(ii) (entered into force 1 September 1972).

208 *OST,* opened for signature 27 January 1967, 610 UNTS 205, art VI (entered into force 10 October 1967).

209 *Articles on Responsibility of States for Internationally Wrongful Acts*, GA Res 56/83, UN GAOR, 56th sess, Supp No 10, UN Doc A/Res/56/83 (2001), art 1 ('*State Responsibility Articles*').

legal obligations, those being primary obligations of space law.'[210] Article VI therefore requires States to ensure that the acts of their nationals are conducted in accordance with international law, and imposes international responsibility on States for any violation of international law by their private entities.[211]

Although the satellite collision was highly unfortunate, it did not involve any breach of international law and therefore did not entail Usurpia's international responsibility. As discussed above in section 1, Usurpia and New Satelsat acted at all times in conformity with Usurpia's international obligations under the *OST, Registration Convention* and *GLITSO*.

Although Article VI of the *OST* does not specify the precise manner in which States are to supervise the activities of their non-governmental activities in outer space, Usurpia took appropriate steps to supervise the activities of New Satelsat in accordance with established State practice under Article VI. The UTA issued a license for the operation of Satelsat-18, in line with the practice of other spacefaring States which require the space activities of their private entities to be licensed.[212] Moreover, Usurpia rightly commenced notification to the UN Secretary General of the satellite's changed status.[213]

Even if New Satelsat's speedy relocation of the satellite was not in conformity with the Usurpian license, this would not constitute a breach of international law, as no legal regime for the regulation of satellite manoeuvring or space traffic has yet been formulated.[214] In 2004, DIRECTV, a private company, relocated one of its satellites from a US orbital slot to an orbital position which had been assigned to Canada, without authorisation from the appropriate US authority. Although the company was fined under US domestic law for breaching internal regulations, it was never alleged that either the US or Canada had breached international law.[215] Similarly, the entire process of New Satelsat's relocation of Satelsat-18 did not breach any of Usurpia's international obligations. Thus, Usurpia is not internationally responsible or liable for the satellite collision.

1.8 Even if Usurpia is liable for the collision, the losses claimed by Landia are not compensable under either the *Liability Convention* or general international law

The *Liability Convention* expressly limits compensable damage in Article I(a) to 'loss of life, personal injury or other impairment of health or loss of or damage to property.'[216] The use of the term 'means' rather than 'includes' in Article I(a) suggests that the definition of damage was intended to be exhaustive, and confined to direct damage.[217]

Indirect damage which does not 'flow directly and immediately from the act, but only from some of the consequences or results of such act' is therefore not recoverable under the

---

210 Frans von der Dunk, 'Liability versus Responsibility in Space Law: Misconception or Misconstruction' (1991) *Proceedings of the 34th Colloquium on the Law of Outer Space* 363, 366.

211 Armel Kerrest, 'Remarks on the Responsibility and Liability for Damages Other Than Those Caused by the Fall of a Space Object' (1997) *Proceedings of the 40th Colloquium on the Law of Outer Space* 134, 139; Ricky J Lee, 'Liability Arising From Article VI of the Outer Space Treaty: States, Domestic Law and Private Operators' (2005) *Proceedings of the 48th Colloquium on the Law of Outer Space* 216, 217.

212 *Compromis* ¶13; See examples of Australian, Russian, South African, Swedish, Ukrainian, British and US licensing laws pursuant to Article VI of the *OST* described in *Review of the concept of the 'launching State,'* UN DOC A/AC.105/768 (2002), 5-6.

213 *Compromis* ¶24.

214 Pamela L Meredith, 'Spacecraft Motion Management (SMM): Institutional and Legal Frameworks' (1992) *Proceeding of the 35th Colloquium on the Law of Outer Space* 102.

215 *In the Matter of DIRECTV, Inc, Before the Federal Communications Commission* (2004) <http://hraunfoss.fcc.gov/edocs_public/attachmatch/FCC-04-138A1.pdf> at 21 July 2008.

216 *Liability Convention,* opened for signature 29 March 1972, 961 UNTS 187, art I(a) (entered into force 1 September 1972).

217 Hurwitz, above n 199, 19.

Liability Convention.[218] This is supported by the *travaux préparatoires* of the *Liability Convention* which indicate that negotiating States specifically considered the issue of indirect damages, and most countries, including the United States and the USSR, opposed the inclusion of reference to indirect damages in the Convention.[219]

Also, in the negotiation of Article VII of the *OST*, which preceded the full elaboration of the *Liability Convention*, the US delegate stated that 'any reasonable interpretation of that clause would mean physical damage' and that the article 'pertains only to physical, non-electronic damages that space activities may cause to the citizens or property of a signatory State.'[220] Moreover, subsequent space law instruments address liability issues by explicitly providing for the recovery of indirect damage.[221] This supports the proposition that when States intend to include compensation for indirect damage, they do so explicitly.

It is acknowledged that some commentators have attempted to use the 1978 Cosmos 954 incident, in which a Soviet Union satellite deposited nuclear debris on Canadian territory, to demonstrate that indirect damage is recoverable under the *Liability Convention.*[222] However, this example does not support any particular interpretation of the *Liability Convention*, let alone one which includes the recovery of indirect damage. Although Canada's claim against the Soviet Union for cleanup costs was based on the *Liability Convention,* the final payment by the USSR did not represent its acceptance of liability under the *Liability Convention.*[223] Rather, under the settlement reached, the USSR paid half the original amount claimed by Canada which it categorised as an 'ex gratia' sum.[224]

The expansion of the definition of 'damage' is, moreover, not supported by the writings of highly qualified publicists. Even given the considerations of today's commercialised space industry, Professor Bin Cheng has stated that the definition of damage in Article I does not need to be widened.[225] Thus, any claim for indirect damage arising from the satellite collision falls outside the scope of the *Liability Convention.*

In any event, the satellite collision was not the legal cause of Landia's economic loss. Space law, like general international law,

---

218 Stephen Gorove, 'Some Comments on the Convention on International Liability for Damage Caused by Space Objects' (1973) *Proceedings of the 16th Colloquium on the Law of Outer Space* 253, 255.

219 Hurwitz above n 199, 16; Carl Q Christol, 'International Liability for Damage Caused by Space Objects' (1980) 74 *American Journal of International Law* 346, 361; W F Foster, 'The Convention on International Liability for Damage caused by Space Objects' (1972) 10 *Canadian Yearbook of International Law* 137, 157-158.

220 Christol, above n 219, 354.

221 For example, the multilateral Space Station Agreement includes indirect damage in the meaning of 'damage' in art 16(c)(4); *Agreement among the Government of Canada, the Governments of the Member States of the European Space Agency, the Government of Japan, the Government of the Russian Federation and the Government of the United States concerning cooperation on a civil international space station* (1998) ; <ftp://ftp.hw.nasa.gov/pub/pao/reports/1998/IGA.html>at 23 March 2008; while the *Principles Relevant to the Use of Nuclear Power Sources in Outer Space* states that compensation for nuclear damage shall include certain expressly stated indirect expenses; *Principles Relevant to the Use of Nuclear Power Sources in Outer Space*, GA Res 47/68, UN GAOR 47th sess, Supp 20, UN Doc A/47/20 (1992), Principle 9(3).

222 See eg, discussion in Christol, above n 219 , 362.

223 Steven Freeland, 'There's A Satellite In My Backyard! – Mir And The Convention On International Liability For Damage Caused By Space Objects' (2001) 24(2) *University of NSW Law Journal.*

224 Ibid, ¶40.

225 Cited in Maureen Williams, 'Review of Space Law Treaties in View of Commercial Activities' (Paper presented at the International Law Association London Conference: Space Law Committee, London, 25-29 July 2000) 13-16.

requires the claimant to establish causation.[226] Articles II and III of the *Liability Convention* limit the scope of compensation to damage which is 'caused by' a space object. The connection which must be established is that of 'proximate causation,' a general principle of international law,[227] which is supported in international case law[228] and by the most highly qualified publicists.[229] The rule of 'proximate causation' limits recoverable damage to that which arises as a normal consequence of an act and is reasonably foreseeable.

In particular, it has been held that a State does not proximately cause foreseeable damage to entities which have contracted with the party which is actually injured.[230] For example, in the *Life Insurance Claims* case, insurance companies were unsuccessful in their claims against Germany for the premature maturation of life insurance contracts, as it was held that the acts of Germany in taking the lives of the insured did not 'proximately cause' the economic loss to the insurance companies.[231] So too, Usurpia cannot be held liable to compensate Landia simply because the destruction of SpaceStar caused it to lose any entitlement it had under a contract with Orbitsat. A fortiori this rule applies in the present case where Landia's contract with Orbitsat was provisional, and even if SpaceStar had not been destroyed, Landia might have lost the benefits under that contract due to its inability to procure the necessary finances.[232] For this Court to decide otherwise would constitute an unreasonable extension of space liability to every party which is contractually connected to the entity which is actually injured, even in situations when this connection is tenuous.

Finally, Landia's $2 billion worth of predicted economic loss due to disruption to its telecommunications infrastructure is not a recoverable head of damage under general principles of international law.[233] Landia has not yet suffered $2 billion worth of indirect economic loss, but only predicts that this will occur. International law refuses the recovery of future loss where this is based upon speculation or contingency.[234] In the *Trail Smelter Arbitration*, residents whose businesses were affected by the metal smelter's fumes were refused compensation for their 'reduced economic status' as this was held to be 'too indirect, remote and uncertain to be appraised.'[235] More recently, damages for prospective loss were denied in the *Gabčíkovo-Nagymaros Case* as they were not sufficiently certain.[236] Landia should not be able to claim for disruption to its economic welfare due to its loss of telecommunications services from SpaceStar as this loss is heavily contingent on Landia having obtained the financial assistance to actually pay for the lease, at five times the cost of its previous satellite services.[237]

1.9 Even if Usurpia is liable, Landia's compensation should be limited due to its failure to mitigate its loss.

The rule that an injured party is required to mitigate its damage is well-accepted in

226 Foster, above n 219, 155; Christol, above n 219, 361.

227 To which recourse may be had pursuant to Article XII of the *Liability Convention.*

228 *Administrative Decision No II, (US v Germany)* (1923) 7 RIAA 23, 29-30; *Trail Smelter Arbitration* (1938) 3 RIAA 1905, 1963; see also discussion of relevant cases in Bin Cheng, *General Principles of Law as Applied by International Courts and Tribunals* (1953) 241-56.

229 Cheng, above n 97; Hurwitz, above n 199, 28; Christol, above n 161, 109.

230 *Dickson Car Wheel Co (Mexico v United States)* (1931) 4 RIAA 681.

231 *Life Insurance Claims (US v Germany)* (1924) 7 RIAA 91, 116.

232 *Compromis* ¶15.

233 Which also condition the recovery of compensation under Article XII the *Liability Convention.*

234 *Alabama Claims (United States-Great Britain) Claims Arbitration* cited in Bruce Hurwitz, 'Reflections on the Cosmos 954 Incident' (1989) *Proceedings of the 32nd Colloquium on the Law of Outer Space* 348, 356.

235 *Trail Smelter Arbitration* (1938) 3 RIAA 1905, 1931.

236 *Gabčíkovo-Nagymaros Project (Hungary v Slovakia)(Judgment)* [1997] ICJ Rep 7.

237 *Compromis* ¶15.

international law.[238] The principle is also relevant to compensatory awards under the *Liability Convention*, as Article XII requires that compensation be determined 'in accordance with international law and the principles of justice and equity'. The ILC's commentaries to the *State Responsibility Articles* states that 'even the wholly innocent victim… is expected to act reasonably when confronted by the injury.'[239] When confronted with the fact that it would no longer receive the services of Satelsat-18, and the prospect that it would not be able to afford the expensive replacement services of SpaceStar, Landia should have accepted the alternative, albeit inferior service from Satelsat to limit the damage it would sustain. Although this option admittedly would have required modification of Landia's ground infrastructure at some expense, this is likely to have been far less costly than entering into the SpaceStar lease at five times the previous cost of its satellite services. Further, as there is evidence that Landia may have obtained monetary assistance to satisfy the cost of the Orbitsat lease, it is entirely reasonable to assume that it could have obtained similar assistance to modify its infrastructure. Accordingly, any damage for which Usurpia is liable should be reduced to the extent that Landia failed to mitigate its loss by accepting the proposed alternative Satelsat service.

2. Concordia is not entitled to compensation from Usurpia for the loss of Satelsat-18 as a result of the collision

2.1 Concordia is precluded from claiming compensation under the *Liability Convention*

Concordia may not claim compensation from Usurpia under the *Liability Convention* for two reasons. First, as discussed above in section 2.1, Usurpia is not a launching State of either satellite involved in the collision. Second, Concordia's claim does not fall within the parameters of Article III of the *Liability Convention* which relates to compensation for damage caused in outer space. Article III of the *Liability Convention* provides that a launching State will be liable only if its space object causes damage to a space object of '*another* launching State.'[240] That is, where damage occurs in outer space, a claim can be made only if the collision is between space objects of different launching States. Professor Hurwitz affirms that where a collision occurs between space objects owned by different private entities but launched by the same State, Article III is not applicable.[241] In this case, as Concordia was the launching State of both Satelsat-18 and SpaceStar, its claim falls outside the scope of the *Liability Convention*.

2.2 In any event, Concordia may not present a claim for compensation as its nationals have not suffered damage

Article VIII of the *Liability Convention* provides that only 'a State which suffers damage, or whose natural or juridical persons suffer damage, may present…a claim for compensation for damage.'[242] It is submitted that neither Concordia nor its national, Satelsat, suffered damage.

In determining a State's eligibility to present a claim under Article VIII of the *Liability Convention*, it is necessary to consider the general principles of international law relating to the nationality of claims.[243] The *Barcelona Traction Case* established the principle that 'a State of nationality of shareholders in a corporation shall not be entitled to exercise diplomatic protection in respect of such shareholders in the case of injury

238 *Gabčíkovo-Nagymaros Project (Hungary v Slovakia)*[1997] ICJ Rep 7, 80; James Crawford, *The International Law Commission's Articles on State Responsibility: Introduction, Text and Commentaries* (2002) 228.

239 Crawford, above n 238.

240 *Liability Convention,* opened for signature 29 March 1972, 961 UNTS 187, art III (entered into force 1 September 1972).

241 See Hurwitz, above n 199, 33.

242 *Liability Convention,* opened for signature 29 March 1972, 961 UNTS 187, art VIII (entered into force 1 September 1972).

243 Article XII of *Liability Convention* provides for reference to general principles of international law in the determination of compensation claims under the Convention.

to the corporation.'[244] This Court recently confirmed that this rule extends to circumstances where the injured corporation is incorporated in the defendant State.[245]

In the present case, Concordia seeks compensation for the loss of Satelsat-18. However, at the time of the collision, the satellite was owned by New Satelsat, a company incorporated in Usurpia. As such, the destruction of Satelsat-18 did not directly damage any Concordian interest. The only possible basis for Concordia's claim is that Satelsat incurred loss by virtue of its shareholding in New Satelsat. However, based on the authorities analysed above, Concordia's claim on behalf of Satelsat as New Satelsat's shareholder cannot succeed.

3. Concordia is not entitled to indemnification from Usurpia for any compensation it owes to Landia as a result of the collision

3.1 The *Liability Convention* does not apply

Article V of the *Liability Convention* provides that 'a launching State which has paid compensation for damage shall have the right to present a claim for indemnification to other participants in the joint launching.'[246] As discussed above in section 2.1, Usurpia is not the launching state of either satellite. Therefore, Article V does not apply, and Usurpia is not obliged to indemnify Concordia for any compensation it might owe Landia.

3.2 In any event, Usurpia is not liable to indemnify Concordia under the *Liability Convention* as it was not at fault

Even if this Court decides that Usurpia is a launching State of Satelsat-18, Usurpia is not liable to indemnify Concordia under Article V as it was not at fault in the collision. Joint launching States are liable to one another for compensation paid to a third State to the extent to which they were at fault.[247]

Fault is not defined in the *Liability Convention*. However, in international law the principle of 'fault' refers not to culpability or malice, but rather a failure to comply with a legal duty or obligation.[248] There is no duty at international law to conduct satellite relocations at a particular speed or for the relocation to follow a particular course or trajectory.[249] As the most highly qualified publicists have noted, even if States were to follow directives as to the spacing of their satellite objects meticulously, they would still lack the ability to predict the occurrence of a collision.[250] There is no evidence that Usurpia was negligent or reckless in its relocation of the satellite. Thus, Usurpia was not at fault in the collision and does not attract liability to indemnify Concordia under the *Liability Convention*.

3.3 This result is not unfair to Concordia

There is no injustice in Concordia bearing absolute liability under the Liability Convention for any damage caused to Landia. As discussed above in section 2.1, through domestic legislation, spacefaring countries[251] routinely require private satellite operators to indemnify them against any international liability they might incur under the Liability Convention in order to 'bridge the gap between

[244] *Barcelona Traction, Light and Power Company Limited* (Belgium v Spain) (Second Phase) [1970] ICJ Rep 3, 34; This has been codified in Article 11 of the *Draft Articles of Diplomatic Protection, International Law Commission, Report of the 55th session,* UN GAOR 58th sess, Supp 10, UN Doc A/58/10 (2003).

[245] *Case Concerning Ahmadou Sadio Diallo (Republic of Guinea v Democratic Republic of the Congo)(Preliminary Objections)* (2007) ICJ ¶76-¶95.

[246] *Liability Convention,* opened for signature 29 March 1972, 961 UNTS 187, art V (entered into force 1 September 1972).

[247] Kerrest, above n 187, 312.

[248] Cheng, above n 97, 218.

[249] Meredith, above n 214.

[250] Howard A Baker, 'Liability for Damage Caused in Outer Space by Space Refuse' (1988) 12 *Annals of Air & Space Law* 183 at 192.

[251] See examples of Australian, Japanese, Russian, South African, Swedish, British and US indemnification laws described in *Review of the concept of the 'launching State,'* UN DOC A/AC.105/768 (2002), 10-12.

the entity legally liable and the entity actually guilty'.[252] Professor Kerrest has remarked that 'as far as private entities' activities are concerned, the launching State's obligation under the Liability Convention is, in fact, more an obligation to control and to guarantee [the disbursement of liability] than an obligation to pay for the damage.'[253] In line with this practice, Concordia would have (or indeed should have) entered into appropriate agreements or enacted relevant legislation to ensure that Satelsat and any of its successors in title would assume the burden of Concordia's liability as a launching State of Satelsat-18.

SUBMISSIONS TO THE COURT

For the foregoing reasons, the Kingdom of Usurpia, Respondent, respectfully requests the Court to adjudge and declare that:

5. Usurpia's decision to license Satelsat-18 and permit it to be deployed at an Usurpian orbital location was consistent with applicable principles of international law;

6. Landia is not entitled to compensation from Usurpia as a result of the collision between Satelsat-18 and SpaceStar satellites;

7. Concordia is not entitled to compensation for the loss of Satelsat-18;

8. Concordia is not entitled to indemnification from Usurpia for any financial obligation owed to Landia, as a result of the collision between Satelsat-18 and SpaceStar.

---

252 Frans von der Dunk 'Commercial Space Activities, an Inventory of Liability, an Inventory of Problems' (1994) *Proceedings of the 37th Colloquium of the Law of Outer Space* 161,164.

253 Kerrest, above n 187, 311.

AUTHOR INDEX